D0164452

Mathematics

A Discrete Introduction

Second Edition

Edward R. Scheinerman
Department of Applied Mathematics and Statistics
The Johns Hopkins University

THOMSON
BROOKS/COLE™

Australia • Canada • Mexico • Singapore • Spain
United Kingdom • United States

THOMSON
BROOKS/COLE

Mathematics: A Discrete Introduction, Second Edition
Edward R. Scheinerman

Publisher: *Bob Pirtle*
Assistant Editor: *Stacy Green*
Editorial Assistant: *Katherine Cook*
Technology Project Manager: *Earl Perry*
Executive Marketing Manager: *Tom Ziolkowski*
Marketing Communications Manager: *Bryan Vann*
Project Manager, Editorial Production: *Kelsey McGee*
Art Director: *Vernon Boes*
Print Buyer: *Doreen Suruki*
Permissions Editor: *Joohee Lee*
Production Service: *Matrix Productions/ Merrill Peterson*
Text Designer: *Kim Rokusek*
Photo Researcher: *Pat Quest*
Copy Editor: *Connie Day*
Cover Designer: *Jeanne Calabrese*
Cover Image: *Albert Kocourek*
Cover Printer: *Phoenix Color Corp*
Compositor: *Interactive Composition Corporation*
Printer: *R.R. Donnelley, Crawfordsville*

Printed in the United States of America
1 2 3 4 5 6 7 09 08 07 06 05

Library of Congress Control Number: 2005922132

Student Edition: ISBN 0-534-39898-7

Thomson Higher Education
10 Davis Drive
Belmont, CA 94002-3098
USA

To
David, Karen, Matthew, Zachary, and Alanna
and
Robert, Suzanne, Calida, and Olivia

Contents

To the Student

Welcome.

This book is an introduction to *mathematics*. In particular, it is an introduction to *discrete* mathematics. What do these terms—*discrete* and *mathematics*—mean?

Continuous versus *discrete* mathematics.

The world of mathematics can be divided roughly into two realms: the *continuous* and the *discrete*. The difference is illustrated nicely by wristwatches. Continuous mathematics corresponds to analog watches—the kind with separate hour, minute, and second hands. The hands move smoothly over time. From an analog watch perspective, between 12:02 P.M. and 12:03 P.M. there are infinitely many possible different times as the second hand sweeps around the watch face. Continuous mathematics studies concepts that are infinite in scope, in which one object blends smoothly into the next. The real-number system lies at the core of continuous mathematics, and—just as on the watch—between any two real numbers, there is an infinity of real numbers. Continuous mathematics provides excellent models and tools for analyzing real-word phenomena that change smoothly over time, including the motion of planets around the sun and the flow of blood through the body.

Discrete mathematics, on the other hand, is comparable to a digital watch. On a digital watch there are only *finitely* many possible different times between 12:02 P.M. and 12:03 P.M. A digital watch does not acknowledge split seconds! There is no time between 12:02:03 and 12:02:04. The watch leaps from one time to the next. A digital watch can show only finitely many different times, and the transition from one time to the next is sharp and unambiguous. Just as the real numbers play a central role in continuous mathematics, *integers* are the primary tool of discrete mathematics. Discrete mathematics provides excellent models and tools for analyzing real-world phenomena that change abruptly and that lie clearly in one state or another. Discrete mathematics is the tool of choice in a host of applications, from computers to telephone call routing and from personnel assignments to genetics.

What is mathematics? A more sophisticated answer is that mathematics is the study of sets, functions, and concepts built on these fundamental notions.

Let us turn to a harder question: What is mathematics? A reasonable answer is that mathematics is the study of numbers and shapes. The particular word in this definition we would like to clarify is *study*. How do mathematicians approach their work?

Every field has its own criteria for success. In medicine, success is healing and the relief of suffering. In science, the success of a theory is determined through experimentation. Success in art is the creation of beauty. Lawyers are successful when they argue cases before juries and convince the jurors of their clients' cases. Players in professional sports are judged by whether they win or lose. And success in business is profit.

What is successful mathematics? Many people lump mathematics together with science. This is plausible, because mathematics is incredibly useful for science, but of the various fields just described, mathematics has less to do with science than it does with law and art!

Mathematical success is measured by *proof*. A *proof* is an essay in which an assertion, such as "There are infinitely many prime numbers," is incontrovertibly shown to be correct. Mathematical statements and proofs are, first and foremost, judged in terms of their correctness. Other, secondary criteria are also important. Mathematicians are concerned with creating beautiful mathematics. And mathematics is often judged in terms of its utility; mathematical concepts and techniques are enormously useful in solving real-world problems.

Proof writing.

One of the principal aims of this book is to teach you, the student, how to write proofs. Long after you complete this course in discrete mathematics, you may find that you do not need to know how many k-element subsets an n-element set has or how Fermat's Little Theorem can be used as a test for primality. Proof writing, by contrast, will always serve you well. It trains us to think clearly and present our case logically.

Many students find proof writing frightening and difficult. They might freeze after writing the word *proof* on their paper, unsure what to do next. The antidote to this proof phobia can be found in the pages of this book! We demystify the proof-writing process by decoding the idiosyncrasies of mathematical English and by providing *proof templates*. The proof templates, scattered throughout this book, provide the structure (and boilerplate language) for the most common varieties of mathematical proofs. Do you need to prove that two sets are equal? See Proof Template 5! Trying to show that a function is one-to-one? Consult Proof Template 20!

Proof templates.

How to Read a Mathematics Book

Reading a mathematics book is an active process. You should have a pad of paper and a pencil handy as you read. Work out the examples and create examples of your own. Before you read the proofs of the theorems in this book, try to write your own proof. Then, if you get stuck, read the proof in the book.

One of the marvelous features of mathematics is that you need not (perhaps, should not!) trust the author. If a physics book refers to an experimental result, it might be difficult or prohibitively expensive for you to do the experiment yourself. If a history book describes some events, it might be highly impractical to consult the original sources (which may be in a language you do not understand). But with mathematics, all is before you to verify. Have a reasonable attitude of doubt as you read; demand of yourself to verify the material presented. Mathematics is not so much about the truths it espouses but about how those truths are established. Be an active participant in the process.

One way to be an active participant is to work on the hundreds of exercises found in this text. If you run into difficulty, you may be helped by the many hints and occasional answers in Appendix A. However, I hope you will not treat this book as just a collection of problems with some stuff thrown in to keep the publisher happy. I tried hard to make the exposition clear and useful to students. If you find it otherwise, please let me know. I hope to improve this book continually, so send your comments to me by email at `ers@jhu.edu` or by conventional letter to Professor Edward Scheinerman, Department of Applied Mathematics and Statistics, The Johns Hopkins University, Baltimore, Maryland 21218, USA. Thank you.

I hope you enjoy.

Exercises

1. On a digital watch there are only finitely many different times that can be displayed. How many different times can be displayed on a digital watch that shows hours, minutes, and seconds and that distinguishes between A.M. and P.M.?
2. An ice cream shop sells ten different flavors of ice cream. You order a two-scoop sundae. In how many ways can you choose the flavors for the sundae if the two scoops in the sundae are different flavors?

To the Instructor

Please also read the "To the Student."

Why do we teach discrete mathematics? I think there are two good reasons. First, discrete mathematics is useful, especially to students whose interests lie in computer science and engineering, as well as those who plan to study probability, statistics, operations research, and other areas of modern applied mathematics. But I believe there is a second, more important reason to teach discrete mathematics. Discrete mathematics is an excellent venue for teaching students to write proofs.

Thus this book has two primary objectives:

- to teach students fundamental concepts in discrete mathematics (from counting to basic cryptography to graph theory) and
- to teach students proof-writing skills.

Audience and Prerequisites

This text is designed for an introductory-level course on discrete mathematics. The aim is to introduce students to the world of mathematics through the ideas and topics of discrete mathematics.

A course based on this text requires only core high school mathematics: algebra and geometry. No calculus is presupposed or necessary.

Serving the computer science/engineering student.

Discrete mathematics courses are taken by nearly all computer science and computer engineering students. Consequently, some discrete mathematics courses focus on topics such as logic circuits, finite state automata, Turing machines, algorithms, and so on. Although these are interesting, important topics, there is more that a computer scientist/engineer should know. We take a broader approach. All of the material in this book is directly applicable to computer science and engineering, but it is presented from a mathematician's perspective. As college instructors, our job is to educate students, not just to train them. We serve our computer science and engineering students better by giving them a broader approach, by exposing them to different ideas and perspectives, and, above all, by helping them to think and write clearly. To be sure, in this book you will find algorithms and their analysis, but the emphasis is on mathematics.

Topics Covered and Navigating the Sections

The topics covered in this book include

- the nature of mathematics (definition, theorem, proof, and counterexample),
- basic logic,
- lists and sets,

- relations and partitions,
- advanced proof techniques,
- recurrence relations,
- functions and their properties,
- permutations and symmetry,
- discrete probability theory,
- number theory,
- group theory,
- public-key cryptography,
- graph theory, and
- partially ordered sets.

Furthermore, enumeration (counting) and proof writing are developed throughout the text. Please consult the table of contents for more detail.

Each section of this book corresponds (roughly) to one classroom lecture. Some sections do not require this much attention, and others require two lectures.

There is enough material in this book for a year-long course in discrete mathematics. If you are teaching a year-long sequence, you can cover all the sections.

A semester course based on this text can be roughly divided into two halves. In the first half, core concepts are covered. This core consists of Sections 2 through 23 (optionally omitting Sections 17 and 18).

From there, the choice of topics depends on the needs and interests of the students. Sample course outlines are given below. The interdependence of the various sections is depicted in the following diagram.

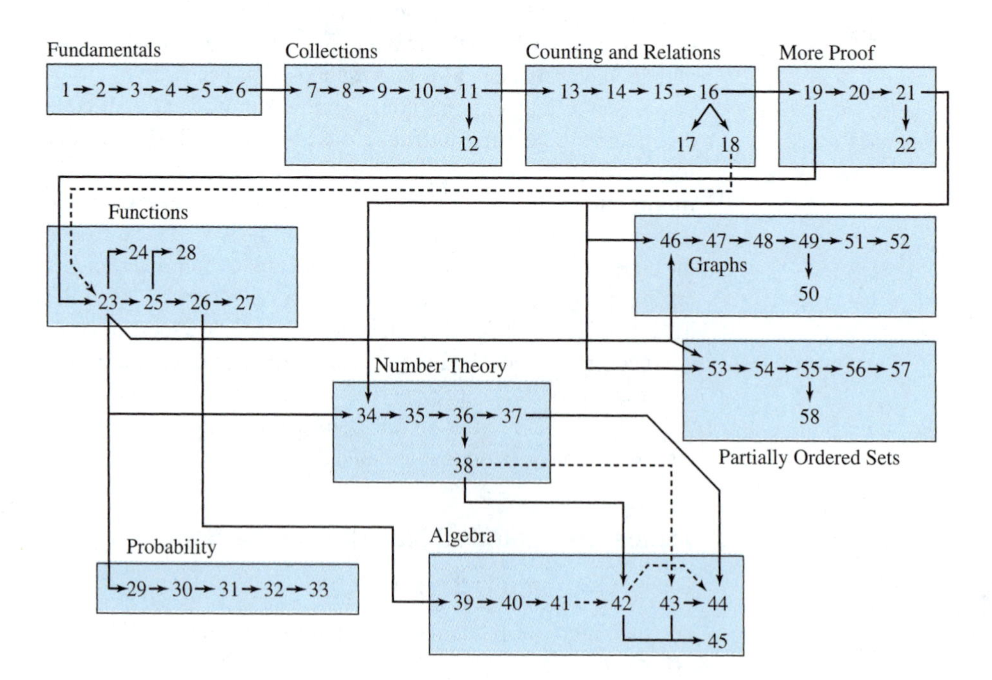

Sample Course Outlines

Thanks to its plentiful selection of topics, this book can serve a variety of discrete mathematics courses. The following outlines provide some ideas on how to structure a course based on this book.

- **Computer science/engineering focus:** Cover sections 1–16, 19–23, 28, 29–33, 34–36, 46–49, and 51. This plan covers the core material, special computer science notation, discrete probability, essential number theory, and graph theory.
- **Abstract algebra focus:** Cover sections 1–16, 19–27, and 34–45. This plan covers the core material, permutations and symmetry, number theory, group theory, and cryptography.
- **Discrete structures focus:** Cover sections 1–26, 46–56, and 58. This plan includes the core material, inclusion-exclusion, multisets, permutations, graph theory, and partially ordered sets.
- **Broad focus:** Cover sections 1–16, 19–23, 25–26, 34–38, 42–45, and 46–52. This plan covers the core material, permutations, number theory, cryptography, and graph theory. It most closely resembles the course I teach at Johns Hopkins.

Special Features

- **Proof templates:** Many students find proof writing difficult. When presented with a task such as proving two sets are equal, they have trouble structuring their proof and don't know what to write first. (See Proof Template 5 on page 51.) The proof templates appearing throughout this book give students the basic skeleton of the proof as well as boilerplate language. A list of the proof templates appears on the inside front cover.
- **Growing proofs:** Experienced mathematicians can write proofs sentence by sentence in proper order. They are able to do so because they can see the entire proof in their minds before they begin. Novice mathematicians (our students) often cannot see the whole proof before they begin. It is difficult for a student to learn how to write a proof simply by studying completed examples. I instruct students to begin their proofs by first writing the first sentence and next writing the last sentence. We then work the proof from both ends until we (ideally) meet in the middle.

  This approach is presented in the text through ever-expanding proofs in which the new sentences appear in color. See, for example, the proof of Proposition 11.11 (pages 69–73).
- **Mathspeak:** Mathematicians write well. We are concerned with expressing our ideas clearly and precisely. However, we change the meaning of some words (e.g., *injection* and *group*) to suit our needs. We invent new words (e.g., *poset* and *bijection*), and we change the part of speech of others (e.g., we use the noun *maximum* and the preposition *onto* as adjectives). I point out and explain many of the idiosyncrasies of mathematical English in marginal notes flagged with the term *Mathspeak*.

- **Hints:** Appendix A contains an extensive collection of hints (and some answers). It is often difficult to give hints that point a student in the correct direction without revealing the full answer. Some hints may give away too much, and others may be cryptic, but on balance, students will find this section enormously helpful. They should be instructed to consult this section only after mounting a hearty first attack on the problems.
- **Answers:** An *Instructor's Guide and Solutions* book is available from Brooks/Cole. Not only does this supplement give solutions to all the problems, it also gives helpful tips for teaching each of the sections.
- **Self tests:** Every chapter ends with a self test for students. Complete answers appear in Appendix B. These problems are of varying degrees of difficulty. Instructors may wish to specify which problems students should attempt in case not all sections of a chapter have been covered in class.

What's New in This Second Edition

In addition to correcting various errors (thank you to all those who wrote!), the following new features have been added:

- Self tests: These are described at the end of the previous section.
- A new example proof in Section 4: A number of instructors remarked that the first statements proved (sum of evens is even and transitivity of divisibility) are too simplistic. A new example has been added that is moderately more complicated.
- Section 12 is entirely new and gives a more thorough introduction to combinatorial proof via two nontrivial examples.
- Section 21 on induction has been expanded and made essentially independent of Section 20 on proof by smallest counterexample.
- Section 22 on recurrence relations is entirely new. We develop methods (with full supporting theory) to solve first- and second-order homogeneous constant coefficient recurrence relations. First-order recurrence relations are solved in both the homogeneous and nonhomogeneous cases, whereas the second-order equations are solved only in the homogeneous case (but the more general case is explored in an exercise).

  We also show how to find the formula for the nth term of a sequence of numbers if that sequence is generated by a polynomial function of n.
- Section 26 includes a new proof that two decompositions of a permutation into transpositions must have the same parity. The new proof avoids the tedious consideration of inversions in a permutation and is based on T. L. Bartlow, "An historical note on the parity of permutations," *American Mathematical Monthly* **79** (1972) 766–769 and S. Nelson, "Defining the sign of a permutation," *American Mathematical Monthly* **94** (1987) 543–545.
- There is a new opening section that describes the pleasure of doing mathematics.

Acknowledgments

This New Edition

I have many people to thank for their help in the preparation of this second edition.

My colleagues at Harvey Mudd College, Professor Arthur Benjamin and Andrew Bernoff, have used preliminary drafts of this second edition in their classrooms and have provided valuable feedback. A number of their students sent me comments and suggestions; many thanks to Jon Azose, Alan Davidson, Rachel Harris, Christopher Kain, John McCullough, and Hadley Watson.

For a number of years, my colleagues at Johns Hopkins University have been teaching our discrete mathematics course using this text. I especially want to thank Donniell Fishkind and Fred Torcaso for their helpful comments and encouragement.

It has been a pleasure working with Bob Pirtle, my editor at Brooks/Cole. I greatly value his support, encouragement, patience, and flexibility.

Brooks/Cole arranged for independent reviewers to provide feedback on this revision. Their comments were valuable and helped improve this new edition. Many thanks to Mike Daven (Mount Saint Mary College), Przemo Kranz (University of Mississippi), Jeff Johannes (The State University of New York Geneseo), and Michael Sullivan (San Diego State University).

The beautiful cover photograph is by my friend and former neighbor (and bridge partner) Albert Kocourek. This glorious image, entitled *New Wharf*, was taken in Maryland on the eastern shore of the Chesapeake Bay. Thank you, Al! More of Al's artwork can be seen on his website, `www.albertkocourek.com`. Thanks also to Jeanne Calabrese for the beautiful design of the cover.

The first edition had a number of errors. I greatly appreciate feedback from various students and instructors for bringing these mistakes to my attention. In particular, I thank Seema Aggarwal, Ben Babcock, Richard Belshoff, Kent Donnelly, Usit Duongsaa, Donniell Fishkind, George Huang, Sandi Klavzar, Peter Landweber, George Mackiw, Ryan Mansfield, Gary Morris, Evan O'Dea, Levi Ortiz, Russ Rutledge, Rachel Scheinerman, Karen Seyffarth, Douglas Shier, and Kimberly Tucker.

From the First Edition

These acknowledgments appeared in the first edition of this book; I still owe the individuals mentioned below a debt of gratitude.

During academic year 1998–99, students at Harvey Mudd College, Loyola College in Maryland, and the Johns Hopkins University used a preliminary version of this text. I am grateful to George Mackiw (Loyola) and Greg Levin (Harvey Mudd) for test-piloting this text. They provided me with many helpful comments, corrections, and suggestions.

I would especially like to thank the many students at these various institutions who had to endure typo-ridden first drafts. They offered many valuable suggestions that improved the text. In particular, I received helpful comments from all of the following:

Harvey Mudd: Jesse Abrams, Rob Adams, Gillian Allen, Matt Brubeck, Zeke Burgess, Nate Chessin, Jocelyn Chew, Brandon Duncan, Adam Fischer, Brad Forrest, Jon Erickson, Cecilia Giddings, Joshdan Griffin, David Herman, Doug Honma, Eric Huang, Keith Ito, Masashi Ito, Leslie Joe, Mike Lauzon, Colin Little, Dale Lovell, Steven Matthews, Laura Mecurio, Elizabeth Millan, Joel Miller, Greg Mulert, Bryce Nichols, Lizz Norton, Jordan Parker, Niccole Parker, Jane Pratt, Katie Ray, Star Roth, Mike Schubmehl, Roy Shea, Josh Smallman, Virginia Stoll, Alex Teoh, Jay Trautman, Richard Trinh, Kim Wallmark, Zach Walters, Titus Winters, Kevin Wong, Matthew Wong, Nigel Wright, Andrew Yamashita, Steve Yan, and Jason Yelinek.

Loyola: Richard Barley and Deborah Kunder.

Johns Hopkins: Adam Cannon, William Chang, Lara Diamond, Elias Fenton, Eric Hecht, Jacqueline Huang, Brian Iacoviello, Mark Schwager, David Tucker, Aaron Whittier, and Hani Yasmin.

Art Benjamin (Harvey Mudd College) contributed a collection of problems he uses when he teaches discrete mathematics; many of these problems appear in this text. Many years ago, Art was my teaching assistant when I first taught discrete mathematics. His help in developing that course undoubtedly has an echo in this book.

Thanks to Ran Libeskind-Hadas (also from Harvey Mudd) for contributing his collection of problems.

I had many enjoyable philosophical discussions with Mike Bridgland (Center for Computer Sciences) and Paul Tanenbaum (Army Research Laboratory). They kept me logically honest and gave excellent advice on how to structure my approach. Paul carefully read through an early draft of the book and made many helpful suggestions.

Thanks to Laura Tateosian, who drew the cartoon for the hint to Exercise 47.7.

Brooks/Cole arranged for an early version of this book to be reviewed by various mathematicians. I thank the following individuals for their helpful suggestions and comments: Douglas Burke (University of Nevada–Las Vegas), Joseph Gallian (University of Minnesota), John Gimbel (University of Alaska–Fairbanks), Henry Gould (West Virginia University), Arthur Hobbs (Texas A&M University), and George MacKiw (Loyola College in Maryland).

Lara Diamond painstakingly read through every sentence, uncovering numerous mathematical errors; I appreciate this tremendous support. Thank you, Lara.

I would like to believe that with so many people looking over my shoulder, all the errors have been found, but this is ridiculous. I am sure I have made many more errors than these people could find. This leaves some more for you, my reader, to find. Please tell me about them. (Send email to `ers@jhu.edu`.)

I am lucky to work with wonderful colleagues and graduate students in the Department of Applied Mathematics and Statistics at Johns Hopkins. In one way or another, they all have influenced me and my teaching and in this way contributed to this book. I thank them all and would like to add particular mention to these.

Bob Serfling was department chair when I first came to Hopkins; he empowered and trusted me to develop the discrete mathematics curriculum for the department. Over more than a decade, I have received tremendous support, encouragement, and advice from my current department chair, John Wierman. And Lenore Cowen not only contributed her enthusiasm, but also read over various portions of the text and made helpful suggestions.

Thanks also to Gary Ostedt, Carole Benedict, and their colleagues at Brooks/Cole. It was a pleasure working with them. Gary's enthusiasm for this project often exceeded my own. Carole was my main point of contact with Brooks/Cole and was always helpful, reliable, and cheerful.

Finally, thanks (and hugs and kisses) to my wife, Amy, and to our children, Rachel, Danny, Naomi, and Jonah, for their patience, support, and love throughout the writing of this book.

Edward R. Scheinerman

CHAPTER

1 Fundamentals

The cornerstones of mathematics are definition, theorem, and proof. *Definitions* specify precisely the concepts in which we are interested, *theorems* assert exactly what is true about these concepts, and *proofs* irrefutably demonstrate the truth of these assertions.

Before we get started, though, we ask a question: Why?

1 Joy

Why?

Please also read the *To the Student* preface, where we briefly address the questions: What is mathematics, and what is discrete mathematics? We also give important advice on how to read a mathematics book.

Before we roll up our sleeves and get to work in earnest, I want to share with you a few thoughts on the question: Why study mathematics?

Mathematics is incredibly useful. Mathematics is central to every facet of modern technology: the discovery of new drugs, the scheduling of airlines, the reliability of communication, the encoding of music and movies on CDs and DVDs, the efficiency of automobile engines, and on and on. Its reach extends far beyond the technical sciences. Mathematics is also central to all the social sciences, from understanding the fluctuations of the economy to the modeling of social networks in schools or businesses. Every branch of the fine arts—including literature, music, sculpture, painting, and theater—has also benefited from (or been inspired by) mathematics.

Because mathematics is both flexible (new mathematics is invented daily) and rigorous (we can incontrovertibly prove our assertions are correct), it is the finest analytic tool humans have developed.

The unparalleled success of mathematics as a tool for solving problems in science, engineering, society, and the arts is reason enough to engage actively this wonderful subject. We mathematicians are immensely proud of the accomplishments that are fueled by mathematical analysis. However, for many of us, this is not the primary motivation in studying mathematics.

The Agony and the Ecstasy

Why do mathematicians devote their lives to the study of mathematics? For most of us, it is because of the joy we experience when doing mathematics.

Mathematics is difficult for everyone. No matter what level of accomplishment or skill in this subject you (or your instructor) have attained, there is always a harder, more frustrating problem waiting around the bend. Demoralizing? Hardly! The greater the challenge, the greater the sense of accomplishment we experience when the challenge has been met. The best part of mathematics is the joy we experience in practicing this art.

Most art forms can be enjoyed by spectators. I can delight in a concert performed by talented musicians, be awestruck by a beautiful painting, or be deeply moved by literature. Mathematics, however, releases its emotional surge only for those who actually do the work.

I want you to feel the joy, too. So at the end of this brief section there is a single problem for you to tackle. In order for you to experience the joy, **do not under any circumstances let anyone help you solve this problem.** I hope that when you first look at the problem, you do not immediately see the solution but, rather, have to struggle with it for a while. Don't feel bad: I've shown this problem to extremely talented mathematicians who did not see the solution right away. Keep working and thinking—the solution will come. My hope is that when you solve this puzzle, it will bring a smile to your face. Here's the puzzle:

Conversely, if you have solved this problem, do not offer your assistance to others; you don't want to spoil their fun.

1 Exercise

1.1. Simplify the following algebraic expression:

$$(x-a)(x-b)(x-c)\cdots(x-z).$$

2 Definition

Mathematics exists only in people's minds. There is no such "thing" as the number 6. You can draw the symbol for the number 6 on a piece of paper, but you can't physically hold a 6 in your hands. Numbers, like all other mathematical objects, are purely conceptual.

Mathematical objects come into existence by definitions. For example, a number is called *prime* or *even* provided it satisfies precise, unambiguous conditions. These highly specific conditions are the definition for the concept. In this way, we are acting like legislators, laying down specific criteria such as eligibility for a government program. The difference is that laws may allow for some ambiguity, whereas a mathematical definition must be absolutely clear.

Let's take a look at an example.

Definition 2.1 **(Even)** An integer is called *even* provided it is divisible by two.

In a definition, the word(s) being defined are set in *italic* type.

Clear? Not entirely. The problem is that this definition contains terms that we have not yet defined, in particular *integer* and *divisible*. If we wish to be terribly fussy, we can complain that we haven't defined the term *two*. Each of these terms—*integer*, *divisible*, and *two*—can be defined in terms of simpler concepts, but this is a game we cannot entirely win. If every term is defined in terms of simpler terms, we will be chasing definitions forever. Eventually we must come to a point where we say, "This term is undefined, but we think we understand what it means."

The situation is like building a house. Each part of the house is built up from previous parts. Before roofing and siding, we must build the frame. Before the frame goes up, there must be a foundation. As house builders, we think of pouring the foundation as the first step, but this is not really the first step. We also have to own the land and run electricity and water to the property. For there to be water, there must be wells and pipes laid in the ground. STOP! We have descended to a level in the process that really has little to do with building a house. Yes, utilities are vital to home construction, but it is not our job, as home builders, to worry about what sorts of transformers are used at the electric substation!

Let us return to mathematics and Definition 2.1. It is possible for us to define the terms *integer*, *two*, and *divisible* in terms of more basic concepts. It takes a great deal of work to define integers, multiplication, and so forth in terms of simpler concepts. What are we to do? Ideally, we should begin from the most basic mathematical object of all—the *set*—and work our way up to the integers. Although this is a worthwhile activity, in this book we build our mathematical house assuming the foundation has already been formed.

The symbol $\mathbb{Z}$ stands for the integers. This symbol is easy to draw, but often people do a poor job. Why? They fall into the following trap: They first draw a Z and then try to add an extra slash. That doesn't work! Instead, make a 7 and then an interlocking, upside-down 7 to draw $\mathbb{Z}$.

Where shall we begin? What may we assume? In this book, we take the integers as our starting point. The *integers* are the positive whole numbers, the negative whole numbers, and zero. That is, the set of integers, denoted by the letter $\mathbb{Z}$, is

$$\mathbb{Z} = \{\ldots, -3, -2, -1, 0, 1, 2, 3, \ldots\}.$$

We also assume that we know how to add, subtract, and multiply, and we need not prove basic number facts such as $3 \times 2 = 6$. We assume the basic algebraic properties of addition, subtraction, and multiplication and basic facts about order relations ($<$, $\leq$, $>$, and $\geq$). See Appendix D for more details on what you may assume.

Thus, in Definition 2.1, we need not define *integer* or *two*. However, we still need to define what we mean by *divisible*. To underscore the fact that we have not made this clear yet, consider the question: Is 3 divisible by 2? We want to say that the answer to this question is no, but perhaps the answer is yes since $3 \div 2$ is $1\frac{1}{2}$. So it is possible to divide 3 by 2 if we allow fractions. Note further that in the previous paragraph we were granted basic properties of addition, subtraction, and multiplication, but not—and conspicuous by its absence—division. Thus we need a careful definition of *divisible*.

Definition 2.2 **(Divisible)** Let a and b be integers. We say that a is *divisible* by b provided there is an integer c such that $bc = a$. We also say b *divides* a, or b is a *factor* of a, or b is a *divisor* of a. The notation for this is $b|a$.

This definition introduces various terms (*divisible*, *factor*, *divisor*, and *divides*) as well as the notation $b|a$. Let's look at an example.

Example 2.3 Is 12 divisible by 4? To answer this question, we examine the definition. It says that $a = 12$ is divisible by $b = 4$ if we can find an integer c so that $4c = 12$. Of course, there is such an integer, namely, $c = 3$.

In this situation, we also say that 4 divides 12 or, equivalently, that 4 is a factor of 12. We also say 4 is a divisor of 12.

The notation to express this fact is $4|12$.

On the other hand, 12 is not divisible by 5 because there is no integer x for which $5x = 12$; thus $5|12$ is false.

Now Definition 2.1 is ready to use. The number 12 is even because $2|12$, and we know $2|12$ because $2 \times 6 = 12$. On the other hand, 13 is not even, because 13 is not divisible by 2; there is no integer x for which $2x = 13$. Note that we did not say that 13 is odd because we have yet to define the term *odd*. Of course, we know that 13 is an odd number, but we simply have not "created" odd numbers yet by specifying a definition for them. All we can say at this point is that 13 is not even. That being the case, let us define the term *odd*.

Definition 2.4 **(Odd)** An integer a is called *odd* provided there is an integer x such that $a = 2x + 1$.

Thus 13 is odd because we can choose $x = 6$ in the definition to give $13 = 2 \times 6 + 1$. Note that the definition gives a clear, unambiguous criterion for whether or not an integer is odd.

Please note carefully what the definition of *odd* does not say: It does not say that an integer is odd provided it is not even. This, of course, is true, and we prove it in a subsequent chapter. "Every integer is odd or even but not both" is a fact that we *prove*.

Here is a definition for another familiar concept.

Definition 2.5 **(Prime)** An integer p is called *prime* provided that $p > 1$ and the only positive divisors of p are 1 and p.

For example, 11 is prime because it satisfies both conditions in the definition: First, 11 is greater than 1, and second, the only positive divisors of 11 are 1 and 11.

Is 1 a prime? No. To see why, take $p = 1$ and see if p satisfies the definition of primality. There are two conditions: First we must have $p > 1$, and second, the only positive divisors of p are 1 and p. The second condition is satisfied: the only divisors of 1 are 1 and itself. However, $p = 1$ does not satisfy the first condition because $1 > 1$ is false. Therefore, 1 is not a prime.

We have answered the question: Is 1 a prime? The reason why 1 isn't prime is that the definition was specifically designed to make 1 nonprime! However, the real "why question" we would like to answer is: Why did we write Definition 2.5 to exclude 1?

I will attempt to answer this question in a moment, but there is an important philosophical point that needs to be underscored. The decision to exclude the number 1 in the definition was deliberate and conscious. In effect, the reason 1 is not prime is "because I said so!" In principle, you could define the word *prime* differently and allow the number 1 to be prime. The main problem with your using a different definition for prime is that the concept of a *prime number* is well established in the mathematical community. If it were useful to you to allow 1 as a prime in your work, you ought to choose a different term for your concept, such as *relaxed prime* or *alternative prime*.

Now, let us address the question: Why did we write Definition 2.5 to exclude 1? The idea is that the prime numbers should form the "building blocks" of multiplication. Later, we prove the fact that every positive integer can be factored in a unique fashion into prime numbers. For example, 12 can be factored as $12 = 2 \times 2 \times 3$. There is no other way to factor 12 down to primes (other than rearranging the order of the factors). The prime factors of 12 are precisely 2, 2, and 3. Were we to allow 1 as a prime number, then we could also factor 12 down to "primes" as $12 = 1 \times 2 \times 2 \times 3$, a different factorization.

Since we have defined prime numbers, it is appropriate to define composite numbers.

Definition 2.6 **(Composite)** A positive integer a is called *composite* provided there is an integer b such that $1 < b < a$ and $b|a$.

For example, the number 25 is composite because it satisfies the condition of the definition: There is a number b with $1 < b < 25$ and $b|25$; indeed, $b = 5$ is the only such number.

Similarly, the number 360 is composite. In this case, there are several numbers b that satisfy $1 < b < 360$ and $b|360$.

Prime numbers are not composite. If p is prime, then, by definition, there can be no divisor of p between 1 and p (read Definition 2.5 carefully).

Furthermore, the number 1 is not composite. (Clearly, there is no number b with $1 < b < 1$.) Poor number 1! It is neither prime nor composite! (There is, however, a special term that is applied to the number 1—the number 1 is called a *unit*.)

Recap

In this section, we introduced the concept of a mathematical definition. Definitions typically have the form "An object X is called *the term being defined* provided it satisfies *specific conditions*." We presented the integers $\mathbb{Z}$ and defined the terms *divisible*, *odd*, *even*, *prime*, and *composite*.

2 Exercises

2.1. Please determine which of the following are true and which are false; use Definition 2.2 to explain your answers.

a. $3|100$.

b. $3|99$.

c. $-3|3$.

d. $-5 \mid -5$.
e. $-2 \mid -7$.
f. $0 \mid 4$.
g. $4 \mid 0$.
h. $0 \mid 0$.

2.2. Here is a possible alternative to Definition 2.2: We say that a is *divisible* by b provided $\frac{a}{b}$ is an integer. Explain why this alternative definition is different from Definition 2.2.

Here, *different* means that Definition 2.2 and the alternative definition specify *different concepts*. So, to answer this question, you should find integers a and b such that a is divisible by b according to one definition, but a is not divisible by b according to the other definition.

2.3. None of the following numbers is prime. Explain why they fail to satisfy Definition 2.5. Which of these numbers is composite?
a. 21.
b. 0.
c. π.
d. $\frac{1}{2}$.
e. -2.
f. -1.

The symbol $\mathbb{N}$ stands for the natural numbers.

2.4. The *natural numbers* are the nonnegative integers; that is,

$$\mathbb{N} = \{0, 1, 2, 3, \ldots\}.$$

Use the concept of natural numbers to create definitions for the following relations about integers: *less than* ($<$), *less than or equal to* ($\leq$), *greater than* ($>$), and *greater than or equal to* ($\geq$).

Note: Many authors define the natural numbers to be just the positive integers; for them, zero is not a natural number. To me, this seems unnatural ☺. The concepts *positive integers* and *nonnegative integers* are unambiguous and universally recognized among mathematicians. The term *natural number*, however, is not 100% standardized.

The symbol $\mathbb{Q}$ stands for the rational numbers.

2.5. A *rational number* is a number formed by dividing two integers a/b where $b \neq 0$. The set of all rational numbers is denoted $\mathbb{Q}$.

Explain why every integer is a rational number, but not all rational numbers are integers.

2.6. Define what it means for an integer to be a *perfect square*. For example, the integers 0, 1, 4, 9, and 16 are perfect squares. Your definition should begin

An integer x is called a *perfect square* provided. . .

2.7. This problem involves basic geometry. Suppose the concept of distance between points in the plane is already defined. Write a careful definition for one point to be *between* two other points. Your definition should begin

Suppose A, B, C are points in the plane. We say that C is *between* A and B provided. . . .

Note: Since you are crafting this definition, you have a bit of flexibility. Consider the possibility that the point C might be the same as the point A or

B, or even that A and B might be the same point. Personally, if A and C were the same point, I would say that C is between A and B (regardless of where B may lie), but you may choose to design your definition to exclude this possibility. Whichever way you decide is fine, but be sure your definition does what you intend.

Note further: You do not need the concept of collinearity to define *between*. Once you have defined *between*, please use the notion of between to define what it means for three points to be collinear. Your definition should begin

> Suppose A, B, C are points in the plane. We say that they are collinear provided. . . .

Note even further: Now if, say, A and B are the same point, you certainly want your definition to imply that A, B, and C are collinear.

2.8. Discrete mathematicians especially enjoy *counting problems*: problems that ask *how many*. Here we consider the question: How many positive divisors does a number have? For example, 6 has four positive divisors: 1, 2, 3, and 6.

How many positive divisors does each of the following have?

a. 8.
b. 32.
c. 2^n where n is a positive integer.
d. 10.
e. 100.
f. 1,000,000.
g. 10^n where n is a positive integer.
h. $30 = 2 \times 3 \times 5$.
i. $42 = 2 \times 3 \times 7$. (Why do 30 and 42 have the same number of positive divisors?)
j. $2310 = 2 \times 3 \times 5 \times 7 \times 11$.
k. $1 \times 2 \times 3 \times 4 \times 5 \times 6 \times 7 \times 8$.
l. 0.

2.9. An integer n is called *perfect* provided it equals the sum of all its divisors that are both positive and less than n. For example, 28 is perfect because the positive divisors of 28 are 1, 2, 4, 7, 14, and 28. Note that $1+2+4+7+14 = 28$.

a. There is a perfect number smaller than 28. Find it.
b. Write a computer program to find the next perfect number after 28.

2.10. *At a Little League game there are three umpires. One is an engineer, one is a physicist, and one is a mathematician. There is a close play at home plate, but all three umpires agree the runner is out.*

Furious, the father of the runner screams at the umpires, "Why did you call her out?!"

The engineer replies, "She's out because I call them as they are."
The physicist replies, "She's out because I call them as I see them."
The mathematician replies, "She's out because I called her out."

Explain the mathematician's point of view.

3 Theorem

A *theorem* is a declarative statement about mathematics for which there is a proof.

The notion of proof is the subject of the next section—indeed, it is a central theme of this book. Suffice it to say for now that a *proof* is an essay that incontrovertibly shows that a statement is true.

In this section we focus on the notion of a theorem. Reiterating, a *theorem* is a declarative statement about mathematics for which there is a proof.

What is a declarative statement? In everyday English we utter many types of sentences. Some sentences are questions: Where is the newspaper? Other sentences are commands: Come to a complete stop. And perhaps the most common sort of sentence is a *declarative statement*—a sentence that expresses an idea about how something is, such as: It's going to rain tomorrow or The Yankees won last night.

Practitioners of every discipline make declarative statements about their subject matter. The economist says, "If the supply of a commodity decreases, then its price will increase." The physicist asserts, "When an object is dropped near the surface of the earth, it accelerates at a rate of 9.8 meter/sec^2."

Mathematicians also make statements that we believe are true about mathematics. Such statements fall into three categories:

- Statements we know to be true because we can prove them—we call these *theorems*.
- Statements whose truth we cannot ascertain—we call these *conjectures*.
- Statements that are false—we call these *mistakes*!

Please be sure to check your own work for nonsensical sentences. This type of mistake is all too common. Think about every word and symbol you write. Ask yourself, what does this term mean? Do the expressions on the left and right sides of your equations represent objects of the same type?

There is one more category of mathematical statements. Consider the sentence "The square root of a triangle is a circle." Since the operation of extracting a square root applies to numbers, not to geometric figures, the sentence doesn't make sense. We therefore call such statements *nonsense*!

The Nature of Truth

To say that a statement is *true* asserts that the statement is correct and can be trusted. However, the nature of truth is much stricter in mathematics than in any other discipline. For example, consider the following well-known meteorological fact: "In July, the weather in Baltimore is hot and humid." Let me assure you, from personal experience, that this statement is true! Does this mean that every day in every July is hot and humid? No, of course not. It is not reasonable to expect such a rigid interpretation of a general statement about the weather.

Consider the physicist's statement just presented: "When an object is dropped near the surface of the earth, it accelerates at a rate of 9.8 meter/sec^2." This statement is also true and is expressed with greater precision than our assertion about the climate in Baltimore. But this physics "law" is not absolutely correct. First, the value 9.8 is an approximation. Second, the term *near* is vague. From a galactic perspective, the moon is "near" the earth, but that is not the meaning of *near* that we intend. We can clarify *near* to mean "within 100 meters of the surface of the earth," but this leaves us with a problem. Even at an altitude of 100 meters, gravity is slightly less than at the surface. Worse yet, gravity at the surface is not constant; the gravitational pull at the top of Mount Everest is a bit smaller than the pull at sea level!

Despite these various objections and qualifications, the claim that objects dropped near the surface of the earth accelerate at a rate of 9.8 meter/sec^2 is true. As climatologists or physicists, we learn the limitations of our notion of truth. Most statements are limited in scope, and we learn that their truth is not meant to be considered absolute and universal.

However, in mathematics the word *true* is meant to be considered absolute, unconditional, and without exception.

Let us consider an example. Perhaps the most celebrated theorem in geometry is the following classical result of Pythagoras.

Theorem 3.1 **(Pythagorean)** If a and b are the lengths of the legs of a right triangle and c is the length of the hypotenuse, then

$$a^2 + b^2 = c^2.$$

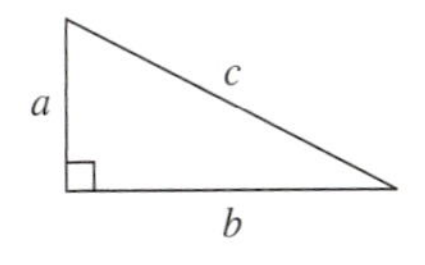

The relation $a^2 + b^2 = c^2$ holds for the legs and hypotenuse of every right triangle, absolutely and without exception! We know this because we can prove this theorem (more on proofs later).

Is the Pythagorean Theorem really absolutely true? We might wonder: If we draw a right triangle on a piece of paper and measure the lengths of the sides down to a billionth of an inch, would we have exactly $a^2 + b^2 = c^2$? Probably not, because a drawing of a right triangle is not a right triangle! A drawing is a helpful visual aid for understanding a mathematical concept, but a drawing is just ink on paper. A "real" right triangle exists only in our minds.

On the other hand, consider the statement, "Prime numbers are odd." Is this statement true? No. The number 2 is prime but not odd. Therefore, the statement is false. We might like to say it is nearly true since all prime numbers except 2 are odd. Indeed, there are far more exceptions to the rule "July days in Baltimore are hot and humid" (a sentence regarded to be true) than there are to "Prime numbers are odd."

Mathematicians have adopted the convention that a statement is called *true* provided it is absolutely true without exception. A statement that is not absolutely true in this strict way is called *false*.

An engineer, a physicist, and a mathematician are taking a train ride through Scotland. They happen to notice some black sheep on a hillside.

"Look," shouts the engineer. "Sheep in this part of Scotland are black!"

"Really," retorts the physicist. "You mustn't jump to conclusions. All we can say is that in this part of Scotland there are some black sheep."

"Well, at least on one side," mutters the mathematician.

If-Then

Mathematicians use the English language in a slightly different way than ordinary speakers. We give certain words special meanings that are different from that of standard usage. Mathematicians take standard English words and use them as

Consider the mathematical and the ordinary usage of the word *prime*. When an economist says that the prime interest rate is now 8%, we are not upset that 8 is not a prime number!

technical terms. We give words such as *set*, *group*, and *graph* new meanings. We also invent our own words, such as *bijection* and *poset*. (All these words are defined later in this book.)

Not only do mathematicians expropriate nouns and adjectives and give them new meanings, we also subtly change the meaning of common words, such as *or*, for our own purposes. Although we may be guilty of fracturing standard usage, we are highly consistent in how we do it. I call such altered usage of standard English *mathspeak*, and the most important example of mathspeak is the if-then construction.

In the statement "If A, then B," condition A is called the *hypothesis* and condition B is called the *conclusion*.

The vast majority of theorems can be expressed in the form "If A, then B." For example, the theorem "The sum of two even integers is even" can be rephrased "If x and y are even integers, then $x + y$ is also even."

In casual conversation, an if-then statement can have various interpretations. For example, I might say to my daughter, "If you mow the lawn, then I will pay you $10." If she does the work, she will expect to be paid. She certainly wouldn't object if I gave her $10 when she didn't mow the lawn, but she wouldn't expect it. Only one consequence is promised.

On the other hand, if I say to my son, "If you don't finish your lima beans, then you won't get dessert," he understands that unless he finishes his vegetables, no sweets will follow. But he also understands that if he does finish his lima beans, then he will get dessert. In this case two consequences are promised: one in the event he finishes his lima beans and one in the event he doesn't.

The mathematical use of if-then is akin to that of "If you mow the lawn, then I will pay you $10." The statement "If A, then B" means: Every time condition A is true, condition B must be true as well. Consider the sentence "If x and y are even, then $x + y$ is even." All this sentence promises is that when x and y are both even, it must also be the case that $x + y$ is even. (The sentence does not rule out the possibility of $x + y$ being even despite x or y not being even. Indeed, if x and y are both odd, we know that $x + y$ is also even.)

In the statement "If A, then B," we might have condition A true or false, and we might have condition B true or false. Let us summarize this in a chart. If the statement "If A, then B" is true, we have the following.

Condition A	**Condition B**	
True	True	Possible
True	False	Impossible
False	True	Possible
False	False	Possible

All that is promised is that whenever A is true, B must be true as well. If A is not true, then no claim about B is asserted by "If A, then B."

Here is an example. Imagine I am a politician running for office, and I announce in public, "If I am elected, then I will lower taxes." Under what circumstances would you call me a liar?

- Suppose I am elected and I lower taxes. Certainly you would not call me a liar—I kept my promise.

- Suppose I am elected and I do not lower taxes. Now you have every right to call me a liar—I have broken my promise.
- Suppose I am not elected, but somehow (say, through active lobbying) I manage to get taxes lowered. You certainly would not call me a liar—I have not broken my promise.
- Finally, suppose I am not elected and taxes are not lowered. Again, you would not accuse me of lying—I promised to lower taxes only if I were elected.

The only circumstance under which "If (A) I am elected, then (B) I will lower taxes" is a lie is when A is true and B is false.

In summary, the statement "If A, then B" promises that condition B is true whenever A is true but makes no claim about B when A is false.

Alternative wordings for "If A, then B."

If-then statements pervade all of mathematics. It would be tiresome to use the same phrases over and over in mathematical writing. Consequently, there is an assortment of alternative ways to express "If A, then B." All of the following express exactly the same statement as "If A, then B."

- "A implies B." This can also be expressed in passive voice: "B is implied by A."
- "Whenever A, we have B." Also: "B, whenever A."
- "A is sufficient for B." Also: "A is a sufficient condition for B."
  This is an example of mathspeak. The word *sufficient* can carry, in standard English, the connotation of being "just enough." No such connotation should be ascribed here. The meaning is "Once A is true, then B must be true as well."
- "In order for B to hold, it is enough that we have A."
- "B is necessary for A."
  This is another example of mathspeak. The way to understand this wording is as follows: In order for A to be true, it is *necessarily* the case that B is also true.
- "A, only if B."
  The meaning is that A can happen *only if* B happens as well.
- "$A \implies B$."
  The special arrow symbol $\implies$ is pronounced "implies."
- "$B \impliedby A$".
  The arrow $\impliedby$ is pronounced "is implied by."

If and Only If

The vast majority of theorems are—or can readily be expressed—in the if-then form. Some theorems go one step further; they are of the form "If A then B, and if B then A." For example, we know the following is true:

If an integer x is even, then $x + 1$ is odd, and if $x + 1$ is odd, then x is even.

This statement is verbose. There are concise ways to express statements of the form "A implies B and B implies A" in which we do not have to write out the conditions A and B twice each. The key phrase is *if and only if*. The statement

"If A then B, and if B then A" can be rewritten as "A if and only if B." The example just given is more comfortably written as follows:

An integer x is even if and only if $x + 1$ is odd.

What does an if-and-only-if statement mean? Consider the statement "A if and only if B." Conditions A and B may each be either true or false, so there are four possibilities that we can summarize in a chart. If the statement "A if and only if B" is true, we have

Condition A	**Condition B**	
True	True	Possible
True	False	Impossible
False	True	Impossible
False	False	Possible

It is impossible for condition A to be true while B is false, because $A \implies B$. Likewise, it is impossible for condition B to be true while A is false, because $B \implies A$. Thus the two conditions A and B must be both true or both false.

Let's revisit the example statement.

An integer x is even if and only if $x + 1$ is odd.

Condition A is "x is even" and condition B is "$x + 1$ is odd." For some integers (e.g., $x = 6$), conditions A and B are both true (6 is even and 7 is odd), but for other integers (e.g., $x = 9$), both conditions A and B are false (9 is not even and 10 is not odd).

Alternative wordings for "A if and only if B."

Just as there are many ways to express an if-then statement, so too are there several ways to express an if-and-only-if statement.

- "A iff B."
  Because the phrase "if and only if" occurs so frequently, the abbreviation "iff" is often used.
- "A is necessary and sufficient for B."
- "A is equivalent to B".
  The reason for the word *equivalent* is that condition A holds under exactly the same circumstances under which condition B holds.
- "$A \iff B$".
  The symbol $\iff$ is an amalgamation of the symbols $\impliedby$ and $\implies$.

And, Or, and Not

Mathematicians use the words *and*, *or*, and *not* in very precise ways. The mathematical usage of *and* and *not* is essentially the same as that of standard English. The usage of *or* is more idiosyncratic.

Mathematical use of *and*.

The statement "A and B" means that both statements A and B are true. For example, "Every integer whose ones digit is 0 is divisible by 2 *and* by 5." This means that a number that ends in a zero, such as 230, is divisible both by 2 and by 5. The use of *and* can be summarized in the following chart.

A	B	A **and** B
True	True	True
True	False	False
False	True	False
False	False	False

Mathematical use of *not*.

The statement "not A" is true if and only if A is false. For example, the statement "All primes are odd" is false. Thus the statement "Not all primes are odd" is true. Again, we can summarize the use of *not* in a chart.

A	**not** A
True	False
False	True

Mathematical use of *or*.

Thus the mathematical usage of *and* and *not* corresponds closely with standard English. The use of *or*, however, does not. In standard English, *or* often suggests a choice of one option or the other, but not both. Consider the question "Tonight, when we go out for dinner, would you like to have pizza or Chinese food?" The implication is that we'll dine on one or the other, but not both.

In contradistinction, the mathematical *or* allows the possibility of *both*. The statement "A or B" means that A is true, or B is true, or both A and B are true. For example, consider the following:

Suppose x and y are integers with the property that $x|y$ and $y|x$. Then $x = y$ or $x = -y$.

The conclusion of this result says that we may have any one of the following:

- $x = y$ but not $x = -y$ (e.g., take $x = 3$ and $y = 3$).
- $x = -y$ but not $x = y$ (e.g., take $x = -5$ and $y = 5$).
- $x = y$ *and* $x = -y$, which is possible only when $x = 0$ and $y = 0$.

Here is a chart for *or* statements.

A	B	A **or** B
True	True	True
True	False	True
False	True	True
False	False	False

What Theorems Are Called

The word *theorem* should not be confused with the word *theory*. A *theorem* is a specific statement that can be proved. A *theory* is a broader assembly of ideas on a particular issue.

Some theorems are more important or more interesting than others. There are alternative nouns that mathematicians use in place of *theorem*. Each has a slightly different connotation. The word *theorem* carries the connotation of importance and generality. The Pythagorean Theorem certainly deserves to be called a *theorem*. The statement "The square of an even integer is also even" is also a theorem, but perhaps it doesn't deserve such a profound name. And the statement "$6 + 3 = 9$" is technically a theorem but does not merit such a prestigious appellation.

Here we list words that are alternatives to *theorem* and offer a guide to their usage.

Result A modest, generic word for a theorem. There is an air of humility in calling your theorem merely a "result." Both important and unimportant theorems can be called results.

Fact A very minor theorem. The statement "$6 + 3 = 9$" is a fact.

Proposition A minor theorem. A proposition is more important or more general than a fact but not as prestigious as a theorem.

Lemma A theorem whose main purpose is to help prove another, more important theorem. Some theorems have complicated proofs. Often one can break the job of proving a complicated theorem down into smaller parts. The lemmas are the parts, or tools, used to build the more complicated proof.

Corollary A result with a short proof whose main step is the use of another, previously proved theorem.

Claim Similar to lemma. A claim is a theorem whose statement usually appears inside the proof of a theorem. The purpose of a claim is to help organize key steps in a proof. Also, the statement of a claim may involve terms that make sense only in the context of the proof.

Vacuous Truth

What are we to think of an if-then statement in which the hypothesis is impossible? Consider the following:

Statement 3.2 **(Vacuous)** If an integer is both a perfect square and prime, then it is negative.

Is this statement true or false?

The statement is not nonsense. The terms *perfect square* (see Exercise 2.6), *prime*, and *negative* properly apply to integers.

We might be tempted to say that the statement is false because square numbers and prime numbers cannot be negative. However, for a statement of the form "If A, then B" to be declared *false*, we need to find an instance in which clause A is true and clause B is false. In the case of Statement 3.2, condition A is impossible; there are no numbers that are both a perfect square and prime. So we can never find an integer that renders condition A true and condition B false. Therefore, Statement 3.2 is true!

Statements of the form "If A, then B" in which condition A is impossible are called *vacuous*, and mathematicians consider such statements true because they have no exceptions.

Recap

This section introduced the notion of a *theorem*: a declarative statement about mathematics that has a proof. We discussed the absolute nature of the word *true* in mathematics. We discussed extensively the if-then and if-and-only-if forms of theorems, as well as alternative language to express such results. We clarified the way in which mathematicians use the words *and*, *or*, and *not*. We presented a number of synonyms for *theorem* and explained their connotations. Finally, we discussed vacuous if-then statements and noted that mathematicians regard such statements as true.

3 Exercises

3.1. Each of the following statements can be recast in the if-then form. Please rewrite each of the following sentences in the form "If A, then B."

a. The product of an odd integer and an even integer is even.
b. The square of an odd integer is odd.
c. The square of a prime number is not prime.
d. The product of two negative integers is negative. (This, of course, is false.)

The statement "If B, then A" is called the *converse* of the statement "If A, then B."

3.2. It is a common mistake to confuse the following two statements:

a. If A, then B.
b. If B, then A.

Find two conditions A and B such that statement (a) is true but statement (b) is false.

3.3. Consider the two statements

a. If A, then B.
b. (not A) or B.

Under what circumstances are these statements true? When are they false? Explain why these statements are, in essence, identical.

The statement "If (not B), then (not A)" is called the *contrapositive* of the statement "If A, then B."

3.4. Consider the two statements

a. If A, then B.
b. If (not B), then (not A).

Under what circumstances are these statements true? When are they false? Explain why these statements are, in essence, identical.

3.5. Consider the two statements

a. A iff B.
b. (not A) iff (not B).

Under what circumstances are these statements true? Under what circumstances are they false? Explain why these statements are, in essence, identical.

3.6. Consider an equilateral triangle whose side lengths are $a = b = c = 1$. Notice that in this case $a^2 + b^2 \neq c^2$. Explain why this is not a violation of the Pythagorean Theorem.

A side of a spherical triangle is an arc of a great circle of the sphere on which it is drawn.

3.7. Explain how to draw a triangle on the surface of a sphere that has three right angles. Do the legs and hypotenuse of such a right triangle satisfy the condition $a^2 + b^2 = c^2$? Explain why this is not a violation of the Pythagorean Theorem.

3.8. Consider the sentence "A line is the shortest distance between two points." Strictly speaking, this sentence is nonsense.

Find two errors with this sentence and rewrite it properly.

3.9. Consider the following rather grotesque claim: "If you pick a guinea pig up by its tail, then its eyes will pop out." Is this true?

3.10. What are the two plurals of the word *lemma*?

4 Proof

We create mathematical concepts via definitions. We then posit assertions about mathematical notions, and then we try to prove our ideas are correct.

What is a *proof*?

In science, truth is borne out through experimentation. In law, truth is ascertained by a trial and decided by a judge and/or jury. In sports, the truth is the ruling of referees to the best of their ability. In mathematics, we have *proof*.

Truth in mathematics is not demonstrated through experimentation. This is not to say that experimentation is irrelevant for mathematics—quite the contrary! Trying out ideas and examples helps us to formulate statements we believe to be true (conjectures); we then try to prove these statements (thereby converting conjectures to theorems).

For example, recall the statement "All prime numbers are odd." If we start listing the prime numbers from 3, we find hundreds and thousands of prime numbers, and they are all odd! Does this mean all prime numbers are odd? Of course not! We simply missed the number 2.

Let us consider a far less obvious example.

Conjecture 4.1 **(Goldbach)** Every even integer greater than two is the sum of two primes.

Let's see that this statement is true for the first few even numbers. We have

$$\begin{array}{cccc} 4 = 2+2 & 6 = 3+3 & 8 = 3+5 & 10 = 3+7 \\ 12 = 5+7 & 14 = 7+7 & 16 = 11+5 & 18 = 11+7. \end{array}$$

One could write a computer program to verify that the first few billion even numbers (starting with 4) are each the sum of two primes. Does this imply Goldbach's Conjecture is true? No! The numerical evidence makes the conjecture believable, but it does not prove that it is true. To date, no proof has been found for Goldbach's Conjecture, so we simply do not know whether it is true or false.

Mathspeak!

A proof is often called an *argument*. In standard English, the word *argument* carries a connotation of disagreement or controversy. No such negative connotation should be associated with a mathematical argument. Indeed, mathematicians are honored when their proofs are called "beautiful arguments."

A proof is an essay that incontrovertibly shows that a statement is true. Mathematical proofs are highly structured and are written in a rather stylized manner. Certain key phrases and logical constructions appear frequently in proofs. In this and subsequent sections, we show how proofs are written.

The theorems we prove in this section are all rather simple. Indeed, you won't learn any facts about numbers you probably didn't already know quite well. The point in this section is not to learn new information about numbers; the point is to learn how to write proofs. So without further ado, let's start writing proofs!

We prove the following:

Proposition 4.2 The sum of two even integers is even.

I will write the proof here in full, and then we will discuss how this proof was created. In this proof, I have numbered each sentence so we can examine the proof piece by piece. Normally we would write this short proof out in a single paragraph and not number the sentences.

Proof (of Proposition 4.2)

1. We show that if x and y are even integers, then $x + y$ is an even integer.
2. Let x and y be even integers.
3. Since x is even, we know by Definition 2.1 that x is divisible by 2 (i.e., $2|x$).
4. Likewise, since y is even, $2|y$.
5. Since $2|x$, we know, by Definition 2.2, that there is an integer a such that $x = 2a$.
6. Likewise, since $2|y$, there is an integer b such that $y = 2b$.
7. Observe that $x + y = 2a + 2b = 2(a + b)$.
8. Therefore there is an integer c (namely, $a + b$) such that $x + y = 2c$.
9. Therefore (Definition 2.2) $2|(x + y)$.
10. Therefore (Definition 2.1) $x + y$ is even. ■

Let us examine exactly how this proof was written.

Convert the statement to if-then form.

- The first step is to convert the statement of the proposition into the if-then form.

  The statement reads, "The sum of two even integers is even."

  We convert the statement into if-then form as follows:

  "If x and y are even integers, then $x + y$ is an even integer."

  Note that we introduced letters (x and y) to name the two even integers. These letters come in handy in the proof.

  Observe that the first sentence of the proof spells out the proposition in if-then form.

  Sentence 1 announces the structure of this proof. The hypothesis (the "if" part) tells the reader that we will assume that x and y are even integers, and the conclusion (the "then" part) tells the reader that we are working to prove that $x + y$ is even.

  Sentence 1 can be regarded as a preamble to the proof. The proof starts in earnest at sentence 2.

Write the first and last sentences using the hypothesis and conclusion of the statement.

- The next step is to write the very beginning and the very *end* of the proof.

  The hypothesis of sentence 1 tells us what to write next. It says, "... if x and y are even integers...," so we simply write, "Let x and y be even integers." (Sentence 2)

  Immediately after we write the first sentence, we write the very last sentence of the proof. The last sentence of the proof is a rewrite of the conclusion of the if-then form of the statement.

  "Therefore, $x + y$ is even." (Sentence 10)

  The basic skeleton of the proof has been constructed. We know where we begin (x and y are even), and we know where we are heading ($x + y$ is even).

Unravel definitions.

- The next step is to unravel definitions. We do this at both ends of the proof.

  Sentence 2 tells us that x is even. What does this mean? To find out, we check (or we remember) the definition of the word *even*. (Take a quick look at Definition 2.1 on page 2.) It says that an integer is even provided it is divisible by 2. So we know that x is divisible by 2, and we can also write that as $2|x$; this gives sentence 3.

Sentence 4 does the same job as sentence 3. Since the reasoning in sentence 4 is identical to that of sentence 3, we use the word *likewise* to flag this parallel construction.

We now unravel the definition of *divisible*. We consult Definition 2.2 to learn that $2|x$ means there is an integer—we need to give that integer a name and we call it a—such that $x = 2a$. So sentence 5 just unravels sentence 3. Similarly (*likewise*!) sentence 6 unravels the fact that $2|y$ (sentence 4), and we know we have an integer b such that $y = 2b$.

At this point, we are stuck. We have unraveled all the definitions at the beginning of the proof, so now we return to the end of the proof and work backward!

We are still in the "unravel definitions" phase of writing this proof. The last sentence of the proof says, "Therefore $x + y$ is even." How do we prove an integer is even? We turn to the definition of *even*, and we see that we need to prove that $x + y$ is divisible by 2. So we know that the penultimate sentence (number 9) should say that $x + y$ is divisible by 2.

How do we get to sentence 9? To show that an integer (namely, $x + y$) is divisible by 2, we need to show there is an integer—let's call it c—such that $(x + y) = 2c$. This gives sentence 8.

Now we have unraveled definitions from both ends of the proof. Let's pause a moment to see what we have. The proof (written more tersely here) reads:

> We show that if x and y are even integers, then $x + y$ is an even integer.
>
> Let x and y be even integers. By definition of *even*, we know that $2|x$ and $2|y$. By definition of *divisibility*, we know there are integers a and b such that $x = 2a$ and $y = 2b$.
>
> ⋮
>
> Therefore there is an integer c such that $x + y = 2c$; hence $2|(x + y)$, and therefore $x + y$ is even.

What do we know? What do we need? Make the ends meet.

- The next step is to think. What do we know and what do we need?

  We know $x = 2a$ and $y = 2b$. We need an integer c such that $x + y = 2c$. So in this case, it is easy to see that we can take $c = a + b$ because the sum of two integers is an integer. We fill in the middle of the proof with sentence 7 and we are finished! To celebrate, and to mark the end of the proof, we append an end-of-proof symbol to the end of the proof: ■

This middle step—which was quite easy—is actually the hardest part of the proof. The translation of the statement of the proposition into if-then form and the unraveling of definitions are routine; once you have written several proofs, you will find these steps are easily produced. The hard part comes when you try to make ends meet!

The proof of Proposition 4.2 is the most basic type of proof; it is called a *direct* proof. The steps in writing a direct proof of an if-then statement are summarized in Proof Template 1.

Proof Template 1 Direct proof of an if-then theorem.

- Write the first sentence(s) of the proof by restating the hypothesis of the result. Invent suitable notation (e.g., assign letters to stand for variables).
- Write the last sentence(s) of the proof by restating the conclusion of the result.
- Unravel the definitions, working forward from the beginning of the proof and backward from the end of the proof.
- Figure out what you know and what you need. Try to forge a link between the two halves of your argument.

Let's use the direct proof technique to prove another result.

Proposition 4.3 Let a, b, and c be integers. If $a|b$ and $b|c$, then $a|c$.

The first step in creating the proof of this proposition is to write the first and last sentences based on the hypothesis and conclusion. This gives

> Suppose a, b, and c are integers with $a|b$ and $b|c$.
>
> ...
>
> Therefore $a|c$. ■

Next we unravel the definition of divisibility.

> Suppose a, b, and c are integers with $a|b$ and $b|c$. Since $a|b$, there is an integer x such that $b = ax$. Likewise there is an integer y such that $c = by$.
>
> ...
>
> Therefore there is an integer z such that $c = az$. Therefore $a|c$. ■

We have unraveled the definitions. Let's consider what we have and what we need.

We have a, b, c, x, and y such that: $b = ax$ and $c = by$.

We want to find z such that: $c = az$.

Now we have to think, but fortunately the problem is not hard. Since $b = ax$, we can substitute ax for b in $c = by$ and get $c = axy$. So the z we need is $z = xy$. We can use this to finish the proof of Proposition 4.3.

> Suppose a, b, and c are integers with $a|b$ and $b|c$. Since $a|b$, there is an integer x such that $b = ax$. Likewise there is an integer y such that $c = by$. Let $z = xy$. Then $az = a(xy) = (ax)y = by = c$.
>
> Therefore there is an integer z such that $c = az$. Therefore $a|c$. ■

A More Involved Proof

Propositions 4.2 and 4.3 are rather simple and not particularly interesting. Here we develop a more interesting proposition and its proof.

One of the most intriguing and most difficult issues in mathematics is the pattern of prime and composite numbers. Here is one pattern for you to consider. Pick a posititive integer, cube it, and then add one. Some examples:

$$3^3 + 1 = 27 + 1 = 28,$$
$$4^3 + 1 = 64 + 1 = 65,$$
$$5^3 + 1 = 125 + 1 = 126, \text{ and}$$
$$6^3 + 1 = 216 + 1 = 217.$$

Notice that the results are all composite. (Note that $217 = 7 \times 31$.) Try a few more examples on your own.

Let us try to convert this observation into a proposition for us to prove. Here's a first (but incorrect) draft: "If x is an integer, then $x^3 + 1$ is composite." This is a good start, but when we examine Definition 2.6, we note that the term *composite* applies only to positive integers. If x is negative, then $x^3 + 1$ is either negative or zero.

Fortunately, it's easy to repair the draft statement; here is a second version: "If x is a positive integer, then $x^3 + 1$ is composite." This looks better, but we're in trouble already when $x = 1$ because, in this case, $x^3 + 1 = 1^3 + 1 = 2$, which is prime. This makes us worry about the entire idea, but we note that when $x = 2$, $x^3 + 1 = 2^3 + 1 = 9$, which is composite, and we can try many other examples with $x > 1$ and always meet with success. The case $x = 1$ turns out to be the only exception, and this leads us to a third (and correct) version of the proposition we wish to prove.

Proposition 4.4 Let x be an integer. If $x > 1$, then $x^3 + 1$ is composite.

Let's write down the basic outline of the proof.

> Let x be an integer and suppose $x > 1$.
>
> ...
>
> Therefore $x^3 + 1$ is composite. ■

To reach the conclusion that $x^3 + 1$ is composite, we need to find a factor of $x^3 + 1$ that is strictly between 1 and $x^3 + 1$. With luck, the word *factor* makes us think about factoring the polynomial $x^3 + 1$ as a polynomial. Recall from basic algebra that

$$x^3 + 1 = (x + 1)(x^2 - x + 1).$$

This is the "Aha!" insight we need. Both $x + 1$ and $x^2 - x + 1$ are factors of $x^3 + 1$. For example, when $x = 6$, the factors $x + 1$ and $x^2 - x + 1$ evaluate to 7 and 31, respectively. Let's add this insight to our proof.

You might have the following concern: "I forgot that $x^3 + 1$ factors. How would I ever come up with this proof?" One idea is to look for patterns in the factors. We saw that $6^3 + 1 = 7 \times 31$, so $6^3 + 1$ is divisible by 7. Trying more examples, you may notice that $7^3 + 1$ is divisible by 8, $8^3 + 1$ is divisible by 9, $9^3 + 1$ is divisible by 10, and so on. With luck, that will help you realize that $x^3 + 1$ is divisible by $x + 1$, and then you can complete the factorization $x^3 + 1 = (x + 1) \times ?$.

> Let x be an integer and suppose $x > 1$. Note that $x^3+1 = (x+1)(x^2-x+1)$.
>
> ...
>
> Since $x + 1$ is a divisor of $x^3 + 1$, we have that $x^3 + 1$ is composite. ■

To correctly say that $x + 1$ is a divisor of $x^3 + 1$, we need the fact that both $x + 1$ and $x^2 - x + 1$ are integers. This is clear, because x itself is an integer. Let's be sure we include this detail in our proof.

> Let x be an integer and suppose $x > 1$. Note that $x^3+1 = (x+1)(x^2-x+1)$. Because x is an integer, both $x + 1$ and $x^2 - x + 1$ are integers. Therefore $(x + 1)|(x^3 + 1)$.
>
> ...
>
> Since $x + 1$ is a divisor of $x^3 + 1$, we have that $x^3 + 1$ is composite. ■

The proof isn't quite finished yet. Consult Definition 2.6; we need that the divisor be strictly between 1 and $x^3 + 1$, and we have not proved that yet. So let's figure out what we need to do. We must prove

$$1 < x + 1 < x^3 + 1.$$

The first part is easy. Since $x > 1$, adding 1 to both sides gives

$$x + 1 > 1 + 1 = 2 > 1.$$

Showing that $x+1 < x^3+1$ is only slightly more difficult. Working backward, to show $x + 1 < x^3 + 1$, it will be enough if we can prove that $x < x^3$. Notice that since $x > 1$, multiplying both sides by x gives $x^2 > x$, and since $x > 1$, we have $x^2 > 1$. Multiplying both sides of this by x gives $x^3 > x$.

Let's take these ideas and add them to the proof.

> Let x be an integer and suppose $x > 1$. Note that $x^3+1 = (x+1)(x^2-x+1)$. Because x is an integer, both $x + 1$ and $x^2 - x + 1$ are integers. Therefore $(x + 1)|(x^3 + 1)$.
>
> Since $x > 1$, we have $x + 1 > 1 + 1 = 2 > 1$.
>
> Also $x > 1$ implies $x^2 > x$, and since $x > 1$, we have $x^2 > 1$. Multiplying both sides by x again yields $x^3 > x$. Adding 1 to both sides gives $x^3 + 1 > x + 1$.
>
> Thus $x + 1$ is an integer with $1 < x + 1 < x^3 + 1$.
>
> Since $x + 1$ is a divisor of $x^3 + 1$ and $1 < x + 1 < x^3 + 1$, we have that $x^3 + 1$ is composite. ■

Proving If-and-Only-If Theorems

The basic technique for proving a statement of the form "A iff B" is to prove two if-then statements. We prove both "If A, then B" and "If B, then A." Here is an example:

Proposition 4.5 Let x be an integer. Then x is even if and only if $x + 1$ is odd.

The basic skeleton of the proof is as follows:

> Let x be an integer.
> ($\Rightarrow$) Suppose x is even. ... Therefore $x + 1$ is odd.
> ($\Leftarrow$) Suppose $x + 1$ is odd. ... Therefore x is even. ■

Notice that we flag the two sections of the proof with the symbols ($\Rightarrow$) and ($\Leftarrow$). This lets the reader know which section of the proof is which.

Now we unravel the definitions at the front of each part of the proof. (Recall the definition of *odd*; see Definition 2.4 on page 4.)

> Let x be an integer.
> ($\Rightarrow$) Suppose x is even. This means that $2|x$. Hence there is an integer a such that $x = 2a$. ... Therefore $x + 1$ is odd.
> ($\Leftarrow$) Suppose $x + 1$ is odd. So there is an integer b such that $x + 1 = 2b + 1$. ... Therefore x is even. ■

The next steps are clear. In the first part of the proof, we have $x = 2a$, and we want to prove $x + 1$ is odd. We just add 1 to both sides of $x = 2a$ to get $x + 1 = 2a + 1$, and that shows that $x + 1$ is odd.

In the second part of the proof, we know $x + 1 = 2b + 1$, and we want to prove that x is even. We subtract 1 from both sides and we are finished.

> Let x be an integer.
> ($\Rightarrow$) Suppose x is even. This means that $2|x$. Hence there is an integer a such that $x = 2a$. Adding 1 to both sides gives $x + 1 = 2a + 1$. By the definition of *odd*, $x + 1$ is odd.
> ($\Leftarrow$) Suppose $x + 1$ is odd. So there is an integer b such that $x + 1 = 2b + 1$. Subtracting 1 from both sides gives $x = 2b$. This shows that $2|x$ and therefore x is even. ■

Proof Template 2 shows the basic method for proving an if-and-only-if theorem.

Proof Template 2 Direct proof of an if-and-only-if theorem.

To prove a statement of the form "A iff B":

- ($\Rightarrow$) Prove "If A, then B."
- ($\Leftarrow$) Prove "If B, then A."

When is it safe to skip steps?

As you become more comfortable writing proofs, you may find yourself getting bored writing the same steps over and over again. We have seen the sequence (1) x is even, so (2) x is divisible by 2, so (3) there is an integer a such that $x = 2a$ several times already. You may be tempted to skip step (2) and just write "x is even, so there is an integer a such that $x = 2a$." The decision about skipping steps requires some careful judgment, but here are some guidelines.

- Would it be easy (and perhaps boring) for you to fill in the missing steps? Are the missing steps obvious? If you answer yes, then omit the steps.
- Does the same sequence of steps appear several times in your proof(s), but the sequence of steps is not very easy to reconstruct? Here you have two choices:
  - Write the sequence of steps out once, and the next time the same sequence appears, use an expression such as "as we saw before" or "likewise."
  - Alternatively, if the consequence of the sequence of steps can be described in a statement, first prove that statement, calling it a *lemma*. Then invoke (refer to) your lemma whenever you need to repeat those steps.
- When in doubt, write it out.

Let us illustrate the idea of explicitly separating off a portion of a proof into a lemma. Consider the following statement.

Proposition 4.6 Let a, b, c, and d be integers. If $a|b$, $b|c$, and $c|d$, then $a|d$.

Here is the proof as suggested by Proof Template 1.

> Let a, b, c, and d be integers with $a|b$, $b|c$, and $c|d$.
> Since $a|b$, there is an integer x such that $ax = b$.
> Since $b|c$, there is an integer y such that $by = c$.
> Since $c|d$, there is an integer z such that $cz = d$.
> Note that $a(xyz) = (ax)(yz) = b(yz) = (by)z = cz = d$.
> Therefore there is an integer $w = xyz$ such that $aw = d$.
> Therefore $a|d$. ■

There is nothing wrong with this proof, but there is a simpler, less verbose way to handle it. We have already shown that $a|b,\ b|c \Rightarrow a|c$ in Proposition 4.3. Let us use this proposition to prove Proposition 4.6.

Here is the alternative proof.

> Let a, b, c, and d be integers with $a|b$, $b|c$, and $c|d$.
> Since $a|b$ and $b|c$, by Proposition 4.3 we have $a|c$.
> Now, since $a|c$ and $c|d$, again by Proposition 4.3 we have $a|d$. ■

The key idea was to use Proposition 4.3 twice. Once it was applied to a, b, and c to get $a|c$. When we have $a|c$, we can use Proposition 4.3 again on the integers a, c, and d to finish the proof.

Proposition 4.3 serves as a lemma in the proof of Proposition 4.6.

Proving Equations and Inequalities

The basic algebraic manipulations you already know are valid steps in a proof. It is not necessary for you to prove that $x + x = 2x$ or that $x^2 - y^2 = (x-y)(x+y)$. In your proofs, feel free to use standard algebraic steps without detailed comment.

However, even these simple facts can be proved using the fundamental properties of numbers and operation (see Appendix D). We show how here, simply to illustrate that algebraic manipulations can be justified in terms of more basic principles.

For $x + x = 2x$:

$$\begin{aligned} x + x &= 1 \cdot x + 1 \cdot x && \text{1 is the identity element for multiplication} \\ &= (1+1)x && \text{distributive property} \\ &= 2x && \text{because } 1 + 1 = 2. \end{aligned}$$

For $(x-y)(x+y) = x^2 - y^2$:

$$\begin{aligned} (x-y)(x+y) &= x(x+y) - y(x+y) && \text{distributive property} \\ &= x^2 + xy - yx - y^2 && \text{distributive property} \\ &= x^2 + xy - xy - y^2 && \text{commutative property for multiplication} \\ &= x^2 + 1xy - 1xy - y^2 && \text{1 is the identity element for multiplication} \\ &= x^2 + (1-1)xy - y^2 && \text{distributive property} \\ &= x^2 + 0xy - y^2 && \text{because } 1 - 1 = 0 \\ &= x^2 + 0 - y^2 && \text{because anything multiplied by 0 is 0} \\ &= x^2 - y^2 && \text{0 is the identity element for addition.} \end{aligned}$$

Working with inequalities may be less familiar, but the basic steps are the same. For example, suppose you are asked to prove the following statement: If $x > 2$ then $x^2 > x + 1$. Here is a proof:

We need to comment that x is positive because multiplying both sides of an inequality by a negative number reverses the inequality.

Proof. We are given that $x > 2$. Since x is positive, multiplying both sides by x gives $x^2 > 2x$. So we have

$$\begin{aligned} x^2 &> 2x \\ &= x + x \\ &> x + 2 && \text{because } x > 2 \\ &> x + 1 && \text{because } 2 > 1. \end{aligned}$$

Therefore $x^2 > x + 1$. ■

Recap

We introduced the concept of proof and presented the basic technique of writing a direct proof for an if-then statement. For if-and-only-if statements, we apply this basic technique to both the forward ($\Rightarrow$) and the backward ($\Leftarrow$) implications.

4 Exercises

4.1. Prove that the sum of two odd integers is even.

4.2. Prove that the sum of an odd integer and an even integer is odd.

4.3. Prove that the product of two even integers is even.

4.4. Prove that the product of an even integer and an odd integer is even.

4.5. Prove that the product of two odd integers is odd.

4.6. Suppose a, b, and c are integers. Prove that if $a|b$ and $a|c$, then $a|(b+c)$.

4.7. Suppose a, b, and c are integers. Prove that if $a|b$, then $a|(bc)$.

4.8. Suppose a, b, d, x, and y are integers. Prove that if $d|a$ and $d|b$, then $d|(ax+by)$.

4.9. Suppose a, b, c, and d are integers. Prove that if $a|b$ and $c|d$, then $(ac)|(bd)$.

4.10. Let x be an integer. Prove that x is odd if and only if $x+1$ is even.

4.11. Let x be an integer. Prove that $0|x$ if and only if $x=0$.

4.12. Let a and b be integers. Prove that $a<b$ if and only if $a \le b-1$.

4.13. Prove that an integer is odd if and only if it is the sum of two consecutive integers.

4.14. Suppose you are asked to prove a statement of the form "If A or B, then C." Explain why you need to prove (a) "If A, then C" and also (b) "If B, then C." Why is it not enough to prove only one of (a) and (b)?

4.15. Suppose you are asked to prove a statement of the form "A iff B." The standard method is to prove both $A \Rightarrow B$ and $B \Rightarrow A$.

Consider the following alternative proof strategy: Prove both $A \Rightarrow B$ and (not A) $\Rightarrow$ (not B). Explain why this would give a valid proof.

5 Counterexample

In the previous section, we developed the notion of proof: a technique to demonstrate irrefutably that a given statement is true. Not all statements about mathematics are true! Given a statement, how do we show that it is false? Disproving false statements is usually simpler than proving theorems. The typical way to disprove an if-then statement is to create a *counterexample*. Consider the statement "If A, then B." A counterexample to such a statement would be an instance where A is true but B is false.

For example, consider the statement "If x is a prime, then x is odd." This statement is false. To prove that it is false, we just have to give an example of an integer that is prime but is not odd. The integer 2 has the requisite properties.

Let's consider another false statement.

Statement 5.1 **(false)** Let a and b be integers. If $a|b$ and $b|a$, then $a=b$.

This statement appears plausible. It seems that if $a|b$, then $a \leq b$, and if $b|a$, then $b \leq a$, and so $a = b$. But this reasoning is incorrect.

To disprove Statement 5.1, we need to find integers a and b that, on the one hand, satisfy $a|b$ and $b|a$ but, on the other hand, do not satisfy $a = b$.

Here is a counterexample: Take $a = 5$ and $b = -5$. To check that this is a counterexample, we simply note that, on the one hand, $5|-5$ and $-5|5$ but, on the other hand, $5 \neq -5$.

Proof Template 3 Refuting a false if-then statement via a counterexample.

To disprove a statement of the form "If A, then B":

Find an instance where A is true but B is false.

Refuting false statements is usually easier than proving true statements. However, finding counterexamples can be tricky. To create a counterexample, I recommend you try creating several instances where the hypothesis of the statement is true and check each to see whether or not the conclusion holds. All it takes is one counterexample to disprove a statement.

Unfortunately, it is easy to get stuck in a thinking rut. For Statement 5.1, we might consider $3|3$ and $4|4$ and $5|5$ and never think about making one number positive and the other negative.

Try to break out of such a rut by creating strange examples. Don't forget about the number 0 (which acts strangely) and negative numbers. Of course, following that advice, we might still be stuck in the rut $0|0$, $-1|-1$, $-2|-2$, and so on.

A strategy for finding counterexamples.

Here is a strategy for finding counterexamples. Begin by trying to prove the statement; when you get stuck, try to figure out what the problem is and look there to build a counterexample.

Let's apply this technique to Statement 5.1. We start, as usual, by converting the hypothesis and conclusion of the statement into the beginning and end of the proof.

> Let a and b be integers with $a|b$ and $b|a$. ... Therefore $a = b$. ■

Next we unravel definitions.

> Let a and b be integers with $a|b$ and $b|a$. Since $a|b$, there is an integer x such that $b = ax$. Since $b|a$, there is an integer y such that $a = by$. ... Therefore $a = b$. ■

Now we ask: What do we know? What do we need? We know

$$b = ax \qquad \text{and} \qquad a = by$$

and we want to show $a = b$. To get there, we can try to show that $x = y = 1$. Let's try to solve for x or y.

Since we have two expressions in terms of a and b, we can try substituting one in the other. We use the fact that $b = ax$ to eliminate b from $a = by$. We get

$$a = by \quad \Rightarrow \quad a = (ax)y \quad \Rightarrow \quad a = (xy)a.$$

It now looks quite tempting to divide both sides of the last equation by a, but we need to worry that perhaps, $a = 0$. Let's ignore the possibility of $a = 0$ for just a moment and go ahead and write $xy = 1$. We see that we have two integers whose product is 1. And we realize at this point that there are two ways that can happen: either $1 = 1 \times 1$ or $1 = -1 \times -1$. So although we know $xy = 1$, we can't conclude that $x = y = 1$ and finish the proof. We're stuck and now we consider the possibility that Statement 5.1 is false. We ask: What if $x = y = -1$? We see that this would imply that $a = -b$; for example, $a = 5$ and $b = -5$. And then we realize that in such a case, $a|b$ and $b|a$ but $a \neq b$. We have found a counterexample. Do we need to go back to our worry that perhaps $a = 0$? No! We have refuted the statement with our counterexample. The attempted proof served only to help us find a counterexample.

Recap

This section showed how to disprove an if-then statement by finding an example that satisfies the hypothesis of the statement but not the conclusion.

5 Exercises

5.1. Disprove: If a and b are integers with $a|b$, then $a \leq b$.

5.2. Disprove: If a and b are nonnegative integers with $a|b$, then $a \leq b$.

Note: A counterexample to this statement would also be a counterexample for the previous problem, but not necessarily vice versa.

5.3. Disprove: If a, b, and c are positive integers with $a|(bc)$, then $a|b$ or $a|c$.

5.4. Disprove: If a, b, and c are positive integers, then $a^{(b^c)} = (a^b)^c$.

5.5. Consider the polynomial $n^2 + n + 41$. Calculate the value of this polynomial for $n = 1, 2, 3, \ldots, 10$. Notice that all the numbers you computed are prime.

Disprove: If n is a positive integer, then $n^2 + n + 41$ is prime.

5.6. What does it mean for an if-and-only-if statement to be false? What properties should a counterexample for an if-and-only-if statement have?

5.7. Disprove: An integer x is positive if and only if $x + 1$ is positive.

5.8. Disprove: Two right triangles have the same area if and only if the lengths of their hypotenuses are the same.

5.9. Disprove: A positive integer is composite if and only if it has two different prime factors.

6 Boolean Algebra

Algebra is useful for reasoning about *numbers*. An algebraic relationship, such as $x^2 - y^2 = (x - y)(x + y)$, describes a general relationship that holds for any numbers x and y.

In a similar way, Boolean algebra provides a framework for dealing with *statements*. We begin with basic statements, such as "x is prime," and combine them using connectives such as *if-then*, *and*, *or*, *not*, and so on.

For example, in Section 3 you were asked (see Exercise 3.3) to explain why the statements "If A, then B" and "(not A) or B" mean essentially the same thing. In this section, we present a simple method for showing that such sentences have the same meaning.

In an ordinary algebraic expression, such as $3x - 4$, letters stand for numbers, and the operations are the familiar ones of addition, subtraction, multiplication, and so forth. The value of the expression $3x - 4$ depends on the number x. When $x = 1$, the value of the expression is -1, and if $x = 10$, the value is 26.

Variables stand for TRUE and FALSE.

Boolean algebra also has expressions containing letters and operations. Letters (variables) in a Boolean expression do not stand for numbers. Rather, they stand for the values TRUE and FALSE. Thus letters in a Boolean algebraic expression can only have two values!

The basic operations of Boolean algebra are $\wedge$, $\vee$, and $\neg$. These operations are also present in many computer languages. Since computer keyboards typically do not have these symbols, the symbols & (for $\wedge$), | (for $\vee$), and ~ (for $\neg$) are often used instead.

There are several operations we can perform on the values TRUE and FALSE. The most basic operations are called *and* (symbol: $\wedge$), *or* (symbol: $\vee$), and *not* (symbol: $\neg$).

We begin with $\wedge$. To define $\wedge$, we need to define the value of $x \wedge y$ for all possible values of x and y. Since there are only two possible values for each of x and y, this is not hard. Without further ado, here is the definition of the operation $\wedge$.

$$\text{TRUE} \wedge \text{TRUE} = \text{TRUE}$$
$$\text{TRUE} \wedge \text{FALSE} = \text{FALSE}$$
$$\text{FALSE} \wedge \text{TRUE} = \text{FALSE}$$
$$\text{FALSE} \wedge \text{FALSE} = \text{FALSE}.$$

In other words, the value of the expression $x \wedge y$ is TRUE when both x and y are TRUE and is FALSE otherwise. A convenient way to write all this is in a *truth table*, which is a chart showing the value of a Boolean expression depending on the values of the variables. Here is a truth table for the operation $\wedge$.

x	y	$x \wedge y$
TRUE	TRUE	TRUE
TRUE	FALSE	FALSE
FALSE	TRUE	FALSE
FALSE	FALSE	FALSE

The definition of the operation $\wedge$ is designed to mirror exactly the mathematical use of the English word *and*. Similarly, the Boolean operation $\vee$ is designed to mirror exactly the mathematical use of the English word *or*. Here is the definition of $\vee$.

$$\text{TRUE} \vee \text{TRUE} = \text{TRUE}$$
$$\text{TRUE} \vee \text{FALSE} = \text{TRUE}$$
$$\text{FALSE} \vee \text{TRUE} = \text{TRUE}$$
$$\text{FALSE} \vee \text{FALSE} = \text{FALSE}.$$

In other words, the value of the expression $x \vee y$ is TRUE in all cases except when both x and y are FALSE. We summarize this in a truth table.

x	y	$x \vee y$
TRUE	TRUE	TRUE
TRUE	FALSE	TRUE
FALSE	TRUE	TRUE
FALSE	FALSE	FALSE

The third operation, $\neg$, is designed to reproduce the mathematical use of the English word *not*:

$$\neg\text{TRUE} = \text{FALSE}$$
$$\neg\text{FALSE} = \text{TRUE}.$$

In truth table form, $\neg$ is as follows:

x	$\neg x$
TRUE	FALSE
FALSE	TRUE

Ordinary algebraic expressions (e.g., $3 \times 2 - 4$) may combine several operations. Likewise we can combine the Boolean operations. For example, consider

$$\text{TRUE} \wedge ((\neg\text{FALSE}) \vee \text{FALSE}).$$

Let us calculate the value of this expression step by step:

$$\begin{aligned} \text{TRUE} \wedge ((\neg\text{FALSE}) \vee \text{FALSE}) &= \text{TRUE} \wedge (\text{TRUE} \vee \text{FALSE}) \\ &= \text{TRUE} \wedge \text{TRUE} \\ &= \text{TRUE}. \end{aligned}$$

In algebra we learn how to manipulate formulas so we can derive identities such as

$$(x + y)^2 = x^2 + 2xy + y^2.$$

In Boolean algebra we are interested in deriving similar identities. Let us begin with a simple example:

$$x \wedge y = y \wedge x.$$

What does this mean? The ordinary algebraic identity $(x + y)^2 = x^2 + 2xy + y^2$ means that once we choose (numeric) values for x and y, the two expressions $(x + y)^2$ and $x^2 + 2xy + y^2$ must be equal. Similarly, the identity $x \wedge y = y \wedge x$ means that once we choose (truth) values for x and y, the results $x \wedge y$ and $y \wedge x$ must be the same.

Now it would be ridiculous to try to prove an identity such as $(x + y)^2 = x^2 + 2xy + y^2$ by trying to substitute all possible values for x and y—there are infinitely many possibilities! However, it is not hard to try all the possibilities to prove a Boolean algebraic identity. In the case of $x \wedge y = y \wedge x$, there are only four possibilities. Let us summarize these in a truth table.

x	y	$x \wedge y$	$y \wedge x$
TRUE	TRUE	TRUE	TRUE
TRUE	FALSE	FALSE	FALSE
FALSE	TRUE	FALSE	FALSE
FALSE	FALSE	FALSE	FALSE

By running through all possible combinations of values for x and y, we have a *proof* that $x \wedge y = y \wedge x$.

Logical equivalence.

When two Boolean expressions, such as $x \wedge y$ and $y \wedge x$, are equal for all possible values of their variables, we call these expressions *logically equivalent*. The simplest method to show that two Boolean expressions are logically equivalent is to run through all the possible values for the variables in the two expressions and to check that the results are the same in every case.

Let us consider a more interesting example.

Proposition 6.1 The Boolean expressions $\neg(x \wedge y)$ and $(\neg x) \vee (\neg y)$ are logically equivalent.

Proof. To show this is true, we construct a truth table for both expressions. To save space, we write T for TRUE and F for FALSE.

x	y	$x \wedge y$	$\neg(x \wedge y)$	$\neg x$	$\neg y$	$(\neg x) \vee (\neg y)$
T	T	T	F	F	F	F
T	F	F	T	F	T	T
F	T	F	T	T	F	T
F	F	F	T	T	T	T

The important thing to notice is that the columns for $\neg(x \wedge y)$ and $(\neg x) \vee (\neg y)$ are exactly the same. Therefore, no matter how we choose the values for x and y, the expressions $\neg(x \wedge y)$ and $(\neg x) \vee (\neg y)$ evaluate to the same truth value. Therefore the expressions $\neg(x \wedge y)$ and $(\neg x) \vee (\neg y)$ are logically equivalent. ■

Proof Template 4 Truth table proof of logical equivalence.

To show that two Boolean expressions are logically equivalent:

Construct a truth table showing the values of the two expressions for all possible values of the variables.

Check to see that the two Boolean expressions always have the same value.

Proofs by means of truth tables are easy but dull. The following result summarizes the basic algebraic properties of the operations $\wedge$, $\vee$, and $\neg$. In several cases, we give names for the properties.

Theorem 6.2

- $x \wedge y = y \wedge x$ and $x \vee y = y \vee x$. (Commutative properties)
- $(x \wedge y) \wedge z = x \wedge (y \wedge z)$ and $(x \vee y) \vee z = x \vee (y \vee z)$. (Associative properties)
- $x \wedge \text{TRUE} = x$ and $x \vee \text{FALSE} = x$. (Identity elements)
- $\neg(\neg x) = x$.
- $x \wedge x = x$ and $x \vee x = x$.
- $x \wedge (y \vee z) = (x \wedge y) \vee (x \wedge z)$ and $x \vee (y \wedge z) = (x \vee y) \wedge (x \vee z)$. (Distributive properties)
- $x \wedge (\neg x) = \text{FALSE}$ and $x \vee (\neg x) = \text{TRUE}$.
- $\neg(x \wedge y) = (\neg x) \vee (\neg y)$ and $\neg(x \vee y) = (\neg x) \wedge (\neg y)$. (DeMorgan's Laws)

All of these logical equivalences are easily proved via truth tables. In some of these identities, there is only one variable (e.g., $x \wedge \neg x = \text{FALSE}$); in this case, there would be only two rows in the truth table (one for $x = \text{TRUE}$ and one for $x = \text{FALSE}$). In the cases where there are three variables, there are eight rows in the truth table as (x, y, z) take on the possible values (T, T, T), (T, T, F), (T, F, T), (T, F, F), (F, T, T), (F, T, F), (F, F, T), and (F, F, F).

More Operations

The operations $\wedge$, $\vee$, and $\neg$ were created to replicate mathematicians' use of the words *and*, *or*, and *not*. We now introduce two more operations, $\rightarrow$ and $\leftrightarrow$, designed to model statements of the form "If A, then B" and "A if and only if B," respectively. The simplest way to define these is through truth tables.

x	y	$x \rightarrow y$
TRUE	TRUE	TRUE
TRUE	FALSE	FALSE
FALSE	TRUE	TRUE
FALSE	FALSE	TRUE

and

x	y	$x \leftrightarrow y$
TRUE	TRUE	TRUE
TRUE	FALSE	FALSE
FALSE	TRUE	FALSE
FALSE	FALSE	TRUE

The expression $x \rightarrow y$ models an if-then statement. We have $x \rightarrow y = \text{TRUE}$ except when $x = \text{TRUE}$ and $y = \text{FALSE}$. Likewise the statement "If A, then B" is true unless there is an instance in which A is true but B is false. Indeed, the arrow $\rightarrow$ reminds us of the implication arrow $\Rightarrow$.

Similarly, the expression $x \leftrightarrow y$ models the statement "A if and only if B." The expression $x \leftrightarrow y$ is true provided x and y are either both true or both false. Likewise the statement "$A \iff B$" is true provided that in every instance A and B are both true or both false.

Let us revisit the issue that the statements "If A, then B" and "(not A) or B" mean the same thing (see Exercise 3.3).

Proposition 6.3 The expressions $x \rightarrow y$ and $(\neg x) \vee y$ are logically equivalent.

Proof. We construct a truth table for both expressions.

x	y	$x \to y$	$\neg x$	y	$(\neg x) \vee y$
TRUE	TRUE	TRUE	FALSE	TRUE	TRUE
TRUE	FALSE	FALSE	FALSE	FALSE	FALSE
FALSE	TRUE	TRUE	TRUE	TRUE	TRUE
FALSE	FALSE	TRUE	TRUE	FALSE	TRUE

The columns for $x \to y$ and $(\neg x) \vee y$ are the same, and therefore these expressions are logically equivalent. ■

Proposition 6.3 shows how the operation $\to$ can be reexpressed just in terms of the basic operations $\vee$ and $\neg$. Similarly, the operation $\leftrightarrow$ can be expressed in terms of the basic operations $\wedge$, $\vee$, and $\neg$ (see Exercise 6.14).

Recap

This section presented Boolean algebra as "arithmetic" with the values TRUE and FALSE. The basic operations are $\wedge$, $\vee$, and $\neg$. Two Boolean expressions are logically equivalent provided they always yield the same values when we substitute for their variables. We can prove Boolean expressions are logically equivalent using truth tables. We concluded this section by defining the operations $\to$ and $\leftrightarrow$.

6 Exercises

6.1. Do the following calculations:

a. TRUE $\wedge$ TRUE $\wedge$ TRUE $\wedge$ TRUE $\wedge$ FALSE. F

b. ($\neg$TRUE) $\vee$ TRUE.

c. $\neg$(TRUE $\vee$ TRUE). F

d. (TRUE $\vee$ TRUE) $\wedge$ FALSE. F ?

e. TRUE $\vee$ (TRUE $\wedge$ FALSE).

The point of the last four is that the order in which you do the operations matters! Compare the expressions in (b)–(c) and (d)–(e) and note that they are the same except for the placement of the parentheses.

Now rethink your answer to (a). Does your answer to (a) depend on the order in which you do the operations?

6.2. Prove by use of truth tables as many parts of Theorem 6.2 as you can tolerate.

6.3. Prove: $(x \wedge y) \vee (x \wedge \neg y)$ is logically equivalent to x.

Exercise 4 shows that an if-then statement is logically equivalent to its contrapositive.

6.4. Prove: $x \to y$ is logically equivalent to $(\neg y) \to (\neg x)$.

6.5. Prove: $x \leftrightarrow y$ is logically equivalent to $(\neg x) \leftrightarrow (\neg y)$.

6.6. Prove: $x \leftrightarrow y$ is logically equivalent to $(x \to y) \wedge (y \to x)$.

6.7. Prove: $x \leftrightarrow y$ is logically equivalent to $(x \to y) \wedge ((\neg x) \to (\neg y))$.

6.8. Prove: $(x \vee y) \to z$ is logically equivalent to $(x \to z) \wedge (y \to z)$.

6.9. Suppose we have two Boolean expressions that involve ten variables. To prove that these two expressions are logically equivalent, we construct a truth table. How many rows (besides the "header" row) would this table have?

An if-then statement is not logically equivalent to its converse.

6.10. How would you disprove a logical equivalence? Show the following:

a. $x \to y$ is not logically equivalent to $y \to x$.

b. $x \to y$ is not logically equivalent to $x \leftrightarrow y$.

c. $x \vee y$ is not logically equivalent to $(x \wedge \neg y) \vee ((\neg x) \wedge y)$.

6.11. A *tautology* is a Boolean expression that evaluates to TRUE for all possible values of its variables. For example, the expression $x \vee \neg x$ is TRUE both when $x =$ TRUE and when $x =$ FALSE. Thus $x \vee \neg x$ is a tautology.

Explain how to use a truth table to prove that a Boolean expression is a tautology and prove that the following are tautologies.

a. $(x \vee y) \vee (x \vee \neg y)$.

b. $(x \wedge (x \to y)) \to y$.

c. $(\neg(\neg x)) \leftrightarrow x$.

d. $x \to x$.

e. $((x \to y) \wedge (y \to z)) \to (x \to z)$.

f. FALSE $\to x$.

6.12. A *contradiction* is a Boolean expression that evaluates to FALSE for all possible values of its variables. For example, $x \wedge \neg x$ is a contradiction.

Prove that the following are contradictions:

a. $(x \vee y) \wedge (x \vee \neg y) \wedge \neg x$.

b. $x \wedge (x \to y) \wedge (\neg y)$.

c. $(x \to y) \wedge ((\neg x) \to y) \wedge \neg y$.

6.13. Suppose A and B are Boolean expressions—that is, A and B are formulas involving variables (x, y, z, etc.) and Boolean operations ($\wedge$, $\vee$, $\neg$, etc.).

Prove: A is logically equivalent to B if and only if $A \leftrightarrow B$ is a tautology.

6.14. The expression $x \to y$ can be rewritten in terms of only the basic operations $\wedge$, $\vee$, and $\neg$; that is, $x \to y = (\neg x) \vee y$.

Find an expression that is logically equivalent to $x \leftrightarrow y$ and uses only the basic operations $\wedge$, $\vee$, and $\neg$ (and prove that you are correct).

The phrase *exclusive or* is sometimes written as *xor*.

6.15. Here is another Boolean operation called *exclusive or*; it is denoted by the symbol $\veebar$. It is defined in the following table.

x	y	$x \veebar y$
TRUE	TRUE	FALSE
TRUE	FALSE	TRUE
FALSE	TRUE	TRUE
FALSE	FALSE	FALSE

Please do the following:

a. Prove that $\veebar$ obeys the commutative and associative properties; that is, prove the logical equivalences $x \veebar y = y \veebar x$ and $(x \veebar y) \veebar z = x \veebar (y \veebar z)$.

b. Prove that $x \veebar y$ is logically equivalent to $(x \wedge \neg y) \vee ((\neg x) \wedge y)$. (Thus $\veebar$ can be expressed in terms of the basic operations $\wedge$, $\vee$, and $\neg$.)

c. Prove that $x \veebar y$ is logically equivalent to $(x \vee y) \wedge (\neg(x \wedge y))$. (This is another way that $\veebar$ can be expressed in terms of $\wedge$, $\vee$, and $\neg$.)

d. Explain why the operation $\veebar$ is called *exclusive or*.

A binary operation is an operation that combines two values. The operation $\neg$ is not binary because it works on just one value at a time; it is called *unary*.

6.16. We have discussed several binary Boolean operations: $\wedge$, $\vee$, $\rightarrow$, $\leftrightarrow$, and (in the previous problem) $\underline{\vee}$. How many different binary Boolean operations can there be? In other words, in how many different ways can we complete the following chart?

x	y	$x * y$
TRUE	TRUE	?
TRUE	FALSE	?
FALSE	TRUE	?
FALSE	FALSE	?

There aren't too many possibilities, and, in worst case, you can try writing out all of them. Be sure to organize your list carefully so you don't miss any or accidentally list the same operation twice.

6.17. We have seen that the operations $\rightarrow$, $\leftrightarrow$, and $\underline{\vee}$ can be reexpressed in terms of the basic operations $\wedge$, $\vee$, and $\neg$. Show that all binary Boolean operations (see the previous problem) can be expressed in terms of these basic three.

6.18. Prove that $x \vee y$ can be reexpressed in terms of just $\wedge$ and $\neg$ so all binary Boolean operations can be reduced to just two basic operations.

Nand.

6.19. Here is yet another Boolean operation called *nand*; it is denoted by the symbol $\overline{\wedge}$. We define $x \,\overline{\wedge}\, y$ to be $\neg(x \wedge y)$.
Please do the following:

a. Construct a truth table for $\overline{\wedge}$.

b. Is the operation $\overline{\wedge}$ commutative? Associative?

c. Show how the operations $x \wedge y$ and $\neg x$ can be reexpressed just in terms of $\overline{\wedge}$.

d. Conclude that all binary Boolean operations can be reexpressed just in terms of $\overline{\wedge}$ alone.

Chapter 1 Self Test

1. True or false: Every positive integer is either prime or composite. Explain your answer.

2. Find all integers x for which $x \mid (x+2)$. You do not need to prove your answer.

3. Let a and b be positive integers. Explain why the notation $a \mid b + 1$ can be interpreted only as $a \mid (b+1)$ and not as $(a \mid b) + 1$.

It is not known whether every perfect number is even, but it is conjectured that there are no odd perfect numbers.

4. Write the following statement in if-then form: "Every perfect integer is even."

5. What is the converse of the statement "If you love me, then you will marry me."

6. Determine which of the following statements are true and which are false. You should base your reply on your common knowledge of mathematics; you do not need to prove your answers.

a. Every integer is positive or negative.

b. Every integer is even and odd.

c. If x is an integer and $x > 2$ and x is prime, then x is odd.

d. Let x and y be integers. We have $x^2 = y^2$ if and only if $x = y$.

e. The sides of a triangle are all congruent to each other if and only if its three angles are all 60°.

f. If an integer x satisfies $x = x + 1$, then $x = 6$.

7. Consider the following statement (which you are not expected to understand): "If a matroid is graphic, then it is representable."

Write the first and last lines of a direct proof of this statement. It is customary to use the letter M to stand for a matroid.

8. The following statement is false: If x, y, and z are integers and $x > y$, then $xz > yz$. Please do the following:

a. Find a counterexample.

b. Modify the hypothesis of the statement by adding a condition concerning z so that the edited statement is true.

9. Prove or disprove the following statements:

a. Let a, b, c be integers. If $a|c$ and $b|c$, then $(a + b)|c$.

b. Let a, b, c be integers. If $a|b$, then $(ac)|(bc)$.

10. Consider the following proposition. Let N be a two-digit number and let M be the number formed from N by reversing N's digits. Now compare N^2 and M^2. The digits of M^2 are precisely those of N^2, but reversed. For example:

$$\begin{array}{ll} 10^2 = 100 & 01^2 = 001 \\ 11^2 = 121 & 11^2 = 121 \\ 12^2 = 144 & 21^2 = 441 \\ 13^2 = 169 & 31^2 = 961 \end{array}$$

and so on.

Here is a proof of the proposition:

Proof. Since N is a two-digit number, we can write $N = 10a + b$ where a and b are the digits of N. Since M is formed from N by reversing digits, $M = 10b + a$.

Note that $N^2 = (10a + b)^2 = 100a^2 + 20ab + b^2 = (a^2) \times 100 + (2ab) \times 10 + (b^2) \times 1$, so the digits of N^2 are, in order, $a^2, 2ab, b^2$.

Likewise, $M^2 = (10b + a)^2 = (b^2) \times 100 + (2ab) \times 10 + (a^2) \times 1$, so the digits of M^2 are, in order, $b^2, 2ab, a^2$, exactly the reverse of N^2. ■

Your job: Show that the proposition is false and explain why the proof is invalid.

11. Suppose we are asked to prove the following identity:

$$x(x + y - 1) - y(x + 1) = x(x - 1) - y.$$

The identity is true (i.e., the equation is valid for all real numbers x and y).

The following "proof" is incorrect. Explain why.

Proof. We begin with

$$x(x + y - 1) - y(x + 1) = x(x - 1) - y$$

and expand the terms (using the distributive property)

$$x^2 + xy - x - yx - y = x^2 - x - y.$$

We cancel the terms x^2, $-x$, and $-y$ from both sides to give

$$xy - yx = 0,$$

and finally xy and $-yx$ cancel to give

$$0 = 0,$$

which is correct. ■

12. Are the Boolean expressions $x \to \neg y$ and $\neg(x \to y)$ logically equivalent? Justify your answer.

13. Is the Boolean expression $(x \to y) \vee (x \to \neg y)$ a tautology? Justify your answer.

14. Prove that the sum of any three consecutive integers is divisible by three.

15. In the previous problem you were asked to prove that the sum of any three consecutive integers is divisible by three. Note, however, that the sum of any four consecutive integers is never divisible by four. For example, $10 + 11 + 12 + 13 = 46$, which is not divisible by four.

For which positive integers a is the sum of a consecutive integers divisible by a? That is, complete the following sentence to give a true statement:

Let a be a positive integer. The sum of a consecutive integers is divisible by a if and only if . . .

You need not prove your conjecture.

16. Let a be an integer. Prove: If $a \geq 3$, then $a^2 > 2a + 1$.

17. Suppose a is a perfect square and $a \geq 9$. Prove that $a - 1$ is composite.

See Exercise 2.6 and its solution on page 487 for the definition of *perfect square*.

18. Consider the following definition:

A pair of positive integers, x and y, are called *square mates* if their sum, $x + y$, is a perfect square. (The concept of square mates was contrived just for this test, problems 18 to 20.)

For example, 4 and 5 are square mates because $4 + 5 = 9 = 3^2$. Likewise, 8 and 8 are square mates because $8 + 8 = 16 = 4^2$. However, 3 and 8 are not square mates.

Explain why 10 and -1 are not square mates.

19. Let x be a positive integer. Prove that there is an integer y that is greater than x such that x and y are square mates.

20. Prove that if x is an integer and $x \geq 5$, then x has a square mate y with $y < x$.

You may use the following fact in your proof. If x is a positive integer, then x lies between two consecutive perfect squares; that is, there is a positive integer a such that $a^2 \leq x < (a + 1)^2$.

CHAPTER

2 Collections

This chapter deals with collections. We consider two types of collections: ordered collections (lists) and unordered collections (sets).

7 Lists

A *list* is an ordered sequence of objects. We write lists by starting with an open parenthesis, following with the elements of the list separated by commas, and finishing with a close parenthesis. For example, $(1, 2, \mathbb{Z})$ is a list whose first element is the number 1, whose second element is the number 2, and whose third element is the set of integers.

The order in which elements appear in a list is significant. The list $(1, 2, 3)$ is not the same as the list $(3, 2, 1)$.

Elements in a list might be repeated, as in $(3, 3, 2)$.

The number of elements in a list is called its *length*. For example, the list $(1, 1, 2, 1)$ is a list of length four.

A list of length two has a special name; it is called an *ordered pair*.

A list of length zero is called the *empty list* and is denoted $()$.

What it means for two lists to be equal.

Two lists are *equal* provided they have the same length, and elements in the corresponding positions on the two lists are equal. Lists (a, b, c) and (x, y, z) are equal iff $a = x$, $b = y$, and $c = z$.

Mathspeak!

Another word mathematicians use for lists is *tuple*. A list of n elements is known as an n-tuple.

Lists are all-pervasive in mathematics and beyond. A point in the plane is often specified by an ordered pair of real numbers (x, y). A natural number, when written in standard notation, is a list of digits; you can think of the number 172 as the list $(1, 7, 2)$. An English word is a list of letters. An identifier in a computer program is a list of letters and digits (where the first element of the list is a letter).

Counting Two-element Lists

In this section, we address questions of the form "How many lists can we make?"

Example 7.1 Suppose we wish to make a two-element list where the entries in the list may be any of the digits 1, 2, 3, and 4. How many such lists are possible?

The most direct approach to answering this question is to write out all the possibilities.

$$\begin{array}{cccc} (1,1) & (1,2) & (1,3) & (1,4) \\ (2,1) & (2,2) & (2,3) & (2,4) \\ (3,1) & (3,2) & (3,3) & (3,4) \\ (4,1) & (4,2) & (4,3) & (4,4) \end{array}$$

There are 16 such lists.

We organized the lists in a manner that ensures we have neither repeated nor omitted a list. The first row of the chart contains all the possible lists that begin with 1, the second row those that begin with 2, and so on. Thus there are $4 \times 4 = 16$ length-two lists whose elements are any one of the digits 1 through 4.

Let's generalize this example a little bit. Suppose we wish to know the number of two-element lists where there are n possible choices for each entry in the list. We may assume the possible elements are the integers 1 through n. As before, we organize all the possible lists into a chart.

Mathspeak!

The mathematical use of the word *choice* is strange. If a restaurant has a menu with only one entrée, the mathematician would say that this menu offers one choice. The rest of the world would probably say that the menu offers no choices! The mathematical use of the word *choice* is similar to *option*.

$$\begin{array}{cccc} (1,1) & (1,2) & \cdots & (1,n) \\ (2,1) & (2,2) & \cdots & (2,n) \\ \vdots & \vdots & \ddots & \vdots \\ (n,1) & (n,2) & \cdots & (n,n) \end{array}$$

The first row contains all the lists that begin with 1, the second those that begin with 2, and so forth. There are n rows in all. Each row has exactly n lists. Therefore there are $n \times n = n^2$ possible lists.

When a list is formed, the options for the second position may be different from the options for the first position. Imagine that a meal is a two-element list consisting of an entrée followed by a dessert. The number of possible entrées might be different from the number of possible desserts.

Therefore let us ask: How many two-element lists are possible in which there are n choices for the first element and m choices for the second element? Suppose that the possible entries in the first position of the list are the integers 1 through n, and the possible entries in the second position are 1 through m.

We construct a chart of all the possibilities as before.

$$\begin{array}{cccc} (1,1) & (1,2) & \cdots & (1,m) \\ (2,1) & (2,2) & \cdots & (2,m) \\ \vdots & \vdots & \ddots & \vdots \\ (n,1) & (n,2) & \cdots & (n,m) \end{array}$$

There are n rows (for each possible first choice), and each row contains m entries. Thus the number of possible such lists is

$$\underbrace{m + m + \cdots + m}_{n \text{ times}} = m \times n.$$

Sometimes the elements of a list satisfy special properties. In particular, the choice of the second element might depend on what the first element is. For example, suppose we wish to count the number of two-element lists we can form from the integers 1 through 5, in which the two numbers on the list must be different. For example, we want to count (3, 2) and (2, 5) but not (4, 4). We make a chart of the possible lists.

$$\begin{array}{ccccc} - & (1,2) & (1,3) & (1,4) & (1,5) \\ (2,1) & - & (2,3) & (2,4) & (2,5) \\ (3,1) & (3,2) & - & (3,4) & (3,5) \\ (4,1) & (4,2) & (4,3) & - & (4,5) \\ (5,1) & (5,2) & (5,3) & (5,4) & - \end{array}$$

As before, the first row contains all the possible lists that begin with 1, the second row those lists that start with 2, and so on, so there are 5 rows. Notice that each row contains exactly $5 - 1 = 4$ lists, so the number of lists is $5 \times 4 = 20$.

Let us summarize and generalize what we have learned in a general principle.

Theorem 7.2 **(Multiplication Principle)** Consider two-element lists for which there are n choices for the first element, and for each choice of the first element there are m choices for the second element. Then the number of such lists is nm.

Proof. Construct a chart of all the possible lists. Each row of this chart contains all the two-element lists that begin with a particular element. Since there are n choices for the first element, there are n rows in the chart. Since, for each choice of the first element, there are m choices for the second element, we know that every row of the chart has m entries. Therefore the number of lists is

$$\underbrace{m + m + \cdots + m}_{n \text{ times}} = n \times m. \qquad \blacksquare$$

Let us consider some examples.

Example 7.3 A person's initials are the two-element list consisting of the initial letters of their first and last names. For example, the author's initials are ES. In how many ways can we form a person's initials? In how many ways can we form initials where the two letters are different?

The first question asks for the number of two-element lists where there are 26 choices for each element. There are 26^2 such lists.

The second question asks for the number of two-element lists where there are 26 choices for the first element and, for each choice of first element, 25 choices for the second element. Thus there are 26×25 such lists.

Another way to answer the second question in Example 7.3 is as follows: There are 26^2 ways to form initials (repetitions allowed). Of these, there are 26 "bad" sets of initials in which there is a repetition, namely, AA, BB, CC, ..., ZZ. The remaining lists are the ones we want to count, so there are $26^2 - 26$ possibilities. Since $26 \times 25 = 26 \times (26 - 1) = 26^2 - 26$, the two answers agree.

Please note that we reported the answers to these questions as 26^2 and 26×25, and not as 676 and 650. Although the latter pair of answers are correct, the answers 26^2 and 26×25 are preferred because they retain the essence of the reasoning used to derive them. Furthermore, the conversion of 26^2 and 26×25 to 676 and 650, respectively, is not interesting and can be done easily by anyone with a calculator.

Example 7.4 A club has ten members. The members wish to elect a president and to elect someone else as a vice president. In how many ways can these posts be filled?

We recast this question as a list-counting problem. How many two-element lists of people can be formed in which the two people in the list are selected from a collection of ten candidates and the same person may not be selected twice?

There are ten choices for the first element of the list. For each choice of the first element (for each president), there are nine possible choices for the second element of the list (vice president). By the Multiplication Principle, there are 10×9 possibilities.

Longer Lists

Let us explore how to use the Multiplication Principle to count longer lists.

Consider the following problem. How many lists of three elements can we make using the numbers 1, 2, 3, 4, and 5? Let us write out all the possibilities. Here is a way we might organize our work:

(1,1,1) (1,1,2) (1,1,3) (1,1,4) (1,1,5)
(1,2,1) (1,2,2) (1,2,3) (1,2,4) (1,2,5)
(1,3,1) (1,3,2) (1,3,3) (1,3,4) (1,3,5)
(1,4,1) (1,4,2) (1,4,3) (1,4,4) (1,4,5)
(1,5,1) (1,5,2) (1,5,3) (1,5,4) (1,5,5)
(2,1,1) (2,1,2) (2,1,3) (2,1,4) (2,1,5)
(2,2,1) (2,2,2) (2,2,3) (2,2,4) (2,2,5)
and so forth until
(5,5,1) (5,5,2) (5,5,3) (5,5,4) (5,5,5)

The first line of this chart contains all lists that begin $(1, 1, \ldots)$. The second line is all lists that begin $(1, 2, \ldots)$ and so forth. Clearly, each line has five lists. The question becomes:

How many lines are there in this chart?

This is a problem we have already solved! Notice that each line of the chart begins, effectively, with a different two-element list; the number of two-element lists where each element is one of five possible values is 5×5, so this chart has 5×5 lines. Therefore, since each line of the chart has five entries, the number of three-element lists is $(5 \times 5) \times 5 = 5^3$.

Suppose A and B are lists. Their *concatenation* is the new list formed by listing first the elements of A and then the elements of B. The concatenation of the lists $(1, 2, 1)$ and $(1, 3, 5)$ is the list $(1, 2, 1, 1, 3, 5)$.

We can think of a three-element list as the concatenation of a two-element list and a one-element list. In this problem, there are 25 possible two-element lists to occupy the front of the three-element list, and for each choice of the front part, there are five choices for the back.

Next, let us count three-element lists whose elements are the integers 1 through 5 in which no number is repeated. As before, we make a chart.

$$
\begin{array}{ccc}
(1,2,3) & (1,2,4) & (1,2,5) \\
(1,3,2) & (1,3,4) & (1,3,5) \\
(1,4,2) & (1,4,3) & (1,4,5) \\
(1,5,2) & (1,5,3) & (1,5,4) \\
(2,1,3) & (2,1,4) & (2,1,5) \\
 & \textit{and so forth until} & \\
(5,4,1) & (5,4,2) & (5,4,3)
\end{array}
$$

The first line of the chart contains all the lists that begin $(1, 2, \ldots)$. (There can be no lines that begin $(1, 1, \ldots)$ because repetitions are disallowed.) The second line contains all lists that begin $(1, 3, \ldots)$, and so on. Each line of the chart contains just three lists; once we have chosen the first and second elements of the list (from a world of only five choices), there are exactly three ways to finish the list. So, as before, the question becomes: How many lines are on this chart? And as before, this is a problem we have already solved!

The first two elements of the list form, unto themselves, a two-element list with each element chosen from a list of five possible objects and without repetition. So, by the Multiplication Principle, there are 5×4 lines on the chart. Since each line has three elements, there is a total of $5 \times 4 \times 3$ possible lists in all.

These three-element lists are a concatenation of a two-element list (20 choices), and, for each two-element list, a one-element list (3 choices), giving a total of 20×3 lists.

We extend the Multiplication Principle to count longer lists. Consider length-three lists. Suppose we have a choices for the first element of the list, and for each choice of first element, there are b choices for the second element, and for each choice of first and second elements, there are c choices for the third element. Thus, in all, there are abc possible lists. To see why, imagine that the three-element list consists of two parts: the initial two elements and the final element. There are ab ways to fill in the first two elements (by the Multiplication Principle!) and there are c ways to complete the last element once the first two are specified. So, by the Multiplication Principle again, there are $(ab)c$ ways to make the lists. The extension of these ideas to lists of length-four or more is analogous.

A useful way to think about list-counting problems is to make a diagram with boxes. Each box stands for a position in the list, so if the length of the list is four, there will be four boxes on the list. We write the number of possible entries in each box. The number of possible lists is computed by multiplying these numbers together.

Example 7.5 Let us revisit Example 7.4. We have a club with ten members. We want to elect an executive board consisting of a president, a vice president, a secretary, and a treasurer. In how many ways can we do this (assuming no member of the club can fill two offices)? We draw the following diagram.

Pres.	V.P.	Sec.	Treas.
10	9	8	7

This shows there are ten choices for president. Once the president is selected, there are are nine choices for vice president, so there are 10×9 ways to fill in the first two elements of the list. Once these are filled, there are eight ways to fill in the next element of the list (secretary), so there are $(10 \times 9) \times 8$ ways to complete the first three slots. Finally, once the first three offices are filled, there are seven ways to select a treasurer, so there are $(10 \times 9 \times 8) \times 7$ ways to select the entire slate of officers.

Two particular list-making problems recur often enough to warrant special attention. These problems both involve making a list of length k in which each element of the list is selected from among n possibilities. In the first problem, we count all such lists; in the second problem, we count those without repeated elements.

When repetitions are allowed, we have n choices for the first element of the list, n choices for the second element of the list, and so on, and n choices for the last element of the list. All told, there are

The number of lists of length k where there are n possible entries in each position of the list and repetitions are allowed.

$$\underbrace{n \times n \times \cdots \times n}_{k \text{ times}} = n^k \tag{1}$$

possible lists.

Now suppose we fill in the length-k list with n possible values, but in this case, repetition is not allowed. There are n ways to select the first element of the list. Once this is done, there are $n-1$ choices for the second element of the list. There are $n-2$ ways to fill in position three, $n-3$ ways to fill in position four, and so on, and finally, there are $n-(k-1) = n-k+1$ ways to fill in position k. Therefore, the number of ways to make a list of length k where the elements are chosen from a pool of n possibilities and no two elements on the list may be the same is

Lists without repetitions are sometimes called *permutations*. However, in this book, the word *permutation* has another meaning described later.

The number of lists of length k where the elements are chosen from a pool of n possibilities and no two elements on the list are the same.

$$n \times [n-1] \times [n-2] \times \cdots \times [n-(k-1)]. \tag{2}$$

This formula is correct, but there is a minor mistake in our reasoning! How many length-six lists can we make where each element of the list is one of the digits 1, 2, 3, or 4 and repetition is not allowed? The answer, obviously, is zero; you cannot make a length-six list using only four possible elements and not repeat an element! What does the formula give? Equation (2) says the number of such lists is

$$4 \times 3 \times 2 \times 1 \times 0 \times -1$$

which equals 0. However, the reasoning behind the formula breaks down. Although it is true that there are 4, 3, 2, 1, and 0 choices for positions one through five, it does not make any sense to say there are -1 choices for the last position! Formula (2) gives the correct answer, but the reasoning used to arrive there needs to be rechecked.

In this paragraph, we use Exercise 4.12: If $a, b \in \mathbb{Z}$, then $a < b \iff a \leq b - 1$.

If the number of elements from which we select entries in the list, n, is less than the length of the list, k, no repetition-free list is possible. But since $n < k$, we know that $n - k < 0$ and so $n - k + 1 < 1$. Since $n - k + 1$ is an integer, we know that $n - k + 1 \leq 0$. Therefore, in the product $n \times (n-1) \times \cdots \times (n-k+1)$, we know that at least one of the factors is zero. Therefore the whole expression evaluates to zero, which is what we wanted!

On the other hand, if $n \geq k$, our reasoning makes sense (all the numbers are positive), and the formula in (2) gives the correct answer.

Because the expression $n(n-1)(n-2)\cdots(n-k+1)$ occurs fairly often, there is a special notation for it. The notation is

$$(n)_k = n(n-1)(n-2)\cdots(n-k+1).$$

The special notation for $n(n-1)\cdots(n-k+1)$ is $(n)_k$. An alternative notation, still in use on some calculators, is ${}_nP_k$.

This notation is called *falling factorial*. We summarize our results on lists with or without repetition concisely using this notation.

Theorem 7.6 The number of lists of length k whose elements are chosen from a pool of n possible elements

$$= \begin{cases} n^k & \text{if repetitions are permitted} \\ (n)_k & \text{if repetitions are forbidden.} \end{cases}$$

I do not recommend that you memorize this result because it is too easy to get confused between the meanings of n and k. Rather, rederive it in your mind when you need it. Imagine the k boxes written out in front of you, put the appropriate numbers in the boxes, and multiply.

Recap

This section deals with counting lists of objects. The central tool is the Multiplication Principle. A general formula is developed for counting length-k lists of elements selected from a pool of n elements either with or without repetitions.

7 Exercises

7.1. A *bit string* is a list of 0s and 1s. How many length-k bit strings can be made?

7.2. Airports have names, but they also have three-letter codes. For example, the airport serving Baltimore is BWI, and the code YYY is for the airport in Mont Joli, Québec, Canada. How many different airport codes are possible?

7.3. A car's ventilation system has various controls. The fan control has four settings: off, low, medium, and high. The air stream can be set to come out at the floor, through the vents, or through the defroster. The air conditioning button can be either on or off. The temperature control can be set to cold, cool, warm, or hot. And finally, the recirculate button can be either on or off.

In how many different ways can these various controls be set?

Note: Several of these settings result in the same effect since nothing happens if the fan control is off. However, the problem asks for the number of different settings of the controls, not the number of different ventilation effects possible.

7.4. My compact disc player has space for 5 CDs; there are five trays numbered 1 through 5 into which I load the CDs. I own 100 CDs.

a. In how many ways can the CD player be loaded if all five trays are filled with CDs?

b. In how many ways can the CD player be loaded if only one CD is placed in the machine?

7.5. You own three different rings. You wear all three rings, but no two of the rings are on the same finger, nor are any of them on your thumbs. In how many ways can you wear your rings? (Assume any ring will fit on any finger.)

7.6. In how many ways can a black rook and a white rook be placed on different squares of a chess board such that neither is attacking the other? (In other words, they cannot be in the same row or the same column of the chess board. A standard chess board is 8×8.)

7.7. License plates in a certain state consist of six characters: The first three characters are uppercase letters (A–Z), and the last three characters are digits (0–9).

a. How many license plates are possible?

b. How many license plates are possible if no character may be repeated on the same plate?

7.8. A telephone number (in the United States and Canada) is a ten-digit number whose first digit cannot be a 0 or a 1. How many telephone numbers are possible?

7.9. A U.S. Social Security number is a nine-digit number. The first digit(s) may be 0.

a. How many Social Security numbers are available?

b. How many of these are even?

c. How many have all of their digits even?

d. How many read the same backward and forward (e.g., 122979221)?

e. How many have none of their digits equal to 8?

f. How many have at least one digit equal to 8?

g. How many have exactly one 8?

The word *character* means a letter or a digit.

7.10. A computer operating system allows files to be named using any combination of uppercase letters (A–Z) and digits (0–9), but the number of characters in the file name is at most eight (and there has to be at least one character in the file name). For example, `X23`, `W`, `4AA`, and `ABCD1234` are valid file names, but `W-23` and `WONDERFUL` are not valid (the first has an improper character, and the second is too long).

How many different file names are possible in this system?

7.11. How many five-digit numbers are there that do not have two consecutive digits the same? For example, you would count 12104 and 12397 but not 6321 (it is not five digits) or 43356 (it has two consecutive 3s).

Note: The first digit may not be a zero.

7.12. A padlock has the digits 0 through 9 arranged in a circle on its face. A combination for this padlock is four digits long. Because of the internal mechanics of the lock, no pair of consecutive numbers in the combination can be the same or one place apart on the face. For example 0-2-7-1 is a valid combination, but neither 0-4-4-7 (repeated digit 4) nor 3-0-9-5 (adjacent digits 0-9) are permitted. How many combinations are possible?

7.13. A bookshelf contains 20 books. In how many different orders can these books be arranged on the shelf?

7.14. A class contains ten boys and ten girls. In how many different ways can they stand in a line if they must alternate in gender (no two boys and no two girls are standing next to one another)?

7.15. Four cards are drawn from a standard deck of 52 cards. In how many ways can this be done if the cards are all of different values (e.g., no two 5s or two jacks) and all of different suits? (For this problem, the order in which the cards are drawn matters, so drawing A♤-K♡-3♢-6♧ is not the same as drawing 6♧-K♡-3♢-A♤ even though the same cards are selected.)

8 Factorial

In Section 7, we counted lists of elements of various lengths in which we were either allowed or forbidden to repeat elements. A special case of this problem is to count the number of length-n lists chosen from a pool of n objects in which repetition is forbidden. In other words, we want to arrange n objects into a list, using each object exactly once. By Theorem 7.6, the number of such lists is

$$(n)_n = n(n-1)(n-2)\cdots(n-n+1) = n(n-1)(n-2)\cdots(1).$$

The quantity $(n)_n$ occurs frequently in mathematics and has a special name and notation; it is called *n factorial* and is written $n!$. For example, $5! = 5 \times 4 \times 3 \times 2 \times 1 = 120$.

Two special cases of the factorial function require special attention.

First, let us consider $1!$. This is the result of multiplying all the numbers starting from 1 all the way down to, well, 1. The answer is 1. Just in case this isn't clear, let's return to the list-counting application. In how many ways can we make a length-1 list where there is only one possible element to fill the first (and only!) position? Obviously, there is only one possible list. So $1! = 1$.

The other special case is $0!$.

Much Ado About 0!

0! is 1. Students' reactions to this statement typically range from "That doesn't make sense" to "That's wrong!" There seems to be an overwhelming urge to evaluate 0! as 0.

Because of this confusion, I feel I owe you a clear and unambiguous explanation of why $0! = 1$. Here it is: Because I said so!

No, that wasn't a terribly satisfying answer, and I will endeavor to do a better job in a moment, but the simple fact is that mathematicians have defined 0! to be 1, and we are all in agreement on this point. Just as we declared (via our definition) that the number 1 is not prime, we can also declare $0! = 1$. Mathematics is a human invention, and as long as we are consistent, we can set things up pretty much however we please.

So now the burden falls on me to explain why it is a good idea to have $0! = 1$ and a bad idea for it to be 0, $\sqrt{17}$, or anything else.

To begin, let us rethink the list-counting problem. The number 0! ought to be the answer to the following problem:

In how many ways can we make a length-0 list whose elements come from a pool of 0 elements in which there is no repetition?

It is tempting to say that no such list is possible, but this is not correct. There is a list whose length is zero: the empty list ( ). The empty list has zero length, and (vacuously!) its elements satisfy the conditions of the problem. So the answer to the problem is $0! = 1$.

Here is another explanation why $0! = 1$. Consider the equation

$$n! = n \times (n-1)! \tag{3}$$

For example, $5! = 5 \times (4 \times 3 \times 2 \times 1) = 5 \times 4!$. Equation (3) makes sense for $n = 2$ since $2! = 2 \times 1! = 2 \times 1$. The question becomes: Does Equation (3) make sense for $n = 1$? If we want Equation (3) to work when $n = 1$, we need $1! = 1 \times 0!$. This forces us to choose $0! = 1$.

Here is another explanation why $0! = 1$. We can think of $n!$ as the result of multiplying n numbers together. For example, 5! is the result of multiplying the numbers on the list (5, 4, 3, 2, 1). What should it mean to multiply together the numbers on the empty list ( )? Let me try to convince you that the sensible answer is 1. We begin by considering what it means to add the numbers on the empty list.

Alice and Bob are to add the numbers on the list (2, 3, 3, 5, 4). The answer should be 17.

Alice and Bob work in a number factory and are given a list of numbers to add. They are both quite adept at addition, so they decide to break the list in two. Alice will add her numbers, Bob will add his numbers, and then they will add their results to get the final answer. This is a sensible procedure, and they ask Carlos to break the list in two for them.

Carlos gives Alice the list (2, 3, 3, 5, 4) and Bob the list ( ). Alice adds her numbers and gets 17. What should Bob say?

Carlos, perhaps because he is feeling mischievous, decides to give Alice all of the numbers and Bob none of the numbers. Alice receives the full list and Bob receives the empty list. Alice adds her numbers as usual, but what is Bob to report as the sum of the numbers on his list? If Bob gives any answer other than 0, the final answer to the problem will be incorrect. The only sensible thing for Bob to say is that his list—the empty list—sums to 0.

The sum of the numbers in the empty list is 0.

Alice and Bob are to multiply the numbers on the list (2, 3, 3, 5, 4). The answer should be 360.

Now, all three of them have received a promotion and are working on multiplication. Their multiplication procedure is the same as their addition procedure. They are asked to multiply lists of numbers. When they receive a list, they ask Carlos to break the list into two parts. Alice multiplies the numbers on her list, and Bob multiplies the numbers on his. They then multiply together their individual results to get the final answer.

Carlos gives Alice the list ( ) and Bob the list (2, 3, 3, 5, 4). Bob multiplies his numbers and gets 360. What should Alice say?

But of course Carlos decides to have some fun and gives all the numbers to Bob; to Alice, he gives the empty list. Bob reports the product of his numbers as usual. What should Alice say? What is the product of the numbers on ( )? If she says 0, then when her answer is multiplied by Bob's answer, the final result will be 0, and this is likely to be the wrong answer. Indeed, the only sensible reply that Alice can give is 1.

The product of the numbers in the empty list is 1. Since 0! "asks" you to multiply together a list containing no numbers, the sensible answer is 1.

This reasoning is akin to taking $2^0 = 1$.

The final reason why we declare $0! = 1$ is that as we move on, other formulas work out better if we take $0! = 1$. If we did not set $0! = 1$, these other results would have to treat 0 as a special case, different from other natural numbers.

Product Notation

Here is another way to write $n!$:

$$n! = \prod_{k=1}^{n} k.$$

What does this mean? The symbol Π is the uppercase form of the Greek letter pi (π), and it stands for *product* (i.e., multiply). This notation is similar to using Σ for summation.

The letter k is called a *dummy variable* and is a place holder that ranges from the lower value (written below the Π symbol) to the upper value (written at the top). The variable k takes on the values $1, 2, \ldots, n$.

To the right of the Π symbol are the quantities we multiply. In this case, it is simple: We just multiply the values of k as k goes from 1 to n; that is, we multiply

$$1 \times 2 \times \cdots \times n.$$

The expression on the right of the Π symbol can be more complex. For example, consider the product

$$\prod_{k=1}^{5} (2k + 3).$$

This specifies that we multiply together the various values of $(2k + 3)$ for $k = 1, 2, 3, 4, 5$. In other words,

$$\prod_{k=1}^{5}(2k+3) = 5 \times 7 \times 9 \times 11 \times 13.$$

The expression on the right of the Π can be simpler. For example,

$$\prod_{k=1}^{n} 2$$

is a fancy way to write 2^n.

Consider the following way of writing $0!$:

$$\prod_{k=1}^{0} k.$$

This means that k starts at 1 and goes up to 0. Since there is no possible value of k with $1 \le k \le 0$, there are no terms to multiply. Therefore the product is empty and evaluates to 1.

Recap

In this section, we introduced factorial, discussed why $0! = 1$, and presented product notation.

8 Exercises

8.1. There are six different French books, eight different Russian books, and five different Spanish books.

a. In how many different ways can these books be arranged on a bookshelf?

b. In how many different ways can these books be arranged on a bookshelf if all books in the same language are grouped together?

8.2. Give an Alice-and-Bob discussion about what it means to add (and to multiply) a list of numbers that only contains one number.

8.3. Consider the formula

$$(n)_k = \frac{n!}{(n-k)!}.$$

This formula is mostly correct. For what values of n and k is it correct? Prove the formula is correct under a suitable hypothesis; that is, this problem asks you to find and prove a theorem of the form "If (conditions on n and k), then $(n)_k = n!/(n-k)!$."

8.4. Evaluate $\frac{100!}{98!}$ without calculating $100!$ or $98!$.

8.5. Order the following integers from least to greatest: 2^{100}, 100^2, 100^{100}, $100!$, 10^{10}.

8.6. The Scottish mathematician James Stirling found an approximation formula for $n!$. Stirling's formula is

$$n! \approx \sqrt{2\pi n}\, n^n e^{-n}$$

where $\pi = 3.14159\ldots$ and $e = 2.71828\ldots$. (Scientific calculators have a key that computes e^x; this key might be labeled $\boxed{\exp x}$.)

Compute $n!$ and Stirling's approximation to $n!$ for $n = 10, 20, 30, 40, 50$. What is the relative error in the approximations?

8.7. Calculate the following products:

a. $\prod_{k=1}^{4}(2k+1)$.

b. $\prod_{k=-3}^{4} k$.

c. $\prod_{k=1}^{n} \frac{k+1}{k}$, where n is a positive integer.

d. $\prod_{k=1}^{n} \frac{1}{k}$, where n is a positive integer.

8.8. When 100! is written out in full, it equals

$$100! = 9332621\ldots000000.$$

Without using a computer, determine the number of 0 digits at the end of this number.

8.9. Prove that all of the following numbers are composite: $1000!+2$, $1000!+3$, $1000!+4, \ldots, 1000!+1002$.

The point of this problem is to present a long list of consecutive numbers, all of which are composite.

8.10. Can factorial be extended to negative integers? On the basis of Equation (3), what value would you assign to $(-1)!$?

8.11. *This problem is only for those who have studied calculus.* Evaluate the following integral for $n = 0, 1, 2, 3, 4$:

$$\int_0^\infty x^n e^{-x}\,dx.$$

Note: The case $n = 0$ is easiest. Do the remaining values of n in order (first 1, then 2, etc.) and use integration by parts.

What is the value of this integral for an arbitrary natural number n?

Extra for experts: Evaluate the integral with $n = \frac{1}{2}$.

9 Sets I: Introduction, Subsets

A set is a repetition-free, unordered collection of objects. A given object either is a member of a set or it is not—an object cannot be in a set "more than once." There is no order to the members of a set. The simplest way to specify a set is to list its elements between curly braces. For example, $\{2, 3, \frac{1}{2}\}$ is a set with exactly three members: the integers 2 and 3, and the rational number $\frac{1}{2}$. No other objects are in this set. All of the following sets are the same:

$$\{2, 3, \tfrac{1}{2}\} \qquad \{3, \tfrac{1}{2}, 2\} \qquad \{2, 2, 3, \tfrac{1}{2}\}.$$

It does not matter in what order we list the objects, nor does it matter if we repeat an object. All that matters is what objects are members of the set and what objects are not. In this example, exactly three objects are members of the set; no other objects are members.

Earlier, we introduced three special sets of numbers. These sets are $\mathbb{Z}$ (the integers), $\mathbb{N}$ (the natural numbers), and $\mathbb{Q}$ (the rational numbers).

An object that belongs to a set is called an *element* of the set.

Membership in a set is denoted with the symbol $\in$. The notation $x \in A$ means that the object x is a member of the set A. For example, $2 \in \{2, 3, \frac{1}{2}\}$ is true, but $5 \in \{2, 3, \frac{1}{2}\}$ is false. In the latter case, we can write $5 \notin \{2, 3, \frac{1}{2}\}$; the notation $x \notin A$ means x is not an element of A.

When read aloud, $\in$ is pronounced "is a member of" or "is an element of" or "is in." Often mathematicians write, "If $x \in \mathbb{Z}$, then. . . ." This means exactly the same thing as "If x is an integer, then. . . ."

However, the $\in$ symbol can also stand for "be a member of" or "be in." For example, if we write "Let $x \in \mathbb{Z}$," we mean "Let x be a member of $\mathbb{Z}$" or, more prosaically, "Let x be an integer."

Absolute value bars around a set stand for the *cardinality* or *size* of the set (i.e., the number of elements in that set).

The number of elements in a set A is denoted $|A|$. The *cardinality* of A is simply the number of objects in the set. The cardinality of the set $\{2, 3, \frac{1}{2}\}$ is 3. The cardinality of $\mathbb{Z}$ is infinite. We also call $|A|$ the *size* of the set A.

A set is called *finite* if its cardinality is an integer (i.e., is finite). Otherwise, it is called *infinite*.

The *empty set* is the set with no members. The empty set may be denoted $\{\ \}$, but it is better to use the special symbol $\emptyset$. The statement "$x \in \emptyset$" is false regardless of what object x might represent. The cardinality of the empty set is zero (i.e., $|\emptyset| = 0$).

Please note that the symbol $\emptyset$ is not the same as the Greek letter phi: ϕ or Φ.

Set-builder notation.

There are two principal ways of specifying a set. The most direct way is to list the elements of the set between curly braces, as in $\{3, 4, 9\}$. This notation is appropriate for small sets. More often, *set-builder notation* is used. The form of this notation is

$$\{\text{dummy variable} : \text{conditions}\}.$$

For example, consider

$$\{x : x \in \mathbb{Z},\ x \geq 0\}.$$

This is the set of all objects x that satisfy two conditions: (1) $x \in \mathbb{Z}$ (i.e., x must be an integer) and (2) $x \geq 0$ (i.e., x is nonnegative). In other words, this set is $\mathbb{N}$, the natural numbers.

An alternative way of writing set-builder notation is

$$\{\text{dummy variable} \in \text{set} : \text{conditions}\}.$$

This is the set of all objects drawn from the set mentioned and subject to the conditions specified. For example,

$$\{x \in \mathbb{Z} : 2|x\}$$

is the set of all integers that are divisible by 2 (i.e., the set of even integers).

It is often tempting to write a set by establishing a pattern to the elements and then using three dots (...) to indicate that the pattern continues. For example, we might write $\{1, 2, 3, \ldots, 100\}$ to denote the set of integers from 1 to 100 inclusive. In this case, the notation is clear, but it would be better to write $\{x \in \mathbb{Z} : 1 \le x \le 100\}$.

Here is another example, which is less clear: $\{3, 5, 7, \ldots\}$. What is intended? We have to guess whether we mean the set of odd integers greater than 1 or the set of odd primes. Use the "..." notation sparingly and only when there is absolutely no chance of confusion.

Equality of Sets

What does it mean for two sets to be *equal*? It means that the two sets have exactly the same elements. To prove that sets A and B are equal, one shows that every element of A is also an element of B, and vice versa.

Proof Template 5 Proving two sets are equal.

Let A and B be the sets. To show $A = B$, we have the following template:

- Suppose $x \in A$. ... Therefore $x \in B$.
- Suppose $x \in B$. ... Therefore $x \in A$.

Therefore $A = B$. ■

Let us illustrate the use of Proof Template 5 on a simple statement.

Proposition 9.1 The following two sets are equal:

$$E = \{x \in \mathbb{Z} : x \text{ is even}\}, \quad \text{and}$$
$$F = \{z \in \mathbb{Z} : z = a + b \text{ where } a \text{ and } b \text{ are both odd}\}.$$

In other words, the set F is the set of all integers that can be written as the sum of two odd numbers. Using the template, the proof looks like this:

> Let $E = \{x \in \mathbb{Z} : x \text{ is even}\}$ and $F = \{z \in \mathbb{Z} : z = a+b \text{ where } a \text{ and } b \text{ are both odd}\}$. We seek to prove that $E = F$.
>
> Suppose $x \in E$. ... Therefore $x \in F$.
>
> Suppose $x \in F$. ... Therefore $x \in E$. ■

Start with the first half by unraveling definitions.

> Let $E = \{x \in \mathbb{Z} : x \text{ is even}\}$ and $F = \{z \in \mathbb{Z} : z = a + b$ where a and b are both odd$\}$. We seek to prove that $E = F$.
>
> Suppose $x \in E$. Therefore x is even, and hence divisible by 2, so $x = 2y$ for some integer y. ... Therefore x is the sum of two odd numbers and so $x \in F$.
>
> Suppose $x \in F$. ... Therefore $x \in E$. ■

We have that $x = 2y$, and we want x as the sum of two odd numbers. Here's a simple way to do this: $2y + 1$ is odd (see Definition 2.4) and so is -1 (because $-1 = 2 \times (-1) + 1$). So we can write

$$x = 2y = (2y + 1) + (-1).$$

Let's fold these ideas into the proof.

> Let $E = \{x \in \mathbb{Z} : x \text{ is even}\}$ and $F = \{z \in \mathbb{Z} : z = a + b$ where a and b are both odd$\}$. We seek to prove that $E = F$.
>
> Suppose $x \in E$. Therefore x is even, and hence divisible by 2, so $x = 2y$ for some integer y. Note that $2y + 1$ and -1 are both odd, and since $x = 2y = (2y + 1) + (-1)$, we see that x is the sum of two odd numbers. Therefore $x \in F$.
>
> Suppose $x \in F$. ... Therefore $x \in E$. ■

The second part of the proof was already considered in Exercise 4.1 (and the solution to that exercise can be found in Appendix A). So we simply refer to that previously worked problem to complete the proof.

> Let $E = \{x \in \mathbb{Z} : x \text{ is even}\}$ and $F = \{z \in \mathbb{Z} : z = a + b$ where a and b are both odd$\}$. We seek to prove that $E = F$.
>
> Suppose $x \in E$. Therefore x is even, and hence divisible by 2, so $x = 2y$ for some integer y. Note that $2y + 1$ and -1 are both odd, and since $x = 2y = (2y + 1) + (-1)$, we see that x is the sum of two odd numbers. Therefore $x \in F$.
>
> Suppose $x \in F$. Therefore x is the sum of two odd numbers. As we showed in Exercise 4.1, x must be even and so $x \in E$. ■

Note that Proposition 9.1 can be rewritten as follows: *An integer is even if and only if it can be expressed as the sum of two odd numbers.*

Subset

Next we define *subset*.

Definition 9.2 **(Subset)** Suppose A and B are sets. We say that A is a *subset* of B provided every element of A is also an element of B. The notation $A \subseteq B$ means A is a subset of B.

For example, $\{1, 2, 3\}$ is a subset of $\{1, 2, 3, 4\}$. For any set A, we have $A \subseteq A$ because every element of A is (of course) in A.

Furthermore, for any set A, we have $\emptyset \subseteq A$. This is because every element of $\emptyset$ is in A—since there are no elements in $\emptyset$, there are no elements of $\emptyset$ that fail to be in A. This is an example of a vacuous statement, but a useful one.

The symbol $\subset$ is often used for subset as well, but we do not use it in this book. We prefer $\subseteq$ because it looks more like $\leq$, and we want to emphasize that a set is always a subset of itself. (The symbol $\subseteq$ is a hybrid of the symbols $\subset$ and $=$.) If we want to rule out the equality of the two sets, we may say that A is a *strict* or *proper* subset of B; this means that $A \subseteq B$ and $A \neq B$. It would be tempting to let $\subset$ denote proper subset (because it looks like $<$), but the use of $\subset$ to mean ordinary subset has not completely fallen out of fashion in the mathematics community. We avoid controversy by not using the symbol $\subset$.

$\subseteq$ and $\in$ have related but different meanings. They cannot be interchanged!

It is important to distinguish between $\in$ and $\subseteq$. The notation $x \in A$ means that x is an element (or member) of A. The notation $A \subseteq B$ means that every element of A is also an element of B. Thus $\emptyset \subseteq \{1, 2, 3\}$ is true, but $\emptyset \in \{1, 2, 3\}$ is false.

The difference between $\in$ and $\subset$ is analogous to the difference between x and $\{x\}$. The symbol x refers to some object (a number or whatever), and the notation $\{x\}$ means the set whose one and only element is the object x. It is always correct to write $x \in \{x\}$, but it is incorrect to write $x = \{x\}$ or $x \subseteq \{x\}$. (Well, it *usually* is incorrect to write $x \subseteq \{x\}$; see Exercise 9.9.)

To prove that one set is a subset of another, we need to show that every element of the first set is also a member of the second set.

Proposition 9.3 Let x be anything and let A be a set; then $x \in A$ if and only if $\{x\} \subseteq A$.

Proof. Let x be any object and let A be a set.

($\Rightarrow$) Suppose that $x \in A$. We want to show $\{x\} \subseteq A$. To do this, we need to show that every element of $\{x\}$ is also an element of A. But the only element of $\{x\}$ is x, and we are given that $x \in A$. Therefore $\{x\} \subseteq A$.

($\Leftarrow$) Suppose that $\{x\} \subseteq A$. This means that every element of the first set ($\{x\}$) is also member of the second set (A). But the only element of $\{x\}$ is certainly x and so $x \in A$. ■

The general method for showing that one set is a subset of another is outlined in Proof Template 6.

Proof Template 6 Proving one set is a subset of another.

To show $A \subseteq B$:

Let $x \in A$. ... Therefore $x \in B$. Therefore $A \subseteq B$. ■

We illustrate the use of Proof Template 6 using the following concept.

Definition 9.4 **(Pythagorean Triple)** A list of three integers (a, b, c) is called a *Pythagorean triple* provided $a^2 + b^2 = c^2$.

Please note that $(\sqrt{2}, \sqrt{3}, \sqrt{5})$ is not a Pythagorean triple because the numbers in the list are not integers; the term *Pythagorean triple* only applies to lists of integers.

For example, $(3, 4, 5)$ is a Pythagorean triple because $3^2 + 4^2 = 5^2$. Pythagorean triples are so named because they are the lengths of the sides of a right triangle.

Proposition 9.5 Let P be the set of Pythagorean triples; that is,

$$P = \{(a, b, c) : a, b, c \in \mathbb{Z} \quad \text{and} \quad a^2 + b^2 = c^2\}$$

and let T be the set

$$T = \{(p, q, r) : p = x^2 - y^2,\ q = 2xy,\ \text{and } r = x^2 + y^2 \text{ where } x, y \in \mathbb{Z}\}.$$

Then $T \subseteq P$.

For example, if we let $x = 3$ and $y = 2$ and we calculate

$$p = x^2 - y^2 = 9 - 4 = 5, \qquad q = 2xy = 12, \qquad r = x^2 + y^2 = 9 + 4 = 13$$

we find that $(5, 12, 13) \in T$. Proposition 9.5 asserts that $T \subseteq P$, which implies $(5, 12, 13) \in T$. Indeed, this is correct since

$$5^2 + 12^2 = 25 + 144 = 169 = 13^2.$$

We now develop the proof of Proposition 9.5 by utilizing Proof Template 6.

> Let P and T be as in the statement of Proposition 9.5.
> Let $(p, q, r) \in T$. ... Therefore $(p, q, r) \in P$. ■

Unravel the meaning of $(p, q, r) \in T$.

> Let P and T be as in the statement of Proposition 9.5.
> Let $(p, q, r) \in T$. Therefore there are integers x and y such that $p = x^2 - y^2$, $q = 2xy$, and $r = x^2 + y^2$. ... Therefore $(p, q, r) \in P$. ■

To verify that $(p, q, r) \in P$, we simply have to check that all three are integers (which is clear) and that $p^2 + q^2 = r^2$. We can write p, q, and r in terms of x and y, so the problem reduces to an algebraic computation. We finish the proof.

Let P and T be as in the statement of Proposition 9.5.

Let $(p, q, r) \in T$. Therefore there are integers x and y such that $p = x^2 - y^2$, $q = 2xy$, and $r = x^2 + y^2$. Note that p, q, and r are integers because x and y are integers. We calculate

$$\begin{aligned} p^2 + q^2 &= (x^2 - y^2)^2 + (2xy)^2 \\ &= (x^4 - 2x^2y^2 + y^4) + 4x^2y^2 \\ &= x^4 + 2x^2y^2 + y^4 \\ &= (x^2 + y^2)^2 = r^2. \end{aligned}$$

Therefore (p, q, r) is a Pythagorean triple and so $(p, q, r) \in P$. ■

The symbols $\in$ and $\subseteq$ may be written backward: $\ni$ and $\supseteq$. The notation $A \ni x$ means exactly the same thing as $x \in A$. The symbol $\ni$ can be read, "contains the element." The notation $B \supseteq A$ means exactly the same thing as $A \subseteq B$. We say that B is a *superset* of A.

(We also say that B contains A and A is contained in B, but the word *contains* can be a bit ambiguous. If we say "B contains A," we generally mean that $B \supseteq A$, but it might mean $B \ni A$. We avoid this term unless the meaning is utterly clear from context.)

Counting Subsets

How many subsets does a set have? Let us consider an example.

Example 9.6 How many subsets does $A = \{1, 2, 3\}$ have?

The easiest way to do this is to list all the possibilities. Since $|A| = 3$, a subset of A can have anywhere from zero to three elements. Let's write down all the possibilities organized this way.

Number of elements	Subsets	Number
0	$\emptyset$	1
1	$\{1\}, \{2\}, \{3\}$	3
2	$\{1, 2\}, \{1, 3\}, \{2, 3\}$	3
3	$\{1, 2, 3\}$	1
	Total:	8

Therefore, there are eight subsets of $\{1, 2, 3\}$.

There is another way to analyze this problem. Each element of the set $\{1, 2, 3\}$ either is a member of or is not a member of a subset. Look at the following diagram.

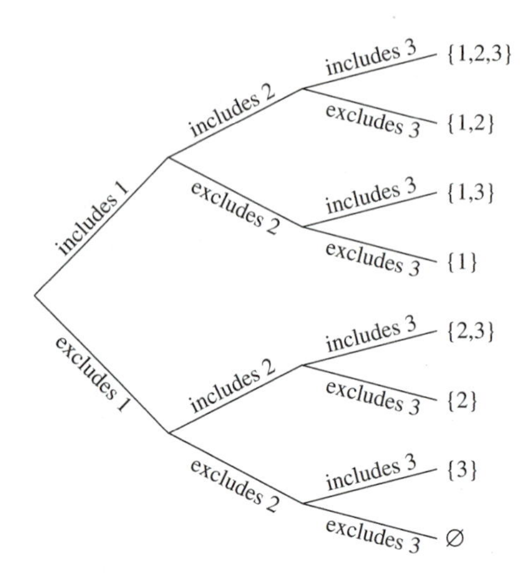

For each element, we have two choices: to include or not to include that element in the subset. We can "ask" each element if it "wants" to be in the subset. The list of answers uniquely determines the subset. So if we ask elements 1, 2, and 3 in turn if they are in the subset and the answers we receive are (yes, yes, no), then the subset is $\{1, 2\}$.

The problem of counting subsets of $\{1, 2, 3\}$ reduces to the problem of counting lists, and we know how to count lists! The number of lists of length three where each entry on the list is either "yes" or "no" is $2 \times 2 \times 2 = 8$.

This list-counting method gives us the solution to the general problem.

Theorem 9.7 Let A be a finite set. The number of subsets of A is $2^{|A|}$.

Proof. Let A be a finite set and let $n = |A|$. Let the n elements of A be named $a_1, a_2, \ldots, a_n$. To each subset B of A we can associate a list of length n; each element of the list is one of the words "yes" or "no." The kth element of the list is "yes" precisely when $a_k \in B$. This establishes a correspondence between length-n yes-no lists and subsets of A. Observe that each subset of A gives a yes-no list, and every yes-no list determines a different subset of A. Therefore the number of subsets of A is exactly the same as the number of length-n yes-no lists. The number of such lists is 2^n, so the number of subsets of A is 2^n where $n = |A|$. ■

This style of proof is called a *bijective* proof. To show that two counting problems have the same answer, we establish a one-to-one correspondence between the two sets we want to count. If we know the answer to one of the counting problems, then we know the answer to the other.

Power Set

A set can be an element of another set. For example, $\{1, 2, \{3, 4\}\}$ is a set with three elements: the number 1, the number 2, and the set $\{3, 4\}$. A special example of this is called the *power set* of a set.

Definition 9.8 **(Power set)** Let A be a set. The *power set* of A is the set of all subsets of A.

For example, the power set of $\{1, 2, 3\}$ is the set

$$\{\emptyset, \{1\}, \{2\}, \{3\}, \{1, 2\}, \{1, 3\}, \{2, 3\}, \{1, 2, 3\}\}.$$

The power set of A is denoted 2^A.

Theorem 9.7 tells us that if a set A has n elements, its power set contains 2^n elements (the subsets of A). As a mnemonic, the notation for the power set of A is 2^A. This is a special notation; there is no general meaning for raising a number to a power that is a set. The only case in which this makes sense is writing the set as a superscript on the number 2; the meaning of the notation is the power set of A. This notation was created so that we would have

$$|2^A| = 2^{|A|}$$

for any finite set A. The left side of this equation is the cardinality of the power set of A; the right side is 2 raised to the cardinality of A. On the left, the superscript on 2 is a set, so the notation means power set; on the right, the superscript on 2 is a number, so the notation means ordinary exponentiation.

Recap

In this section, we introduced the concept of a set and the notation $x \in A$. We presented the set-builder notation $\{x \in A : \ldots\}$. We discussed the concepts of empty set ($\emptyset$), subset ($\subseteq$), and superset ($\supseteq$). We distinguished between finite and infinite sets and presented the notation $|A|$ for the cardinality of A. We considered the problem of counting the number of subsets of a finite set and defined the power set of a set, 2^A.

---

9 Exercises

9.1. Write out the following sets by listing their elements between curly braces.

a. $\{x \in \mathbb{N} : x \le 10 \text{ and } 3|x\}$.
b. $\{x \in \mathbb{Z} : x \text{ is prime, and } 2|x\}$.
c. $\{x \in \mathbb{Z} : x^2 = 4\}$.
d. $\{x \in \mathbb{Z} : x^2 = 5\}$.
e. $2^{\emptyset}$.
f. $\{x \in \mathbb{Z} : 10|x \text{ and } x|100\}$.
g. $\{x : x \subseteq \{1, 2, 3, 4, 5\} \text{ and } |x| \le 1\}$.

9.2. Find the cardinality of the following sets.

a. $\{x \in \mathbb{Z} : |x| \le 10\}$.
b. $\{x \in \mathbb{Z} : 1 \le x^2 \le 2\}$.
c. $\{x \in \mathbb{Z} : x \in \emptyset\}$.
d. $\{x \in \mathbb{Z} : \emptyset \in x\}$.

e. $\{x \in \mathbb{Z} : \emptyset \subseteq \{x\}\}$.
f. $2^{2^{\{1,2,3\}}}$.
g. $\{x \in 2^{\{1,2,3,4\}} : |x| = 1\}$.
h. $\{\{1, 2\}, \{3, 4, 5\}\}$.

9.3. Complete each of the following by writing either $\in$ or $\subseteq$ in place of the $\bigcirc$.
a. $2 \bigcirc \{1, 2, 3\}$.
b. $\{2\} \bigcirc \{1, 2, 3\}$.
c. $\{2\} \bigcirc \{\{1\}, \{2\}, \{3\}\}$.
d. $\emptyset \bigcirc \{1, 2, 3\}$.
e. $\mathbb{N} \bigcirc \mathbb{Z}$.
f. $\{2\} \bigcirc \mathbb{Z}$.
g. $\{2\} \bigcirc 2^{\mathbb{Z}}$.

9.4. Let A and B be sets. Prove that $A = B$ if and only if $A \subseteq B$ and $B \subseteq A$.
(This gives a slightly different proof strategy for showing two sets are equal; compare to Proof Template 5.)

9.5. Let $A = \{x \in \mathbb{Z} : 4|x\}$ and let $B = \{x \in \mathbb{Z} : 2|x\}$. Prove that $A \subseteq B$.

9.6. Generalize the previous problem. Let a and b be integers and let $A = \{x \in \mathbb{Z} : a|x\}$ and $B = \{x \in \mathbb{Z} : b|x\}$. Find and prove a necessary and sufficient condition for $A \subseteq B$. In other words, given the notation developed, find and prove a theorem of the form "$A \subseteq B$ if and only if *some condition involving a and b*."

9.7. Let $C = \{x \in \mathbb{Z} : x|12\}$ and let $D = \{x \in \mathbb{Z} : x|36\}$. Prove that $C \subseteq D$.

9.8. Generalize the previous problem. Let c and d be integers and let $C = \{x \in \mathbb{Z} : x|c\}$ and $D = \{x \in \mathbb{Z} : x|d\}$. Find and prove a necessary and sufficient condition for $C \subseteq D$.

9.9. Give an example of an object x that makes the sentence $x \subseteq \{x\}$ true.

9.10. Please refer to Proposition 9.5, in which we proved that $T \subseteq P$. Show that $T \neq P$.

10 Quantifiers

There are certain phrases that appear frequently in theorems, and the purpose of this section is to clarify and formalize those phrases. At first glance, these phrases are simple, but we'll do our best to try to make them complicated. The expressions are *there is* and *every*.

There Is

Consider a sentence such as the following:

There is a natural number that is prime and even.

The general form of this sentence is "There is an object x, a member of set A, that has the following properties." The example sentence can be rewritten to adhere more strictly to this form as follows:

There is an x, a member of $\mathbb{N}$, such that x is prime and even.

The meaning of the sentence is, we hope, clear. It says that at least one element in $\mathbb{N}$ has the required properties. In this case, there is only one possible x (the number 2), but the phrase *there is* does not rule out the possibility that there can be more than one object with the requisite properties.

The phrase *there exists* is synonymous with *there is*.

Because the phrase *there is* occurs so often, mathematicians have developed a formal notation for statements of the form "There is an x in set A such that . . . ." We write a backward, uppercase E (i.e., $\exists$) that we pronounce *there is* or *there exists*. The general form for using this notation is

$$\exists x \in A, \text{ assertions about } x.$$

This is read, "There is an x, a member of the set A, for which the assertions hold." The sentence "There is a natural number that is prime and even" would be written

$$\exists x \in \mathbb{N}, \ x \text{ is prime and even.}$$

The letter x is a dummy variable—simply a placeholder. It is similar to the index of summation in Σ notation.

At times, we abbreviate the statement "$\exists x \in A$, assertions about x" to "$\exists x$, assertions about x" when context makes it clear what sort of object x ought to be.

The backward E is called the *existential quantifier*.

To prove a statement of the form "$\exists x \in A$, assertions about x," we have to show that some element in A satisfies the assertions. The general form for such a proof is given in Proof Template 7.

Proof Template 7 Proving existential statements.

To prove $\exists x \in A$, assertions about x:

> Let x be (give an explicit example) . . . (Show that x satisfies the assertions.) . . . Therefore x satisfies the required assertions. ■

Proving an existential statement is akin to finding a counterexample. We simply have to find one object with the required properties.

Example 10.1 Here is a proof (very short!) that there is an integer that is even and prime.

Statement: $\exists x \in \mathbb{Z}, \ x$ is even and x is prime.

Proof. Consider the integer 2. Clearly 2 is even and 2 is prime. ■

For All

The other phrase we consider in this section is *every*, as in "Every integer is even or odd." There are alternative phrases we use in place of *every*, including *all*, *each*, and *any*. All of the following sentences mean the same thing:

- *Every* integer is either even or odd.
- *All* integers are either even or odd.

- *Each* integer is either even or odd.
- Let x be *any* integer. Then x is even or odd.

In all cases, we mean that the condition applies to all integers without exception.

There is a fancy notation for these types of sentences. Just as we used the backward E for *there is*, we use an upside-down A ($\forall$) as a notation for *all*. The general form for this notation is

$$\forall x \in A, \text{ assertions about } x.$$

This means that all elements of the set A satisfy the assertions, as in

$$\forall x \in \mathbb{Z}, \ x \text{ is odd or } x \text{ is even.}$$

When the context makes clear what sort of object x is, the notation may be shortened to "$\forall x$, assertions about x."

The upside-down A is called the *universal quantifier*.

To prove an "all" theorem, we need to show that every element of the set satisfies the required assertions. The general form for this sort of proof is given in Proof Template 8.

Proof Template 8 Proving universal statements.

To prove $\forall x \in A$, assertions about x:

Let x be any element of A.... (Show that x satisfies the assertions using only the fact that $x \in A$ and no further assumptions on x.) ... Therefore x satisfies the required assertions. ■

Example 10.2 To prove: Every integer that is divisible by 6 is even.

More formally, let $A = \{x \in \mathbb{Z} : 6|x\}$. Then the statement we want to prove is

$$\forall x \in A, \ x \text{ is even.}$$

Proof. Let $x \in A$; that is, x is an integer that is divisible by 6. This means there is an integer y such that $x = 6y$, which can be rewritten $x = (2 \cdot 3)y = 2(3y)$. Therefore x is divisible by 2 and therefore even. ■

Mathspeak!

Mathematicians use the word *arbitrary* in a slightly nonstandard way. When we say that x is an arbitrary element of a set A, we mean that x might be any element of A, and one should not assume anything about x other than it is an element of A. To say x is an arbitrary even number means that x is even, but we make no further assumptions about x.

Note that this proof is not really any different from proving an ordinary if-then: "If x is divisible by 6, then x is even." The point we are trying to stress is that in the proof, we assume that x is an arbitrary element of A and then move on to show that x satisfies the condition.

Negating Quantified Statements

Consider the statements

- There is no integer that is both even and odd.
- Not all integers are prime.

Symbolically, these can be written

- $\neg(\exists x \in \mathbb{Z},\ x \text{ is even and } x \text{ is odd})$.
- $\neg(\forall x \in \mathbb{Z},\ x \text{ is prime})$.

In both cases, we have negated a quantified statement. What do these negations mean?

Let us first consider a statement of the form

$$\neg(\exists x \in A,\ \text{assertions about } x).$$

This means that none of the elements of A satisfy the assertions, and this is equivalent to saying that *all* of the elements of A fail to satisfy the assertions. In other words, the following two sentences are equivalent:

$$\neg(\exists x \in A,\ \text{assertions about } x)$$
$$\forall x \in A,\ \neg(\text{assertions about } x).$$

For example, the statement "There is no integer that is both even and odd" says the same thing as "Every integer is not both even and odd."

Next we consider the negation of universal statements. Consider a statement of the form

$$\neg(\forall x \in A,\ \text{assertions about } x).$$

This means that not all of the elements of x have the requisite assertions (i.e., some don't). Thus the following two statements are equivalent:

$$\neg(\forall x \in A,\ \text{assertions about } x)$$
$$\exists x \in A,\ \neg(\text{assertions about } x).$$

For example, the statement "Not all integers are prime" is equivalent to the statement "There is an integer that is not prime."

The mnemonic I use to remember these equivalences is

$$\neg\forall \ldots = \exists\neg \ldots \qquad \text{and} \qquad \neg\exists \ldots = \forall\neg \ldots.$$

When the $\neg$ "moves" inside the quantifier, it toggles the quantifier between $\forall$ and $\exists$.

Combining Quantifiers

Quantified statements can become difficult and confusing when there are two (or more!) quantifiers in the same statement. For example, consider the following statements about integers:

- For every x, there is a y such that $x + y = 0$.
- There is a y such that for every x, we have $x + y = 0$.

In symbols, these statements are written

- $\forall x,\ \exists y,\ x + y = 0$.
- $\exists y,\ \forall x,\ x + y = 0$.

What do these mean?

The first sentence makes a claim about an arbitrary integer x. It says that no matter what x is, something is true—namely, we can find an integer y such that $x + y = 0$. Let's say $x = 12$. Can we find a y such that $x + y = 0$? Of course! We just want $y = -12$. Say $x = -53$. Can we find a y such that $x + y = 0$? Yes! Take $y = 53$. Notice that the y that satisfies $x = 12$ is different from the y that satisfies $x = -53$. The statement just requires that no matter how we pick x ($\forall x$), we can find a y ($\exists y$) such that $x + y = 0$. And this is a true statement. Here is the proof:

Let x be any integer. Let y be the integer $-x$. Then $x + y = x + (-x) = 0$. ■

Since the overall statement begins $\forall x$, we begin the proof by considering an arbitrary integer x. We now have to prove something about this number x—namely, we can find a number y such that $x + y = 0$. The choice for y is obvious, just take $y = -x$. The statement $\forall x,\ \exists y,\ x + y = 0$ is true.

Now let us examine the similar statement

$$\exists y,\ \forall x,\ x + y = 0.$$

This sentence is similar to the previous sentence; the only difference is the order of the quantifiers. This sentence alleges that there is an integer y with a certain property—namely, no matter what number we add to y ($\forall x$), we get 0 ($x + y = 0$). This sentence is blatantly false! There is no such integer y. No matter what integer y you might think of, we can always find an integer x such that $x + y$ is not zero.

The statements $\forall x,\ \exists y,\ x + y = 0$ and $\exists y,\ \forall x,\ x + y = 0$ are made a bit clearer through the use of parentheses. They may be rewritten as follows:

$$\forall x,\ (\exists y,\ x + y = 0)$$
$$\exists y,\ (\forall x,\ x + y = 0).$$

These additional parentheses are not strictly necessary, but if they make these statements clearer to you, please feel free to use them.

In general, the two sentences

$$\forall x,\ \exists y,\ \text{assertions about } x \text{ and } y$$
$$\exists y,\ \forall x,\ \text{assertions about } x \text{ and } y$$

are not equivalent to one another.

Recap

We analyzed statements of the form "For all ..." and "There exists ..." and introduced the formal quantifier notation for them. We presented basic proof templates for such sentences. We examined the negation of quantified sentences, and we studied statements with more than one quantifier.

10 Exercises

10.1. Write the following sentences using the quantifier notation (i.e., use the symbols ∃ and/or ∀). *Note*: We do not claim these statements are true, so please do not try to prove them!

a. Every integer is prime.
b. There is an integer that is neither prime nor composite.
c. There is an integer whose square is 2.
d. All integers are divisible by 5.
e. Some integer is divisible by 7.
f. The square of any integer is nonnegative.
g. For every integer x, there is an integer y such that $xy = 1$.
h. There are an integer x and an integer y such that $x/y = 10$.
i. There is an integer that, when multiplied by any integer, always gives the result 0.
j. No matter what integer you choose, there is always another integer that is larger.
k. Everybody loves somebody sometime.

10.2. Write the negation of each of the sentences in the previous problem. You should "move" the negation all the way inside the quantifiers. Give your answer in English and symbolically. For example, the negation of part (a) would be "There is an integer that is not prime" (English) and "$\exists x \in \mathbb{Z}$, x is not prime" (symbolic).

10.3. What does the sentence "Everyone is not invited to my party" mean?

Presumably the meaning of this sentence is not what the speaker intended. Rewrite this sentence to give the intended meaning.

10.4. *True or False*: Please label each of the following sentences about integers as either true or false. (You do not need to prove your assertions.)

a. $\forall x,\ \forall y,\ x + y = 0$.
b. $\forall x,\ \exists y,\ x + y = 0$.
c. $\exists x,\ \forall y,\ x + y = 0$.
d. $\exists x,\ \exists y,\ x + y = 0$.
e. $\forall x,\ \forall y,\ xy = 0$.
f. $\forall x,\ \exists y,\ xy = 0$.
g. $\exists x,\ \forall y,\ xy = 0$.
h. $\exists x,\ \exists y,\ xy = 0$.

10.5. For each of the following sentences, write the negation of the sentence, but place the ¬ symbol as far to the right as possible. Then rewrite the negation in English.

For example, for the sentence

$$\forall x \in \mathbb{Z},\ x \text{ is odd}$$

the negation would be

$$\exists x \in \mathbb{Z},\ \neg(x \text{ is odd}),$$

which in English is "There is an integer that is not odd."

a. $\forall x \in \mathbb{Z},\ x < 0$.
b. $\exists x \in \mathbb{Z},\ x = x + 1$.
c. $\exists x \in \mathbb{N},\ x > 10$.

d. $\forall x \in \mathbb{N}, x + x = 2x$.
e. $\exists x \in \mathbb{Z}, \forall y \in \mathbb{Z}, x > y$.
f. $\forall x \in \mathbb{Z}, \forall y \in \mathbb{Z}, x = y$.
g. $\forall x \in \mathbb{Z}, \exists y \in \mathbb{Z}, x + y = 0$.

10.6. Do the following two statements mean the same thing?

$$\forall x,\ \forall y,\ \text{assertions about } x \text{ and } y$$

$$\forall y,\ \forall x,\ \text{assertions about } x \text{ and } y$$

Explain.

Likewise, do the following two statements mean the same thing?

$$\exists x,\ \exists y,\ \text{assertions about } x \text{ and } y$$

$$\exists y,\ \exists x,\ \text{assertions about } x \text{ and } y$$

Explain.

11 Sets II: Operations

Just as numbers can be added or multiplied, and truth values can be combined with $\wedge$ and $\vee$, there are various operations we perform on sets. In this section, we discuss several set operations.

Union and Intersection

The most basic set operations are *union* and *intersection*.

Definition 11.1 **(Union and intersection)** Let A and B be sets.

The *union* of A and B is the set of all elements that are in A or B. The union of A and B is denoted $A \cup B$.

The *intersection* of A and B is the set of all elements that are in both A and B. The intersection of A and B is denoted $A \cap B$.

In symbols, we can write this as follows:

$$A \cup B = \{x : x \in A \text{ or } x \in B\}, \quad \text{and}$$

$$A \cap B = \{x : x \in A \text{ and } x \in B\}.$$

Example 11.2 Suppose $A = \{1, 2, 3, 4\}$ and $B = \{3, 4, 5, 6\}$. Then $A \cup B = \{1, 2, 3, 4, 5, 6\}$ and $A \cap B = \{3, 4\}$.

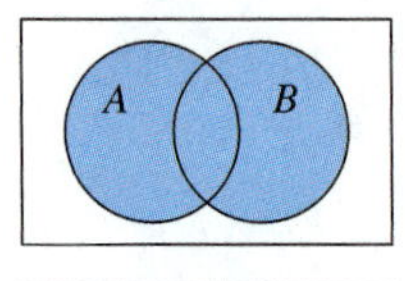

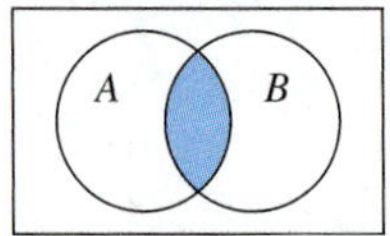

It is useful to have a mental picture of union and intersection. A *Venn diagram* depicts sets as circles or other shapes. In the figure, the shaded region in the upper diagram is $A \cup B$, and the shaded region in the lower diagram is $A \cap B$.

The operations of $\cup$ and $\cap$ obey various algebraic properties. We list some of them here.

Theorem 11.3 Let A, B, and C denote sets. The following are true:

- $A \cup B = B \cup A$ and $A \cap B = B \cap A$. (Commutative properties)
- $A \cup (B \cup C) = (A \cup B) \cup C$ and $A \cap (B \cap C) = (A \cap B) \cap C$. (Associative properties)
- $A \cup \emptyset = A$ and $A \cap \emptyset = \emptyset$.
- $A \cup (B \cap C) = (A \cup B) \cap (A \cup C)$ and $A \cap (B \cup C) = (A \cap B) \cup (A \cap C)$. (Distributive properties)

Proof. Most of the proof is left as Exercise 11.2. Theorem 6.2 is extremely useful in proving this result.

Here we prove the associative property for union. You may use this as a template for proving the other parts of this theorem.

Let A, B, and C be sets. We have the following:

$$\begin{aligned}
A \cup (B \cup C) &= \{x : (x \in A) \vee (x \in B \cup C)\} && \text{definition of union} \\
&= \{x : (x \in A) \vee ((x \in B) \vee (x \in C))\} && \text{definition of union} \\
&= \{x : ((x \in A) \vee (x \in B)) \vee (x \in C)\} && \text{associative property of } \vee \\
&= \{x : (x \in A \cup B) \vee (x \in C)\} && \text{definition of union} \\
&= (A \cup B) \cup C && \text{definition of union.} \quad \blacksquare
\end{aligned}$$

How did we think up this proof? We used the technique of writing the beginning and end of the proof and working toward the middle. Imagine a long sheet of paper. On the left, we write $A \cup (B \cup C) = \ldots$; on the right, we write $\ldots = (A \cup B) \cup C$. On the left, we unravel the definition of $\cup$ for the first $\cup$, obtaining $A \cup (B \cup C) = \{x : (x \in A) \vee (x \in B \cup C)\}$. We unravel the definition of $\cup$ again (this time on the $B \cup C$) to transform the set into

$$\{x : (x \in A) \vee ((x \in B) \vee (x \in C))\}.$$

Meanwhile, we do the same thing on the right. We unravel the second $\cup$ in $(A \cup B) \cup C$ to yield $\{x : (x \in A \cup B) \vee (x \in C)\}$ and then unravel $A \cup B$ to get $\{x : ((x \in A) \vee (x \in B)) \vee (x \in C)\}$.

Now we ask: What do we have and what do we want? On the left, we have

$$\{x : (x \in A) \vee ((x \in B) \vee (x \in C))\}$$

and on the right, we need

$$\{x : ((x \in A) \vee (x \in B)) \vee (x \in C)\}.$$

Finally, we stare at these two sets and realize that the conditions after the colon are logically equivalent (by Theorem 6.2) and we have our proof.

Venn diagrams are also useful for visualizing why these properties hold. For example, the following diagrams illustrate the distributive property $A \cup (B \cap C) = (A \cup B) \cap (A \cup C)$.

First examine the top row of pictures. On the left, we see the set A highlighted; in the center, the region for $B \cap C$ is shaded; and finally, on the right, we show $A \cup (B \cap C)$.

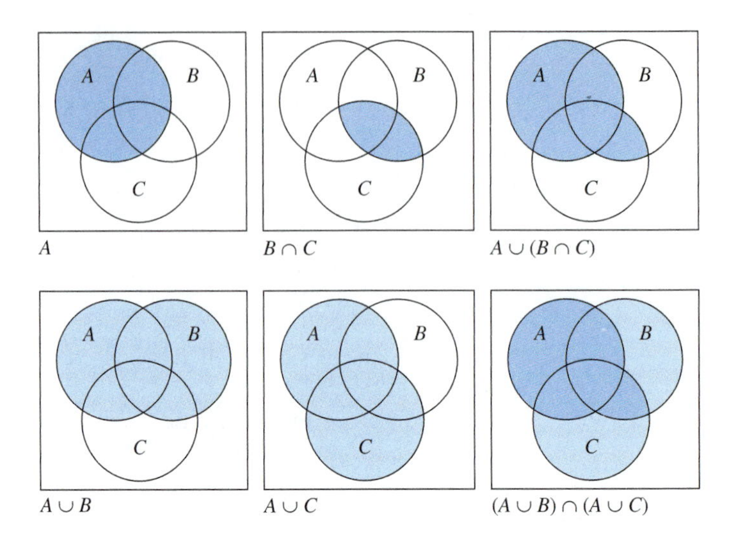

A $\quad B \cap C$ $\quad A \cup (B \cap C)$

$A \cup B$ $\quad A \cup C$ $\quad (A \cup B) \cap (A \cup C)$

Next examine the bottom row. The left and center pictures show $A \cup B$ and $A \cup C$ highlighted, respectively. The rightmost picture superimposes the first two, and the darkened region shows $(A \cup B) \cap (A \cup C)$.

Notice that exactly the same two shapes on the right panels (top and bottom) are dark, illustrating that $A \cup (B \cap C) = (A \cup B) \cap (A \cup C)$.

The Size of a Union

Suppose A and B are finite sets. There is a simple relationship between the quantities $|A|$, $|B|$, $|A \cup B|$, and $|A \cap B|$.

Proposition 11.4 Let A and B be finite sets. Then

$$|A| + |B| = |A \cup B| + |A \cap B|.$$

Proof. Imagine we assign labels to every object. We attach a label A to objects in the set A, and we attach a label B to objects in B.

Question: How many labels have we assigned?

On the one hand, the answer to this question is $|A| + |B|$ because we assign $|A|$ labels to the objects in A and $|B|$ labels to the objects in B.

On the other hand, we have assigned at least one label to the elements in $|A \cup B|$. So $|A \cup B|$ counts the number of objects that get at least one label. Elements in $A \cap B$ receive two labels. Thus $|A \cup B| + |A \cap B|$ counts all elements that receive a label and double counts those elements that receive two labels. This gives the number of labels.

Since these two quantities, $|A| + |B|$ and $|A \cup B| + |A \cap B|$, answer the same question, they must be equal. ■

This proof is an example of a *combinatorial proof.* Typically a combinatorial proof is used to demonstrate that an equation (such as the one in Proposition 11.4) is true. We do this by creating a question and then arguing that both sides of the equation give a correct answer to the question. It then follows, since both sides are correct answers, that the two sides of the alleged equation must be equal. This technique is summarized in Proof Template 9.

Proof Template 9 Combinatorial proof.

To prove an equation of the form LHS = RHS:

Pose a question of the form, "In how many ways . . . ?"
On the one hand, argue why LHS is a correct answer to the question.
On the other hand, argue why RHS is a correct answer.

Therefore LHS = RHS. ■

Finding the correct question to ask can be difficult. Writing combinatorial proofs is akin to playing the television game *Jeopardy!*. You are given the answer (indeed, two answers) to a counting question; your job is to find a question whose answers are the two sides of the equation you are trying to prove.

We shall do more combinatorial proofs, but for now, let us return to Proposition 11.4. One useful way to rewrite this result is as follows:

Basic inclusion-exclusion.

$$|A \cup B| = |A| + |B| - |A \cap B|. \tag{4}$$

This is a special case of a counting method called *inclusion-exclusion*. It can be interpreted as follows: Suppose we want to count the number of things that have one property or another. Imagine that set A contains those things that have the one property and set B contains those that have the other. Then the set $A \cup B$ contains those things that have one property or the other, and we can count those things by calculating $|A| + |B| - |A \cap B|$. This is useful when calculating $|A|$, $|B|$, and $|A \cap B|$ is easier than calculating $|A \cup B|$. We develop the concept of inclusion-exclusion more extensively in Section 18.

Example 11.5 How many integers in the range 1 to 1000 (inclusive) are divisible by 2 or by 5? Let

$$A = \{x \in \mathbb{Z} : 1 \leq x \leq 1000 \text{ and } 2|x\}, \qquad \text{and}$$
$$B = \{x \in \mathbb{Z} : 1 \leq x \leq 1000 \text{ and } 5|x\}.$$

The problem asks for $|A \cup B|$.

It is not hard to see that $|A| = 500$ and $|B| = 200$. Now $A \cap B$ are those numbers (in the range from 1 to 1000) that are divisible by both 2 and 5. Now an integer is divisible by both 2 and 5 if and only if it is divisible by 10 (this can be shown rigorously using ideas developed in Section 38; see Exercise 38.3), so

$$A \cap B = \{x \in \mathbb{Z} : 1 \leq x \leq 1000 \text{ and } 10|x\}$$

and it follows that $|A \cap B| = 100$. Finally, we have

$$|A \cup B| = |A| + |B| - |A \cap B| = 500 + 200 - 100 = 600.$$

There are 600 integers in the range 1 to 1000 that are divisible by 2 or by 5.

In case $A \cap B = \emptyset$, Equation (4) simplifies to $|A \cup B| = |A| + |B|$. In words, if two sets have no elements in common, then the size of their union equals the sum of their sizes. There is a special term for sets with no elements in common.

Definition 11.6 **(Disjoint, pairwise disjoint)** Let A and B be sets. We call A and B *disjoint* provided $A \cap B = \emptyset$.

Let $A_1, A_2, \ldots, A_n$ be a collection of sets. These sets are called *pairwise disjoint* provided $A_i \cap A_j = \emptyset$ whenever $i \neq j$. In other words, they are pairwise disjoint provided no two of them have an element in common.

Example 11.7 Let $A = \{1, 2, 3\}$, $B = \{4, 5, 6\}$, and $C = \{7, 8, 9\}$. These sets are pairwise disjoint because $A \cap B = A \cap C = B \cap C = \emptyset$.

However, let $X = \{1, 2, 3\}$, $Y = \{4, 5, 6, 7\}$, and $Z = \{7, 8, 9, 10\}$. This collection of sets is not pairwise disjoint because $Y \cap Z \neq \emptyset$ (all other pairwise intersections are empty).

Corollary 11.8 **(Addition Principle)** Let A and B be finite sets. If A and B are disjoint, then $|A \cup B| = |A| + |B|$.

Corollary 11.8 follows immediately from Proposition 11.4. There is an extension of the Addition Principle to more than two sets.

If $A_1, A_2, \ldots, A_n$ are pairwise disjoint sets, then

$$|A_1 \cup A_2 \cup \cdots \cup A_n| = |A_1| + |A_2| + \cdots + |A_n|.$$

This can be shown formally using the methods from Section 20; see Exercise 20.9.

A fancy way to write this is

$$\left| \bigcup_{k=1}^{n} A_k \right| = \sum_{k=1}^{n} |A_k|.$$

The big $\bigcup$ is analogous to the $\sum$ and $\prod$ symbols. It means, as k goes from 1 to n (the lower and upper values), take the union of the expression to the right (in this case A_k). So the big $\bigcup$ notation is just a shorthand for $A_1 \cup A_2 \cup \cdots \cup A_n$. This is surrounded by vertical bars, so we want the size of that set. On the right, we see an ordinary summation symbol telling us to add up the cardinalities of A_1, $A_2, \ldots, A_n$.

Difference and Symmetric Difference

Definition 11.9 **(Set difference)** Let A and B be sets. The *set difference*, $A - B$, is the set of all elements of A that are not in B:

$$A - B = \{x : x \in A \text{ and } x \notin B\}.$$

The *symmetric difference* of A and B, denoted $A \Delta B$, is the set of all elements in A but not B or in B but not A. That is,

$$A \Delta B = (A - B) \cup (B - A).$$

Example 11.10 Suppose $A = \{1, 2, 3, 4\}$ and $B = \{3, 4, 5, 6\}$. Then $A - B = \{1, 2\}$, $B - A = \{5, 6\}$, and $A \Delta B = \{1, 2, 5, 6\}$.

The figures show Venn diagram for these operations.

In general, the sets $A - B$ and $B - A$ are different (but see Exercise 11.14).

Here is another way to express symmetric difference:

Proposition 11.11 Let A and B be sets. Then

$$A \Delta B = (A \cup B) - (A \cap B).$$

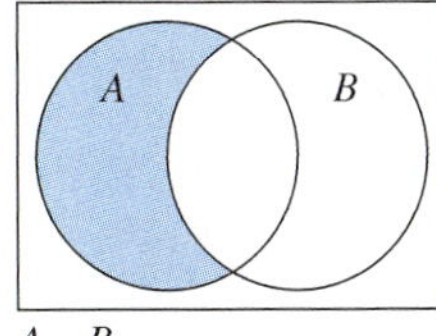

$A - B$

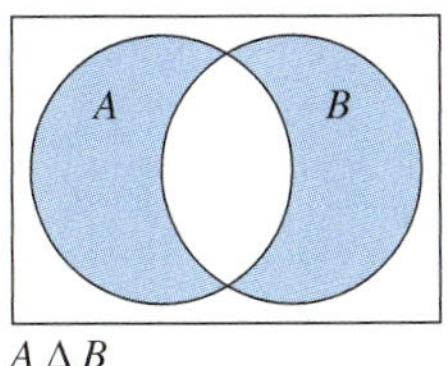

$A \Delta B$

Let us illustrate the various proof techniques by developing the proof of Proposition 11.11 step by step. The proposition asks us to prove that two sets are equal, namely, $A \Delta B$ and $(A \cup B) - (A \cap B)$. We use Proof Template 5 to form the skeleton of the proof.

> Let A and B be sets.
>
> **(1)** Suppose $x \in A \Delta B$. . . . Therefore $x \in (A \cup B) - (A \cap B)$.
> **(2)** Suppose $x \in (A \cup B) - (A \cap B)$. . . . Therefore $x \in A \Delta B$.
>
> Therefore $A \Delta B = (A \cup B) - (A \cap B)$. ■

We begin with part (1) of the proof. We unravel definitions from both ends. We know that $x \in A \Delta B$. By definition of Δ, this means $x \in (A - B) \cup (B - A)$. The proof now reads as follows:

> Let A and B be sets.
>
> **(1)** Suppose $x \in A \Delta B$. Thus $x \in (A - B) \cup (B - A)$. . . . Therefore $x \in (A \cup B) - (A \cap B)$.
> **(2)** Suppose $x \in (A \cup B) - (A \cap B)$. . . . Therefore $x \in A \Delta B$.
>
> Therefore $A \Delta B = (A \cup B) - (A \cap B)$. ■

Now we know that $x \in (A - B) \cup (B - A)$. What does this mean? By definition of union, it means that $x \in (A - B)$ or $x \in (B - A)$. We have to consider both possibilities since we don't know in which of these sets x lies. This means that part (1) of the proof breaks into cases depending on whether $x \in A - B$ or $x \in B - A$. In both cases, we need to show that $x \in (A \cup B) - (A \cap B)$.

Let A and B be sets.

(1) Suppose $x \in A \,\Delta\, B$. Thus $x \in (A - B) \cup (B - A)$. This means either $x \in A - B$ or $x \in B - A$. We consider both cases.
- Suppose $x \in A - B$.... Therefore $x \in (A \cup B) - (A \cap B)$.
- Suppose $x \in B - A$.... Therefore $x \in (A \cup B) - (A \cap B)$.

...Therefore $x \in (A \cup B) - (A \cap B)$.

(2) Suppose $x \in (A \cup B) - (A \cap B)$.... Therefore $x \in A \,\Delta\, B$.

Therefore $A \,\Delta\, B = (A \cup B) - (A \cap B)$. ■

Let's focus on the first case, $x \in A - B$. This means that $x \in A$ and $x \notin B$. We put that in.

Let A and B be sets.

(1) Suppose $x \in A \,\Delta\, B$. Thus $x \in (A - B) \cup (B - A)$. This means either $x \in A - B$ or $x \in B - A$. We consider both cases.
- Suppose $x \in A - B$. So $x \in A$ and $x \notin B$. ...Therefore $x \in (A \cup B) - (A \cap B)$.
- Suppose $x \in B - A$.... Therefore $x \in (A \cup B) - (A \cap B)$.

...Therefore $x \in (A \cup B) - (A \cap B)$.

(2) Suppose $x \in (A \cup B) - (A \cap B)$.... Therefore $x \in A \,\Delta\, B$.

Therefore $A \,\Delta\, B = (A \cup B) - (A \cap B)$. ■

We appear to be stuck. We have unraveled definitions down to $x \in A$ and $x \notin B$. To proceed, we work backward from our goal; we want to show that $x \in (A \cup B) - (A \cap B)$. To do that, we need to show that $x \in A \cup B$ and $x \notin A \cap B$. We add this language to the proof.

Let A and B be sets.

(1) Suppose $x \in A \,\Delta\, B$. Thus $x \in (A - B) \cup (B - A)$. This means either $x \in A - B$ or $x \in B - A$. We consider both cases.
- Suppose $x \in A - B$. So $x \in A$ and $x \notin B$....
  Thus $x \in A \cup B$, but $x \notin A \cap B$. Therefore $x \in (A \cup B) - (A \cap B)$.
- Suppose $x \in B - A$.... Therefore $x \in (A \cup B) - (A \cap B)$.

...Therefore $x \in (A \cup B) - (A \cap B)$.

(2) Suppose $x \in (A \cup B) - (A \cap B)$.... Therefore $x \in A \,\Delta\, B$.

Therefore $A \,\Delta\, B = (A \cup B) - (A \cap B)$. ■

Now the two parts of this proof are moving closer together. Let's record what we know and what we want.

We already know:	$x \in A$ and $x \notin B$.
We want to show:	$x \in A \cup B$ and $x \notin A \cap B$.

The gap is now easy to close! Since we know $x \in A$, certainly x is in A or B (we just said it's in A!), so $x \in A \cup B$. Since $x \notin B$, x is not in both A and B (we just said it's not in B!), so $x \notin A \cap B$. We add this to the proof.

Let A and B be sets.

(1) Suppose $x \in A \,\Delta\, B$. Thus $x \in (A - B) \cup (B - A)$. This means either $x \in A - B$ or $x \in B - A$. We consider both cases.

- Suppose $x \in A - B$. So $x \in A$ and $x \notin B$. Since $x \in A$, we have $x \in A \cup B$. Since $x \notin B$, we have $x \notin A \cap B$. Thus $x \in A \cup B$, but $x \notin A \cap B$. Therefore $x \in (A \cup B) - (A \cap B)$.
- Suppose $x \in B - A$. ... Therefore $x \in (A \cup B) - (A \cap B)$.

... Therefore $x \in (A \cup B) - (A \cap B)$.

(2) Suppose $x \in (A \cup B) - (A \cap B)$. ... Therefore $x \in A \,\Delta\, B$.

Therefore $A \,\Delta\, B = (A \cup B) - (A \cap B)$. ■

We can now return to the second case of part (1) of the proof: "Suppose $x \in B - A$. ... Therefore $x \in (A \cup B) - (A \cap B)$." We have good news! This case looks just like the previous case, except we have A and B switched around. The argument in this case is going to proceed exactly as before. Since the steps are (essentially) the same, we don't really have to write them out. (If you are not 100% certain that the steps in this second case are exactly the same as before, I urge you to write out this portion of the proof for yourself using the previous case as a guide.) We can now complete part (1) of the proof.

Let A and B be sets.

(1) Suppose $x \in A \,\Delta\, B$. Thus $x \in (A - B) \cup (B - A)$. This means either $x \in A - B$ or $x \in B - A$. We consider both cases.

- Suppose $x \in A - B$. So $x \in A$ and $x \notin B$. Since $x \in A$, we have $x \in A \cup B$. Since $x \notin B$, we have $x \notin A \cap B$. Thus $x \in A \cup B$, but $x \notin A \cap B$. Therefore $x \in (A \cup B) - (A \cap B)$.
- Suppose $x \in B - A$. By the same argument as above, we have $x \in (A \cup B) - (A \cap B)$.

Therefore $x \in (A \cup B) - (A \cap B)$.

(2) Suppose $x \in (A \cup B) - (A \cap B)$. ... Therefore $x \in A \,\Delta\, B$.

Therefore $A \,\Delta\, B = (A \cup B) - (A \cap B)$. ■

Now we are ready to work on part (2). We begin by unraveling $x \in (A \cup B) - (A \cap B)$. This means that $x \in A \cup B$, but $x \notin A \cap B$ (by the definition of set difference).

> Let A and B be sets.
>
> **(1)** Suppose $x \in A \,\Delta\, B$. Thus $x \in (A - B) \cup (B - A)$. This means either $x \in A - B$ or $x \in B - A$. We consider both cases.
> - Suppose $x \in A - B$. So $x \in A$ and $x \notin B$. Since $x \in A$, we have $x \in A \cup B$. Since $x \notin B$, we have $x \notin A \cap B$. Thus $x \in A \cup B$, but $x \notin A \cap B$. Therefore $x \in (A \cup B) - (A \cap B)$.
> - Suppose $x \in B - A$. By the same argument as above, we have $x \in (A \cup B) - (A \cap B)$.
>
> Therefore $x \in (A \cup B) - (A \cap B)$.
>
> **(2)** Suppose $x \in (A \cup B) - (A \cap B)$. Thus $x \in A \cup B$ and $x \notin A \cap B$. ... Therefore $x \in A \,\Delta\, B$.
>
> Therefore $A \,\Delta\, B = (A \cup B) - (A \cap B)$. ■

Now let's work backward from the end of part (2). We want to show $x \in A \Delta B$, so we need to show $x \in (A - B) \cup (B - A)$.

> Let A and B be sets.
>
> **(1)** Suppose $x \in A \,\Delta\, B$. Thus $x \in (A - B) \cup (B - A)$. This means either $x \in A - B$ or $x \in B - A$. We consider both cases.
> - Suppose $x \in A - B$. So $x \in A$ and $x \notin B$. Since $x \in A$, we have $x \in A \cup B$. Since $x \notin B$, we have $x \notin A \cap B$. Thus $x \in A \cup B$, but $x \notin A \cap B$. Therefore $x \in (A \cup B) - (A \cap B)$.
> - Suppose $x \in B - A$. By the same argument as above, we have $x \in (A \cup B) - (A \cap B)$.
>
> Therefore $x \in (A \cup B) - (A \cap B)$.
>
> **(2)** Suppose $x \in (A \cup B) - (A \cap B)$. Thus $x \in A \cup B$ and $x \notin A \cap B$. ... So $x \in (A - B) \cup (B - A)$. Therefore $x \in A \,\Delta\, B$.
>
> Therefore $A \,\Delta\, B = (A \cup B) - (A \cap B)$. ■

To show $x \in (A - B) \cup (B - A)$, we need to show that either $x \in A - B$ or $x \in B - A$. Let's pause and write down what we know and what we want.

We already know:	$x \in A \cup B$ and $x \notin A \cap B$.
We want to show:	$x \in A - B$ or $x \in B - A$.

What we know says: x is in A or B but not both. In other words, either x is in A, in which case it's not in B, or x is in B, in which case it's not in A. In other words, $x \in A - B$ or $x \in B - A$, and that's what we want to show! Let's work this into the proof.

Let A and B be sets.

(1) Suppose $x \in A \Delta B$. Thus $x \in (A - B) \cup (B - A)$. This means either $x \in A - B$ or $x \in B - A$. We consider both cases.

- Suppose $x \in A - B$. So $x \in A$ and $x \notin B$. Since $x \in A$, we have $x \in A \cup B$. Since $x \notin B$, we have $x \notin A \cap B$. Thus $x \in A \cup B$, but $x \notin A \cap B$. Therefore $x \in (A \cup B) - (A \cap B)$.
- Suppose $x \in B - A$. By the same argument as above, we have $x \in (A \cup B) - (A \cap B)$.

Therefore $x \in (A \cup B) - (A \cap B)$.

(2) Suppose $x \in (A \cup B) - (A \cap B)$. Thus $x \in A \cup B$ and $x \notin A \cap B$. This means that x is in A or B but not both. Thus either x is in A but not B or x is in B but not A. That is, $x \in (A - B)$ or $x \in (B - A)$. So $x \in (A - B) \cup (B - A)$. Therefore $x \in A \Delta B$.

Therefore $A \Delta B = (A \cup B) - (A \cap B)$. ■

And this completes the proof.

More properties of difference and symmetric difference are developed in the exercises. One particularly worthwhile result, however, is the following:

Proposition 11.12 **(DeMorgan's Laws)** Let A, B, and C be sets. Then

$$A - (B \cup C) = (A - B) \cap (A - C) \quad \text{and} \quad A - (B \cap C) = (A - B) \cup (A - C).$$

The proof is left to you (Exercise 11.15).

Cartesian Product

We close this section with one more set operation.

Definition 11.13 **(Cartesian product)** Let A and B be sets. The *Cartesian product* of A and B, denoted $A \times B$, is the set of all ordered pairs (two-element lists) formed by taking an element from A together with an element from B in all possible ways. That is,

$$A \times B = \{(a, b) : a \in A, b \in B\}.$$

Example 11.14 Suppose $A = \{1, 2, 3\}$ and $B = \{3, 4, 5\}$. Then

$$A \times B = \{(1,3), (1,4), (1,5), (2,3), (2,4), (2,5), (3,3), (3,4), (3,5)\}, \quad \text{and}$$
$$B \times A = \{(3,1), (3,2), (3,3), (4,1), (4,2), (4,3), (5,1), (5,2), (5,3)\}.$$

Notice that for the sets in Example 11.14, $A \times B \neq B \times A$, so Cartesian product of sets is not a commutative operation.

In what sense does Cartesian product "multiply" the sets? Why do we use a times sign $\times$ to denote this operation? Notice, in the example, that the two sets both had three elements, and their product had $3 \times 3 = 9$ elements. In general, we have the following:

Proposition 11.15 Let A and B be finite sets. Then $|A \times B| = |A| \times |B|$.

The proof is left for Exercise 11.24.

Recap

In this section we discussed the following set operations:

- Union: $A \cup B$ is the set of all elements in A or B (or both).
- Intersection: $A \cap B$ is the set of all elements in both A and B.
- Set difference: $A - B$ is the set of all elements in A but not B.
- Symmetric difference: $A \Delta B$ is the set of all elements in A or B, but not both.
- Cartesian product: $A \times B$ is the set of all ordered pairs of the form (a, b) where $a \in A$ and $b \in B$.

11 Exercises

11.1. Let $A = \{1, 2, 3, 4, 5\}$ and let $B = \{4, 5, 6, 7\}$. Please compute:
- **a.** $A \cup B$.
- **b.** $A \cap B$.
- **c.** $A - B$.
- **d.** $B - A$.
- **e.** $A \Delta B$.
- **f.** $A \times B$.
- **g.** $B \times A$.

11.2. Prove Theorem 11.3.

11.3. Earlier we presented a Venn diagram illustration of the distributive property $A \cup (B \cap C) = (A \cup B) \cap (A \cup C)$. Please give a Venn diagram illustration of the other distributive property, $A \cap (B \cup C) = (A \cap B) \cup (A \cap C)$.

11.4. Is a Venn diagram illustration a proof? (This is a philosophical question.)

11.5. Suppose A, B, and C are sets with $A \cap B \cap C = \emptyset$. Prove or disprove: $|A \cup B \cup C| = |A| + |B| + |C|$.

11.6. Suppose A, B, and C are pairwise disjoint sets. Prove or disprove: $|A \cup B \cup C| = |A| + |B| + |C|$.

11.7. Let A and B be sets. Prove or disprove: $A \cup B = A \cap B$ if and only if $A = B$.

11.8. Let A and B be sets. Prove or disprove: $|A \Delta B| = |A| + |B| - |A \cap B|$.

11.9. Let A and B be sets. Prove or disprove: $|A \Delta B| = |A - B| + |B - A|$.

11.10. Let A be a set. Prove: $A - \emptyset = A$ and $\emptyset - A = \emptyset$.

11.11. Let A be a set. Prove: $A \Delta A = \emptyset$ and $A \Delta \emptyset = A$.

11.12. Prove that $A \subseteq B$ if and only if $A - B = \emptyset$.

11.13. Let A and B be nonempty sets. Prove: $A \times B = B \times A$ if and only if $A = B$.
Why do we need the condition that A and B are nonempty?

11.14. State and prove necessary and sufficient conditions for $A - B = B - A$. In other words, create a theorem of the form "Let A and B be sets. We have $A - B = B - A$ if and only if (a condition on A and B)." Then prove your result.

11.15. Give a standard proof of Proposition 11.12 and illustrate it with a Venn diagram.

11.16. *True or False*: For each of the following statements, determine whether the statement is true or false and then prove your assertion. That is, for each true statement, please supply a proof, and for each false statement, present a counterexample (with explanation).

In the following, A, B, and C denote sets.

a. $A - (B - C) = (A - B) - C$.
b. $(A - B) - C = (A - C) - B$.
c. $(A \cup B) - C = (A - C) \cap (B - C)$.
d. If $A = B - C$, then $B = A \cup C$.
e. If $B = A \cup C$, then $A = B - C$.
f. $|A - B| = |A| - |B|$.
g. $(A - B) \cup B = A$.
h. $(A \cup B) - B = A$.

Set complement.

11.17. Let A be a set. The *complement* of A, denoted $\overline{A}$, is the set of all objects that are not in A. STOP! This definition needs some amending. Taken literally, the complement of the set $\{1, 2, 3\}$ includes the number -5, the ordered pair $(3, 4)$, and the sun, moon, and stars! After all, it says "... *all objects* that are not in A." This is not what is intended.

When mathematicians speak of set complements, they usually have some overarching set in mind. For example, during a given proof or discussion about the integers, if A is a set containing just integers, $\overline{A}$ stands for the set containing all integers not in A.

If U (for "universe") is the set of all objects under consideration and $A \subseteq U$, then the complement of A is the set of all objects in U that are not in A. In other words, $\overline{A} = U - A$. Thus $\overline{\emptyset} = U$.

Prove the following about set complements. Here the letters A, B, and C denote subsets of a universe set U.

a. $A = B$ if and only if $\overline{A} = \overline{B}$.
b. $\overline{\overline{A}} = A$.
c. $\overline{A \cup B \cup C} = \overline{A} \cap \overline{B} \cap \overline{C}$.

The notation $\overline{A}$ is handy, but it can be ambiguous. Unless it is perfectly clear what the "universe" set U should be, it is better to use the set difference notation rather than complement notation.

The notation $U - A$ is much clearer than $\overline{A}$.

11.18. Design a four-set Venn diagram. Notice that the three-set Venn diagram we have been using has eight regions (including the region surrounding the four circles) corresponding to the eight possible memberships an object might have. An object might be in or not in A, in or not in B, and in or not in C.

Explain why this gives eight possibilities.

Your Venn diagram should show four sets, A, B, C, and D. How many regions should your diagram have?

On your Venn diagram, shade in the set $A \,\Delta\, B \,\Delta\, C \,\Delta\, D$.

Note: Your diagram does not have to use circles to demark sets. Indeed, it is impossible to create a Venn diagram for four sets using circles! You need to use other shapes.

An expanded version of inclusion-exclusion.

11.19. Let A, B, and C be sets. Prove that

$$|A \cup B \cup C| = |A| + |B| + |C| - |A \cap B| - |A \cap C| - |B \cap C| + |A \cap B \cap C|.$$

The connection between set operations and Boolean algebra.

11.20. There is an intimate connection between set concepts and Boolean algebra concepts. The symbols $\wedge$ and $\vee$ are pointy versions of $\cap$ and $\cup$, respectively. This is more than a coincidence. Consider:

$$x \in A \cap B \iff (x \in A) \wedge (x \in B)$$
$$x \in A \cup B \iff (x \in A) \vee (x \in B)$$

Find similar relations between the set-theoretic notions of $\subseteq$ and Δ and notions from Boolean algebra.

11.21. Prove that symmetric difference is a commutative operation; that is, for sets A and B, we have $A \Delta B = B \Delta A$.

11.22. Prove that symmetric difference is an associative operation; that is, for any sets A, B, and C, we have $A \Delta (B \Delta C) = (A \Delta B) \Delta C$.

11.23. Give a Venn diagram illustration of $A \Delta (B \Delta C) = (A \Delta B) \Delta C$.

11.24. Prove Proposition 11.15.

11.25. Let A, B, and C denote sets. Prove the following:

a. $A \times (B \cup C) = (A \times B) \cup (A \times C)$.

b. $A \times (B \cap C) = (A \times B) \cap (A \times C)$.

c. $A \times (B - C) = (A \times B) - (A \times C)$.

d. $A \times (B \Delta C) = (A \times B) \Delta (A \times C)$.

12 Combinatorial Proof: Two Examples

In Section 11 we introduced the concept of combinatorial proof of equations. This technique works by showing that both sides of an equation are answers to a common question. This method was used to prove Proposition 11.4 (for finite sets A and B we have $|A| + |B| = |A \cup B| + |A \cap B|$). See Proof Template 9.

In this section we give two examples that further illustrate this technique. One is based on a set-counting problem and the other on a list-counting problem.

Proposition 12.1 Let n be a positive integer. Then

$$2^0 + 2^1 + \cdots + 2^{n-1} = 2^n - 1.$$

For example, $2^0 + 2^1 + 2^2 + 2^3 + 2^4 = 1 + 2 + 4 + 8 + 16 = 31 = 2^5 - 1$. We seek a question to which both sides of the equation give a correct answer.

The right-hand side is simpler, so let us begin there. The 2^n term answers the question "How many subsets does an n-element set have?" However, the term is $2^n - 1$, not 2^n. We can modify the question to rule out all but one of the subsets. Which subset should we ignore? A natural choice is to skip the empty set. The rephrased question is "How many nonempty subsets does an n-element set have?" Now it is clear that the right-hand side of the equation, $2^n - 1$, is a correct answer. But what of the left?

The left-hand side is a long sum, with each term of the form 2^j. This is a hint that we are considering several subset-counting problems. Somehow, the question of how many nonempty subsets an n-element set has must be broken down into disjoint cases (each a subset-counting problem unto itself) and then combined to give the full answer.

We know we are counting nonempty subsets of an n-element set. For the sake of specificity, suppose the set is $\{1, 2, \ldots, n\}$. Let's start writing down the nonempty subsets of this set. It's natural to start with $\{1\}$. Next we write down $\{1, 2\}$ and $\{2\}$—these are the sets whose largest element is 2. Next we write down the sets whose largest element is 3. Let's organize this into a chart.

Largest element	Subsets of $\{1, 2, \ldots, n\}$
1	$\{1\}$
2	$\{2\}, \{1, 2\}$
3	$\{3\}, \{1, 3\}, \{2, 3\}, \{1, 2, 3\}$
4	$\{4\}, \{1, 4\}, \{2, 4\}, \{1, 2, 4\}, \ldots, \{1, 2, 3, 4\}$
$\vdots$	$\vdots$
n	$\{n\}, \{1, n\}, \{2, n\}, \{1, 2, n\}, \ldots, \{1, 2, 3, \ldots, n\}$

We neglected to write out all the subsets on line 4 of the chart. How many are there? The sets on this line must contain 4 (since that's the largest element). The other elements of these sets are chosen from among 1, 2, and 3. Because there are $2^3 = 8$ possible ways to form a subset of $\{1, 2, 3\}$, there must be 8 sets on this line. Please take a moment to verify this for yourself by completing line 4 of the chart.

Now skip to the last line of the chart. How many subsets of $\{1, 2, \ldots, n\}$ have largest element n? We must include n together with any subset of $\{1, 2, \ldots, n-1\}$, for a total of 2^{n-1} choices.

Notice that every nonempty subset of $\{1, 2, \ldots, n\}$ must appear exactly once in the chart. Totaling the row sizes gives

$$1 + 2 + 4 + 8 + \cdots + 2^{n-1}.$$

Aha! This is precisely the lefthand side of the equation we seek to prove.

Armed with these insights, we are ready to write the proof.

Proof (of Proposition 12.1)

Let n be a positive integer, and let $N = \{1, 2, \ldots, n\}$. How many nonempty subsets does N have?

Answer 1: Since N has 2^n subsets, when we disregard the empty set, we see that N has $2^n - 1$ nonempty subsets.

Answer 2: We consider the number of subsets of N whose largest element is j where $1 \le j \le n$. Such subsets must be of the form $\{\ldots, j\}$ where the other elements are chosen from $\{1, \ldots, j-1\}$. Since this latter set has 2^{j-1} subsets, N has 2^{j-1} subsets whose largest element is j. Summing these answers over all j gives

$$2^0 + 2^1 + 2^2 + \cdots + 2^{n-1}$$

nonempty subsets of N.

Since answers 1 and 2 are both correct solutions to the same counting problem, we have

$$2^0 + 2^1 + 2^2 + \cdots + 2^{n-1} = 2^n. \qquad ■$$

We now turn to a second example.

Proposition 12.2 Let n be a positive integer. Then

$$1 \cdot 1! + 2 \cdot 2! + \cdots + n \cdot n! = (n+1)! - 1.$$

For example, with $n = 4$, observe that

$$\begin{aligned} 1 \cdot 1! + 2 \cdot 2! + 3 \cdot 3! + 4 \cdot 4! &= 1 \cdot 1 + 2 \cdot 2 + 3 \cdot 6 + 4 \cdot 24 \\ &= 1 + 4 + 18 + 96 \\ &= 119 = 120 - 1 = 5! - 1. \end{aligned}$$

The key to proving Proposition 12.2 is to find a question to which both sides of the equation give a correct answer. As with the first example, the righthand side is simpler, so we begin there.

The $(n+1)!$ term reminds us of counting lists without replacement. Specifically, it answers the question "How many lists can we form using the elements of $\{1, 2, \ldots, n+1\}$ in which every element is used exactly once?" Because the righthand side also includes a -1 term, we need to discard one of these lists. Which? A natural choice is to skip the list $(1, 2, 3, \ldots, n+1)$; this is the only list in which every element j appears in position j for every $j = 1, 2, \ldots, n$. In every other list, some element j is not in the jth position on this list. Alternatively, the discarded list is the only one in which the elements appear in increasing order.

We therefore consider the question "How many lists can we form using the elements of $\{1, 2, \ldots, n+1\}$ in which every element appears exactly once and in which the elements do not appear in increasing order?"

Clearly $(n+1)! - 1$ is one solution to this problem; we need to show that the lefthand side is also a correct answer. If the elements in the list are not in increasing order, then some element, say k, will not be in position k. We can organize this counting problem by considering where this first happens.

Let us consider the case $n = 4$. We form a chart containing all length-5 repetition-free lists we can form from the elements of $\{1, 2, 3, 4, 5\}$ that are not in increasing order. We organize the chart by considering the first time slot k is not element k. For example, when $k = 3$ the lists are $12\underline{4}35$, $12\underline{4}53$, $12\underline{5}34$, and

12$\underline{5}$43 since the entries in positions 1 and 2 are elements 1 and 2, respectively, but entry 3 is not 3. (We have omitted the commas and parentheses for the sake of clarity.)

The chart for $n = 4$ follows.

k	first "misplaced" element at position k	
1	21345 21354 21435 21453 21534 21534	23145 23154 23415 23154 23514 23541
	24135 24153 24315 24351 24513 24531	25134 25143 25314 25341 25413 25431
	31245 31254 31425 31452 31524 31542	32145 32154 32415 32451 32514 32541
	34125 34152 34215 34251 34512 34521	35124 35142 35214 35241 35412 35421
	41235 41253 41325 41352 41523 41532	42135 42153 42315 42351 42513 42531
	43125 43152 43215 43251 43512 43521	45123 45132 45213 45231 45312 45321
	51234 51243 51324 51342 51423 51432	52134 52143 52314 52341 52413 52431
	53124 53142 53214 53241 53412 53421	54123 54132 54213 54231 54312 54321
2	13245 13254 13425 13452 13524 13542	
	14235 14253 14325 14352 14523 14532	
	15234 15243 15324 15342 15423 15432	
3	12435 12453 12534 12543	
4	12354	
5	—	

Notice that row 5 of the chart is empty; why? This row should contain all repetition-free lists in which the first slot k that does not contain element k is $k = 5$. Such a list must be of the form $(1, 2, 3, 4, ?)$, but then there is no valid way to fill in the last position.

Next, count the number of lists in each portion of the chart. Working from the bottom, there are $1 + 4 + 18 + 96 = 119$ lists (all $5! = 120$ except the list $(1, 2, 3, 4, 5)$). The sum $1 + 4 + 18 + 96$ should be familiar; it is precisely $1 \cdot 1! + 2 \cdot 2! + 3 \cdot 3! + 4 \cdot 4!$. Of course, this is not a coincidence. Consider the first row of the chart. The lists in this row must not begin with a 1 but may begin with any element of $\{2, 3, 4, 5\}$; there are 4 choices for the first element. Once the first element is chosen, the remaining four elements in the lists may be chosen in any way we like. Since there are 4 elements remaining (after selecting the first), these 4 elements can be arranged in $4!$ ways. Thus, by the Multiplication Principle, there are $4 \cdot 4!$ lists in which the first element is not 1.

The same analysis works for the second row. Lists on this row must begin with a 1, and then the second element must not be a 2. There are 3 choices for the second element because we must choose it from $\{3, 4, 5\}$. Once the second element has been selected, the remaining three elements may be arranged in any way we wish, and there are $3!$ ways to do so. Thus the second row of the chart contains $3 \cdot 3! = 18$ lists.

We are ready to complete the proof.

Proof (of Proposition 12.2)

Let n be a positive integer. We ask, "How many repetition-free lists can we form using all the elements in $\{1, 2, \ldots, n + 1\}$ in which the elements do not appear in increasing order?"

Answer 1: There are $(n+1)!$ repetition-free lists, and in only one such list do the elements appear in order, namely $(1, 2, \ldots, n, n+1)$. Thus the answer to the question is $(n+1)! - 1$.

Answer 2: Let j be an integer between 1 and n, inclusive. Let us consider those lists in which the first $j-1$ elements are $1, 2, \ldots, j-1$, respectively, but for which the jth element is not j. How many such lists are there? For element j there are $n+1-j$ choices because elements 1 through $j-1$ have already been chosen and we may not use element j. The remaining $n+1-j$ elements may fill in the remaining slots on the list in any order, giving $(n+1-j)!$ possibilities. By the Multiplication Principle, there are $(n+1-j) \cdot (n+1-j)!$ such lists. Summing over $j = 1, 2, \ldots, n$ gives

$$n \cdot n! + (n-1) \cdot (n-1)! + \cdots + 3 \cdot 3! + 2 \cdot 2! + 1 \cdot 1!.$$

Since answers 1 and 2 are both correct solutions to the same counting problem, we have

$$1 \cdot 1! + 2 \cdot 2! + \cdots + n \cdot n! = (n+1)! - 1.$$

■

Recap

In this section we illustrated the concept of combinatorial proof by applying the technique to demonstrate two identities.

12 Exercises

12.1. Give an alternative proof of Proposition 12.1 in which you use list counting instead of subset counting.

12.2. Let n be a positive integer. Use algebra to simplify the following expression:

$$(x-1)(1 + x + x^2 + \cdots + x^{n-1}).$$

Use this to give another proof of Proposition 12.1.

12.3. Substituting $x = 3$ into your expression in the previous problem yields

$$2 \cdot 3^0 + 2 \cdot 3^1 + 2 \cdot 3^2 + \cdots + 2 \cdot 3^{n-1} = 3^n - 1.$$

Prove this equation combinatorially.

Next, substitute $x = 10$ and illustrate the result using ordinary base-10 numbers.

12.4. Let a and b be positive integers with $a > b$. Give a combinatorial proof of the identity $(a+b)(a-b) = a^2 - b^2$.

12.5. Let n be a positive integer. Give a combinatorial proof that $n^2 = n(n-1) + n$.

Chapter 2 Self Test

1. The call sign for a radio station in the United States is a list of three or four letters, such as WJHU or WJZ. The first letter must be a W or a K, and there is no restriction on the other letters. In how many ways can the call sign of a radio station be formed?

2. In how many ways can we make a list of three integers (a, b, c) where $0 \le a, b, c \le 9$ and $a + b + c$ is even?

3. In how many ways can we make a list of three integers (a, b, c) where $0 \le a, b, c \le 9$ and abc is even?

4. Without the use of any computational aid, simplify the following expression:

$$\frac{20!}{17! \cdot 3!}$$

5. In how many ways can we arrange a standard deck of 52 cards so that all cards in a given suit appear contiguously (e.g., first all the spades appear, then all the diamonds, then all the hearts, and then all the clubs)?

6. Ten married couples are waiting in line to enter a restaurant. Husbands and wives stand next to each other, but either one might be ahead of the other. How many such arrangements are possible?

7. Evaluate the following:

$$\prod_{k=0}^{100} \frac{k^2}{k+1}.$$

8. Let $A = \{x \in \mathbb{Z} : |x| < 10\}$. Evaluate $|A|$.

9. Let $A = \{1, 2, \{3, 4\}\}$. Which of the following are true and which false? No proof is required.

a. $1 \in A$.
b. $\{1\} \in A$.
c. $3 \in A$.
d. $\{3\} \in A$.
e. $\{3\} \subseteq A$.

10. Let A and B be finite sets. Determine whether the following statements are true or false. Justify your answer with a proof or counterexample, as appropriate.

a. $2^{A \cap B} = 2^A \cap 2^B$.
b. $2^{A \cup B} = 2^A \cup 2^B$.
c. $2^{A \Delta B} = 2^A \Delta 2^B$.

11. Let A be a set. Which of the following are true and which false?

a. $x \in A$ iff $x \in 2^A$.
b. $T \subseteq A$ iff $T \in 2^A$.
c. $x \in A$ iff $\{x\} \in 2^A$.
d. $\{x\} \in A$ iff $\{\{x\}\} \in 2^A$.

12. Which of the following statements about integers are true and which false? No proof is required.

a. $\forall x, \forall y, x > y$.
b. $\exists x, \forall y, x > y$.
c. $\forall x, \exists y, x > y$.
d. $\exists x, \exists y, x > y$.

13. Let $p(x, y)$ stand for a sentence about two integers, x and y. For example, $p(x, y)$ could mean "$x - y$ is a perfect square."

Assume the statement $\forall x, \exists y, p(x, y)$ is true. Which of the following statements about integers must also be true?

a. $\forall x, \exists y, \neg p(x, y)$.

b. $\neg(\exists x, \forall y, \neg p(x, y))$.

c. $\exists x, \exists y, p(x, y)$.

14. Let A and B be sets and suppose $A \times B = \{(1, 2), (1, 3), (2, 2), (2, 3)\}$. Find $A \cup B$, $A \cap B$, and $A - B$.

15. Let A, B, and C denote sets. Prove that $(A \cup B) - C = (A - C) \cup (B - C)$ and give a Venn diagram illustration.

16. Suppose A and B are finite sets. Given that $|A| = 10$, $|A \cup B| = 15$, and $|A \cap B| = 3$, determine $|B|$.

17. Let A and B be sets. Create an expression that evaluates to $A \cap B$ and uses only the operations union and set difference. That is, find a formula that uses only the symbols $A, B, \cup, -$, and parentheses; this formula should equal $A \cap B$ for all sets A and B.

18. Let n be a positive integer. Give a combinatorial proof of the identity

$$n^3 = n(n-1)(n-2) + 3n(n-1) + n.$$

CHAPTER

3 Counting and Relations

13 Relations

Mathematics is teeming with relations. Intuitively, a relation is a comparison between two objects. The two objects either are or are not related according to some rule. For example, less than ($<$) is a relation defined on integers. Some pairs of numbers, such as $(2, 8)$, satisfy the less-than relation (since $2 < 8$), but other pairs of numbers do not, such as $(10, 3)$ (since $10 \not< 3$).

There are other relations defined on the integers, such as divisibility, greater than, equality, and so on. Furthermore, there are relations on other sorts of objects. We can ask whether a pair of sets satisfies the $\subseteq$ relation or whether a pair of triangles satisfies the is-congruent-to relation.

Typically we use relations to study objects. For example, the is-congruent-to relation is a central tool in geometry in the study of triangles. In this section, we take a different point of view. Our purpose is to study the relations themselves.

What is a relation? The precise definition follows. Beware! At first glance, it may seem utterly perplexing and bear little resemblance to what you understand relations, such as $\leq$, to be. Rest assured that we will explain this definition thoroughly.

Definition 13.1 **(Relation)** A *relation* is a set of ordered pairs.

A set of ordered pairs??? Yes, we mean a set of two-element lists. For example, $R = \{(1, 2), (1, 3), (3, 0)\}$ is a relation, though not a particularly interesting one. This seems to have little to do with familiar relations such as $<$ and $\subseteq$ and $|$.

In truth, when mathematicians think about relations, we rarely think about them as sets of ordered pairs. We think of a relation R as a "test." If x and y are related by R—if they pass the test—then we write $x\ R\ y$. Otherwise, if they are not related by the relation R, we put a slash through the relation symbol, as in $x \neq y$ or $A \not\subseteq B$ (A is not a subset of B).

How can we understand Definition 13.1 in this way? The set of ordered pairs is a complete listing of all pairs of objects that "satisfy" the relation.

Let's return to the example $R = \{(1, 2), (1, 3), (3, 0)\}$. This says that, for the relation R, 1 is related to 2, 1 is related to 3, and 3 is related to 0, and for any other objects x, y, it is not the case that x is related to y. We can write,

$$(1, 2) \in R, \quad (1, 3) \in R, \quad (3, 0) \in R, \quad (5, 6) \notin R$$

and this means that $(1, 2)$, $(1, 3)$, and $(3, 0)$ are related by R, but $(5, 6)$ is not. Although this is a formally correct way to express these facts, it is not how mathematicians write. We would rather write,

$$1\,R\,2, \quad 1\,R\,3, \quad 3\,R\,0, \quad 5\,\not R\,6.$$

$x\,R\,y \iff (x, y) \in R.$

In other words, the symbols $x\ R\ y$ mean $(x, y) \in R$. Read aloud, $x\ R\ y$ can be spoken "x is related by the relation R to y," or, if everyone knows what relation is under consideration at the moment, we can simply say, "x is related to y."

The familiar relations of mathematics can be thought of in these terms. For example, the less-than-or-equal-to relation on the set of integers can be written as follows:

$$\{(x, y) : x, y \in \mathbb{Z} \text{ and } y - x \in \mathbb{N}\}.$$

This says that (x, y) is in the relation provided $y - x \in \mathbb{N}$—that is, provided $y - x$ is a nonnegative integer, which in turn is equivalent to $x \leq y$.

Let's reiterate the two salient points:

- A relation R is a set of ordered pairs (x, y); we include an ordered pair in R just when (x, y) "satisfies" the relation R. Any set of ordered pairs constitutes a relation, and a relation does not have to be specified by a general "rule" or special principle.
- Even though relations are sets of ordered pairs, we usually do not write $(x, y) \in R$. Rather, we write $x\ R\ y$ and say, "x is related to y by the relation R."

Next we extend Definition 13.1 a bit.

Definition 13.2 **(Relation on, between sets)** Let R be a relation and let A and B be sets. We say R is a *relation on* A provided $R \subseteq A \times A$, and we say R is a *relation from* A *to* B provided $R \subseteq A \times B$.

Example 13.3 Let $A = \{1, 2, 3, 4\}$ and $B = \{4, 5, 6, 7\}$. Let

$$\begin{aligned} R &= \{(1, 1), (2, 2), (3, 3), (4, 4)\}, \\ S &= \{(1, 2), (3, 2)\}, \\ T &= \{(1, 4), (1, 5), (4, 7)\}, \\ U &= \{(4, 4), (5, 2), (6, 2), (7, 3)\}, \text{ and} \\ V &= \{(1, 7), (7, 1)\}. \end{aligned}$$

All of these are relations.

- R is a relation on A. Note that it is the equality relation on A.
- S is a relation on A. Note that element 4 is never mentioned.

- T is a relation from A to B. Note that elements $2, 3 \in A$, and $6 \in B$ are never mentioned.
- U is a relation from B to A. Note that $1 \in A$ is never mentioned.
- V is a relation, but it is neither a relation from A to B nor a relation from B to A.

Since, formally, a relation is a set, all the various set operations apply to relations. For example, if R is a relation and A is a set, then $R \cap (A \times A)$ is the relation R *restricted* to the set A. [We can also consider $R \cap (A \times B)$, in which case we have restricted R to be a relation from A to B.]

Here is another operation we can perform on relations.

Definition 13.4 **(Inverse relation)** Let R be a relation. The *inverse* of R, denoted R^{-1}, is the relation formed by reversing the order of all the ordered pairs in R.

In symbols,

$$R^{-1} = \{(x, y) : (y, x) \in R\}.$$

Example 13.5 Let

$$R = \{(1, 5), (2, 6), (3, 7), (3, 8)\}.$$

Then

$$R^{-1} = \{(5, 1), (6, 2), (7, 3), (8, 3)\}.$$

If R is a relation on A, so is R^{-1}. If R is a relation from A to B, then R^{-1} is a relation from B to A.

Note that writing $1/R$ is nonsense. To form the inverse of a relation simply means to reverse all the ordered pairs in the relation; it has nothing to do with division. The -1 superscript is a convenient notation. We have not defined a general operation of raising a relation to a power.

Since the inverse operation reverses the ordered pairs in a relation, it is clear that $(R^{-1})^{-1} = R$. Here are a formal statement and proof.

Proposition 13.6 Let R be a relation. Then $(R^{-1})^{-1} = R$.

Note that R, R^{-1}, and $(R^{-1})^{-1}$ are all sets. Thus, to prove that $(R^{-1})^{-1} = R$, we use Proof Template 5.

Proof. Suppose $(x, y) \in R$. Then $(y, x) \in R^{-1}$ and thus $(x, y) \in (R^{-1})^{-1}$.

Now suppose $(x, y) \in (R^{-1})^{-1}$. Then $(y, x) \in R^{-1}$ and so $(x, y) \in R$.

We have shown that $(x, y) \in R \iff (x, y) \in (R^{-1})^{-1}$; therefore $R = (R^{-1})^{-1}$. ■

Properties of Relations

We introduce special terms to describe relations.

Definition 13.7 **(Properties of relations)** Let R be a relation defined on a set A.

- If for all $x \in A$ we have $x \, R \, x$, we call R *reflexive*.
- If for all $x \in A$ we have $x \not R \, x$, we call R *irreflexive*.
- If for all $x, y \in A$ we have $x \, R \, y \implies y \, R \, x$, we call R *symmetric*.
- If for all $x, y \in A$ we have $(x \, R \, y \wedge y \, R \, x) \implies x = y$, we call R *antisymmetric*.
- If for all $x, y, z \in A$ we have $(x \, R \, y \wedge y \, R \, z) \implies x \, R \, z$, we call R *transitive*.

We present several examples to illustrate this vocabulary.

Example 13.8 Consider the relation $=$ (equality) on the integers. It is reflexive (any integer is equal to itself), symmetric (if $x = y$, then $y = x$), and transitive (if $x = y$ and $y = z$, then we must have $x = z$).

The relation $=$ is antisymmetric, though this is not an interesting example of antisymmetry. See the subsequent examples.

However, the relation $=$ is not irreflexive (which would say that $x \neq x$ for all $x \in \mathbb{Z}$).

Example 13.9 Consider the relation $\leq$ (less than or equal to) on the integers. Note that $\leq$ is reflexive because for any integer x, it is true that $x \leq x$. It is also transitive, since $x \leq y$ and $y \leq z$ imply that $x \leq z$.

The relation $\leq$ is not symmetric because that would mean that $x \leq y \implies y \leq x$. This is false; for example, $3 \leq 9$, but $9 \not\leq 3$.

However, $\leq$ is antisymmetric: If we know $x \leq y$ and $y \leq x$, it must be the case that $x = y$.

Finally, $\leq$ is not irreflexive; for example, $5 \leq 5$.

Example 13.10 Consider the relation $<$ (strict less than) on the integers. Note that $<$ is not reflexive because, for example, $3 < 3$ is false. Further, $<$ is irreflexive because $x < x$ is never true.

The relation $<$ is not symmetric because $x < y$ does not imply $y < x$; for example, $0 < 5$ but $5 \not< 0$.

The relation $<$ is antisymmetric, but it fulfills the condition vacuously. The condition states

$$(x < y \text{ and } y < x) \implies x = y.$$

However, it is impossible to have both $x < y$ and $y < x$, so the hypothesis of this if-then statement can never be satisfied. Therefore it is true.

Finally, $<$ is transitive.

Example 13.11 Consider the relation | (divides) on the natural numbers. Note that | is antisymmetric because, if x and y are natural numbers with $x|y$ and $y|x$, then $x = y$.

However, the relation | on the integers is not antisymmetric. For example, $3|-3$ and $-3|3$, but $3 \neq -3$.

Also notice that | is not symmetric (e.g., $3|9$, but 9 does not divide 3).

The properties in Definition 13.7 depend on the context of the relation. The | (divides) relation on the integers is different from the | relation when restricted to the natural numbers.

This example also shows that a relation can be neither symmetric nor antisymmetric.

The terms in Definition 13.7, such as *reflexive,* are attributes of a relation R defined on a set A. Consider the relation $R = \{(1, 1), (1, 2), (2, 2), (2, 3), (3, 3)\}$. We ask: Is R reflexive? This question does not have a definitive answer. If we think of R as a relation on the set $\{1, 2, 3\}$, then the answer is yes. However, we can also consider R as a relation on all of $\mathbb{Z}$; in this context, the answer is no. One can only say that a relation R is reflexive if we are presented with the set A on which R is a relation. In most cases, the set A will either be explicitly mentioned or be obvious from context.

Recap

We introduced the notion of a relation in both the intuitive sense as a "condition" and in the formal sense as a set of ordered pairs. We presented the concept of an inverse relation and defined the following properties of relations: reflexive, irreflexive, symmetric, antisymmetric, and transitive.

13 Exercises

13.1. For each of the following relations defined on the set $\{1, 2, 3, 4, 5\}$, determine whether the relation is reflexive, irreflexive, symmetric, antisymmetric, and/or transitive.

- **a.** $R = \{(1, 1), (2, 2), (3, 3), (4, 4), (5, 5)\}$.
- **b.** $R = \{(1, 2), (2, 3), (3, 4), (4, 5)\}$.
- **c.** $R = \{(1, 1), (1, 2), (1, 3), (1, 4), (1, 5)\}$.
- **d.** $R = \{(1, 1), (1, 2), (2, 1), (3, 4), (4, 3)\}$.
- **e.** $R = \{1, 2, 3, 4, 5\} \times \{1, 2, 3, 4, 5\}$.

13.2. Let us say that two integers are *near* one another provided their difference is 2 or smaller (i.e., the numbers are at most 2 apart). For example, 3 is near to 5, 10 is near to 9, but 8 is not near to 4. Let R stand for this is-near-to relation. Please do the following:

- **a.** Write down R as a set of ordered pairs. Your answer should look like this:

$$R = \{(x, y) : \ldots\}.$$

- **b.** Prove or disprove: R is reflexive.
- **c.** Prove or disprove: R is irreflexive.
- **d.** Prove or disprove: R is symmetric.

e. Prove or disprove: R is antisymmetric.
f. Prove or disprove: R is transitive.

13.3. For each of the following relations, find R^{-1}.
a. $R = \{(1, 2), (2, 3), (3, 4)\}$.
b. $R = \{(1, 1), (2, 2), (3, 3)\}$.
c. $R = \{(x, y) : x, y \in \mathbb{Z},\ x - y = 1\}$.
d. $R = \{(x, y) : x, y \in \mathbb{N},\ x|y\}$.
e. $R = \{(x, y) : x, y \in \mathbb{Z},\ xy > 0\}$.

13.4. Suppose that R and S are relations and $R = S^{-1}$. Prove that $S = R^{-1}$.

13.5. Let R be a relation on a set A. Prove or disprove: If R is antisymmetric, then R is irreflexive.

13.6. Let R be the relation has-the-same-size-as defined on all finite subsets of $\mathbb{Z}$ (i.e., $A\,R\,B$ iff $|A| = |B|$). Which of the five properties (reflexive, irreflexive, symmetric, antisymmetric, transitive) does R have? Prove your answers.

13.7. Consider the relation $\subseteq$ on $2^{\mathbb{Z}}$ (i.e., the is-a-subset-of relation defined on all sets of integers). Which of the properties in Definition 13.7 does $\subseteq$ have? Prove your answers.

13.8. What is $\leq^{-1}$?

13.9. The property *irreflexive* is not the same as being not reflexive. To illustrate this, please do the following:
a. Give an example of a relation on a set that is neither reflexive nor irreflexive.
b. Give an example of a relation on a set that is both reflexive and irreflexive.
Part (a) is not too hard, but for (b), you will need to create a rather strange example.

13.10. A fancy way to say R is symmetric is $R = R^{-1}$. Prove this (i.e., prove that a relation R is symmetric if and only if $R = R^{-1}$).

13.11. Prove: A relation R on a set A is antisymmetric if and only if

$$R \cap R^{-1} \subseteq \{(a, a) : a \in A\}.$$

13.12. Give an example of a relation on a set that is both symmetric and transitive but not reflexive.

Explain what is wrong with the following "proof."

> *Statement*: If R is symmetric and transitive, then R is reflexive.
> *"Proof"*: Suppose R is symmetric and transitive. Symmetric means that $x\ R\ y$ implies $y\ R\ x$. We apply transitivity to $x\ R\ y$ and $y\ R\ x$ to give $x\ R\ x$. Therefore R is reflexive. ■

13.13. *Drawing pictures of relations*. Pictures of mathematical objects are wonderful aids in understanding concepts. There is a nice way to draw a picture of a relation on a set or of a relation from one set to another.

To draw a picture of a relation R on a set A, we make a diagram in which each element of A is represented by a dot. If $a\ R\ b$, then we draw an arrow from dot a to dot b. If it should happen that b is also related to a, we draw another arrow from b to a. And if $a\ R\ a$, then we draw a looping arrow from a to itself.

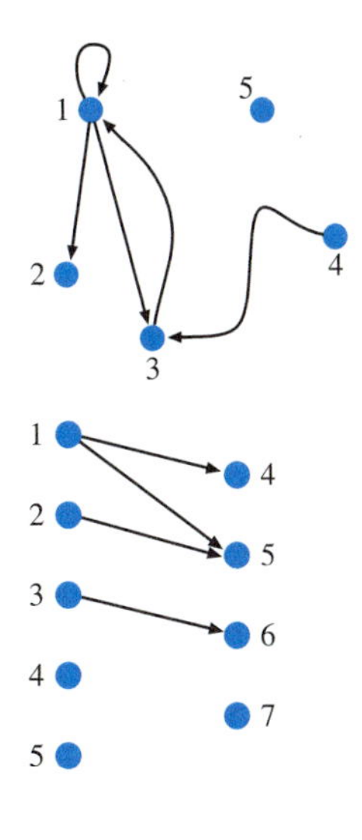

For example, let $A = \{1, 2, 3, 4, 5\}$ and $R = \{(1, 1), (1, 2), (1, 3), (4, 3), (3, 1)\}$. A picture of the relation R on A is given in the first figure.

To draw a picture of a relation from A to B, we draw two collections of dots. The first collection of dots corresponds to the elements in A, and we place these on the left side of the figure. The dots for B go on the right. We then draw an arrow from $a \in A$ to $b \in B$ just when (a, b) is in the relation.

For example, let $A = \{1, 2, 3, 4, 5\}$ and $B = \{4, 5, 6, 7\}$ and let S be the relation $\{(1, 4), (1, 5), (2, 5), (3, 6)\}$. A picture of the relation S is given in the second figure.

Please draw pictures of the following relations.

a. Let $A = \{a \in \mathbb{N} : a|10\}$ and let R be the relation $|$ (divides) restricted to A.

b. Let $A = \{1, 2, 3, 4, 5\}$ and let R be the less-than relation restricted to A.

c. Let $A = \{1, 2, 3, 4, 5\}$ and let R be the relation $=$ (equals) restricted to A.

d. Let $A = \{1, 2, 3, 4, 5\}$ and let $B = \{2, 3, 4, 5\}$. Let R be the relation $\geq$ (greater than or equal to) from A to B.

e. Let $A = \{-1, -2, -3, -4, -5\}$ and let $B = \{1, 2, 3, 4, 5\}$ and let $R = \{(a, b) : a \in A,\ b \in B,\ a|b\}$.

14 Equivalence Relations

As we proceed with our study of discrete mathematics, we shall encounter various relations. Certain relations bear a strong resemblance to the relation *equality*. A good example (from geometry) is the is-congruent-to relation (often denoted by $\cong$) on the set of triangles. Roughly speaking, triangles are congruent if they have exactly the same shape. Congruent triangles are not equal (i.e., they might be in different parts of the plane), but in a sense, they act like equal triangles. Why? What is special about $\cong$ that it acts like equality?

Of the five properties listed in Definition 13.7, $\cong$ is reflexive, symmetric, and transitive (but it is neither irreflexive nor antisymmetric). Relations with these three properties are akin to equality and are given a special name.

Definition 14.1 **(Equivalence relation)** Let R be a relation on a set A. We say R is an *equivalence relation* provided it is reflexive, symmetric, and transitive.

Example 14.2 Consider the has-the-same-size-as relation on finite sets (see Exercise 13.6): For finite sets of integers A and B, let $A\ R\ B$ provided $|A| = |B|$. Note that R is reflexive, symmetric, and transitive and therefore is an equivalence relation.

It is not the case that two sets with the same size are the same. For example, $\{1, 2, 3\}\ R\ \{2, 3, 4\}$, but $\{1, 2, 3\} \neq \{2, 3, 4\}$. Nonetheless, sets related by R are "like" each other in that they share a common property: their size.

The following equivalence relation plays a central role in number theory.

Definition 14.3 **(Congruence modulo *n*)** Let n be a positive integer. We say that integers x and y are *congruent modulo* n, and we write

$$x \equiv y \pmod{n}$$

provided $n|(x-y)$.

In other words, $x \equiv y \pmod{n}$ if and only if x and y differ by a multiple of n.

Example 14.4

$3 \equiv 13 \pmod{5}$ because $3 - 13 = -10$ is a multiple of 5.
$4 \equiv 4 \pmod{5}$ because $4 - 4 = 0$ is a multiple of 5.
$16 \not\equiv 3 \pmod{5}$ because $16 - 3 = 13$ is not a multiple of 5.

Congruence of numbers (modulo n) is different from congruence of geometric figures. They are both equivalence relations. Unfortunately, mathematicians do use the same word with different meanings. We try, however, to make sure the meaning is always clear from context.

We often abbreviate the word *modulo* to just *mod*. If the integer n is known and unchanging throughout the discussion, we may omit the (mod n) on the right. Also, the (mod n) is often shortened to just (n).

The simplest case for this definition is when $n = 1$. In this case, we have $x \equiv y$ provided the integer $x - y$ is divisible by 1. However, all integers are divisible by 1, so any two integers are congruent modulo 1. This is not interesting.

The next case is when $n = 2$. Two numbers are congruent mod 2 provided their difference is divisible by 2 (i.e., they differ by an even number). For example,

$$3 \equiv 15 \pmod{2}, \qquad 0 \equiv -14 \pmod{2}, \qquad \text{and} \qquad 3 \equiv 3 \pmod{2}.$$

However,

$$3 \not\equiv 12 \pmod{2} \qquad \text{and} \qquad -1 \not\equiv 0 \pmod{2}.$$

Two numbers that are both even or both odd are said to have the same *parity*.

Please notice that two numbers are congruent mod 2 iff they are both even or both odd (we prove this later).

Theorem 14.5 Let n be a positive integer. The is-congruent-to-mod-n relation is an equivalence relation on the set of integers.

The proof of this result is not hard if we use the proof techniques we have developed. Our goal is to prove that a relation is an equivalence relation. This means the proof should look like this:

> Let n be a positive integer and let $\equiv$ denote congruence mod n. We need to show that $\equiv$ is reflexive, symmetric, and transitive.
>
> - Claim: $\equiv$ is reflexive. ... Thus $\equiv$ is reflexive.
> - Claim: $\equiv$ is symmetric. ... Thus $\equiv$ is symmetric.
> - Claim: $\equiv$ is transitive. ... Thus $\equiv$ is transitive.
>
> Therefore $\equiv$ is an equivalence relation. ■

Note that the proof breaks into three parts corresponding to the three conditions in Definition 14.1. Each section is announced with the word *claim*. A *claim* is a statement we plan to prove during the course of a proof. This helps the reader know what's coming next and why.

We can now start unraveling each part of the proof. For example, to show that $\equiv$ is reflexive, we have to show $\forall x \in \mathbb{Z}$, $x \equiv x$ (see Definition 13.7). Let's put that into the proof.

> Let n be a positive integer and let $\equiv$ denote congruence mod n. We need to show that $\equiv$ is reflexive, symmetric, and transitive.
>
> - Claim: $\equiv$ is reflexive. Let x be an arbitrary integer, ... Therefore $x \equiv x$. Thus $\equiv$ is reflexive.
> - Claim: $\equiv$ is symmetric, ... Thus $\equiv$ is symmetric.
> - Claim: $\equiv$ is transitive, ... Thus $\equiv$ is transitive.
>
> Therefore $\equiv$ is an equivalence relation. ■

Now we want to prove $x \equiv x$. What does this mean? It means $n|(x-x)$—that is, $n|0$—and this is obvious! Clearly 0 is a multiple of n since $n \cdot 0 = 0$. We add this to the proof:

> Let n be a positive integer and let $\equiv$ denote congruence mod n. We need to show that $\equiv$ is reflexive, symmetric, and transitive.
>
> - Claim: $\equiv$ is reflexive. Let x be an arbitrary integer. Since $0 \cdot n = 0$, we have $n|0$, which we can rewrite as $n|(x-x)$. Therefore $x \equiv x$. Thus $\equiv$ is reflexive.
> - Claim: $\equiv$ is symmetric. ... Thus $\equiv$ is symmetric.
> - Claim: $\equiv$ is transitive. ... Thus $\equiv$ is transitive.
>
> Therefore $\equiv$ is an equivalence relation. ■

Now we tackle the symmetry of $\equiv$. To show symmetry, we consult Definition 13.7 to see that we must prove $x \equiv y \implies y \equiv x$. This is an if-then statement, so we write:

> Let n be a positive integer and let $\equiv$ denote congruence mod n. We need to show that $\equiv$ is reflexive, symmetric, and transitive.
>
> - Claim: $\equiv$ is reflexive. Let x be an arbitrary integer. Since $0 \cdot n = 0$, we have $n|0$, which we can rewrite as $n|(x-x)$. Therefore $x \equiv x$. Thus $\equiv$ is reflexive.
> - Claim: $\equiv$ is symmetric. Let x and y be integers and suppose $x \equiv y$. ... Therefore $y \equiv x$. Thus $\equiv$ is symmetric.
> - Claim: $\equiv$ is transitive. ... Thus $\equiv$ is transitive.
>
> Therefore $\equiv$ is an equivalence relation. ■

Next we unravel definitions.

> Let n be a positive integer and let $\equiv$ denote congruence mod n. We need to show that $\equiv$ is reflexive, symmetric, and transitive.
>
> - Claim: $\equiv$ is reflexive. Let x be an arbitrary integer. Since $0 \cdot n = 0$, we have $n|0$, which we can rewrite as $n|(x - x)$. Therefore $x \equiv x$. Thus $\equiv$ is reflexive.
> - Claim: $\equiv$ is symmetric. Let x and y be integers and suppose $x \equiv y$. This means that $n|(x - y)$. ... And so $n|(y - x)$. Therefore $y \equiv x$. Thus $\equiv$ is symmetric.
> - Claim: $\equiv$ is transitive. ... Thus $\equiv$ is transitive.
>
> Therefore $\equiv$ is an equivalence relation. ■

We're nearly done. We know $n|(x - y)$. We want $n|(y - x)$. We can unravel the definition of divisibility and complete this section of the proof. (Alternatively, we can use Exercise 4.7.)

> Let n be a positive integer and let $\equiv$ denote congruence mod n. We need to show that $\equiv$ is reflexive, symmetric, and transitive.
>
> - Claim: $\equiv$ is reflexive. Let x be an arbitrary integer. Since $0 \cdot n = 0$, we have $n|0$, which we can rewrite as $n|(x - x)$. Therefore $x \equiv x$. Thus $\equiv$ is reflexive.
> - Claim: $\equiv$ is symmetric. Let x and y be integers and suppose $x \equiv y$. This means that $n|(x - y)$. So there is an integer k such that $(x - y) = kn$. But then $(y - x) = (-k)n$. And so $n|(y - x)$. Therefore $y \equiv x$. Thus $\equiv$ is symmetric.
> - Claim: $\equiv$ is transitive. ... Thus $\equiv$ is transitive.
>
> Therefore $\equiv$ is an equivalence relation. ■

The proof of the third section nearly writes itself and we leave it to you (Exercise 14.4).

Equivalence Classes

We noted earlier that two numbers are congruent mod 2 if and only if they are either (1) both odd or (2) both even. (We have not proved this yet; we will. See Corollary 34.5.)

We have two classes of numbers: odd and even. Any two odd numbers are congruent modulo 2 (this you can prove), and any two even numbers are congruent modulo 2. The two classes are disjoint (have no elements in common) and, taken together, contain all the integers.

Similarly, let R denote the has-the-same-size-as relation on the finite subsets of $\mathbb{Z}$. We noted that R is an equivalence relation. Notice that we can categorize finite subsets of $\mathbb{Z}$ according to their cardinality. There is just one finite subset of $\mathbb{Z}$ that has cardinality zero—namely, the empty set. The only set related by R to $\emptyset$ is $\emptyset$. Next, there are the subsets of size one:

$$\ldots, \{-2\}, \{-1\}, \{0\}, \{1\}, \{2\}, \ldots$$

These are all R-related to one another but not to other sets. There is also the class of all subsets of $\mathbb{Z}$ of size two; again, these are related to one another but not to any other sets.

This decomposition of a set by an equivalence relation is an important idea we now formalize.

Definition 14.6 **(Equivalence class)** Let R be an equivalence relation on a set A and let $a \in A$. The *equivalence class of* a, denoted $[a]$, is the set of all elements of A related (by R) to a; that is,

$$[a] = \{x \in A : x \mathrel{R} a\}.$$

Example 14.7 Consider the equivalence relation congruence mod 2. What is [1]? By definition,

$$[1] = \{x \in \mathbb{Z} : x \equiv 1 \pmod 2\}.$$

This is the set of all integers x such that $2|(x-1)$ (i.e., $x - 1 = 2k$ for some k), so $x = 2k + 1$ (i.e., x is odd)! The set [1] is the set of odd numbers.

It's not hard to see (you should prove) that [0] is the set of even numbers.

Consider [3]. You should also prove that [3] is the set of odd numbers, so $[1] = [3]$. (See Exercise 14.6.)

The equivalence relation congruence mod 2 has only two equivalence classes: the set of odd integers [1] and the set of even integers [0].

Example 14.8 Let R be the has-the-same-size-as relation defined on the set of finite subsets of $\mathbb{Z}$. What is $[\emptyset]$? By definition,

$$[\emptyset] = \{A \subseteq \mathbb{Z} : |A| = 0\} = \{\emptyset\}$$

since $\emptyset$ is the only set of cardinality zero.

What is $[\{2, 4, 6, 8\}]$? The set of all finite subsets of $\mathbb{Z}$ related to $\{2, 4, 6, 8\}$ are exactly those of size 4:

$$[\{2, 4, 6, 8\}] = \{A \subseteq \mathbb{Z} : |A| = 4\}.$$

The relation R separates the set of finite subsets of $\mathbb{Z}$ into infinitely many equivalence classes (one for each element of $\mathbb{N}$). Every class contains sets that are related to each other but not to anything not in that class.

We now present several propositions describing the salient features of equivalence classes.

Proposition 14.9 Let R be an equivalence relation on a set A and let $a \in A$. Then $a \in [a]$.

Proof. Note that $[a] = \{x \in A : x \; R \; a\}$. To show that $a \in [a]$, we just need to show that $a \; R \; a$, and that is true by definition (R is reflexive). ■

One consequence of Proposition 14.9 is that equivalence classes are not empty. A second consequence is that the union of all the equivalence classes is A (see Exercise 14.7).

Proposition 14.10 Let R be an equivalence relation on a set A and let $a, b \in A$. Then $a \; R \; b$ if and only if $[a] = [b]$.

Proof. ($\Rightarrow$) Suppose $a \; R \; b$. We need to show that the sets $[a]$ and $[b]$ are the same (see Proof Template 5).

Suppose $x \in [a]$. This means that $x \; R \; a$. Since $a \; R \; b$, we have (by transitivity) $x \; R \; b$. Therefore $x \in [b]$.

On the other hand, suppose $y \in [b]$. This means that $y \; R \; b$. We are given $a \; R \; b$, and this implies $b \; R \; a$ (symmetry). By transitivity (applied to $y \; R \; b$ and $b \; R \; a$), we have $y \; R \; a$. Therefore $y \in [a]$.

Hence $[a] = [b]$.

($\Leftarrow$) Suppose $[a] = [b]$. We know (Proposition 14.9) that $a \in [a]$. But $[a] = [b]$, so $a \in [b]$. Therefore $a \; R \; b$. ■

Proposition 14.11 Let R be an equivalence relation on a set A and let $a, x, y \in A$. If $x, y \in [a]$, then $x \; R \; y$.

You are asked to prove Proposition 14.11 in Exercise 14.9.

Proposition 14.12 Let R be an equivalence relation on A and suppose $[a] \cap [b] \neq \emptyset$. Then $[a] = [b]$.

Before we work on the proof of this result, let us understand clearly what it is telling us. It says that either two equivalence classes have nothing in common or else (if they do have a common element) they are identical. In other words, equivalence classes must be pairwise disjoint.

Now we develop the proof of Proposition 14.12. This proposition asks us to prove that two sets ($[a]$ and $[b]$) are the same. We could use Proof Template 5, and the proof would not be too hard to do (you can try this for yourself).

However, please notice that Proposition 14.10 gives us a necessary and sufficient condition to prove that two equivalence classes are the same. To show that $[a] = [b]$, it is enough to show $a \; R \; b$. The proof skeleton is as follows:

> Let R be an equivalence relation on A and suppose $[a]$ and $[b]$ are equivalence classes with $[a] \cap [b] \neq \emptyset$. ... Therefore $a \; R \; b$. By Proposition 14.10, we therefore have $[a] = [b]$. ■

Now we need to unravel the fact that $[a] \cap [b] \neq \emptyset$. The fact that two sets have a nonempty intersection means there is some element that is in both.

> Let R be an equivalence relation on A and suppose $[a]$ and $[b]$ are equivalence classes with $[a] \cap [b] \neq \emptyset$. Hence there is an $x \in [a] \cap [b]$—that is, an element x with $x \in [a]$ and $x \in [b]$. ... Therefore $a\ R\ b$. By Proposition 14.10, we therefore have $[a] = [b]$. ■

We can now unravel the facts $x \in [a]$ and $x \in [b]$ to give $x\ R\ a$ and $x\ R\ b$ (by Definition 14.6).

> Let R be an equivalence relation on A and suppose $[a]$ and $[b]$ are equivalence classes with $[a] \cap [b] \neq \emptyset$. Hence there is an $x \in [a] \cap [b]$—that is, an element x with $x \in [a]$ and $x \in [b]$. So $x\ R\ a$ and $x\ R\ b$. ... Therefore $a\ R\ b$. By Proposition 14.10, we therefore have $[a] = [b]$. ■

Now we are almost finished.

We know:	$x\ R\ a$ and $x\ R\ b$.
We want:	$a\ R\ b$.

We can switch $x\ R\ a$ to $a\ R\ x$ (by symmetry) and then use transitivity on $a\ R\ x$ and $x\ R\ b$ to get $a\ R\ b$, completing the proof.

> Let R be an equivalence relation on A and suppose $[a]$ and $[b]$ are equivalence classes with $[a] \cap [b] \neq \emptyset$. Hence there is an $x \in [a] \cap [b]$—that is, an element x with $x \in [a]$ and $x \in [b]$. So $x\ R\ a$ and $x\ R\ b$. Since $x\ R\ a$, we have $a\ R\ x$ (symmetry), and since $a\ R\ x$ and $x\ R\ b$, we have (transitivity) $a\ R\ b$. By Proposition 14.10, we therefore have $[a] = [b]$. ■

The proof is finished.

Let us reiterate some of what we have learned.

Corollary 14.13 Let R be an equivalence relation on a set A. The equivalence classes of R are nonempty, pairwise disjoint subsets of A whose union is A.

Recap

An equivalence relation is a relation on a set that is reflexive, symmetric, and transitive. We discussed the important equivalence relation congruence modulo n on $\mathbb{Z}$. We developed the notion of equivalence classes and discussed various properties of equivalence classes.

14 Exercises

14.1. Which of the following are equivalence relations?

a. $R = \{(1, 1), (1, 2), (2, 1), (2, 2), (3, 3)\}$ on the set $\{1, 2, 3\}$.

b. $R = \{(1, 2), (2, 3), (3, 1)\}$ on the set $\{1, 2, 3\}$.

c. $|$ on $\mathbb{Z}$.

d. $\leq$ on $\mathbb{Z}$.

e. $\{1, 2, 3\} \times \{1, 2, 3\}$ on the set $\{1, 2, 3\}$.

f. $\{1, 2, 3\} \times \{1, 2, 3\}$ on the set $\{1, 2, 3, 4\}$.

g. Is-an-anagram-of on the set of English words. (For example, STOP is an anagram of POTS because we can form one from the other by rearranging its letters.)

14.2. Prove that if x and y are both odd, then $x \equiv y \pmod 2$.

Prove that if x and y are both even, then $x \equiv y \pmod 2$.

14.3. Prove: If a is an integer, then $a \equiv -a \pmod 2$.

14.4. Complete the proof of Theorem 14.5; that is, prove that congruence modulo n is transitive.

14.5. For each equivalence relation, find the requested equivalence class.

a. $R = \{(1, 1), (1, 2), (2, 1), (2, 2), (3, 3), (4, 4)\}$ on $\{1, 2, 3, 4\}$. Find $[1]$.

b. $R = \{(1, 1), (1, 2), (2, 1), (2, 2), (3, 3), (4, 4)\}$ on $\{1, 2, 3, 4\}$. Find $[4]$.

c. R is has-the-same-tens-digit-as on the set $\{x \in \mathbb{Z} : 100 < x < 200\}$. Find $[123]$.

d. R is has-the-same-parents-as on the set of all human beings. Find [you].

e. R is has-the-same-birthday-as on the set of all human beings. Find [you].

f. R is has-the-same-size-as on $2^{\{1,2,3,4,5\}}$. Find $[\{1, 3\}]$.

14.6. Please refer to the Example 14.7, in which we discussed the congruence modulo 2 relation on the integers. For that relation, prove that $[1] = [3]$.

14.7. Let R be an equivalence relation on a set A. Prove that the union of all of R's equivalence classes is A.

In symbols this is

$$\bigcup_{a \in A} [a] = A.$$

The big $\bigcup$ notation on the left is worthy of comment. It is akin to the notation developed in Section 9. There, however, we had an index that ran between two integers, as in

$$\bigcup_{k=1}^{n} (\text{sets depending on } k)$$

The dummy variable is k, and we take a union of sets that depend on k as k ranges over the integers $1, 2, \ldots, n$.

The situation here is slightly different. The dummy variable is not necessarily an integer. The notation is of the form

$$\bigcup_{a \in A} (\text{sets depending on } a).$$

This means we take the union over all possible (sets depending on a) as a ranges over the various members of A.

Notice that in this problem the union may be redundant. It is possible for $[a] = [a']$ where a and a' are different members of A. For example, if R is congruence mod 2 and $A = \mathbb{Z}$, then

$$\bigcup_{a\in\mathbb{Z}}[a] = \cdots \cup [-2] \cup [-1] \cup [0] \cup [1] \cup [2] \cup \cdots = [0] \cup [1] = \mathbb{Z}$$

because $\cdots = [-2] = [0] = [2] = \cdots$ and $\cdots = [-3] = [-1] = [1] = [3] = \cdots$.

14.8. Suppose R is an equivalence relation on a set A and suppose $a, b \in A$.
Prove: $a \in [b] \iff b \in [a]$.

14.9. Prove Proposition 14.11.

14.10. Let R and S be equivalence relations on a set A. Prove that $R = S$ if and only if the equivalence classes of R are the same as the equivalence classes of S.

14.11. Please refer to Exercise 13.13 on drawing pictures of relations.
Let $A = \{1, 2, 3, \ldots, 10\}$. Do the following:

a. Draw three pictures of different equivalence relations on A.
b. For each equivalence relation, list all of its equivalence classes.
c. Describe what equivalence relations "look like."

14.12. Here is another way to draw a picture of an equivalence relation: Draw the equivalence classes. For example, consider the following equivalence relation on $A = \{1, 2, 3, 4, 5, 6\}$:

$$R = \{(1, 1), (1, 2), (2, 1), (2, 2), (3, 3), (4, 4), (4, 5), (4, 6), (5, 4), (5, 5), (5, 6), (6, 4), (6, 5), (6, 6)\}.$$

The equivalence classes of this relation on A are

$$[1] = [2] = \{1, 2\}, \quad [3] = \{3\}, \quad \text{and} \quad [4] = [5] = [6] = \{4, 5, 6\}.$$

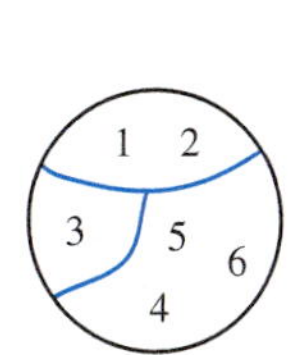

The picture of the relation R, rather than showing relation arrows, simply shows the equivalence classes of A. The elements of A are enclosed in a circle, and we subdivide the circle into regions to show the equivalence classes. By Corollary 14.13, we know that the equivalence classes of R are nonempty, are pairwise disjoint, and contain all the elements of A. So in the picture, the regions are nonoverlapping, and every element of A ends up in exactly one region.

For each of the equivalence relations you found in the previous problem, draw a diagram of the equivalence classes.

14.13. There is only one possible equivalence relation on a one-element set: If $A = \{1\}$, then $R = \{(1, 1)\}$ is the only possible equivalence relation.

There are exactly two possible equivalence relations on a two-element set: If $A = \{1, 2\}$, then $R_1 = \{(1, 1), (2, 2)\}$ and $R_2 = \{(1, 1), (1, 2), (2, 1), (2, 2)\}$ are the only equivalence relations on A.

How many different equivalence relations are possible on a three-element set? ... on a four-element set?

14.14. Describe the equivalence classes for the is-similar-to relation on the set of all triangles.

15 Partitions

We ended the previous section with Corollary 14.13. Let us repeat that result here.

Let R be an equivalence relation on a set A. The equivalence classes of R are nonempty, pairwise disjoint subsets of A whose union is A.

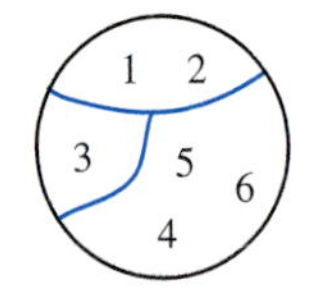

This corollary is illustrated nicely by the diagrams you drew in Exercise 14.12. The equivalence classes of R are drawn as separate regions inside a circle containing the elements of A.

The technical language for this property is that the equivalence classes of R *partition* A.

Definition 15.1 **(Partition)** Let A be a set. A *partition of* (or *on*) A is a set of nonempty, pairwise disjoint sets whose union is A.

There are four key points in this definition, and we shall examine them closely in an example. The four points are

The parts of a partition are also called *blocks*.

- A partition is a set of sets; each member of a partition is a subset of A. The members of the partition are called *parts*.
- The parts of a partition are nonempty. The empty set is never a part of a partition.
- The parts of a partition are pairwise disjoint. No two parts of a partition may have an element in common.
- The union of the parts is the original set.

Example 15.2 Let $A = \{1, 2, 3, 4, 5, 6\}$ and let

$$\mathcal{P} = \{\{1, 2\}, \{3\}, \{4, 5, 6\}\}.$$

We often use a fancy letter $\mathcal{P}$ to denote a partition. We do this because $\mathcal{P}$ is a set of sets. The elements of $\mathcal{P}$ are subsets of A. This hierarchy of letters—lowercase, uppercase, fancy—is a useful convention for distinguishing elements, sets, and sets of sets, respectively.

This is a partition of A into three parts. The three parts are $\{1, 2\}$, $\{3\}$, and $\{4, 5, 6\}$. These three sets are (1) nonempty, (2) they are pairwise disjoint, and (3) their union is A.

The partition $\{\{1, 2\}, \{3\}, \{4, 5, 6\}\}$ is not the only partition of $A = \{1, 2, 3, 4, 5, 6\}$. Here are two more that are worthy of note:

$$\{\{1, 2, 3, 4, 5, 6\}\} \qquad \text{and} \qquad \{\{1\}, \{2\}, \{3\}, \{4\}, \{5\}, \{6\}\}.$$

The first is a partition of A into just one part containing all the elements of A, and the second is a partition of A into six parts, each containing just one element.

Corollary 14.13 can be restated as follows:

Let R be an equivalence relation on a set A. The equivalence classes of R form a partition of the set A.

Forming an equivalence relation from a partition.

Given an equivalence relation on a set, the equivalence classes of that relation form a partition of the set. We start with an equivalence relation, and we form a partition. We can also go the other way; given a partition, there is a natural way to construct an equivalence relation.

Let $\mathcal{P}$ be a partition of a set A. We use $\mathcal{P}$ to form a relation on A. We call this relation the is-in-the-same-part-as relation and denote it by $\stackrel{\mathcal{P}}{\equiv}$. It is defined as follows. Let $a, b \in A$. Then

$$a \stackrel{\mathcal{P}}{\equiv} b \iff \exists P \in \mathcal{P},\ a, b \in P.$$

In words, a and b are related by $\stackrel{\mathcal{P}}{\equiv}$ provided there is some part of the partition $\mathcal{P}$ that contains both a and b.

Proposition 15.3 Let A be a set and let $\mathcal{P}$ be a partition on A. The relation $\stackrel{\mathcal{P}}{\equiv}$ is an equivalence relation on A.

Proof. To show that $\stackrel{\mathcal{P}}{\equiv}$ is an equivalence relation, we must show that it is (1) reflexive, (2) symmetric, and (3) transitive.

- Claim: $\stackrel{\mathcal{P}}{\equiv}$ is reflexive.

  Let a be an arbitrary element of A. Since $\mathcal{P}$ is a partition, there must be a part $P \in \mathcal{P}$ that contains a (the union of the parts is A). We have $a \stackrel{\mathcal{P}}{\equiv} a$, since $a, a \in P \in \mathcal{P}$.
- Claim: $\stackrel{\mathcal{P}}{\equiv}$ is symmetric.

  Suppose $a \stackrel{\mathcal{P}}{\equiv} b$ for $a, b \in A$. This means there is a $P \in \mathcal{P}$ such that $a, b \in P$. Since b and a are in the same part of $\mathcal{P}$, we have $b \stackrel{\mathcal{P}}{\equiv} a$.
- Claim: $\stackrel{\mathcal{P}}{\equiv}$ is transitive. (This step is more interesting.)

  Let $a, b, c \in A$ and suppose $a \stackrel{\mathcal{P}}{\equiv} b$ and $b \stackrel{\mathcal{P}}{\equiv} c$. Since $a \stackrel{\mathcal{P}}{\equiv} b$, there is a part $P \in \mathcal{P}$ containing both a and b. Since $b \stackrel{\mathcal{P}}{\equiv} c$, there is a part $Q \in \mathcal{P}$ with $b, c \in Q$. Notice that b is in both P and Q. Thus parts P and Q have a common element. Since parts of a partition must be pairwise disjoint, it must be the case that $P = Q$. Therefore all three of a, b, c are together in the same part of $\mathcal{P}$. Since a, c are in a common part of $\mathcal{P}$, we have $a \stackrel{\mathcal{P}}{\equiv} c$. ■

We have confirmed that $\stackrel{\mathcal{P}}{\equiv}$ is an equivalence relation on A. What are its equivalence classes?

Proposition 15.4 Let $\mathcal{P}$ be a partition on a set A and let $\stackrel{\mathcal{P}}{\equiv}$ be the is-in-the-same-part-as relation. The equivalence classes of $\stackrel{\mathcal{P}}{\equiv}$ are exactly the parts of $\mathcal{P}$.

We leave the proof for you (Exercise 15.5).

The salient point here is that equivalence relations and partitions are flip sides of the same mathematical coin. Given a partition, we can form the in-the-same-part-as equivalence relation. Given an equivalence relation, we can form the partition into equivalence classes.

Counting Classes/Parts

In discrete mathematics, we often encounter counting problems of the form "In how many different ways can . . . " The word on which we wish to focus is *different*.

For example, in how many different ways can the letters in the word HELLO be rearranged? The difficult part of this problem is the repeated L. So let us begin with an easier word.

Example 15.5 In how many ways can the letters in the word WORD be rearranged? A word is simply a list of letters. We have a list of four possible letters, and we want to count lists using each of them exactly once. This is a problem we have already solved (see Sections 7 and 8). The answer is $4! = 24$. Here they are:

WORD	WODR	WROD	WRDO	WDOR	WDRO
OWRD	OWDR	ORWD	ORDW	ODWR	ODRW
RWOD	RWDO	ROWD	RODW	RDWO	RDOW
DWOR	DWRO	DOWR	DORW	DRWO	DROW

Anagrams of HELLO.

Let us return to the problem of counting the number of ways it is possible to rearrange the letters in the word HELLO. If there were no repeated letters, then the answer would be $5! = 120$. Imagine for a moment that the two Ls are different letters. Let us write one larger than the other: HELLO. If we were to write down all 120 different ways to rearrange the letters in HELLO, we would have a chart that looks like this:

HELLO	HELOL	HELLO	HELOL	HEOLL	HEOLL
HLELO	HLEOL	HLLEO	HLLOE	HLOLE	HLOEL
		many lines omitted			
LLHEO	LLHOE	LLEHO	LLEOH	LLOHE	LLOEH
LLHEO	LLHOE	LLEHO	LLEOH	LLOHE	LLOEH

Now we shrink the large Ls back to their proper size. When we do, we can no longer distinguish between HELLO and HELLO, or between LEHLO and LEHLO.

I hope at this point it is clear that the answer to the counting problem is 60: There are 120 entries in the chart (from HELLO to LLOEH), and each rearrangement of HELLO appears exactly twice on the chart.

Let's think about this by using equivalence relations and partitions. The set A is the set of all 120 different rearrangements of HELLO. Suppose a and b are elements of A (anagrams of HELLO). Define a relation R with $a\ R\ b$ provided that a and b give the same rearrangement of HELLO when we shrink the large L to a small L. For example, (HELOL) R (HELOL).

Is R an equivalence relation? Clearly R is reflexive, symmetric, and transitive (if in doubt, think this out), and so, yes, R is an equivalence relation. The equivalence classes of R are all the different ways of rearranging HELLO that look the same when we shrink the large L. For example,

$$[\text{HLEOL}] = \{\text{HLEOL}, \text{HLEOL}\}$$

since HLEOL and HLEOL both give HLEOL when we shrink the big L.

Here is the important point: The number of ways to rearrange the letters in HELLO is exactly the same as the number of equivalence classes of R.

Now let's do the arithmetic: There are 120 different ways to rearrange the letters in HELLo (i.e., $|A| = 120$). The relation R partitions the set A into a certain number of equivalence classes. Each equivalence class has exactly two elements in it. So all told, there are $120 \div 2 = 60$ different equivalence classes. Hence there are 60 different ways to rearrange HELLO.

Anagrams of AARDVARK.

Let us consider another example. How many different ways can we rearrange the letters in the word AARDVARK? This eight-letter word features two Rs and three As. Let us use two styles of R (say, R and R) and three styles of A (say, a, A, and A), so the word is AARDVaRK.

Let X be the set of all rearrangements of AARDVaRK. We consider two spellings to be related by relation R if they are the same once their letters are restored to normal size. Clearly R is an equivalence relation on X, and we want to count the number of equivalence classes.

The problem becomes: How large are the equivalence classes? Let us consider the size of the equivalence class [RADaKRAV]. These are all the rearrangements that become RADAKRAV when their letters are all the same style. How many are there? This is a list-counting problem! We want to count the number of lists wherein the entries on the list satisfy the following restrictions:

- Elements 3, 5, and 8 of the list must be D, K, and V.
- Elements 1 and 6 must be one each of two different styles of R.
- Elements 2, 4, and 7 must be one each of three different styles of A.

See the figure.

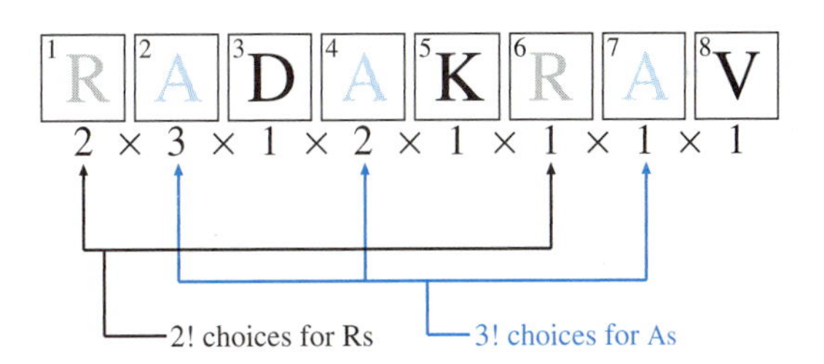

The letters R and A in the figure are dimmed to show that their final form is to be determined.

Now let's count how many ways we can build this list. There are two choices for the first position (we can user either R). There are three choices for the second position (we can use any A). There is only one choice for position 3 (it must be D). Now, given the choices thus far, there are only two choices for position 4 (the first A has already been selected, and so there are only two choices of A left at this point). For each of the remaining positions, there is only one choice (the K and V are predetermined, and we are down to only one choice each on the remaining A and R).

Therefore, the number of rearrangements of AARDVaRK in [RADaKRAV] is $2 \times 3 \times 1 \times 2 \times 1 \times 1 \times 1 \times 1 = 3! \times 2! = 12$.

And now for a critical comment: All equivalence classes have the same size! No matter how we rearrange the letters in AARDVaRK, the analysis we just conducted remains the same. Regardless of where the As may fall, there will be exactly 3! ways to fill them in, and regardless of where the Rs are, there are 2! ways to select their styles. And there is only one choice each for the style of D, K, and V. So all of the equivalence classes have size twelve.

Therefore the number of rearrangements of AARDVARK is

$$\frac{8!}{3!2!} = \frac{40320}{12} = 3360.$$

It is worth summarizing the central idea of this counting technique in an official statement.

Theorem 15.6 **(Counting equivalence classes)** Let R be an equivalence relation on a finite set A. If all the equivalence classes of R have the same size, m, then the number of equivalence classes is $|A|/m$.

There is an important hypothesis in this result: The equivalence classes must all be the same size. This does not always happen.

Example 15.7 Let $A = 2^{\{1,2,3,4\}}$—that is, the set of all subsets of $\{1, 2, 3, 4\}$. Let R be the has-the-same-size-as relation. This relation partitions A into five parts (subsets of size 0 through 4). The sizes of these equivalence classes are not all the same. For example, $[\emptyset]$ contains only $\emptyset$, so that class has size 1. However, $[\{1\}] = \{\{1\}, \{2\}, \{3\}, \{4\}\}$, so this class contains four members of A. Here is a full chart.

Equivalence class	Size of the class
$[\emptyset]$	1
$[\{1\}]$	4
$[\{1, 2\}]$	6
$[\{1, 2, 3\}]$	4
$[\{1, 2, 3, 4\}]$	1

Recap

A partition of a set A is a set of nonempty, pairwise disjoint subsets of A whose union is A. We explored the connection between partitions and equivalence relations. We applied these ideas to counting problems where we seek to count the number of equivalence classes when all the equivalence classes have the same size.

15 Exercises

15.1. There are only two possible partitions of the set $\{1, 2\}$. They are $\{\{1\}, \{2\}\}$ and $\{\{1, 2\}\}$. Find all possible partitions of $\{1, 2, 3\}$ and of $\{1, 2, 3, 4\}$.

15.2. How many different anagrams (including nonsensical words) can be made from each of the following?

a. STAPLE

b. DISCRETE

c. MATHEMATICS

d. SUCCESS

e. MISSISSIPPI

15.3. How many different anagrams (including nonsensical words) can be made from SUCCESS if we require that the first and last letters must both be S.

15.4. How many different anagrams (including nonsensical words) can be made from FACETIOUSLY if we require that all six vowels must remain in alphabetical order (but not necessarily contiguous with each other).

15.5. Prove Proposition 15.4.

15.6. Prove Theorem 15.6. You may assume the generalized Addition Principle (see after Corollary 11.8).

15.7. Twelve people join hands for a circle dance. In how many ways can they do this?

15.8. *Continued from the previous problem.* Suppose six of these people are men, and the other six are women. In how many ways can they join hands for a circle dance, assuming they alternate in gender around the circle?

15.9. You wish to make a necklace with 20 different beads. In how many different ways can you do this?

22	4	5	20	23
16	3	8	7	14
21	1	25	9	15
6	12	11	2	24
19	10	17	13	18

20	4	5	22	23
7	3	8	16	14
9	1	25	21	15
2	12	11	6	24
13	10	17	19	18

15.10. The integers 1 through 25 are arranged in a 5×5 array (we use each number from 1 to 25 exactly once). All that matters is which numbers are in each column and how they are arranged in the columns. It does not matter in what order the columns appear. (See the figure. The two arrays shown should be considered to be the same.)

How many different such arrays can be formed?

15.11. Twenty people are to be divided into two teams with ten players on each team. In how many ways can this be done?

15.12. One hundred people are to be divided into ten discussion groups with ten people in each group. In how many ways can this be done?

15.13. How many different partitions with exactly two parts can be made of the set $\{1, 2, 3, 4\}$?

Answer the same question for the set $\{1, 2, 3, \ldots, 100\}$.

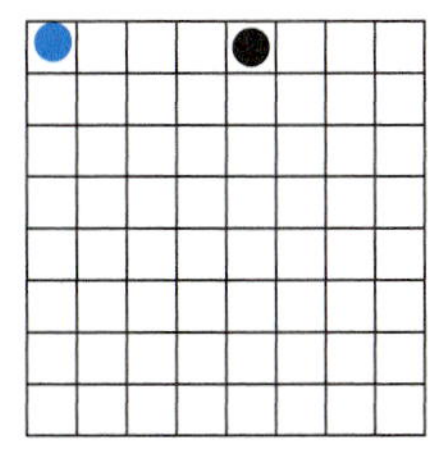

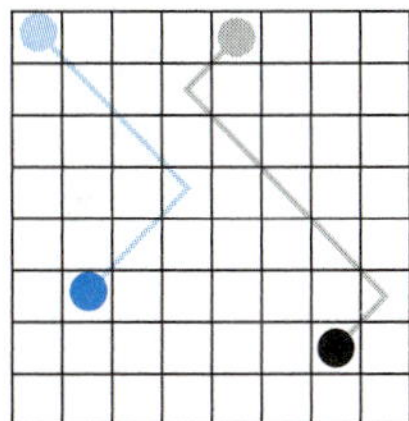

15.14. Two different coins are placed on squares of a standard 8×8 chess board; they may both be placed on the same square.

Let us call two arrangements of these coins on the chess board equivalent if we can move the coins diagonally to get from one arrangement to another. For example, the two positions shown on the two boards in the figure are equivalent.

How many different (inequivalent) ways can the coins be placed on the chess board?

15.15. Please redo the previous problem, this time assuming the coins are identical.

15.16. Let A be a set and let $\mathcal{P}$ be a partition of A. Is it possible to have $A = \mathcal{P}$?

16 Binomial Coefficients

The notation $\binom{n}{k}$ is pronounced "n choose k." Another form of this notation, still in use on some calculators, is ${}_nC_k$. Occasionally people write $C(n, k)$. An alternative way to express $\binom{n}{k}$ is as the number of "combinations" of n things taken k at a time. The word *combinatorics* (a term that refers to counting problems in discrete mathematics) comes from "combinations." I dislike the use of the word "combinations" and believe it is clearer to say $\binom{n}{k}$ stands for the number of k-element subsets of an n-element set.

We ended the previous section with Example 15.7, in which we counted the number of equivalence classes of the has-the-same-size-as relation on the set of subsets of $\{1, 2, 3, 4\}$. We found five different equivalence classes (corresponding to the five integers from 0 to 4), and these equivalence classes have various sizes. Their sizes are, in order, 1, 4, 6, 4, and 1. These numbers may be familiar to you. Observe:

$$(x+y)^4 = 1x^4 + 4x^3y + 6x^2y^2 + 4xy^3 + 1y^4.$$

These numbers are the coefficients of $(x + y)^4$ after it is expanded. You may also recognize these numbers as the fourth row of Pascal's triangle. In this section, we explore these numbers in detail.

The central problem we consider in this section is the following:

How many subsets of size k does an n-element set have?

There is a special notation for the answer to this question: $\binom{n}{k}$.

Definition 16.1 **(Binomial coefficient)** Let $n, k \in \mathbb{N}$. The symbol $\binom{n}{k}$ denotes the number of k-element subsets of an n-element set.

We call the number $\binom{n}{k}$ a *binomial coefficient*. The reason for this nomenclature is that the numbers $\binom{n}{k}$ are the coefficients of binomial $(x + y)^n$. This is explained more thoroughly below.

Example 16.2 Evaluate $\binom{5}{0}$.

Solution: We need to count the number of subsets of a five-element set that have zero elements. The only possible such set is $\emptyset$, so the answer is $\binom{5}{0} = 1$.

Clearly there is nothing special about the number 5 in this example. The number of zero-element subsets of any set is always 1. So we have, for all $n \in \mathbb{N}$,

$$\binom{n}{0} = 1.$$

Example 16.3 Evaluate $\binom{5}{1}$.

Solution: This asks for the number of one-element subsets of a five-element set. For example, consider the five-element set $\{1, 2, 3, 4, 5\}$. The one-element subsets are $\{1\}$, $\{2\}$, $\{3\}$, $\{4\}$, and $\{5\}$, so $\binom{5}{1} = 5$. The number of one-element subsets of an n-element set is exactly n:

$$\binom{n}{1} = n.$$

Example 16.4 Evaluate $\binom{5}{2}$.

Solution: The symbol $\binom{5}{2}$ stands for the number of two-element subsets of a five-element set. The simplest thing to do is to list all the possibilities.

$$\begin{array}{llll} \{1,2\} & \{1,3\} & \{1,4\} & \{1,5\} \\ \{2,3\} & \{2,4\} & \{2,5\} & \\ \{3,4\} & \{3,5\} & & \\ \{4,5\} & & & \end{array}$$

Therefore, there are 10 two-element subsets of a five-element set, so $\binom{5}{2} = 4 + 3 + 2 + 1 = 10$.

There is an interesting pattern in Example 16.4. Let us try to generalize it. Suppose we want to know the number of two-element subsets of an n-element set. Let's say that the n-element set is $\{1, 2, 3, \ldots, n\}$. We can make a chart as in the example. The first row of the chart lists the two-element subsets whose smaller element is 1. The second row lists those two-element subsets whose smaller element is 2, and so on, and the last row of the chart lists the (one and only) two-element subset whose smaller element is $n - 1$ (i.e., $\{n - 1, n\}$).

Notice that our chart exhausts all the possibilities (the smaller element must be one of the numbers from 1 to $n - 1$), and no duplication takes place (subsets on different rows of the chart have different smaller elements).

The number of sets in the first row of this hypothetical chart is $n - 1$, because once we decide that the smaller element is 1, the subset looks like this: $\{1, \_\}$. The second element must be larger than 1, and so it is chosen from $\{2, \ldots, n\}$; there are $n - 1$ ways to complete the set $\{1, \_\}$.

The number of sets in the second row of this chart is $n - 2$. All subsets in this row look like this: $\{2, \_\}$. The second element needs to be chosen from the numbers 3 to n, so there are $n - 2$ ways to complete this set.

In general, the number of sets in row k of this hypothetical chart is $n - k$. Subsets on this row look like $\{k, \_\}$, the second element of the set needs to be an integer from $k + 1$ to n, and there are $n - k$ possibilities.

This discussion is the proof of the following result.

Proposition 16.5 Let n be an integer with $n \geq 2$. Then

$$\binom{n}{2} = 1 + 2 + 3 + \cdots + (n - 1) = \sum_{k=1}^{n-1} k.$$

So far we have evaluated $\binom{5}{0}$, $\binom{5}{1}$, and $\binom{5}{2}$. Let us continue this exploration.

Example 16.6 Evaluate $\binom{5}{3}$.

Solution: We simply list the three-element subsets of $\{1, 2, 3, 4, 5\}$:

$$\begin{array}{lllll} \{1,2,3\} & \{1,2,4\} & \{1,2,5\} & \{1,3,4\} & \{1,3,5\} \\ \{1,4,5\} & \{2,3,4\} & \{2,3,5\} & \{2,4,5\} & \{3,4,5\} \end{array}$$

There are ten such sets, so $\binom{5}{3} = 10$.

Notice that $\binom{5}{2} = \binom{5}{3} = 10$. This equality is not a coincidence. Let's see why these numbers are equal. The idea is to find a natural way to match up the two-element subsets of $\{1, 2, 3, 4, 5\}$ with the three-element subsets. We want a one-to-one correspondence between these two kinds of sets. Of course, we could just list them down two columns of a chart, but this is not necessarily "natural." The idea is to take the complement (see Exercise 11.17) of a two-element subset to form a three-element subset, or vice versa. We do this here:

This is an example of a *bijective* proof.

The concept of *set complement* is developed in Exercise 11.17.

A	$\overline{A}$	A	$\overline{A}$
$\{1, 2\}$	$\{3, 4, 5\}$	$\{2, 4\}$	$\{1, 3, 5\}$
$\{1, 3\}$	$\{2, 4, 5\}$	$\{2, 5\}$	$\{1, 3, 4\}$
$\{1, 4\}$	$\{2, 3, 5\}$	$\{3, 4\}$	$\{1, 2, 5\}$
$\{1, 5\}$	$\{2, 3, 4\}$	$\{3, 5\}$	$\{1, 2, 4\}$
$\{2, 3\}$	$\{1, 4, 5\}$	$\{4, 5\}$	$\{1, 2, 3\}$

Each two-element subset A is paired up with $\{1, 2, 3, 4, 5\} - A$ (which we denote $\overline{A}$ since $\{1, 2, 3, 4, 5\}$ is the "universe" we are considering at the moment).

This pairing, $A \leftrightarrow \overline{A}$, is a one-to-one correspondence between the two-element and three-element subsets of $\{1, 2, 3, 4, 5\}$. If A_1 and A_2 are two different two-element subsets, then $\overline{A_1}$ and $\overline{A_2}$ are two different three-element subsets. Every two-element subset is paired up with exactly one three-element subset, and no sets are left unpaired. This thoroughly explains why $\binom{5}{2} = \binom{5}{3}$ and gives us an avenue for generalization.

We might guess $\binom{n}{2} = \binom{n}{3}$, but this is not right. Let's apply our complement analysis to $\binom{n}{2}$ and see what we learn. Let A be a two-element subset of $\{1, 2, \ldots, n\}$. In this context, $\overline{A}$ means $\{1, 2, \ldots, n\} - A$. The pairing $A \leftrightarrow \overline{A}$ does not pair up two-element and three-element subsets. The complement of a two-element subset would be an $(n - 2)$-element subset of $\{1, 2, \ldots, n\}$. Aha! Now we have the correct result: $\binom{n}{2} = \binom{n}{n-2}$.

We can push this analysis further. Instead of forming the complement of the two-element subsets of $\{1, 2, \ldots, n\}$, we can form the complements of subsets of another size. What are the complements of the k-element subsets of $\{1, 2, \ldots, n\}$? They are precisely the $(n - k)$-element subsets. Furthermore, the correspondence $A \leftrightarrow \overline{A}$ gives a one-to-one pairing of the k-element and $(n - k)$-element subsets of $\{1, 2, \ldots, n\}$. This implies that the number of k- and $(n - k)$-element subsets of an n-element set must be the same. We have shown the following:

Proposition 16.7 Let $n, k \in \mathbb{N}$ with $0 \le k \le n$. Then

$$\binom{n}{k} = \binom{n}{n-k}.$$

Here is another way to think about this result. Imagine a class with n children. The teacher has k identical candy bars to give to exactly k of the children. In how many ways can the candy bars be distributed? The answer is $\binom{n}{k}$ because we are selecting a set of k lucky children to get candy. But the pessimistic view is also interesting. We can think about selecting the unfortunate children who will not be receiving candy. There are $n-k$ children who do not get candy, and we can select that subset of the class in $\binom{n}{n-k}$ ways. Since the two counting problems are clearly the same, we must have $\binom{n}{k} = \binom{n}{n-k}$.

Thus far we have evaluated $\binom{5}{0}$, $\binom{5}{1}$, $\binom{5}{2}$, and $\binom{5}{3}$. Let us continue. We can use Proposition 16.7 to evaluate $\binom{5}{4}$; the proposition says that

$$\binom{5}{4} = \binom{5}{5-4} = \binom{5}{1}$$

and we already know that $\binom{5}{1} = 5$. So $\binom{5}{4} = 5$.

Next is $\binom{5}{5}$. We can use Proposition 16.7 and reason $\binom{5}{5} = \binom{5}{5-5} = \binom{5}{0} = 1$, or we can realize that there can be only one five-element subset of a five-element set—namely, the whole set!

Next comes $\binom{5}{6}$. We can try to use Proposition 16.7, but we run into a snag. We write

$$\binom{5}{6} = \binom{5}{5-6} = \binom{5}{-1}$$

but we don't know what $\binom{5}{-1}$ is. Actually, the situation is worse: $\binom{5}{-1}$ is nonsense. It does not make sense to ask for the number of subsets of a five-element set that have -1 elements; it does not make sense to consider sets with a negative number of elements! (This is why we included the hypothesis $0 \le k \le n$ in the statement of Proposition 16.7.)

However, a set *can* have six elements, so $\binom{5}{6}$ is not nonsense; it is simply zero. A five-element set cannot have any six-element subsets, so $\binom{5}{6} = 0$. Similarly, $\binom{5}{7} = \binom{5}{8} = \cdots = 0$.

Let us summarize what we know so far:

- We have evaluated $\binom{5}{k}$ for all natural numbers k. The values are 1, 5, 10, 10, 5, 1, 0, 0, ..., for $k = 0, 1, 2, \ldots$, respectively.
- We have $\binom{n}{0} = 1$ and $\binom{n}{1} = n$.
- We have $\binom{n}{2} = 1 + 2 + \cdots + (n-1)$.
- We have $\binom{n}{k} = \binom{n}{n-k}$.
- If $k > n$, $\binom{n}{k} = 0$.

Calculating $\binom{n}{k}$

Thus far we have calculated various values of $\binom{n}{k}$, but our work has been *ad hoc*. We do not have a general method for obtaining these values. We found that the nonzero values of $\binom{5}{k}$ are

$$1, 5, 10, 10, 5, 1.$$

If we expand $(x+y)^5$, we get

$$\begin{aligned}(x+y)^5 &= 1x^5 + 5x^4y + 10x^3y^2 + 10x^2y^3 + 5xy^4 + 1y^5 \\ &= \binom{5}{0}x^5 + \binom{5}{1}x^4y + \binom{5}{2}x^3y^2 + \binom{5}{3}x^2y^3 + \binom{5}{4}xy^4 + \binom{5}{5}y^5.\end{aligned}$$

This suggests a way to calculate $\binom{n}{k}$: Expand $(x+y)^n$ and $\binom{n}{k}$ is the coefficient of $x^{n-k}y^k$. This is marvelous! Let's prove it.

Theorem 16.8 **(Binomial)** Let $n \in \mathbb{N}$. Then

$$(x+y)^n = \sum_{k=0}^{n} \binom{n}{k} x^{n-k} y^k.$$

This result explains why $\binom{n}{k}$ is called a *binomial coefficient*. The numbers $\binom{n}{k}$ are the coefficients that appear in the expansion of $(x+y)^n$.

Proof. The key to proving the Binomial Theorem is to think about how we multiply polynomials. When we multiply $(x+y)^2$, we calculate as follows:

$$(x+y)^2 = (x+y)(x+y) = xx + xy + yx + yy$$

and then we collect like terms to get $x^2 + 2xy + y^2$.

The procedure for $(x+y)^n$ is much the same. We write out n factors of $(x+y)$:

$$\underbrace{(x+y)}_{1}\underbrace{(x+y)}_{2}\underbrace{(x+y)}_{3}\cdots\underbrace{(x+y)}_{n}.$$

We then form all possible terms by taking either an x or a y from factors 1, 2, 3, …, n. This is like making lists (see Section 7). We are forming all possible n-element lists where each element is either an x or a y. For example,

$$(x+y)(x+y)(x+y) = xxx + xxy + xyx + xyy + yxx + yxy + yyx + yyy.$$

The next step is to collect like terms. In the example $(x+y)^3$ there is one term with three xs and no ys, three terms with two xs and one y, three terms with one x and two ys, and one term with no xs and three ys. This gives

$$(x+y)^3 = 1x^3 + 3x^2y + 3xy^2 + 1y^3.$$

The question now becomes: How many terms in $(x+y)^n$ have precisely k ys (and $n-k$ xs)? Let us think of this as a list-counting question. We want to count the number of n-element lists with precisely $n-k$ xs and k ys. And we know what we want the answer to be: $\binom{n}{k}$. We need to justify this answer.

We can specify all the lists with k ys (and $n-k$ xs) by reporting the positions of the ys (and the xs fill in the remaining positions). For example, if $n = 10$ and we say that the set of y positions is $\{2, 3, 7\}$, then we know we are speaking of the term (list) $xyyxxxyxxx$. We could make a chart: On the left of the chart would be all the lists with k ys and $n-k$ xs, and on the right we would write the set of y positions for each list. The right column of the chart would simply be the k-element subsets of $\{1, 2, \ldots, n\}$. Aha! The number of lists with k ys and $n-k$ xs is exactly the same as the number of k-element subsets of $\{1, 2, \ldots, n\}$. Therefore the number of $x^{n-k}y^k$ terms we collect is $\binom{n}{k}$. And this completes the proof! ■

Example 16.9 Expand $(x+y)^5$ and find all the terms with two ys and three xs. Pair these terms up with the two-element subsets of $\{1, 2, 3, 4, 5\}$.

Solution:

$yyxxx \leftrightarrow \{1, 2\}$	$xyxyx \leftrightarrow \{2, 4\}$
$yxyxx \leftrightarrow \{1, 3\}$	$xyxxy \leftrightarrow \{2, 5\}$
$yxxyx \leftrightarrow \{1, 4\}$	$xxyyx \leftrightarrow \{3, 4\}$
$yxxxy \leftrightarrow \{1, 5\}$	$xxyxy \leftrightarrow \{3, 5\}$
$xyyxx \leftrightarrow \{2, 3\}$	$xxxyy \leftrightarrow \{4, 5\}$

We now have a procedure to calculate, say, $\binom{20}{10}$. All we have to do is expand out $(x+y)^{20}$ and find the coefficient of $x^{10}y^{10}$. To do that, we just write down all the terms from $xxx\cdots xx$ to $yyy\cdots yy$ and collect like terms. There are only $2^{20} = 1{,}048{,}576$ terms. Sounds like fun!

No? You are right. This is not a good way to find $\binom{20}{10}$. It is no better than writing out all the possible ten-element subsets of $\{1, 2, \ldots, 20\}$. And there are a lot of them. How many? We don't know! That's what we're trying to find out. We need another method (see also Exercise 16.29).

Pascal's Triangle

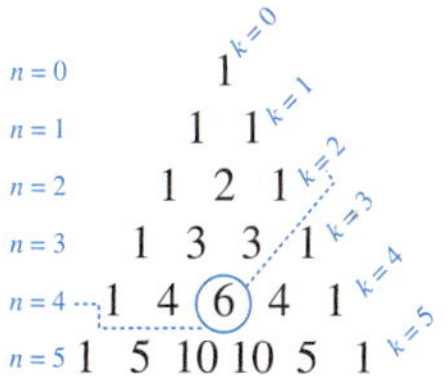

$n=0$ 1 ($k=0$)
$n=1$ 1 1 ($k=1$)
$n=2$ 1 2 1 ($k=2$)
$n=3$ 1 3 3 1 ($k=3$)
$n=4$ 1 4 (6) 4 1 ($k=4$)
$n=5$ 1 5 10 10 5 1 ($k=5$)

Recall from your algebra class that the coefficients of $(x+y)^n$ form the nth row of Pascal's triangle. The figure shows Pascal's triangle. The entry in row $n = 4$ and diagonal $k = 2$ is $\binom{4}{2} = 6$, as shown (we count the rows and diagonals starting from 0).

How is Pascal's triangle generated? Here is a complete description:

- The zeroth row of Pascal's triangle contains just the single number 1.
- Each successive row contains one more number than its predecessor.
- The first and last number in every row is 1.
- An intermediate number in any row is formed by adding the two numbers just to its left and just to its right in the previous row. For example, the first 10 in row $n = 5$ (and diagonal $k = 2$) is formed by adding the 4 to its upper left (at $n = 4, k = 1$) and the 6 to its upper right (at $n = 4, k = 2$ as shown circled in the figure).

How do we know that Pascal's triangle generates the binomial coefficients? How do we know that the entry in row n and column k is $\binom{n}{k}$?

To see why this works, we need to show that the binomial coefficients follow the same four rules we just listed.

In other words, we form a triangle containing $\binom{0}{0}$ on the zeroth row; $\binom{1}{0}$, $\binom{1}{1}$ on the first row, $\binom{2}{0}$, $\binom{2}{1}$, $\binom{2}{2}$ on the second row, and so on. We then need to prove that this triangle of binomial coefficients is generated by exactly the same four rules as Pascal's triangle! This is three-fourths easy plus one-fourth tricky. Here we go.

- *The zeroth row of the binomial coefficient triangle contains the single number* 1.

  This is easy: The zeroth row of the binomial coefficients triangle is $\binom{0}{0} = 1$.
- *Each successive row contains one more number than its predecessor.*

This is easy: Row n of the binomial coefficient triangle contains exactly $n+1$ numbers: $\binom{n}{0}, \binom{n}{1}, \ldots, \binom{n}{n}$.

- *The first and last number in every row is* 1.

  This is easy: The first and last numbers in row n of the binomial coefficient triangle are $\binom{n}{0} = \binom{n}{n} = 1$.

- *The intermediate number in any row is formed by adding the two numbers just to its left and just to its right in the previous row.*

  This is tricky! The first thing we need to do is write down a careful statement of what we need to prove about binomial coefficients. We need an *intermediate number in any row*. This means we do not need to worry about $\binom{n}{0}$ or $\binom{n}{n}$; we already know those are 1. An intermediate number in row n would be $\binom{n}{k}$ with $0 < k < n$.

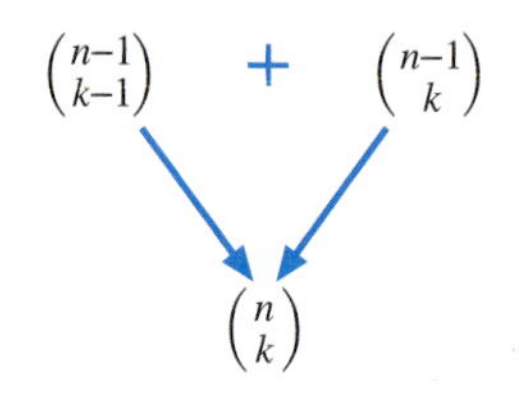

  What are the numbers just above $\binom{n}{k}$? To find the upper left neighbor, we move up to row $n-1$ and up to diagonal $k-1$. So the number to the upper left is $\binom{n-1}{k-1}$. To find the upper right neighbor, we move up to row $n-1$ but stay on diagonal k. So the number to the upper right is $\binom{n-1}{k}$.

  We need to prove the following:

Theorem 16.10 **(Pascal's Identity)** Let n and k be integers with $0 < k < n$. Then

$$\binom{n}{k} = \binom{n-1}{k-1} + \binom{n-1}{k}.$$

How can we prove this? We don't have a formula for $\binom{n}{k}$. The idea is to use combinatorial proof (see Proof Template 9). We need to ask a question and then prove that the left and right sides of the equation in Theorem 16.10 both give correct answers to this question. What question has these answers? There is a clear question to which the left-hand side gives an answer. The question is: How many k-element subsets does an n-element set have?

Proof. To prove $\binom{n}{k} = \binom{n-1}{k-1} + \binom{n-1}{k}$, we consider the question: How many k-element subsets does the set $\{1, 2, 3, \ldots, n\}$ have?

- Answer 1: $\binom{n}{k}$, by definition.

Now we need another answer. The right-hand side of the equation gives us some hints. It contains the numbers $n-1$, $k-1$, and k. It is telling us to pick either $k-1$ or k elements from an $(n-1)$-element set. But we have been thinking about an n-element set, so let's throw away one of the elements; let's say that element n is a "weirdo." The right-hand side is telling us to pick either $k-1$ or k elements from among the normal elements $1, 2, \ldots, n-1$. If we only pick $k-1$ elements, that doesn't make a full k-element subset—in this case, we can add the weirdo to the $(k-1)$-element subset. Or we pick k elements from the normal elements. Now we have a full k-element subset, and no room is left for the weirdo.

We now have all the ideas in place; let's express them clearly.

Let n be called the "weird" element of $\{1, 2, \ldots, n\}$. When we form a k-element subset of $\{1, 2, \ldots, n\}$, there are two possibilities. Either we have a subset that includes the weirdo, or we have a subset that does not include the weirdo—these mutually exclusive possibilities cover all cases.

If we put the weird element in the subset, then we have $\binom{n-1}{k-1}$ choices for how to complete the subset because we must choose $k-1$ elements from $\{1, 2, \ldots, n-1\}$.

If we do not put the weird element in the subset, then we have $\binom{n-1}{k}$ ways to make the subset because we must choose all k elements from $\{1, 2, \ldots, n-1\}$.

Thus we have another answer.

- Answer 2: $\binom{n-1}{k-1} + \binom{n-1}{k}$.

Since Answer 1 and Answer 2 are correct answers to the same question, they must be equal, and we are finished. ■

Example 16.11 We show that $\binom{6}{2} = \binom{5}{1} + \binom{5}{2}$ by listing all the two-element subsets of $\{1, 2, 3, 4, 5, 6\}$.

There are $\binom{5}{1} = 5$ two-element subsets that include the weirdo 6:

$$\{1, 6\}\ \{2, 6\}\ \{3, 6\}\ \{4, 6\}\ \{5, 6\}$$

and there are $\binom{5}{2} = 10$ two-element subsets that do not include 6:

$$\{1, 2\}\ \{1, 3\}\ \{1, 4\}\ \{1, 5\}\ \{2, 3\}$$
$$\{2, 4\}\ \{2, 5\}\ \{3, 4\}\ \{3, 5\}\ \{4, 5\}.$$

We now want to calculate $\binom{20}{10}$. The technique we could follow is to generate Pascal's triangle down to the 20th row and look up the entry on diagonal 10. How much work would this be? The 20th row of Pascal's triangle contains 21 numbers. The previous row contains 20, and the one before that has 19. There are only $1 + 2 + 3 + \cdots + 21 = 231$ numbers. We get most of them by simple addition and we need to do about 200 addition problems. (We can be more efficient; see Exercise 16.30.) If you were to implement this procedure on a computer, you would not need to save all 210 numbers. You would only need to save about 40. Once you have calculated a row of Pascal's triangle, you can discard the previous row. So at any time, you would only keep the previous row and the current row. And if you are clever, you can save even more memory.

In any case, if you follow this procedure, you will find that $\binom{20}{10} = 184{,}756$.

A Formula for $\binom{n}{k}$

The technique of generating Pascal's triangle to calculate binomial coefficients is a good one. We can calculate $\binom{20}{10}$ by performing roughly 200 addition problems instead of sifting through a million terms in a polynomial (see also Exercise 16.29).

There is something a bit unsatisfying about this answer. We like formulas! We want a nice way to express $\binom{n}{k}$ in a simple expression using familiar operations. We have an expression for $\binom{n}{2}$: Proposition 16.5 says

$$\binom{n}{2} = 1 + 2 + 3 + \cdots + (n-1).$$

This is not bad, but it suggests that we still need to do a lot of addition to get the answer. There is, however, a nice trick for simplifying this sum. Write the integers 1 through $n - 1$ forward and backward, and then add:

$$\begin{array}{rcccccccccccccc} \binom{n}{2} = & 1 & + & 2 & + & 3 & + & \cdots & + & n-2 & + & n-1 & \\ +\binom{n}{2} = & n-1 & + & n-2 & + & n-3 & + & \cdots & + & 2 & + & 1 & \\ \hline 2\binom{n}{2} = & n & + & n & + & n & + & \cdots & + & n & + & n & = n(n-1) \end{array}$$

and therefore

$$\binom{n}{2} = \frac{n(n-1)}{2}.$$

This equation is a special case of a more general result. Here is another way to count k-element subsets of an n-element set.

Let us begin by counting all k-element lists, without repetition, whose elements are selected from an n-element set. This is a problem we have already solved (see Section 7)! The number of such lists is $(n)_k$.

For example, there are $(5)_3 = 5 \cdot 4 \cdot 3 = 60$ three-element, repetition-free lists we can form from the members of $\{1, 2, 3, 4, 5\}$:

All the entries in a single row of this chart expresses the same three-element subset in six different ways. Since this chart has 60 entries, the number of three-element subsets of $\{1, 2, 3, 4, 5\}$ is $60 \div 6 = 10$.

123	132	213	231	312	321
124	142	214	241	412	421
125	152	215	251	512	521
		and so on, until			
345	354	435	453	534	543

Notice how we have organized our chart. All lists on the same row contain exactly the same elements, just in different orders. Let us define a relation R on these lists. The relation is "has-the-same-elements-as"—two lists are related by R just when their elements are the same (but their orders might be different). Clearly R is an equivalence relation. Each row of the chart gives an equivalence class. We want to count the equivalence classes. There are 60 elements of the set (all three-element lists). Each equivalence class contains six lists. Therefore the number of equivalence classes is $\frac{60}{6} = 10 = \binom{5}{3}$ by Theorem 15.6.

Let's repeat this analysis for the general problem. We want to count the number of k-element subsets of $\{1, 2, \ldots, n\}$. Instead, we consider the k-element, repetition-free lists we can form from $\{1, 2, \ldots, n\}$. We declare two of these lists equivalent if they contain the same members. Finally, we compute the number of equivalence classes to calculate $\binom{n}{k}$.

The reason why each list is equivalent to $(k)_k = k!$ lists also follows from Theorem 7.6; we want to know how many length-k, repetition-free lists we can form using k elements.

The number of k-element, repetition-free lists we can form from $\{1, 2, \ldots, n\}$ is a problem we already solved (Theorem 7.6); there are $(n)_k$ such lists.

Therefore the number of equivalence classes is $(n)_k/k! = \binom{n}{k}$. We can rewrite $(n)_k$ as $n!/(n-k)!$ (provided $k \le n$), and we have the following result.

Theorem 16.12 **(Formula for $\binom{n}{k}$)** Let n and k be integers with $0 \le k \le n$. Then

$$\binom{n}{k} = \frac{n!}{k!(n-k)!}.$$

We have found a "formula" for $\binom{n}{k}$. Are we happy? Perhaps. If we want to compute $\binom{20}{10}$, what does this theorem tell us to do? It asks us to calculate

$$\binom{20}{10} = \frac{20 \times 19 \times 18 \times \cdots \times 3 \times 2 \times 1}{10 \times 9 \times 8 \times \cdots \times 2 \times 1 \times 10 \times 9 \times 8 \times \cdots \times 2 \times 1}.$$

This entails about 40 multiplications and 1 division. Also, the intermediate results (the numerator and denominator) are very large (more digits than most calculators can handle).

Of course, we can cancel some terms between the numerator and the denominator to speed things up. The last ten terms of the numerator are $10 \times \cdots \times 1$, and that cancels out one of the 10!s in the denominator. So now the problem reduces to

$$\binom{20}{10} = \frac{20 \times 19 \times 18 \times \cdots \times 11}{10 \times 9 \times 8 \times \cdots \times 1}.$$

We can hunt for more cancellations, but now it requires us to think about the numbers involved. The cancellation of one 10! in the denominator was mindless; we could build that easily into a computer program. Other cancellations may be tricky to find. If we're doing this on a computer, we may as well just do the remaining multiplications and final division, which would be

$$\frac{670442572800}{3628800} = 184756.$$

Recap

This section dealt entirely with the binomial coefficient $\binom{n}{k}$, the number of k-element subsets of an n-element set. We proved the Binomial Theorem, we showed that the binomial coefficients are the entries in Pascal's triangle, and we developed a formula to express $\binom{n}{k}$ in terms of factorials.

16 Exercises

16.1. Mixed Matched Marvin has a drawer full of 30 different socks (no two are the same). He reaches in and grabs two. In how many different ways can he do this? Now he puts them on his feet (presumably, one on the left and the other on the right). In how many different ways can he do that?

16.2. Twenty people attend a party. If everyone shakes everyone else's hand exactly once, how many handshakes take place?

16.3. **a.** How many n-digit binary (0,1) sequences contain exactly k 1s?

b. How many n-digit ternary (0,1,2) sequences contain exactly k 1s?

16.4. Fifty runners compete in a 10K race. How many different outcomes are possible?

To make this problem tractable, assume that there are no ties.

The answer to this question depends on what we are judging. Find different answers to this question depending on the context.

a. We want to know in what place every runner finished.

b. The race is a qualifying race, and we just want to pick the ten fastest runners.

c. The race is an Olympic final event, and we care only about who gets the gold, silver, and bronze medals.

16.5. Write out all the three- and four-element subsets of $\{1, 2, 3, 4, 5, 6, 7\}$ in two columns. Pair each three-element subset with its complement. Your chart should have 35 rows.

16.6. A special type of door lock has a panel with five buttons labeled with the digits 1 through 5. This lock is opened by a sequence of three actions. Each action consists of either pressing one of the buttons or pressing a pair of them simultaneously.

For example, 12-4-3 is a possible combination. The combination 12-4-3 is the same as 21-4-3 because both the 12 and the 21 simply mean to press buttons 1 and 2 simultaneously.

a. How many combinations are possible?

b. How many combinations are possible if no digit is repeated in the combination?

16.7. In how many different ways can we partition an n-element set into two parts if one part has four elements and the other part has all the remaining elements?

16.8. Look down the middle column of Pascal's triangle. Notice that, except for the very top 1, all these numbers are even. Why?

16.9. Use Theorem 16.12 to prove Proposition 16.7.

16.10. Prove that the sum of the numbers in the nth row of Pascal's triangle is 2^n.

One easy way to do this is to substitute $x = y = 1$ into the Binomial Theorem (Theorem 16.8).

However, please give a combinatorial proof. That is, prove that

$$2^n = \sum_{k=0}^{n} \binom{n}{k}$$

by finding a question that is correctly answered by both sides of this equation.

16.11. Use the Binomial Theorem (Theorem 16.8) to prove

$$\binom{n}{0} - \binom{n}{1} + \binom{n}{2} - \binom{n}{3} + \cdots \pm \binom{n}{n} = 0$$

provided $n > 0$.

Move all the negative terms over to the right-hand side to give

$$\binom{n}{0} + \binom{n}{2} + \binom{n}{4} + \cdots = \binom{n}{1} + \binom{n}{3} + \binom{n}{5} + \cdots.$$

Give a combinatorial description of what this means and convert it into a combinatorial proof. Use the "weirdo" method.

16.12. Consider the following formula:

$$k\binom{n}{k} = n\binom{n-1}{k-1}.$$

Give two different proofs. One proof should use the factorial formula for $\binom{n}{k}$ (Theorem 16.12). The other proof should be combinatorial; develop a question that both sides of the equation answer.

16.13. Let $n \geq k \geq m \geq 0$ be integers. Consider the following formula:

$$\binom{n}{k}\binom{k}{m} = \binom{n}{m}\binom{n-m}{k-m}.$$

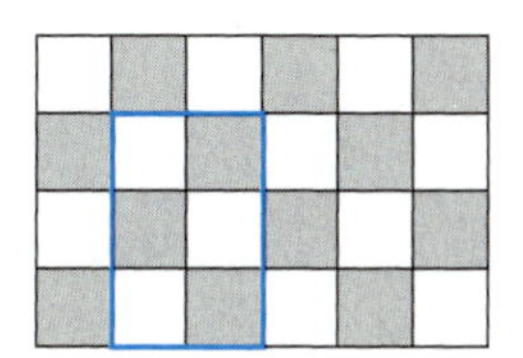

Give two different proofs. One proof should use the factorial formula for $\binom{n}{k}$ (Theorem 16.12). The other proof should be combinatorial. Try to develop a question that both sides of the equation answer.

16.14. How many rectangles can be formed from an $m \times n$ chess board? For example, for a 2×2 chess board, there are nine possible rectangles.

16.15. Let n be a natural number. Give a combinatorial proof of the following:

$$\binom{2n+2}{n+1} = \binom{2n}{n+1} + 2\binom{2n}{n} + \binom{2n}{n-1}.$$

16.16. Use Stirling's formula (see Exercise 8.6) to develop an approximation formula for $\binom{2n}{n}$. Without using Stirling's formula, give a direct proof that $\binom{2n}{n} \le 4^n$.

16.17. Use the factorial formula for $\binom{n}{k}$ (Theorem 16.12) to prove Pascal's Identity (Theorem 16.10).

16.18. Prove

$$\binom{n}{3} = \binom{2}{2} + \binom{3}{2} + \binom{4}{2} + \cdots + \binom{n-1}{2}.$$

Hint: Mimic the argument for Proposition 16.5.

16.19. *Continued from the previous problem.* Proposition 16.5 says $\binom{n}{2} = 1 + 2 + \cdots + (n-1)$. Make a large copy of Pascal's triangle and mark the numbers $\binom{7}{2}$, 6, 5, 4, 3, 2, and 1. You have several choices—do this "right." What's the pattern?

The previous exercise asks you to prove $\binom{n}{3} = \binom{2}{2} + \binom{3}{2} + \binom{4}{2} + \cdots + \binom{n-1}{2}$. On a large copy of Pascal's triangle, mark the numbers $\binom{7}{3}$, $\binom{6}{2}$, $\binom{5}{2}$, $\binom{4}{2}$, $\binom{3}{2}$, and $\binom{2}{2}$. What's the pattern?

Now generalize these formulas and prove your assertion.

16.20. Give a geometric and an algebraic proof that

$$1 + 2 + 3 + \cdots + (n-1) + n + (n-1) + (n-2) + \cdots + 2 + 1 = n^2.$$

16.21. Prove: $\binom{n}{0}\binom{n}{n} + \binom{n}{1}\binom{n}{n-1} + \binom{n}{2}\binom{n}{n-2} + \cdots + \binom{n}{n-1}\binom{n}{1} + \binom{n}{n}\binom{n}{0} = \binom{2n}{n}$.

16.22. How many Social Security numbers (see Exercise 7.9) have their nine digits in strictly increasing order?

The following series of problems introduce the concept of *multinomial coefficients*.

16.23. The binomial coefficient $\binom{n}{k}$ is the number of k-element subsets of an n-element set. Here is another way to think of $\binom{n}{k}$. Let A be an n-element set and suppose we have a supply of labels; we have k labels that say "good" and $n - k$ labels that say "bad." In how many ways can we affix exactly one label to each element of A?

16.24. Let A be an n-element set. Suppose we have three types of labels to assign to the elements of A. We can call these labels "good," "bad," and "ugly" or give them less interesting names such as "Type 1," "Type 2," and "Type 3."

Let $a, b, c \in \mathbb{N}$. Define the symbol $\binom{n}{a\ b\ c}$ to be the number of ways to label the elements of an n-element set with three types of labels in which we give exactly a of the elements labels of Type 1, b of the elements labels of Type 2, and c of the elements labels of Type 3.

Evaluate the following from first principles:

a. $\binom{3}{1\ 1\ 1}$.

b. $\binom{10}{1\ 2\ 5}$.

c. $\binom{5}{0\ 5\ 0}$.

d. $\binom{10}{7\ 3\ 0}$.

e. $\binom{10}{5\ 2\ 3} - \binom{10}{2\ 3\ 5}$.

16.25. Let $n, a, b, c \in \mathbb{N}$ with $a + b + c = n$. Please prove the following:

a. $\binom{n}{a\ b\ c} = \binom{n}{a}\binom{n-a}{b}$.

b. $\binom{n}{a\ b\ c} = \frac{n!}{a!b!c!}$.

c. If $a + b + c \neq n$, then $\binom{n}{a\ b\ c} = 0$.

16.26. Let $n \in \mathbb{N}$. Prove

$$(x + y + z)^n = \sum_{a+b+c=n} \binom{n}{a\ b\ c} x^a y^b z^c$$

where the sum is over all natural numbers a, b, c with $a + b + c = n$.

16.27. A poker hand consists of 5 cards chosen from a standard deck of 52 cards. How many different poker hands are possible?

If you divide the answers to this problem by $\binom{52}{5}$ (the answer to the previous problem), you will have the *probability* that a randomly selected poker hand is of the sort described. The concept of probability is developed in Chapter 6.

16.28. *Poker continued.* There are a variety of special hands that one can be dealt in poker. For each of the following types of hands, count the number of hands that have that type.

a. Four of a kind: The hand contains four cards of the same numerical value (e.g., four jacks) and another card.

b. Three of a kind: The hand contains three cards of the same numerical value and two other cards with two other numerical values.

c. Flush: The hand contains five cards all of the same suit.

d. Full house: The hand contains three cards of one value and two cards of another value.

e. Straight: The five cards have consecutive numerical values, such as 7-8-9-10-jack. Treat ace as being higher than king but not less than 2. The suits are irrelevant.

f. Straight flush: The hand is both a straight and a flush.

16.29. It is silly to compute $(x + y)^{20}$ by expanding it to a million terms and then collecting like terms. A much better way is to calculate $(x + y)^2$ and collect like terms. Then multiply that result by $(x + y)$ and collect like terms to give $(x + y)^3$. Now multiply that again by $(x + y)$ and so on until you reach $(x + y)^{20}$. Compare this method to the method of generating all of Pascal's triangle down to the 20th row.

16.30. To compute $\binom{n}{k}$ by generating Pascal's triangle, it is not necessary to generate the entire triangle down to row n; you need only the part of the triangle in a 90° wedge above $\binom{n}{k}$.

Estimate how many addition problems you would need to perform to calculate $\binom{100}{30}$ by this method. How many addition problems would you need to perform if you were to compute the entire Pascal's triangle down to row 30?

16.31. Use a computer to print out a very large copy of Pascal's triangle, but with a twist. Instead of printing the number, print a dot if the number is odd and leave the location blank if the number is even. Produce at least 64 rows.

Note that the computer doesn't actually need to compute the entries in Pascal's triangle; it needs only to calculate their parity. (Explain.) What do you see?

17 Counting Multisets

We have considered two kinds of counting problems: lists and sets. The list-counting problems (see Section 7) come in two flavors: we either allow or forbid repetition of the members of the lists. The number of lists of length k whose members are drawn from an n-element set is either n^k (if repetition is allowed) or $(n)_k$ (if repetition is forbidden).

Subsets as unordered lists.

Sets may be thought of as unordered lists (i.e., lists of elements where the order of the members does not matter). As we saw in Section 16, the number of unordered lists of length k whose members are drawn without repetition from an n-element set is $\binom{n}{k}$. This is a set-counting problem.

The goal of this section is to count the number of unordered lists of length k whose elements are drawn from an n-element set with repetition allowed. It is difficult, however, to express this idea in the language of sets. We need the more general concept of *multiset*.

Multisets

A given object either is or is not in a set. An element cannot be in a set "twice." The following sets are all identical:

$$\{1, 2, 3\} = \{3, 1, 2\} = \{1, 1, 2, 2, 3, 3\} = \{1, 2, 3, 1, 2, 3, 1, 1, 1, 1\}.$$

A *multiset* is a generalization of a set. A multiset is, like a set, an unordered collection of elements. However, in a multiset, an object may be considered to be in the multiset more than once.

There is no standard notation for multisets. Our notation $\langle \cdots \rangle$ is not widely used. The delimiters $\langle$ and $\rangle$ are called *angle brackets* and should not be confused with the less-than $<$ and greater-than $>$ symbols. Some mathematicians simply use curly braces $\{\cdots\}$ for both sets and multisets.

In this book, we write a multiset as follows: $\langle 1, 2, 3, 3 \rangle$. This multiset contains four elements: the element 1, the element 2, and the element 3 counted twice. We say that element 3 has *multiplicity* equal to 2 in the multiset $\langle 1, 2, 3, 3 \rangle$. The *multiplicity* of an element is the number of times it is a member of the multiset.

Two multisets are the same provided they contain the same elements with the same multiplicities. For example, $\langle 1, 2, 3, 3 \rangle = \langle 3, 1, 3, 2 \rangle$, but $\langle 1, 2, 3, 3 \rangle \neq \langle 1, 2, 3, 3, 3 \rangle$.

The *cardinality* of a multiset is the sum of the multiplicities of its elements. In other words, it is the number of elements in the multiset where we take into account the number of times each element is present. The notation is the same as for sets. If M is a multiset, then $|M|$ denotes its cardinality. For example, $|\langle 1, 2, 3, 3 \rangle| = 4$.

The notation $\left(\!\binom{n}{k}\!\right)$ is pronounced "n multichoose k." The doubled parentheses remind us that we may include elements more than once.

The counting problem we consider is: How many k-element multisets can we form by choosing elements from an n-element set? In other words, how many unordered length-k lists can we form using the elements $\{1, 2, \ldots, n\}$ with repetition allowed?

Just as we defined $\binom{n}{k}$ to represent the answer to a set-counting problem, we have a special notation for the answer to this multiset-counting problem.

Definition 17.1 Let $n, k \in \mathbb{N}$. The symbol $\left(\!\binom{n}{k}\!\right)$ denotes the number of multisets with cardinality equal to k whose elements belong to an n-element set such as $\{1, 2, \ldots, n\}$.

Example 17.2 Let n be a positive integer. Evaluate $\left(\!\binom{n}{1}\!\right)$.

Solution: This asks for the number of one-element multisets whose elements are selected from $\{1, 2, \ldots, n\}$. The multisets are

$$\langle 1 \rangle, \quad \langle 2 \rangle, \quad \ldots, \quad \langle n \rangle$$

and so $\left(\!\binom{n}{1}\!\right) = n$.

Example 17.3 Let k be a positive integer. Evaluate $\left(\!\binom{1}{k}\!\right)$.

Solution: This asks for the number of k-element multisets whose elements are selected from $\{1\}$. Since there is only one possible member of the multiset, and the multiset has cardinality k, the only possibility is

$$\langle 1, 1, \ldots, 1 \rangle$$

and so $\left(\!\binom{1}{k}\!\right) = 1$.

Example 17.4 Evaluate $\left(\!\binom{2}{2}\!\right)$.

Solution: We need to count the number of two-element multisets whose elements are selected from the set $\{1, 2\}$. We simply list all the possibilities. They are

$$\langle 1, 1 \rangle, \quad \langle 1, 2 \rangle, \quad \text{and,} \quad \langle 2, 2 \rangle.$$

Therefore $\left(\!\binom{2}{2}\!\right) = 3$.

In general, consider $\left(\!\binom{2}{k}\!\right)$. We need to form a k-element multiset using only the elements 1 and 2. We can decide how many 1s are in the multiset (anywhere from 0 to k, giving $k + 1$ possibilities), and then the remaining elements of the multiset must be 2s. Therefore $\left(\!\binom{2}{k}\!\right) = k + 1$.

Example 17.5 Evaluate $\left(\!\binom{3}{3}\!\right)$.

Solution: We need to count the number of three-element multisets whose elements are selected from the set $\{1, 2, 3\}$. We list all the possibilities. They are

$$\begin{array}{ccccc} \langle 1,1,1 \rangle & \langle 1,1,2 \rangle & \langle 1,1,3 \rangle & \langle 1,2,2 \rangle & \langle 1,2,3 \rangle \\ \langle 1,3,3 \rangle & \langle 2,2,2 \rangle & \langle 2,2,3 \rangle & \langle 2,3,3 \rangle & \langle 3,3,3 \rangle \end{array}$$

Therefore $\left(\!\binom{3}{3}\!\right) = 10$.

Formulas for $\left(\!\binom{n}{k}\!\right)$

In the foregoing examples, we calculated $\left(\!\binom{n}{k}\!\right)$ by explicitly listing all possible multisets. This, of course, is not practical if we want to calculate $\left(\!\binom{n}{k}\!\right)$ for large values of n and k. We need a better way to perform this computation.

For ordinary binomial coefficients, we have two methods to calculate $\binom{n}{k}$. We can generate Pascal's triangle using the relation $\binom{n}{k} = \binom{n-1}{k} + \binom{n-1}{k-1}$ or we can use the formula $\binom{n}{k} = \frac{n!}{k!(n-k)!}$.

Let's look for patterns in the values of $\left(\!\binom{n}{k}\!\right)$. Here is a table of values of $\left(\!\binom{n}{k}\!\right)$ for $0 \le n, k \le 6$.

		k						
		0	1	2	3	4	5	6
n	0	1	0	0	0	0	0	0
	1	1	1	1	1	1	1	1
	2	1	2	3	4	5	6	7
	3	1	3	6	10	15	21	28
	4	1	4	10	20	35	56	84
	5	1	5	15	35	70	126	210
	6	1	6	21	56	126	252	462

In Pascal's triangle, we found that the value of $\binom{n}{k}$ can be computed by adding two values in the previous row. Does a similar relationship hold here?

Look at the value 56 in row $n = 6$ and column $k = 3$. The number just above this 56 is 35. Is 21 next to 35 so we can get 56 by adding 21 and 35? There is no 21 in row 5, but just to the left of the 56 in row 6 there is a 21.

Examine other numbers in this chart. Each is the sum of the number just above and just to the left. The number to the left of $\left(\!\binom{n}{k}\!\right)$ is $\left(\!\binom{n}{k-1}\!\right)$ and number above is $\left(\!\binom{n-1}{k}\!\right)$.

We have observed the following:

Proposition 17.6 Let n, k be positive integers. Then

$$\left(\!\binom{n}{k}\!\right) = \left(\!\binom{n-1}{k}\!\right) + \left(\!\binom{n}{k-1}\!\right).$$

The proof of this result is similar to that of Theorem 16.10. I recommend you reread that proof now. The essential idea of that proof and the one we are about to present is to consider a weird element. We count [multi]sets of size k that either include or exclude the weirdo.

Proof. We use a combinatorial proof to prove this result (see Proof Template 9). We ask a question that we expect will be answered by both sides of the equation:

> How many multisets of size k can we form using the elements $\{1, 2, \ldots, n\}$?

A simple answer to this question is $\left(\!\binom{n}{k}\!\right)$.

For a second answer, we analyze the meanings of $\left(\!\binom{n-1}{k}\!\right)$ and $\left(\!\binom{n}{k-1}\!\right)$.

The first has an easy interpretation. The number $\left(\!\binom{n-1}{k}\!\right)$ is the number of k-element multisets using the members of $\{1, 2, \ldots, n\}$ in which we *never* use element n.

How should we interpret $\left(\!\binom{n}{k-1}\!\right)$? What we want to say is that this represents the number of k-element multisets using the members of $\{1, 2, \ldots, n\}$ in which we must use element n. To see why this is true, suppose we must use element n when forming a k-element multiset. So we throw element n into the multiset. Now we are free to complete this multiset in any way we wish. We need to pick $k - 1$ more elements from $\{1, 2, \ldots, n\}$; the number of ways to do that is precisely $\left(\!\binom{n}{k-1}\!\right)$.

Since element n either is or is not in the multiset, we have $\left(\!\binom{n}{k}\!\right) = \left(\!\binom{n-1}{k}\!\right) + \left(\!\binom{n}{k-1}\!\right)$. ■

Example 17.7 We illustrate the proof of Proposition 17.6 by considering $\left(\!\binom{3}{4}\!\right) = \left(\!\binom{2}{4}\!\right) + \left(\!\binom{3}{3}\!\right)$.

We list all the multisets of size 4 we can form using the elements $\{1, 2, 3\}$.

First, we list all the multisets of size 4 we can form from the elements in $\{1, 2, 3\}$ that do not use element 3. In other words, we want all the multisets of size 4 we can form that use just elements $\{1, 2\}$. There are $\left(\!\binom{2}{4}\!\right) = 5$ of them. They are

$$\langle 1,1,1,1\rangle\ \langle 1,1,1,2\rangle\ \langle 1,1,2,2\rangle\ \langle 1,2,2,2\rangle\ \langle 2,2,2,2\rangle$$

Second, we list all the multisets of size 4 that include the element 3 (at least once). They are

$$\langle 1,1,1,3\rangle\ \langle 1,1,2,3\rangle\ \langle 1,1,3,3\rangle\ \langle 1,2,2,3\rangle\ \langle 1,2,3,3\rangle$$
$$\langle 1,3,3,3\rangle\ \langle 2,2,2,3\rangle\ \langle 2,2,3,3\rangle\ \langle 2,3,3,3\rangle\ \langle 3,3,3,3\rangle$$

Notice that if we ignore the mandatory 3 (in color), we have listed all the three-element multisets we can form from the elements in $\{1, 2, 3\}$. There are $\left(\!\binom{3}{3}\!\right) = 10$ of them.

This result, $\left(\!\binom{n}{k}\!\right) = \left(\!\binom{n-1}{k}\!\right) + \left(\!\binom{n}{k-1}\!\right)$, and its proof are quite similar to Theorem 16.10, $\binom{n}{k} = \binom{n-1}{k-1} + \binom{n-1}{k}$. The table of $\left(\!\binom{n}{k}\!\right)$ values is similar to Pascal's triangle in another way. If we read the table of $\left(\!\binom{n}{k}\!\right)$ values diagonally from the lower-left corner to the upper-right corner, we read off the values

$$1 \quad 5 \quad 10 \quad 10 \quad 5 \quad 1$$

and this is the fifth row of Pascal's triangle. We can write this as follows:

$$\begin{array}{cccccc} 1 & 5 & 10 & 10 & 5 & 1 \\ \updownarrow & \updownarrow & \updownarrow & \updownarrow & \updownarrow & \updownarrow \\ \left(\!\binom{6}{0}\!\right) & \left(\!\binom{5}{1}\!\right) & \left(\!\binom{4}{2}\!\right) & \left(\!\binom{3}{3}\!\right) & \left(\!\binom{2}{4}\!\right) & \left(\!\binom{1}{5}\!\right) \\ \updownarrow & \updownarrow & \updownarrow & \updownarrow & \updownarrow & \updownarrow \\ \binom{5}{0} & \binom{5}{1} & \binom{5}{2} & \binom{5}{3} & \binom{5}{4} & \binom{5}{5} \end{array}$$

Observe that $\left(\!\binom{n}{k}\!\right) = \binom{?}{k}$. What number should we fill in for the question mark? A bit of guesswork and we see that $? = n + k - 1$ fits the pattern we observed. For example, $\left(\!\binom{4}{2}\!\right) = \binom{5}{2} = \binom{4+2-1}{2}$.

We assert the following:

Theorem 17.8 Let $n, k \in \mathbb{N}$. Then

$$\left(\!\!\binom{n}{k}\!\!\right) = \binom{n+k-1}{k}.$$

Proof. The idea of this proof is to develop a way to encode multisets and then count their encodings. To find $\left(\!\binom{n}{k}\!\right)$, we list all (encodings of) the k-element multisets we can form using the integers 1 through n. Before we present the encoding scheme, we need to deal with the special case $n = 0$.

If both $n = 0$ and $k = 0$, then $\left(\!\binom{0}{0}\!\right) = 1$ (the empty multiset). However, the formula gives $\binom{0+0-1}{0} = \binom{-1}{0}$. Although this is nonsense (it is not possible to have a set with -1 elements), it is possible to extend the definition of $\binom{n}{k}$ to allow the upper index, n, to be any real number; see Exercise 17.10. In the extended definition, $\binom{-1}{0} = 1$ as desired.

If $n = 0$ and $k > 0$, then $\left(\!\binom{n}{k}\!\right) = 0$ (there are no multisets of cardinality k whose elements are chosen from the empty set). In this case, $\binom{n+k-1}{k} = \binom{k-1}{k} = 0$, as required.

Hence, from this point on, we may assume n is a positive integer. We now present the scheme for encoding multisets as lists.

Suppose, for the moment, that $n = 5$ and the multiset is $M = \langle 1, 1, 1, 2, 3, 3, 5 \rangle$. We encode this multiset with a sequence of stars $*$ and bars $|$. We have a star for each element and a bar to make separate compartments for the elements. For this multiset, the stars-and-bars encoding is as follows:

$$\langle 1, 1, 1, 2, 3, 3, 5 \rangle \quad \longleftrightarrow \quad *** \,|\, * \,|\, ** \,|\,|\, *$$

The first three $*$s stand for the three 1s in M. Then there is a $|$ to mark the end of the 1s section. Next there is a single $*$ to denote the single 2 in M, and another $|$ to signal the end of the 2s. Two more $*$s follow for the two 3s in the multiset. Now notice that we have two $|$s in a row. Since there are no 4s in M, there are no $*$s in this compartment. Finally, the last $*$ is for the single 5 in M.

In the general case, let M be a k-element multiset formed using the integers 1 through n. Its stars-and-bars notation contains exactly k $*$s (one for each element of M) and exactly $n - 1$ $|$s (to separate n different compartments).

This one-to-one pairing of multisets and stars-and-bars encodings is an example of a bijective proof.

Notice that given any sequence of k $*$s and $n - 1$ $|$s, we can recover a unique multiset of cardinality k whose elements are chosen from the integers 1 through n. Thus there is a one-to-one correspondence between k-element multisets of integers chosen from $\{1, 2, \ldots, n\}$ and lists of stars and bars with k $*$s and $n - 1$ $|$s. The good news is that it is easy to count the number of such stars-and-bars lists.

Each stars-and-bars list contains exactly $n + k - 1$ symbols, of which exactly k are $*$s. The number of such lists is $\binom{n+k-1}{k}$ because we can think of choosing exactly k positions on the length-$(n + k - 1)$ list to be $*$s. In other words, there are $n + k - 1$ positions on this list. We want to select a k-element subset of those $n + k - 1$ positions in all possible ways. There are $\binom{n+k-1}{k}$ ways to do this.

Therefore $\left(\!\binom{n}{k}\!\right) = \binom{n+k-1}{k}$. ■

Example 17.9 In Example 17.5, we explicitly listed all possible size-three multisets formed using the integers 1, 2, and 3. Here we list them with their stars-and-bars notation.

Multiset	Stars-and-bars	Subset
$\langle 1,1,1\rangle$	$*{*}*\,\vert\,\vert$	$\{1,2,3\}$
$\langle 1,1,2\rangle$	$**\,\vert\,*\,\vert$	$\{1,2,4\}$
$\langle 1,1,3\rangle$	$**\,\vert\,\vert\,*$	$\{1,2,5\}$
$\langle 1,2,2\rangle$	$*\,\vert\,**\,\vert$	$\{1,3,4\}$
$\langle 1,2,3\rangle$	$*\,\vert\,*\,\vert\,*$	$\{1,3,5\}$
$\langle 1,3,3\rangle$	$*\,\vert\,\vert\,**$	$\{1,4,5\}$
$\langle 2,2,2\rangle$	$\vert\,***\,\vert$	$\{2,3,4\}$
$\langle 2,2,3\rangle$	$\vert\,**\,\vert\,*$	$\{2,3,5\}$
$\langle 2,3,3\rangle$	$\vert\,*\,\vert\,**$	$\{2,4,5\}$
$\langle 3,3,3\rangle$	$\vert\,\vert\,***$	$\{3,4,5\}$

The column labeled *Subset* shows which of the five positions in the stars-and-bars encoding are occupied by *s. Notice that the $\left(\!\binom{3}{3}\!\right)$ multisets correspond to the $\binom{5}{3}$ subsets. Thus $\left(\!\binom{3}{3}\!\right) = \binom{3+3-1}{3} = \binom{5}{3}$.

Recap

In this section, we considered the following counting problem: How many k-element multisets can we form whose elements are selected from $\{1, 2, \ldots, n\}$? We denoted the answer by $\left(\!\binom{n}{k}\!\right)$. We proved various properties of $\left(\!\binom{n}{k}\!\right)$, most notably that

$$\left(\!\binom{n}{k}\!\right) = \binom{n+k-1}{k}.$$

We have studied four counting problems: counting lists (with or without repetitions), counting subsets, and counting multisets. The answers to these four counting problems are summarized in the following chart.

Counting collections

	Repetition allowed	Repetition forbidden
Ordered	n^k	$(n)_k$
Unordered	$\left(\!\binom{n}{k}\!\right)$	$\binom{n}{k}$

Size of collection: k
Size of universe: n

17 Exercises

17.1. Evaluate $\left(\!\binom{3}{2}\!\right)$ and $\left(\!\binom{2}{3}\!\right)$ by explicitly listing all possible multisets of the appropriate size. Check that your answers agree with the formula in Theorem 17.8.

17.2. Give a stars-and-bars representation for all the sets you found in the previous problem.

17.3. Let n be a positive integer. Evaluate the following from first principles (i.e., don't use Proposition 17.6).

a. $\left(\!\binom{0}{n}\!\right)$.
b. $\left(\!\binom{n}{0}\!\right)$.
c. $\left(\!\binom{0}{0}\!\right)$.
Explain your answers.

17.4. What multiset is encoded by the stars-and-bars notation $*|||***$?

17.5. Express $\left(\!\binom{n}{k}\!\right)$ using factorial notation.

17.6. Prove:

$$\left(\!\binom{n}{k}\!\right) = \left(\!\binom{k+1}{n-1}\!\right).$$

17.7. Let $\left[\!\left[{n \atop k}\right]\!\right]$ denote the number of multisets of cardinality k we can form choosing the elements in $\{1, 2, 3, \ldots, n\}$ with the added condition that we *must* use each of these n elements at least once in the multiset.
a. Evaluate from first principles, $\left[\!\left[{n \atop n}\right]\!\right]$.
b. Prove: $\left[\!\left[{n \atop k}\right]\!\right] = \left(\!\binom{n}{k-n}\!\right)$.

17.8. Let n, k be positive integers. Prove:

$$\left(\!\binom{n}{k}\!\right) = \left(\!\binom{n-1}{0}\!\right) + \left(\!\binom{n-1}{1}\!\right) + \left(\!\binom{n-1}{2}\!\right) + \cdots + \left(\!\binom{n-1}{k}\!\right).$$

17.9. Let n, k be positive integers. Prove:

$$\left(\!\binom{n}{k}\!\right) = \left(\!\binom{1}{k-1}\!\right) + \left(\!\binom{2}{k-1}\!\right) + \cdots + \left(\!\binom{n}{k-1}\!\right).$$

17.10. Let x be a positive integer. We can write

$$\binom{x}{2} = \frac{x(x-1)}{2} = \frac{1}{2}x^2 - \frac{1}{2}x.$$

In this way, we can think of $\binom{x}{2}$ as a *polynomial* in x. Thus, although it does not make sense as a counting problem, we can write $\binom{\frac{1}{3}}{2}$ and this evaluates to $\frac{1}{2}(\frac{1}{3})^2 - \frac{1}{2}(\frac{1}{3}) = -\frac{1}{9}$.
a. Write $\left(\!\binom{x}{2}\!\right)$ as a polynomial in x.
b. As silly as it looks, evaluate

$$\left(\!\binom{-\frac{1}{2}}{2}\!\right).$$

c. Write $\binom{x}{3}$ and $\left(\!\binom{x}{3}\!\right)$ as polynomials in x.
d. Let $k \in \mathbb{N}$. Find (and prove) a relationship between the polynomials $\binom{x}{k}$ and $\left(\!\binom{x}{k}\!\right)$.

18 Inclusion-Exclusion

In Section 11 we learned that for finite sets A and B, we have $|A| + |B| = |A \cup B| + |A \cap B|$. We can rewrite this as

$$|A \cup B| = |A| + |B| - |A \cap B|$$

[see Proposition 11.4 and Equation (4)]. The equation expresses the size of a union of two sets in terms of the sizes of the individual sets and their intersection. In Exercise 11.19, you were asked to extend this result to three sets A, B, and C—that is, to prove that

$$
\begin{aligned}
|A \cup B \cup C| = |A| + |B| + |C| \\
- |A \cap B| - |A \cap C| - |B \cap C| \\
+ |A \cap B \cap C|.
\end{aligned}
$$

Again, the size of the union is expressed in terms of the sizes of the individual sets and their various intersections. These equations are called *inclusion-exclusion* formulas.

In this section, we prove a general inclusion-exclusion formula.

Theorem 18.1 **(Inclusion-Exclusion)** Let $A_1, A_2, \ldots, A_n$ be finite sets. Then

$$
\begin{aligned}
|A_1 \cup A_2 \cup \cdots \cup A_n| &= |A_1| + |A_2| + \cdots + |A_n| \\
&- |A_1 \cap A_2| - |A_1 \cap A_3| - \cdots - |A_{n-1} \cap A_n| \\
&+ |A_1 \cap A_2 \cap A_3| + |A_1 \cap A_2 \cap A_4| + \cdots \\
&+ |A_{n-2} \cap A_{n-1} \cap A_n| \\
&- \cdots + \cdots\cdots\cdots \\
&\pm |A_1 \cap A_2 \cap \cdots \cap A_n|.
\end{aligned}
$$

To find the size of a union, we add the sizes of the individual sets (inclusion), subtract the sizes of all the pairwise intersections (exclusion), add the sizes of all the three-way intersections (inclusion), and so on.

The idea is that when we add up all the sizes of the individual sets, we have added too much because some elements may be in more than one set. So we subtract off the sizes of the pairwise intersections to compensate, but now we may have subtracted too much. Thus we correct back by adding in the sizes of the triple intersections, but this overcounts, so we have to subtract, and so on. Amazingly, at the end, everything is in perfect balance (we prove this in a moment).

The repeated use of ellipsis ($\cdots$) in the formula is unfortunate, but it is difficult to express this formula using the notations we have thus far developed. For four sets (A through D) the formula is

$$
\begin{aligned}
|A \cup B \cup C \cup D| &= |A| + |B| + |C| + |D| \\
&- |A \cap B| - |A \cap C| - |A \cap D| - |B \cap C| \\
&- |B \cap D| - |C \cap D| \\
&+ |A \cap B \cap C| + |A \cap B \cap D| + |A \cap C \cap D| \\
&+ |B \cap C \cap D| \\
&- |A \cap B \cap C \cap D|.
\end{aligned}
$$

Example 18.2 At an art academy, there are 43 students taking ceramics, 57 students taking painting, and 29 students taking sculpture. There are 10 students in both ceramics and painting, 5 in both painting and sculpture, 5 in both ceramics and sculpture, and

2 taking all three courses. How many students are taking at least one course at the art academy?

Solution: Let C, P, and S denote the sets of students taking ceramics, painting, and sculpture, respectively. We want to calculate $|C \cup P \cup S|$. We apply inclusion-exclusion:

$$\begin{aligned}|C \cup P \cup S| &= |C| + |P| + |S| - |C \cap P| - |C \cap S| - |P \cap S| + |C \cap P \cap S| \\ &= 43 + 57 + 29 - 10 - 5 - 5 + 2 = 111.\end{aligned}$$

Proof (of Theorem 18.1)

Let the n sets be $A_1, A_2, \ldots, A_n$ and let the elements in their union be named x_1, $x_2, \ldots, x_m$. We create a large chart. The rows of this chart are labeled by the elements x_1 through x_m. The chart has $2^n - 1$ columns that correspond to all the terms on the right-hand side of the inclusion-exclusion formula. The first n columns are labeled A_1 through A_n. The next $\binom{n}{2}$ columns are labeled by all the pairwise intersections from $A_1 \cap A_2$ through $A_{n-1} \cap A_n$. The next $\binom{n}{3}$ columns are labeled by the triple intersections, and so on.

The entries in this chart either are blank or contain a + or − sign. The entries depend on the row label (element) and column label (set). If the element is not in the set, the entry in that position is blank. If the element is a member of the set, we put a + sign when the column label is an intersection of an odd number of sets or else a − sign when the column label is an intersection of an even number of sets. For the three sets in the Venn diagram in the figure and their elements, the chart would be:

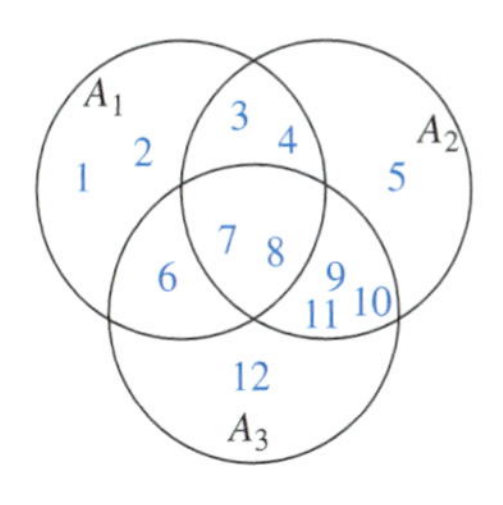

El't	A_1	A_2	A_3	$A_1 \cap A_2$	$A_1 \cap A_3$	$A_2 \cap A_3$	$A_1 \cap A_2 \cap A_3$
1	+						
2	+						
3	+	+		−			
4	+	+		−			
5		+					
6	+		+		−		
7	+	+	+	−	−	−	+
8	+	+	+	−	−	−	+
9		+	+			−	
10		+	+			−	
11		+	+			−	
12			+				

There are three things to notice about this chart.

- First, the number of marks in each column is the cardinality of that column's set; we make a mark in a column just for that set's elements. In the example, there are five marks in the $A_2 \cap A_3$ column (corresponding to elements 7, 8, 9, 10, and 11).

- Second, the sign of the mark ($+$ or $-$) corresponds to whether we are adding or subtracting that set's cardinality in the inclusion-exclusion formula. Thus, if we add 1 for every $+$ sign in the chart and subtract 1 for every $-$ sign, we get precisely the right-hand side of the inclusion-exclusion formula.
- Third, look at the number of $+$s and $-$s in each row. In the example, notice that there is always one more $+$ than $-$. If we can prove this always works, we will be finished because then the net effect of all the $+$s and $-$s is to count 1 for each element in the union of the sets $A_1 \cup A_2 \cup \cdots \cup A_n$. So, if we can prove this works in general, we have completed the proof.

The problem now reduces to proving that every row has exactly one more $+$ than $-$.

Let x be an element of $A_1 \cup A_2 \cup \cdots \cup A_n$. It is in some (perhaps all) of the A_i. Let us say it is in exactly k of them (with $1 \le k \le n$). Let us calculate how many $+$s and $-$s are in x's row.

In the columns indexed by single sets, there will be k $+$s; let's write $\binom{k}{1}$ in place of k (you will see why in a moment).

In the columns indexed by pairwise intersections, there will be $\binom{k}{2}$ $-$s. This is because x is in k of the A_is, and the number of pairs of sets to which x belongs is $\binom{k}{2}$.

In the columns indexed by triple intersections, there will be $\binom{k}{3}$ $+$s.

In general, in the columns indexed by j-fold intersections, there will be $\binom{k}{j}$ marks. The marks are $+$ if j is odd and $-$ if j is even. Thus

$$\text{the number of } +\text{s is } \binom{k}{1} + \binom{k}{3} + \binom{k}{5} + \cdots, \text{ and}$$

$$\text{the number of } -\text{s is } \binom{k}{2} + \binom{k}{4} + \binom{k}{6} + \cdots.$$

Note that these sums do not go on forever; they include only those binomial coefficients whose lower index does not exceed k. Also note that the term $\binom{k}{0}$ is absent.

In Exercise 16.11, you proved

$$\binom{k}{0} - \binom{k}{1} + \binom{k}{2} - \cdots \pm \binom{k}{k} = 0$$

or, equivalently,

$$\binom{k}{0} + \underbrace{\binom{k}{2} + \binom{k}{4} + \binom{k}{6} + \cdots}_{\text{number of } - \text{ signs}} = \underbrace{\binom{k}{1} + \binom{k}{3} + \binom{k}{5} + \cdots}_{\text{number of } + \text{ signs}}.$$

We therefore see that the number of $+$s is exactly $\binom{k}{0} = 1$ more than the number of $-$s in x's row. ■

How to Use Inclusion-Exclusion

Inclusion-exclusion takes one counting problem (How many elements are in $A_1 \cup \cdots \cup A_n$?) and replaces it with $2^n - 1$ new counting problems (How many elements are in the various intersections?). Nevertheless, inclusion-exclusion makes certain counting problems easier. Here is an example.

Example 18.3 **(A list-counting problem)** The number of length-k lists whose elements are chosen from the set $\{1, 2, \ldots, n\}$ is n^k. How many of these lists use all of the elements in $\{1, 2, \ldots, n\}$ at least once?

For example, for $n = 3$ and $k = 3$, there are $3^3 = 27$ length-three lists using the elements in $\{1, 2, 3\}$. Of these, the following six lists use all of the elements 1, 2, and 3:

$$123 \quad 132 \quad 213 \quad 231 \quad 312 \quad 321.$$

Here is how to use inclusion-exclusion to solve this problem. We begin by letting U (for universe) be the set of all length-k lists whose elements are chosen from $\{1, 2, \ldots, n\}$. Thus $|U| = n^k$. We call some of these lists "good"—these are the ones that contain all the elements of $\{1, 2, \ldots, n\}$. And we call some of the lists "bad"—these are the ones that miss one or more of the elements in $\{1, 2, \ldots, n\}$. If we can count the number of bad lists, we'll be finished because

It is convenient to use # to stand for "number of."

$$\text{\# good lists} = n^k - \text{\# bad lists}. \tag{5}$$

Now a list might be bad because it fails to contain the number 1. Or it might be bad if it misses the number 2, and so on. There are n different elements in $\{1, 2, \ldots, n\}$, and there are n different ways a list might be bad. Let B_1 be the set of all lists in U that do not contain the element 1, let B_2 be the set of all lists in U that do not contain the element 2, ..., and let B_n be the set of all lists in U that do not contain the element n. The set

$$B_1 \cup B_2 \cup \cdots \cup B_n$$

contains precisely all the bad lists; what we want to do is calculate the size of this union. This is a job for inclusion-exclusion! To calculate the size of this union, we need to calculate the sizes of each of the sets B_i and all possible intersections, and then invoke Theorem 18.1.

$B_1 = \{222, 223, 232, 233, 322, 323, 332, 333\}$.

To begin, we calculate the size of B_1. This is the number of length-k lists whose elements are chosen from $\{1, 2, \ldots, n\}$ with the added condition that the element 1 is never used. In other words, $|B_1|$ is the number of length-k lists whose elements are chosen from $\{2, 3, \ldots, n\}$ (notice we deleted element 1). Thus we have $n - 1$ choices for each position on the list, so $|B_1| = (n-1)^k$.

$B_2 = \{111, 113, 131, 133, 311, 313, 331, 333\}$.

What about $|B_2|$? The analysis is exactly the same as for $|B_1|$. The number of length-k lists that do not use element 2 is the number of length-k lists whose elements are chosen from $\{1, 3, 4, \ldots, n\}$ (we deleted 2). So $|B_2| = (n-1)^k$.

Indeed, for every j, $|B_j| = (n-1)^k$. The first part of the inclusion-exclusion formula now gives

$$\begin{aligned} |B_1 \cup \cdots \cup B_n| &= |B_1| + \cdots + |B_n| - \cdots\cdots \\ &= n(n-1)^k - \cdots\cdots \end{aligned}$$

$B_1 \cap B_2 = \{333\}$.

Now we continue to the second row of terms in Theorem 18.1. These are all the terms of the form $|B_i \cap B_j|$. We begin with $|B_1 \cap B_2|$. This is the number of lists that (1) do not include the element 1 and (2) do not include the element 2. In other words, $|B_1 \cap B_2|$ equals the number of length-k lists whose elements are chosen from the set $\{3, 4, \ldots, n\}$. The number of these lists is $|B_1 \cap B_2| = (n-2)^k$.

$B_1 \cap B_3 = \{222\}$.

What about $|B_1 \cap B_3|$? The analysis is exactly the same as before. These lists avoid the elements 1 and 3, so they are drawn from an $(n-2)$-element set. Thus $|B_1 \cap B_3| = (n-2)^k$. Indeed, all terms in the second row of the inclusion-exclusion formula give $(n-2)^k$.

The question that remains is: How many terms are on the second row? We want to pick all possible pairs of sets from B_1 through B_n and there are $\binom{n}{2}$ such pairs. Thus far, we have

$$\begin{aligned}|B_1 \cup \cdots \cup B_n| &= |B_1| + \cdots + |B_n| - |B_1 \cap B_2| - \cdots + \cdots\cdots \\ &= n(n-1)^k - \binom{n}{2}(n-2)^k + \cdots\cdots.\end{aligned}$$

Let's think about the triple intersections before we do the general case. How many lists are in $B_1 \cap B_2 \cap B_3$? This is the number of length-k lists that avoid all three of the elements 1, 2, and 3. In other words, these are the length-k lists whose elements are drawn from $\{4, \ldots, n\}$. The number of such lists is $(n-3)^k$. Of course, this analysis applies to any triple intersection. How many triple intersections are there? There are $\binom{n}{3}$. So we now have

$$|B_1 \cup \cdots \cup B_n| = n(n-1)^k - \binom{n}{2}(n-2)^k + \binom{n}{3}(n-3)^k - \cdots\cdots.$$

The pattern should be emerging. To make the pattern look better, replace the first n by $\binom{n}{1}$ in the above equation. We expect the next term to be $-\binom{n}{4}(n-4)^k$.

To make sure the pattern we see is correct, let us think about the size of a j-fold intersection of the B sets. How many elements are in $B_1 \cap B_2 \cap \cdots \cap B_j$? These are the length-$k$ lists that avoid all elements from 1 to j; that is, they draw their elements from $\{j+1, \ldots, n\}$ (a set of size $n-j$). So $|B_1 \cap B_2 \cap \cdots \cap B_j| = (n-j)^k$. Of course, all j-fold intersections work exactly like this. How many j-fold intersections are there? There are $\binom{n}{j}$. Thus the jth term in the inclusion-exclusion is $\pm\binom{n}{j}(n-j)^k$. The sign is positive when j is odd and negative when j is even.

The last term in the example is $B_1 \cap B_2 \cap B_3 = \emptyset$.

As a sanity check, let us make sure this formula applies to $|B_1 \cap \cdots \cap B_n|$, the last term in the inclusion-exclusion. This is the number of lists of length k that contain none of the elements 1 through n. If we can't use any of the elements, we certainly can't make any lists. The size of this set is zero. Our formula for this term is $\pm\binom{n}{0}(n-n)^k$, which, of course, is 0.

We now have

$$|B_1 \cup \cdots \cup B_n| = \binom{n}{1}(n-1)^k - \binom{n}{2}(n-2)^k + \binom{n}{3}(n-3)^k - \cdots \pm \binom{n}{n}(n-n)^k$$

which can be rewritten using $\sum$ notation as

$$|B_1 \cup \cdots \cup B_n| = \sum_{j=1}^{n}(-1)^{j+1}\binom{n}{j}(n-j)^k.$$

The $(-1)^{j+1}$ term is a device that gives a plus sign when j is odd and a minus sign when j is even.

We have nearly answered the question from Example 18.3. The set $B_1 \cup \cdots \cup B_n$ counts the number of bad lists; we want the number of good lists. We simply substitute into Equation (5) to get

$$\begin{aligned}
\#\text{ good lists} &= n^k - \#\text{ bad lists} \\
&= n^k - \left[\binom{n}{1}(n-1)^k - \binom{n}{2}(n-2)^k \right. \\
&\quad \left. + \binom{n}{3}(n-3)^k - \cdots \pm \binom{n}{n}(n-n)^k\right] \\
&= n^k - \binom{n}{1}(n-1)^k + \binom{n}{2}(n-2)^k \\
&\quad - \binom{n}{3}(n-3)^k + \cdots \mp \binom{n}{n}(n-n)^k \\
&= \binom{n}{0}n^k - \binom{n}{1}(n-1)^k + \binom{n}{2}(n-2)^k \\
&\quad - \binom{n}{3}(n-3)^k + \cdots \mp \binom{n}{n}(n-n)^k \\
&= \sum_{j=0}^{n}(-1)^j\binom{n}{j}(n-j)^k
\end{aligned}$$

answering the question from Example 18.3.

Example 18.4 is known as the *hat-check problem*. The story is that n people go to the theater and check their hats with a deranged clerk. The clerk hands the hats back to the patrons at random. The problem is: What is the probability that none of the patrons get their own hat back? The answer to this probability question is the answer to Example 18.4 divided by $n!$.

Derangements

We illustrate the method of Proof Template 10 on the following classical problem.

Example 18.4 **(Counting derangements)** There are $n!$ ways to make lists of length n using the elements of $\{1, 2, \ldots, n\}$ without repetition. Such a list is called a *derangement* if the number j does not occupy position j of the list for any $j = 1, 2, \ldots, n$. How many derangements are there?

For example, if $n = 8$, the lists $(8, 7, 6, 5, 4, 3, 2, 1)$ and $(6, 5, 7, 8, 1, 2, 3, 4)$ are derangements but $(3, 5, 1, 4, 8, 6, 7)$ and $(2, 1, 4, 3, 8, 6, 7, 5)$ are not.

Proof Template 10 Using inclusion-exclusion.

Counting with inclusion-exclusion:

- Classify the objects as either "good" (the ones you want to count) or "bad" (the ones you don't want to count).
- Decide whether you want to count the good objects directly or to count the bad objects and subtract from the total.
- Cast the counting problem as the size of a union of sets. Each set describes one way the objects might be "good" or "bad."
- Use inclusion-exclusion (Theorem 18.1).

Example 18.5 The derangements of $\{1, 2, 3, 4\}$ are

$$\begin{array}{ccc} 2143 & 2341 & 2413 \\ 3142 & 3412 & 3421 \\ 4123 & 4312 & 4321 \end{array}$$

There are $n!$ lists under consideration. The "good" lists are the derangements. The "bad" lists are the lists in which one (or more) element j of $\{1, 2, \ldots, n\}$ appears at position j of the list.

We count the number of bad lists and subtract from $n!$ to count the good lists.

We count the number of bad lists by counting a union. There are n ways in which a list might be bad: 1 might be in position 1, 2 might be in position 2, and so forth, and n might be in position n. So we define the following sets:

$$\begin{aligned} B_1 &= \{\text{lists with 1 in position 1}\} \\ B_2 &= \{\text{lists with 2 in position 2}\} \\ &\vdots \\ B_n &= \{\text{lists with } n \text{ in position } n\}. \end{aligned}$$

Our goal is to count $|B_1 \cup \cdots \cup B_n|$ and finally to subtract from $n!$. To compute the size of a union, we use inclusion-exclusion.

$B_1 = \{1234, 1243, 1324, 1342, 1423, 1432\}$.

We first calculate $|B_1|$. This is the number of lists with 1 in position 1; the other $n - 1$ elements may be anywhere. There are $(n - 1)!$ such lists. Likewise, $|B_2| = (n - 1)!$ because element 2 must be in position 2, but the other $n - 1$ elements may be anywhere. We have

$$\begin{aligned} |B_1 \cup \cdots \cup B_n| &= |B_1| + \cdots + |B_n| - \cdots\cdots \\ &= n(n-1)! - \cdots\cdots. \end{aligned}$$

$B_1 \cap B_2 = \{1234, 1243\}$.

Next consider $|B_1 \cap B_2|$. These are the lists in which 1 must be in position 1, 2 must be in position 2, and the remaining $n - 2$ elements may be anywhere. There are $(n - 2)!$ such lists. Indeed, for any $i \neq j$, we have $|B_i \cap B_j| = (n - 2)!$ since element i goes in position i, element j goes in position j, and the remaining $n - 2$ elements may go anywhere they want. There are $\binom{n}{2}$ pairwise intersections, and they all have size $(n - 2)!$. This gives

$$\begin{aligned} |B_1 \cup \cdots \cup B_n| &= |B_1| + \cdots + |B_n| - |B_1 \cap B_2| - \cdots + \cdots\cdots \\ &= n(n-1)! - \binom{n}{2}(n-2)! + \cdots\cdots. \end{aligned}$$

The $\binom{n}{3}$ triple intersections all work the same, too. The size of $B_1 \cap B_2 \cap B_3$ is $(n - 3)!$ because elements 1, 2, and 3 must go into their respective positions, but the remaining $n - 3$ elements go wherever they please. So far we have

$$|B_1 \cup \cdots \cup B_n| = n(n-1)! - \binom{n}{2}(n-2)! + \binom{n}{3}(n-3)! - \cdots.$$

If we rewrite the first n as $\binom{n}{1}$, this becomes

$$|B_1 \cup \cdots \cup B_n| = \binom{n}{1}(n-1)! - \binom{n}{2}(n-2)! + \binom{n}{3}(n-3)! - \cdots.$$

The pattern is emerging. To see that this works, let us consider the k-fold intersections such as $|B_1 \cap B_2 \cap \cdots \cap B_k|$. There are $\binom{n}{k}$ terms of this form. Each evaluates to $(n-k)!$ because k of the elements/positions on the list are determined, and the remaining $n-k$ elements can go wherever they wish. Thus we have

$$|B_1 \cup \cdots \cup B_n| = \binom{n}{1}(n-1)! - \binom{n}{2}(n-2)! + \binom{n}{3}(n-3)! - \cdots \pm \binom{n}{n}(n-n)!.$$

$B_1 \cap B_2 \cap B_3 \cap B_4 = \{1234\}$.

Note the last term is $\binom{n}{n}0! = 1$. To see this is correct, note that this is the size of $B_1 \cap \cdots \cap B_n$. This is the set of lists in which 1 must be in position 1, 2 must be in position 2, and so on, and n must be in position n. There is exactly one such list—namely, $(1, 2, 3, \ldots, n)$.

Finally, we subtract $|B_1 \cup \cdots \cup B_n|$ from $n!$ to get the number of derangements. This is

$$n! - \left[\binom{n}{1}(n-1)! - \binom{n}{2}(n-2)! + \binom{n}{3}(n-3)! - \cdots \pm \binom{n}{n}(n-n)!\right]$$

which equals

$$\binom{n}{0}n! - \binom{n}{1}(n-1)! + \binom{n}{2}(n-2)! - \binom{n}{3}(n-3)! + \cdots \mp \binom{n}{n}(n-n)!$$

or, in Σ notation,

$$\#\text{ derangements} = \sum_{k=0}^{n}(-1)^k\binom{n}{k}(n-k)!.$$

We can simplify this answer. Recall that

$$\binom{n}{k} = \frac{n!}{k!(n-k)!}$$

(see Theorem 16.12). Therefore

$$\begin{aligned}\#\text{ derangements} &= \sum_{k=0}^{n}(-1)^k\binom{n}{k}(n-k)! = \sum_{k=0}^{n}(-1)^k\frac{n!}{k!(n-k)!}(n-k)! \\ &= \sum_{k=0}^{n}(-1)^k\frac{n!}{k!}.\end{aligned}$$

Finally, we can factor out the $n!$ from all the terms and just have

$$\#\text{ derangements} = n!\sum_{k=0}^{n}\frac{(-1)^k}{k!}.$$

A Ghastly Formula

The inclusion-exclusion formula is

$$\begin{aligned}|A_1 \cup A_2 \cup \cdots \cup A_n| = {} & |A_1| + |A_2| + \cdots + |A_n| \\ & - |A_1 \cap A_2| - |A_1 \cap A_3| - \cdots - |A_{n-1} \cap A_n| \\ & + |A_1 \cap A_2 \cap A_3| + |A_1 \cap A_2 \cap A_4| + \cdots \\ & + |A_{n-2} \cap A_{n-1} \cap A_n| \\ & - \cdots + \cdots\cdots\cdots \\ & \pm |A_1 \cap A_2 \cap \cdots \cap A_n|.\end{aligned}$$

Can this be rewritten without resorting to use of ellipsis $(\cdots)$? Here we reduce the formula so that it contains only a single ellipsis. You decide whether this is better.

$$\left|\bigcup_{k=1}^{n} A_k\right| = \sum_{k=1}^{n} (-1)^{k+1} \sum_{1 \le a_1 < \cdots < a_k \le n} \left|\bigcap_{j=1}^{k} A_{a_j}\right|.$$

Can you invent a notation that does not require even one ellipsis?

Recap

We extended the simple formula $|A \cup B| = |A| + |B| - |A \cap B|$ to deal with the size of the union of many sets in terms of the sizes of their various intersections. We then showed how to apply inclusion-exclusion to some complicated counting problems.

18 Exercises

18.1. There are four large groups of people, each with 1000 members. Any two of these groups have 100 members in common. Any three of these groups have 10 members in common. And there is 1 person in all four groups. All together, how many people are in these groups?

18.2. Let A, B, and C be finite sets. Prove or disprove: If $|A \cup B \cup C| = |A| + |B| + |C|$, then A, B, and C must be pairwise disjoint.

18.3. How many five-letter "words" can you make in which no two consecutive letters are the same? A "word" may be any list of the standard 26 letters, so WENJW is a word you would count, but NUTTY is not.

Here is an easy solution: By the list-counting methods of Section 7, the answer is $26 \times 25 \times 25 \times 25 \times 25 = 26 \times 25^4$.

Give a hard solution using inclusion-exclusion, and then show that the two answers are the same.

18.4. This problem asks you to give two proofs for

$$9^n = \sum_{k=0}^{n} (-1)^k \binom{n}{k} 10^{n-k}.$$

a. The first proof should use the binomial theorem (see Theorem 16.8).
b. The second should be a combinatorial proof using inclusion-exclusion.

18.5. How many six-digit numbers do not have three consecutive digits the same? (For this problem, you may consider six-digit numbers whose initial digits might be 0. Thus you should count 012345 and 001122, but not 000987 or 122234.)

18.6. Note the following: $|A \cap B| = |A| + |B| - |A \cup B|$. Find a general formula for the size of the intersection of several finite sets in terms of the sizes of their unions.

Chapter 3 Self Test

1. Let R be the relation on the set of all human beings (not just those in your family) defined by $x\ R\ y$ if and only if x is a parent of y.

a. If x is you, describe the set of people $\{y : x\ R\ y\}$.

b. If y is you, describe the set of people $\{x : x\ R\ y\}$.

c. Determine which of the following properties is satisfied by R: reflexive, irreflexive, symmetric, antisymmetric, transitive.

d. Describe R^{-1}.

2. Which of the following relations R defined on the set of all human beings (not just those in your family) are equivalence relations?

a. $x\ R\ y$ provided x and y have the same mother.

b. $x\ R\ y$ provided x and y have the same mother and the same father.

c. $x\ R\ y$ provided x and y have at least one parent in common.

3. Let $A = \{1, 2, 3, 4\}$. How many different relations on A are there?

4. Let x and y be integers. Suppose $x \equiv y \pmod{10}$ and $x \equiv y \pmod{11}$. Do these imply that $x = y$?

5. Let $R = \{(x, y) : x, y \in \mathbb{Z} \text{ and } |x| = |y|\}$.

a. Prove that R is an equivalence relation on the integers.

b. Find the equivalence classes $[5]$, $[-2]$, and $[0]$.

6. Let $A = \{1, 2, 3\}$, $B = \{4, 5\}$, and $R = (A \times A) \cup (B \times B)$. Note that R is an equivalence relation on $A \cup B$. Find all the equivalence classes of R.

7. Let $A = \{1, 2, 3, 4, 5\}$ and define an equivalence relation R on 2^A by $X\ R\ Y$ if and only if $|X| = |Y|$. How many equivalence classes does R have?

8. Let $\mathcal{P} = \{N, Z, P\}$ be a partition of the integers, $\mathbb{Z}$ defined by

- $N = \{x \in \mathbb{Z} : x < 0\}$,
- $Z = \{0\}$, and
- $P = \{x \in \mathbb{Z} : x > 0\}$.

Describe the equivalence relation $\stackrel{\mathcal{P}}{\equiv}$. Your answer should be of the following form: "Suppose x and y are integers. Then $x \stackrel{\mathcal{P}}{\equiv} y$ if and only if ...."

9. Ten married couples are seated around a large circular table. In how many different ways can they do this, assuming husbands and wives sit next to one another? Please note that if everyone moves one (or more) places to the left, the arrangement is not considered to be different.

10. The letters in the word ELECTRICITY are scrambled to make two, possibly nonsensical words (e.g., TREEL CICTY). How many such anagrams are possible?

11. Two children are playing tic-tac-toe. In how many ways can the first two moves be made?

One possible answer is $9 \times 8 = 72$ since there are 9 locations for the first player to mark X and, for each such choice, 8 locations for the second player to mark O.

However, because of symmetry, some of these opening pairs of moves are the same. For example, if the first player chooses a corner square and the second player chooses the center, it doesn't really matter which corner the first player chose.

Taking this into account, in how many distinct ways can the first two moves be made?

12. There are 21 students in a chemistry class. The students must pair up to work as lab partners, but, of course, one student will be left over to work alone. In how many ways can the students be paired up?

13. Let $A = \{1, 2, 3, \ldots, 100\}$. How many 10-element subsets of A consist of only odd numbers?

14. The expression $(x + 2)^{50}$ is expanded. What is the coefficient of x^{17}?

15. Let n be a positive integer. Simplify the following expression:

$$n + (n + 1) + (n + 2) + \cdots + (2n).$$

16. In a school of 200 children, 15 students are chosen to be on the school's math team, and of those, 2 students are chosen to be co-captains. In how many ways can this be done?

17. Let n and k be positive integers with $k + 2 \le n$. Prove the identity

$$\binom{n+2}{k+2} = \binom{n}{k} + 2\binom{n}{k+1} + \binom{n}{k+2}$$

by the following two methods: combinatorially and by use of Pascal's Identity (Theorem 16.10).

18. A pizza restaurant features ten different kinds of toppings. When you order a quadruple pie, you get to pick four toppings on your pizza.

a. How many different quadruple pizzas can be made if the four toppings must be different?

b. How many different quadruple pizzas can be made if toppings may be repeated (e.g., onions, olives, and double mushrooms, or triple anchovies and garlic).

19. Let n be a positive integer. How many multisets can be made using the numbers 1 through n, where each is used at most three times? Be sure to justify your answer.

For example, if $n = 5$, then we would count $\langle 1, 2, 2, 3\rangle$ and $\langle 1, 2, 3, 4, 4, 4, 5\rangle$, but we would not count $\langle 1, 2, 4, 4, 4, 4, 4, 4, 4\rangle$ (too many 4s) or $\langle 3, 4, 6\rangle$ (6 is not in the range from 1 to n).

20. The squares of a 4×4 checkerboard are colored black or white. Use inclusion-exclusion to find the number of ways the checkerboard can be colored so that no row is entirely one color.

Explain why your expression simplifies to 14^4.

CHAPTER

4 More Proof

Thus far we have used primarily one proof technique known as *direct* proof. In this method, we work from hypothesis to conclusion, showing how each statement follows from previous statements. The central idea is to unravel definitions and bridge the gap from what we have to what we want.

We are now ready for, and need, more sophisticated proof methods. In this chapter, we present two powerful methods: *proof by contradiction* and *proof by induction* (and its variant *proof by smallest counterexample*).

19 Contradiction

Most theorems can be expressed in the if-then form. The usual way to prove "If A, then B" is to assume the conditions listed in A and then work to prove the conditions in B (see Proof Template 1). In this section, we present two alternatives to the direct proof method.

Proof by Contrapositive

The statement "If A, then B" is logically equivalent to the statement "If (not B), then (not A)." The statement "If (not B), then (not A)" is called the *contrapositive* of "If A, then B."

Why are a statement and its contrapositive logically equivalent? For "If A, then B" to be true, it must be the case that whenever A is true, B must also be true. If it ever should happen that B is false, then it must have been the case that A was false. In other words, if B is false, then A must be false. Thus we have "If (not B), then (not A)."

Here's another explanation. We know that "If A, then B" is logically equivalent to "(not A) or B" (see Exercise 3.3). By the same reasoning, "If (not B), then (not A)" is equivalent to "(not (not B)) or (not A)," but "not (not B)" is the same as B, so this becomes "B or (not A)," which is equivalent to "(not A) or B." In symbols,

$$a \to b \quad = \quad (\neg a) \vee b \quad = \quad (\neg(\neg b)) \vee (\neg a) \quad = \quad (\neg b) \to (\neg a).$$

If these explanations are difficult to follow, here is a mechanical way to proceed. We build a truth table for $a \to b$ and $(\neg b) \to (\neg a)$ and see the same results.

a	b	$a \to b$	$\neg b$	$\neg a$	$(\neg b) \to (\neg a)$
T	T	T	F	F	T
T	F	F	T	F	F
F	T	T	F	T	T
F	F	T	T	T	T

The bottom line is this: To prove "If A, then B," it is acceptable to prove "If (not B), then (not A)." This is outlined in Proof Template 11.

Proof Template 11 Proof by contrapositive

To prove "If A, then B": Assume (not B) and work to prove (not A).

Let's work through an example.

Proposition 19.1 Let R be an equivalence relation on a set A and let $a, b \in A$. If $a \not R\, b$, then $[a] \cap [b] = \emptyset$.

We have essentially proved this already (see Proposition 14.12). Our purpose here is to illustrate proof by contrapositive. We set up the proof using Proof Template 11.

> Let R be an equivalence relation on a set A and let $a, b \in A$. We prove the contrapositive of the statement.
>
> Suppose $[a] \cap [b] \neq \emptyset$. . . . Therefore $a \, R \, b$. ■

The key point to observe is that we suppose the opposite of the conclusion (not $[a] \cap [b] = \emptyset$) and work toward proving the opposite of the hypothesis (not $a \not R\, b$; i.e., $a \, R \, b$).

Notice that we alerted our reader that we are not using direct proof by announcing that we are going to prove the contrapositive.

To continue the proof, we observe that $[a] \cap [b] \neq \emptyset$ means there is an element in both $[a]$ and $[b]$. We put this into the proof.

> Let R be an equivalence relation on a set A and let $a, b \in A$. We prove the contrapositive of the statement.
>
> Suppose $[a] \cap [b] \neq \emptyset$. Thus there is an $x \in [a] \cap [b]$; that is, $x \in [a]$ and $x \in [b]$. . . . Therefore $a \, R \, b$. ■

We use the definition of equivalence class to finish.

> Let R be an equivalence relation on a set A and let $a, b \in A$. We prove the contrapositive of the statement.
>
> Suppose $[a] \cap [b] \neq \emptyset$. Thus there is an $x \in [a] \cap [b]$; that is, $x \in [a]$ and $x \in [b]$. Hence $x \; R \; a$ and $x \; R \; b$. By symmetry $a \; R \; x$, and since $x \; R \; b$, by transitivity we have $a \; R \; b$. ■

Is there an advantage to proof by contrapositive? Yes. Try proving Proposition 19.1 by direct proof. We would assume $a \not R \, b$ and try to show $[a] \cap [b] = \emptyset$. How would we unravel the hypothesis $a \not R \, b$? How do we show that two sets have nothing in common? We don't have good ways of accomplishing these tasks; a direct proof here looks hard. By switching to the contrapositive, we have conditions that are easier for us to use.

Reductio Ad Absurdum

Proof by contradiction is also called *indirect proof*.

One mistake. Here is another way to think about proof by contradiction. We assume A and (not B) and then follow with valid reasoning until we reach an impossible situation. This means there must be a mistake. If all our reasoning is valid, and since we are allowed to assume A, the mistake *must* have been in supposing (not B). Since (not B) is the mistake, we must have B.

Proof by contrapositive is an alternative to direct proof. If you can't find a direct proof, try proving the contrapositive. Wouldn't it be nice if there were a proof technique that combined both direct proof and proof by contrapositive? There is! It is called *proof by contradiction* or, in Latin, *reductio ad absurdum*. Here is how it works.

We want to prove "If A, then B." To do this, we show that it is impossible for A to be true while B is false. In other words, we want to show that "A and (not B)" is impossible.

How do we prove that something is impossible? We suppose the impossible thing is true and prove that this supposition leads to an absurd conclusion. If a statement implies something clearly wrong, then that statement must have been false!

To prove "If A, then B," we make two assumptions. We assume the hypothesis A and we assume the opposite of the conclusion; that is, we assume (not B). From these two assumptions, we try to reach a clearly false statement. The general outline is given in Proof Template 12.

Proof Template 12 Proof by contradiction.

To prove "If A, then B":

We assume the conditions in A.
Suppose, for the sake of contradiction, not B.
Argue until we reach a contradiction.
$\Rightarrow\Leftarrow$ ■

(The symbol $\Rightarrow\Leftarrow$ is an abbreviation for the following: Thus we have reached a contradiction. Therefore the supposition (not B) must be false. Hence B is true.)

Let us present a formal description of proof by contradiction and then give an example.

We want to prove a statement of the form "If A, then B." To do this, we assume A and (not B) and show this implies something false. Symbolically, we want to show $a \to b$. To do this, we prove $(a \wedge \neg b) \to \text{FALSE}$. These two are logically equivalent.

Proposition 19.2 The Boolean formulas $a \to b$ and $(a \wedge \neg b) \to \text{FALSE}$ are logically equivalent.

Proof. To see that these two are logically equivalent, we build a truth table.

a	b	$a \to b$	$a \wedge \neg b$	$(a \wedge \neg b) \to \text{FALSE}$
T	T	T	F	T
T	F	F	T	F
F	T	T	F	T
F	F	T	F	T

Therefore $a \to b = (a \wedge \neg b) \to \text{FALSE}$. ■

Let's apply this method to prove the following:

Proposition 19.3 No integer is both even and odd.

Reexpressed in if-then form, Proposition 19.3 reads, "If x is an integer, then x is not both even and odd."

Let's set up a proof by contradiction.

> Let x be an integer.
>
> Suppose, for the sake of contradiction, that x is both even and odd.
>
> ...
>
> That is impossible. Thus we have reached a contradiction, so our supposition (that x is both even and odd) is false. Therefore x is not both even and odd, and the proposition is proved. ■

Several comments are in order:

- The first sentence gives the hypothesis (let x be an integer).
- The second sentence serves two purposes.

  First, it announces to the reader that this is going to be a proof by contradiction using the phrase "for the sake of contradiction."

  Second, it supposes the opposite of the conclusion. The supposition is that x is both even and odd.
- The next sentence reads, "That is impossible." We don't know what the antecedent to "That" is! What is impossible? We don't know yet! As the proof develops, we hope to run into a contradiction.

- Given that we have reached a contradiction, here is how we finish the proof. We say that the supposition is impossible because it leads to an absurd statement. Therefore the supposition (not B) must be false. Hence the conclusion (B) must be true.

  The last few sentences of a proof by contradiction are almost always the same. Mathematicians use a special symbol to abbreviate a lot of words. The symbol is $\Rightarrow\Leftarrow$. The image is that two implications are crashing into one another.

  The symbol $\Rightarrow\Leftarrow$ is an abbreviation for "Thus we have reached a contradiction; therefore the supposition is false."

  The supposition is that which we have supposed—namely, (not B).

We don't know (yet) what contradiction we might reach. Let's just continue working with what we have. We know that x is both even and odd, so we unravel.

> Let x be an integer.
>
> Suppose, for the sake of contradiction, that x is both even and odd.
>
> Since x is even, we know $2|x$; that is, there is an integer a such that $x = 2a$.
>
> Since x is odd, we know that there is an integer b such that $x = 2b + 1$.
>
> ...
>
> $\Rightarrow\Leftarrow$ Therefore x is not both even and odd, and the proposition is proved. ■

No contradiction yet. The definitions are completely unraveled. What we have to work with is $x = 2a = 2b + 1$ where a and b are integers. Somehow, we need to manipulate these into something false. Let's try dividing the equation $x = 2a = 2b + 1$ through by 2 to give $\frac{x}{2} = a = b + \frac{1}{2}$, and this says that one integer is just $\frac{1}{2}$ bigger than another (i.e., $a - b = \frac{1}{2}$), but $a - b$ is an integer and $\frac{1}{2}$ is not! A number $(a - b)$ cannot be both an integer and not an integer! That's a contradiction. Hurray!! Let's put it into the proof. (Notice we didn't use $\frac{x}{2}$ in the contradiction, so we can simplify this a bit.)

> Let x be an integer.
>
> Suppose, for the sake of contradiction, that x is both even and odd.
>
> Since x is even, we know $2|x$; that is, there is an integer a such that $x = 2a$.
>
> Since x is odd, we know that there is an integer b such that $x = 2b + 1$.
>
> Therefore $2a = 2b + 1$. Dividing both sides by 2 gives $a = b + \frac{1}{2}$ so $a - b = \frac{1}{2}$. Note that $a - b$ is an integer (since a and b are integers) but $\frac{1}{2}$ is not an integer. $\Rightarrow\Leftarrow$ Therefore x is *not* both even and odd, and the proposition is proved. ■

This completes the proof. We did not know when we began this proof that the absurdity we would reach is that $\frac{1}{2}$ is an integer. This is typical in a proof by contradiction; we begin with A and (not B) and see where the implications lead.

Proposition 19.3 can also be expressed as follows. Let

$$X = \{x \in \mathbb{Z} : x \text{ is even}\}, \text{ and}$$
$$Y = \{x \in \mathbb{Z} : x \text{ is odd}\}.$$

Then $X \cap Y = \emptyset$.

Proof by contradiction is usually the best technique for showing that a set is empty. This is worth codifying in a proof template.

Proof Template 13 Proving that a set is empty.

To prove a set is empty:

Assume the set is nonempty and argue to a contradiction.

Proof Template 13 is appropriate to prove statements of the form "There is no object that satisfies conditions."

Contradiction is also the proof technique of choice when proving *uniqueness* statements. Such statements assert that there can be only one object that satisfies the given conditions.

Mathspeak!

You would think that mathematicians, of all people, would use the word *two* correctly. So it may come as a surprise that when mathematicians say "two" they sometimes mean "one or two." Here is an example. Consider the following statement: Every positive even integer is the sum of two odd positive integers. Mathematicians consider this statement to be true despite the fact that the only way to write 2 as the sum of two positive odd numbers is $2 = 1 + 1$. The two odd numbers in this case are 1 and 1. The two numbers just happen to be the same.

The phrase "Let x and y be two integers . . . " allows for the integers x and y to be the same. This is the convention, albeit a slightly dangerous one. It would be better simply to write, "Let x and y be integers . . . ."

Occasionally we truly wish to eliminate the possibility that $x = y$. In this case, we write, "Let x and y be two different integers . . . " or "Let x and y be two distinct integers . . . ."

Proof Template 14 Proving uniqueness.

To prove there is at most one object that satisfies conditions:

Proof: Suppose there are two different objects, x and y, that satisfy conditions.

Argue to a contradiction.

Often the contradiction in a uniqueness proof is that the two allegedly different objects are in fact the same. Here is a simple example.

Proposition 19.4 Let a and b be numbers with $a \neq 0$. There is at most one number x with $ax + b = 0$.

Proof. Suppose there are two different numbers x and y such that $ax + b = 0$ and $ay + b = 0$. This gives $ax + b = ay + b$. Subtracting b from both sides gives $ax = ay$. Since $a \neq 0$, we can divide both sides by a to give $x = y$.$\Rightarrow\Leftarrow$ ■

A Matter of Style

Proof by contradiction of "If A, then B" is often easier than direct proof because there are more conditions available. Instead of starting with only condition A and trying to demonstrate condition B, we start with both A and (not B) and hunt for a contradiction. This gives us more material with which to work.

Sometimes, when you elect to write a proof by contradiction, you may discover that proof by contradiction was not really required and a simpler sort of proof is possible. A proof is a proof, and you should be happy to have found a correct proof. Nonetheless, a simpler way to present your argument is always preferable. Here is how to tell when you can simplify a proof of "If A, then B."

- You assumed A and (not B). You used only the hypothesis A, and the contradiction you reached was B and (not B).

  In this case, you really have a direct proof and you can remove the extraneous proof-by-contradiction apparatus.
- You assumed A and not B. You used only the supposition (not B), and the contradiction you reached was A and (not A).

  In this case, you really have a proof by contrapositive. Rewrite it in that form.

Recap

We introduced two new proof techniques for statements of the form "If A, then B." In a proof by contrapositive, we assume (not B) and work to prove (not A). In a proof by contradiction, we assume both A and (not B) and work to produce a contradiction.

19 Exercises

19.1. Please state the contrapositive of each of the following statements:

a. If x is odd, then x^2 is odd.

b. If p is prime, then $2^p - 2$ is divisible by p.

c. If x is nonzero, then x^2 is positive.

d. If the diagonals of a parallelogram are perpendicular, then the parallelogram is a rhombus.

e. If the battery is fully charged, the car will start.

f. If A or B, then C.

19.2. What is the contrapositive of the contrapositive of an if-then statement?

19.3. A statement of the form "A if and only if B" is usually proved in two parts: one part to show $A \Rightarrow B$ and another to show $B \Rightarrow A$.

Explain why the following is also an acceptable structure for a proof. First prove $A \Rightarrow B$ and then prove $\neg A \Rightarrow \neg B$.

19.4. For each of the following statements, write the first sentences of a proof by contradiction (you should not attempt to complete the proofs). Please use the phrase "for the sake of contradiction."

a. If $A \subseteq B$ and $B \subseteq C$, then $A \subseteq C$.

b. The sum of two negative integers is a negative integer.

c. If the square of a rational number is an integer, then the rational number must also be an integer.

d. If the sum of two primes is prime, then one of the primes must be 2.

e. A line cannot intersect all three sides of a triangle.

f. Distinct circles intersect in at most two points.

g. There are infinitely many primes.

19.5. Prove by contradiction that consecutive integers cannot be both even.

19.6. Prove by contradiction that consecutive integers cannot be both odd.

19.7. Prove by contradiction: If the sum of two primes is prime, then one of the primes must be 2.

You may assume that every integer is either even or odd, but never both.

19.8. Let A and B be sets. Prove by contradiction that $(A - B) \cap (B - A) = \emptyset$.

19.9. Let A and B be sets. Prove $A \cap B = \emptyset$ if and only if $(A \times B) \cap (B \times A) = \emptyset$.

19.10. Prove the converse of the Addition Principle (Corollary 11.8). The *converse* of a statement "If A, then B" is the statement "If B, then A." In other words, your job is to prove the following:

Let A and B be finite sets. If $|A \cup B| = |A| + |B|$, then $A \cap B = \emptyset$.

19.11. Let A be a subset of the integers.

a. Write a careful definition for the *smallest element* of A.

b. Let E be the set of even integers; that is, $E = \{x \in \mathbb{Z} : 2|x\}$. Prove by contradiction that E has no smallest element.

c. Prove that if $A \subseteq \mathbb{Z}$ has a smallest element, it is unique.

20 Smallest Counterexample

Proof by contradiction as proof by lack of counterexample.

In Section 19 we developed the method of proof by contradiction. Here is another way we can think about this technique.

We want to prove a result of the form "If A, then B." Let's suppose this result were false. If that were the case, there would be a *counterexample* to the statement. That is, there would be an instance where A is true and B is false. We then analyze that alleged counterexample and produce a contradiction. Since the supposition that there is a counterexample leads to an absurd conclusion (a contradiction), that supposition must be wrong; there is no counterexample. Since there is no counterexample, the result must be true.

For example, we showed that no integer could be both even and odd. We can rephrase the argument as follows:

> Suppose the statement "No integer is both even and odd" were false. Then there would be a counterexample; let's say x were such an integer (i.e., x is both even and odd). Since x is even, there is an integer a such that $x = 2a$. Since x is odd, there is an integer b such that $x = 2b + 1$. Thus $2a = 2b + 1$, which implies $a - b = \frac{1}{2}$. Since a and b are integers, so is $a - b$, $\Rightarrow\Leftarrow$ ($\frac{1}{2}$ is not an integer). ■

In this section, we extend this idea by considering *smallest* counterexamples. It's a little idea that wields enormous power. The essence of the idea is that we not only consider an alleged counterexample to an if-then result, we consider a smallest counterexample. This needs to be done carefully, and we explore this idea at length.

We have not yet proved a fact you know well: Every integer is even or odd. We have shown that no integer can be both even and odd, but we have not yet ruled out the possibility that some integer is neither. It is sensible to try to prove this by contradiction. We would structure the proof as follows:

> Suppose, for the sake of contradiction, that there were an integer x that is neither even nor odd. ... $\Rightarrow\Leftarrow$ Therefore every integer is either even or odd. ■

Next we could unravel definitions as follows:

> Suppose, for the sake of contradiction, that there were an integer x that is neither even nor odd. So there is no integer a with $x = 2a$ and there is no integer b with $x = 2b + 1$. ... $\Rightarrow\Leftarrow$ Therefore every integer is either even or odd. ■

And now we're stuck. What do we do next? We need a new idea. The new idea is to consider a *smallest counterexample*. We begin with a restricted version of what we are trying to prove.

Proposition 20.1 Every natural number is either even or odd.

Note that we are just proving that every *natural number* (member of $\mathbb{N}$) is either even or odd; we'll extend this to all integers later. (The reason for this restriction is presented later.)

We begin the proof using the idea of smallest counterexample.

> Suppose, for the sake of contradiction, that not all natural numbers are even or odd. Then there is a smallest natural number, x, that is neither even nor odd. ... $\Rightarrow\Leftarrow$ ■

Why did we restrict the scope of Proposition 20.1 to natural numbers? If we were trying to prove that every integer is either even or odd, we could not rule out the possibility that there might be infinitely many counterexamples, marching off to $-\infty$. Then we could not sensibly talk about the smallest counterexample. It is akin to talking about the smallest odd integer; there is no such thing! The odd numbers descend forever $-3, -5, -7, \ldots$; there is no smallest odd integer.

On the other hand, the natural numbers do not descend forever; they "stop" at zero. It makes sense to speak of the smallest odd natural number, namely 1.

This is why we proved Proposition 20.1 only for natural numbers. We extend this result to all integers after we complete the proof.

Let us return to the proof. We add the next sentence to the proof, and let me warn you that the next sentence has an error! Read the sentence carefully and try to find the mistake.

> Suppose, for the sake of contradiction, that not all natural numbers are even or odd. Then there is a smallest natural number, x, that is neither even nor odd. Since $x - 1 < x$, we see that $x - 1$ is a smaller natural number and therefore is not a counterexample to Proposition 20.1.
>
> $\ldots \Rightarrow\Leftarrow$ ■

Do you see the problem? It is subtle. Let's dissect the new sentence.

- Since $x - 1 < x \ldots$. No problem here. Obviously $x - 1 < x$.
- $\ldots x - 1 \ldots$ is not a counterexample to Proposition 20.1. No problem here either. We know x is the smallest counterexample. Because $x - 1$ is smaller than x, it is not a counterexample to Proposition 20.1.

  Where is the problem?
- ...natural number.... How do we know $x - 1$ is a natural number? Here's the mistake. We do not know that $x - 1$ is a natural number because we have not ruled out the possibility that $x = 0$.

Now it is not hard to rule out $x = 0$; we simply haven't done it yet. Let's take care of this seemingly minor point.

> Suppose, for the sake of contradiction, that not all natural numbers are even or odd. Then there is a smallest natural number, x, that is neither even nor odd.
>
> We know $x \neq 0$ because 0 is even. Therefore $x \geq 1$.
>
> Since $0 \leq x - 1 < x$, we see that $x - 1$ is a smaller natural number and therefore is not a counterexample to Proposition 20.1.
>
> $\ldots \Rightarrow\Leftarrow$ ■

We can now continue the proof. We know that $x - 1 \in \mathbb{N}$ and $x - 1$ is not a counterexample to the proposition. What does this mean? It means that since $x - 1$ is a natural number, it must be either even or odd. We don't know which of these might be true, so we consider both possibilities.

> Suppose, for the sake of contradiction, that not all natural numbers are even or odd. Then there is a smallest natural number, x, that is neither even nor odd.
>
> We know $x \neq 0$ because 0 is even. Therefore $x \geq 1$.
>
> Since $0 \leq x - 1 < x$, we see that $x - 1$ is a smaller natural number and therefore is not a counterexample to Proposition 20.1.
>
> Therefore $x - 1$ is either even or odd. We consider both possibilities.
>
> **(1)** Suppose $x - 1$ is odd....
> **(2)** Suppose $x - 1$ is even....
>
> $\ldots \Rightarrow\Leftarrow$ ■

Now we unravel definitions. In case (1), $x - 1$ is odd, so $x - 1 = 2a + 1$ for some integer a. In case (2), $x - 1$ is even, so $x - 1 = 2b$ for some integer b.

> Suppose, for the sake of contradiction, that not all natural numbers are even or odd. Then there is a smallest natural number, x, that is neither even nor odd.
>
> We know $x \neq 0$ because 0 is even. Therefore $x \geq 1$.
>
> Since $0 \leq x - 1 < x$, we see that $x - 1$ is a smaller natural number and therefore is not a counterexample to Proposition 20.1.
>
> Therefore $x - 1$ is either even or odd. We consider both possibilities.
>
> **(1)** Suppose $x - 1$ is odd. Therefore $x - 1 = 2a + 1$ for some integer a. ...
> **(2)** Suppose $x - 1$ is even. Therefore $x - 1 = 2b$ for some integer b. ...
>
> ... $\Rightarrow\Leftarrow$ ■

In case (1), we have $x - 1 = 2a + 1$, so $x = 2a + 2 = 2(a + 1)$, so x is even; this is a contradiction to the fact that x is neither even nor odd. In case (2), we get a similar contradiction.

> Suppose, for the sake of contradiction, that not all natural numbers are even or odd. Then there is a smallest natural number, x, that is neither even nor odd.
>
> We know $x \neq 0$ because 0 is even. Therefore $x \geq 1$.
>
> Since $0 \leq x - 1 < x$, we see that $x - 1$ is a smaller natural number and therefore is not a counterexample to Proposition 20.1.
>
> Therefore $x - 1$ is either even or odd. We consider both possibilities.
>
> **(1)** Suppose $x - 1$ is odd. Therefore $x - 1 = 2a + 1$ for some integer a. Thus $x = 2a + 2 = 2(a + 1)$, so x is even $\Rightarrow\Leftarrow$ (x is neither even nor odd).
> **(2)** Suppose $x - 1$ is even. Therefore $x - 1 = 2b$ for some integer b. Thus $x = 2b + 1$, so x is odd $\Rightarrow\Leftarrow$ (x is neither even nor odd).
>
> In every case, we have a contradiction, so the supposition is false and the proposition is proved. ■

Let us summarize the main points of this proof.

- It is a proof by contradiction.
- We consider a smallest counterexample to the result.
- We need to treat the *very* smallest possibility as a special case.
- We descend to a smaller case for which the theorem is true and work back.

Before we present another example, let us finish the job we set out to accomplish.

Corollary 20.2 Every integer is either even or odd.

The key idea is that either $x \geq 0$ (in which case we are finished by Proposition 20.1) or else $x < 0$ (in which case $-x \in \mathbb{N}$, and again we can use Proposition 20.1).

Proof. Let x be any integer.

If $x \geq 0$, then $x \in \mathbb{N}$, so by Proposition 20.1, x is either even or odd.

Otherwise, $x < 0$. In this case $-x > 0$, so $-x$ is either even or odd.

- If $-x$ is even, then $-x = 2a$ for some integer a. But then $x = -2a = 2(-a)$, so x is even.
- If $-x$ is odd, then $-x = 2b + 1$ for some integer b. From this we have $x = -2b - 1 = 2(-b - 1) + 1$, so x is odd.

In every case, x is either even or odd. ■

Proof Template 15 gives the general form of this technique.

Proof Template 15 Proof by smallest counterexample.

First, let x be a smallest counterexample to the result we are trying to prove. It must be clear that there can be such an x.

Second, rule out x being the *very* smallest possibility. This (usually easy) step is called the *basis* step.

Third, consider an instance x' of the result that is "just" smaller than x. Use the fact that the result for x' is true but the result for x is false to reach a contradiction $\Rightarrow\Leftarrow$.

Conclude that the result is true. ■

Here is another proposition we prove using the smallest-counterexample method.

Proposition 20.3 Let n be a positive integer. The sum of the first n odd natural numbers is n^2.

The first n odd natural numbers are $1, 3, 5, \ldots, 2n-1$. The proposition claims that

$$1 + 3 + 5 + \cdots + (2n - 1) = n^2$$

or, in $\sum$ notation,

$$\sum_{k=1}^{n}(2k - 1) = n^2.$$

For example, with $n = 5$ we have $1 + 3 + 5 + 7 + 9 = 25 = 5^2$.

Proof. Suppose Proposition 20.3 is false. This means that there is a smallest positive integer x for which the statement is false (i.e., the sum of the first x odd numbers is not x^2); that is,

$$1 + 3 + 5 + \cdots + (2x - 1) \neq x^2. \tag{6}$$

Note that $x \neq 1$ because the sum of the first 1 odd numbers is $1 = 1^2$. (This is the basis step.)

So $x > 1$. Since x is the smallest number for which Proposition 20.3 fails and since $x > 1$, the sum of the first $x - 1$ odd numbers must equal $(x - 1)^2$; that is,

$$1 + 3 + 5 + \cdots + [2(x - 1) - 1] = (x - 1)^2. \tag{7}$$

(So far this proof has been on "autopilot." We are simply using Proof Template 15.)

Notice that the left-hand side of (7) is one term short of the sum of the first x odd numbers. We add one more term to both sides of this equation to give

$$1 + 3 + 5 + \cdots + [2(x - 1) - 1] + (2x - 1) = (x - 1)^2 + (2x - 1).$$

The right-hand side can be algebraically expanded; thus

$$\begin{aligned} 1 + 3 + 5 + \cdots + [2(x - 1) - 1] + (2x - 1) &= (x - 1)^2 + (2x - 1) \\ &= (x^2 - 2x + 1) + (2x - 1) \\ &= x^2 \end{aligned}$$

contradicting (6).⇒⇐ ■

The absolute importance of the basis step.

In the two proofs we have considered thus far, there is a basis step. In the proof that all natural numbers are either even or odd, we first checked that 0 was not a counterexample. In the proof that the sum of the first n odd numbers is n^2, we first checked that 1 was not a counterexample. These steps are important. They show that the immediate smaller case of the result still makes sense. Perhaps the best way to convince you that this basis step is absolutely essential is to show how we can prove an erroneous result if we omit it.

Statement 20.4 **(false)** Every natural number is both even and odd.

Obviously Statement 20.4 is false! Here we give a bogus proof using the smallest-counterexample method, but omitting the basis step.

Proof. Suppose Proposition 20.4 is false. Then there is a smallest natural number x that is not both even and odd. Consider $x - 1$. Since $x - 1 < x$, $x - 1$ is not a counterexample to Proposition 20.4. Therefore $x - 1$ is both even and odd.

Since $x - 1$ is even, $x - 1 = 2a$ for some integer a, and so $x = 2a + 1$, so x is odd.

Since $x - 1$ is odd, $x - 1 = 2b + 1$ for some integer b, and so $x = 2b + 2 = 2(b + 1)$, so x is even.

Thus x is both even and odd, but x is not both even and odd.$\Rightarrow\Leftarrow$ ■

The proof is 99% correct. Where is the mistake? The error is in the sentence "Therefore $x - 1$ is both even and odd." It is correct that $x - 1$ is not a counterexample, but we do not know that $x - 1$ is a natural number. We do not know this because we have not ruled out the possibility that $x - 1 = -1$ (i.e., $x = 0$). Of course, no natural number is both even and odd. So the smallest natural number that is not both even and odd is zero (the exact problem case!).

Well-Ordering

Let us take a closer look at the proof-by-smallest-counterexample technique. We saw that it was appropriate to apply this technique to showing that all natural numbers are either even or odd, but the method is invalid for integers. The difference is that the integers contain infinitely descending negative numbers. However, consider the following statement and its bogus proof.

Statement 20.5 **(false)** Every nonnegative rational number is an integer.

Recall that a *rational number* is any number that can be expressed as a fraction a/b where $a, b \in \mathbb{Z}$ and $b \neq 0$. This statement is asserting that numbers such as $\frac{1}{4}$ are integers. Ridiculous! Notice, however, that the statement is restricted to nonnegative rational numbers; this is analogous to Proposition 20.1, which was restricted to nonnegative integers.

Let's look at the "proof."

Proof. Suppose Statement 20.5 were false. Let x be a smallest counterexample.

Notice that $x = 0$ is not a counterexample because 0 is an integer. (This is the basis step.)

Since x is a nonnegative rational, so is $x/2$. Furthermore, since $x \neq 0$, we know that $x/2 < x$, so $x/2$ is smaller than the smallest counterexample, x. Therefore $x/2$ is not a counterexample, so $x/2$ is an integer. Now $x = 2(x/2)$, and 2 times an integer is an integer; therefore x is an integer.$\Rightarrow\Leftarrow$ ■

What is wrong with this proof? It looks like we followed Proof Template 15, and we even remembered to do a basis step (we considered $x = 0$).

The problem is in the sentence "Let x be a smallest counterexample." There are infinitely many counterexamples to Statement 20.5, including $\frac{1}{2}, \frac{1}{3}, \frac{1}{4}, \frac{1}{5}, \ldots$. These form an infinite descent of counterexamples, and so there can be no smallest counterexample!

We need to worry that we do not make subtle mistakes like the "proof" of Statement 20.5 when we use the proof-by-smallest-counterexample technique. The central issue is: When can we be certain to find a smallest counterexample?

The guiding principle is the following.

Statement 20.6 **(Well-Ordering Principle)** Every nonempty set of natural numbers contains a least element.

Example 20.7 Let $P = \{x \in \mathbb{N} : x \text{ is prime}\}$. This set is a nonempty subset of the natural numbers. By the Well-Ordering Principle, P contains a least element. Of course, the least element in P is 2.

Example 20.8 Consider the set

$$X = \{x \in \mathbb{N} : x \text{ is even and odd}\}.$$

We know that this set is empty because we have shown that no natural number is both even and odd (Proposition 20.1). But for the sake of contradiction, we suppose that $X \neq \emptyset$; then, by the Well-Ordering Principle, X would contain a smallest element. This is the central idea in the proof of Proposition 20.1.

The term *well-ordered* applies to an ordered set (i.e., a set X with a $<$ relation). The set X is called *well-ordered* if every nonempty subset of X contains a least element.

Example 20.9 In contradistinction, consider the set

$$Y = \{y \in \mathbb{Q} : y \geq 0, \ y \notin \mathbb{Z}\}.$$

The bogus proof of Statement 20.5 sought a least element of Y. We subsequently realized that Y has no least element, and that was the error in our "proof." The Well-Ordering Principle applies to $\mathbb{N}$, but not to $\mathbb{Q}$.

The Well-Ordering Principle is an *axiom* of the natural numbers.

Notice that we called the Well-Ordering Principle a *statement;* we did not call it a *theorem*. Why? The reason harks back to the beginning of this book. We could, but did not, define exactly what the integers are. Were we to go through the difficult task of writing a careful definition of the integers, we would begin by defining the natural numbers. The natural numbers are defined to be a set of "objects" that satisfy certain conditions; these defining conditions are called *axioms*. One of these defining axioms is the Well-Ordering Principle. So the natural numbers obey the Well-Ordering Principle by definition. There are other ways to define integers and natural numbers, and in those contexts one can prove the Well-Ordering Principle. If you are intrigued about how all this is done, I recommend you take a course in foundations of mathematics (such a course might be called Logic and Set Theory).

In any case, our approach has been to assume fundamental properties of the integers; we take the Well-Ordering Principle to be one of those fundamental properties.

The Well-Ordering Principle explains why the smallest-counterexample technique works to prove that natural numbers cannot be both even and odd, but it does not work to prove that nonnegative rationals are integers.

Proof Template 16 gives an alternative to Proof Template 15 that explicitly uses the Well-Ordering Principle.

Proof Template 16 Proof by the Well-Ordering Principle.

To prove a statement about natural numbers:

Proof. Suppose, for the sake of contradiction, that the statement is false. Let $X \subseteq \mathbb{N}$ be the set of counterexamples to the statement. (I like the letter X for eXceptions.) Since we have supposed the statement is false, $X \neq \emptyset$. By the Well-Ordering Principle, X contains a least element, x.

(Basis step.) We know that $x \neq 0$ because *show that the result holds for* 0; *this is usually easy.*

Consider $x - 1$. Since $x > 0$, we know that $x - 1 \in \mathbb{N}$ and the statement is true for $x - 1$ (because $x - 1 < x$). *From here we argue to a contradiction, often that x both is and is not a counterexample to the statement.*$\Rightarrow\Leftarrow$ ■

Here is an example of how to use Proof Template 16.

Proposition 20.10 Let $n \in \mathbb{N}$. If $a \neq 0$ and $a \neq 1$, then

$$a^0 + a^1 + a^2 + \cdots + a^n = \frac{a^{n+1} - 1}{a - 1}. \tag{8}$$

In fancy notation, we want to prove

$$\sum_{k=0}^{n} a^k = \frac{a^{n+1} - 1}{a - 1}.$$

We rule out $a = 1$ because the right-hand side would be $\frac{0}{0}$. We also rule out $a = 0$ to avoid worrying about 0^0. If we take $0^0 = 1$, then the formula still works.

Proof. We prove Proposition 20.10 using the Well-Ordering Principle.

Suppose, for the sake of contradiction, that Proposition 20.10 were false. Let X be the set of counterexamples—that is, those integers n for which Equation (8) does not hold. Hence

$$X = \left\{ n \in \mathbb{N} : \sum_{k=0}^{n} a^k \neq \frac{a^{n+1} - 1}{a - 1} \right\}.$$

As we have supposed that the proposition is false, there must be a counterexample, so $X \neq \emptyset$.

Since X is a nonempty subset of $\mathbb{N}$, by the Well-Ordering Principle, it contains a least element x.

Note that for $n = 0$, Equation (8) reduces to

$$1 = \frac{a^1 - 1}{a - 1}$$

and this is true. This means that $n = 0$ is not a counterexample to the proposition. Thus $x \neq 0$. (This is the basis step.)

Therefore $x > 0$. Now $x - 1 \in \mathbb{N}$ and $x - 1 \notin X$ because $x - 1$ is smaller than the least element of X. Therefore the proposition holds for $n = x - 1$, so we have

$$a^0 + a^1 + a^2 + \cdots + a^{x-1} = \frac{a^x - 1}{a - 1}.$$

We add a^x to both sides of this equation to get

$$a^0 + a^1 + a^2 + \cdots + a^{x-1} + a^x = \frac{a^x - 1}{a - 1} + a^x. \tag{9}$$

Putting the right-hand side of Equation (9) over a common denominator gives

$$\begin{aligned} \frac{a^x - 1}{a - 1} + a^x &= \frac{a^x - 1}{a - 1} + a^x \left(\frac{a - 1}{a - 1}\right) \\ &= \frac{a^x - 1 + a^{x+1} - a^x}{a - 1} \\ &= \frac{a^{x+1} - 1}{a - 1} \end{aligned}$$

and so

$$a^0 + a^1 + a^2 + \cdots + a^x = \frac{a^{x+1} - 1}{a - 1}.$$

This shows that x satisfies the proposition and is therefore not a counterexample, contradicting $x \in X$.⇒⇐ ■

Proof Template 16 is more rigidly specified than Proof Template 15. Often you will need to modify Proof Template 16 to suit a particular situation. For example, consider the following:

Proposition 20.11 For all integers $n \geq 5$, we have $2^n > n^2$.

Notice that the inequality $2^n > n^2$ is not true for a few small values of n:

n	0	1	2	3	4	5
2^n	1	2	4	8	16	32
n^2	0	1	4	9	16	25

Thus Proposition 20.11 does not apply to all of $\mathbb{N}$. We need to modify Proof Template 16 slightly. Here is the proof of Proposition 20.11:

Proof. Suppose, for the sake of contradiction, Proposition 20.11 were false. Let X be the set of counterexamples; that is,

$$X = \{n \in \mathbb{Z} : n \geq 5,\ 2^n \not> n^2\}.$$

Since our supposition is that the proposition is false, we have $X \neq \emptyset$. By the Well-Ordering Principle, X contains a least element x.

We claim that $x \neq 5$. Note that $2^5 = 32 > 25 = 5^2$, so 5 is not a counterexample to the proposition (i.e., $x \notin X$), and hence $x \neq 5$. Thus $x \geq 6$.

Now consider $x - 1$. Since $x \geq 6$, we have $x - 1 \geq 5$. Since x is the least element of X, we know that the proposition is true for $n = x - 1$; that is,

$$2^{x-1} > (x-1)^2. \tag{10}$$

We know $2^{x-1} = \frac{1}{2} \cdot 2^x$ and $(x-1)^2 = x^2 - 2x + 1$, so Equation (10) can be rewritten as

$$\frac{1}{2} \cdot 2^x > x^2 - 2x + 1.$$

Multiplying both sides by 2 gives

$$2^x > 2x^2 - 4x + 2. \tag{11}$$

We will be finished once we can prove

$$2x^2 - 4x + 2 \geq x^2. \tag{12}$$

To prove Equation (12), we just need to prove

$$x^2 - 4x + 4 \geq 2. \tag{13}$$

We got Equation (13) from Equation (12) by adding $2 - x^2$ to both sides. Notice that Equation (13) can be rewritten

$$(x-2)^2 \geq 2. \tag{14}$$

So we have reduced the problem to proving Equation (14), and to prove that, it certainly is enough to prove

$$x - 2 \geq 2. \tag{15}$$

and that's true because $x \geq 6$ (all we need is $x \geq 4$). ■

The only modification to Proof Template 16 is that the basis case was $x = 5$ instead of $x = 0$.

We present another example where we need to modify slightly the Well-Ordering Principle method. This example involves the following celebrated sequence of numbers.

Definition 20.12 **(Fibonacci numbers)** The *Fibonacci numbers* are the list of integers $(1, 1, 2, 3, 5, 8, \ldots) = (F_0, F_1, F_2, \ldots)$ where

$$F_0 = 1,$$
$$F_1 = 1, \text{ and}$$
$$F_n = F_{n-1} + F_{n-2}, \text{ for } n \geq 2.$$

In words, the Fibonacci numbers are the sequence that begins 1, 1, 2, 3, 5, 8, ... and each successive term is produced by adding the two previous terms. We label these numbers F_n (starting with F_0).

Proposition 20.13 For all $n \in \mathbb{N}$, we have $F_n \le 1.7^n$.

Proof. Suppose, for the sake of contradiction, that Proposition 20.13 were false. Let X be the set of counterexamples; that is,

$$X = \{n \in \mathbb{N} : F_n \not\le 1.7^n\}.$$

Since we have supposed that the proposition is false, we know that $X \neq \emptyset$. Thus, by the Well-Ordering Principle, X contains a least element x.

Observe that $x \neq 0$ because $F_0 = 1 = 1.7^0$ and $x \neq 1$ because $F_1 = 1 \le 1.7^1$.

Notice that we have considered two basis cases: $x \neq 0$ and $x \neq 1$. Why? We explain in just a moment.

Thus $x \ge 2$. Now we know that

$$F_x = F_{x-1} + F_{x-2} \tag{16}$$

and we know, since $x - 1$ and $x - 2$ are natural numbers less than x, that

$$F_{x-2} \le 1.7^{x-2} \qquad \text{and} \qquad F_{x-1} \le 1.7^{x-1}. \tag{17}$$

This is why! We want to use the fact that the proposition is true for $x-1$ and $x-2$ in the proof. We cannot do this unless we are sure that $x - 1$ and $x - 2$ are natural numbers; that is why we must rule out both $x = 0$ and $x = 1$.

Combining Equations (16) and (17), we have

$$\begin{aligned} F_x &= F_{x-1} + F_{x-2} \\ &\le 1.7^{x-1} + 1.7^{x-2} \\ &= 1.7^{x-2}(1.7 + 1) \\ &= 1.7^{x-2}(2.7) \\ &< 1.7^{x-2}(2.89) \\ &= 1.7^{x-2}(1.7^2) \\ &= 1.7^x. \end{aligned}$$

(The trick was recognizing $2.7 < 2.89 = 1.7^2$.)

Therefore Proposition 20.13 is true for $n = x$, contradicting $x \in X$.$\Rightarrow\Leftarrow$ ■

Recap

In this section, we extended the proof-by-contradiction method to proof by smallest counterexample. We refined this method by explicit use of the Well-Ordering Principle. We underscored the vital importance of the (usually easy) basis case.

20 Exercises

20.1. What is the smallest positive real number?

20.2. Prove by the techniques of this section that $1+2+3+\cdots+n = \frac{1}{2}(n)(n+1)$ for all positive integers n.

20.3. Prove by the techniques of this section that $n < 2^n$ for all $n \in \mathbb{N}$.

20.4. Prove by the techniques of this section that $n! \le n^n$ for all positive integers n.

20.5. The inequality $F_n > 1.6^n$ is true once n is big enough. Do some calculations to find out from what value n this inequality holds. Prove your assertion.

20.6. Calculate the sum of the first n Fibonacci numbers for $n = 0, 1, 2, \ldots, 5$. In other words, calculate

$$F_0 + F_1 + \cdots + F_n$$

for several values of n.

Formulate a conjecture about these sums and prove it.

20.7. Criticize the following statement and proof:

Statement. All natural numbers are divisible by 3.

Proof. Suppose, for the sake of contradiction, the statement were false. Let X be the set of counterexamples (i.e., $X = \{x \in \mathbb{N} : x \text{ is not divisible by } 3\}$). The supposition that the statement is false means that $X \neq \emptyset$. Since X is a nonempty set of natural numbers, it contains a least element x.

Note that $0 \notin X$ because 0 is divisible by 3. So $x \neq 3$.

Now consider $x - 3$. Since $x - 3 < x$, it is not a counterexample to the statement. Therefore $x - 3$ is divisible by 3; that is, there is an integer a such that $x - 3 = 3a$. So $x = 3a + 3 = 3(a + 1)$ and x is divisible by 3, contradicting $x \in X$.$\Rightarrow\Leftarrow$ ■

20.8. In Section 16 we discussed that Pascal's triangle and the triangle of binomial coefficients are the same, and we explained why. Rewrite that discussion as a careful proof using the method of smallest counterexample. Your proof should contain a sentence akin to "Consider the first row where Pascal's triangle and the binomial coefficient triangle are not the same."

20.9. Prove the generalized Addition Principle by use of the Well-Ordering Principle. That is, please prove the following:

Suppose $A_1, A_2, \ldots, A_n$ are pairwise disjoint finite sets. Then

$$|A_1 \cup A_2 \cup \cdots \cup A_n| = |A_1| + |A_2| + \cdots + |A_n|.$$

And Finally

Theorem 20.14 **(Interesting)** Every natural number is interesting.

Proof. Suppose, for the sake of contradiction, that Theorem 20.14 were false. Let X be the set of counterexamples (i.e., X is the set of those natural numbers that are *not* interesting). Because we have supposed the theorem to be false, we have $X \neq \emptyset$. By the Well-Ordering Principle, let x be the smallest element of X.

Of course, 0 is an interesting number: It is the identity element for addition, it is the first natural number, any number multiplied by 0 is 0, and so on. So $x \neq 0$. Similarly, $x \neq 1$ because 1 is the only unit in $\mathbb{N}$, it is the identity element for multiplication, and so on. And $x \neq 2$ because 2 is the only even prime. These are interesting numbers!

What is x? It is the first natural number that isn't interesting. That makes it very interesting! $\Rightarrow\Leftarrow$ ■

21 Induction

In this section, we present an alternative to proof by smallest counterexample. This method is called *mathematical induction,* or *induction* for short.

Mathspeak!

In standard English, the word *induction* refers to drawing general conclusions from examining several particular facts. For example, the general principle that the sun always rises in the east follows by induction from the observations that every sunrise ever seen has been in the east. This, of course, does not prove the sun will rise in the east tomorrow, but even a mathematician would not bet against it! The mathematician's use of the word *induction* is quite different and is explained in this section.

The Induction Machine

Imagine: Sitting before you is a statement to be proved. Rather than prove it yourself, suppose you could build a machine to prove it for you. Although some progress has been made by computer scientists to create theorem-proving programs, the dream of a personal theorem-proving robot is still the stuff of science fiction.

Nevertheless, some statements can be proved by an imaginary theorem-proving machine. Let us illustrate with an example.

Proposition 21.1 Let n be a positive integer. The sum of the first n odd natural numbers is n^2.

(This is Proposition 20.3, repeated here for our reconsideration.)

We can think of Proposition 21.1 as an assertion that infinitely many equations are true:

$$\begin{aligned} 1 &= 1^2 \\ 1+3 &= 2^2 \\ 1+3+5 &= 3^2 \\ 1+3+5+7 &= 4^2 \\ &\vdots \end{aligned}$$

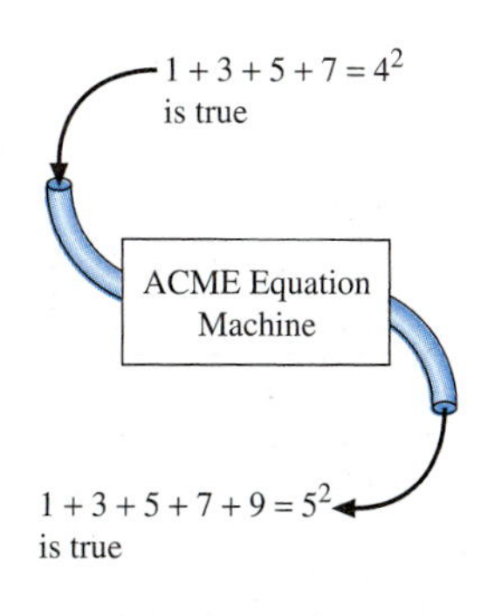

It is neither difficult nor particularly interesting to verify any one of these equations; we just need to add some numbers and check that we get the promised answer.

We could write a computer program to check these equations, but we cannot wait for the program to run forever to verify the entire list. Instead, we are going to build a different sort of machine. Here is how the machine works.

We give the machine one of the equations that has already been proved, say $1 + 3 + 5 = 3^2$. The machine takes this equation and uses it to prove the next equation on the list, say $1 + 3 + 5 + 7 = 4^2$. That's all the machine does. When we give the machine one equation, it uses that equation to prove the next equation on the list.

Suppose such a machine has been built and is ready to work. We drop in $1+3+5+7=4^2$ and out pops $1+3+5+7+9=5^2$. Then we push in $1+3+5+7+9=5^2$ and out comes $1+3+5+7+9+11=6^2$. Amazing! But it gets tiring feeding the machine these equations, so let's attach a pipe from the "out" tube of the machine around to the "in" tube of the machine. As verified equations pop out of the machine, they are immediately shuttled over to the machine's intake to produce the next equation, and the whole cycle repeats *ad infinitum*.

Our machine is all ready to work. To start it off, we put in the first equation, $1=1^2$, switch on the machine and let it run. Out pops $1+3=2^2$, and then $1+3+5=3^2$, and so on. Marvelous!

Would such a machine be able to prove Proposition 21.1? Won't we need to wait forever for the machine to prove all the equations? Certainly the machine is fun to watch, but who has all eternity to wait?

We need one more idea. Suppose we could prove that the machine is 100% reliable. Whenever one equation on the list is fed into the machine, we are absolutely guaranteed that the machine will verify the next equation on the list. If we had such a guarantee, then we would know that every equation on the list will eventually be proved, so they all must be correct.

Let's see how this is possible. The machine takes an equation that has already been proved, say $1+3+5+7=4^2$. The machine is now required to prove that $1+3+5+7+9=5^2$. The machine could simply add up 1, 3, 5, 7, and 9 to get 25 and then check that $25=5^2$. But that is rather inefficient. The machine already knows that $1+3+5+7=4^2$, so it is faster and simpler to add 9 to both sides of the equation: $1+3+5+7+9=4^2+9$. Now the machine just has to calculate $4^2+9=16+9=25=5^2$.

Here are the blueprints for the machine:

1. The machine receives an equation of the form
$$1+3+5+\cdots+(2k-1)=k^2$$
through its intake tube.
Note: We are allowed to insert only equations that have already been proved, so we trust that this particular equation is correct.
2. The next odd number after $2k-1$ is $(2k-1)+2=2k+1$. The machine adds $2k+1$ to both sides of the equation. The equation now looks like this:
$$1+3+5+\cdots+(2k-1)+(2k+1)=k^2+(2k+1).$$
3. The machine calculates $k^2+(2k+1)$ and checks to see whether it equals $(k+1)^2$. If so, it is happy and ejects the newly proved equation
$$1+3+5+\cdots+(2k-1)+(2k+1)=(k+1)^2$$
through its output tube.

To be sure this machine is reliable, we need to check that whenever we feed the machine a valid equation, the machine will always verify that the next equation on this list is valid.

As we examine the inner workings of the machine carefully, the only place the machine's gears might jam is when it checks whether $k^2 + (2k + 1)$ is equal to $(k + 1)^2$. If we can be sure that step always works, then we can have complete confidence in the machine. Of course, we know from basic algebra that $k^2+2k+1 = (k + 1)^2$, and so we know with complete certainty that this machine will perform its job flawlessly!

The proof boils down to this. It is easy to check the first equation: $1 = 1^2$. We now imagine this equation being fed into the machine (which we proved is flawless) and the machine will prove all the equations on the list. We don't need to wait for the machine to run forever; we know that every equation on the list is going to be proved. Therefore, Proposition 21.1 must be true.

Theoretical Underpinnings

The essence of proof by mathematical induction is embedded in the metaphor of the equation-proving machine. The method is embodied in the following theorem.

Theorem 21.2 **(Principle of Mathematical Induction)** Let A be a set of natural numbers. If

- $0 \in A$, and
- $\forall k \in \mathbb{N},\ k \in A \Longrightarrow k + 1 \in A$,

then $A = \mathbb{N}$.

The two conditions say that (1) 0 is in the set A, and (2) whenever a natural number k is in A, it must be the case that $k + 1$ is also in A. The only way these two conditions can be met is if A is the full set of natural numbers.

First we prove this result, and then we explain how to use it as the central tool of a proof technique.

Proof. Suppose, for the sake of contradiction, that $A \neq \mathbb{N}$. Let $X = \mathbb{N} - A$ (i.e., X is the set of natural numbers not in A). Our supposition that $A \neq \mathbb{N}$ means there is some natural number not in A (i.e., $X \neq \emptyset$).

Since X is a nonempty set of natural numbers, we know that X contains a least element x (Well-Ordering Principle). So x is the smallest natural number not in A.

Note that $x \neq 0$ because we are given that $0 \in A$, so $0 \notin X$. Therefore $x \geq 1$. Thus $x - 1 \geq 0$, so $x - 1 \in \mathbb{N}$. Furthermore, since x is the smallest element not in A, we have $x - 1 \in A$.

Now the second condition of the theorem says that whenever a natural number is in A, so is the next larger natural number. Since $x - 1 \in A$, we know that $(x - 1) + 1 = x$ is in A. But $x \notin A$.$\Rightarrow\Leftarrow$ ■

Proof by Induction

We can use Theorem 21.2 as a proof technique. The general kind of statement we prove by induction can be expressed in the form "Every natural number has a certain property." For example, consider the following:

Proposition 21.3 Let n be a natural number. Then

$$0^2 + 1^2 + 2^2 + \cdots + n^2 = \frac{(2n+1)(n+1)(n)}{6}. \tag{18}$$

The overall outline of the proof is summarized in Proof Template 17. We use this method to prove Proposition 21.3.

Proof Template 17 Proof by induction.

To prove every natural number has *some property*.

Proof.

- Let A be the set of natural numbers for which the result is true.
- Prove that $0 \in A$. This is called the *basis step*. It is usually easy.
- Prove that if $k \in A$, *then* $k + 1 \in A$. This is called the *inductive step*. To do this, we
  - Assume that the result is true for $n = k$. This is called the *induction hypothesis*.
  - Use the induction hypothesis to prove the result is true for $n = k + 1$.
- We invoke Theorem 21.2 to conclude $A = \mathbb{N}$.
- Therefore the result is true for all natural numbers. ■

Proof (of Proposition 21.3)

We prove this result by induction on n. Let A be the set of natural numbers for which Proposition 21.3 is true—that is, those n for which Equation (18) holds.

- **Basis step:** Note that the theorem is true for $n = 0$ because both sides of Equation (18) evaluate to 0.
- **Induction hypothesis:** Suppose the result is true for $n = k$; that is, we may assume

$$0^2 + 1^2 + 2^2 + \cdots + k^2 = \frac{(2k+1)(k+1)(k)}{6}. \tag{19}$$

- Now we need to prove that Equation (18) holds for $n = k + 1$; that is, we need to prove

$$0^2 + 1^2 + 2^2 + \cdots + k^2 + (k+1)^2 = \frac{[2(k+1)+1][(k+1)+1][k+1]}{6}. \tag{20}$$

- To prove Equation (20) from Equation (19), we add $(k + 1)^2$ to both sides of Equation (19):

$$0^2 + 1^2 + 2^2 + \cdots + k^2 + (k+1)^2 = \frac{(2k+1)(k+1)(k)}{6} + (k+1)^2. \tag{21}$$

To complete the proof, we need to show that the right-hand side of Equation (20) equals the right-hand side of Equation (21); that is, we have to prove

$$\frac{(2k+1)(k+1)(k)}{6} + (k+1)^2 = \frac{[2(k+1)+1][(k+1)+1][k+1]}{6}. \quad (22)$$

The verification of Equation (22) is a simple, if mildly painful, algebra exercise that we leave to you (Exercise 21.2).

- We have shown $0 \in A$ and $k \in A \implies (k+1) \in A$. Therefore, by induction (Theorem 21.2), we know that $A = \mathbb{N}$; that is, the proposition is true for all natural numbers. ■

This proof can be described using the machine metaphor. We want to prove all of the following equations:

$$\begin{aligned}
0^2 &= \frac{(2\cdot 0+1)(0+1)(0)}{6} \\
0^2+1^2 &= \frac{(2\cdot 1+1)(1+1)(1)}{6} \\
0^2+1^2+2^2 &= \frac{(2\cdot 2+1)(2+1)(2)}{6} \\
0^2+1^2+2^2+3^2 &= \frac{(2\cdot 3+1)(3+1)(3)}{6} \\
0^2+1^2+2^2+3^2+4^2 &= \frac{(2\cdot 4+1)(4+1)(4)}{6} \\
&\vdots
\end{aligned}$$

So we build a machine that accepts one of these equations in its input tube; the equation entering the machine is assumed to have been proved already. The machine then uses that known equation to verify the next equation on the list. Suppose we know that the machine is absolutely reliable, and whenever one equation is fed into the machine, the next equation on the list will emerge from the machine as verified.

So if we can prove that the machine is completely reliable, all we need to do is feed in the first equation on the list and let the machine churn through the rest. Our job reduces to this: Prove the first equation (which is easy), design the machine, and prove it works.

The design of the machine is not particularly difficult. It simply adds the next term in the long sum to both sides of the equation and checks for equality.

The challenging part is to verify that the machine will always work. For this, we must have to check an algebraic identity, namely

$$\frac{(2k+1)(k+1)(k)}{6} + (k+1)^2 = \frac{[2(k+1)+1][(k+1)+1][k+1]}{6}.$$

In the proof of Proposition 21.3, we explicitly referred to the set A of all natural numbers for which the result is true. As you become more comfortable with proofs by induction, you can omit explicit mention of this set. The important steps in a proof by induction are these:

- Prove the basis case; that is, prove the result is true for $n = 0$.
- Assume the induction hypothesis; that is, assume the result for $n = k$.
- Use the induction hypothesis to prove the next case (i.e., for $n = k + 1$).

Note that in proving the case $n = k + 1$, you should use the fact that the result is true in case $n = k$. If you do not use the induction hypothesis, then either (1) you can write a simpler proof of the result without induction or (2) you have made a mistake.

The basis case is always essential and, thankfully, usually easy. If the result you wish to prove does not cover all natural numbers—say, it covers just the positive integers—then the basis step may begin at a value other than 0.

The induction hypothesis is a seemingly magical tool that makes proving theorems easier. To prove the case $n = k + 1$, not only may you assume the hypotheses of the theorem, but you also may assume the induction hypothesis; this gives you more with which to work.

Proving Equations and Inequalities

Proof by induction takes practice. One common application of this technique is to prove equations and inequalities. Here we present some examples for you to study. You will find that the general outlines of the proofs are the same; the only difference is in some of the algebra. The first two examples are results also proved in Section 12 by the combinatorial method (see Propositions 12.1 and 12.2).

Proposition 21.4 Let n be a positive integer. Then

$$2^0 + 2^1 + \cdots + 2^{n-1} = 2^n - 1.$$

Note that this induction proof begins with $n = 1$ because the Proposition is asserted for positive integers.

Proof. We prove this by induction on n.

Basis step: The case $n = 1$ is true because both sides of the equation, 2^0 and $2^1 - 1$, evaluate to 1.

Induction hypothesis: Suppose the result is true when $n = k$; that is, we assume

$$2^0 + 2^1 + \cdots + 2^{k-1} = 2^k - 1. \tag{23}$$

We must prove that the Proposition is true when $n = k + 1$; that is, we must use Equation (23) to prove

$$2^0 + 2^1 + \cdots + 2^{(k+1)-1} = 2^{k+1} - 1. \tag{24}$$

Note that the left-hand side of Equation (24) can be formed from the left-hand side of Equation (23) by adding the term 2^k. So we add 2^k to both sides of Equation (23) to get

$$2^0 + 2^1 + \cdots + 2^{k-1} + 2^k = 2^k - 1 + 2^k. \tag{25}$$

We need to show that the right-hand side of Equation (25) equals the right-hand side of Equation (24). Fortunately, this is easy:

$$2^k - 1 + 2^k = 2 \cdot 2^k - 1 = 2^{k+1} - 1. \quad (26)$$

Using Equations (24) and (26) gives

$$2^0 + 2^1 + \cdots + 2^{(k+1)-1} = 2^{k+1} - 1$$

which is what we needed to show. ■

As our comfort and confidence in writing proofs by induction grow, we can be a bit terser. The next proof is written in a more compact style.

Proposition 21.5 Let n be a positive integer. Then

$$1 \cdot 1! + 2 \cdot 2! + \cdots + n \cdot n! = (n+1)! - 1.$$

Proof. We prove the result by induction on n.

Basis case: The Proposition is true in the case $n = 1$, because both sides of the equation, $1! \cdot 1$ and $2! - 1$, evaluate to 1.

Induction hypothesis: Suppose the Proposition is true in case $n = k$; that is, we have that

$$1 \cdot 1! + 2 \cdot 2! + \cdots + k \cdot k! = (k+1)! - 1. \quad (27)$$

We need to prove the Proposition for the case $n = k + 1$. To this end, we add $(k+1) \cdot (k+1)!$ to both sides of Equation (27) to give

$$1 \cdot 1! + 2 \cdot 2! + \cdots + k \cdot k! + (k+1) \cdot (k+1)! = (k+1)! - 1 + (k+1) \cdot (k+1)!. \quad (28)$$

The right-hand side of Equation (28) can be manipulated as follows:

$$\begin{aligned} (k+1)! - 1 + (k+1) \cdot (k+1)! &= (1 + k + 1) \cdot (k+1)! - 1 \\ &= (k+2) \cdot (k+1)! - 1 \\ &= (k+2)! - 1 = [(k+1) + 1]! - 1. \end{aligned}$$

Substituting this into Equation (28) gives

$$1 \cdot 1! + 2 \cdot 2! + \cdots + k \cdot k! + (k+1) \cdot (k+1)! = [(k+1) + 1]! - 1.$$ ■

Inequalities can be proved by induction as well. Here is a simple example whose proof is a bit terser still.

Proposition 21.6 Let n be a natural number. Then

$$10^0 + 10^1 + \cdots + 10^n < 10^{n+1}.$$

Proof. The proof is by induction on n. The basis case, when $n = 0$, is clear because $10^0 < 10^1$.

Assume (induction hypothesis) that the result holds for $n = k$; that is, we have

$$10^0 + 10^1 + \cdots + 10^k < 10^{k+1}.$$

To show that the Proposition is true when $n = k + 1$, we add 10^{k+1} to both sides and find

$$\begin{aligned} 10^0 + 10^1 + \cdots + 10^k + 10^{k+1} &< 10^{k+1} + 10^{k+1} \\ &= 2 \cdot 10^k < 10 \cdot 10^k = 10^{k+1}. \end{aligned}$$

Therefore the result holds when $n = k + 1$. ■

Other Examples

With a bit of practice, proving equations and inequalities by induction will become routine. Generally, we manipulate both sides of the given equation (assumed by the induction hypothesis, $n = k$) to demonstrate the next equation ($n = k + 1$). However, other kinds of results can be proved by induction. For example, consider the following:

Proposition 21.7 Let n be a natural number. Then $4^n - 1$ is divisible by 3.

Proof. The proof is by induction on n. The basis case, $n = 0$, is clear since $4^0 - 1 = 1 - 1 = 0$ is divisible by 3.

Suppose (induction hypothesis) that the Proposition is true for $n = k$; that is, $4^k - 1$ is divisible by 3. We must show that $4^{k+1} - 1$ is also divisible by 3.

Note that $4^{k+1} - 1 = 4 \cdot 4^k - 1 = 4(4^k - 1) + 4 - 1 = 4(4^k - 1) + 3$. Since $4^k - 1$ and 3 are both divisible by 3, it follows that $4(4^k - 1) + 3$ is divisible by 3; hence $4^{k+1} - 1$ is divisible by 3. ■

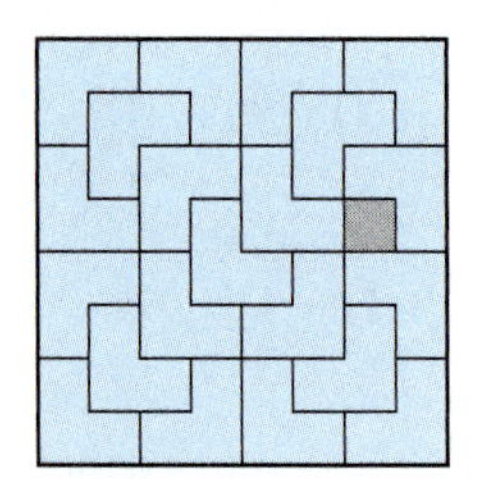

The next example involves some geometry. We wish to cover a chess board with special tiles called *L-shaped triominoes,* or *L-triominoes* for short. These are tiles formed from three 1×1 squares joined at their edges to form an L shape.

It is not possible to tile a standard 8×8 chess board with L-triominoes because there are 64 squares on the chess board and 64 is not divisible by 3. However, it is possible to cover all but one square of the chess board, and such a tiling is shown in the figure.

Is it possible to tile larger chess boards? A $2^n \times 2^n$ chess board has 4^n squares, so, applying Proposition 21.7, we know that $4^n - 1$ is divisible by 3. Hence there is a hope that we may be able to cover all but one of the squares.

Proposition 21.8 Let n be a positive integer. For every square on a $2^n \times 2^n$ chess board, there is a tiling by L-triominoes of the remaining $4^n - 1$ squares.

Proof. The proof is by induction on n. The basis case, $n = 1$, is obvious since placing an L-triomino on a 2×2 chess board covers all but one of the squares, and by rotating the triomino we can select which square is missed.

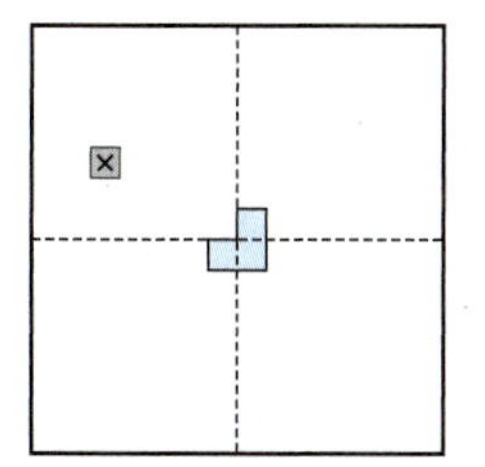

Suppose (induction hypothesis) that the Proposition has been proved for $n = k$.

We are given a $2^{k+1} \times 2^{k+1}$ chess board with one square selected. Divide the board into four $2^n \times 2^n$ subboards (as shown); the selected square must lie in one of these subboards. Place an L-triomino overlapping three corners from the remaining subboards as shown in the diagram.

We now have four $2^k \times 2^k$ subboards each with one square that does not need to be covered. By induction, the remaining squares in the subboards can be tiled by L-triominoes. ■

Strong Induction

Here is a variation on Theorem 21.2.

Theorem 21.9 **(Principle of Mathematical Induction—strong version)** Let A be a set of natural numbers. If

- $0 \in A$ and
- for all $k \in \mathbb{N}$, if $0, 1, 2, \ldots, k \in A$, then $k + 1 \in A$

then $A = \mathbb{N}$.

The proof of this theorem is left to you (see Exercise 21.14).

Why is this called *strong* induction? Suppose you are using induction to prove a proposition. In both standard and strong induction, you begin by showing the basis case ($0 \in A$). In standard induction, you assume the induction hypothesis ($k \in A$; i.e., the proposition is true for $n = k$) and then use that to prove $k + 1 \in A$ (i.e., the proposition is true for $n = k + 1$). Strong induction gives you a stronger induction hypothesis. In strong induction, you may assume $0, 1, 2, \ldots, k \in A$ (the proposition is true for all n from 0 to k) and use that to prove $k + 1 \in A$ (the proposition is true for $n = k + 1$).

This method is outlined in Proof Template 18.

Proof Template 18 Proof by strong induction.

To prove every natural number has *some property:*

Proof.

- Let A be the set of natural numbers for which the result is true.
- Prove that $0 \in A$. This is called the *basis step*. It is usually easy.
- Prove that *if* $0, 1, 2, \ldots, k \in A$, *then* $k+1 \in A$. This is called the *inductive step*. To do this, we
  - Assume that the result is true for $n = 0, 1, 2, \ldots, k$. This is called the *strong induction hypothesis*.
  - Use the strong induction hypothesis to prove the result is true for $n = k + 1$.
- Invoke Theorem 21.9 to conclude $A = \mathbb{N}$.
- Therefore the result is true for all natural numbers. ■

Let us see how to use strong induction and why it gives us more flexibility than standard induction. We illustrate proof by strong induction on a geometry problem.

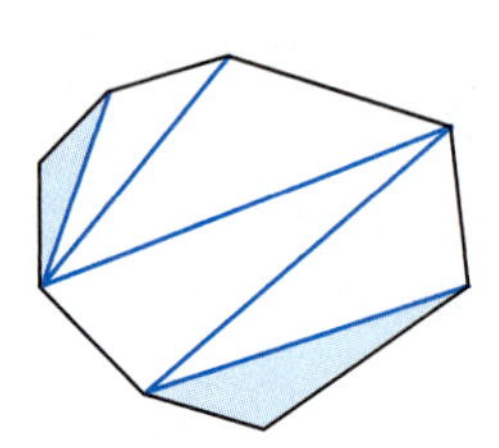

Let P be a polygon in the plane. To *triangulate* a polygon is to draw diagonals through the interior of the polygon so that (1) the diagonals do not cross each other and (2) every region created is a triangle (see the figure). Notice that we have shaded two of the triangles. These triangles are called *exterior* triangles because two of their three sides are on the exterior of the original polygon.

We prove the following result using strong induction.

Proposition 21.10 If a polygon with four or more sides is triangulated, then at least two of the triangles formed are exterior.

Proof. Let n denote the number of sides of the polygon. We prove Proposition 21.10 by strong induction on n.

Basis case: Since this result makes sense only for $n \geq 4$, the basis case is $n = 4$. The only way to triangulate a quadrilateral is to draw in one of the two possible diagonals. Either way, the two triangles formed must be exterior.

Strong induction hypothesis: Suppose Proposition 21.10 has been proved for all polygons on $n = 4, 5, \ldots, k$ sides.

Let P be any triangulated polygon with $k + 1$ sides. We must prove that at least two of its triangles are exterior.

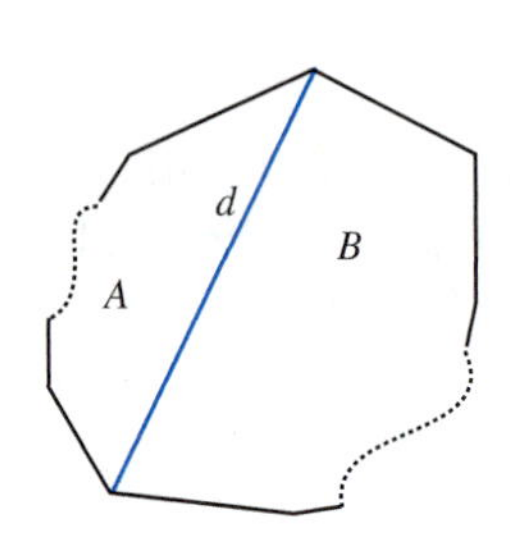

Let d be one of the diagonals. This diagonal separates P into two polygons A and B where (this is the key comment) A and B are triangulated polygons with fewer sides than P. It is possible that one or both of A and B are triangles themselves. We consider the cases where neither, one, or both A and B are triangles.

- *If A is not a triangle*: Then, since A has at least four, but at most k sides, by strong induction we know that two or more of A's triangles are exterior. Now we need to worry: Are the exterior triangles of A really exterior triangles of P? Not necessarily. If one of A's exterior triangles uses the diagonal d, then it is not an exterior triangle of P. Nonetheless, the other exterior triangle of A can't also use the diagonal d, and so at least one exterior triangle of A is also an exterior triangle of P.
- *If B is not a triangle*: As in the previous case, B contributes at least one exterior triangle to P.
- *If A is a triangle*: Then A is an exterior triangle of P.
- *If B is a triangle*: Then B is an exterior triangle of P.

In every case, both A and B contribute at least one exterior triangle to P, and so P has at least two exterior triangles. ■

Strong induction helped us enormously in this proof. When we considered the diagonal d, we did not know the number of sides of the two polygons A and B. All we knew for sure was that they had fewer sides than P. To use ordinary induction, we would need to have chosen a diagonal such that A had k sides and B had three;

in other words, we would have to select B to be an exterior triangle. The problem is that we had not yet proved that a triangulated polygon has an exterior triangle!

Curiously, it is harder to prove that a triangulated polygon has one exterior triangle than to prove that a triangulated polygon has two exterior triangles! See Exercise 21.13.

Strong induction gives more flexibility than standard induction because the induction hypothesis lets you assume more. It is probably best not to write your proof in the style of strong induction when standard induction suffices. In the cases where you need to use strong induction, you also have proof by smallest counterexample as an alternative.

A More Complicated Example

Fibonacci numbers were introduced in Definition 20.12.

We prove the following result by strong induction. The hard part of this example is keeping track of the many binomial coefficients. The overall structure of the proof is no different from the proof of Proposition 21.10. We follow Proof Template 18.

Proposition 21.11 Let $n \in \mathbb{Z}$ and let F_n denote the nth Fibonacci number. Then

$$\binom{n}{0} + \binom{n-1}{1} + \binom{n-2}{2} + \cdots + \binom{0}{n} = F_n. \tag{29}$$

Note that the last several terms in the sum are all zero. Eventually the lower index in the binomial coefficient will exceed the upper index, and all terms from that point on are zero. For example,

$$\begin{aligned}\tbinom{7}{0} + \tbinom{6}{1} + \tbinom{5}{2} + \tbinom{4}{3} + \tbinom{3}{4} + \tbinom{2}{5} + \tbinom{1}{6} + \tbinom{0}{7} &= 1+6+10+4+0+0+0+0 \\ &= 21 = F_7.\end{aligned}$$

In fancy notation,

$$\sum_{j=0}^{n} \binom{n-j}{j} = F_n.$$

Before we present the formal proof of Proposition 21.11, let us look to see why this might be true and why we need strong induction.

In general, to prove that some expression gives a Fibonacci number, we use the fact that $F_n = F_{n-1} + F_{n-2}$. If we know that the expression works for F_{n-1} and F_{n-2}, then we can add the appropriate expressions and hope we get F_n. In ordinary induction, we can only assume the immediate smaller case of the result; here we need the two previous values, and strong induction allows us to do this.

Let's see how we can apply this to Proposition 21.11 by examining the case $n = 8$. We want to prove

$$F_8 = \binom{8}{0} + \binom{7}{1} + \cdots + \binom{4}{4}.$$

We do this by assuming

$$F_6 = \binom{6}{0} + \binom{5}{1} + \binom{4}{2} + \binom{3}{3} \quad \text{and}$$

$$F_7 = \binom{7}{0} + \binom{6}{1} + \binom{5}{2} + \binom{4}{3}.$$

We want to add these equations because $F_8 = F_7 + F_6$. The idea is to interleave the terms from the two expressions:

$$F_7 + F_6 = \binom{7}{0} + \binom{6}{0} + \binom{6}{1} + \binom{5}{1} + \binom{5}{2} + \binom{4}{2} + \binom{4}{3} + \binom{3}{3}$$

Now we can use Pascal's identity (Theorem 16.10) to combine pairs of terms:

$$\binom{6}{0} + \binom{6}{1} = \binom{7}{1} \qquad \binom{5}{1} + \binom{5}{2} = \binom{6}{2} \qquad \binom{4}{2} + \binom{4}{3} = \binom{5}{3}$$

We can therefore combine every other term to get

$$\begin{aligned} F_7 + F_6 &= \binom{7}{0} + \left[\binom{6}{0} + \binom{6}{1}\right] + \left[\binom{5}{1} + \binom{5}{2}\right] + \left[\binom{4}{2} + \binom{4}{3}\right] + \binom{3}{3} \\ &= \binom{7}{0} + \binom{7}{1} + \binom{6}{2} + \binom{5}{3} + \binom{3}{3}. \end{aligned}$$

We are nearly finished. Notice that the $\binom{7}{0}$ term should be $\binom{8}{0}$ and the $\binom{3}{3}$ term should be $\binom{4}{4}$. The good news is that these terms both equal 1, so we can replace what we have by what we want to finish this example:

$$\begin{aligned} F_7 + F_6 &= \binom{7}{0} + \binom{6}{0} + \binom{6}{1} + \binom{5}{1} + \binom{5}{2} + \binom{4}{2} + \binom{4}{3} + \binom{3}{3} \\ &= \binom{7}{0} + \binom{7}{1} + \binom{6}{2} + \binom{5}{3} + \binom{3}{3} \\ &= \binom{8}{0} + \binom{7}{1} + \binom{6}{2} + \binom{5}{3} + \binom{4}{4}. \end{aligned}$$

The case $F_9 = F_8 + F_7$ is similar, but there are some minor differences. It is important that you write out the steps for this case yourself before reading the proof. Be sure you see what the differences are between these two cases.

Proof (of Proposition 21.11)

We use strong induction.

Basis case: The result is true for $n = 0$; Equation (29) reduces to $\binom{0}{0} = 1 = F_1$, which is true. Notice that the result is also true for $n = 1$ since $\binom{1}{0} + \binom{0}{1} = 1 + 0 = 1 = F_1$.

Strong induction hypothesis: Proposition 21.11 is true for all values of n from 0 to k. (We may also assume $k \geq 1$ since we have already proved the result for $n = 0$ and $n = 1$.)

We seek to prove Equation (29) in the case $n = k + 1$; that is, we want to prove

$$F_{k+1} = \binom{k+1}{0} + \binom{k}{1} + \binom{k-1}{2} + \cdots.$$

By the strong induction hypothesis, we know the following two equations are true:

$$F_{k-1} = \binom{k-1}{0} + \binom{k-2}{1} + \binom{k-3}{2} + \cdots$$
$$F_k = \binom{k}{0} + \binom{k-1}{1} + \binom{k-2}{2} + \cdots.$$

We add these two lines to get

$$\begin{aligned} F_{k+1} &= F_k + F_{k-1} \\ &= \binom{k}{0} + \binom{k-1}{0} + \binom{k-1}{1} + \binom{k-2}{1} + \binom{k-2}{2} + \binom{k-3}{2} + \cdots. \end{aligned}$$

The next step is to combine terms with the same upper index using Pascal's identity (Theorem 16.10). First, we are going to worry about where this long sum ends.

In the case k is even, the sum ends

$$F_{k+1} = \cdots + \binom{\frac{k}{2}+1}{\frac{k}{2}-2} + \binom{\frac{k}{2}+1}{\frac{k}{2}-1} + \binom{\frac{k}{2}}{\frac{k}{2}-1} + \binom{\frac{k}{2}}{\frac{k}{2}}$$

and in the case k is odd, it ends

$$F_{k+1} = \cdots + \binom{\frac{1}{2}(k-1)+1}{\frac{1}{2}(k-1)-1} + \binom{\frac{1}{2}(k-1)+1}{\frac{1}{2}(k-1)} + \binom{\frac{1}{2}(k-1)}{\frac{1}{2}(k-1)}.$$

Now we apply Pascal's identity, combining those pairs of terms with the same upper entry (each black term and the color term that follows).

In the case k is even, we have

$$\begin{aligned} F_{k+1} &= \binom{k}{0} + \left[\binom{k}{1} + \binom{k-1}{2} + \cdots + \binom{\frac{k}{2}+2}{\frac{k}{2}-1} + \binom{\frac{k}{2}+1}{\frac{k}{2}}\right] \\ &= \binom{k+1}{0} + \left[\binom{k}{1} + \binom{k-1}{2} + \cdots + \binom{\frac{k}{2}+2}{\frac{k}{2}-1} + \binom{\frac{k}{2}+1}{\frac{k}{2}}\right] \end{aligned}$$

and in the case k is odd, we have

$$\begin{aligned} F_{k+1} &= \binom{k}{0} + \left[\binom{k}{1} + \binom{k-1}{2} + \cdots + \binom{\frac{1}{2}(k-1)+2}{\frac{1}{2}(k-1)}\right] + \binom{\frac{1}{2}(k-1)}{\frac{1}{2}(k-1)} \\ &= \binom{k+1}{0} + \left[\binom{k}{1} + \binom{k-1}{2} + \cdots + \binom{\frac{1}{2}(k-1)+2}{\frac{1}{2}(k-1)}\right] + \binom{\frac{1}{2}(k+1)}{\frac{1}{2}(k+1)}. \end{aligned}$$

In both cases, we have verified Equation (29) with $n = k + 1$, completing the proof. ■

The most difficult part of this proof was dealing with the upper and lower indices of the binomial coefficients.

A Matter of Style

Proof by induction and proof by smallest counterexample are usually interchangeable. I prefer, however, proof by smallest counterexample. This is mostly a stylistic preference, but there is a mathematical reason to prefer the smallest-counterexample technique. When mathematicians try to prove statements, they may believe that the statement is true, but they don't *know*—until they have a proof—whether or not the statement is true. We often alternate between trying to prove the statement and trying to find a counterexample. One way to do both activities simultaneously is to try to deduce properties a smallest counterexample might have. In this way, we either reach a contradiction (and then we have a proof of the statement) or we learn enough about how the counterexample should behave to construct a counterexample.

Recap

Proof by induction is an alternative method to proof by smallest counterexample. The first step in a proof by induction is to prove a basis case (often that the result you want to prove is true for $n = 0$). In standard induction, we make an induction hypothesis (the proposition is true when $n = k$) and use it to prove the next case (the proposition is true when $n = k + 1$). Strong induction is similar, but the strong-induction hypothesis is that the proposition is true for $n = 0, 1, 2, \ldots, k$.

Any result you prove by induction (standard or strong) can just as well be proved using the smallest-counterexample method. Induction proofs are more popular.

21 Exercises

21.1. Induction is often likened to climbing a ladder. If you can master the following two skills, then you can climb a ladder: (1) get your foot on the first rung and (2) advance from one rung to the next.

Explain why both parts (1) and (2) are necessary, and explain what this has to do with induction.

21.2. Prove Equation (22).

21.3. Prove the following by induction. In each case, n is a positive integer.

a. $1 + 4 + 7 + \cdots + (3n - 2) = \frac{n(3n-1)}{2}$.

b. $1^3 + 2^3 + \cdots + n^3 = \frac{n^2(n+1)^2}{4}$.

c. $9 + 9 \times 10 + 9 \times 100 + \cdots + 9 \times 10^{n-1} = 10^n - 1$.

d. $\frac{1}{1\cdot 2} + \frac{1}{2\cdot 3} + \cdots + \frac{1}{n(n+1)} = 1 - \frac{1}{n+1}$.

21.4. Prove the following by induction. In each case, n is a positive integer.

a. $2^n \le 2^{n+1} - 2^{n-1} - 1$.

b. $(1 - \frac{1}{2})(1 - \frac{1}{4})(1 - \frac{1}{8}) \cdots (1 - \frac{1}{2^n}) \ge \frac{1}{4} + \frac{1}{2^{n+1}}$.

c. $1 + \frac{1}{2} + \frac{1}{3} + \frac{1}{4} + \cdots + \frac{1}{2^n} \ge 1 + \frac{n}{2}$.

21.5. A group of people stand in line to purchase movie tickets. The first person in line is a woman and the last person in line is a man. Use proof by induction to show that somewhere in the line a woman is directly in front of a man.

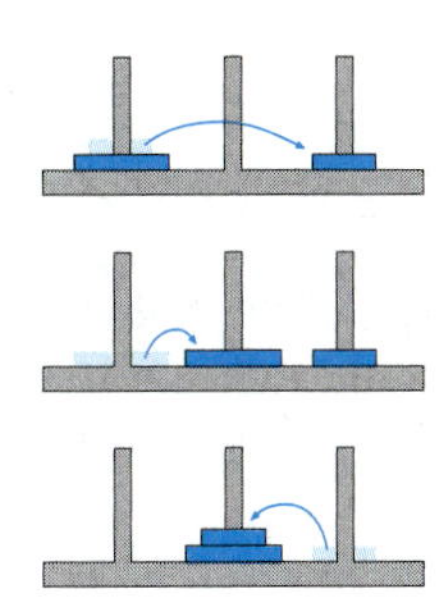

21.6. The *Tower of Hanoi* is a puzzle consisting of a board with three dowels and a collection of n disks of n different sizes (radii). The disks have holes drilled through their centers so that they can fit on the dowels on the board. Initially, all the disks are on the first dowel and are arranged in size order (from the largest on the bottom to the smallest on the top).

The object is to move all the disks to another dowel in as few moves as possible. Each move consists of taking the top disk off one of the stacks and placing it on another stack, with the added condition that you may not place a larger disk atop a smaller one. The figure shows how to solve the Tower of Hanoi in three moves when $n = 2$.

Prove: For every positive integer n, the Tower of Hanoi puzzle (with n disks) can be solved in $2^n - 1$ moves.

21.7. Let $A_1, A_2, \ldots, A_n$ be sets (where $n \geq 2$). Suppose for any two sets A_i and A_j either $A_i \subseteq A_j$ or $A_j \subseteq A_i$.

Prove by induction that one of these n sets is a subset of all of them.

The intimate connection between recursive definition and proof by induction.

21.8. May a word be used in its own definition? Generally, the answer is no. However, in Definition 20.12, we defined the Fibonacci numbers as the sequence $F_0, F_1, F_2, \ldots$ by setting $F_0 = 1$, $F_1 = 1$, and for $n \geq 2$, $F_n = F_{n-1} + F_{n-2}$. Notice that we defined Fibonacci numbers in terms of themselves! This works because we have defined F_n in terms of previously defined Fibonacci numbers. This type of definition is called a *recursive* definition.

Recursive definitions bear a strong resemblance to proofs by induction. There are typically one or a few basis cases, and then the rest of the definition refers back to smaller cases (this is like the inductive step in a proof by induction).

Induction is the proof technique of choice to prove statements about recursively defined concepts.

The following sequences of numbers are recursively defined. Answer the questions asked.

a. Let $a_0 = 1$ and, for $n > 0$, let $a_n = 2a_{n-1} + 1$. The first few terms of the sequence $a_0, a_1, a_2, a_3, \ldots$ are $1, 3, 7, 15, \ldots$.
What are the next three terms?
Prove: $a_n = 2^{n+1} - 1$.

b. Let $b_0 = 1$ and, for $n > 0$, let $b_n = 3b_{n-1} - 1$.
What are the first five terms of the sequence $b_0, b_1, b_2, \ldots$?
Prove: $b_n = \frac{3^n+1}{2}$.

c. Let $c_0 = 3$ and, for $n > 0$, let $c_n = c_{n-1} + n$.
What are the first five terms of the sequence $c_0, c_1, c_2, \ldots$?
Prove: $c_n = \frac{n^2+n+6}{2}$.

d. Let $d_0 = 2$, $d_1 = 5$ and, for $n > 1$, let $d_n = 5d_{n-1} - 6d_{n-2}$.
Why did we give two basis definitions?
What are the first five terms of the sequence $d_0, d_1, d_2, \ldots$?
Prove: $d_n = 2^n + 3^n$.

e. Let $e_0 = 1$, $e_1 = 4$ and, for $n > 1$, let $e_n = 4(e_{n-1} - e_{n-2})$.
What are the first five terms of the sequence $e_0, e_1, e_2, \ldots$?
Prove: $e_n = (n + 1)2^n$.

f. Let F_n denote the nth Fibonacci number. Prove:

$$F_n = \frac{\left(\frac{1+\sqrt{5}}{2}\right)^{n+1} - \left(\frac{1-\sqrt{5}}{2}\right)^{n+1}}{\sqrt{5}}.$$

21.9. A flagpole is n feet tall. On this pole we display flags of the following types: red flags that are 1 foot tall, blue flags that are 2 feet tall, and green flags that are 2 feet tall. The sum of the heights of the flags is exactly n feet.

Prove that there are $\frac{2}{3}2^n + \frac{1}{3}(-1)^n$ ways to display the flags.

21.10. Prove that every positive integer can be expressed as the sum of distinct Fibonacci numbers.

For example, $20 = 2+5+13$ where 2, 5, 13 are, of course, Fibonacci numbers. Although we can write $20 = 2+5+5+8$, this does not illustrate the result because we have used 5 twice.

21.11. Consider the following computer program.

```
function findMax(array, first, last) {
   if (first == last) return array[first];
   mid = first + (last-first)/2;
   a = findMax(array,first,mid);
   b = findMax(array,mid+1,last);
   if (a<b) return b;
   return a;
}
```

Here `array` is an array of integers. All other variables are integers. We assume that `first` and `last` are between 1 and the number of elements in `array` and that `first` $\le$ `last`.

The purpose of this program is to find the largest value in the array between two indices; that is, it should return the largest value of `array[first]`, `array[first+1]`, ..., `array[last]`.

Your job: Prove that this program fulfills this task.

[*Technical note*: If `last-first` is odd, then `(last-first)/2` is rounded down to the nearest integer. For example, if `first` is 7 and `last` is 20, then `(last-first)/2` is 6.]

21.12. Consider the following computer program.

```
function lookUp(array, first, last, key) {
   mid = first + (last-first)/2;
   if (array[mid] == key) return mid;
   if (array[mid] > key) return lookUp(array,first,mid-1,key);
   return lookUp(array,mid+1,last,key);
}
```

Here `array` is an array of integers; all other variables represent integers. The values stored in `array` are sorted; that is, we know that

$$\texttt{array[1]} < \texttt{array[2]} < \texttt{array[3]} < \cdots.$$

We also know that $1 \le$ `first` $\le$ `last` and that there is some index j between `first` and `last` for which `array[`j`]` is equal to `key`.

Prove that this program finds that index j.

21.13. Try to prove, using strong or standard induction, that a triangulated polygon has at least one exterior triangle.

What goes wrong when you try to do your proof?

The harder theorem ("... has at least two exterior triangles") is easier to prove than the easier theorem ("... has at least one exterior triangle"). This phenomenon is known as induction loading.

21.14. Prove Theorem 21.9.

21.15. Prove, using strong induction, that every natural number can be expressed as the sum of distinct powers of 2. For example, $21 = 2^4 + 2^2 + 2^0$.

21.16. We showed how to prove the Principle of Mathematical Induction (Theorem 21.2) by use of the Well-Ordering Principle. Now do the opposite. Use induction to prove the Well-Ordering Principle (Statement 20.6).

22 Recurrence Relations

Proposition 21.3 gives a formula for the sum of the squares of the natural numbers up to n:

$$0^2 + 1^2 + 2^2 + \cdots + n^2 = \frac{(2n+1)(n+1)(n)}{6}.$$

How did we derive this formula?

In Exercise 21.8d you were told that a sequence of numbers, $d_0, d_1, d_2, d_3, \ldots$ satisfies the conditions $d_0 = 2$, $d_1 = 5$, and $d_n = 5d_{n-1} - 6d_{n-2}$ and you were asked to prove that $d_n = 2^n + 3^n$. More dramatically, in the same problem, you were asked to prove the following complicated expression for the nth Fibonacci number:

$$F_n = \frac{\left(\frac{1+\sqrt{5}}{2}\right)^{n+1} - \left(\frac{1-\sqrt{5}}{2}\right)^{n+1}}{\sqrt{5}}.$$

How did we create these formulas?

In this section we present methods for solving a *recurrence relation*: a formula that specifies how each term of a sequence is produced from earlier terms.

For example, consider a sequence $a_0, a_1, a_2, \ldots$ defined by

$$a_n = 3a_{n-1} + 4a_{n-2}, \quad a_0 = 3, \quad a_1 = 2.$$

We can now compute a_2 in terms of a_0 and a_1, and then a_3 in terms of a_2 and a_1, and so on:

$$\begin{aligned} a_2 &= 3a_1 + 4a_0 = 3 \times 2 + 4 \times 3 = 18 \\ a_3 &= 3a_2 + 4a_1 = 3 \times 18 + 4 \times 2 = 62 \\ a_4 &= 3a_3 + 4a_2 = 3 \times 62 + 4 \times 18 = 258. \end{aligned}$$

Our goal is to have a simple method to convert the recurrence relation into an explicit formula for the nth term of the sequence. In this case, $a_n = 4^n + 2 \cdot (-1)^n$.

First-Order Recurrence Relations

The recurrence relations with which we begin are called *first order* because a_n can be expressed just in terms of the immediate previous element of the sequence, a_{n-1}. Because the first term of the sequence is a_0, it is not meaningful to speak of the term a_{-1}. Therefore, the recurrence relation holds only for $n \geq 1$. The value of a_0 must be given separately.

The simplest recurrence relation is $a_n = a_{n-1}$. Each term is exactly equal to the one before it, so every term is equal to the initial term, a_0.

Let's try something only slightly more difficult. Consider the recurrence relation $a_n = 2a_{n-1}$. Here, every term is twice as large as the previous term. We also need to give the initial term—say, $a_0 = 5$. Then the sequence is 5, 10, 20, 40, 80, 160, ... It's easy to write down a formula for the nth term of this sequence: $a_n = 5 \times 2^n$.

More generally, if the recurrence relation is

$$a_n = sa_{n-1}$$

then each term is just s times the previous term. Given a_0, then nth term of this sequence is

$$a_n = a_0 s^n.$$

Let's consider a more complicated example. Suppose we define a sequence by

$$a_n = 2a_{n-1} + 3, \quad a_0 = 1.$$

When we calculate the first several terms of this, sequence we find the following values:

$$1, \quad 5, \quad 13, \quad 29, \quad 61, \quad 125, \quad 253, \quad 509, \quad \ldots$$

Because the recurrence relation involves doubling each term, we might suspect that powers of 2 are present in the formula. With this in mind, if we stare at the sequence of values, we might realize that each term is 3 less than a power of 2. We can rewrite the sequence like this:

$$4-3, \quad 8-3, \quad 16-3, \quad 32-3, \quad 64-3, \quad 128-3, \quad 256-3, \quad 512-3, \quad \ldots$$

With this, we obtain $a_n = 4 \times 2^n - 3$.

Unfortunately, "stare and hope you recognize" is not a guaranteed procedure. Let's try to analyze this recurrence relation again in a more systematic fashion.

We begin with the recurrence $a_n = 2a_{n-1} + 3$ but leave the initial term a_0 unspecified for the moment. We derive an expression for a_1 in terms of a_0 using the recurrence relation:

$$a_1 = 2a_0 + 3.$$

Next, let's find an expression for a_2. We know that $a_2 = 2a_1 + 3$, and we have an expression for a_1 in terms of a_0. Combining these, we get

$$a_2 = 2a_1 + 3 = 2(2a_0 + 3) + 3 = 4a_0 + 9.$$

Now that we have a_2, we work out an expression for a_3 in terms of a_0:

$$a_3 = 2a_2 + 3 = 2(4a_0 + 9) + 3 = 16a_0 + 21.$$

Here are the first several terms:

$$\begin{aligned}
a_0 &= a_0 \\
a_1 &= 2a_0 + 3 \\
a_2 &= 4a_0 + 9 \\
a_3 &= 8a_0 + 21 \\
a_4 &= 16a_0 + 45 \\
a_5 &= 32a_0 + 93 \\
a_6 &= 64a_0 + 189.
\end{aligned}$$

One part of this pattern is obvious: a_n can be written as $2^n a_0$ plus something. It's the "plus something" that is still a mystery. We can try staring at the extra terms $0, 3, 9, 21, 45, 93, 189, \ldots$ in the hope of finding a pattern, but we don't want to resort to that. Instead, let's trace out how the term $+189$ was created in a_6. We calculated a_6 from a_5:

$$a_6 = 2a_5 + 3 = 2(32a_0 + 93) + 3$$

so the $+189$ term comes from $2 \times 93 + 3$. Where did the 93 term come from? Let's trace these terms back to the beginning:

$$\begin{aligned}
189 &= 2 \times 93 + 3 \\
&= 2 \times (2 \times 45 + 3) + 3 \\
&= 2 \times (2 \times (2 \times 21 + 3) + 3) + 3 \\
&= 2 \times (2 \times (2 \times (2 \times 9 + 3) + 3) + 3) + 3 \\
&= 2 \times (2 \times (2 \times (2 \times (2 \times 3 + 3) + 3) + 3) + 3) + 3.
\end{aligned}$$

Now let's rewrite the last term as follows:

$$\begin{aligned}
&2 \times (2 \times (2 \times (2 \times (2 \times 3 + 3) + 3) + 3) + 3) + 3 \\
&= 2^5 \times 3 + 2^4 \times 3 + 2^3 \times 3 + 2^2 \times 3 + 2^1 \times 3 + 2^0 \times 3 \\
&= (2^5 + 2^4 + 2^3 + 2^2 + 2^1 + 2^0) \times 3 \\
&= (2^6 - 1) \times 3 = 63 \times 3 = 189
\end{aligned}$$

Based on what we have learned, we predict a_7 to be

$$a_7 = 128a_0 + (2^7 - 1) \times 3 = 2^7(a_0 + 3) - 3 = 128a_0 + 381$$

and this is correct.

We are now ready to conjecture the solution to the recurrence relation $a_n = 2a_{n-1} + 3$. It is

$$a_n = (a_0 + 3)\,2^n - 3.$$

Once we have the formula in hand, it is easy to prove it is correct using induction. However, we don't want to go through all that work every time we need to solve a recurrence relation; we want a much simpler method. We seek a ready-made answer to a recurrence relation of the form

$$a_n = sa_{n-1} + t$$

where s and t are given numbers. Based on our experience with the recurrence $a_n = 2a_{n-1} + 3$, we are in a position to guess that the formula for a_n will be of the following form:

$$a_n = (\text{a number}) \times s^n + (\text{a number}).$$

Let's see that this is correct by finding a_1, a_2, etc., in terms of a_0:

$$\begin{aligned}
a_0 &= a_0 \\
a_1 &= sa_0 + t \\
a_2 &= sa_1 + t = s(sa_0 + t) + t = s^2 a_0 + (s+1)t \\
a_3 &= sa_2 + t = s\left(s^2 a_0 + (s+1)t\right) + t = s^3 a_0 + (s^2 + s + 1)t \\
a_4 &= sa_3 + t = s\left(s^3 a_0 + (s^2 + s + 1)t\right) = s^4 a_0 + (s^3 + s^2 + s + 1)t.
\end{aligned}$$

Continuing with this pattern, we see that

$$a_n = s^n a_0 + (s^{n-1} + s^{n-2} + \cdots + s + 1)t.$$

We can simplify this by noticing that $s^{n-1} + s^{n-2} + \cdots + s + 1$ is a geometric series whose sum is

$$\frac{s^n - 1}{s - 1}$$

provided $s \neq 1$ (a case with which we will deal separately). We can now write

$$a_n = a_0 s^n + \left(\frac{s^n - 1}{s - 1}\right) t$$

or, collecting the s^n terms, we have

$$a_n = \left(a_0 + \frac{t}{s-1}\right) s^n - \frac{t}{s-1}. \tag{30}$$

Despite the precise nature of Equation (30), I prefer expressing the answer as in the following result because it is easier to remember and just as useful.

Proposition 22.1 All solutions to the recurrence relation $a_n = sa_{n-1} + t$ where $s \neq 1$ have the form

$$a_n = c_1 s^n + c_2$$

where c_1 and c_2 are specific numbers.

Let's see how to apply Proposition 22.1.

Example 22.2 Solve the recurrence $a_n = 5a_{n-1} + 3$ where $a_0 = 1$.

Solution: We have $a_n = c_1 5^n + c_2$. We need to find c_1 and c_2. Note that

$$\begin{aligned}
a_0 &= 1 = c_1 + c_2 \\
a_1 &= 8 = 5c_1 + c_2.
\end{aligned}$$

Solving these equations, we find $c_1 = \frac{7}{4}$ and $c_2 = -\frac{3}{4}$, and so

$$a_n = \frac{7}{4} \cdot 5^n - \frac{3}{4}.$$

We have a small bit of unfinished business: the case $s = 1$. Fortunately this case is easy. The recurrence relation is of the form

$$a_n = a_{n-1} + t$$

where t is some number. It's easy to write down the first few terms of this sequence and see the result:

$$\begin{aligned}
a_0 &= a_0 \\
a_1 &= a_0 + t \\
a_2 &= a_1 + t = (a_0 + t) + t = a_0 + 2t \\
a_3 &= a_2 + t = (a_0 + 2t) + t = a_0 + 3t \\
a_4 &= a_3 + t = (a_0 + 3t) + t = a_0 + 4t.
\end{aligned}$$

See the pattern? In retrospect, it's pretty obvious.

Proposition 22.3 The solution to the recurrence relation $a_n = a_{n-1} + t$ is

$$a_n = a_0 + nt.$$

Second-Order Recurrence Relations

In a second-order recurrence relation, a_n is specified in terms of a_{n-1} and a_{n-2}. Since the sequence begins with a_0, the recurrence relation is valid for $n \geq 2$. The values of a_0 and a_1 must be given separately.

A second-order recurrence relation gives each term of a sequence in terms of the previous two terms. Consider, for example, the recurrence

$$a_n = 5a_{n-1} - 6a_{n-2}. \tag{31}$$

(This is the recurrence from Exercise 21.8d.) Let us ignore the fact that we already know a solution to this recurrence and do some creative guesswork. A first-order recurrence, $a_n = sa_{n-1}$ has a solution that's just powers of s. Perhaps such a solution is available for Equation (31). We can try $a_n = 5^n$ or perhaps $a_n = 6^n$, but let's hedge our bets and guess a solution of the form $a_n = r^n$ for some number r. We'll substitute this into Equation (31) and hope for the best. Here goes:

$$a_n = 5a_{n-1} - 6a_{n-2} \qquad \Rightarrow \qquad r^n = 5r^{n-1} - 6r^{n-2}$$

Dividing this through by r^{n-2} gives

$$r^2 = 5r - 6$$

a simple quadratic equation. We can solve this as follows:

$$r^2 = 5r - 6 \qquad \Rightarrow \qquad 0 = r^2 - 5r + 6 = (r-2)(r-3) \qquad \Rightarrow \qquad r = 2, 3.$$

This suggests that both 2^n and 3^n are solutions to Equation (31). To see that this is correct, we simply have to check whether 2^n (or 3^n) works in the recurrence. That is, we have to check whether $2^n = 5 \cdot 2^{n-1} - 6 \cdot 2^{n-2}$ (and likewise for 3^n). Here are the proofs:

$$\begin{aligned} 5 \cdot 2^{n-1} - 6 \cdot 2^{n-2} &= 5 \cdot 2^{n-1} - 3 \cdot 2 \cdot 2^{n-2} \\ &= 5 \cdot 2^{n-1} - 3 \cdot 2^{n-1} \\ &= (5-3) \cdot 2^{n-1} = 2^n \end{aligned}$$

$$\begin{aligned} 5 \cdot 3^{n-1} - 6 \cdot 3^{n-2} &= 5 \cdot 3^{n-1} - 2 \cdot 3 \cdot 3^{n-2} \\ &= 5 \cdot 3^{n-1} - 2 \cdot 3^{n-1} \\ &= (5-2) \cdot 3^{n-1} = 3^n. \end{aligned}$$

We have shown that 2^n and 3^n are solutions to Equation (31). Are there other solutions? Here are two interesting observations.

First, if a_n is a solution to Equation (31), so is ca_n where c is any specific number. To see why, we calculate

$$ca_n = c\,(5a_{n-1} - 6a_{n-2}) = 5(ca_{n-1}) - 6(ca_{n-2}).$$

Since 2^n is a solution to (31), so is $5 \cdot 2^n$.

Second, if a_n and a'_n are both solutions to Equation (31), then so is $a_n + a'_n$. To see why, we calculate:

$$a_n + a'_n = (5a_{n-1} - 6a_{n-2}) + (5a'_{n-1} - 6a'_{n-2}) = 5(a_{n-1} + a'_{n-1}) - 6(a_{n-2} + a'_{n-2}).$$

Since 2^n and 3^n are solutions to Equation (31), so is $2^n + 3^n$.

Based on this analysis, any expression of the form $c_1 2^n + c_2 3^n$ is a solution to Equation (31). Are there any others? The answer is no; let's see why.

We are given that $a_n = 5a_{n-1} - 6a_{n-2}$. Once we have set specific values for a_0 and a_1, $a_2, a_3, a_4, \ldots$ are all determined. If we are given a_0 and a_1, we can set up the equations

$$\begin{aligned} a_0 &= c_1 2^0 + c_2 3^0 = c_1 + c_2 \\ a_1 &= c_1 2^1 + c_2 3^1 = 2c_1 + 3c_2 \end{aligned}$$

and solve these for c_1, c_2 to get

$$\begin{aligned} c_1 &= 3a_0 - a_1 \\ c_2 &= -2a_0 + a_1. \end{aligned}$$

Thus, any solution to Equation (31) can be expressed as

$$a_n = (3a_0 - a_1)\,2^n + (-2a_0 + a_1)\,3^n.$$

Encouraged by this success, we are prepared to tackle the general problem

$$a_n = s_1 a_{n-1} + s_2 a_{n-2} \tag{32}$$

where s_1 and s_2 are given numbers.

There is a rough edge in this calculation; since we are dividing by r^{n-2} this analysis is faulty in the case $r = 0$. However, this is not a problem because we check our work in a moment by a different method.

We guess a solution of the form $a_n = r^n$, substitute into Equation (32), and hope for the best:

$$\begin{aligned} a_n &= s_1 a_{n-1} + s_2 a_{n-2} \\ r^n &= s_1 r^{n-1} + s_2 r^{n-2} \\ \Rightarrow \qquad r^2 &= s_1 r + s_2 \end{aligned}$$

so the r we seek is a root of the quadratic equation $x^2 - s_1 x - s_2 = 0$. Let's record this as a proposition.

Proposition 22.4 Let s_1, s_2 be given numbers and suppose r is a root of the quadratic equation $x^2 - s_1 x - s_2 = 0$. Then $a_n = r^n$ is a solution to the recurrence relation $a_n = s_1 a_{n-1} + s_2 a_{n-2}$.

Proof. Let r be a root of $x^2 - s_1 x - s_2 = 0$ and observe

$$\begin{aligned} s_1 r^{n-1} + s_2 r^{n-2} &= r^{n-2}(s_1 r + s_2) \\ &= r^{n-2} r^2 \qquad \text{because } r^2 = s_1 r + s_2 \\ &= r^n. \end{aligned}$$

Therefore r^n satisfies the recurrence $a_n = s_1 a_{n-1} + s_2 a_{n-2}$. ■

We're now in a good position to derive the general solution to Equation (32). As we saw with Equation (31), if a_n is a solution to (32), then so is any constant multiple of a_n—that is, ca_n. Also, if a_n and a_n' are two solutions to (32), then so is their sum $a_n + a_n'$.

Therefore, if r_1 and r_2 are roots of the polynomial $x^2 - s_1 x - s_2 = 0$, then

$$a_n = c_1 r_1^n + c_2 r_2^n$$

is a solution to Equation (32).

Are these all the possible solutions? The answer is yes in most cases. Let's see what works and where we run into some trouble.

The expression $c_1 r_1^n + c_2 r_2^n$ gives all solutions to (32) provided it can produce a_0 and a_1; if we can choose c_1 and c_2 so that

$$\begin{aligned} a_0 &= c_1 r_1^0 + c_2 r_2^0 = c_1 + c_2 \\ a_1 &= c_1 r_1^1 + c_2 r_2^1 = r_1 c_1 + r_2 c_2 \end{aligned}$$

then every possible sequence that satisfies (32) is of the form $c_1 r_1^n + c_2 r_2^n$. So all we have to do is solve those equations for c_1 and c_2. When we do, we get this:

$$c_1 = \frac{a_1 - a_0 r_2}{r_1 - r_2} \qquad \text{and} \qquad c_2 = \frac{-a_1 + a_0 r_1}{r_1 - r_2}.$$

All is well unless $r_1 = r_2$; we'll deal with this difficulty in a moment. First, let's write down what we know so far.

Theorem 22.5 Let s_1, s_2 be numbers and let r_1, r_2 be roots of the equation $x^2 - s_1 x - s_2 = 0$. If $r_1 \neq r_2$, then every solution to the recurrence

$$a_n = s_1 a_{n-1} + s_2 a_{n-2}$$

is of the form

$$a_n = c_1 r_1^n + c_2 r_2^n.$$

Example 22.6 Find the solution to the recurrence relation

$$a_n = 3a_{n-1} + 4a_{n-2}, \quad a_0 = 3, \quad a_1 = 2.$$

Solution: Using Theorem 22.5, we find the roots of the quadratic equation $x^2 - 3x - 4 = 0$. This polynomial factors $x^2 - 3x - 4 = (x-4)(x+1)$ so the roots of the equation are $r_1 = 4$ and $r_2 = -1$. Therefore a_n has the form $a_n = c_1 4^n + c_2(-1)^n$.

To find c_1 and c_2, we note that

$$\begin{aligned} a_0 &= c_1 4^0 + c_2(-1)^0 & \Rightarrow \quad & 3 = c_1 + c_2 \\ a_1 &= c_1 4^1 + c_2(-1)^1 & \Rightarrow \quad & 2 = 4c_1 - c_2 \end{aligned}$$

Solving these gives

$$c_1 = 1 \quad \text{and} \quad c_2 = 2.$$

Therefore $a_n = 4^n + 2 \cdot (-1)^n$.

Example 22.7 The Fibonacci numbers are defined by the recurrence relation $F_n = F_{n-1} + F_{n-2}$. Using Theorem 22.5, we solve the quadratic equations $x^2 - x - 1 = 0$ whose roots are $(1 \pm \sqrt{5})/2$. Therefore there is a formula for F_n of the form

$$F_n = c_1 \left(\frac{1+\sqrt{5}}{2} \right)^n + c_2 \left(\frac{1-\sqrt{5}}{2} \right)^n.$$

We can work out the values of c_1 and c_2 based on the given values of F_0 and F_1.

Example 22.8 Solve the recurrence relation

$$a_n = 2a_{n-1} - 2a_{n-2} \qquad \text{where } a_0 = 1 \text{ and } a_1 = 3.$$

Solution: The associated quadratic equation is $x^2 - 2x + 2 = 0$, which, by the quadratic formula, has two complex roots: $1 \pm i$. Do not panic. There is nothing in the work we did that required the numbers involved to be real. We now just seek a formula of the form $a_n = c_1(1+i)^n + c_2(1-i)^n$. Examining a_0 and a_1, we have

$$\begin{aligned} a_0 &= 1 = c_1 + c_2 \\ a_1 &= 3 = (1+i)c_1 + (1-i)c_2. \end{aligned}$$

Solving these gives $c_1 = \frac{1}{2} - i$ and $c_2 = \frac{1}{2} + i$. Therefore $a_n = (\frac{1}{2} - i)(1+i)^n + (\frac{1}{2} + i)(1-i)^n$.

The Case of the Repeated Root

We now consider the recurrence relations in which the associated polynomial $x^2 - s_1 x - s_2$ has only one root. We begin with the following recurrence relation:

$$a_n = 4a_{n-1} - 4a_{n-2} \tag{33}$$

with $a_0 = 1$ and $a_1 = 3$. The first few values of a_n are 1, 3, 8, 20, 48, 112, 256, and 576.

The quadratic equation associated with this recurrence relation is $x^2 - 4x + 4 = 0$, which factors $(x-2)(x-2)$. So the only root is $r = 2$. We might hope that the formula for a_n takes the form $a_n = c2^n$, but this is incorrect. Consider the first two terms:

$$a_0 = 1 = c2^0 \quad \text{and} \quad a_1 = 3 = c2^1.$$

The first equation implies $c = 1$ and the second implies $c = \frac{3}{2}$.

We need a new idea. We hope that 2^n is involved in the formula, so we try a different approach. Let us guess a formula of the form

$$a_n = c(n)2^n$$

where we can think of $c(n)$ as a "changing" coefficient. Let's work out the first few values of $c(n)$ based on the values of a_n we already calculated:

$$\begin{aligned}
a_0 &= 1 = c(0)2^0 & \Rightarrow \quad & c(0) = 1 \\
a_1 &= 3 = c(1)2^1 & \Rightarrow \quad & c(1) = \frac{3}{2} \\
a_2 &= 8 = c(2)2^2 & \Rightarrow \quad & c(2) = 2 \\
a_3 &= 20 = c(3)2^3 & \Rightarrow \quad & c(3) = \frac{5}{2} \\
a_4 &= 48 = c(4)2^4 & \Rightarrow \quad & c(4) = 4 \\
a_5 &= 112 = c(5)2^5 & \Rightarrow \quad & c(5) = \frac{7}{2}
\end{aligned}$$

The "changing" coefficient $c(n)$ works out to something simple: $c(n) = 1 + \frac{1}{2}n$. We therefore conjecture that $a_n = (1 + \frac{1}{2}n)2^n$.

Please note that the solution has the following form: $a_n = c_1 2^n + c_2 n 2^n$. Let's show that all sequences of this form satisfy the recurrence relation in (33):

$$\begin{aligned}
4a_{n-1} - 4a_{n-2} &= 4\left(c_1 2^{n-1} + c_2(n-1)2^{n-1}\right) - 4\left(c_1 2^{n-2} + c_2(n-2)2^{n-2}\right) \\
&= [2c_1 2^n - c_1 2^n] + [2c_2 n 2^n - c_2 n 2^n] + [-4 \cdot 2^{n-1} + 8 \cdot 2^{n-2}] \\
&= c_1 2^n + c_2 n 2^n + 0 = a_n.
\end{aligned}$$

So every sequence of the form $a_n = c_1 2^n + c_2 n 2^n$ is a solution to Equation (33). Have we found all solutions? Yes we have, because we can choose c_1 and c_2 to match any initial conditions a_0 and a_1; here's how. We solve

$$\begin{aligned}
a_0 &= c_1 2^0 + c_2 \cdot 0 \cdot 2^0 \\
a_1 &= c_1 2^1 + c_2 \cdot 1 \cdot 2^1
\end{aligned}$$

which gives

$$c_1 = a_0 \quad \text{and} \quad c_2 = -a_0 + \frac{1}{2}a_1.$$

Since the formula $a_n = 2^n + \frac{1}{2}n2^n$ is of the form $c_1 2^n + c_2 n 2^n$, we know it satisfies the recurrence (33). Substituting $n = 0$ and $n = 1$ in the formula gives the correct values of a_0 and a_1 (namely, 1 and 3), it follows that we have found the solution to Equation (33).

Inspired by this success, we assert and prove the following statement. Notice the requirement that $r \neq 0$; we'll treat the case $r = 0$ as a special case.

Theorem 22.9 Let s_1, s_2 be numbers so that the quadratic equation $x^2 - s_1 x - s_2 = 0$ has exactly one root, $r \neq 0$. Then every solution to the recurrence relation

$$a_n = s_1 a_{n-1} + s_2 a_{n-2}$$

is of the form

$$a_n = c_1 r^n + c_2 n r^n.$$

Proof. Since the quadratic equation has a single (repeated) root, it must be of the form $(x - r)(x - r) = x^2 - 2rx + r^2$. Thus the recurrence must be $a_n = 2ra_{n-1} - r^2 a_{n-2}$.

To prove the result, we show that a_n satisfies the recurrence and that c_1, c_2 can be chosen so as to produce all possible a_0, a_1.

To see that a_n satisfies the recurrence, we calculate as follows:

$$\begin{aligned}
2ra_{n-1} - r^2 a_{n-2} &= 2r(c_1 r^{n-1} + c_2(n-1)r^{n-1}) - r^2(c_1 r^{n-2} + c_2(n-2)r^{n-2}) \\
&= (2c_1 r^n - c_1 r^n) + (2c_2(n-1)r^n - c_2(n-2)r^n) \\
&= c_1 r^n + c_2 n r^n = a_n.
\end{aligned}$$

To see that we can choose c_1, c_2 to produce all possible a_0, a_1, we simply solve

$$\begin{aligned}
a_0 &= c_1 r^0 + c_2 \cdot 0 \cdot r^0 = c_1 \\
a_1 &= c_1 r^1 + c_2 \cdot 1 \cdot r = r(c_1 + c_2).
\end{aligned}$$

So long as $r \neq 0$, we can solve these. They yield

$$c_1 = a_0 \quad \text{and} \quad c_2 = \frac{a_0 r - a_1}{r}.$$

■

Finally, in case $r = 0$, the recurrence is simply $a_n = 0$, which means that all terms are zero.

Sequences Generated by Polynomials

We began this section by recalling Proposition 21.3, which gives a formula for the sum of the squares of the natural numbers up to n:

$$0^2 + 1^2 + 2^2 + \cdots + n^2 = \frac{(2n+1)(n+1)(n)}{6}.$$

Notice that the formula for the sum of the first n squares is a polynomial expression. In Exercise 21.3b you were asked to show that the sum of the first n cubes is $n^2(n+1)^2/4$—another polynomial expression. Proving these by induction is relatively routine, but how can we figure out the formulas in the first place?

The difference operator Δ should not be confused with the symmetric difference operation, also denoted by Δ. The difference operator converts a sequence of numbers into a new sequence of numbers, whereas the symmetric difference operation takes a pair of sets and returns another set.

Good news: We will now develop a simple method for detecting whether a sequence of numbers is generated by a polynomial expression and, if so, for determining the polynomial that created the numbers.

The key is the *difference operator*. Let $a_0, a_1, a_2, \ldots$ be a sequence of numbers. From this sequence we form a new sequence

$$a_1 - a_0, a_2 - a_1, a_3 - a_2, \ldots$$

in which each term is the difference of two consecutive terms of the original sequence. We denote this new sequence as Δa. That is, Δa is the sequence whose nth term is $\Delta a_n = a_{n+1} - a_n$. We call Δ the *difference operator*.

Example 22.10 Let a be the sequence $0, 2, 7, 15, 26, 40, 57, \ldots$. The sequence Δa is 2, 5, 8, 11, 14, 17. This is easier to see if we write the sequence a on one row and Δa on a second row with Δa_n written between a_n and a_{n+1}.

$$\begin{array}{lccccccccccccc} a: & 0 & & 2 & & 7 & & 15 & & 26 & & 40 & & 57 \\ \Delta a: & & 2 & & 5 & & 8 & & 11 & & 14 & & 17 & \end{array}$$

If the sequence a_n is given by a polynomial expression, then we can use that expression to find a formula for Δa. For example, if $a_n = n^3 - 5n + 1$, then

$$\begin{aligned} \Delta a_n &= a_{n+1} - a_n \\ &= [(n+1)^3 - 5(n+1) + 1] - [n^3 - 5n + 1] \\ &= n^3 + 3n^2 + 3n + 1 - 5n - 5 + 1 - n^3 + 5n - 1 \\ &= 3n^2 + 3n - 4. \end{aligned}$$

The *degree* of a polynomial expression is the largest exponent appearing in the expression. For example, $3n^5 - n^2 + 10$ is a degree-5 polynomial in n.

Notice that the difference operator converted a degree-3 polynomial formula, $n^3 - 5n + 1$, into a degree-2 polynomial.

Proposition 22.11 Let a be a sequence of numbers in which a_n is given by a degree-d polynomial in n where $d \geq 1$. Then Δa is a sequence given by a polynomial of degree $d - 1$.

Proof. Suppose a_n is given by a polynomial of degree d. That is, we can write

$$a_n = c_d n^d + c_{d-1} n^{d-1} + \cdots + c_1 n + c_0$$

where $c_d \neq 0$ and $d \geq 1$. We now calculate Δa_n:

$$\begin{aligned} \Delta a_n &= a_{n+1} - a_n \\ &= [c_d(n+1)^d + c_{d-1}(n+1)^{d-1} + \cdots + c_1(n+1) + c_0] \\ &\quad - [c_d n^d + c_{d-1} n^{d-1} + \cdots + c_1 n + c_0] \\ &= [c_d(n+1)^d - c_d n^d] + [c_{d-1}(n+1)^{d-1} - c_{d-1} n^{d-1}] \\ &\quad + \cdots + [c_1(n+1) - c_1 n] + [c_0 - c_0]. \end{aligned}$$

Each term on the last line is of the form $c_j(n+1)^j - c_j n^j$. We expand the $(n+1)^j$ term using the Binomial Theorem (Theorem 16.8) to give

$$\begin{aligned} c_j(n+1)^j - c_j n^j &= c_j\left[n^j + \binom{j}{1}n^{j-1} + \binom{j}{2}n^{j-2} + \cdots + \binom{j}{j}n^0\right] - c_j n^j \\ &= c_j\left[\binom{j}{1}n^{j-1} + \binom{j}{2}n^{j-2} + \cdots + \binom{j}{j}\right]. \end{aligned}$$

Notice that $c_j(n+1)^j - c_j n^j$ is a polynomial of degree $j-1$. Thus, if we look at the full expression for Δa_n, we see that the first term $c_d(n+1)^d - c_d n^d$ is a polynomial of degree $d-1$ (because $c_d \neq 0$) and none of the subsequent terms can cancel the n^{d-1} term because they all have degree less than $d-1$. Therefore Δa_n is given by a polynomial of degree $d-1$. ■

If a is given by a polynomial of degree d, then Δa is given by a polynomial of degree $d-1$. This implies that $\Delta(\Delta a)$ is given by a polynomial of degree $d-2$, and so on. Instead of $\Delta(\Delta a)$, we write $\Delta^2 a$. In general, $\Delta^k a$ is $\Delta(\Delta^{k-1}a)$ and $\Delta^1 a$ is just Δa.

What happens if we apply Δ repeatedly to a polynomially generated sequence? Each subsequent sequence is a polynomial of one lower degree until we reach a polynomial of degree zero—which is just a constant. If we apply Δ one more time, we arrive at the all-zero sequence!

Corollary 22.12 If a sequence a is generated by a polynomial of degree d, then $\Delta^{d+1}a$ is the all-zeros sequence.

Example 22.13 The sequence $0, 2, 7, 15, 26, 40, 57, \ldots$ from Example 22.10 is generated by a polynomial. Repeatedly applying Δ to this sequence gives this:

a:	0		2		7		15		26		40		57
Δa:		2		5		8		11		14		17	
$\Delta^2 a$:			3		3		3		3		3		
$\Delta^3 a$:				0		0		0		0			

Corollary 22.12 tells us that if a_n is given by a polynomial expression, then repeated applications of Δ will reduce this sequence to all zeros. We now seek to prove the converse; that is, if there is a positive integer k such that $\Delta^k a_n$ is the all-zeros sequence, then a_n is given by a polynomial formula. Furthermore, we develop a simple method for deducing the polynomial that generates a_n.

Our first tool is the following simple proposition.

Proposition 22.14 Let a, b, and c be sequences of numbers and let s be a number.

(1) If, for all n, $c_n = a_n + b_n$, then $\Delta c_n = \Delta a_n + \Delta b_n$.
(2) If, for all n, $b_n = s a_n$, then $\Delta b_n = s\Delta a_n$.

For those who have studied linear algebra. If we think of a sequence as a vector (with infinitely many components), then Proposition 22.14 says that Δ is a linear transformation.

This proposition can be written more succinctly as follows: $\Delta(a_n + b_n) = \Delta a_n + \Delta b_n$ and $\Delta(sa_n) = s\Delta a_n$.

Proof. Suppose first that for all n, $c_n = a_n + b_n$. Then

$$\begin{aligned}\Delta c_n &= c_{n+1} - c_n \\ &= (a_{n+1} + b_{n+1}) - (a_n + b_n) \\ &= (a_{n+1} - a_n) + (b_{n+1} - b_n) \\ &= \Delta a_n + \Delta b_n.\end{aligned}$$

Next, suppose that $b_n = sa_n$. Then

$$\Delta b_n = b_{n+1} - b_n = sa_{n+1} - sa_n = s\,(a_{n+1} - a_n) = s\Delta a_n. \qquad \blacksquare$$

The next step is to understand how Δ treats some particular polynomial sequences. We start with a specific example.

Not only can $\binom{n}{3}$ be expressed as a polynomial in n, but the same is true for all $\binom{n}{k}$ (where k is a positive integer). Using Theorem 16.12, when $n \geq k$, write $\binom{n}{k}$ as

$$\frac{n(n-1)(n-2)\cdots(n-k+1)}{k!}.$$

For the case $0 \leq n < k$, observe that both $\binom{n}{k}$ and the polynomial evaluate to zero. Thus for every positive integer k, $\binom{n}{k}$ can be written as a polynomial of degree k.

Let a be the sequence whose nth term is $a_n = \binom{n}{3}$. For example, $a_5 = \binom{5}{3} = 10$. By Theorem 16.12, we can write

$$\begin{aligned}a_n = \binom{n}{3} &= \frac{n!}{(n-3)!3!} \\ &= \frac{n(n-1)(n-2)(n-3)(n-4)\cdots(2)(1)}{(n-3)(n-4)\cdots(2)(1)\cdot 3!} \\ &= \frac{1}{6}n(n-1)(n-2)\end{aligned}$$

which is a polynomial. This formula is correct, but there is a minor error. The formula $\binom{n}{k} = \frac{n!}{(n-k)!k!}$ applies only when $0 \leq k \leq n$. The first few terms of the sequence, a_0, a_1, a_2, are $\binom{0}{3}$, $\binom{1}{3}$, and $\binom{2}{3}$. All of these evaluate to zero, but Theorem 16.12 does not apply to them. Fortunately, the polynomial expression $\frac{1}{6}n(n-1)(n-2)$ also evaluates to zero for $n = 0, 1, 2$, so the formula $a_n = \frac{1}{6}n(n-1)(n-2)$ is correct for all values of n.

Now let's calculate Δa_n, $\Delta^2 a_n$, and so on, until we reach the all-zeros sequence (which, by Corollary 22.12, should be by $\Delta^4 a_n$).

a_n:	0		0		0		1		4		10		20		35		56
Δa_n:		0		0		1		3		6		10		15		21	
$\Delta^2 a_n$:			0		1		2		3		4		5		6		
$\Delta^3 a_n$:				1		1		1		1		1		1			
$\Delta^4 a_n$:					0		0		0		0		0				

Please note that every row of this table begins with a zero except for row $\Delta^3 a_n$, which begins with a one.

Since $a_n = \binom{n}{3}$ is a polynomial of degree 3, we know that Δa_n is a polynomial of degree 2. Let's work this out algebraically:

$$\begin{aligned}
\Delta a_n = \Delta\binom{n}{3} &= \binom{n+1}{3} - \binom{n}{3} \\
&= \frac{1}{6}(n+1)(n)(n-1) - \frac{1}{6}n(n-1)(n-2) \\
&= \frac{(n^3-n)-(n^3-3n^2+2n)}{6} = \frac{3n^2-3n}{6} \\
&= \frac{1}{2}n(n-1) = \binom{n}{2}.
\end{aligned}$$

Having discovered that $\Delta\binom{n}{3} = \Delta\binom{n}{2}$, we wonder whether there is an easier way to prove this (there is) and whether this generalizes (it does).

We seek a quick way to prove that $\Delta\binom{n}{3} = \binom{n}{2}$. This can be rewritten $\binom{n+1}{3} - \binom{n}{3} = \binom{n}{2}$, which can be rearranged to $\binom{n}{2} + \binom{n}{3} = \binom{n+1}{3}$. This follows directly from Pascal's Identity (Theorem 16.10).

Seeing that $\Delta\binom{n}{3} = \binom{n}{2}$, it's not a bold leap to guess that $\Delta\binom{n}{4} = \binom{n}{3}$, or in general $\Delta\binom{n}{k} = \binom{n}{k-1}$. The proof is essentially a direct application of Pascal's Identity (with a bit of care in the case $n < k$).

Proposition 22.15 Let k be a positive integer and let $a_n = \binom{n}{k}$ for all $n \geq 0$. Then $\Delta a_n = \binom{n}{k-1}$.

Proof. We need to show that $\Delta\binom{n}{k} = \binom{n}{k-1}$ for all $n \geq 0$. This is equivalent to $\binom{n+1}{k} - \binom{n}{k} = \binom{n}{k-1}$ which in turn is the same as

$$\binom{n+1}{k} = \binom{n}{k} + \binom{n}{k-1}. \tag{34}$$

By Pascal's Identity (Theorem 16.10), Equation (34) holds whenever $0 < k < n+1$, so we need only concern ourselves with the case $n+1 \leq k$ (i.e., $n \leq k-1$).

In the case $n < k-1$, all three terms, $\binom{n+1}{k}$, $\binom{n}{k}$, and $\binom{n}{k-1}$, equal zero, so (34) holds.

In the case $n = k-1$, we have $\binom{n+1}{k} = \binom{k}{k} = 1$, $\binom{n}{k} = \binom{k-1}{k} = 0$, and $\binom{n}{k-1} = \binom{k-1}{k-1} = 1$, and (34) reduces to $1 = 0 + 1$. ■

Earlier we noted that for $a_n = \binom{n}{3}$, we have that $\Delta^j a_0 = 0$ for all j except $j = 3$, and $\Delta^3 a_0 = 1$. This generalizes. Let k be a positive integer and let $a_n = \binom{n}{k}$. Because a_n is expressible as a degree-k polynomial, $\Delta^{k+1} a_n = 0$ for all n. Using Proposition 22.15, we have that $a_0 = \Delta a_0 = \Delta^2 a_0 = \cdots = \Delta^{k-1} a_0 = 0$ but $\Delta^k a_k = 1$; see Exercise 22.5.

Thus, for the sequence $a_n = \binom{n}{k}$, we know (1) that $\Delta^{k+1} a_n = 0$ for all n, (2) the value of a_0, and (3) the value of $\Delta^j a_0$ for $1 \leq j < k$. We claim that these three facts uniquely determine the sequence a_n. Here is a careful statement of that assertion.

Proposition 22.16 Let a and b be sequences of numbers and let k be a positive integer. Suppose that

- $\Delta^k a_n$ and $\Delta^k b_n$ are zero for all n,
- $a_0 = b_0$, and
- $\Delta^j a_0 = \Delta^j b_0$ for all $1 \le j < k$.

Then $a_n = b_n$ for all n.

Proof. The proof is by induction on k.

The basis case is when $k = 1$. In this case we are given that $\Delta a_n = \Delta b_n = 0$ for all n. This means that $a_{n+1} - a_n = 0$ for all n, which implies that $a_{n+1} = a_n$ for all n. In other words, all terms in a_n are identical. Likewise for b_n. Since we also are given that $a_0 = b_0$, the two sequences are the same.

Now suppose (induction hypothesis) that the Proposition has been proved for the case $k = \ell$. We seek to prove the result in the case $k = \ell + 1$. To that end, let a and b be sequences such that

- $\Delta^{\ell+1} a_n = \Delta^{\ell+1} b_n = 0$ for all n,
- $a_0 = b_0$, and
- $\Delta^j a_0 = \Delta^j b_0$ for all $1 \le j < \ell + 1$.

Consider the sequences $a'_n = \Delta a_n$ and $b'_n = \Delta b_n$. By our hypotheses we see that $\Delta^\ell a'_n = \Delta^\ell b'_n = 0$ for all n, $a'_0 = b'_0$, and $\Delta^j a'_0 = \Delta^j b'_0$ for all $1 \le j < \ell$. Therefore, by induction, a' and b' are identical (i.e., $a'_n = b'_n$ for all n).

We now prove that $a_n = b_n$ for all n. Suppose, for the sake of contradiction, that a and b were different sequences. Choose m to be the smallest subscript so that $a_m \ne b_m$. Note that $m \ne 0$ because we are given $a_0 = b_0$; thus $m > 0$. Thus we know $a_{m-1} = b_{m-1}$. We also know that $a'_{m-1} = b'_{m-1}$; here is why:

$$\begin{aligned} a'_{m-1} &= \Delta a_{m-1} = a_m - a_{m-1} \\ = b'_{m-1} &= \Delta b_{m-1} = b_m - b_{m-1} \end{aligned}$$

$$\begin{aligned} a_m - a_{m-1} &= b_m - b_{m-1} \\ a_m - b_m &= a_{m-1} - b_{m-1} = 0 \\ \therefore \quad a_m &= b_m \qquad \Rightarrow\Leftarrow \end{aligned}$$

Thus $a_n = b_n$ for all n. ■

We are now ready to present our main result about sequences generated by polynomial expressions.

Theorem 22.17 Let $a_0, a_1, a_2, \ldots$ be a sequence of numbers. The terms a_n can be expressed as polynomial expressions in n if and only if there is a nonnegative integer k such that for all $n \ge 0$ we have $\Delta^{k+1} a_n = 0$. In this case,

$$a_n = a_0 \binom{n}{0} + (\Delta a_0)\binom{n}{1} + (\Delta^2 a_0)\binom{n}{2} + \cdots + (\Delta^k a_0)\binom{n}{k}.$$

Proof. One half of the if-and-only-if statement has already been proved: If a_n is given by a polynomial of degree d, then $\Delta^{d+1}a_n = 0$ for all n (Corollary 22.12).

Suppose now that a is a sequence of numbers and that there is a natural number k such that for all n, $\Delta^{k+1}a_n = 0$. We prove that a_n is given by a polynomial expression by showing that a_n is equal to

$$b_n = a_0\binom{n}{0} + (\Delta a_0)\binom{n}{1} + (\Delta^2 a_0)\binom{n}{2} + \cdots + (\Delta^k a_0)\binom{n}{k}.$$

To show that $a_n = b_n$ for all n, we apply Proposition 22.16; that is, we need to prove

(1) $\Delta^{k+1}a_n = \Delta^{k+1}b_n = 0$ for all n,
(2) $a_0 = b_0$, and
(3) $\Delta^j a_0 = \Delta^j b_0$ for all $1 \le j \le k$.

We tackle each in turn.

To show (1), note that $\Delta^{k+1}a_n = 0$ for all n by hypothesis. Notice that b_n is a polynomial of degree k, and so $\Delta^{k+1}b_n = 0$ for all n as well (by Corollary 22.12).

It is easy to verify (2) by substituting $n = 0$ into the expression for b_n; every term except the first evaluates to zero, and the first term is $a_0\binom{0}{0} = a_0$.

Finally, we need to prove (3). The notation can become confusing as we calculate $\Delta^j b_n$—there will be too many Δs crawling around the page! To make our work easier to read, we let

$$c_0 = a_0,\ c_1 = \Delta a_0,\ c_2 = \Delta^2 a_0, \ldots,\ c_k = \Delta^k a_0$$

and so we can rewrite b_n as

$$b_n = c_0\binom{n}{0} + c_1\binom{n}{1} + c_2\binom{n}{2} + \cdots + c_k\binom{n}{k}.$$

Now, to calculate $\Delta^j b_n$ we apply Proposition 22.14, Proposition 22.15, and Corollary 22.12:

$$\begin{aligned}
\Delta^j b_n &= \Delta^j\left[c_0\binom{n}{0} + c_1\binom{n}{1} + c_2\binom{n}{2} + \cdots + c_k\binom{n}{k}\right] \\
&= c_0\Delta^j\binom{n}{0} + c_1\Delta^j\binom{n}{1} + c_2\Delta^j\binom{n}{2} + \cdots + c_k\Delta^j\binom{n}{k} \\
&= 0 + \cdots + 0 + c_j\Delta^j\binom{n}{j} + c_{j+1}\Delta^j\binom{n}{j+1} + \cdots + c_k\Delta^j\binom{n}{k} \\
&= c_j\binom{n}{0} + c_{j+1}\binom{n}{1} + \cdots + c_k\binom{n}{k-j}.
\end{aligned}$$

We substitute $n = 0$ into this, which gives

$$\Delta^j b_0 = c_j + 0 + \cdots + 0 = \Delta^j a_0$$

and this completes the proof. ■

Example 22.18 We return to the sequence presented in Examples 22.10 and 22.13: 0, 2, 7, 15, 26, 40, 57, ... We calculated successive differences and found this:

a:	0		2		7		15		26		40		57		
Δa:		2		5		8		11		14		17			
$\Delta^2 a$:			3		3		3		3		3				
$\Delta^3 a$:				0		0		0		0					

By Theorem 22.17,

$$a_n = 0\binom{n}{0} + 2\binom{n}{1} + 3\binom{n}{2} = 0 + 2 \cdot n + 3 \cdot \frac{n(n-1)}{2} = \frac{n(3n+1)}{2}.$$

Example 22.19 Let us derive the following formula from Proposition 21.3:

$$0^2 + 1^2 + 2^2 + \cdots + n^2 = \frac{(2n+1)(n+1)(n)}{6}.$$

Let $a_n = 0^2 + 1^2 + \cdots + n^2$. Computing successive differences, we have

a_n:	0		1		5		14		30		55		91		140
Δa_n:		1		4		9		16		25		36		49	
$\Delta^2 a_n$:			3		5		7		9		11		13		
$\Delta^3 a_n$:				2		2		2		2		2			
$\Delta^4 a_n$:					0		0		0		0				

Therefore

$$\begin{aligned} a_n &= 0\binom{n}{0} + 1\binom{n}{1} + 3\binom{n}{2} + 2\binom{n}{3} \\ &= 0 + n + \frac{3}{2}n(n-1) + \frac{2}{6}n(n-1)(n-2) \\ &= \frac{2n^3 + 3n^2 + n}{6} = \frac{(2n+1)(n+1)(n)}{6}. \end{aligned}$$

Recap

A recurrence relation for a sequence of numbers is an equation that expresses an element of the sequence in terms of earlier elements. We analyzed first-order recurrence relations of the form $a_n = sa_{n-1} + t$ and second-order recurrence relations of the form $a_n = s_1 a_{n-1} + s_2 a_{n-2}$:

- The recurrence $a_n = sa_{n-1} + t$ has the following solution: If $s \neq 1$, then $a_n = c_1 s^n + c_2$ where c_1, c_2 are specific numbers.
- The solution to the recurrence $a_n = s_1 a_{n-1} + s_2 a_{n-2}$ depends on the roots r_1, r_2 of the quadratic equation $x^2 - s_1 x - s_2 = 0$. If $r_1 \neq r_2$, then $a_n = c_1 r_1^n + c_2 r_2^n$ but if $r_1 = r_2 = r$, then $a_n = c_1 r^n + c_2 n r^n$.

We introduced the difference operator, $\Delta a_n = a_{n+1} - a_n$. The sequence of numbers a_n is generated by a polynomial expression of degree d if and only if $\Delta^{d+1} a_n$ is zero for all n. In this case we can write $a_n = a_0\binom{n}{0} + (\Delta a_0)\binom{n}{1} + (\Delta^2 a_0)\binom{n}{2} + \cdots + (\Delta^d a_0)\binom{n}{d}$.

22 Exercises

22.1. For each of the following recurrence relations, calculate the first six terms of the sequence (that is, a_0 through a_5). You do not need to find a formula for a_n.

a. $a_n = 2a_{n-1} + 2$, $a_0 = 1$.
b. $a_n = a_{n-1} + 3$, $a_0 = 5$.
c. $a_n = a_{n-1} + 2a_{n-2}$, $a_0 = 0$, $a_1 = 1$.
d. $a_n = 3a_{n-1} - 5a_{n-2}$, $a_0 = 0$, $a_2 = 0$.
e. $a_n = a_{n-1} + a_{n-2} + 1$, $a_0 = a_1 = 1$.
f. $a_n = a_{n-1} + n$, $a_0 = 1$.

22.2. Solve each of the following recurrence relations by giving an explicit formula for a_n. For each, please calculate a_9.

a. $a_n = \frac{2}{3}a_{n-1}$, $a_0 = 4$.
b. $a_n = 10a_{n-1}$, $a_0 = 3$.
c. $a_n = -a_{n-1}$, $a_0 = 5$.
d. $a_n = 1.2a_{n-1}$, $a_0 = 0$.
e. $a_n = 3a_{n-1} - 1$, $a_0 = 10$.
f. $a_n = 4 - 2a_{n-1}$, $a_0 = 0$.
g. $a_n = a_{n-1} + 3$, $a_0 = 0$.
h. $a_n = 2a_{n-1} + 2$, $a_0 = 0$.
i. $a_n = 8a_{n-1} - 15a_{n-2}$, $a_0 = 1$, $a_1 = 4$.
j. $a_n = a_{n-1} + 6a_{n-2}$, $a_0 = 4$, $a_1 = 4$.
k. $a_n = 4a_{n-1} - 3a_{n-2}$, $a_0 = 1$, $a_1 = 2$.
l. $a_n = -6a_{n-1} - 9a_{n-2}$, $a_0 = 3$, $a_3 = 6$.
m. $a_n = 2a_{n-1} - a_{n-2}$, $a_0 = 5$, $a_1 = 1$.
n. $a_n = -2a_{n-1} - a_{n-2}$, $a_0 = 5$, $a_1 = 1$.
o. $a_n = 2a_{n-1} + 2a_n$, $a_0 = 3$, $a_1 = 3$.
p. $a_n = 2a_{n-1} - 5a_{n-2}$, $a_0 = 2$, $a_1 = 3$.

22.3. Each of the following sequences is generated by a polynomial expression. For each, find the polynomial expression that gives a_n.

a. 1, 6, 17, 34, 57, 86, 121, 162, 209, 262, ...
b. 6, 5, 6, 9, 14, 21, 30, 41, 54, 69, ...
c. 4, 4, 10, 28, 64, 124, 214, 340, 508, 724, ...
d. 5, 16, 41, 116, 301, 680, 1361, 2476, 4181, 6656, ...

22.4. Explain why the notation Δa_n has implicit parentheses $(\Delta a)_n$ and why $\Delta(a_n)$ is not correct.

22.5. Let k be a positive integer and let $a_n = \binom{n}{k}$. Prove that $a_0 = \Delta a_0 = \Delta^2 a_0 = \cdots = \Delta^{k-1} a_0 = 0$ and that $\Delta^k a_0 = 1$.

22.6. Suppose that the sequence a satisfies the recurrence $a_n = a_{n-1} + 12a_{n-2}$ and that $a_0 = 6$ and $a_5 = 4877$. Find an expression for a_n.

22.7. Find a polynomial formula for $1^4 + 2^4 + 3^4 + \cdots + n^4$.

22.8. Let t be a positive integer. Prove that $1^t + 2^t + 3^t + \cdots + n^t$ can be written as a polynomial expression.

22.9. Some so-called intelligence tests often include problems in which a series of numbers is presented and the subject is required to find the next term of the sequence. For example, the sequence might begin 1, 2, 4, 8. No doubt the examiner is looking for 16 as the next term.

Show how to "outsmart" the intelligence test by finding a polynomial expression (of degree 3) for a_n such that $a_0 = 1$, $a_1 = 2$, $a_2 = 4$, $a_3 = 8$, but $a_4 = 15$.

22.10. Let s be a real number with $s \neq 0$. Find a sequence a such that $a_n = s\Delta a_n$ and $a_0 = 1$.

22.11. Solve the equation $\Delta^2 a_n = -a_n$ with $a_0 = a_1 = 2$.

22.12. Find two different sequences a and b for which $\Delta a_n = \Delta b_n$ for all n.

22.13. The second-order recurrence relations we solved were of the form $a_n = s_1 a_{n-1} + s_2 a_{n-2}$. In this problem we extend this to relations of the form $a_n = s_1 a_{n-1} + s_2 a_{n-2} + t$. Typically (but not always) the solution to such a relation is of the form $a_n = c_1 r_1^n + c_2 r_2^n + c_3$ where c_1, c_2, c_3 are specific numbers, and r_1, r_2 are roots of the associated quadratic equation $x^2 - s_1 x - s_2 = 0$. However, if one of these roots is 1, or if the roots are equal to each other, another form of solution is required.

Please solve the following recurrence relations. In the cases where the standard form does not apply, try to work out an appropriate alternative form, but if you get stuck, please consult the Hints (Appendix A).

a. $a_n = 5a_{n-1} - 6a_{n-2} + 2$, $a_0 = 1$, $a_1 = 2$.
b. $a_n = 4a_{n-1} + 5a_{n-2} + 4$, $a_0 = 2$, $a_1 = 3$.
c. $a_n = 2a_{n-1} + 4a_{n-2} + 6$, $a_0 = a_1 = 4$.
d. $a_n = 3a_{n-1} - 2a_{n-2} + 5$, $a_0 = a_1 = 3$.
e. $a_n = 6a_{n-1} - 9a_{n-2} - 2$, $a_0 = -1$, $a_1 = 4$.
f. $a_n = 2a_{n-1} - a_{n-2} + 2$, $a_0 = 4$, $a_1 = 2$.

22.14. Extrapolate from Theorems 22.5 and 22.9 to solve the following third-order recurrence relations.

a. $a_n = 4a_{n-1} - a_{n-2} - 6a_{n-3}$, $a_0 = 8$, $a_1 = 3$, and $a_2 = 27$.
b. $a_n = 2a_{n-1} + 2a_{n-2} - 4a_{n-3}$, $a_0 = 11$, $a_1 = 10$, and $a_2 = 32$.
c. $a_n = -a_{n-1} + 8a_{n-2} + 12a_{n-3}$, $a_0 = 6$, $a_1 = 19$, and $a_2 = 25$.
d. $a_n = 6a_{n-1} - 12a_{n-2} + 8a_{n-3}$, $a_0 = 3$, $a_1 = 2$, and $a_2 = 36$.

22.15. Suppose you wish to generate elements of a recurrence relation using a computer program. It is tempting to write such a program recursively.

For example, consider the recurrence $a_n = 3a_{n-1} - 2a_{n-2}$, $a_0 = 1$, $a_1 = 5$. Here is a program to calculate the values a_n:

```
procedure get_term(n)
   if (n < 0)
      print 'Illegal argument'
      exit
   end

   if (n == 0)
      return 1
   end
```

```
    if (n == 1)
        return 5
    end
    return 3*get_term(n-1) - 2*get_term(n-2)
end
```

Although this program is easy to understand, it is extremely inefficient. Explain why.

In particular, let b_n be the number of times this routine is called when it calculates a_n. Find a recurrence—and solve it!—for b_n.

22.16. There are many types of recurrence relations that are of different forms from those presented in this section. Try your hand at finding a formula for a_n for these:

a. $a_n = na_{n-1}$, $a_0 = 1$.
b. $a_n = a_{n-1}^2$, $a_0 = 2$.
c. $a_n = a_0 + a_1 + a_2 + \cdots + a_{n-1}$, $a_0 = 1$.
d. $a_n = na_0 + (n-1)a_1 + (n-2)a_2 + \cdots + 2a_{n-2} + 1a_{n-1}$, $a_0 = 1$.
e. $a_n = 3.9a_{n-1}(1 - a_{n-1})$, $a_0 = \frac{1}{2}$.

Chapter 4 Self Test

1. Prove that the equation $x^2 + 1 = 0$ does not have any real solutions.
2. Prove that the sum of any four consecutive integers is not divisible by 4.
3. Let a and b be positive integers. Prove: If $a|b$ and $b|a$, then $a = b$.
4. Which of the following sets are well-ordered?
   a. The set of all even integers.
   b. The set of all primes.
   c. $\{-100, -99, -98, \ldots, 98, 99, 100\}$.
   d. $\emptyset$.
   e. The negative integers.
   f. $\{\pi, \pi^2, \pi^3, \pi^4, \ldots\}$ where π is the familiar real number $3.14159\ldots$.
5. Let n be a positive integer. Prove that

$$1 + 4 + 7 + \cdots + (3n - 2) = \frac{3n^2 - n}{2}.$$

6. Let n be a natural number. Prove that

$$0! + 1! + 2! + \cdots + n! \le (n + 1)!.$$

7. Suppose $a_0 = 1$ and $a_n = 4a_{n-1} - 1$ when $n \ge 1$. Prove that for all natural numbers n, we have $a_n = (2 \cdot 4^n + 1)/3$.
8. Prove by induction: If $n \in \mathbb{N}$, then $n < 2^n$.
9. Consider the following proposition.

Let P be a finite set of (three or more) points in the plane and suppose any three points in P are collinear. Then all the points in P must lie on a common line.

Prove this two ways: by contradiction and by induction.

10. Let n be a positive integer. Prove that

$$\sqrt{1} + \sqrt{2} + \cdots + \sqrt{n} \leq n\sqrt{n}.$$

11. Prove the Binomial Theorem (Theorem 16.8) by induction. That is, if n is a natural number, then

$$(x+y)^n = \sum_{j=0}^{n} \binom{n}{j} x^j y^{n-j}.$$

12. Let n be a positive integer and suppose n distinct lines are drawn in the plane. No two of these lines are parallel, and no three of these lines intersect at a common point. Prove that these lines divide the plane into $\binom{n}{0} + \binom{n}{1} + \binom{n}{2}$ regions.

13. Let F_n denote the nth Fibonacci number (see Definition 20.12). Prove that for all natural numbers n, we have

$$F_n + 2F_{n+1} = F_{n+4} - F_{n+2}.$$

14. Let F_n denote the nth Fibonacci number. If n is a natural number, then 1 is the only positive divisor of both F_n and F_{n+1} (i.e., if $d > 0$, $d|F_n$, and $d|F_{n+1}$, then $d = 1$).

15. A horizontal stripe is to be tiled. The tiles come in two shapes: 1×1 rectangles and 1×2 rectangles. The 1×1 tiles are available in two colors (white and dark blue), and the 1×2 tiles are available in three colors (white, light blue, and dark blue). For a positive integer n, let a_n denote the number of different ways to tile an n-long stripe using these tiles. The figure shows one possible tiling with $n = 11$.

a. Show that for $n \geq 2$, $a_n = 2a_{n-1} + 3a_{n-2}$.

b. Prove that $a_n = (3^{n+1} + (-1)^n)/4$.

16. Let n be a positive integer. Prove there is a unique pair of nonnegative integers a, b such that $n = 2^a b$ and b is odd.

17. Let A be a nonempty finite set of positive integers. Suppose that for any two elements $r, s \in A$, we have $r|s$ or $s|r$. (In symbols, $\forall r \in A$, $\forall s \in A$, $(r|s$ or $s|r)$.)

a. Prove that A contains an element t with the property that for all $a \in A$, $a|t$. (In symbols, $\exists t \in A$, $\forall a \in A$, $a|t$.)

b. Furthermore, prove that t is unique (i.e., there is only one element of A that is a multiple of all elements of A).

c. Finally, give an example to show that uniqueness does not hold if we do not assume that all the elements of A are positive.

18. For each of the following recurrence relations, find a formula for the nth term, a_n.

a. $a_n = 2a_{n-1} + 15a_{n-2}$, $a_0 = 4$, $a_1 = 0$.

b. $a_n = 2a_{n-1} + 15$, $a_0 = 4$, $a_1 = 0$.
c. $a_n = 12a_{n-1} - 36a_{n-2}$, $a_0 = 1$, $a_1 = 2$.

19. The following sequence of numbers is generated by a polynomial expression. Find the polynomial. (The first term is a_0; you should find a polynomial expression for a_n.)

The sequence is

$$5, 26, 67, 146, 281, 490, 791, 1202, 1741, 2426, 3275, \ldots.$$

CHAPTER

5 Functions

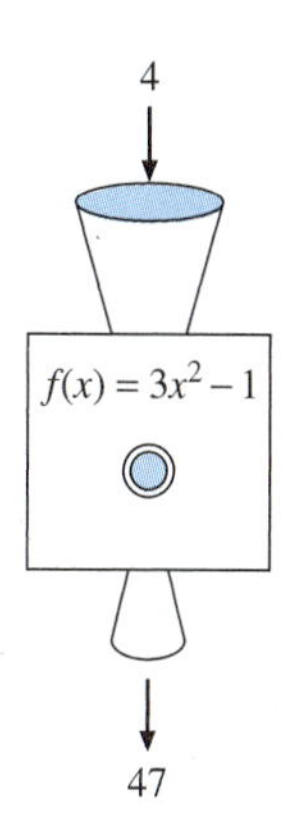

The concept of *function* is central to mathematics. Intuitively, a function can be thought of as a machine. You put a number into the machine, push a button, and out comes an answer. A key property of being a function is consistency. Every time we put a specific number—say, 4—into the machine, the same answer emerges. We illustrate this in the figure. Here the function takes an integer x as input and returns the value $3x^2 - 1$. Thus every time the number 4 is entered into the machine, the answer 47 is produced.

Note that the function in the figure operates on numbers. It would not make sense to try to put a triangle down the hopper of this machine! However, we can create a function whose inputs are triangles and whose outputs are numbers. For example, we can define f to be the function whose inputs are triangles, and for each triangle entered into the function, the output is the area of the triangle.

The "mechanism" in the function "machine" need not be dictated by an algebraic formula. All that is required is that we carefully specify the allowable inputs and, for each allowable input, the corresponding output. This is often done with an algebraic expression, but there are other ways to specify a function.

In this chapter, we take a careful look at functions. We begin with a precise definition.

23 Functions

Intuitively, a function is a "rule" or "mechanism" that transforms one quantity into another. For example, the function $f(x) = x^2 + 4$ takes an integer x and transforms it into the integer $x^2 + 4$. The function $g(x) = |x|$ takes the integer x and returns x if $x \geq 0$ and $-x$ if $x < 0$.

In this section, we develop a more abstract and rigorous view of functions. Functions are special types of relations (please review Section 13).

Recall that a *relation* is simply a set of ordered pairs. Just as this definition of a relation was at first counterintuitive, the precise definition of a function may at first seem strange.

Definition 23.1 **(Function)** A relation f is called a *function* provided $(a, b) \in f$ and $(a, c) \in f$ imply $b = c$.

Stated in a negative fashion, a relation f is not a function if there exist a, b, c with $(a, b) \in f$ and $(a, c) \in f$, and $b \neq c$.

Example 23.2 Let

$$f = \{(1, 2), (2, 3), (3, 1), (4, 7)\} \quad \text{and}$$
$$g = \{(1, 2), (1, 3), (4, 7)\}.$$

The relation f is a function, but the relation g is not because $(1, 2), (1, 3) \in g$ and $2 \neq 3$.

Mathspeak!

Mathematicians often use the word *map* as a synonym for *function*. In addition to saying "f of 1 equals 2," we also say "f *maps* 1 to 2." And there is a notation for this. We write $1 \mapsto 2$. The special arrow $\mapsto$ means $f(1) = 2$. The function f is not explicitly mentioned in the notation $1 \mapsto 2$; when we use the $\mapsto$ notation, we need to be certain that the reader knows what function is being discussed.

When expressed as a set of ordered pairs, functions do not look like rules for transforming one object into another, but let us look closer. The ordered pairs in f associate "input" values (the first elements in the lists in f) with "output" values (the second elements in the lists). In Example 23.2, the function f associates the input value 1 with the output value 2, because $(1, 2) \in f$. The reason why g is not a function is that for the input value 1, there are two different output values: 2 and 3. What makes f a function is that for each input there can be at most one output.

Mathematicians rarely use the notation $(1, 2) \in f$, even though this is formally correct. Instead, we use the $f(\cdot)$ notation.

Definition 23.3 **(Function notation)** Let f be a function and let a be an object. The notation $f(a)$ is defined provided there exists an object b such that $(a, b) \in f$. In this case, $f(a)$ equals b. Otherwise [there is no ordered pair of the form $(a, \_) \in f$], the notation $f(a)$ is undefined. The symbols $f(a)$ are pronounced "f of a."

For the function f from Example 23.2, we have

$$f(1) = 2 \qquad f(2) = 3 \qquad f(3) = 1 \qquad f(4) = 7$$

but for any other object x, $f(x)$ is undefined. The reason why we don't call g a function becomes clearer. What is $g(1)$? Since both $(1, 2)$ and $(1, 3) \in g$, the notation $g(1)$ does not specify an unambiguous value.

Example 23.4 **Problem:** Express the integer function $f(x) = x^2$ as a set of ordered pairs.
Solution: We might write this out using ellipses:

$$f = \{\ldots, (-3, 9), (-2, 4), (-1, 1), (0, 0), (1, 1), (2, 4), (3, 9), \ldots\}$$

but it is much clearer if we use set-builder notation:

$$f = \{(x, y) : x, y \in \mathbb{Z},\ y = x^2\}.$$

It is often clearer to write, "Let f be the function defined for an integer x by $f(x) = x^2$" than to write out f as a set of ordered pairs as in the example.

The set-of-ordered-pairs notation for a function is similar to writing a function as a chart:

x	$f(x)$
$\vdots$	$\vdots$
-3	9
-2	4
-1	1
0	0
1	1
2	4
3	9
$\vdots$	$\vdots$

Domain and Image

The sets of allowable inputs and possible outputs of a function have special names.

Definition 23.5 **(Domain, image)** Let f be a function. The set of all possible first elements of the ordered pairs in f is called the *domain* of f and is denoted dom f. The set of all possible second elements of the ordered pairs in f is called the *image* of f and is denoted im f.

We have avoided using the word *range*. Students are often taught that the word *range* means the same thing as our word *image*. The mathematician's use of the word *range* is different from that commonly taught in high school. We avoid confusion simply by not using this word.

In other notation,

$$\text{dom } f = \{a : \exists b,\ (a, b) \in f\} \qquad \text{and} \qquad \text{im } f = \{b : \exists a,\ (a, b) \in f\}.$$

Alternatively, we can write

$$\text{dom } f = \{a : f(a) \text{ is defined}\} \qquad \text{and} \qquad \text{im } f = \{b : b = f(a) \text{ for some } a\}.$$

Example 23.6 Let $f = \{(1, 2), (2, 3), (3, 1), (4, 7)\}$. (This is the function from Example 23.2.) Then

$$\text{dom } f = \{1, 2, 3, 4\} \qquad \text{and} \qquad \text{im } f = \{1, 2, 3, 7\}.$$

Example 23.7 Let f be the function from Example 23.4; that is,

$$f = \{(x, y) : x, y \in \mathbb{Z},\ y = x^2\}.$$

The domain of f is the set of all integers, and the image of f is the set of all perfect squares.

Next we introduce a special notation for functions.

Definition 23.8 ($f : A \to B$) Let f be a function and let A and B be sets. We say that f is a *function from A to B* provided dom $f = A$ and im $f \subseteq B$. In this case, we write $f : A \to B$. We also say that f is a *mapping from A to B*.

The notation $f : A \to B$ is read aloud "f is a function from A to B." The notation $f : A \to B$ makes three promises: First, f is a function. Second, $\text{dom } f = A$. And third, $\text{im } f \subseteq B$.

Mathspeak!

The notation $f : A \to B$ can be an entire sentence, an independent clause, or a noun phrase. In a theorem, we might write, "If $f : A \to B$, then ...." In this case, we would pronounce the symbols as "If f is a function from A to B...."

However, we may also write, "Let $f : A \to B$ ...." In this case, we would read the symbols as "Let f be a function from A to B...."

Example 23.9 Consider the sine function. This function is defined for every real number and returns a real value. The domain of the sine function is all real numbers, and the image is the set $[-1, 1] = \{x \in \mathbb{R} : -1 \le x \le 1\}$. We can write $\sin : \mathbb{R} \to \mathbb{R}$ because $\text{dom } \sin = \mathbb{R}$ and $\text{im } \sin \subseteq \mathbb{R}$. It would also be correct to write $\sin : \mathbb{R} \to [-1, 1]$.

To prove that $f : A \to B$ (i.e., to prove that f is a function from A to B), use Proof Template 19.

Proof Template 19 To show $f : A \to B$.

To prove that f is a function from a set A to a set B:

- Prove that f is a function.
- Prove that $\text{dom } f = A$.
- Prove that $\text{im } f \subseteq B$.

Pictures of Functions

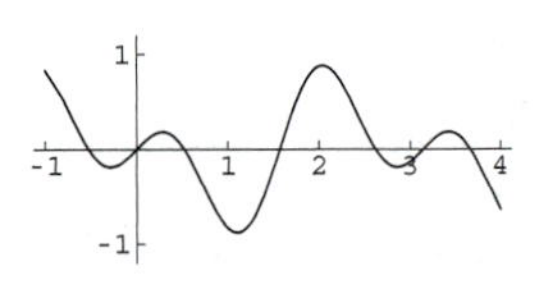

Graphs provide an excellent way to visualize functions whose inputs and outputs are real numbers. For example, the figure shows the graph of the function $f(x) = \sin x \cos 3x$. To draw the graph of a function, we plot a point in the plane at coordinates $(x, f(x))$ for every $x \in \text{dom } f$.

Formally, the *graph* of a function is the set $\{(x, y) : y = f(x)\}$. What is interesting is that this set is the function! The function f is the set of all ordered pairs (x, y) for which $y = f(x)$. So to speak of "the graph of a function" is redundant! This is not bad. When we use the word *graph* in this context, we are conjuring up a geometric view of the function.

Mathspeak!

Later in this book we use the word *graph* in an entirely different way. Here the word *graph* refers to the diagram used to depict the relation between one quantity (x) and another ($y = f(x)$).

Graphs are helpful tools for understanding functions to and from the real numbers. To verify that a picture represents a function, we can apply the *vertical line test*: Every vertical line in the plane may intersect the graph of a function in at most one point. A vertical line may not hit the graph twice; otherwise we would have two different points (x, y_1) and (x, y_2), both on the graph of the function. This would mean that both $(x, y_1), (x, y_2) \in f$ with $y_1 \neq y_2$. And this is forbidden by the definition of function.

In discrete mathematics, we are particularly interested in considering functions to and from finite sets (or $\mathbb{N}$ or $\mathbb{Z}$). In such cases, traditional graphs of functions are either not helpful or nonsensical. For example, let A be a finite set. We can consider the function $f : 2^A \to \mathbb{N}$ defined by $f(x) = |x|$. (Alert: The vertical value bars in this context do not mean absolute value!) To each subset x of A, the function f assigns its size. There is no practical way to draw this as a graph on coordinate axes.

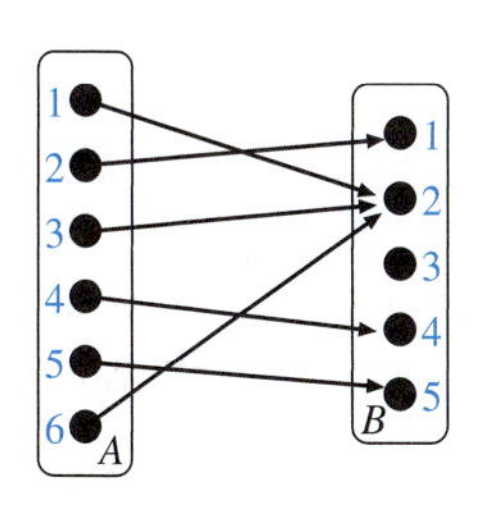

We have an alternative way to draw pictures of functions $f : A \to B$ where A and B are finite sets. Let $A = \{1, 2, 3, 4, 5, 6\}$ and $B = \{1, 2, 3, 4, 5\}$ and consider the function $f : A \to B$ defined by

$$f = \{(1, 2), (2, 1), (3, 2), (4, 4), (5, 5), (6, 2)\}.$$

A picture of f is created by drawing two sets of dots: one for A on the left and one for B on the right. We draw an arrow from a dot $a \in A$ to a dot $b \in B$ just when $(a, b) \in f$—that is, when $f(a) = b$. From the picture, it is easy to see that $\operatorname{im} f = \{1, 2, 4, 5\}$.

Now consider g defined by

$$g = \{(1, 3), (2, 1), (2, 4), (3, 2), (4, 4), (5, 5)\}.$$

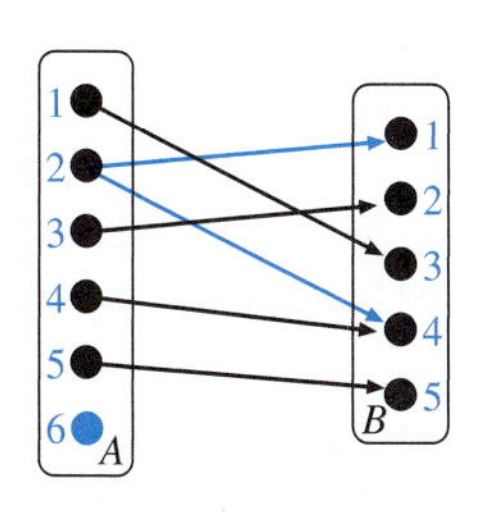

Is g a function from $A = \{1, 2, 3, 4, 5, 6\}$ to $B = \{1, 2, 3, 4, 5\}$? There are two reasons why $g : A \to B$ is false.

First, $6 \in A$ but $6 \notin \operatorname{dom} g$. Thus $\operatorname{dom} g \neq A$. You can see this in the picture: There are no arrows emanating from element 6.

Second, g is not a function (from any set to any set). Notice that $(2, 1), (2, 4) \in g$, which violates Definition 23.1. You can see this in the picture as well: There are two arrows emanating from element 2.

If f is a function from A to B ($f : A \to B$), its picture satisfies the following: Every dot on the left (in A) has exactly one arrow leaving it, ending at the right (in B).

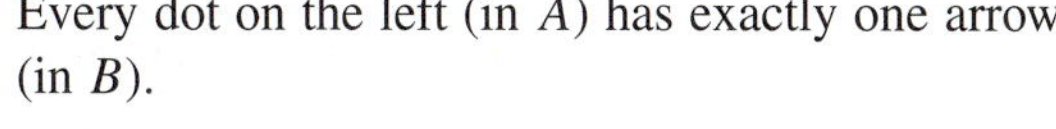

An alternative way to count functions is to count charts. In how many ways can we replace the question marks in the following chart with elements from B?

x	$f(x)$
1	?
2	?
$\vdots$	$\vdots$
a	?

The right-hand column is a length-a list of elements chosen from the b-element set B. There are b^a ways to complete this chart.

Counting Functions

Let A and B be finite sets. How many functions from A to B are there? Without loss of generality, we can choose A to be the set $\{1, 2, \ldots, a\}$ and B to be the set $\{1, 2, \ldots, b\}$. Every function $f : A \to B$ can be written out as

$$f = \{(1, ?), (2, ?), (3, ?), \ldots, (a, ?)\}$$

where the ? entries are elements from B. In how many ways can we replace the ?s with elements in B? There are b choices for the element ? in $(1, ?)$, and for each such choice, there are b choices for the ? in $(2, ?)$, etc., and finally b choices for the ? in $(a, ?)$ given all the previous choices. Thus, all told, there are b^a choices. We have shown the following:

Proposition 23.10 Let A and B be finite sets with $|A| = a$ and $|B| = b$. The number of functions from A to B is b^a.

Example 23.11 Let $A = \{1, 2, 3\}$ and $B = \{4, 5\}$. Find all functions $f : A \to B$.

Solution: Proposition 23.10 tells that there are $2^3 = 8$ such functions. They are

$$\begin{array}{ll} \{(1,4),(2,4),(3,4)\} & \{(1,5),(2,4),(3,4)\} \\ \{(1,4),(2,4),(3,5)\} & \{(1,5),(2,4),(3,5)\} \\ \{(1,4),(2,5),(3,4)\} & \{(1,5),(2,5),(3,4)\} \\ \{(1,4),(2,5),(3,5)\} & \{(1,5),(2,5),(3,5)\}. \end{array}$$

In Section 9 we introduced the notation 2^A for the set of all subsets of A. This notation was a mnemonic for remembering that the number of subsets of an a-element set is 2^a. Similarly, there is a special notation for the set of all functions from A to B. The notation is B^A. This is a mnemonic for Proposition 23.10, because we can write

The notation B^A stands for the set of all functions $f : A \to B$.

$$|B^A| = |B|^{|A|}.$$

In this book, we do not use this notation. Furthermore, people often find it confusing. It is tempting to pronounce the symbols B^A as "B to the A," whereas the notation means the set of functions from A to B.

Inverse Functions

A function is a special type of relation. Recall that in Section 13 we defined the inverse of a relation R, denoted R^{-1}, to be the relation formed from R by reversing all its ordered pairs.

Since a function, f, is a relation, we may also consider f^{-1}. The problem we consider here is: If f is a function from A to B, is f^{-1} a function from B to A?

Example 23.12 Let $A = \{0, 1, 2, 3, 4\}$ and $B = \{5, 6, 7, 8, 9\}$. Let $f : A \to B$ be defined by

$$f = \{(0, 5), (1, 7), (2, 8), (3, 9), (4, 7)\},$$

so

$$f^{-1} = \{(5, 0), (7, 1), (8, 2), (9, 3), (7, 4)\}.$$

Is f^{-1} a function from B to A? The answer is no for two reasons. First, f^{-1} is not a function. Note that both $(7, 1)$ and $(7, 4)$ are in f^{-1}. Second, $\operatorname{dom} f^{-1} = \{5, 7, 8, 9\} \neq B$. See the figure.

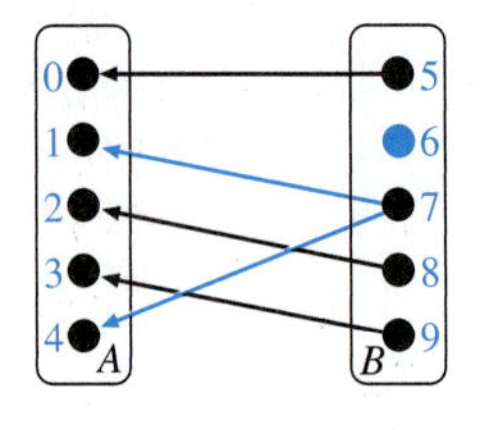

In this example, f^{-1} is not a function. Let us examine why. Consulting Definition 23.1, we observe that for f^{-1} to be a function, it must, first, be a relation. This is not an issue; since f is a relation, so is f^{-1}. Second, whenever $(a, b), (a, c) \in f^{-1}$, we must have $b = c$. Restating this in terms of f, whenever $(b, a), (c, a) \in f$, we must have $b = c$. This is what went wrong in Example 23.12; we had $(1, 7), (4, 7) \in f$, but $1 \neq 4$.

Pictorially, f^{-1} is not a function because there are two f-arrows entering element 7 on the right.

Let us formalize this condition as a definition.

Definition 23.13 **(One-to-one)** A function f is called *one-to-one* provided that, whenever (x, b), $(y, b) \in f$, we must have $x = y$. In other words, if $x \neq y$, then $f(x) \neq f(y)$.

Mathspeak!

The term *one-to-one* is often written as 1:1. Another word for a one-to-one function is *injection*.

The function in Example 23.12 is not one-to-one because $f(1) = f(4)$ but $1 \neq 4$. Compare closely Definitions 23.13 (one-to-one) and 23.1 (function). The conditions are quite similar.

Proposition 23.14 Let f be a function. The inverse relation f^{-1} is a function if and only if f is one-to-one.

The proof is left to you (Exercise 23.10). While you are at it, also prove the following:

Proposition 23.15 Let f be a function and suppose f^{-1} is also a function. Then $\operatorname{dom} f = \operatorname{im} f^{-1}$ and $\operatorname{im} f = \operatorname{dom} f^{-1}$.

It is common to want to prove that a function is one-to-one. Proof Template 20 gives strategies for proving that a function is one-to-one.

Proof Template 20 Proving a function is one-to-one.

To show that f is one-to-one:

Direct method: Suppose $f(x) = f(y)$. ... Therefore $x = y$. Therefore f is one-to-one. ■

Contrapositive method: Suppose $x \neq y$. ... Therefore $f(x) \neq f(y)$. Therefore f is one-to-one. ■

Contradiction method: Suppose $f(x) = f(y)$ but $x \neq y$. ... $\Rightarrow\Leftarrow$ Therefore f is one-to-one. ■

Example 23.16 Let $f : \mathbb{Z} \to \mathbb{Z}$ by $f(x) = 3x + 4$. Prove that f is one-to-one.

Proof. Suppose $f(x) = f(y)$. Then $3x + 4 = 3y + 4$. Subtracting 4 from both sides gives $3x = 3y$. Dividing both sides by 3 gives $x = y$. Therefore f is one-to-one. ■

On the other hand, to prove that a function is not one-to-one typically requires us to present a counterexample—that is, a pair of objects x and y with $x \neq y$ but $f(x) = f(y)$.

Example 23.17 Let $f : \mathbb{Z} \to \mathbb{Z}$ by $f(x) = x^2$. Prove that f is not one-to-one.

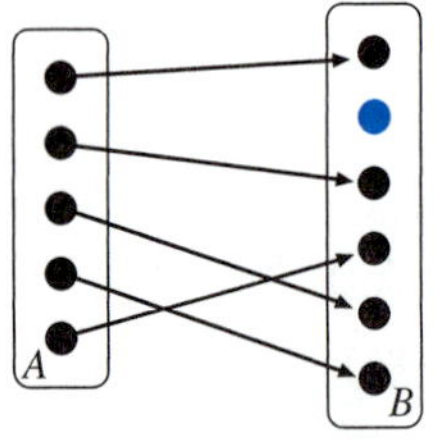

Proof. Notice that $f(3) = f(-3) = 9$, but $3 \neq -3$. Therefore f is not one-to-one. ■

Mathspeak!

In standard English, the word *onto* is a preposition. In mathematical English, we use *onto* as an adjective. Another word for an onto function is *surjection*.

For the inverse of a function also to be a function, it is necessary and sufficient that the function be one-to-one. Now we consider a more focused question. Let $f : A \to B$. We want to know when f^{-1} is a function from B to A. Recall that we had two difficulties in Example 23.12. We have dealt with the first difficulty: f^{-1} needs to be a function. The second difficulty was that there was an element in B that had no incoming arrow.

Consider the function $f : A \to B$ shown in the figure. Clearly f is one-to-one, so f^{-1} is a function. However, f^{-1} is not a function from B to A because there is an element $b \in B$ for which $f^{-1}(b)$ is undefined. For $f^{-1} : B \to A$, there must be an f-arrow pointing to every element of B. Here is the careful way to say this:

Definition 23.18 **(Onto)** Let $f : A \to B$. We say that f is *onto* B provided that for every $b \in B$ there is an $a \in A$ so that $f(a) = b$. In other words, $\operatorname{im} f = B$.

The sentence "$f : A \to B$ is onto" is a promise that the following are true. First, f is a function. Second, $\operatorname{dom} f = A$. And third, $\operatorname{im} f = B$ (see Exercise 23.7).

Example 23.19 Let $A = \{1, 2, 3, 4, 5, 6\}$ and $B = \{7, 8, 9, 10\}$. Let

$$f = \{(1,7), (2,7), (3,8), (4,9), (5,9), (6,10)\}, \quad \text{and}$$
$$g = \{(1,7), (2,7), (3,7), (4,9), (5,9), (6,10)\}.$$

Note that $f : A \to B$ is onto because for each element b of B, we can find one or more elements $a \in A$ such that $f(a) = b$. It is also easy to check that $\operatorname{im} f = B$.

However, $g : A \to B$ is not onto. Note that $8 \in B$, but there is no $a \in A$ with $g(a) = 8$. Also, $\operatorname{im} g = \{7, 9, 10\} \neq B$.

The condition that $f : A \to B$ is onto can be expressed using the quantifiers $\exists$ and $\forall$ as

$$\forall b \in B, \exists a \in A, \ f(a) = b.$$

The condition that f is not onto can be expressed

$$\exists b \in B, \forall a \in A, \ f(a) \neq b.$$

These ways of thinking about onto functions are formalized in Proof Template 21.

Proof Template 21 Proving a function is onto.

To show $f : A \to B$ is onto:
Direct method: Let b be an arbitrary element of B. Explain how to find/construct an element $a \in A$ such that $f(a) = b$. Therefore f is onto. ■

Set method: Show that the sets B and $\operatorname{im} f$ are equal. ■

Example 23.20 Let $f : \mathbb{Q} \to \mathbb{Q}$ by $f(x) = 3x + 4$. Prove that f is onto $\mathbb{Q}$.

Recall that $\mathbb{Q}$ stands for the set of rational numbers.

Proof. Let $b \in \mathbb{Q}$ be arbitrary. We seek an $a \in \mathbb{Q}$ such that $f(a) = b$. Let $a = \frac{1}{3}(b - 4)$. (Since b is a rational number, so is a.) Notice that

$$f(a) = 3\left[\tfrac{1}{3}(b-4)\right] + 4 = (b-4) + 4 = b.$$

Therefore $f : \mathbb{Q} \to \mathbb{Q}$ is onto. ■

How did we ever "guess" that we should take $a = \frac{1}{3}(b-4)$? We didn't guess; we worked backward!

Let $f : A \to B$. In order for f^{-1} to be a function, it is necessary and sufficient that f be one-to-one. Given that, in order for $f^{-1} : B \to A$, it is necessary for f to be onto B. Otherwise, if f is not onto B, we can find a $b \in B$ such that $f^{-1}(b)$ is undefined.

Theorem 23.21 Let A and B be sets and let $f : A \to B$. The inverse relation f^{-1} is a function from B to A if and only if f is one-to-one and onto B.

Proof. Let $f : A \to B$.

($\Rightarrow$) Suppose f is one-to-one and onto B. We need to prove that $f^{-1} : B \to A$. We use Proof Template 19.

- Since f is one-to-one, we know by Proposition 23.14 that f^{-1} is a function.
- Since f is onto B, $\operatorname{im} f = B$. By Proposition 23.15, $\operatorname{dom} f^{-1} = B$.
- Since the domain of f is A, by Proposition 23.15, $\operatorname{im} f^{-1} = A$.

Therefore $f^{-1} : B \to A$.

($\Leftarrow$) Suppose $f : A \to B$ and $f^{-1} : B \to A$. Since f^{-1} is a function, f is one-to-one (Proposition 23.14). Since $\operatorname{im} f = \operatorname{dom} f^{-1} = B$, we see that f is onto B. ■

A function $f : A \to B$ that is both kinds of "jection"—an injection and a surjection—is called a *bijection*.

A function that is both one-to-one and onto has a special name.

Definition 23.22 **(Bijection)** Let $f : A \to B$. We call f a *bijection* provided it is both one-to-one and onto.

Example 23.23 Let A be the set of even integers and let B be the set of odd integers. The function $f : A \to B$ defined by $f(x) = x + 1$ is a bijection.

Proof. We must prove that f is both one-to-one and onto. To see that f is one-to-one, suppose $f(x) = f(y)$ where x and y are even integers. Thus

$$f(x) = f(y) \quad \Rightarrow \quad x + 1 = y + 1 \quad \Rightarrow \quad x = y.$$

Hence f is one-to-one.

To see that f is onto B, let $b \in B$ (i.e., b is an odd integer). By definition, $b = 2k + 1$ for some integer k. Let $a = 2k$; clearly a is even. Then $f(a) = a + 1 = 2k + 1 = b$, so f is onto. Since f is both one-to-one and onto, f is a bijection. ■

Counting Functions, Again

Let A and B be finite sets with $|A| = a$ and $|B| = b$. How many functions $f : A \to B$ are one-to-one? How many are onto?

Let's look at two easy special cases. If $|A| > |B|$, then f cannot be one-to-one. Why? Consider the function $f : A \to B$ that we hope is one-to-one. Because f is one-to-one, for distinct elements $x, y \in A$, $f(x)$ and $f(y)$ are distinct elements of B. So let's say the first b elements of A are mapped by f to b different elements in B. After that, there are no further elements in B to which we can map elements of A!

On the other hand, if $|A| < |B|$, then f cannot be onto. Why? There aren't enough elements in A to "cover" all the elements in B!

Let's summarize these comments.

Proposition 23.24 **(Pigeonhole Principle)** Let A and B be finite sets and let $f : A \to B$. If $|A| > |B|$, then f is not one-to-one. If $|A| < |B|$, then f is not onto.

Stated in the contrapositive, if $f : A \to B$ is one-to-one, then $|A| \le |B|$, and if $f : A \to B$ is onto, then $|A| \ge |B|$. If f is both, we have the following:

Proposition 23.25 Let A and B be finite sets and let $f : A \to B$. If f is a bijection, then $|A| = |B|$.

Let us return to the problem of counting those functions from an a-element set to a b-element set that are one-to-one and those functions that are onto.

The good news is that we have solved these problems in previous sections of this book!

Counting one-to-one functions.

Consider the problem of counting one-to-one functions. Without loss of generality, suppose $A = \{1, 2, \ldots, a\}$ and $B = \{1, 2, \ldots, b\}$. A one-to-one function from A to B is of the form

$$f = \{(1, ?), (2, ?), (3, ?), \ldots, (a, ?)\}$$

where the ?s are filled in with elements of B without repetition. This is a list-counting problem that we solved in Section 7.

Counting onto functions.

Now consider the problem of counting onto functions. Here we want to fill in the ?s with elements of B so that every element is used at least once. The number of length-a lists whose elements come from B and use all the elements in B at least once was solved in Section 18.

Let us collect what we learned in those sections and summarize them in the following result.

Theorem 23.26 Let A and B be finite sets with $|A| = a$ and $|B| = b$.

(1) The number of functions from A to B is b^a.

(2) If $a \le b$, the number of one-to-one functions $f : A \to B$ is

$$(b)_a = b(b-1)\cdots(b-a+1) = \frac{b!}{(b-a)!}.$$

If $a > b$, the number of such functions is zero.

(3) If $a \ge b$, the number of onto functions $f : A \to B$ is

$$\sum_{j=0}^{b} (-1)^j \binom{b}{j} (b-j)^a.$$

If $a < b$, the number of such functions is zero.

(4) If $a = b$, the number of bijections $f : A \to B$ is $a!$. If $a \neq b$, the number of such functions is zero.

Recap

We introduced the concept of function, as well as the notation $f : A \to B$. We investigated when the inverse relation of a function is itself a function. We studied the properties one-to-one and onto. We counted functions between finite sets.

---

23 Exercises

23.1. For each of the following relations, please answer these questions:

(1) Is it a function? If not, explain why and stop. Otherwise, continue with the remaining questions.

(2) What are its domain and image?

(3) Is the function one-to-one? If not, explain why and stop. Otherwise, answer the remaining question.

(4) What is its inverse function?

a. $\{(1, 2), (3, 4)\}$.

b. $\{(x, y) : x, y \in \mathbb{Z},\ y = 2x\}$.

c. $\{(x, y) : x, y \in \mathbb{Z},\ x + y = 0\}$.

d. $\{(x, y) : x, y \in \mathbb{Z},\ xy = 0\}$.

e. $\{(x, y) : x, y \in \mathbb{Z},\ y = x^2\}$.

f. $\emptyset$.

g. $\{(x, y) : x, y \in \mathbb{Q},\ x^2 + y^2 = 1\}$.

h. $\{(x, y) : x, y \in \mathbb{Z},\ x|y\}$.

i. $\{(x, y) : x, y \in \mathbb{N},\ x|y \text{ and } y|x\}$.

j. $\{(x, y) : x, y \in \mathbb{N},\ \binom{x}{y} = 1\}$.

23.2. Let $A = \{1, 2, 3\}$ and $B = \{4, 5\}$. Write down all functions $f : A \to B$. Indicate which are one-to-one and which are onto B.

23.3. Let $A = \{1, 2\}$ and $B = \{3, 4, 5\}$. Write down all functions $f : A \to B$. Indicate which are one-to-one and which are onto B.

23.4. Let $A = \{1, 2\}$ and $B = \{3, 4\}$. Write down all functions $f : A \to B$. Indicate which are one-to-one and which are onto B.

23.5. For each of the following functions, find $f(2)$.

a. $f = \{(x, y) : x, y \in \mathbb{Z}, x + y = 0\}$.

b. $f = \{(1, 2), (2, 3), (3, 2)\}$.

c. $f : \mathbb{N} \to \mathbb{N}$ by $f(x) = (x + 1)^{(x+1)}$.

d. $f = \{1, 2, 3, 4, 5\} \times \{1\}$.

e. $f : \mathbb{N} \to \mathbb{N}$ by $f(n) = n!$.

23.6. Let $A = \{1, 2, 3, 4\}$ and $B = \{5, 6, 7\}$. Let f be the relation

$$f = \{(1, 5), (2, 5), (3, 6), (?, ?)\}$$

where the two question marks are to be determined by you. Your job is to find replacements for (?, ?) so that each of the following is true. [Three different answers—one for each of (a), (b), and (c)—are expected. The ordered pair (?, ?) should be a member of $A \times B$.]

a. The relation f is not a function.

b. The relation f is a function from A to B but is not onto B.

c. The relation f is a function from A to B and is onto B.

Despite the fact that the phrase "f is onto" does not make sense in isolation, mathematicians often write it. It makes sense if we are thinking about a particular pair of sets A and B with $f : A \to B$. In this context, "f is onto" means "f is onto B."

23.7. Consider the following two sentences about a function f:

a. f is onto.

b. $f : A \to B$ is onto.

Explain why (a) does not make sense but (b) does.

23.8. The sine function is a function to and from the real numbers; that is $\sin : \mathbb{R} \to \mathbb{R}$. The sine function is neither one-to-one nor onto. Yet the arc sine function, $\sin^{-1}$, is known as its inverse function.

Explain.

23.9. For each of the following, determine whether the function is one-to-one, onto, or both. Prove your assertions.

a. $f : \mathbb{Z} \to \mathbb{Z}$ defined by $f(x) = 2x$.

b. $f : \mathbb{Z} \to \mathbb{Z}$ defined by $f(x) = 10 + x$.

c. $f : \mathbb{N} \to \mathbb{N}$ defined by $f(x) = 10 + x$.

d. $f : \mathbb{Z} \to \mathbb{Z}$ defined by

$$f(x) = \begin{cases} \frac{x}{2} & \text{if } x \text{ is even} \\ \frac{x-1}{2} & \text{if } x \text{ is odd.} \end{cases}$$

e. $f : \mathbb{Q} \to \mathbb{Q}$ defined by $f(x) = x^2$.

23.10. Prove Propositions 23.14 and 23.15.

23.11. Let A and B be finite sets and let $f : A \to B$. Prove that any two of the following statements being true implies the third.

a. f is one-to-one.

b. f is onto.

c. $|A| = |B|$.

23.12. Give an example of a set A and a function $f : A \to A$ where f is onto but not one-to-one.

Give an example where f is one-to-one but not onto.

Are your examples contradictions to the previous exercise?

23.13. Suppose $f : A \to B$ is a bijection. Prove that $f^{-1} : B \to A$ is a bijection as well.

23.14. Let A be an n-element set and let $k \in \mathbb{N}$. How many functions $f : A \to \{0, 1\}$ are there for which there are exactly k elements a in A with $f(a) = 1$?

23.15. Let A be an n-element set and let $i, j, k \in \mathbb{N}$ with $i + j + k = n$. How many functions $f : A \to \{0, 1, 2\}$ are there for which there are exactly i elements $a \in A$ with $f(a) = 0$, exactly j elements $a \in A$ with $f(a) = 1$, and exactly k elements $a \in A$ with $f(a) = 2$.

24 The Pigeonhole Principle

Proposition 23.24 is called the *Pigeonhole Principle*. It asserts that if A and B are finite sets and if $|A| > |B|$, then there can be no one-to-one function $f : A \to B$. The reason is clear: There are too many elements in A. What, you might ask, does this result have to do with *pigeons*?

Imagine that we own a flock of pigeons and that the pigeons live in a coop. The pigeon coop is divided into separate compartments called *holes* where the pigeons nest.

Suppose we own p pigeons and our coop has h holes. If $p \le h$, then the coop is large enough so that pigeons do not have to share holes. However, if $p > h$, then there are not enough holes to give every pigeon a private room; some pigeons will have to share quarters.

There are a number of interesting mathematical problems that can be solved by the Pigeonhole Principle. Here we present some examples.

Proposition 24.1 Let $n \in \mathbb{N}$. Then there exist positive integers a and b, with $a \ne b$, such that $n^a - n^b$ is divisible by 10.

For example, if $n = 17$, then we can subtract

$$\begin{array}{rrcr} & 17^6 & = & 24{,}137{,}569 \\ - & 17^2 & = & 289 \\ \hline & & & 24{,}137{,}280 \end{array}$$

which is divisible by 10.

To prove this result, we use the well-known fact that a natural number is divisible by 10 if and only if its last digit is a zero. A more careful approach would use ideas developed in Section 34.

Proof. Consider the 11 natural numbers

$$n^1 \quad n^2 \quad n^3 \quad \cdots \quad n^{11}.$$

The ones digits of these numbers take on values in the set $\{0, 1, 2, \ldots, 9\}$. Since there are only ten possible ones digits, and we have 11 different numbers, two of these numbers (say n^a and n^b) must have the same ones digit. Therefore $n^a - n^b$ is divisible by 10. ■

The next example comes from geometry. Every point in the plane can be expressed in terms of its x- and y-coordinates. A point whose coordinates are both integers is called a *lattice* point. For example, the points $(1, 2)$, $(-3, 8)$, and the origin are lattice points, but $(1.3, 0)$ is not.

Proposition 24.2 Given five distinct lattice points in the plane, at least one of the line segments determined by these points has a lattice point as its midpoint.

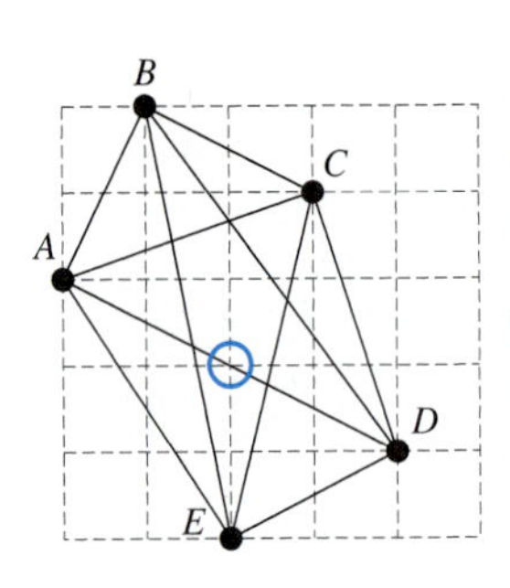

In other words, suppose A, B, C, D, and E are distinct lattice points. There are $\binom{5}{2} = 10$ different line segments we can form whose endpoints are in the set $\{A, B, C, D, E\}$. Proposition 24.2 asserts that the midpoint of one (or more) of these line segments must also be a lattice point. For example, consider the five points in the figure. The midpoint of segment AD is a lattice point.

To prove this result, we recall the midpoint formula from coordinate geometry. Let (a, b) and (c, d) be two points in the plane (not necessarily lattice points). The midpoint of the line segment determined by these points can be found using the following formula:

$$\left(\frac{a+c}{2}, \frac{b+d}{2}\right).$$

Proof (of Proposition 24.2)

We are given five distinct lattice points in the plane. The various coordinates are integers and hence are either even or odd. Given a lattice point's coordinates, we can classify it as one of the following four types:

(even, even) (even, odd) (odd, even) (odd, odd)

depending on the parity of its coordinates. Notice that we have five lattice points, but only four parity categories. Therefore (by the Pigeonhole Principle) two of these points must have the same parity type. Suppose these two points have coordinates (a, b) and (c, d). The midpoint of this segment is at coordinates $\left(\frac{a+c}{2}, \frac{b+d}{2}\right)$. Since a and c have the same parity, $a + c$ is even, and so $\frac{a+c}{2}$ is an integer. Likewise $\frac{b+d}{2}$ is an integer. This proves that the midpoint is a lattice point. ■

The third example concerns sequences of integers. A *sequence* is simply a list. Given a sequence of integers, a *subsequence* is a list formed by deleting elements from the original list and keeping the remaining elements in the same order in which they originally appeared.

For example, the sequence

$$1 \quad 9 \quad 10 \quad 8 \quad 3 \quad 7 \quad 5 \quad 2 \quad 6 \quad 4$$

contains the subsequence

$$9 \quad 8 \quad 6 \quad 4.$$

Notice that the four numbers in the subsequence are in decreasing order, and so we call it a *decreasing* subsequence. Similarly, a subsequence whose elements are in increasing order is called an *increasing* subsequence.

We claim that every sequence of ten distinct integers must contain a subsequence of four elements that is either increasing or decreasing. The sequence above has a decreasing subsequence of length four and also an increasing subsequence of length four (find it). The sequence

$$10 \quad 9 \quad 8 \quad 7 \quad 6 \quad 5 \quad 4 \quad 3 \quad 1 \quad 2$$

has several length-four decreasing subsequences, but no length-four increasing subsequence.

A sequence that is either increasing or decreasing is called *monotone*. Our claim is that every sequence of ten distinct integers must contain a monotone, length-four subsequence. This claim is a special case of a more general result.

Theorem 24.3 **(Erdős-Szekeres)** Let n be a positive integer. Every sequence of $n^2 + 1$ distinct integers must contain a monotone subsequence of length $n + 1$.

Our example (sequences of length ten) is the case $n = 3$ of the Erdős-Szekeres Theorem.

Proof. Let n be a positive integer. Suppose, for the sake of contradiction, that there is a sequence S of $n^2 + 1$ distinct integers that does not contain a monotone subsequence of length $n + 1$. In other words, all the monotone subsequences of S have length n or less.

Let x be an element of the sequence S. We label x with a pair of integers (u_x, d_x). The integer u_x (u for up) is the length of a longest increasing subsequence of S that starts at x. Similarly, d_x (d for down) is the length of a longest decreasing subsequence of S that starts at x.

For example, the sequence

$$1 \quad 9 \quad 10 \quad 8 \quad 3 \quad 7 \quad 5 \quad 2 \quad 6 \quad 4$$

would be labeled as follows:

1	9	10	8	3	7	5	2	6	4
(4,1)	(2,5)	(1,5)	(1,4)	(3,2)	(1,3)	(2,2)	(2,1)	(1,2)	(1,1)

Element 4 is the last element in the sequence, so it gets the label $(1, 1)$—the only sequences starting at 4 have length one. Element 9 has label $(2, 5)$ because the length of a longest increasing subsequence starting at 9 is two: $(9, 10)$. The length of a longest decreasing subsequence starting at 9 is five: $(9, 8, 7, 5, 4)$ or $(9, 8, 7, 6, 4)$.

Returning to the proof, we make the following observations.

- Because there are no monotone subsequences of length $n+1$ (or longer), *the labels on the sequence S use only the integers* 1 *through* n.

  Hence, we use at most n^2 labels (from $(1, 1)$ to (n, n)).
- We claim that *two distinct elements of the sequence cannot have the same label*.

  To see why, suppose x and y are distinct elements of the sequence with x appearing before y. Their labels are (u_x, d_x) and (u_y, d_y). Because the numbers on the list are distinct, either $x < y$ or $x > y$.

  If $x < y$, then we claim $u_x > u_y$: We know there is an increasing subsequence of length u_y starting at y. If we insert x at the beginning of this subsequence, we get an increasing subsequence of length $u_y + 1$. Thus $u_x \geq u_y + 1$, or, equivalently, $u_x > u_y$. Thus x and y have different labels.

  Similarly, if $x > y$, then we have $d_x > d_y$ and we again conclude that x and y have different labels.

However, these two observations lead to a contradiction. There are only n^2 different labels, and S has $n^2 + 1$ elements. By the Pigeonhole Principle, two of the elements must have the same label. However, this contradicts the second observation that no two elements can have the same label.$\Rightarrow\Leftarrow$ Therefore S must have a monotone subsequence of length $n + 1$. ■

Cantor's Theorem

The Pigeonhole Principle asserts that if $|A| > |B|$, there can be no one-to-one function $f : A \to B$. The flip side of this coin is that if $|A| < |B|$, there can be no onto function $f : A \to B$. Therefore, if $f : A \to B$ is both one-to-one and onto, then $|A| = |B|$.

These assertions are meaningful only if A and B are finite sets. Of course, it is possible to find bijections between infinite sets. For example, here is a bijection from $\mathbb{N}$ onto $\mathbb{Z}$. Define $f : \mathbb{N} \to \mathbb{Z}$ by

$$f(n) = \begin{cases} -n/2 & \text{if } n \text{ is even and} \\ (n+1)/2 & \text{if } n \text{ is odd.} \end{cases}$$

It is a bit awkward to see that f is a bijection from $\mathbb{N}$ onto $\mathbb{Z}$ just by staring at these formulas. However, if we compute a few values of f (for some small values of n), the picture snaps into focus.

n	0	1	2	3	4	5	6	7	8	9
$f(n)$	0	1	−1	2	−2	3	−3	4	−4	5

Clearly, f is a one-to-one function (every integer appears at most once in the lower row of the chart) and is onto $\mathbb{Z}$ (every integer is somewhere on the lower row). See Exercise 24.9.

Since there is a bijection from $\mathbb{N}$ to $\mathbb{Z}$, it makes a little bit of sense to write $|\mathbb{N}| = |\mathbb{Z}|$. This means that $\mathbb{N}$ and $\mathbb{Z}$ are "just as infinite." This often strikes people as counterintuitive because $\mathbb{Z}$ ought to be "twice as infinite" as $\mathbb{N}$. However, the bijection shows that we can match up—in a one-to-one fashion—the elements of the two sets.

You might be tempted to reconcile this in your mind by saying $|\mathbb{Z}| = |\mathbb{N}|$ because both are infinite. This is not correct. The notation $|\mathbb{Z}| = |\mathbb{N}|$ should not be used because the sets are infinite; however, the meaning we are trying to convey is that there is a bijection between $\mathbb{N}$ and $\mathbb{Z}$. In this sense, the two infinite sets have the same size despite the fact that $\mathbb{Z}$ superficially appears to be "twice as big" as $\mathbb{N}$.

Is it possible for two infinite sets not to have the same "size"? At first, this seems like a silly question. If the two sets are both infinite, then they are both infinite—end of story! But this doesn't quite answer the question.

It is reasonable to define two sets as having the same size provided there is a bijection between them. In this sense, $\mathbb{N}$ and $\mathbb{Z}$ have the same size. Do all infinite sets have the same size? The surprising answer to this question is no.

We prove that $\mathbb{Z}$ and $2^{\mathbb{Z}}$ (the set of integers and the set of all subsets of the integers) do not have the same size. Here is the general result:

Theorem 24.4 **(Cantor)** Let A be a set. If $f : A \to 2^A$, then f is not onto.

If A is a finite set, this result is easy. If $|A| = a$, then $|2^A| = 2^a$ and we know that $a < 2^a$ (see Exercise 20.3). Since 2^A is a larger set, there can be no onto function $f : A \to 2^A$. This argument, however, applies only to finite sets. Cantor's Theorem applies to all sets.

Proof. Let A be a set and let $f : A \to 2^A$. To show that f is not onto, we must find a $B \in 2^A$ (i.e., $B \subseteq A$) for which there is no $a \in A$ with $f(a) = B$. In other words, B is a set that f "misses." To this end, let

Since $f(x)$ is a set, indeed a subset of A, the condition $x \notin f(x)$ makes sense.

$$B = \{x \in A : x \notin f(x)\}.$$

We claim there is no $a \in A$ with $f(a) = B$.

Suppose, for the sake of contradiction, there is an $a \in A$ such that $f(a) = B$. We ponder: Is $a \in B$?

- If $a \in B$, then, since $B = f(a)$, we have $a \in f(a)$. So, by definition of B, $a \notin f(a)$; that is, $a \notin B$.$\Rightarrow\Leftarrow$
- If $a \notin B = f(a)$, then, by definition of B, $a \in B$.$\Rightarrow\Leftarrow$

Both $a \in B$ and $a \notin B$ lead to contradictions, and hence our supposition [there is an $a \in A$ with $f(a) = B$] is false, and therefore f is not onto. ∎

Example 24.5 We illustrate the proof of Theorem 24.4 with a specific example. Let $A = \{1, 2, 3\}$. Let $f : A \to 2^A$ as defined in the following chart.

a	$f(a)$	$a \in f(a)$?
1	$\{1, 2\}$	yes
2	$\{3\}$	no
3	$\emptyset$	no

Now $B = \{x \in A : x \notin f(x)\}$. Since $1 \in f(1)$, but $2 \notin f(2)$ and $3 \notin f(3)$, we have $B = \{2, 3\}$. Notice that there is no $a \in A$ with $f(a) = B$.

The implication of Cantor's Theorem is that $|\mathbb{Z}| \neq |2^{\mathbb{Z}}|$. In a correct sense $2^{\mathbb{Z}}$ is more infinite than $\mathbb{Z}$. Cantor developed these notions by creating a new set of numbers "beyond" the natural numbers; he called these numbers *transfinite cardinals*. The smallest infinite sets, Cantor proved, have the same size as $\mathbb{N}$. The size of $\mathbb{N}$ is denoted by the transfinite number named $\aleph_0$ (aleph null).

Recap

There cannot be a one-to-one function from a set to a smaller set; this fact is known as the Pigeonhole Principle. We illustrated how this fact can be used in proofs. We also know that there cannot be a function from a set onto a larger set. We showed that for any set A, the set 2^A is larger, even for infinite sets A.

24 Exercises

24.1. Let $(a_1, a_2, a_3, a_4, a_5)$ be a sequence of five distinct integers. We call such a sequence increasing if $a_1 < a_2 < a_3 < a_4 < a_5$ and decreasing if $a_1 > a_2 > a_3 > a_4 > a_5$. Other sequences may have a different pattern of $<$s and $>$s. For the sequence $(1, 5, 2, 3, 4)$ we have $1 < 5 > 2 < 3 < 4$. Different sequences may have the same pattern of $<$s and $>$s between their elements. For example, $(1, 5, 2, 3, 4)$ and $(0, 6, 1, 3, 7)$ have the same pattern of $<$s and $>$s as illustrated here:

$$\begin{array}{ccccccccc} 1 & < & 5 & > & 2 & < & 3 & < & 4 \\ & \updownarrow & & \updownarrow & & \updownarrow & & \updownarrow & \\ 0 & < & 6 & > & 1 & < & 3 & < & 7 \end{array}$$

Given a collection of 17 sequences of five distinct integers, prove that 2 of them have the same pattern of $<$s and $>$s.

24.2. Two Social Security numbers (see Exercise 7.9) *match zeros* if a digit of one number is zero iff the corresponding digit of the other is also zero. For example, the Social Security numbers 120-90-1109 and 430-20-5402 have matching zeros.

Prove: Given a collection of 513 Social Security numbers, there must be two that match zeros.

24.3. Given a set of seven distinct positive integers, prove that there is a pair whose sum or whose difference is a multiple of 10.

You may use the fact that if the ones digit of an integer is 0, then that integer is a multiple of 10.

24.4. Consider a square whose side has length one. Suppose we select five points from this square. Prove that there are two points whose distance is at most $\sqrt{2}/2$.

24.5. Show that Proposition 24.2 is best possible by finding four lattice points in the plane such that none of their midpoints are lattice points.

24.6. Find and prove a generalization of Proposition 24.2 to three dimensions.

24.7. Find a sequence of nine distinct integers that does not contain a monotone subsequence of length four.

Generalize your construction by showing how to construct (for every positive integer n) a sequence of n^2 distinct integers that does not contain a monotone subsequence of length $n+1$.

24.8. Write a computer program that takes as its input a sequence of distinct integers and returns as its output the length of a longest monotone subsequence.

24.9. Let $f : \mathbb{N} \to \mathbb{Z}$ by

$$f(n) = \begin{cases} -n/2 & \text{if } n \text{ is even and} \\ (n+1)/2 & \text{if } n \text{ is odd.} \end{cases}$$

Prove that f is a bijection.

24.10. Let E denote the set of even integers. Find a bijection between E and $\mathbb{Z}$.

25 Composition

Just as there are operations (e.g., $+$ and $\times$) for combining integers and there are operations for combining sets (e.g., $\cup$ and $\cap$), there is a natural operation for combining functions.

Definition 25.1 **(Composition of functions)** Let A, B, and C be sets and let $f : A \to B$ and $g : B \to C$. Then the function $g \circ f$ is a function from A to C defined by

$$(g \circ f)(a) = g[f(a)]$$

where $a \in A$. The function $g \circ f$ is called the *composition* of g and f.

Example 25.2 Let $A = \{1, 2, 3, 4, 5\}$, $B = \{6, 7, 8, 9\}$, and $C = \{10, 11, 12, 13, 14\}$. Let $f : A \to B$ and $g : B \to C$ be defined by

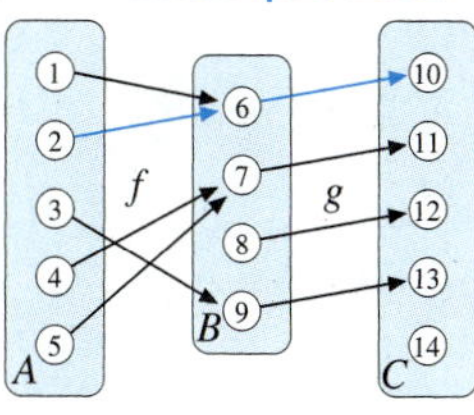

$$f = \{(1,6), (2,6), (3,9), (4,7), (5,7)\}, \quad \text{and}$$
$$g = \{(6,10), (7,11), (8,12), (9,13)\}.$$

Then $(g \circ f)$ is the function

$$(g \circ f) = \{(1,10), (2,10), (3,13), (4,11), (5,11)\}.$$

For example,

$$(g \circ f)(2) = g[f(2)] = g[6] = 10.$$

So $(2, 10) \in g \circ f$; that is, $(g \circ f)(2) = 10$.

Example 25.3 Let $f : \mathbb{Z} \to \mathbb{Z}$ by $f(x) = x^2 + 1$ and $g : \mathbb{Z} \to \mathbb{Z}$ by $g(x) = 2x - 3$. What is $(g \circ f)(4)$?

We calculate $(g \circ f)(4) = g[f(4)] = g(4^2 + 1) = g(17) = 2 \times 17 - 3 = 31$. (See the figure.)

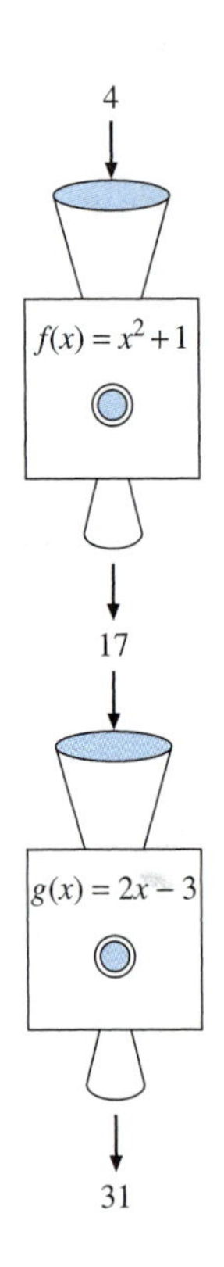

In general,

$$\begin{aligned}(g \circ f)(x) &= g[f(x)] \\ &= g(x^2 + 1) \\ &= 2(x^2 + 1) - 3 \\ &= 2x^2 + 2 - 3 \\ &= 2x^2 - 1.\end{aligned}$$

Some comments:

- The notation $g \circ f$ means that we do f first and then g. It may seem strange that although we evaluate f first, we write its symbol after g. Why? When we apply the function $(g \circ f)$ to an element a, as in

$$(g \circ f)(a)$$

the letter f is closer to a and "hits" it first:

$$(g \circ f)(a) \longrightarrow g[f(a)].$$

- The domain of $g \circ f$ is the same as the domain of f:

$$\text{dom}\,(g \circ f) = \text{dom}\, f.$$

- In order for $g \circ f$ to make sense, every output of f must be an acceptable input to g. Properly said, we need $\text{im}\, f \subseteq \text{dom}\, g$. The requirements $f : A \to B$ and $g : B \to C$ ensure that the functions fit together when we form $g \circ f$.

  For the functions in Example 25.2, $f \circ g$ is undefined because $g(6) = 10$, but $10 \notin \text{dom}\, f$.
- It is possible that $g \circ f$ and $f \circ g$ both make sense (are defined). In this situation, it may be the case that $f \circ g \neq g \circ f$ (are different functions).

Example 25.4 $(g \circ f \neq f \circ g)$

Let $A = \{1, 2, 3, 4, 5, 6\}$ and let $f : A \to A$ and $g : A \to A$ be defined by

$$f = \{(1, 1), (2, 1), (3, 1), (4, 1), (5, 1)\}, \quad \text{and}$$
$$g = \{(1, 5), (2, 4), (3, 3), (4, 2), (5, 1)\}.$$

Then $g \circ f$ and $f \circ g$ are as follows:

$$g \circ f = \{(1, 5), (2, 5), (3, 5), (4, 5), (5, 5)\} \quad \text{and}$$
$$f \circ g = \{(1, 1), (2, 1), (3, 1), (4, 1), (5, 1)\}.$$

Thus $g \circ f \neq f \circ g$.

Example 25.5 Recall the functions f and g from Example 25.3: $f(x) = x^2+1$ and $g(x) = 2x-3$. For these, we have

$$(g \circ f)(4) = g[f(4)] = g(17) = 31 \quad \text{and}$$
$$(f \circ g)(4) = f[g(4)] = f(5) = 26.$$

Therefore $g \circ f \neq f \circ g$.

More generally,

$$\begin{aligned}(g \circ f)(x) &= g[f(x)] = g[x^2 + 1] \\ &= 2[x^2 + 1] - 3 = 2x^2 - 1 \qquad \text{and}\end{aligned}$$

$$\begin{aligned}(f \circ g)(x) &= f[g(x)] = f[2x - 3] \\ &= [2x - 3]^2 + 1 \\ &= 4x^2 - 12x + 10.\end{aligned}$$

Therefore $g \circ f \neq f \circ g$.

Therefore function composition does not satisfy the commutative property. It does, however, satisfy the associative property.

Proposition 25.6 Let A, B, C, and D be sets and let $f : A \to B$, $g : B \to C$, and $h : C \to D$. Then

$$h \circ (g \circ f) = (h \circ g) \circ f.$$

This proposition asserts that two functions, $h \circ (g \circ f)$ and $(h \circ g) \circ f$, are the same function. Before we begin this proof, let us pause to consider: How do we prove two functions are the same? We can go back to basics and recall that functions are relations, and relations in turn are sets of ordered pairs. We can then follow Proof Template 5 to show that the sets are equal.

However, it is simpler if we show that the two functions have the same domain, and for every element in their common domain, they produce the same value. This implies that the two sets are the same (see Exercise 25.2). This is summarized in Proof Template 22.

Proof Template 22 Proving two functions are equal.

Let f and g be functions. To prove $f = g$, do the following:

- Prove that dom f = dom g.
- Prove that for every x in their common domain, $f(x) = g(x)$.

We now proceed with the proof of Proposition 25.6.

Proof. Let $f : A \to B$, $g : B \to C$, and $h : C \to D$. We seek to prove $h \circ (g \circ f) = (h \circ g) \circ f$.

First, we check that the domains of $h \circ (g \circ f)$ and $(h \circ g) \circ f$ are the same. Earlier we noted that dom $(g \circ f)$ = dom f. Applying this fact to the current situation, we have

$$\begin{aligned}\operatorname{dom}[h \circ (g \circ f)] &= \operatorname{dom}(g \circ f) = \operatorname{dom} f = A, \quad \text{and} \\ \operatorname{dom}[(h \circ g) \circ f] &= \operatorname{dom} f = A\end{aligned}$$

so both functions have the same domain, A.

Second, we check that for any $a \in A$, the two functions produce the same value. Let $a \in A$ be arbitrary. We compute

$$\begin{aligned}[h \circ (g \circ f)](a) &= h[(g \circ f)(a)] \\ &= h[g[f(a)]]\end{aligned}$$

and

$$\begin{aligned}[(h \circ g) \circ f](a) &= (h \circ g)[f(a)] \\ &= h[g[f(a)]].\end{aligned}$$

Hence $h \circ (g \circ f) = (h \circ g) \circ f$. ■

Identity Function

The integer 1 is the identity element for multiplication, and $\emptyset$ is the identity element for union. What serves as an identity element for composition? There is no single identity element; instead we have many.

Definition 25.7 **(Identity function)** Let A be a set. The *identity function on* A is the function id_A whose domain is A, and for all $a \in A$, $\mathrm{id}_A(a) = a$. In other words,

$$\mathrm{id}_A = \{(a, a) : a \in A\}.$$

The reason we call id_A the identity function is the following:

Proposition 25.8 Let A and B be sets. Let $f : A \to B$. Then

$$f \circ \mathrm{id}_A = \mathrm{id}_B \circ f = f.$$

Proof. We need to show that the functions $f \circ \mathrm{id}_A$, $\mathrm{id}_B \circ f$, and f are all the same. We use Proof Template 22.

Consider $f \circ \mathrm{id}_A$ and f. We have

$$\mathrm{dom}(f \circ \mathrm{id}_A) = \mathrm{dom}\ \mathrm{id}_A = A = \mathrm{dom}\ f$$

so they have the same domain. Let $a \in A$. We calculate

$$(f \circ \mathrm{id}_A)(a) = f(\mathrm{id}_A(a)) = f(a)$$

so $f \circ \mathrm{id}_A$ and f give the same value for all $a \in A$. Therefore $f \circ \mathrm{id}_A = f$.

The argument that $\mathrm{id}_B \circ f = f$ is nearly the same (see Exercise 25.5). ■

Just as multiplying a rational number by its reciprocal gives 1, composing a function with its inverse gives an identity function.

Proposition 25.9 Let A and B be sets and suppose $f : A \to B$ is one-to-one and onto. Then

$$f \circ f^{-1} = \mathrm{id}_B \quad \text{and} \quad f^{-1} \circ f = \mathrm{id}_A.$$

Please prove this (Exercise 25.6).

Recap

In this section we studied the composition of functions and identity functions.

25 Exercises

25.1. We list several pairs of functions f and g. For each pair, please do the following:

- Determine which of $g \circ f$ and $f \circ g$ is defined.
- If one or both are defined, find the resulting function(s).
- If both are defined, determine whether $g \circ f = f \circ g$.

a. $f = \{(1,2), (2,3), (3,4)\}$ and $g = \{(2,1), (3,1), (4,1)\}$.
b. $f = \{(1,2), (2,3), (3,4)\}$ and $g = \{(2,1), (3,2), (4,3)\}$.
c. $f = \{(1,2), (2,3), (3,4)\}$ and $g = \{(1,2), (2,0), (3,5), (4,3)\}$.
d. $f = \{(1,4), (2,4), (3,3), (4,1)\}$ and $g = \{(1,1), (2,1), (3,4), (4,4)\}$.
e. $f = \{(1,2), (2,3), (3,4), (4,5), (5,1)\}$ and $g = \{(1,3), (2,4), (3,5), (4,1), (5,2)\}$.
f. $f(x) = x^2 - 1$ and $g(x) = x^2 + 1$ (both for all $x \in \mathbb{Z}$).
g. $f(x) = x + 3$ and $g(x) = x - 7$ (both for all $x \in \mathbb{Z}$).
h. $f(x) = 1 - x$ and $g(x) = 2 - x$ (both for all $x \in \mathbb{Q}$).
i. $f(x) = \frac{1}{x}$ for $x \in \mathbb{Q}$ except $x = 0$ and $g(x) = x + 1$ for all $x \in \mathbb{Q}$.
j. $f = \mathrm{id}_A$ and $g = \mathrm{id}_B$ where $A \subseteq B$ but $A \neq B$.

25.2. Consider functions f and g. Prove that $f = g$ (as sets) if and only if $\mathrm{dom}\, f = \mathrm{dom}\, g$ and for every x in their common domain, $f(x) = g(x)$. This justifies Proof Template 22.

25.3. Let A and B be sets. Prove that $A = B$ if and only if $\mathrm{id}_A = \mathrm{id}_B$.

25.4. What is the difference between the identity function defined on a set A and the is-equal-to relation defined on A?

25.5. Complete the proof of Proposition 25.8.

25.6. Prove Proposition 25.9.

25.7. Suppose A and B are sets, and f and g are functions with $f : A \to B$ and $g : B \to A$.

Prove: If $g \circ f = \mathrm{id}_A$ and $f \circ g = \mathrm{id}_B$, then f is invertible and $g = f^{-1}$.
Note: This result is a converse to Proposition 25.9.

25.8. Suppose $f : A \to B$ is a bijection. Explain why the following are *incorrect:*

$$f \circ f^{-1} = \mathrm{id}_A \qquad \text{and} \qquad f^{-1} \circ f = \mathrm{id}_B.$$

25.9. Suppose A, B, and C are sets and $f : A \to B$ and $g : B \to C$. Prove the following:

a. If f and g are one-to-one, so is $g \circ f$.
b. If f and g are onto, so is $g \circ f$.
c. If f and g are bijections, so is $g \circ f$.

25.10. Find a pair of functions f and g, from set A to itself, such that $f \circ g = g \circ f$.
Any of the following will work:

- Choose f and g to be the same function.
- Choose f or g to be id_A.
- Choose $g = f^{-1}$.

Those are too easy. Find another example.

25.11. Let A be a set and f a function with $f : A \to A$.

a. Suppose f is one-to-one. Must f be onto?

b. Suppose f is onto. Must f be one-to-one?

Justify your answers.

25.12. Suppose $f : A \to A$ and $g : A \to A$ are both bijections.

a. Prove or disprove: $g \circ f$ is a bijection from A to itself.

b. Prove or disprove: $(g \circ f)^{-1} = g^{-1} \circ f^{-1}$.

c. Prove or disprove: $(g \circ f)^{-1} = f^{-1} \circ g^{-1}$.

25.13. Let A be a set and let $f : A \to A$. Then $f \circ f$ is also a function from A to itself, as is $f \circ f \circ f$.

Note that $f^{(n)}(x)$ does not mean $[f(x)]^n$. For example, if $f(x) = \frac{1}{2}x + 1$, then $f^{(2)}(x) = f[f(x)] = \frac{1}{2}[\frac{1}{2}x + 1] + 1 = \frac{1}{4}x + \frac{3}{2}$. This is not the same as $[f(x)]^2 = (\frac{1}{2}x + 1)^2 = \frac{1}{4}x^2 + x + 1$.

Let us write $f^{(n)}$ to stand for the n-fold composition of f with itself; that is,

$$f^{(n)} = \underbrace{f \circ f \circ \cdots \circ f}_{n \text{ times}}.$$

Of course, $f^{(1)} = f$.

a. Develop a sensible meaning for $f^{(0)}$.

b. If $f, g : A \to A$, must it be the case that $(g \circ f)^{(2)} = g^{(2)} \circ f^{(2)}$? Prove or disprove.

c. If f is invertible, must it be the case that $(f^{-1})^{(n)} = (f^{(n)})^{-1}$? Prove or disprove.

The following questions are best answered with the aid of a computer.

d. Let $f : \mathbb{R} \to \mathbb{R}$ by $f(x) = 2.8x(1 - x)$. Consider the sequence of values

$$f(\tfrac{1}{2}),\ f^{(2)}(\tfrac{1}{2}),\ f^{(3)}(\tfrac{1}{2}),\ f^{(4)}(\tfrac{1}{2}),\ \ldots.$$

Describe the long-term behavior of these numbers.

e. Let $f : \mathbb{R} \to \mathbb{R}$ by $f(x) = 3.1x(1 - x)$. Consider the sequence of values

$$f(\tfrac{1}{2}),\ f^{(2)}(\tfrac{1}{2}),\ f^{(3)}(\tfrac{1}{2}),\ f^{(4)}(\tfrac{1}{2}),\ \ldots.$$

Describe the long-term behavior of these numbers.

f. Let $f : \mathbb{R} \to \mathbb{R}$ by $f(x) = 3.9x(1 - x)$. Consider the sequence of values

$$f(\tfrac{1}{2}),\ f^{(2)}(\tfrac{1}{2}),\ f^{(3)}(\tfrac{1}{2}),\ f^{(4)}(\tfrac{1}{2}),\ \ldots.$$

Describe the long-term behavior of these numbers.

26 Permutations

Informally, a *permutation* is an ordering of objects. The precise meaning of permutation is the following.

Definition 26.1 **(Permutation)** Let A be a set. A *permutation* on A is a bijection from A to itself.

Example 26.2 Let $A = \{1, 2, 3, 4, 5\}$ and let $f : A \to A$ by

$$f = \{(1, 2), (2, 4), (3, 1), (4, 3), (5, 5)\}.$$

Since f is a one-to-one and onto function (i.e., a bijection) from A to A, it is a permutation.

Notice that because f is a bijection, the list $(f(1), f(2), f(3), f(4), f(5)) = (2, 4, 1, 3, 5)$ is simply a reordering of $(1, 2, 3, 4, 5)$.

Mathematicians use the notation S_n to denote the set of all permutations on any n-element set.

It is customary to use lowercase Greek letters (especially π, σ, and τ) to stand for permutations. Note that in this context, π does not stand for the real number $3.14159\ldots$.

The set of all permutations on $\{1, 2, \ldots, n\}$ has a special notation.

Definition 26.3 (S_n) The set of all permutations on the set $\{1, 2, \ldots, n\}$ is denoted S_n.

In later sections, we refer to S_n as the *symmetric group* on n elements.

The symbol ι is a lowercase Greek *iota*. It looks much like an *i* but does not have a dot. It is called the *identity permutation*.

The following result lists important properties of S_n. One of these properties is that the identity function $\mathrm{id}_{\{1,2,\ldots,n\}}$ is a permutation and therefore in S_n. We usually denote the identity function by the lowercase Greek letter ι.

Proposition 26.4 There are $n!$ permutations in S_n. The set S_n satisfies the following properties.

- $\forall \pi, \sigma \in S_n,\ \pi \circ \sigma \in S_n$.
- $\forall \pi, \sigma, \tau \in S_n,\ \pi \circ (\sigma \circ \tau) = (\pi \circ \sigma) \circ \tau$.
- $\forall \pi \in S_n,\ \pi \circ \iota = \iota \circ \pi = \pi$.
- $\forall \pi \in S_n,\ \pi^{-1} \in S_n$ and $\pi \circ \pi^{-1} = \pi^{-1} \circ \pi = \iota$.

Proof. We have already proved all the assertions in this proposition! The fact that $|S_n| = n!$ comes from Theorem 23.26. The fact that the composition of two permutations is a permutation is a consequence of Exercise 25.9. The equation $\iota \circ \pi = \pi \circ \iota = \pi$ follows from Proposition 25.8. The fact that $\pi \in S_n \implies \pi^{-1} \in S_n$ comes from Exercise 23.13 and the fact that $\pi \circ \pi^{-1} = \pi^{-1} \circ \pi = \iota$ is shown in Proposition 25.9. ■

Cycle Notation

The permutation from Example 26.2 in the form of a chart:

x	$\pi(x)$
1	2
2	4
3	1
4	3
5	5

In Example 26.2, we considered the following permutation in S_5:

$$\pi = \{(1, 2), (2, 4), (3, 1), (4, 3), (5, 5)\}.$$

Writing a function as a list of ordered pairs is correct, but it is not always the most useful notation. Here we consider alternative ways of expressing permutations.

We can express π in chart form as in the figure. Another popular form is to express a permutation as a $2 \times n$ array of integers. The top row contains the integers 1 through n in their usual order, and the bottom row contains $\pi(1)$ through $\pi(n)$:

$$\pi = \begin{bmatrix} 1 & 2 & 3 & 4 & 5 \\ 2 & 4 & 1 & 3 & 5 \end{bmatrix}.$$

Notice that the $2 \times n$ array notation is not significantly different from a chart.

The top row in the array notation is not strictly necessary. We could express the permutation π simply by reporting the bottom row; all the information we need is there. We could write $\pi = [2, 4, 1, 3, 5]$. When n is small (e.g., $n = 5$), this notation is reasonable. However, for a larger value of n (e.g., $n = 200$), it is awkward for human beings to distinguish between the values for $\pi(83)$ and $\pi(84)$. On the other hand, this is a reasonable way to store a permutation in a computer.

An alternative notation for expressing permutations is known as *cycle notation*. The cycle notation for the permutation $\pi = \left[\begin{smallmatrix} 1 & 2 & 3 & 4 & 5 \\ 2 & 4 & 1 & 3 & 5 \end{smallmatrix}\right]$ is

$$\pi = (1, 2, 4, 3)(5).$$

Let us explain what this notation means. The two lists in parentheses, $(1, 2, 4, 3)$ and (5), are called *cycles*. The cycle $(1, 2, 4, 3)$ means that

$$1 \mapsto 2 \mapsto 4 \mapsto 3 \mapsto 1.$$

In other words,

$$\pi(1) = 2, \quad \pi(2) = 4, \quad \pi(4) = 3, \quad \text{and} \quad \pi(3) = 1.$$

Each number k is followed by $\pi(k)$. Taken literally, if we began the cycle with 1, we would go on forever: $(1, 2, 4, 3, 1, 2, 4, 3, 1, 2, 4, 3, 1, \ldots)$. Instead, when we reach the first 3, we write a close parenthesis meaning "return to the start of the cycle." Thus $(1, \ldots, 3)$ means that $\pi(3) = 1$.

What does the lonely (5) mean? It means $\pi(5) = 5$.

Let's continue with a more complicated example.

Example 26.5 Let $\pi = \left[\begin{smallmatrix} 1 & 2 & 3 & 4 & 5 & 6 & 7 & 8 & 9 \\ 2 & 7 & 5 & 6 & 3 & 8 & 1 & 4 & 9 \end{smallmatrix}\right] \in S_9$. Express π in cycle notation.

Solution: Note that $\pi(1) = 2$, $\pi(2) = 7$, and $\pi(7) = 1$ (we have returned to start). So far we have

$$\pi = (1, 2, 7) \cdots.$$

The first element we have not considered is 3. Restarting from 3, we have $\pi(3) = 5$ and $\pi(5) = 3$, so the next cycle is $(3, 5)$. So far we have $\pi = (1, 2, 7)(3, 5) \cdots$.

The next element we have yet to consider is 4. We have $\pi(4) = 6$, $\pi(6) = 8$, and $\pi(8) = 4$ to complete the cycle. The next cycle is $(4, 6, 8)$. Thus far we have $(1, 2, 7)(3, 5)(4, 6, 8) \cdots$.

Finally, we have $\pi(9) = 9$, so the last cycle is just (9). The permutation π in cycle notation is

$$\pi = (1, 2, 7)(3, 5)(4, 6, 8)(9).$$

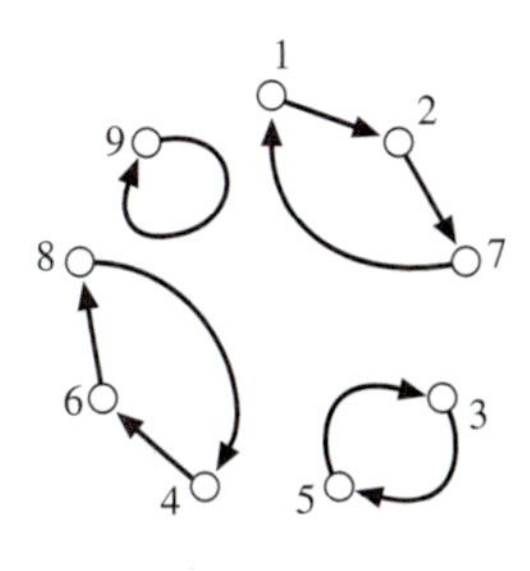

We can draw a picture of a permutation. Let $\pi \in S_n$. We draw a dot for each element of the set $\{1, 2, \ldots, n\}$. We draw an arrow from dot k to dot $\pi(k)$. The figure shows the permutation $\pi = \begin{bmatrix} 1 & 2 & 3 & 4 & 5 & 6 & 7 & 8 & 9 \\ 2 & 7 & 5 & 6 & 3 & 8 & 1 & 4 & 9 \end{bmatrix}$. Notice that each cycle in $(1, 2, 7)(3, 5)(4, 6, 8)(9)$ corresponds to a cycle of arrows in the diagram.

Does the cycle notation method work for all permutations? Is it possible that we begin making a cycle $(1, 5, 2, 9, \ldots)$ and the first repetition is not to the first element of the cycle. In other words, could we run into a situation such as

$$\pi(1) = 5 \quad \pi(5) = 2 \quad \pi(2) = 9 \quad \pi(9) = 5?$$

In the diagram, we would have a chain of arrows starting at 1, going to 5, then 2, then 9, but then back to 5 rather than 1. Might this happen? No. Notice that in this case we would have $\pi(1) = \pi(9) = 5$, contradicting the fact that π is one-to-one.

More formally, let $\pi \in S_n$. Consider the sequence

$$1, \pi(1), (\pi \circ \pi)(1), (\pi \circ \pi \circ \pi)(1), \ldots$$

which we can rewrite

$$1, \pi(1), \pi^{(2)}(1), \pi^{(3)}(1), \ldots$$

(see Exercise 25.13). This is a sequence of integers in the finite set $\{1, 2, \ldots, n\}$, so eventually this sequence must repeat itself. Let's say that the *first* repeat is at $\pi^{(k)}(1)$. [It is possible that the first repeat is at $k = 1$—that is, that $\pi(1) = 1$.] We want to conclude that $\pi^{(k)}(1) = 1$. Suppose, for the sake of contradiction, that $\pi^{(k)}(1) \neq 1$. In this case, we have

$$\pi^{(k)}(1) = \pi^{(j)}(1) \tag{35}$$

where $0 < j < k$. Because this is the first repeat, we have

$$\pi^{(k-1)}(1) \neq \pi^{(j-1)}(1). \tag{36}$$

Since π is one-to-one, applying π to both sides of Equation (36) yields

$$\pi^{(k)}(1) \neq \pi^{(j)}(1)$$

contradicting (35). Therefore the first repeat must go back to element 1.

The cycle starting at element 1 might not include all the elements of $\{1, 2, \ldots, n\}$. In this case, we can restart with an as-yet-unconsidered element and start building a new cycle.

Is it possible that this new cycle "runs into" an existing cycle? For this to happen, we would have two arrows pointing to the same dot, a violation of the fact that π is one-to-one. More formally, if the element s is not an element of the cycle $(t, \pi(t), \pi^{(2)}(t), \ldots)$, is it possible that $\pi^{(k)}(s)$ is an element of the cycle? If so, there is an element c on the cycle with the property that there are two different elements a and b with $\pi(a) = \pi(b) = c$, contradicting the fact that π is one-to-one.

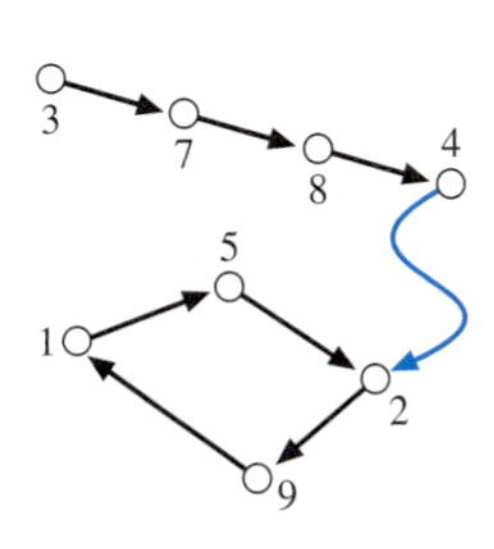

Therefore we can write π as a collection of *pairwise disjoint* cycles; that is, no two of the cycles have a common element.

We can say more. Is it possible to write the same permutation as a collection of disjoint cycles in two different ways? At first glance, the answer is yes.

For example,

$$\pi = (1, 2, 7)(3, 5)(4, 6, 8)(9) = (5, 3)(6, 8, 4)(9)(7, 1, 2);$$

both represent the permutation $\pi = \begin{bmatrix} 1 & 2 & 3 & 4 & 5 & 6 & 7 & 8 & 9 \\ 2 & 7 & 5 & 6 & 3 & 8 & 1 & 4 & 9 \end{bmatrix}$. However, on closer inspection, we see that the two representations of π have the same cycles; the cycles $(1, 2, 7)$ and $(7, 1, 2)$ both say the same thing—namely, $\pi(1) = 2$, $\pi(2) = 7$, $\pi(7) = 1$.

There is only one way to write π as a collection of disjoint cycles. Suppose, for the sake of contradiction, that we had two ways to write π. Then an element, say element 1, would be listed in one cycle in the first representation and in a different cycle in the second representation. However, if we consider the sequence,

$$1,\ \pi(1),\ \pi^{(2)}(1),\ \pi^{(3)}(1), \ldots$$

we see that the two different cycles predict two different sequences. This is nonsense because the sequence is solely dependent on π and not on the notation in which we write it!

We summarize what we have discussed in the following result:

Theorem 26.6 Every permutation of a finite set can be expressed as a collection of pairwise disjoint cycles. Furthermore, this representation is unique up to rearranging the cycles and the cyclic order of the elements within cycles.

Calculations with Permutations

The cycle notation is handy for doing pencil-and-paper calculations with permutations. Here we show how to compute the inverse of a permutation and the composition of two permutations. Let us begin with calculating π^{-1}.

If π maps $a \mapsto b$, then π^{-1} maps $b \mapsto a$. Thus if $(a, b, c, \ldots)$ is a cycle of π, then $(\ldots, c, b, a)$ is a cycle of π^{-1}.

Example 26.7 **(Inverting π)** Let $\pi = (1, 2, 7, 9, 8)(5, 6, 3)(4) \in S_9$. Calculate π^{-1}.

Solution: $\pi^{-1} = (8, 9, 7, 2, 1)(3, 6, 5)(4)$.

To check that this is correct, let k be any element in $\{1, 2, \ldots, 9\}$. If $\pi(k) = j$ (if j follows k in a cycle in π), check that $\pi^{-1}(j) = k$ (then k follows j in a cycle of π^{-1}).

Let us explore how to compute the composition of two permutations. For example, let $\pi, \sigma \in S_9$ be given by

$$\pi = (1, 3, 5)(4, 6)(2, 7, 8, 9), \quad \text{and}$$
$$\sigma = (1, 4, 7, 9)(2, 3)(5)(6, 8).$$

We compute $\pi \circ \sigma$. To do this, we calculate $(\pi \circ \sigma)(k)$ for all $k \in \{1, 2, \ldots, 9\}$. We begin with $(\pi \circ \sigma)(1)$. This can be written out as

$$[\underbrace{(1, 3, 5)(4, 6)(2, 7, 8, 9)}_{\pi}] \circ [\underbrace{(1, 4, 7, 9)(2, 3)(5)(6, 8)}_{\sigma}](1).$$

Notice that σ acts on 1 first and sends $1 \mapsto 4$.

The problem reduces to computing $\pi(4)$; that is,

$$[(1,3,5)(4,6)(2,7,8,9)](4)$$

and we see that π sends $4 \mapsto 6$. Thus $(\pi \circ \sigma)(1) = \pi(4) = 6$, and we can write

$$\pi \circ \sigma = (1, 6, \ldots .$$

To continue the cycle, we calculate $(\pi \circ \sigma)(6)$. We have

$$\begin{aligned} &[\underbrace{(1,3,5)(4,6)(2,7,8,9)}_{\pi}] \circ [\underbrace{(1,4,7,9)(2,3)(5)(6,8)}_{\sigma}](6) \\ &= [\underbrace{(1,3,5)(4,6)(2,7,8,9)}_{\pi}](8) = 9. \end{aligned}$$

So $\pi \circ \sigma$ maps $6 \mapsto 9$. Now we have

$$\pi \circ \sigma = (1, 6, 9, \ldots$$

Next we compute $(\pi \circ \sigma)(9) = \pi(1) = 3$, so $\pi \circ \sigma = (1, 6, 9, 3, \ldots$. Continuing in this fashion, we get

$$1 \mapsto 6 \mapsto 9 \mapsto 3 \mapsto 7 \mapsto 2 \mapsto 5 \mapsto 1$$

and we have completed a cycle! Thus $(1, 6, 9, 3, 7, 2, 5)$ is a cycle of $\pi \circ \sigma$. Notice that 4 is not on this cycle, so we start over computing $(\pi \circ \sigma)(4)$. We find

$$[\underbrace{(1,3,5)(4,6)(2,7,8,9)}_{\pi}] \circ [\underbrace{(1,4,7,9)(2,3)(5)(6,8)}_{\sigma}](4) = 8$$

so $4 \mapsto 8$. The second cycle in $\pi \circ \sigma$ begins $(4, 8, \ldots$. Now we calculate $(\pi \circ \sigma)(8) = 4$, so the entire cycle is simply $(4, 8)$. The two cycles $(1, 6, 9, 3, 7, 2, 5)$ and $(4, 8)$ exhaust all the elements of $\{1, 2, \ldots, 9\}$, and so we are finished. We have found

$$\pi \circ \sigma = (1, 6, 9, 3, 7, 2, 5)(4, 8).$$

Transpositions

The simplest permutation is the identity permutation ι; it satisfies $\iota(x) = x$ for every x in its domain. The identity permutation maps every element to itself.

The next simplest type of permutation is called a *transposition*. Transpositions map almost all elements to themselves, except that they exchange one pair of elements. For example,

$$\tau = (1)(2)(3,6)(4)(5)(7)(8)(9) \in S_9$$

is a transposition. Here is a formal definition:

Definition 26.8 **(Transposition)** A permutation $\tau \in S_n$ is called a *transposition* provided

- there exist $i, j \in \{1, 2, \ldots, n\}$ with $i \neq j$ so that $\tau(i) = j$ and $\tau(j) = i$, and
- for all $k \in \{1, 2, \ldots, n\}$ with $k \neq i$ and $k \neq j$, we have $\tau(k) = k$.

When written in cycle notation, the vast majority of the cycles are singletons. It is more convenient not to write out all these 1-cycles and to write just $\tau = (3, 6)$ instead of the verbose $\tau = (1)(2)(3,6)(4)(5)(7)(8)(9)$.

There is a nice trick for converting a cycle into a composition of transpositions.

Example 26.9 Let $\pi = (1, 2, 3, 4, 5)$. Write π as the composition of transpositions.
Solution: $(1, 2, 3, 4, 5) = (1, 5) \circ (1, 4) \circ (1, 3) \circ (1, 2)$.

To see that this is correct, let $\pi = (1, 5) \circ (1, 4) \circ (1, 3) \circ (1, 2)$ and calculate $\pi(1)$, $\pi(2)$, $\pi(3)$, $\pi(4)$, and $\pi(5)$. Look at how the elements 1 through 5 pass (from right to left) through the transpositions. For example, $1 \mapsto 2$ by $(1, 2)$, then $2 \mapsto 2$ by $(1, 3)$, then $2 \mapsto 2$ by $(1, 4)$, and finally $2 \mapsto 2$ by $(1, 5)$. So overall, $1 \mapsto 2$. Here is how all the elements are handled as they pass through $(1, 5) \circ (1, 4) \circ (1, 3) \circ (1, 2)$:

$$1 \mapsto 2 \mapsto 2 \mapsto 2 \mapsto 2$$
$$2 \mapsto 1 \mapsto 3 \mapsto 3 \mapsto 3$$
$$3 \mapsto 3 \mapsto 1 \mapsto 4 \mapsto 4$$
$$4 \mapsto 4 \mapsto 4 \mapsto 1 \mapsto 5$$
$$5 \mapsto 5 \mapsto 5 \mapsto 5 \mapsto 1$$

so overall $\pi = (1, 2, 3, 4, 5)$.

Example 26.10 Let $\pi = (1, 2, 3, 4, 5)(6, 7, 8)(9)(10, 11)$. Write π as the composition of transpositions.
Solution: $\pi = [(1, 5) \circ (1, 4) \circ (1, 3) \circ (1, 2)] \circ [(6, 8) \circ (6, 7)] \circ (10, 11)$. (The brackets are unnecessary; their purpose is to show how the answer was obtained.)

Let π be any permutation. Write π as a collection of disjoint cycles. Using the technique from Example 26.9, we can rewrite each of its cycles as a composition of transpositions. Because the cycles are disjoint, there is no effect of one cycle on another. Thus we can simply string together the transpositions for the various cycles into one long composition of cycles.

What about the identity permutation ι? Can it also be represented as the composition of transpositions? Yes. We can write $\iota = (1, 2) \circ (1, 2)$. Or we can say that ι is the result of composing together a list of no permutations (this is akin to an empty product—see Section 8).

Let us summarize what we have shown here.

Theorem 26.11 Let π be any permutation on a finite set. Then π can be expressed as the composition of transpositions defined on that set.

The decomposition (great word to use here!) of a permutation into transpositions is not unique. For example, we can write

$$\begin{aligned}(1, 2, 3, 4) &= (1, 4) \circ (1, 3) \circ (1, 2)\\ &= (1, 2) \circ (2, 3) \circ (3, 4)\\ &= (1, 2) \circ (1, 4) \circ (2, 3) \circ (1, 4) \circ (3, 4).\end{aligned}$$

These ways of writing $(1, 2, 3, 4)$ are not simple rearrangements of one another. We see that they do not even have the same length. However, they do have something in common. In all three cases, we used an odd number of transpositions.

Theorem 26.12 Let $\pi \in S_n$. Let π be decomposed into transpositions as

$$\pi = \tau_1 \circ \tau_2 \circ \cdots \circ \tau_a \quad \text{and}$$
$$\pi = \sigma_1 \circ \sigma_2 \circ \cdots \circ \sigma_b.$$

Then a and b have the same parity; that is, they are both odd or both even.

The key to proving this theorem is to prove a special case first.

Lemma 26.13 If the identity permutation is written as a composition of transpositions, then that composition must use an even number of transpositions. That is, if

$$\iota = \tau_1 \circ \tau_2 \circ \cdots \circ \tau_a,$$

where the τs are transpositions, then a must be even.

Before we prove this lemma, we show how to use it to prove Theorem 26.12.

Proof (of Theorem 26.12)

Let π be a permutation decomposed into transpositions as

$$\pi = \tau_1 \circ \tau_2 \circ \cdots \circ \tau_a, \quad \text{and}$$
$$\pi = \sigma_1 \circ \sigma_2 \circ \cdots \circ \sigma_b.$$

Note that we can write π^{-1} as (see Exercise 26.11)

$$\pi^{-1} = \sigma_b \circ \sigma_{b-1} \circ \cdots \circ \sigma_2 \circ \sigma_1$$

and so

$$\iota = \pi \circ \pi^{-1} = \tau_1 \circ \tau_2 \circ \cdots \circ \tau_a \circ \sigma_b \circ \sigma_{b-1} \circ \cdots \circ \sigma_2 \circ \sigma_1.$$

This is a decomposition of ι into $a + b$ transpositions. Hence $a + b$ is even, and so a and b have the same parity. ■

Our job now reduces to proving Lemma 26.13. To do this, we introduce the concept of an *inversion* in a permutation.

Definition 26.14 **(Inversion in a permutation)** Let $\pi \in S_n$ and let $i, j \in \{1, 2, \ldots, n\}$ with $i < j$. The pair i, j is called an *inversion* in π if $\pi(i) > \pi(j)$.

It is easier to understand inversions when the permutation is written in $2 \times n$ array form. Let

$$\pi = \begin{bmatrix} 1 & 2 & 3 & 4 & 5 \\ 4 & 2 & 1 & 5 & 3 \end{bmatrix}.$$

Here is another way to think about inversions. Draw two collections of dots labeled 1 through n on the left and on the right. For each element i on the left, draw a straight arrow from i to $\pi(i)$ on the right. The number of crossings is the number of inversions.

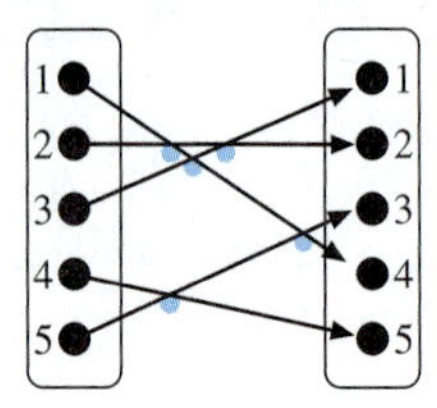

There are $\binom{5}{2} = 10$ ways we can choose a pair of elements $1 \le i < j \le 5$. In the following chart, we list all such pairs i, j and check whether $\pi(i) > \pi(j)$.

i	j	$\pi(i)$	$\pi(j)$	**Inversion?**
1	2	4	2	YES
1	3	4	1	YES
1	4	4	5	no
1	5	4	3	YES
2	3	2	1	YES
2	4	2	5	no
2	5	2	3	no
3	4	1	5	no
3	5	1	3	no
4	5	5	3	YES

Thus π has five inversions. We can also write π as the composition of transpositions:

$$\pi = \begin{bmatrix} 1 & 2 & 3 & 4 & 5 \\ 4 & 2 & 1 & 5 & 3 \end{bmatrix} = (1, 4, 5, 3)(2) = (1, 4)(4, 5)(5, 3).$$

In this decomposition there are three transpositions (odd) and the permutation π has five inversions (also odd).

For a second, more abstract example, we calculate the number of inversions in a transposition $(a, b) \in S_n$. Let us assume $a < b$ so we can write this as

$$(a, b) = \begin{bmatrix} 1 & 2 & \cdots & a-1 & a & a+1 & \cdots & b-1 & b & b+1 & \cdots & n \\ 1 & 2 & \cdots & a-1 & b & a+1 & \cdots & b-1 & a & b+1 & \cdots & n \end{bmatrix}.$$

Let us count the inversions. To begin, the only inversions possible are those that involve a or b. For any i, j (with neither i nor j equal to a or b), the transposition (a, b) does not invert the order of i and j; there are no inversions of this sort.

We now count three types of inversions: those involving only a, those involving only b, and those involving both a and b.

- Inversions involving a but not b.

  Element a has advanced from column a to column b. In so doing, it has skipped past elements $a + 1, a + 2, \ldots, b - 1$ and creates inversions with those elements. It is still in its proper order with respect to all other columns. The number of inversions of this sort is $(b - 1) - (a + 1) + 1 = b - a - 1$.
- Inversions involving b but not a.

  Element b has retreated from column b to column a. In so doing, it has ducked under elements $a + 1, a + 2, \ldots, b - 1$ and creates inversions with those elements. It is still in its proper order with respect to all other columns. The number of inversions of this sort is, again, $(b-1)-(a+1)+1 = b-a-1$.
- Inversions involving both a and b.

  This is just one inversion.

Therefore the total number of inversions is $2(b - a - 1) + 1$, an odd number. The number of inversions involving a but not b equals the number of inversions involving b but not a. Further, all these inversions involve the elements appearing between a and b.

The identity permutation has, of course, 0 (even) inversions. We now return to the goal of showing that any decomposition of ι into transpositions uses an even number of transpositions.

Proof (of Lemma 26.13)

Write ι as a composition of transpositions:

$$\iota = \tau_a \circ \tau_{a-1} \circ \cdots \circ \tau_2 \circ \tau_1.$$

(We have written the τs in reverse order because we want to think of doing τ_1 first, τ_2 second, and so on.)

Our goal is to prove that a is even. Imagine applying the transpositions τ_i one at a time. We begin with a "clean slate"; that is, $\begin{bmatrix} 1 & 2 & \cdots & n \\ 1 & 2 & \cdots & n \end{bmatrix}$.

Now we apply τ_1. As we analyzed earlier, the resulting number of inversions is now odd. We now show that as we apply each τ_i, the number of inversions changes by an odd amount. Since the number of inversions at the start and at the end is zero, and since each transposition increases or decreases the number of inversions by an odd amount, the number of transpositions must be even.

Suppose $\tau_k = (a, b)$ and

$$\tau_{k-1} \circ \cdots \circ \tau_1 = \begin{bmatrix} \cdots & i & \cdots & m & \cdots & j & \cdots \\ \cdots & a & \cdots & x & \cdots & b & \cdots \end{bmatrix}.$$

Now when we apply $\tau_k = (a, b)$, the effect is

$$\tau_k \circ \tau_{k-1} \circ \cdots \circ \tau_1 = \begin{bmatrix} \cdots & i & \cdots & m & \cdots & j & \cdots \\ \cdots & b & \cdots & x & \cdots & a & \cdots \end{bmatrix}.$$

The only change is that a and b are exchanged in the bottom row. What has happened to the number of inversions?

The first thing to note is that for a pair of columns including neither column i nor column j, there is no change. All changes involve column i or j or both.

The second thing to note is that columns to the left of column i and columns to the right of column j are unaffected by the interchange of a and b; these elements do not change their order with respect to these outer columns.

Therefore we only need to pay attention to columns between columns i and j. Let's say that column m is between these ($i < m < j$), and the entry in column m is x. When we exchange a and b, the bottom row changes from $[\cdots a \cdots x \cdots b \cdots]$ to $[\cdots b \cdots x \cdots a \cdots]$.

We break into cases depending on x's size compared to a and b; x can be larger than both a and b, smaller than both a and b, or between a and b.

- If $x > a$ and $x > b$, then there is no change in the number of inversions involving x and a or b. Before applying τ_k, we had a and x inverted, but x and b were in natural order. After applying τ_k, we have x and b inverted, but x and a are in their natural order.
- If $x < a$ and $x < b$, then there is no change in the number of inversions involving x and a or b; the argument is analogous to the case where x is larger than both.

- If $a < x < b$, then upon switching a and b, we gain two inversions involving a and x and involving b and x.
- If $a > x > b$, then upon switching a and b, we lose two inversions.

In every case, the number of inversions either stays the same or changes by two. Thus the number of inversions involving column i or j and a column other than i or j changes by an even amount.

Finally, the exchange of a and b either increases the number of inversions by one (if $a < b$) or decreases the number of inversions by one (if $a > b$).

Thus the cumulative effect of τ_k is to change the number of inversions by an odd amount.

In conclusion, since we begin and end with zero inversions, the number of transpositions in

$$\iota = \tau_a \circ \tau_{a-1} \circ \cdots \circ \tau_2 \circ \tau_1$$

must be even. ■

Theorem 26.12 enables us to separate permutations into two disjoint categories: those that can be expressed as the composition of an even number of transpositions, and those that can be expressed as the composition of an odd number of transpositions.

Definition 26.15 **(Even, odd permutations)** Let π be a permutation on a finite set. We call π *even* provided it can be written as the composition of an even number of transpositions. Otherwise it can be written as the composition of an odd number of transpositions, in which case we call π *odd*.

The *sign* of a permutation is ± 1 depending on whether the permutation is odd or even. The sign of π is 1 if π is even and -1 if π is odd. The sign of π is written $\operatorname{sgn} \pi$.

A Graphical Approach

We close with an alternative approach to understanding even and odd permutations. The ideas we present here yield another proof of Theorem 26.12. We use Theorem 26.6, which asserts that every permutation $\pi \in S_n$ can be expressed as a collection of disjoint cycles in, essentially, only one way.

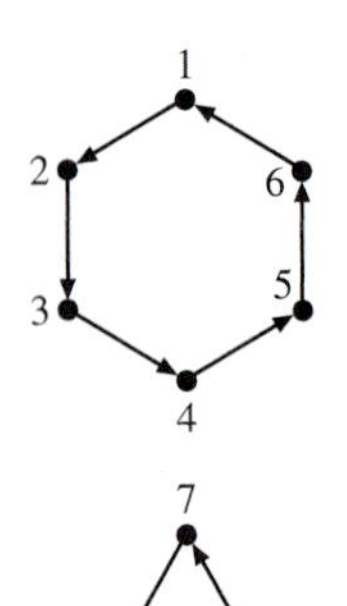

We begin by drawing a picture of the permutation. Given $\pi \in S_n$, we make a figure in which the numbers $1, 2, \ldots, n$ are represented by points, and if $\pi(a) = b$, we draw an arrow from a to b. A picture for the permutation $\pi = (1, 2, 3, 4, 5, 6)(7, 8, 9)$ is shown in the figure. In case $\pi(a) = a$, we draw a looping arrow from a to itself. Each cycle of π corresponds precisely to a closed path in the diagram.

Suppose we compose a permutation π with a transposition τ. What is the effect on the diagram? Suppose $\pi, \tau \in S_n$ and $\tau = (a, b)$ where $a \neq b$ and $a, b \in \{1, 2, \ldots, n\}$. When we express π as disjoint cycles, cycles that contain neither a nor b are the same in π and $\pi \circ \tau$. The only cycles that are affected are ones that contain a or b (or both).

If a and b are in the same cycle, then π is of the form

$$\pi = (p, a, q, \ldots, s, b, t, \ldots, z)(\cdots).$$

Then $\pi \circ (a, b)$ will be of the form

$$\begin{aligned}\pi \circ (a, b) &= (p, a, q, \ldots, s, b, t, \ldots, z)(\cdots) \circ (a, b) \\ &= (p, a, t, \ldots, z)(q, \ldots, s, b)(\cdots).\end{aligned}$$

In other words, the cycle containing a and b in π is split into two cycles in $\pi \circ (a, b)$: one containing a and the other containing b.

The opposite effect occurs when a and b are in different cycles. In this case, π is of the form

$$\pi = (p, a, q, \ldots)(s, b, t, \ldots)(\cdots)$$

and so $\pi \circ (a, b)$ has the form

$$\begin{aligned}\pi \circ (a, b) &= (p, a, q, \ldots)(s, b, t, \ldots)(\cdots) \circ (a, b) \\ &= (p, a, t, \ldots, s, b, q, \ldots)(\cdots).\end{aligned}$$

The cycles containing a and b in π are merged into a single cycle in $\pi \circ (a, b)$.

For example, suppose $\pi = (1, 2, 3, 4, 5)(6, 7, 8, 9)$ and let $\sigma = \pi \circ (4, 7)$. Observe that $\sigma = (1, 2, 3, 4, 8, 9, 6, 7, 5)$. Because 4 and 7 are in separate cycles of π, they are in a common cycle of $\pi \circ (4, 7)$. Conversely, 4 and 7 are in the same cycle of σ but are split into separate cycles in $\sigma \circ (4, 7)$. See the figure.

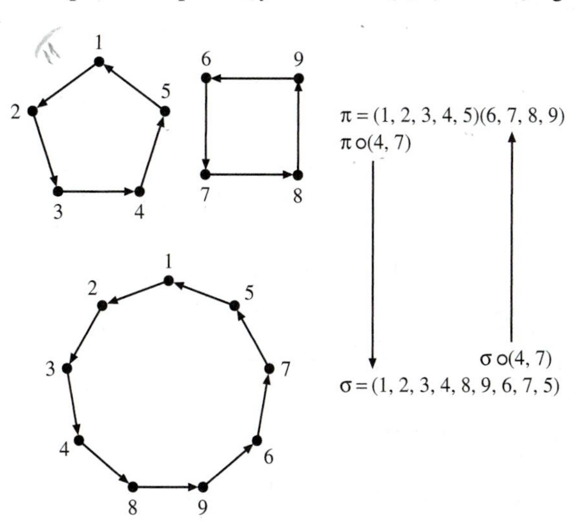

With only a bit more care, these observations can be made into a rigorous proof of the following result.

Proposition 26.16 Let n be a positive integer and $\pi, \tau \in S_n$, and suppose τ is a transposition. Then the number of cycles in the disjoint cycle representations of π and $\pi \circ \tau$ differ by exactly one.

For the remainder of this section, it is convenient to write $c(\pi)$ to stand for the number of cycles in the unique disjoint cycle representation of π. Proposition 26.16 can be expressed as $c(\pi \circ \tau) = c(\pi) \pm 1$.

We now apply Proposition 26.16 to give another proof of Theorem 26.12.

Note that for a transposition $\tau \in S_n$, we have $n - c(\tau) = 1$. Remember that $\tau = (a, b)$ is an abbreviated form of the permutation in which the 1-cycles are not written. For example, in S_6 the transposition $\tau = (3, 5)$ is, when written in full, $(1)(2)(3, 5)(4)(6)$. Therefore $n - c(\tau) = 6 - 5 = 1$.

Proof (of Theorem 26.12)

Suppose $\pi \in S_n$ and

$$\pi = \tau_1 \circ \tau_2 \circ \cdots \circ \tau_a \tag{37}$$

where the τs are transpositions. We claim that $a \equiv n - c(\pi) (\text{mod } 2)$. In other words, the parity of the number of transpositions in Equation (37) equals the parity of $n - c(\pi)$, and so two different decompositions of π into transpositions will both have an even or both have an odd number of terms.

Consider the sequence $\iota, \tau_1, \tau_1 \circ \tau_2, \tau_1 \circ \tau_2 \circ \tau_3, \ldots, \pi$. Each term is formed from the previous by appending the appropriate τ_j. We calculate $n - c(\cdot)$ for each of these permutations; see the following chart.

Permutation σ	$n - c(\sigma)$
ι	0
τ_1	1
$\tau_1 \circ \tau_2$	1 ± 1
$\tau_1 \circ \tau_2 \circ \tau_3$	$1 \pm 1 \pm 1$
$\vdots$	$\vdots$
$\pi = \tau_1 \circ \cdots \circ \tau_a$	$\underbrace{1 \pm 1 \pm 1 \pm \cdots \pm 1}_{a \text{ terms}}$

Note that the parity of the expression $1 \pm 1 \pm 1 \pm \cdots \pm 1$ (with a terms) is exactly the same as the parity of a, and the result follows. ■

This proof of Theorem 26.12 yields the following corollary.

Corollary 26.17 Let n be a positive integer and $\pi \in S_n$. Then $\operatorname{sgn} \pi = (-1)^{n-c(\pi)}$.

Recap

This section dealt with permutations: bijections from a set to itself. We studied properties of composition with respect to the set S_n of all permutations on $\{1, 2, \ldots, n\}$. We showed how to represent permutations in various forms, but we were especially interested in studying permutations in disjoint cycle form. We showed how to represent permutations as compositions of transpositions and discussed even and odd permutations.

26 Exercises

26.1. Consider the permutation $\pi = \begin{bmatrix} 1 & 2 & 3 & 4 & 5 & 6 & 7 & 8 & 9 \\ 2 & 4 & 1 & 6 & 5 & 3 & 8 & 9 & 7 \end{bmatrix}$. Please express π in as many forms as possible, including the following:

a. As a set of ordered pairs. (Never forget: A permutation is a function, and functions are sets of ordered pairs.)
b. As a two-column chart.
c. In cycle notation (disjoint cycle).
d. As the composition of transpositions.
e. As a diagram with two collections of dots for the numbers 1 through 9 (one collection on the left and one collection on the right) with arrows from left to right.
f. As a diagram with one collection of dots for the numbers 1 through 9 with arrows from i to $\pi(i)$ for each $i = 1, 2, \ldots, 9$.

26.2. Please express the following permutations in disjoint cycle form.

a. $\sigma = \begin{bmatrix} 1 & 2 & 3 & 4 & 5 & 6 \\ 2 & 4 & 6 & 1 & 3 & 5 \end{bmatrix}$.

b. $\pi = \begin{bmatrix} 1 & 2 & 3 & 4 & 5 & 6 \\ 2 & 3 & 4 & 5 & 6 & 1 \end{bmatrix}$.

c. $\pi \circ \pi$, where π is the permutation from part (b).
d. π^{-1} where π is the permutation from part (b).
e. $\iota \in S_5$.
f. $(1, 2) \circ (2, 3) \circ (3, 4) \circ (4, 5) \circ (5, 1)$.

26.3. How many permutations in S_n have exactly one cycle?

26.4. How many permutations in S_n do not have a cycle of length one in their disjoint cycle notation?

26.5. Let $\pi, \sigma, \tau \in S_9$ be given by

$$\pi = (1)(2, 3, 4, 5)(6, 7, 8, 9),$$
$$\sigma = (1, 3, 5, 7, 9, 2, 4, 6, 8), \quad \text{and}$$
$$\tau = (1, 9)(2, 8)(3, 5)(4, 6)(7).$$

Please calculate the following:

a. $\pi \circ \sigma$.
b. $\sigma \circ \pi$.
c. $\pi \circ \pi$.
d. π^{-1}.
e. σ^{-1}.
f. $\tau \circ \tau$.
g. τ^{-1}.

26.6. Prove or disprove: For all $\pi, \sigma \in S_n$, $\pi \circ \sigma = \sigma \circ \pi$.

26.7. Prove or disprove: If τ and σ are transpositions, then $\tau \circ \sigma = \sigma \circ \tau$.

26.8. Prove or disprove: For all $\pi, \sigma \in S_n$, $(\pi \circ \sigma)^{-1} = \sigma^{-1} \circ \pi^{-1}$.

26.9. Prove or disprove: For all $\pi, \sigma \in S_n$, $(\pi \circ \sigma)^{-1} = \pi^{-1} \circ \sigma^{-1}$.

26.10. Prove or disprove: A permutation τ is a transposition if and only if $\tau \neq \iota$ and $\tau = \tau^{-1}$.

26.11. Let $\tau_1, \tau_2, \ldots, \tau_a$ be transpositions and suppose

$$\pi = \tau_1 \circ \tau_2 \circ \cdots \circ \tau_a.$$

Prove that

$$\pi^{-1} = \tau_a \circ \tau_{a-1} \circ \cdots \circ \tau_1.$$

26.12. Let $\pi = (1, 2)(3, 4, 5, 6, 7)(8, 9, 10, 11)(12) \in S_{12}$. Find the smallest positive integer k for which

$$\pi^{(k)} = \underbrace{\pi \circ \pi \circ \cdots \circ \pi}_{k \text{ times}} = \iota.$$

Generalize. If a π's disjoint cycles have lengths $n_1, n_2, \ldots, n_t$, what is the smallest integer k so that $\pi^{(k)} = \iota$?

26.13. Although permutations are uniquely expressible as disjoint permutations, there is some choice in the way the permutations can be written. For example,

$$(1, 3, 9, 2)(7)(4, 6, 5, 8) = (7)(2, 1, 3, 9)(5, 8, 4, 6)$$
$$= (6, 5, 8, 4)(3, 9, 2, 1)(7).$$

Devise a standard form for writing permutations as disjoint cycles that makes it easy to check whether two permutations are the same.

26.14. Prove: If $\pi, \sigma \in S_n$ and $\pi \circ \sigma = \sigma$, then $\pi = \iota$.

26.15. Let $\pi, \sigma, \tau \in S_n$ and suppose $\pi \circ \sigma = \pi \circ \tau$. Prove that $\sigma = \tau$.

26.16. For each of the permutations listed, please do the following:

(1) Write the permutation as a composition of transpositions.

(2) Find the number of inversions.

(3) Determine whether the permutation is even or odd.

a. $(1, 2, 3, 4, 5)$.

b. $(1, 3)(2, 4, 5)$.

c. $[(1, 3)(2, 4, 5)]^{-1}$.

d. $\begin{bmatrix} 1 & 2 & 3 & 4 & 5 \\ 2 & 4 & 1 & 3 & 5 \end{bmatrix}$.

26.17. Prove: The number of inversions in a permutation equals the number of inversions in its inverse.

26.18. Prove the following:

a. The composition of two even permutations is even.

b. The composition of two odd permutations is even.

c. The composition of an even permutation and an odd permutation is odd.

d. The inverse of an even permutation is even.

e. The inverse of an odd permutation is odd.

f. For $n > 1$, the number of odd permutations in S_n equals the number of even permutations in S_n.

26.19. Suppose permutation π is written as a disjoint collection of cycles of lengths $n_1, n_2, \ldots, n_t$. Can you determine, just from these numbers, whether π is even or odd?

To answer yes, you need to develop and prove a formula for the parity of a permutation given only its disjoint cycle lengths.

To answer no, you need to find two permutations—one even and one odd—whose disjoint cycles have the same length.

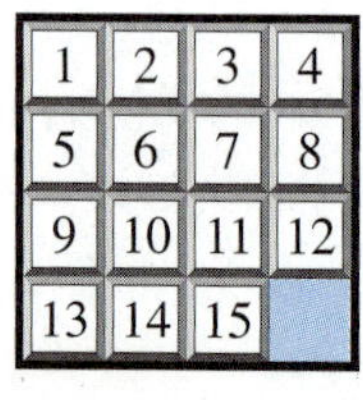

26.20. The Fifteen Puzzle is a 4×4 array of tiles numbered 1 to 15 with one empty space. You move the tiles about this board by sliding a number tile into the empty position. The initial configuration of the puzzle is shown in the upper diagram. To play, you scramble the pieces about randomly and then try to restore the initial configuration.

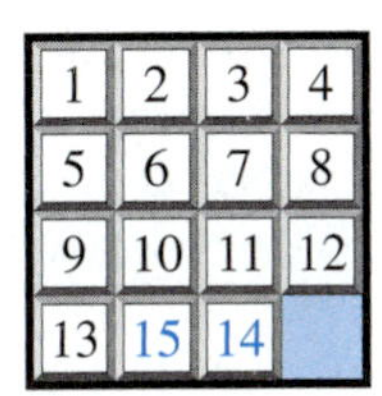

Prove that it is impossible to move the pieces in the puzzle from the initial configuration to a new position in which all numbers are in their original positions, but tiles 14 and 15 are interchanged (shown in the lower figure).

27 Symmetry

In this section, we take a careful look at the concept of *symmetry*. What does it mean to say that an object is *symmetric*? A human face is symmetric because the left half and the right half are mirror images of one another. On the other hand, a human hand is not symmetric.

In mathematics, the word *symmetry* typically refers to geometric figures. We give an informal definition of symmetry here; a precise definition is given later.

A *symmetry* of a figure is a motion that, when applied to an object, results in a figure that looks exactly the same as the original.

For example, consider a square sitting in the plane. If we rotate the square counterclockwise about its center through an angle of 90°, the resulting figure is exactly the same as the original. However, if we rotate the square through an angle of, say, 30°, the resulting figure is not the same as the original. Therefore a 90° rotation is a symmetry of the square, but a 30° rotation is not.

Symmetries of a Square

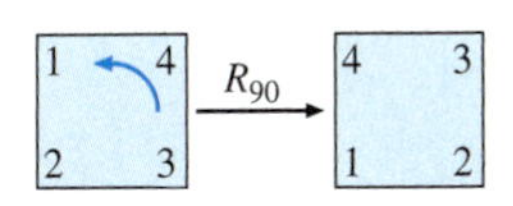

Rotating a square 90° counterclockwise through its center leaves the square unchanged. What are the other motions we can apply to a square that leave it unchanged? To aid us in our analysis, imagine that the numbers 1 through 4 are written in the corners of the square. Since the square looks exactly the same before and after we move it, the labels enable us to see how the square was moved. The figure shows a counterclockwise rotation through 90°; we call this symmetry R_{90}.

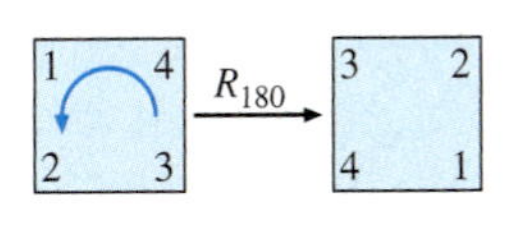

We may also rotate the square counterclockwise through 180°. After this rotation, the square will look exactly the same as before. We call this symmetry R_{180}. We might also rotate the square clockwise through 180°. Even though the physical motion of the square might be different (clockwise versus counterclockwise rotation), the end results are identical. By looking at the corner labels, you can tell that the square was rotated 180°, but you cannot tell whether that rotation was clockwise or counterclockwise. We consider these two motions to be exactly the same; they give the same symmetry of the square.

Next, we can rotate the square through 270° and leave the image unchanged. We call this symmetry R_{270}.

Finally, we can rotate the square through 360° and the result is unchanged. Should we call this R_{360}? Although this is not a bad idea, notice that a 360° rotation has no effect on the labels. It is as if no motion whatsoever was applied to the square. We therefore call this symmetry I, for identity.

If we rotate the square through 450° [Note: $450 = 360 + 90$], it is as if we rotated only through 90°. A rotation through 450° is simply R_{90}.

So far we have found four symmetries: I, R_{90}, R_{180}, and R_{270}. Are there more?

In addition to rotating the square, we can pick the square up, flip it over, and set it back down in the plane. For example, we can flip the square over along a horizontal axis. The result of this motion is shown in the figure. Notice that after this motion, the square looks exactly the same as when it started. We call this symmetry F_H for "flip-horizontal."

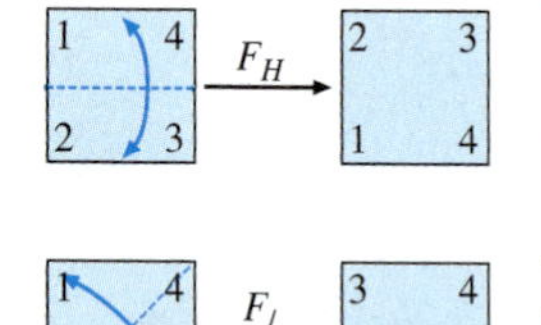

We can also flip the square over along its vertical axis; we call that motion F_V. Please draw a picture of this symmetry yourself.

We can also hold the square by two opposite corners and flip it over along its diagonal. If we hold the upper-right and lower-left corners, the result is as shown in the figure. We call this symmetry $F_/$ for "flip along the / diagonal."

We can also hold the upper-left and lower-right corners firm and flip over along the \ diagonal. We call this symmetry $F_\backslash$.

The eight symmetries we have found thus far are I, R_{90}, R_{180}, R_{270}, F_H, F_V, $F_/$, and $F_\backslash$. The following figure shows all of them.

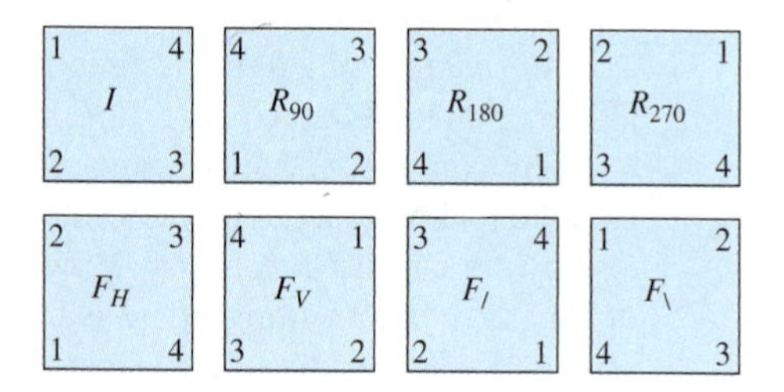

Two questions arise.

- First, have we repeated ourselves? Just as a 360° rotation and the identity symmetry are the same, are (perhaps) two of the above symmetries the same?

  The answer is no. If you look at the labels, you can observe that no two of the squares are labeled the same. The eight symmetries we have found are all different.

- Second, are there other symmetries we didn't think of?

  The labels can help us to see that the answer to this question is also no. Imagine that we pick up the square and lay it back down in its original place (but perhaps rotated and/or flipped). Where does the corner labeled 1 go? We have four choices: It might end up in the northeast, northwest, southeast, or southwest. Once we have decided where corner 1 goes, consider the final resting place of corner 2. We now have only two choices because corner 2 must end up next to (and not opposite) corner 1. Once we have placed corners 1 and 2, the remaining corners are forced into position. Therefore, there are $4 \times 2 = 8$ choices (four choices for corner 1 and, for each such choice, two choices for corner 2). We have found all the symmetries.

Symmetries as Permutations

Sylvia and Steve work in a symmetry factory. One day their boss asks them to rotate the big stone square in the company lobby 90°. Of course, the only way the boss can know that the square has been moved is by the labels on the corners of the square. So rather than move the big, heavy square, they peel the stickers off the corners of the square and reattach them in their new locations.

To perform the rotation R_{90}, they simply move label 1 to position 2, label 2 to position 3, label 3 to position 4, and label 4 to position 1.

The symmetry R_{90} can be expressed as $\begin{bmatrix} 1 & 2 & 3 & 4 \\ 2 & 3 & 4 & 1 \end{bmatrix}$. The first column means that label 1 moves to position 2, the second column means that label 2 moves to position 3, and so on.

Now $\begin{bmatrix} 1 & 2 & 3 & 4 \\ 2 & 3 & 4 & 1 \end{bmatrix}$ is a permutation! We can express this permutation in cycle form as $(1, 2, 3, 4)$. Indeed, all eight symmetries of the square can be expressed in this notation.

Symmetry name	1 2 3 4 go to positions	Cycle form
I	1 2 3 4	$(1)(2)(3)(4)$
R_{90}	2 3 4 1	$(1, 2, 3, 4)$
R_{180}	3 1 4 2	$(1, 3)(2, 4)$
R_{270}	4 1 2 3	$(1, 4, 3, 2)$
F_H	2 1 4 3	$(1, 2)(3, 4)$
F_V	4 3 2 1	$(1, 4)(2, 3)$
$F_/$	3 2 1 4	$(1, 3)(2)(4)$
$F_\backslash$	1 4 3 2	$(1)(2, 4)(3)$

Every day, Steve and Sylvia's boss asks them to reposition the big, heavy square in the lobby. And every day, they just move the stickers around. One day, they switch stickers 1 and 2 and then take a lunch break. Meanwhile, their boss sees that the "symmetry" they performed is $(1, 2)(3)(4)$, and there is no such symmetry of the square. Not all permutations in S_4 correspond to symmetries of the square—just the eight we listed. Sylvia and Steve were summarily sacked for their sham stone square symmetry stratagem!

Combining Symmetries

What happens if we first flip the square horizontally and then rotate it through $90°$? The combined motion looks like this:

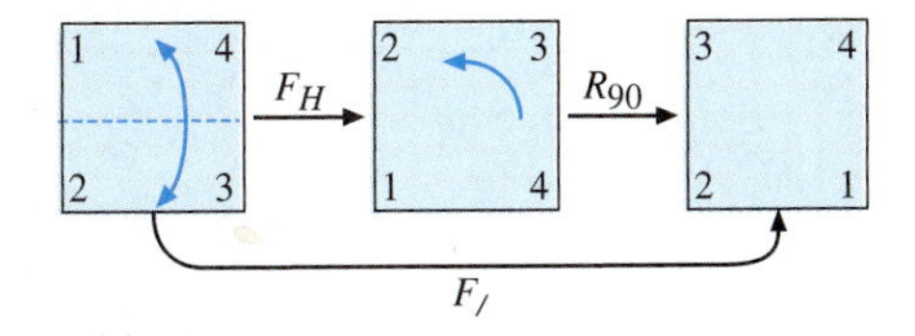

The net effect of combining these two symmetries is a flip along the / diagonal, (i.e., $F_/$). We write this as

$$R_{90} \circ F_H = F_/.$$

This is not a misprint! We did the horizontal flip F_H first and then followed it by the $90°$ rotation R_{90}. Why did we write R_{90} first? We are reusing the function composition symbol $\circ$ in this context. Recall (Section 25) that when we write $g \circ f$, it means we perform function f first and then function g.

Suppose we want to calculate the result of

$$F_H \circ R_{270} \circ F_V.$$

We could draw several pictures or work with a physical model, but there is a better way. We saw above that the symmetries of the square can be thought of as relabeling permutations of its corners. Behold:

$$\begin{aligned} R_{90} \circ F_H &= (1,2,3,4) \circ (1,2)(3,4) \\ &= (1,3)(2)(4) \\ &= F_/. \end{aligned}$$

The first $\circ$ stands for combining symmetries, and the second $\circ$ is permutation composition. Notice, however, that the calculation with permutations gives the correct answer for the symmetries.

Let's think about why this works. We first do F_H, which we can express as $\pi = (1,2)(3,4)$. The effect is to take whatever is in position 1 (label 1) to position 2. Then $\sigma = (1,2,3,4)$ takes whatever is in position 2 (label 1) to position 3. So the net effect is $1 \mapsto 2 \mapsto 3$. The other corners work the same way.

It is a mildly laborious but worthwhile chore to make an 8×8 chart showing the combined effect of each pair of symmetries. Here is the result:

$\circ$	I	R_{90}	R_{180}	R_{270}	F_H	F_V	$F_/$	$F_\backslash$
I	I	R_{90}	R_{180}	R_{270}	F_H	F_V	$F_/$	$F_\backslash$
R_{90}	R_{90}	R_{180}	R_{270}	I	$F_/$	$F_\backslash$	F_V	F_H
R_{180}	R_{180}	R_{270}	I	R_{90}	F_V	F_H	$F_\backslash$	$F_/$
R_{270}	R_{270}	I	R_{90}	R_{180}	$F_\backslash$	$F_/$	F_H	F_V
F_H	F_H	$F_\backslash$	F_V	$F_/$	I	R_{180}	R_{270}	R_{90}
F_V	F_V	$F_/$	F_H	$F_\backslash$	R_{180}	I	R_{90}	R_{270}
$F_/$	$F_/$	F_H	$F_\backslash$	F_V	R_{90}	R_{270}	I	R_{180}
$F_\backslash$	$F_\backslash$	F_V	$F_/$	F_H	R_{270}	R_{90}	R_{180}	I

Some comments:

- The operation $\circ$ is not commutative. Notice that $R_{90} \circ F_H = F_/$ but $F_H \circ R_{90} = F_\backslash$.
- Element I is an identity element for $\circ$.
- Every element has an inverse. For example, $R_{90}^{-1} = R_{270}$ because $R_{90} \circ R_{270} = R_{270} \circ R_{90} = I$.

  It is also interesting to notice that most of the elements are their own inverse.
- The operation $\circ$ is associative. This is not easy to see just from looking at the table. However, it follows from the fact that we can replace symmetries

by permutations and then interpret $\circ$ as composition. Since composition is associative, so is $\circ$ for symmetries.

- Compare these remarks to Proposition 26.4. If we $\circ$ together two symmetries of the square, we get a symmetry of the square. The operation $\circ$ is associative and has an identity element, and every symmetry has an inverse. The operation of composition on the set of all permutations of n elements, S_n, also exhibits these same properties.

Formal Definition of Symmetry

The plane is denoted by the symbol $\mathbb{R}^2$. Why? The notation $\mathbb{R}^2$ is a shorthand way of writing $\mathbb{R} \times \mathbb{R}$—that is, the set of all ordered pairs (x, y) where x and y are real numbers. This corresponds to the representation of points in the plane by two coordinates.

A geometric figure, such as a square, is a set of points in the plane ($\mathbb{R}^2$). For example, the following set is a square:

$$S = \{(x, y) \in \mathbb{R}^2 : -1 \le x \le 1, \text{ and } -1 \le y \le 1\}. \tag{38}$$

The distance between points (a, b) and (c, d) is (by the Pythagorean Theorem)

$$\operatorname{dist}[(a, b), (c, d)] = \sqrt{(a - c)^2 + (b - d)^2}$$

where $\operatorname{dist}[(a, b), (c, d)]$ stands for the distance between the points (a, b) and (c, d).

Definition 27.1 **(Isometry)** Let $f : \mathbb{R}^2 \to \mathbb{R}^2$. We call f an *isometry* provided

$$\forall (a, b), (c, d) \in \mathbb{R}^2, \ \operatorname{dist}[(a, b), (c, d)] = \operatorname{dist}[f(a, b), f(c, d)].$$

A synonym for isometry is a *distance-preserving* function.

Let $X \subseteq \mathbb{R}^2$ (i.e., X is a geometric figure). Let $f : \mathbb{R}^2 \to \mathbb{R}^2$. Now writing $f(X)$ is nonsense because X is a set of points and the domain of f is the set of points in the plane. Nonetheless, $f(X)$ is a useful notation. It means

$$f(X) = \{f(a, b) : (a, b) \in X\}.$$

That is, $f(X)$ is the set we obtain by evaluating f at all the points in X.

We can now say precisely what a symmetry is.

Definition 27.2 **(Symmetry)** Let $X \subseteq \mathbb{R}^2$. A *symmetry* of X is an isometry $f : \mathbb{R}^2 \to \mathbb{R}^2$ such that $f(X) = X$.

Let S be the square in the plane defined by Equation (38). The symmetries of S are

$$\begin{aligned} I(a, b) &= (a, b) & F_H(a, b) &= (a, -b) \\ R_{90}(a, b) &= (-b, a) & F_V(a, b) &= (-a, b) \\ R_{180}(a, b) &= (-a, -b) & F_/(a, b) &= (b, a) \\ R_{270}(a, b) &= (b, -a) & F_\backslash(a, b) &= (-a, -b). \end{aligned}$$

This discussion has been limited to geometric figures in the plane. One can extend all these ideas to three-dimensional space and beyond.

Recap

This section introduced the concept of symmetry, related symmetry to permutations of labels, and explored the operation of combining symmetries. Finally, we gave a technical definition of symmetry.

27 Exercises

27.1. Verify by pictures and by permutation calculation that $F_H \circ R_{90} = F_{\backslash}$.

27.2. Let R be a rectangle that is not a square. Describe the set of symmetries of R and write down the $\circ$ table for this set.

27.3. Which of the symmetries of a square are represented by even permutations? Compare your answer to this exercise to the previous one.

27.4. Let T be an equilateral triangle. Find all the symmetries of T and represent them as permutations of the corners. Compare this to S_3.

27.5. What are the symmetries of a triangle that is isosceles but not equilateral?

27.6. What are the symmetries of a triangle that is not isosceles (all three sides have different lengths)?

27.7. Let P be a regular pentagon. Find all the symmetries of P (give them sensible names) and represent them as permutations of the corners.

27.8. Let Q be a cube in space. How many symmetries does Q have?

a. Show that a correct answer to this question is 24.

b. Show that another correct answer to this question is 48.

c. By Proof Template 9, since 24 and 48 are both answers to the same question, it must be the case that $24 = 48$.

Actually, the question "How many symmetries does Q have?" is a bit ambiguous.

What is different about the second set of 24 symmetries?

d. Represent the 48 symmetries of the cube as permutations of its corners.

27.9. *This problem is only for those who have studied linear algebra.*

Let C be a circle in the plane.

a. Describe the set of all symmetries of C.

b. Show how the symmetries of the circle can be represented by 2×2 matrices A with $\det A = \pm 1$.

c. What is the difference between symmetries whose matrix has determinant 1 and those whose matrix has determinant -1?

d. Does every matrix with determinant ± 1 correspond to a symmetry of the circle?

28 Assorted Notation

Big oh

Pricing items at \$9.99 drives me crazy.

I wish merchants would just sell the item for \$10 and not try to deceive me that the item costs "about" \$9. It's much easier for humans to deal with round, whole numbers, and that is why approximating is a valuable skill.

Just as it is valuable to approximate numbers, it is also useful to express functions in an approximate manner. Consider a complicated function f (defined on the natural numbers) defined by

$$f(n) = 4n^5 - \frac{n(n+1)(n+2)}{3} + 3n^2 - 12.$$

When n is large, the most "important" part is the n^5. In this section, we develop a notation that expresses this idea precisely.

The "big oh" notation expresses the idea that one function is bounded by another. Here is the definition.

Definition 28.1 **(Big oh)** Let f and g be real-valued functions defined on the natural numbers (i.e., $f : \mathbb{N} \to \mathbb{R}$ and $g : \mathbb{N} \to \mathbb{R}$). We say that $f(n)$ is $O(g(n))$ provided there is a positive number M such that, with at most finitely many exceptions,

$$|f(n)| \le M|g(n)|.$$

In other words, $f(n)$ is $O(g(n))$ means that $|f(n)|$ is no greater than a constant multiple of $|g(n)|$ (with, perhaps, a few exceptions).

Example 28.2 Let $f(n) = \binom{n}{2}$. We claim that $f(n)$ is $O(n^2)$. Recall that $\binom{n}{2} = n(n-1)/2$. Thus

$$f(n) = \frac{n(n-1)}{2} \le \frac{n^2}{2}$$

and so $f(n) \le \frac{1}{2}n^2$ for all n. So we can take $M = \frac{1}{2}$ in the definition of big oh and conclude that $f(n)$ is $O(n^2)$.

Example 28.3 Let $f(n) = n(n+5)/2$. We claim that $f(n)$ is $O(n^2)$. Note that, except for $n = 0$, we have

$$\begin{aligned}
\frac{|f(n)|}{|n^2|} &= \frac{f(n)}{n^2} \qquad \text{because } f(n) \ge 0 \text{ for all } n \in \mathbb{N} \\
&= \frac{n(n+5)}{2n^2} \\
&= \frac{n+5}{2n} \\
&= \frac{1}{2} + \frac{5}{2n} \\
&\le \frac{1}{2} + \frac{5}{2} \le 3.
\end{aligned}$$

Thus $|f(n)| \le 3|n^2|$ and so $f(n)$ is $O(n^2)$.

Let us consider a more complicated example. Recall the function we mentioned at the start of this section:

$$f(n) = 4n^5 - \frac{n(n+1)(n+2)}{3} + 3n^2 - 12.$$

We show that this function is $O(n^5)$. To do this, we need to compare $|f(n)|$ and $|n^5|$ where $n \in \mathbb{N}$. Since n is nonnegative, $|n^5| = n^5$. However, because the polynomial defining $f(n)$ has negative coefficients, we need a tool to handle $|f(n)|$.

Proposition 28.4 **(Triangle inequality)** Let a, b be real numbers. Then

$$|a + b| \le |a| + |b|.$$

Proof. We consider four cases depending on whether or not each of a and b is negative.

- If neither a nor b is negative, we have $|a + b| = a + b = |a| + |b|$.
- If $a \ge 0$ but $b < 0$, we have $|a| + |b| = a - b$.

  If $|a+b| = a+b$ (when $a+b \ge 0$) and we have that $|a+b| = a+b < a < a - b = |a| + |b|$.

  Otherwise $|a+b| = -(a+b)$ (when $a+b < 0$), and we have $|a+b| = -a - b = -|a| + |b| < |a| + |b|$.

  In both cases, $|a + b| < |a| + |b|$.
- The case $a < 0$ and $b \ge 0$ is analogous to the preceding case.
- Finally, if a and b are both negative, we have $|a + b| = -(a + b) = (-a) + (-b) = |a| + |b|$.

In all cases, we have $|a + b|$ is either equal to or less than $|a| + |b|$. ■

We return to the analysis of $f(n)$. If we multiply out all the terms in f, we get an expression of the form

$$f(n) = 4n^5 + ?n^3 + ?n^2 + ?n + ?$$

where the question marks represent numbers that I'm too lazy to figure out. Therefore

$$\begin{aligned}|f(n)| &= |4n^5+?n^3+?n^2+?n+?| \\ &\le 4n^5 + |?|n^3 + |?|n^2 + |?|n + |?|.\end{aligned}$$

We divide this expression by n^5 and get

$$\frac{|f(n)|}{|n^5|} = 4 + \frac{|?|}{n^2} + \frac{|?|}{n^3} + \frac{|?|}{n^4} + \frac{|?|}{n^5}.$$

Notice that once n is larger than $|?|$ for all the terms I neglected to calculate, each of the terms with a question mark is less than 1. So I may conclude that, except for finitely many values of n, we have

$$\frac{|f(n)|}{|n^5|} < 4 + 1 + 1 + 1 + 1 = 8.$$

That is, with at most finitely many exceptions, $|f(n)| \le 8|n^5|$ and so $f(n)$ is $O(n^5)$.

Example 28.5 n^2 is $O(n^3)$ but n^3 is not $O(n^2)$.

It is clear that $|n^2| \le |n^3|$ for all $n \in \mathbb{N}$, so n^2 is $O(n^3)$.

However, suppose, for the sake of contradiction, that n^3 is $O(n^2)$. This means there is a constant M so that, except for finitely many $n \in \mathbb{N}$, we have $|n^3| \le M|n^2|$. Since $n \in \mathbb{N}$, we may drop the absolute value bars and divide by n^2 to get $n \le M$ for all but finitely many $n \in \mathbb{N}$, but this is obviously false. Therefore n^3 is not $O(n^2)$.

When we say $f(n)$ is $O(g(n))$, the function $g(n)$ serves as a bound on $|f(n)|$. That is, it says that $|f(n)|$ grows no faster than a multiple of $|g(n)|$. So the function n^2 grows no faster than the function n^3, but not vice versa.

The awful but useful and popular notation $f(n) = O(g(n))$.

For better or worse, mathematicians use the big oh notation in a sloppy way. It is proper to write "$f(n)$ is $O(g(n))$." This means that the function f has a certain property—namely, that its absolute value is bounded by a constant multiple of g. Now it is natural to use the word *is* when we see an equals sign (=). As a result, mathematicians often write the abhorrent $f(n) = O(g(n))$.

Why do we use this terrible notation? It's like the old joke:
A: My uncle is crazy. He thinks he's a chicken!
B: So why don't you send him to a psychiatrist and have him helped??
A: Because we need the eggs!

I deplore this terrible notation. But, of course, I use it all the time. The problem is that $f(n)$ does not equal $O(g(n))$. Rather, $f(n)$ has a certain property that we call $O(g(n))$.

Further, we often write "equations" such as

$$\binom{n}{3} = \frac{n^3}{6} + O(n^2).$$

This means that the function $\binom{n}{3}$ is equal to the function $\frac{n^3}{6}$ plus another function that is $O(n^2)$. This is a handy way to absorb all the less important information about $\binom{n}{3}$ into a "remainder" term. The proper way to express the foregoing "equation" is to say that $\binom{n}{3} - \frac{n^3}{6}$ is $O(n^2)$.

Although we tolerate the $f(n) = O(g(n))$ notation, we adamantly reject writing $O(g(n)) = f(n)$.

On the other end of the spectrum, some mathematicians write $f(n) \in O(g(n))$. This is actually a nice notation. Many mathematicians define the notation $O(g(n))$ to be the set of all functions whose absolute values are bounded by a constant multiple of $|g(n)|$ (with finitely many exceptions). When we write $f(n) \in O(g(n))$, we assert that f is such a function.

Ω and Θ

The big oh notation establishes an upper bound on the growth of $|f(n)|$. Conversely, the Ω (big omega) notation defines a lower bound on its growth.

Definition 28.6 (Ω) Let f and g be real-valued functions defined on the natural numbers (i.e., $f : \mathbb{N} \to \mathbb{R}$ and $g : \mathbb{N} \to \mathbb{R}$). We say that $f(n)$ is $\Omega(g(n))$ provided there is a positive number M such that, with at most finitely many exceptions,

$$|f(n)| \ge M|g(n)|.$$

There is a simple relation between the O and Ω notations.

Proposition 28.7 Let f and g be functions from $\mathbb{N}$ to $\mathbb{R}$. Then $f(n)$ is $O(g(n))$ if and only if $g(n)$ is $\Omega(f(n))$.

Proof. ($\Rightarrow$) Suppose $f(n)$ is $O(g(n))$. Then there is a positive constant M such that $|f(n)| \le M|g(n)|$ for all but finitely many n. Therefore $|g(n)| \ge \frac{1}{M}|f(n)|$ for all but finitely many n, and so $g(n) = \Omega(f(n))$.

($\Leftarrow$) Analogous to the previous argument. ■

Example 28.8 Let $f(n) = n^2 - 3n + 2$. Then $f(n)$ is $\Omega(n^2)$ and $f(n)$ is also $\Omega(n)$, but $f(n)$ is not $\Omega(n^3)$.

The O notation is an upper bound and the Ω is a lower bound. The following notation combines them. The symbol Θ is a Greek capital theta.

Definition 28.9 (Θ) Let f and g be real-valued functions defined on the natural numbers (i.e., $f : \mathbb{N} \to \mathbb{R}$ and $g : \mathbb{N} \to \mathbb{R}$). We say that $f(n)$ is $\Theta(g(n))$ provided there are positive numbers A and B such that, with at most finitely many exceptions,

$$A|g(n)| \le |f(n)| \le B|g(n)|.$$

Example 28.10 Let $f(n) = \binom{n}{3}$. Then $f(n)$ is $\Theta(n^3)$, but $f(n)$ is neither $\Theta(n^2)$ nor $\Theta(n^4)$.

Proposition 28.11 Let f and g be functions from $\mathbb{N}$ to $\mathbb{R}$. Then $f(n)$ is $\Theta(g(n))$ if and only if $f(n)$ is $O(g(n))$ and $f(n)$ is $\Omega(g(n))$.

The proof is left for you (see Exercise 28.4).

The statement that $f(n)$ is $\Theta(g(n))$ says, in effect, that as n gets large, $f(n)$ and $g(n)$ grow at roughly the same rate.

As with the O notation, mathematicians often misuse the Ω and Θ notations, writing "equations" of the form $f(n) = \Omega(g(n))$ and $f(n) = \Theta(g(n))$.

Little oh

This section is only for those who have studied calculus.

The statement that $f(n)$ is $O(g(n))$ says that $f(n)$ does not grow faster than $g(n)$ as n gets large. Sometimes it is useful to say that $f(n)$ grows "much" slower than $g(n)$. For this, we have the "little oh" notation.

Definition 28.12 **(Little oh)** Let f and g be real-valued functions defined on the natural numbers (i.e., $f : \mathbb{N} \to \mathbb{R}$ and $g : \mathbb{N} \to \mathbb{R}$). We say that $f(n)$ is $o(g(n))$ provided

$$\lim_{n\to\infty} \frac{f(n)}{g(n)} = 0.$$

Example 28.13 Let $f(n) = \sqrt{n}$. Then $f(n) = o(n)$. To see why, we calculate

$$\lim_{n\to\infty} \frac{\sqrt{n}}{n} = \lim_{n\to\infty} \frac{1}{\sqrt{n}} = 0.$$

Mathematicians misuse the little oh notation with the same reckless abandon with which they misuse the O, Ω, and Θ notations. You are more likely to see the "equation" $f(n) = o(n^2)$ than the words "$f(n)$ is $o(n^2)$."

Floor and Ceiling

I have n marbles to give to two children. How should I divide them fairly? The answer is to give each child $n/2$ marbles. That is, of course, unless n is odd. Half a marble does neither child any good, so I might as well give one child $(n-1)/2$ and the other child $(n+1)/2$. (To be totally fair, I would flip a coin to decide who gets the extra.)

There is an alternative notation for floor and ceiling. Some mathematicians write $[x]$ to stand for the floor of x and $\{x\}$ to stand for the ceiling of x. The problem with this notation is that square brackets [ ] are used as big parentheses and curly braces {} are used for sets. You may see $[x]$ in some older mathematics books; just remember that it means $\lfloor x \rfloor$.

The "give each child $n/2$ marbles" answer is easier to express than the more elaborate answer that applies when n is odd. Sometimes, the only sensible answer to a problem is an integer, but the algebraic expression we derive does not necessarily evaluate to an integer. It is useful, in many instances, to have a notation for rounding off a noninteger answer to an integral answer.

There are a number of different ways to round off nonintegers. The standard method is to round the quantity to the nearest integer (and to round up if we are midway between). There are, however, two other natural alternatives: We can always round up or we can always round down. These functions have special names and notations.

Definition 28.14 **(Floor and ceiling)** Let x be a real number.

The *floor* of x, denoted $\lfloor x \rfloor$, is the largest integer n such that $n \le x$.

The *ceiling* of x, denoted $\lceil x \rceil$, is the smallest integer n such that $n \ge x$.

In other words, $\lfloor x \rfloor$ is the integer we form from x by rounding down (unless x is already an integer), and $\lceil x \rceil$ is the integer we form from x by rounding up.

Example 28.15 The following illustrate the floor and ceiling functions.

$$\lfloor 3.2 \rfloor = 3 \qquad \lfloor -3.2 \rfloor = -4 \qquad \lfloor 5 \rfloor = 5$$
$$\lceil 3.2 \rceil = 4 \qquad \lceil -3.2 \rceil = -3 \qquad \lceil 5 \rceil = 5$$

Recap

This section introduced the following notation for approximating functions: O, Ω, Θ, and o. We also introduced the floor and ceiling functions for rounding off real numbers to integer values.

28 Exercises

28.1. Prove the following:

a. n^2 is $O(n^4)$.
b. n^2 is $O(1.1^n)$.
c. $(n)_k$ is $O(n^k)$ where k is a fixed, positive integer.
d. $\frac{n+1}{n}$ is $O(1)$.
e. 2^n is $O(3^{n-1})$.
f. $n \sin n$ is $O(n)$.

28.2. *True or false*: Determine whether the following statements are true or false.

a. Suppose $x \in \mathbb{Q}$. Then $x \in \mathbb{Z}$ if and only if $\lceil x \rceil = x$.
b. Suppose $x \in \mathbb{Q}$. Then $x \in \mathbb{Z}$ if and only if $\lceil x \rceil = \lfloor x \rfloor$.
c. Suppose $x, y \in \mathbb{Q}$. Then $\lfloor x + y \rfloor = \lfloor x \rfloor + \lfloor y \rfloor$.
d. Suppose $x, y \in \mathbb{Q}$. Then $\lfloor xy \rfloor = \lfloor x \rfloor \cdot \lfloor y \rfloor$.
e. Suppose $x \in \mathbb{Z}$ and $y \in \mathbb{Q}$. Then $\lfloor x + y \rfloor = x + \lfloor y \rfloor$.
f. Suppose $x \in \mathbb{Q}$. Then $\lfloor x \rfloor$ can be calculated as follows: Write x as a decimal and then drop all the digits to the right of the decimal point.

28.3. Suppose $f(n)$ is $O(g(n))$ and $g(n)$ is $O(h(n))$. Prove that $f(n)$ is $O(h(n))$.

28.4. Prove Proposition 28.11.

28.5. Let a and b be real numbers with $a, b > 1$. Prove that $\log_a n = O(\log_b n)$. Conclude that $\log_a n = \Theta(\log_b n)$.

28.6. Let $p(n)$ be a polynomial of degree d in n. Prove that $p(n)$ is $\Theta(n^d)$.

28.7. Develop an expression (using the floor or ceiling notation) for the ordinary meaning of rounding off a real number x to the nearest integer. Be sure your formula properly handles rounding 3.49 to 3, but 3.5 to 4.

28.8. Develop an expression (using the floor or ceiling notation) for the ones digit of a positive integer. That is, if $n = 326$, then your expression should evaluate to 6.

Chapter 5 Self Test

1. Let $f = \{(1, 2), (2, 3), (3, 4)\}$ and $g = \{(2, 1), (3, 1), (4, 2)\}$. Please answer the following:

a. What is $f(2)$?
b. What is $f(4)$?
c. What is dom f?
d. What is im f?
e. What is f^{-1}?
f. Note that g^{-1} is not a function. Why?
g. What is $g \circ f$?
h. What is $f \circ g$?

2. Suppose A and B are sets and f is a function with $f : A \to B$. Suppose also that $f(a) = b$. Please mark each of the following statements as true or false.

a. $a \in A$.

b. $b \in B$.

c. $\operatorname{dom} f = A$.

d. $\operatorname{im} f = B$.

3. Let $A = \{1, 2, 3\}$ and let $B = \{3, 4, 5, 6\}$.

a. How many functions $f : A \to B$ are there?

b. How many one-to-one functions $f : A \to B$ are there?

c. How many onto functions $f : A \to B$ are there?

4. Suppose $f : A \to B$ is one-to-one and $g : B \to A$ is one-to-one. Must it be the case that f is onto? Justify your answer.

5. Let $f : \mathbb{Z} \to \mathbb{N}$ by $f(x) = |x|$. (a) Is f one-to-one? (b) Is f onto? Prove your answers.

6. Let $f : \mathbb{Z} \to \mathbb{Z}$ by $f(x) = x^3$. (a) Is f one-to-one? (b) Is f onto? Prove your answers.

7. Functions are relations, although it is not customary to consider whether they exhibit properties such as reflexive or antisymmetric. Nevertheless, find a function that is also an equivalence relation on the set $\{1, 2, 3, 4, 5\}$.

8. The squares of a 9×9 chess board are arbitrarily colored black and white. When we examine the 2×2 blocks of squares, we must see repeated patterns (prove this). Indeed, prove that some pattern must be repeated at least four times, as illustrated in the figure.

9. Let $A = \{1, 2, 3, 4, 5\}$ with $f : A \to A$, $g : A \to A$, and $h : A \to A$. We are given the following:

- $f = \{(1, 2), (2, 3), (3, 1), (4, 3), (5, 5)\}$,
- $h = \{(1, 3), (2, 3), (3, 2), (4, 5), (5, 3)\}$, and
- $h = f \circ g$.

Find all possible functions g that satisfy these conditions.

10. Suppose $f, g : \mathbb{R} \to \mathbb{R}$ are defined by

$$f(x) = x^2 + x - 1 \qquad \text{and} \qquad g(x) = 3x + 2.$$

Express, in simplest terms, $(f \circ g)(x) - (g \circ f)(x)$.

11. Let $f, g, h : \mathbb{R} \to \mathbb{R}$ defined by $f(x) = 3x - 4$, $g(x) = ax + b$, and $h(x) = 2x + 1$, where a and b are real numbers. Suppose that $(f \circ g \circ h)(x) = 6x + 5$. Find $(h \circ g \circ f)(x)$.

12. It is standard mathematical convention to consider the expression 0^0 to be undefined. However, from the perspective of a discrete mathematician, there is a natural value to associate with 0^0. What is that value? Justify your answer.

13. Let A be a set. Suppose f and g are functions $f : A \to A$ and $g : A \to A$ with the property that $f \circ g = \mathrm{id}_A$.

Prove or disprove: $f = g^{-1}$.

14. Let π be a permutation of $\{1, 2, 3, \ldots, 9\}$ defined by the 2×9 array $\pi = \begin{bmatrix} 1 & 2 & 3 & 4 & 5 & 6 & 7 & 8 & 9 \\ 3 & 9 & 2 & 6 & 5 & 7 & 4 & 1 & 8 \end{bmatrix}$. Please do the following:

a. Express π as a set of ordered pairs.

b. Express π in cycle notation.

c. Express π^{-1} in cycle notation.

d. Express $\pi \circ \pi$ in cycle notation.

e. Express π as the product of transpositions and determine whether π is an even or odd permutation.

15. Let n be a positive integer and let $\pi \in S_n$. Prove there is a positive integer k such that $\pi^{(k)} = \pi^{-1}$.

Note: $\pi^{(k)} = \pi \circ \pi \circ \cdots \circ \pi$ where π appears on the right k times.

16. Let n be a positive integer and $\pi, \sigma \in S_n$. Evaluate

$$\sum_{k=1}^{n} (\pi(k) - \sigma(k))$$

and explain your answer.

17. Let n be a positive integer and let $\pi \in S_n$.

a. Prove that π can be written in the following form:

$$\pi = (1, x_1) \circ (1, x_2) \circ \cdots \circ (1, x_a)$$

where $1 < x_i \le n$ for all n.

b. If the identity permutation ι is written in the form presented in part (a) of this problem, we know that a must be even. Give such a representation of ι in which some of the transpositions $(1, x)$ appear an odd number of times. (The total number of transpositions must be even, but some of the particular transpositions appear an odd number of times.)

18. Let n be a positive integer and $\pi \in S_n$. Let $x_1, x_2, \ldots, x_n$ be real numbers. Prove that

$$\prod_{1 \le i < j \le n} (x_j - x_i) = (\operatorname{sgn} \pi) \cdot \prod_{1 \le i < j \le n} \left(x_{\pi(j)} - x_{\pi(i)}\right).$$

Note: The products are over all pairs of integers i, j between 1 and n where $i < j$. For example, with $n = 3$, the products are

$$(x_2 - x_1)(x_3 - x_1)(x_3 - x_2) \quad \text{and}$$
$$\left(x_{\pi(2)} - x_{\pi(1)}\right)\left(x_{\pi(3)} - x_{\pi(1)}\right)\left(x_{\pi(3)} - x_{\pi(2)}\right).$$

19. Let T be a tetrahedron (a solid figure with four triangular faces) all of whose sides have the same length.

a. Describe the set of symmetries of T, assuming reflections of the tetrahedron are considered the same.

b. Describe the set of symmetries of T, assuming reflections of the tetrahedron are considered different.

In both cases, the symmetries should be described as permutations of the four vertices (corners) of T, which may be labeled 1, 2, 3, and 4.

20. Let x be a real number and suppose that $\lfloor x \rfloor = \lceil x \rceil$. What can you conclude about x?

21. Show that 2^n is $O(3^n)$, but 3^n is not $O(2^n)$.

CHAPTER 6 Probability

Few things in life are certain. Probability theory provides us with tools for analyzing situations in which events occur at random. Probability theory is used in a wide range of disciplines, including sociology, nuclear physics, genetics, and finance.

It is important to distinguish between mathematical probability theory and its application to problems in the real world. In mathematics, a probability is simply a number associated with some object. In applications, the object is some event or uncertain action, and the number is a measure of how frequent or how likely that event is. Imagine you are prescribed a medication for some disease. Your doctor might tell you that the probability the medication will be effective is 94%. This means that if a large number of patients were to use this drug for this disease, we would expect 94% of them to be cured and the remaining 6% of them would not be cured. In applications, probability is often synonymous with frequency.

There are many ways to write a number. The number 94% is exactly the same as 0.94, which is the same as $\frac{94}{100}$, which is the same as $\frac{47}{50}$. Percentages are convenient ways to express numbers between 0 and 1, but they are no different from fractions or decimal numbers.

Probabilities are real numbers between 0 and 1. An event with probability 1 is certain to occur, and an event with probability 0 is impossible. Probabilities between 0 and 1 reflect the relative likelihood between these two extremes. Unlikely events have probabilities close to 0, and likely events have probabilities close to 1.

In this chapter we introduce fundamental ideas from discrete probability theory. Discrete probability problems are often counting problems recast in the language of probability theory.

29 Sample Space

Consider the toss of a die. We cannot say in advance which of the six sides of the die will land face up; the outcome of this experiment is unpredictable. However, if the die is fair, we can say that all six outcomes are equally likely. Thus, although we cannot predict which of the six sides will emerge on top, we can describe the likelihood of seeing, for example, a 4 when we roll the die.

There are two parts to a sample space: a list of outcomes and an assignment of probabilities to these outcomes.

Mathematicians model the roll of a die using a concept called a *sample space*. A sample space has two parts. First, it contains a list of all the *outcomes* of some experiment. In this case, there are six outcomes: any of faces 1 through 6 might land face up. Second, it quantifies the likelihood of each of these outcomes. In this case, since all six outcomes are equally likely, we give the same numerical score to each result; we call this likelihood score the *probability* of the result. By convention, we require that the sum of the probabilities of the various possible outcomes be 1. Thus we assign probability $\frac{1}{6}$ to each of the six outcomes of the die-rolling experiment.

Defined more carefully, a sample space consists of a set and a function. The *set* is the collection of all possible outcomes of some experiment. The *function* assigns a numerical score to each outcome; this numerical score—called the *probability* of the outcome—is simply a real number between 0 and 1 (inclusive). We also require the sum of the probabilities of all the outcomes to be exactly 1. It is customary to use the letter S for the set of outcomes and the letter P for the function that assigns to each $s \in S$ the probability of that outcome, $P(s)$.

Example 29.1 **(Roll of a die)** Let S be the set of outcomes from the roll of a die. The simplest way to name the outcomes is with the integers 1, 2, 3, 4, 5, and 6, so

$$S = \{1, 2, 3, 4, 5, 6\}.$$

We also have a function $P : S \to \mathbb{R}$ defined by

$$P(1) = \frac{1}{6} \quad P(2) = \frac{1}{6} \quad P(3) = \frac{1}{6}$$
$$P(4) = \frac{1}{6} \quad P(5) = \frac{1}{6} \quad P(6) = \frac{1}{6}.$$

Note that the probabilities are nonnegative real numbers and the sum of the probabilities of all the elements in S is 1.

With this example in mind, we present the definition of a sample space formally:

Definition 29.2 **(Sample space)** A *sample space* is a pair (S, P) where S is a finite, nonempty set and P is a function $P : S \to \mathbb{R}$ such that $P(s) \geq 0$ for all $s \in S$ and

$$\sum_{s \in S} P(s) = 1.$$

The condition $\sum_{s \in S} P(s) = 1$ means that the sum of the probabilities of all the elements in S must be exactly 1.

Example 29.3 **(Spinner)** Consider the spinner shown in the figure. The arrow represents a needle that can be spun around to point to one of the four regions 1, 2, 3, or 4.

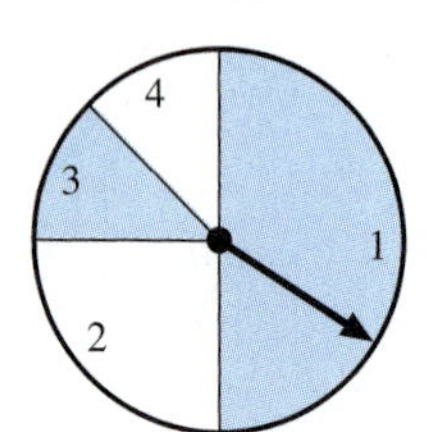

We model this physical device with a sample space. The set of outcomes S contains the names of the four regions; that is,

$$S = \{1, 2, 3, 4\}.$$

The probability function $P : S \to \mathbb{R}$ measures how likely it is for the spinner to land in each of the regions. The likelihood is proportional to the area of the region.

Thus we have

$$P(1) = \frac{1}{2}, \quad P(2) = \frac{1}{4}, \quad P(3) = \frac{1}{8}, \quad P(4) = \frac{1}{8}.$$

We check that

$$\sum_{s \in S} P(S) = P(1) + P(2) + P(3) + P(4) = \frac{1}{2} + \frac{1}{4} + \frac{1}{8} + \frac{1}{8} = 1.$$

Example 29.4 **(Pair of dice)** Two dice are tossed. Die 1 can land in any one of 6 equally likely ways, and the same is true for die 2. We can express the outcome of this experiment as an ordered pair (a, b) where a and b are integers between 1 and 6. Thus there are $6 \times 6 = 36$ possible outcomes for this experiment. We let

$$\begin{aligned} S &= \{1, 2, 3, 4, 5, 6\} \times \{1, 2, 3, 4, 5, 6\} \\ &= \{(1, 1), (1, 2), (1, 3), \ldots, (6, 5), (6, 6)\}. \end{aligned}$$

Each of the 36 possible outcomes of this experiment is equally likely; that is, $P(s) = \frac{1}{36}$ for all $s \in S$.

Note that the fundamental outcomes of rolling a pair of dice are the 36 different ways the pair can land. In the next section, we consider events such as "the sum of the numbers on the dice is eight." Rolling a 6 on the first die and a 2 on the second is an outcome of the dice-rolling experiment. There are several different outcomes in which the two values sum to 8.

Example 29.5 **(Poker hand)** A *hand* of poker is a five-element subset of the standard deck of 52 cards. There are $\binom{52}{5}$ different five-element subsets of a 52-element set. The set S consists of all these different five-element subsets. Since they are all equally likely, we have

$$P(s) = \frac{1}{\binom{52}{5}}$$

for all $s \in S$.

Example 29.6 **(Coin tossing)** A fair coin is tossed five times in a row, and the sequence of HEADS and TAILS is recorded. We model this as a sample space. The set S contains all possible outcomes of this experiment. We denote an outcome as a length-five list of Hs and Ts (where H stands for HEADS and T for TAILS). There are $2^5 = 32$ such lists, and they are all equally likely. Thus

$$S = \{\text{TTTTT}, \text{TTTTH}, \text{TTTHT}, \ldots, \text{HHHHT}, \text{HHHHH}\}$$

and $P(s) = \frac{1}{32}$ for all $s \in S$.

Are the sequences HHHHT and HHHHH equally likely? After seeing a coin turn up HEADS several times in a row, some people have an intuition that the next roll is more likely to be TAILS. They feel that the coin is "ready" to come up TAILS.

This intuition is incorrect, but the reason is physical, not mathematical. The coin is not capable of "remembering" the results it gave for the past several rolls; from the coin's perspective, each roll is a new trial that has nothing to do with the past.

Perhaps a brilliant mechanical engineer could design a coin that could keep track of how it lands; if the coin accumulated a series of HEADS, it would silently shift internal parts to make TAILS more likely on the next roll. Then our model that HHHHT and HHHHH are equally likely would not accurately reflect physical reality.

How can we tell whether our model is accurate? Ultimately, because this is a physical issue and not a mathematical one, at some point we need to rely on physical measurements. We would record each group of five flips to see if all possible length-5 lists came up about $1/32$ of the time.

A sample space (S, P) is a mathematical model of a physical experiment. In its pure form, the sample space (S, P) is simply a set and a function with certain properties. The interpretation of S as a set of outcomes and $P(s)$ as the likelihood of S is an added layer of meaning. This added layer of meaning is what makes probability theory useful. However, we can create sample spaces that have no specific physical interpretation. Here is an example:

Example 29.7 Let $S = \{1, 2, 3, 4, 5, 6\}$ and define $P : S \to \mathbb{R}$ by

$$P(1) = 0.1 \qquad P(2) = 0.4 \qquad P(3) = 0.1$$
$$P(4) = 0 \qquad P(5) = 0.2 \qquad P(6) = 0.2.$$

Note that $\sum_{s \in S} P(s) = 1$.

In this example $P(4) = 0$; this is perfectly acceptable. The interpretation is that outcome 4 is impossible. Thus the set S of outcomes might include results that cannot occur.

Recap

We introduced the concept of a sample space: a pair (S, P) where S is a set and P is a function that assigns to each element in S a nonnegative number called its probability. The sum of the probabilities over all outcomes in S must be exactly 1. In applications, the elements of S represent the fundamental outcomes or results of some experiment.

29 Exercises

29.1. Let (S, P) be the sample space in which $S = \{1, 2, 3, 4\}$ and $P(1) = 0.1$, $P(2) = 0.1$, $P(3) = 0.2$, and $P(4) = x$. Find x.

29.2. Let (S, P) be the sample space in which $S = \{1, 2, 3, 4\}$. Suppose $P(1) = x$, $P(2) = 2x$, $P(3) = 3x$, and $P(4) = 4x$. Find x.

29.3. An experiment is performed in which a coin is flipped and a die is rolled. Describe this experiment as a sample space. Explicitly list all elements of the set S and the value of $P(s)$ for each element of the S.

29.4. *Tetrahedral dice.* A *tetrahedron* is a solid figure with four faces, each of which is an equilateral triangle. We can make dice in the shape of tetrahedra and label their faces with the numbers 1 through 4. When such a die is rolled, the number that lands face down on the table is the result.

a. Create a sample space that represents the toss of a tetrahedral die.

b. Create a sample space that represents the toss of a pair of tetrahedral dice.

29.5. A bag contains 20 marbles. These marbles are identical, except they are labeled with the integers 1 through 20. Five marbles are drawn at random from the bag. There are a few ways to think about this.

a. *Marbles are drawn one at a time without replacement.* Once a marble is drawn, it is not replaced in the bag. We consider all the lists of marbles we might create. (In this case, picking marbles 1, 2, 3, 4, 5 in that order is different from picking marbles 5, 4, 3, 2, 1.)

b. *Marbles are drawn all at once without replacement.* Five marbles are snatched up at once. (In this case, picking marbles 1, 2, 3, 4, 5 and picking marbles 5, 4, 3, 2, 1 are considered the same outcome.)

c. *Marbles are drawn one at a time with replacement.* Once a marble is drawn, it is tossed back into the bag (where it is hopelessly mixed up with the marbles still in the bag). Then the next marble is drawn, tossed back in, and so on. (In this case, picking 1, 1, 2, 3, 5 and picking 1, 2, 1, 3, 5 are different outcomes.)

For each of these interpretations, describe the sample space that models these experiments.

4
3
2
1

29.6. A dart is thrown blindly at the target shown in the figure. The probability that the dart lands in one of the four concentric regions is proportional to the area of the region. The radii of the circles in the figure are 1, 2, 3, and 4 units, respectively. Please note that region 2 consists of just the annular region from radius 1 to 2, and does not the include the enclosed circular region 1.

Let (S, P) be a sample space modeling this situation. The set S consists of four outcomes: hitting region 1, 2, 3, or 4. We can abbreviate that as $S = \{1, 2, 3, 4\}$.

Please find $P(1)$, $P(2)$, $P(3)$, and $P(4)$.

29.7. Give an example of a sample space with three elements in which one of the elements has probability equal to 1.

29.8. Given an example of a sample space in which all of the elements have probability 1.

29.9. Definition 29.2 requires that the set S be nonempty. In fact, this requirement is redundant. Show that if we delete this requirement from the definition, it is, nevertheless, impossible to have a sample space in which the set S is empty.

30 Events

Rolling a 2 is an outcome of the die-rolling experiment. It is a fundamental result of the experiment. Rolling an even number is an event; an event is a set of outcomes.

In this section we extend the scope of the probability function P of a sample space.

Let us return to the die-throwing example (Example 29.1). In this sample space (S, P), the probability function P gives the probability of each of the six possible outcomes of rolling the die.

We might wish to know, for example, the probability that the die will show an even number. There are three ways the die might yield an even result: face 2,

4, and 6. We want to know the probability that the die produces a result in the set $\{2, 4, 6\}$. We call such a set an *event*. The probability of this event is $\frac{1}{2}$. Each of the three outcomes of the die has probability $\frac{1}{6}$, and we simply add them.

We denote the probability of the event $\{2, 4, 6\}$ as $P(\{2, 4, 6\})$. This is a forgivable abuse of notation. The function P is a function defined on the elements of the set S of a sample space. We use the same symbol applied to a subset of S. We define this extended use of the symbol P so that

$$P(\{2, 4, 6\}) = P(2) + P(4) + P(6).$$

Definition 30.1 **(Event)** Let (S, P) be a sample space. An *event* A is a subset of S (i.e., $A \subseteq S$).

The probability of an event A, denoted $P(A)$, is

$$P(A) = \sum_{a \in A} P(a).$$

Example 30.2 **(Pair of dice)** Let (S, P) be the sample space representing the toss of a pair of dice (see Example 29.4). What is the probability that the sum of the numbers on the two dice is 7?

Let A denote the event that the numbers on the dice sum to 7. In other words,

$$A = \{(a, b) \in S : a + b = 7\} = \{(1, 6), (2, 5), (3, 4), (4, 3), (5, 2), (6, 1)\}.$$

The probability of this event is

$$\begin{aligned} P(A) &= P[(1, 6)] + P[(2, 5)] + P[(3, 4)] + P[(4, 3)] + P[(5, 2)] + P[(6, 1)] \\ &= \frac{1}{36} + \frac{1}{36} + \frac{1}{36} + \frac{1}{36} + \frac{1}{36} + \frac{1}{36} = \frac{6}{36} = \frac{1}{6}. \end{aligned}$$

Example 30.3 **(Coin tossing)** Let (S, P) be the sample space that models tossing a coin five times (see Example 29.6). What is the probability that we see exactly one HEAD?

Let A denote the event that exactly one HEAD emerges. We can write this out explicitly as

$$A = \{\text{HTTTT}, \text{THTTT}, \text{TTHTT}, \text{TTTHT}, \text{TTTTH}\}.$$

Note that A contains five outcomes, each of which has probability $\frac{1}{32}$. Therefore $P(A) = \frac{5}{32}$.

What is the probability that exactly two HEADs are shown? Let B be the event that exactly two of the coin flips show HEADs. We can write out the elements of B explicitly, but all we really need to know is how many elements are in B (because all elements of S have the same probability). The size of B is $|B| = \binom{5}{2} = 10$ because we are choosing a two-element subset (the positions of the Hs) from a five-element set (the five positions in the list). Thus $P(B) = \frac{10}{32} = \frac{5}{16}$.

Example 30.4 **(Ten dice)** Ten dice are tossed. What is the probability that none of the dice shows the number 1?

We begin by constructing a sample space (S, P). Let S denote the set of all possible outcomes of this experiment. An outcome of this experiment can be expressed as a length-ten list formed from the symbols 1, 2, 3, 4, 5, and 6. There are 6^{10} such lists and they are all equally likely, so $P(s) = 6^{-10}$ for all $s \in S$.

Let A be the event that none of the dice shows the number 1. Since all elements of S have the same probability, this problem reduces to finding the number of elements in A.

The number of outcomes that do not have the number 1 is the number of lists of length ten whose elements are chosen from the symbols 2, 3, 4, 5, and 6. The number of such lists is 5^{10}. Therefore there are 5^{10} elements in A, all of which have probability 6^{-10}. Therefore

$$P(A) = 5^{10} \times 6^{-10} = \left(\frac{5}{6}\right)^{10} \approx 0.1615.$$

Example 30.5 **(Four of a kind)** Recall the poker hand sample space of Example 29.5. A poker hand is called a *four of a kind* if four of the five cards show the same value (e.g., all 7s or all kings). What is the probability that a poker hand is a four of a kind?

Let A be the event that the poker hand is a four of a kind. Since every poker hand has probability $1 \Big/ \binom{52}{5}$, we simply need to calculate $|A|$. There are 13 choices for which value is repeated four times. Given that value, there are 48 choices for the fifth card. Thus

$$P(A) = \frac{13 \times 48}{\binom{52}{5}} = \frac{1}{4165} \approx 0.00024.$$

Example 30.6 **(Four children)** A couple has four children. Which is more likely: They have two boys and two girls, or they have three of one gender and one of the other?

Let S be the set of all possible lists of genders the couple might have. We can represent the genders of the children as a list of length four drawn from the symbols b and g. There are $2^4 = 16$ such lists, and they are all equally likely.

Let A be the event that the couple has two boys and two girls. Then

$$A = \{ggbb, gbgb, gbbg, bbgg, bgbg, bggb\}$$

so $P(A) = \frac{6}{16} = \frac{3}{8} = 0.375$.

Let B be the event that the couple has three of one gender and one of the other. Thus

$$B = \{gggb, ggbg, gbgg, bggg, bbbg, bbgb, bgbb, gbbb\}$$

so $P(B) = \frac{8}{16} = \frac{1}{2} = 0.5$.

Since $P(B) > P(A)$, we conclude that it is more likely for the couple to have three of one gender and one of the other than for them to have two boys and two girls.

Combining Events

Events are subsets of a probability space. We can use the usual operations of set theory (e.g., union and intersection) to combine events.

Union of events.

Let (S, P) be a sample space. If A and B are events, so is $A \cup B$. We can think of $A \cup B$ as the event that A or B occurs. For example, suppose A is the event that a die shows an even number and B is the event that the die shows a prime number. Then $A \cup B$ is the event that the die shows a number that is even or prime (or both), so $A \cup B = \{2, 4, 6\} \cup \{2, 3, 5\} = \{2, 3, 4, 5, 6\}$. The probability of the event $A \cup B$ is $\frac{5}{6}$.

Intersection of events.

Likewise, $A \cap B$ is the event that represents when both A and B occur. If A is the event that a die shows an even number and B is the event that it shows a prime number, then $A \cap B = \{2, 4, 6\} \cap \{2, 3, 5\} = \{2\}$. The probability of this event is $P(A \cap B) = \frac{1}{6}$.

Difference of events.

The set $A - B$ is the event that A occurs but B does not. For the die-rolling example, $A - B = \{2, 4, 6\} - \{2, 3, 5\} = \{4, 6\}$. The probability of rolling a number that is even but not prime is $P(A - B) = \frac{2}{6}$.

Complement of an event.

Since the set S of a sample space is the "universe" of all outcomes, it is sensible to write $\overline{A}$ to stand for the set $S - A$. The set $\overline{A}$ represents the event when A does not occur. For the die-rolling example, $\overline{A}$ is the event that we do not roll an even number, so $P(\overline{A}) = P(\{1, 3, 5\}) = \frac{3}{6}$.

Can we find $P(A \cup B)$ if we know only $P(A)$ and $P(B)$? The answer is no. Consider these two examples (from rolling a die).

- Let $A = \{2, 4, 6\}$ and $B = \{2, 3, 5\}$. (Event A is rolling an even number and event B is rolling a prime number.) Note that $P(A) = P(B) = \frac{1}{2}$ and $P(A \cup B) = \frac{5}{6}$.
- Let $A = \{2, 4, 6\}$ and let $B = \{1, 3, 5\}$. (Event A is rolling an even number and event B is rolling an odd number.) Note that $P(A) = P(B) = \frac{1}{2}$ and $P(A \cup B) = 1$.

These examples show that knowing $P(A) = P(B) = \frac{1}{2}$ is not enough to determine the value of $P(A \cup B)$.

We can, however, relate the quantities $P(A)$, $P(B)$, $P(A \cup B)$, and $P(A \cap B)$.

Proposition 30.7 Let A and B be events in a sample space (S, P). Then

$$P(A) + P(B) = P(A \cup B) + P(A \cap B).$$

It is interesting to compare this to Proposition 11.4, which asserts that

$$|A| + |B| = |A \cup B| + |A \cap B|.$$

In both cases, the results relate the "sizes" of sets. In the case of Proposition 11.4, we are relating the number of elements in the various sets. In Proposition 30.7, we find the analogous relation among the probabilities of the events.

Proof (of Proposition 30.7)

Consider the two sides of the equation,

$$P(A) + P(B) \qquad \text{and} \qquad P(A \cup B) + P(A \cap B).$$

We can expand these two sides as sums of $P(s)$ for various members of S. The left side is

$$P(A) + P(B) = \sum_{s \in A} P(s) + \sum_{s \in B} P(s)$$

and the right side is

$$P(A \cup B) + P(A \cap B) = \sum_{s \in A \cup B} P(s) + \sum_{s \in A \cap B} P(s).$$

Consider an arbitrary element $s \in S$. There are four possibilities:

- s is in neither A nor B. In this case, the term $P(s)$ does not enter either side of the equation.
- s is in A but not in B. In this case, $P(s)$ enters exactly once into both sides of the equation [once in $P(A)$ and once in $P(A \cup B)$, but not in $P(B)$ or $P(A \cap B)$].
- s is in B but not in A. As before, $P(s)$ enters exactly once into both sides of the equation.
- s is in both A and B. In this case, $P(s)$ appears twice on each side of the equation [once each in $P(A)$ and $P(B)$ and once each in $P(A \cup B)$ and $P(A \cap B)$].

Therefore the two sides of the equation $P(A) + P(B)$ and $P(A \cup B) + P(A \cap B)$ sum exactly the same terms and are therefore equal. ■

Proposition 30.8 Let (S, P) be a sample space and let A and B be events. We have the following:

- If $A \cap B = \emptyset$, then $P(A \cup B) = P(A) + P(B)$.
- $P(A \cup B) \le P(A) + P(B)$.
- $P(S) = 1$.
- $P(\emptyset) = 0$.
- $P(\overline{A}) = 1 - P(A)$.

The proof is left for you (Exercise 30.13). In the first item, events whose intersection is the empty set are called *mutually exclusive*.

The Birthday Problem

Four people are chosen at random. What is the probability that two (or more) of them have the same birthday?

To make this problem more tractable, we make two simplifying assumptions. First, we ignore the possibility that a person might be born on February 29. Second, we assume that it is equally likely that a person is born on any given day of the year; that is, the probability a random person is born on a given day of the year is $\frac{1}{365}$.

We model this problem with a sample space (S, P). The sample space consists of all length-4 lists of days of the year; we can represent these lists as (d_1, d_2, d_3, d_4) where the d_i are integers from 1 to 365. All such lists are equally likely with probability 365^{-4}.

Let A be the event that two (or more) of the people have the same birthday. It is easier to calculate $\overline{A}$, the probability they all have different birthdays. Because the four birthdays must be different, we can choose the first date in 365 ways, the second date in 364 ways, the third in 363, and the last in 362. Therefore

$$P(\overline{A}) = \frac{365 \cdot 364 \cdot 363 \cdot 362}{365^4} = \frac{47831784}{48627125}$$

so

$$P(A) = 1 - P(\overline{A}) = \frac{795341}{48627125} \approx 1.64\%.$$

It is rather unlikely that two of them have the same birthday.

Now suppose that 23 people are chosen at random. What is the probability that some of them have the same birthday? It would seem, since 23 is much smaller than 365, that this is also an unlikely event. However, let us analyze this situation carefully.

Consider the sample space (S, P) where S contains all length-23 lists $(d_1, d_2, \ldots, d_{23})$ where each of the d_i is an integer from 1 to 365. We assign probability 365^{-23} to each of these lists.

Let A be the event that two (or more) of the d_is are equal. As before, it is easier to calculate the probability of $\overline{A}$. The number of length-23 repetition-free lists we can form from 365 different symbols is $(365)_{23}$. Therefore,

$$P(\overline{A}) = \frac{(365)_{23}}{365^{23}} = \frac{365 \cdot 364 \cdots 343}{365^{23}}$$

and so

$$P(A) = 1 - P(\overline{A}) = 1 - \frac{(365)_{23}}{365^{23}}.$$

Using a computer, it is not hard to calculate that $P(A) = 50.73\%$, so it is more likely that two (or more) of the people will have the same birthday than it is that no two of them have the same birthday!

Recap

Let (S, P) be a sample space. An *event* is a subset A of S. The probability of the event A is the sum of the probabilities of the elements of A; that is, $P(A) = \sum_{s \in A} P(s)$. We can combine events with the usual set operations, such as union ($A \cup B$ represents the event that A or B occurs) and intersection ($A \cap B$ is the event that both A and B occur). We investigated the birthday problem.

30 Exercises

30.1. Recall the tetrahedral dice of Exercise 29.4. Suppose a pair of these dice are tossed. The sum of the values we get (face down) can range from $2 = 1+1$ to $8 = 4+4$. Let A_k be the event that the sum of the values of the dice is k. For each value of k from 2 to 8, please do the following:

a. Write down the event A_k by explicitly writing out its elements between curly braces.

b. Calculate $P(A_k)$.

30.2. A coin is flipped four times. Let A be the event that we record an equal number of HEADS and TAILS.

a. Write down the event A by explicitly writing its elements between curly braces.

b. Evaluate $P(A)$.

30.3. A coin is flipped ten times. What is the probability that we record an equal number of HEADS and TAILS?

30.4. A coin is flipped n times. What is the probability that exactly h HEADS emerge?

30.5. Let (S, P) denote the sample space for flipping a coin ten times. Let A denote the event that the results alternate between HEADS and TAILS.

a. Explicitly write down the set A.

b. Evaluate $P(A)$.

30.6. A pair of dice are rolled. Let A denote the event that the sum of the numbers showing is 8.

a. Explicitly write down the set A (as a set of ordered pairs).

b. Evaluate $P(A)$.

30.7. Three dice are rolled. What is the probability that all three dice show even numbers?

30.8. Three dice are rolled. What is the probability that the sum of the numbers showing is even?

30.9. Two dice are rolled. Let A denote the event that the number on the first die is greater than the number on the second die.

a. Explicitly write down A as a set.

b. Evaluate $P(A)$.

30.10. A bag contains ten identically wrapped boxes, but the contents of the boxes have different values (e.g., each contains a different amount of money). Alice and Bob are each going to pick one box from the bag.

Suppose Alice picks first (one of the ten boxes at random) and then Bob picks at random from the remaining boxes.

What is the probability that the contents of Alice's box are more valuable than the contents of Bob's box? Is there an advantage to going first?

30.11. *Nontransitive dice*. In this problem we consider three dice with unusual numbering. Call the three dice 1, 2, and 3. The spots on the three dice are given in the following chart.

Die 1	5	6	7	8	9	18
Die 2	2	3	4	15	16	17
Die 3	1	10	11	12	13	14

A game is played with these dice. Each player gets one of the dice (and the two players have different dice). They each roll their die, and whoever has the higher number wins.

a. If dice 1 and 2 are rolled, what is the probability that die 1 beats die 2?
b. If dice 2 and 3 are rolled, what is the probability that die 2 beats die 3?
c. If dice 3 and 1 are rolled, what is the probability that die 3 beats die 1?
d. Which die is best?

30.12. *More poker hands.*

a. What is the probability that a poker hand is a three of a kind? (A *three of a kind* has three cards of the same value and two other cards of different values, such as three 10s, a 7, and a jack.)
b. What is the probability that a poker hand is a full house? (A *full house* has three cards with one common value and two other cards of another common value, such as three queens and two 4s.)
c. What is the probability that a poker hand has one pair? (*One pair* means two cards have the same value and three other cards have three other values, such as two 9s, a king, an 8, and a 5.)
d. What is the probability that a poker hand has two pairs? (*Two pairs* means two cards have one common value, two more cards have another common value, and a fifth card has yet another value, such as two jacks, two 8s, and a 3.)
e. What is the probability that a poker hand is a flush? (A *flush* means all five cards have the same suit.)

30.13. Prove Proposition 30.8.

30.14. A coin is flipped ten times.

a. What is the probability that there are an equal number of HEADS and TAILS?
b. What is the probability that the first three flips are HEADS?
c. What is the probability that there are an equal number of HEADS and TAILS and the first three flips are HEADS?
d. What is the probability that there are an equal number of HEADS and TAILS or the first three flips are HEADS (or both)?

30.15. Three dice are rolled.

a. What is the probability that none of the dice shows 1?
b. What is the probability that at least one die shows 1?
c. What is the probability that at least one die shows 2?
d. What is the probability that none of the dice shows 1 or 2?
e. What is the probability that at least one die shows 1 or at least one die shows 2 (or both)?
f. What is the probability that at least one die shows 1 and at least one die shows 2?

30.16. Let A and B be events in a sample space. Please prove that

$$P(A \cap B) + P(A \cap \overline{B}) = P(A).$$

30.17. Suppose A and B are events in a sample space. Please prove: If $A \subseteq B$, then $P(A) \le P(B)$.

30.18. Suppose that A and B are events in a sample space and that $P(A) > \frac{1}{2}$ and $P(B) > \frac{1}{2}$. Prove that $P(A \cap B) \neq 0$.

30.19. Suppose $A_1, A_2, \ldots, A_n$ are events in a sample space. Prove that

$$P(A_1 \cup A_2 \cup \cdots \cup A_n) \leq P(A_1) + P(A_2) + \cdots + P(A_n).$$

30.20. Let A be an event in a sample space. Find $P(A \cap \overline{A})$ and give a commonsense interpretation.

30.21. Write a computer program that takes as its input an integer n between 1 and 365 and returns as its output the probability that, among n randomly chosen people, two (or more) have the same birthday.

Use your program to find the least positive integer k such that the probability is greater than 99%.

31 Conditional Probability and Independence

An event is a subset of a sample space. Accordingly, we can apply set-theoretic operations to create new events. For example, if A and B are events, then $A \cap B$ is the event in which both A and B occur.

In this section, we present the concept of one event being *conditional* on another. We illustrate this concept with a nonmathematical example.

Let A represent the event that a student misses the school bus. Let B represent the event that the student's alarm clock malfunctions. Both these events have low probability; $P(A)$ and $P(B)$ are small numbers. However, let us ask, "What is the probability of the student missing the school bus given the fact that the alarm clock malfunctioned?" Now it is likely the student will miss the bus! We denote this probability as $P(A|B)$: This is the probability that event A occurs given that event B occurs.

We can think of $P(A)$ as the frequency (percentage of mornings) with which the student misses the bus. Similarly, $P(B)$ measures how often the alarm clock fails. The conditional probability $P(A|B)$ is the frequency with which the student misses the bus, but only considering the mornings when the alarm clock is broken.

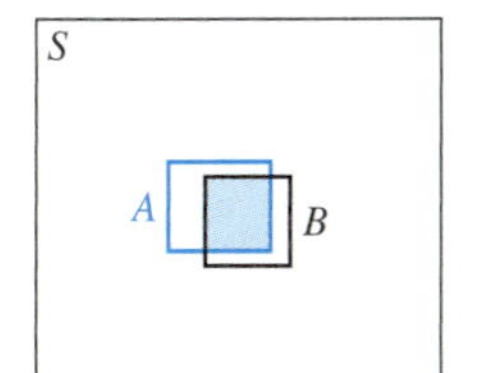

We can illustrate this with a Venn diagram. Since events are sets, we illustrate them as regions in the diagram. The box S represents the entire sample space. Regions A and B represent the two events (missing the bus and alarm clock malfunction). We have drawn boxes A and B relatively small to illustrate the fact that these are infrequent events.

The "universe" box S has area 1, and the smaller rectangles for events A and B have area equal to their probabilities, $P(A)$ and $P(B)$.

Look closely at box B—the alarm clock malfunction event. A large proportion of B's area is overlapped by box A. This overlap region represents those days on which the student misses the bus and the alarm clock fails. Given that the alarm clock has failed, a large proportion of the time the student misses the bus. The overlapping region has area $P(A \cap B)$. What proportion of box B does this overlap region cover? It covers $P(A \cap B)/P(B)$. This ratio, $P(A \cap B)/P(B)$, is fairly close to 1 and represents the frequency with which the student misses the bus on days the alarm clock fails. The *conditional probability* of event A given event B is $P(A|B) = P(A \cap B)/P(B)$.

We consider another example. Let (S, P) be the pair-of-dice sample space (Example 29.4). Consider the events A and B defined by

- Event A: the numbers on the dice sum to 8.
- Event B: the numbers on the dice are both even.

As sets, these can be written as follows:

$$A = \{(2,6), (3,5), (4,4), (5,3), (6,2)\}, \quad \text{and}$$

$$B = \{(2,2), (2,4), (2,6), (4,2), (4,4), (4,6), (6,2), (6,4), (6,6)\}.$$

Therefore we have $P(A) = \frac{5}{36}$ and $P(B) = \frac{9}{36} = \frac{1}{4}$.

We now consider the problem: What is the probability the dice sum to 8 given that both dice show even numbers? Of the nine, equally likely dice rolls in set B, three of them (highlighted in color) sum to 8. Therefore $P(A|B) = \frac{3}{9} = \frac{1}{3}$. Notice that $P(A \cap B) = \frac{3}{36}$ and we have

$$P(A|B) = \frac{P(A \cap B)}{P(B)} = \frac{3/36}{9/36} = \frac{3}{9} = \frac{1}{3}.$$

The conditional probability $P(A|B)$ when $P(B) = 0$ does not make sense for us. This asks for the probability that A occurs given that an impossible event B occurred.

The equation $P(A|B) = P(A \cap B)/P(B)$ is the definition of $P(A|B)$, and we interpret it as the probability of event A given that event B occurred. The only instance in which this definition does not make sense is when $P(B) = 0$.

Definition 31.1 **(Conditional probability)** Let A and B be events in a sample space (S, P) and suppose $P(B) \neq 0$. The *conditional probability* $P(A|B)$, the probability of A given B, is

$$P(A|B) = \frac{P(A \cap B)}{P(B)}.$$

Example 31.2 **(Spinner revisited)** Consider the spinner from Example 29.3 (see the figure). Let A be the event that we spin to a 1 (i.e., $A = \{1\}$) and let B be the event that the pointer ends in a colored region (i.e., $B = \{1, 3\}$). What is the probability that we spin to a 1 given that the pointer ends in a colored region?

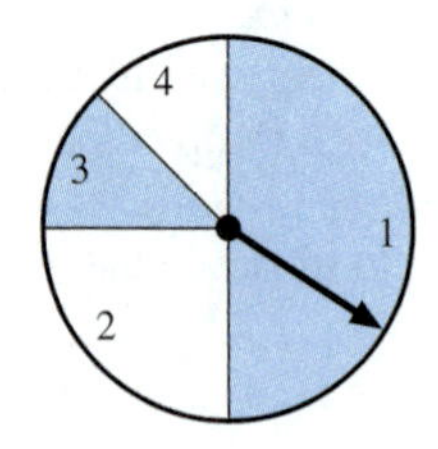

Notice that region 1 consumes $\frac{4}{5}$ of the colored portion of the diagram. We can also calculate

$$P(A|B) = \frac{P(A \cap B)}{P(B)} = \frac{P(\{1\})}{P(\{1,3\})} = \frac{1/2}{5/8} = \frac{4}{5}.$$

Example 31.3 A coin is flipped five times. What is the probability that the first flip is a TAIL given that exactly three HEADS are flipped?

Let A be the event that the first flip is TAILS, and let B be the event that we flip exactly three HEADS. We calculate

$$P(A) = \frac{2^4}{2^5} = \frac{1}{2}, \quad \text{and}$$

$$P(B) = \frac{\binom{5}{3}}{2^5} = \frac{10}{32} = \frac{5}{16}.$$

To calculate $P(A|B)$, we also need to know $P(A \cap B)$. The set $A \cap B$ contains exactly $\binom{4}{3} = 4$ sequences since the first flip must be TAILS and exactly three of the remaining four flips are HEADS. So

$$P(A \cap B) = \frac{4}{32} = \frac{1}{8}.$$

Thus

$$P(A|B) = \frac{P(A \cap B)}{P(B)} = \frac{1/8}{5/16} = \frac{2}{5}.$$

Independence

A coin is flipped five times. What is the probability that the first flip comes up HEADS given that the last flip comes up HEADS?

Let A be the event that the first flip comes up HEADS, and let B be the event that the last flip comes up HEADS. We have

$$P(A) = \frac{2^4}{2^5} = \frac{1}{2}$$
$$P(B) = \frac{2^4}{2^5} = \frac{1}{2}$$
$$P(A \cap B) = \frac{2^3}{2^5} = \frac{1}{4}$$

and therefore

$$P(A|B) = \frac{P(A \cap B)}{P(B)} = \frac{1/4}{1/2} = \frac{1}{2}.$$

Notice that $P(A|B)$ and $P(A)$ are equal. This makes intuitive sense. The probability the first flip comes up HEADS is $\frac{1}{2}$ and has nothing to do with the last flip. We call such events *independent* (a formal definition follows).

This situation is quite different from Example 31.3. In that example, knowing that three HEADS were seen decreases the likelihood that the first flip was a TAIL. Indeed, for that example, $P(A|B) = \frac{2}{5} < \frac{1}{2} = P(A)$.

We work out the consequences of the equation $P(A|B) = P(A)$. This equation can be written

$$P(A|B) = \frac{P(A \cap B)}{P(B)} = P(A)$$

and if we multiply through by $P(B)$, we get

$$P(A \cap B) = P(A)P(B).$$

Now if $P(A) \neq 0$, we can divide both sides by $P(A)$, and we have

$$P(B|A) = \frac{P(A \cap B)}{P(A)} = P(B).$$

We can summarize what we learned in the following proposition.

Proposition 31.4 Let A, B be events in a sample space (S, P) and suppose $P(A)$ and $P(B)$ are both nonzero. Then the following statements are equivalent:

The expression "the following statements are equivalent" means that each implies the other. In other words, we have (1) $\iff$ (2), (1) $\iff$ (3), and (2) $\iff$ (3).

1. $P(A|B) = P(A)$.
2. $P(B|A) = P(B)$.
3. $P(A \cap B) = P(A)P(B)$.

Nearly all the ideas for the proof have been presented. We leave it to you to fill in the details (Exercise 31.5).

We use condition (3) to define the concept of *independent* events.

Definition 31.5 **(Independent events)** Let A and B be events in a sample space. We say that these events are *independent* provided

$$P(A \cap B) = P(A)P(B).$$

Events that are not independent are called *dependent*.

We consider another example. A bag contains twenty balls; ten of the balls are painted red and ten are painted blue. Two balls are drawn from the bag. Let A be the event that the first ball drawn is red, and let B be the event that the second ball is red. Are these events independent?

The question is vague because we have not specified whether or not we replace the first ball before drawing the second. We consider both possibilities.

Suppose we replace the first ball before drawing the second. Then there are 20×20 ways to pick the two balls, of which 10×20 have the property that the first ball is red. Thus $P(A) = \frac{200}{400} = \frac{1}{2}$. Likewise, $P(B) = \frac{1}{2}$. Finally, there are 10×10 ways to draw the balls such that both the first and second balls are red. Therefore $P(A \cap B) = \frac{100}{400} = \frac{1}{4}$. Since

$$P(A \cap B) = \frac{1}{4} = \frac{1}{2} \times \frac{1}{2} = P(A)P(B)$$

we conclude that A and B are independent events. This makes sense because the color we observe on the second draw does not in any way depend on the color seen on the first.

But now suppose we do *not* replace the first ball once it is drawn. The situation is a bit more complicated. There are $20 \times 19 = 380$ different ways to draw one ball and then draw a second from those that remain. There are 10×19 ways to pick a ball such that the first ball is red; hence $P(A) = \frac{190}{380} = \frac{1}{2}$. Similarly, there are 190 ways to pick a ball such that the second ball is red, and we have $P(B) = \frac{1}{2}$. However, there are only 10×9 ways to pick the balls such that both are red. Therefore

$$P(A \cap B) = \frac{90}{380} = \frac{9}{38} \neq \frac{1}{4} = P(A)P(B)$$

and so the events are dependent.

It is instructive to calculate the conditional probabilities in this no-replacement scenario. We have

$$P(B|A) = \frac{P(A \cap B)}{P(A)} = \frac{9/38}{1/2} = \frac{9}{19} \approx 47.4\%$$

so we see that the probability the second ball is red given that the first was red is slightly less than the unconditional probability. This makes sense because once we pick the first ball, and it is red, the proportion of red balls left in the bag is less than half. Indeed, exactly nine of the remaining balls are red, and we have $P(B|A) = \frac{9}{19}$, as we noted before.

Independent Repeated Trials

Recall the spinner from Examples 29.3 and 31.2. Suppose we spin the needle twice. Now, instead of 4 possible outcomes (1, 2, 3, and 4), there are 16 [from (1, 1) through (4, 4)]. What is the probability that we spin a 3 and then we spin a 2?

We cannot express this question in the limited confines of the spinner sample space (S, P) (where $S = \{1, 2, 3, 4\}$). Nonetheless, we can answer the question. The first spin of the spinner and the second spin are independent of one another—the number that comes up on the second spin is not in any way dependent on the first number that appears. If we think of "first spin a 3" and "next spin a 2" as independent events with probabilities $\frac{1}{8}$ and $\frac{1}{4}$, respectively, then the probability that we spin a 3 and then a 2 ought to be $\frac{1}{8} \times \frac{1}{4} = \frac{1}{32}$.

Technical note on Definition 31.6: We have overused the symbol P in this definition. We have two sample spaces under consideration here: (S, P) and (S^n, P). It would be more precise to use different symbols for the two probability functions. A reasonable choice would be to write $P^n(\cdot)$ for the second probability function.

This is an example of *repeated independent trials*. We have a sample space (S, P). Instead of taking a single element $s \in S$ at random from S with probability $P(s)$, we take a sequence of events $s_1, s_2, \ldots, s_n$ each drawn at random from S. We construct a new sample space designed to handle this situation.

Definition 31.6 **(Repeated trials)** Let (S, P) be a sample space and let n be a positive integer. Let S^n denote the set of all length-n lists of elements in S. Then (S^n, P) is the *n-fold repeated-trial sample space* in which

$$P[(s_1, s_2, \ldots, s_n)] = P(s_1)P(s_2)\cdots P(s_n).$$

Example 31.7 **(Dice revisited)** The pair-of-dice sample space (Example 29.4) can be considered a repeated trial on a single die. Let (S, P) be the sample space with $S = \{1, 2, 3, 4, 5, 6\}$ and $P(s) = \frac{1}{6}$ for all $s \in S$. Then (S^2, P) represents the roll-two-dice sample space. The elements of S are all possible results for rolling a pair of dice, from (1, 1) through (6, 6), all with probability $\frac{1}{36}$.

Example 31.8 **(Coin tossing revisited)** In Example 29.6, we consider the sample space representing five flips of a fair coin. We can reformulate this situation as follows: Let (S, P) be the sample space in which $S = \{\text{HEADS}, \text{TAILS}\}$ and $P(s) = \frac{1}{2}$ for both $s \in S$.

The toss-five-times sample space is simply (S^5, P). The set S^5 contains all length-five lists of the symbols HEADS and TAILS. All such lists are equally likely with probability $\frac{1}{32}$.

Example 31.9 **(Tossing an unfair coin)** Imagine a coin that is not fairly balanced; that is, it does not turn up HEADS and TAILS with the same frequencies. We model this with a sample space (S, P) where $S = \{\text{HEADS,TAILS}\}$, but

$$P(\text{HEADS}) = p \qquad \text{and} \qquad P(\text{TAILS}) = 1 - p$$

where p is a number with $0 \le p \le 1$.

If we toss this coin five times, what is the probability that we see (in this order): HEADS, HEADS, TAILS, TAILS, HEADS?

The answer is

$$P(\text{HHTTH}) = P(\text{H})P(\text{H})P(\text{T})P(\text{T})P(\text{H}) = p \cdot p \cdot (1-p) \cdot (1-p) \cdot p.$$

This generalization of coin tossing, in which the coin might not produce HEADS and TAILS with the same frequency, is known as a *Bernoulli trial*. The term *Bernoulli trial* refers to a situation in which there are two possible outcomes, often called SUCCESS and FAILURE. The probability of SUCCESS is p and that of FAILURE is $1 - p$.

The Monty Hall Problem

The following problem is inspired by the old television game show *Let's Make a Deal*. On this show, one lucky contestant was presented with a choice of three doors. Behind exactly one of these doors was a terrific prize; the other doors concealed items of considerably less value. The contestant was asked to choose a door. At this point, the host of the show, Monty Hall, would show the contestant one of the worthless prizes behind one of the other doors. Furthermore, the contestant was offered the opportunity to switch to the other closed door. The problem is: Is it helpful to switch to the other door, or doesn't it matter?

An informal—and incorrect!—analysis of this problem runs as follows. The probability that the prize is behind the door originally picked by the contestant is $\frac{1}{3}$. But now that one door has been revealed, the probability that the valuable prize is behind either of the two remaining doors is $\frac{1}{2}$, so it doesn't matter whether the contestant switches to the other door. The error in this argument is that the contestant knows *more* than the fact that the prize is not behind a certain door. The door the host opens depends on which door the contestant originally chose, and this is not an arbitrary choice.

Let us model this situation with a sample space. Suppose, without loss of generality, the contestant chooses door 1. The prize might be behind door 1, in which case the host will show door 2 or 3. Let us suppose the host is equally likely to pick either. If the prize is behind door 2, then the host will certainly show door 3, and if the prize is behind door 3, then the host will certainly show door 2.

Let us write "P1:S2" to stand for "the prize is behind door 1 and the host shows door 2." With this notation, the four possible occurrences are P1:S2, P1:S3, P2:S3, P3:S2. We model this as a sample space by assigning the following probabilities:

$$P(\text{P1:S2}) = \frac{1}{6}, \quad P(\text{P1:S3}) = \frac{1}{6}, \quad P(\text{P2:S3}) = \frac{1}{3}, \quad P(\text{P3:S2}) = \frac{1}{3}.$$

Suppose that after the contestant picks door 1, the host reveals the worthless item behind door 2. Should the contestant switch to door 3?

Consider the following three events:

- A: the prize is behind door 1; i.e. $A = \{\text{P1:S2}, \text{P1:S3}\}$.
- B: the prize is behind door 3; i.e. $B = \{\text{P3:S2}\}$.
- C: the host reveals door 2; i.e., $C = \{\text{P1:S2}, \text{P3:S2}\}$.

Note that $P(A) = P(B) = \frac{1}{3}$. If the host did not reveal a door, there is no reason to switch.

However, let us calculate $P(A|C)$ and $P(B|C)$. We have

$$P(A \cap C) = P(\{\text{P1:S2}\}) = \frac{1}{6}$$

$$P(C) = P(\{\text{P1:S2, P3:S2}\}) = \frac{1}{6} + \frac{1}{3} = \frac{1}{2}$$

and so $$P(A|C) = \frac{P(A \cap C)}{P(C)} = \frac{1/6}{1/2} = \frac{1}{3}.$$

And we also have

$$P(B \cap C) = P(\{\text{P3:S2}\}) = \frac{1}{3}$$

$$P(C) = P(\{\text{P1:S2, P3:S2}\}) = \frac{1}{6} + \frac{1}{3} = \frac{1}{2}$$

and so $$P(B|C) = \frac{P(B \cap C)}{P(C)} = \frac{1/3}{1/2} = \frac{2}{3}.$$

Therefore it is twice as likely that the contestant will win the big prize by switching doors than by staying with the original choice.

Recap

We introduced the notion of *conditional probability*. If A and B are events [with $P(B) > 0$], then $P(A|B)$ is the probability that A occurs given that B occurs. We define $P(A|B) = P(A \cap B)/P(B)$. We discussed independent events. We say A and B are independent provided $P(A \cap B) = P(A)P(B)$. In the case where B has nonzero probability, this implies that $P(A|B) = P(A)$. We showed how to extend a sample space (S, P) into a repeated-trial sample space (S^n, P). We concluded with an analysis of the Monty Hall problem.

31 Exercises

31.1. Let (S, P) be a sample space with $S = \{1, 2, 3, 4, 5\}$ and

$$P(1) = 0.1, \quad P(2) = 0.1, \quad P(3) = 0.2, \quad P(4) = 0.2, \quad \text{and} \quad P(5) = 0.4.$$

Here we list several pairs of events A and B. In each case, please calculate $P(A|B)$.

a. $A = \{1, 2, 3\}$ and $B = \{2, 3, 4\}$.
b. $A = \{2, 3, 4\}$ and $B = \{1, 2, 3\}$.
c. $A = \{1, 5\}$ and $B = \{1, 2, 5\}$.
d. $A = \{1, 2, 5\}$ and $B = \{1, 5\}$.
e. $A = \{1, 2, 3\}$ and $B = \{1, 2, 3\}$.
f. $A = \{1, 2, 3\}$ and $B = \{4, 5\}$.
g. $A = \emptyset$ and $B = \{1, 3, 5\}$.
h. $A = \{1, 3, 5\}$ and $B = \emptyset$.
i. $A = \{1, 2, 3, 4, 5\}$ and $B = \{1, 3\}$.
j. $A = \{1, 3\}$ and $B = \{1, 2, 3, 4, 5\}$.

31.2. A pair of dice are rolled. What is the probability that neither die shows a 2 given that they sum to 7?

31.3. A pair of dice are rolled. What is the probability that they sum to 7 given that neither die shows a 2?

31.4. A coin is flipped ten times. What is the probability that the first three flips are all HEADS given that an equal number of HEADS and TAILS are flipped?

How does this conditional probability compare with the simple probability that the first three flips are HEADS?

31.5. Prove Proposition 31.4.

31.6. Are disjoint events independent? Please give a proof or a counterexample.

31.7. Let A and B be events in a sample space with $P(A \cap B) \neq 0$. Prove that $P(A|B) = P(B|A)$ if and only if $P(A) = P(B)$.

31.8. Let A and B be events in a sample space for which $P(A) > 0$, $P(B) > 0$, but $P(A \cap B) = 0$. Prove that $P(A|B) = P(B|A)$.

Give an example of two such events with $P(A) \neq P(B)$.

31.9. Let A and B be events in a sample space (S, P) and suppose $0 < P(B) < 1$. Please prove:

$$P(A|B)P(B) + P(A|\overline{B})P(\overline{B}) = P(A).$$

31.10. Let A and B be events with nonzero probability in a sample space.

Suppose $P(A|B) > P(A)$. Must it be the case that $P(B|A) > P(B)$?

Suppose $P(A|B) < P(A)$. Must it be the case that $P(B|A) < P(B)$?

Please prove your answers.

31.11. Let A and B be events in a sample space with $P(B) \neq 0$. Suppose $P(A|B) > 0$. Must it be the case that $P(A) > 0$? (Prove your answer.)

31.12. Let A, B, and C be events in a sample space and suppose $P(A \cap B) \neq 0$. Please prove:

$$P(A \cap B \cap C) = P(A)P(B|A)P(C|A \cap B).$$

31.13. A card is drawn from a well-shuffled standard deck of 52 cards.

a. What is the probability that it is a spade (♤)?

b. What is the probability that it is a king?

c. What is the probability that it is the king of spades?

d. Are the events in parts (a) and (b) independent?

31.14. Two cards are sequentially drawn (without replacement) from a well-shuffled standard deck of 52 cards. Let A be the event that the two cards drawn have the same value (e.g., both 4s) and let B be the event that the first card drawn is an ace. Are these events independent?

31.15. Two cards are sequentially drawn (without replacement) from a well-shuffled standard deck of 52 cards. Let A be the event that the two cards drawn have the same value (e.g., both 4s) and let B be the event that the two cards have the same suit (e.g., both diamonds [♢]). Are these events independent?

31.16. Two cards are sequentially drawn (without replacement) from a well-shuffled standard deck of 52 cards. Let A be the event that the first card drawn is a club (♧) and let B be the event that the second card drawn is also a club. Are these events independent?

31.17. Let A and B be events in a sample space. Prove or disprove the following statements.

a. If A and B are independent, then A and $\overline{B}$ are independent.

b. If A and B are independent, then $\overline{A}$ and $\overline{B}$ are independent.

31.18. Let A and B be events in a sample space. Prove or disprove:

a. If $P(A) = 0$, then A and B are independent.

b. If $P(A) = 1$, then A and B are independent.

31.19. Let A, B, and C be events in a sample space. Prove or disprove:

a. If A and B are independent, and B and C are independent, then A and C are independent.

b. If $P(A \cap B \cap C) = P(A)P(B)P(C)$, then A and B are independent, A and C are independent, and B and C are independent.

c. If A and B are independent, A and C are independent, and B and C are independent, then $P(A \cap B \cap C) = P(A)P(B)P(C)$.

31.20. Recall the spinner sample space (S, P) from Examples 29.3 and 31.2. Write down all the elements in (S^2, P) as well as the value of $P(\cdot)$ for every member of S^2.

31.21. The spinner from Examples 29.3 and 31.2 is spun twice. What is the probability that the sum of the two numbers is 6?

31.22. The spinner from Examples 29.3 and 31.2 is spun five times. What is the probability the number 4 is never spun?

31.23. An unfair coin shows HEADS with probability p and TAILS with probability $1 - p$ (see Example 31.9). Suppose this coin is tossed five times. Let A be the event that HEADS comes up exactly twice.

a. Write down A as a set.

b. Find $P(A)$.

31.24. An unfair coin shows HEADS with probability p and TAILS with probability $1 - p$ (see Example 31.9). Suppose this coin is tossed n times. Let A be the event that HEADS comes up exactly h times. Find $P(A)$.

31.25. An unfair coin shows HEADS with probability p and TAILS with probability $1 - p$ (see Example 31.9). Suppose this coin is tossed twice. Let A be the event that the coin comes up first HEADS and then TAILS, and let B be the event that the coin comes up first TAILS and then HEADS.

a. Calculate $P(A)$.

b. Calculate $P(B)$.

c. Calculate $P(A|A \cup B)$.

d. Calculate $P(B|A \cup B)$.

e. Explain how to use an unfair coin to make a fair decision (choose between two alternatives with equal probability).

31.26. Penelope the Pessimist and Olivia the Optimist are two of ten finalists in a contest. One of these ten finalists will be randomly chosen to receive the grand prize (all finalists have the same chance of winning). Just before the grand prize is awarded, a judge tells eight of the finalists that they have not won the grand prize, and only Penelope and Olivia remain.

Penelope thinks: Even before the judge eliminated the eight contestants, I knew that at least eight of the other people were losers. That I now know that those eight are losers doesn't tell me anything. My chance of winning is still only 10%. What rotten luck!

Olivia thinks: Now that those eight have been eliminated, there are only two of us left in the contest. So now I have a 50% chance of winning. What wonderful luck!

Whose analysis is correct?

31.27. Alice and Bob play the following game. Both players start with a pile of n chips. On each turn, they flip a coin. With probability p, Alice wins the toss and Bob gives her a chip; conversely, with probability $1 - p$, Bob wins the toss and Alice gives him a chip. The game is over when one player (the winner) has all $2n$ chips.

What is the probability that Alice wins this game?

To help you work this out, please do the following:

a. Let a_k denote the probability that Alice wins the game when she has k chips and Bob has $2n - k$. What are the values of a_0 and a_{2n}?

b. Find an expression for a_k in terms of a_{k-1} and a_{k+1}. This expression is valid when $0 < k < 2n$.

c. Using the techniques of Section 22, solve the recurrence relation from part (b) using the boundary conditions you deduced in part (a).

(If you have not studied Section 22, please see the hints in Appendix A.)

d. Your answer to part (c) should be a formula for a_k. Substitute $k = n$ into that formula to find the probability that Alice wins.

In expressing your answers to (b), (c), and (d), it is useful to let $q = 1 - p$.

32 Random Variables

The term *random variable* is, perhaps, one of the greatest misnomers in all of mathematics: A random variable is neither random nor variable! It is a function defined on a sample space. Random variables are used to model quantities whose value is random.

Let (S, P) be a sample space. Although we may be interested in the individual outcomes listed in S, we are often more interested in events. For example, in the pair-of-dice sample space, we may want to know the probability that the numbers on the two dice are different. Or if we flip a coin ten times, we may want to know the probability that we flip an equal number of HEADS and TAILS. We have studied such "compound outcomes"—they are called *events*.

We might not be interested in the specific outcomes in a sample space, but we might be interested in some quantity derived from the outcome. For example, we might want to know the sum of the numbers on two dice. Or we might want to know the number of HEADS observed in ten throws of a fair coin.

In this section we consider the concept of a random variable. A typical random variable associates a number with each outcome in a sample space (S, P). That is, $X(s)$ is a number that depends on $s \in S$. For example, X might represent the number of HEADS observed in ten flips of a coin, and if $s =$ HHTHTTTTHT then $X(s) = 4$.

The proper way to express this idea is to say that X is a *function*. The domain of X is the set S or a sample space (S, P). Each outcome $s \in S$ has a value $X(s)$

that is usually (but not always) a real number. In this case, we have $X : S \to \mathbb{R}$. More generally, a random variable is any function defined on a sample space.

Definition 32.1 **(Random variable)** A *random variable* is a function defined on a probability space; that is, if (S, P) is a sample space, then a random variable is a function $X : S \to V$ (for some set V).

Example 32.2 **(Pair of dice)** Let (S, P) be the pair-of-dice sample space (Example 29.4). Let $X : S \to \mathbb{N}$ be the random variable that gives the sum of the numbers on the two dice. For example,

$$X[(1,2)] = 3, \qquad X[(5,5)] = 10, \qquad \text{and} \qquad X[(6,2)] = 8.$$

Example 32.3 **(Heads minus tails)** Let (S, P) be the sample space representing ten tosses of a fair coin. Let $X : S \to \mathbb{Z}$ be the random variable that gives the number of HEADS minus the number of TAILS. For example,

$$X(\text{HHTHTTTTHT}) = -2.$$

We can also define random variables X_H and X_T as the number of HEADS and the number of TAILS in an outcome. For example,

$$X_H(\text{HHTHTTTTHT}) = 4 \qquad \text{and} \qquad X_T(\text{HHTHTTTTHT}) = 6.$$

Notice that $X = X_H - X_T$. This means that for any $s \in S$, $X(s) = X_H(s) - X_T(s)$.

Example 32.4 Here is an example of a random variable whose values are not numbers. Let (S, P) be the sample space representing ten tosses of a fair coin. For $s \in S$, let $Z(s)$ denote the *set* of positions where HEADS is observed. For example,

$$Z(\text{HHTHTTTTHT}) = \{1, 2, 4, 9\}$$

because the HEADS are in positions 1, 2, 4, and 9. We call Z a *set-valued* random variable because $Z(s)$ is a set.

The random variable X_H from the previous example is closely related to Z. We have $X_H = |Z|$. This means that for all $s \in S$, $X_H(s) = |Z(s)|$.

Random Variables as Events

Let X be a random variable defined on a sample space (S, P). We might like to know the probability that X takes on a particular value v. For example, if we roll a pair of dice, what is the probability that the sum of the numbers is 8? We can express this question in two ways. First, we can let A be the event that the two dice sum to 8; that is, $A = \{(2,6), (3,5), (4,4), (5,3), (6,2)\}$. We then ask: What is $P(A)$? Alternatively, we can define a random variable X to be the sum of the numbers on the dice. We can then ask: What is the probability that $X = 8$? We write this as $P(X = 8)$.

Writing $P(X = 8)$ extends the $P(\cdot)$ notation beyond its previous scope. So far, we allowed two sorts of objects to follow the P. We may write $P(s)$ where s is an element of a sample space, and we may write $P(A)$ where A is an event (i.e., a subset of a sample space).

The way to read the expression $P(X = 8)$ is to interpret the "$X = 8$" as an event. The $X = 8$ is shorthand for the event

$$\{s \in S : X(s) = 8\}.$$

In this case,

$$\begin{aligned} P(X = 8) &= P(\{s \in S : X(s) = 8\}) \\ &= P(\{(2,6),(3,5),(4,4),(5,3),(6,2)\}) = \frac{5}{36}. \end{aligned}$$

What does $P(X \geq 8)$ mean? The "$X \geq 8$" is shorthand for the event $\{s \in S : X(s) \geq 8\}$, so

$$P(X \geq 8) = P(\{s \in S : X(s) \geq 8\}) = \frac{5+4+3+2+1}{36} = \frac{15}{36} = \frac{5}{12}.$$

We can insert even more complicated algebraic expressions involving random variables into the $P(\cdot)$ notation. The notation asks for the probability of an implicit event; the event is the set of all s that satisfy the given expression. For example, recall the random variables X_H and X_T from Example 32.3. (These count the number of HEADS and the number of TAILS, respectively, in ten flips of a fair coin.) We might ask: What is the probability that there are at least four HEADS and at least four TAILS in ten flips of the coin? This question can be expressed in these various ways:

$$\begin{gathered} P(X_H \geq 4 \text{ and } X_T \geq 4) \\ P(X_H \geq 4 \wedge X_T \geq 4) \\ P(X_H \geq 4 \cap X_T \geq 4) \\ P(4 \leq X_H \leq 6). \end{gathered}$$

In every case, we seek the probability of the following event:

$$\{s \in S : X_H(s) \geq 4 \text{ and } X_T(s) \geq 4\}.$$

Incidentally, the answer to this question is

$$P(X_H \geq 4 \wedge X_T \geq 4) = \frac{\binom{10}{4} + \binom{10}{5} + \binom{10}{6}}{2^{10}} = \frac{672}{1024} = \frac{21}{32}.$$

Example 32.5 **(Binomial random variable)** Recall the unfair coin of Example 31.9. Suppose this coin produces HEADS with probability p and TAILS with probability $1 - p$. The coin is flipped n times. Let X denote the number of times that we see HEADS.

Let h be an integer. What is $P(X = h)$?

If $h < 0$ or $h > n$, it is impossible for $X(s) = h$, so $P(X = h) = 0$. Thus we narrow our attention to the case with $0 \leq h \leq n$.

There are exactly $\binom{n}{h}$ sequences of n flips with exactly h HEADS. All of these sequences have the same probability: $p^h(1-p)^{n-h}$. Therefore

$$P(X = h) = \binom{n}{h} p^h (1-p)^{n-h}.$$

We call X a *binomial* random variable for the following reason. Expand the expression $(p+q)^n$ using the binomial theorem. One of the terms in the expansion is $\binom{n}{h} p^h q^{n-h}$. If we set $q = 1-p$, this is exactly $P(X = h)$. See also Exercise 31.24.

Independent Random Variables

Recall the pair-of-dice sample space (Example 29.4). For this sample space, we define two random variables, X_1 and X_2. The value of $X_1(s)$ is the number on the first die and $X_2(s)$ is the number on the second die. For example,

$$X_1[(5,3)] = 5 \qquad \text{and} \qquad X_2[(5,3)] = 3.$$

Finally, let $X = X_1 + X_2$. This means $X(s) = X_1(s) + X_2(s)$; that is, X is the sum of the numbers on the dice. For example, $X[(5,3)] = 8$. Knowledge of X_2 tells us some information about X. For example, if we know that $X_2(s) = 4$, then $X(s) = 4$ is impossible. If we know that $X_2(s) = 4$, then the probability that $X(s) = 5$ is $\frac{1}{6}$ (as opposed to $\frac{4}{36}$). We can express this as $P(X = 5|X_2 = 4) = \frac{1}{6}$. The meaning of $P(X = 5|X_2 = 4)$ is the usual meaning of conditional probability. The events in this case are $X = 5$ and $X_2 = 4$. We can calculate this in the usual way:

$$P(X = 5|X_2 = 4) = \frac{P(X = 5 \text{ and } X_2 = 4)}{P(X_2 = 4)} = \frac{1/36}{1/6} = \frac{1}{6}.$$

However, knowledge of X_2 tells us nothing about X_1. Indeed, if a and b are integers from 1 to 6, we have

$$P(X_1 = a|X_2 = b) = \frac{P(X_1 = a \text{ and } X_2 = b)}{P(X_2 = b)} = \frac{1/36}{1/6} = \frac{1}{6} = P(X_1 = a).$$

We can say even more. Since

$$P(X_1 = a \text{ and } X_2 = b) = \frac{1}{36} = \frac{1}{6} \cdot \frac{1}{6} = P(X_1 = a)P(X_2 = b) \qquad (39)$$

the events "$X_1 = a$" and "$X_2 = b$" are independent. Furthermore, if either a or b is not an integer from 1 to 6, then both sides of Equation (39) are zero. So we have

$$\forall a, b \in \mathbb{Z},\ P(X_1 = a \text{ and } X_2 = b) = P(X_1 = a)P(X_2 = b).$$

The events $X_1 = a$ and $X_2 = b$ are independent for all a and b. This is precisely what it means to say that X_1 and X_2 are *independent* random variables.

Definition 32.6 **(Independent random variables)** Let (S, P) be a sample space and let X and Y be random variables defined on (S, P). We say that X and Y are *independent* if, for all a, b,

$$P(X = a \text{ and } Y = b) = P(X = a)P(Y = b).$$

Let us expand on the phrase "for all a, b" in this definition. The random variables X and Y are functions defined on (S, P). Therefore we may write $X : S \to A$ and $Y : S \to B$ for some sets A and B. It is not possible for X to take on a value outside of A or for Y to take on a value outside of B. So the phrase "for all a, b" can be written more extensively as "for all $a \in A$ and all $b \in B$." We can rewrite the condition in the definition as

$$\forall a \in A, \forall b \in B, \ P(X = a \text{ and } Y = b) = P(X = a)P(Y = b).$$

Recap

A random variable is neither random nor variable. Rather, a random variable is a function defined on a sample space (S, P). That is, for every $s \in S$, the random variable X returns a value $X(s)$. This value is often a number. We expanded the $P(\cdot)$ notation to include events described by random variables; for example, $P(X = 3)$ is the probability of the event $\{s \in S : X(s) = 3\}$. Random variables X and Y are independent if the events $X = a$ and $Y = b$ are independent for all a and b.

32 Exercises

32.1. Let (S, P) be a sample space with $S = \{a, b, c, d\}$ and

$$P(a) = 0.1, \quad P(b) = 0.2, \quad P(c) = 0.3, \quad \text{and} \quad P(d) = 0.4.$$

Define random variables X and Y on this sample space according to the following table.

s	$X(s)$	$Y(s)$
a	1	−1
b	3	3
c	5	6
d	8	10

Please answer the following questions.

a. Write down the event "$X > 3$" as a set of outcomes (i.e., a subset of S) and calculate $P(X > 3)$.

b. Write down the event "Y is odd" as a set of outcomes and calculate $P(Y \text{ is odd})$.

c. Write down the event "$X > Y$" as a set of outcomes and calculate $P(X > Y)$.

d. Write down the event "$X = Y$" as a set of outcomes and calculate $P(X = Y)$.

e. Calculate $P(X = m \text{ and } Y = n)$ for all integers m and n.
Note that for all but finitely many choices of m and n, this probability is zero.

f. Are X and Y independent?

g. Define a new random variable $Z = X + Y$. Find $P(Z = n)$ for all integers n.
Note that for all but finitely values of n, this probability is zero.

32.2. Recall the spinner from Examples 29.3 and 31.2. Suppose a prize of \$10 is awarded for spinning an odd number and \$20 is awarded for spinning an even number.

a. Let X be the random variable that represents the amount of money won in this game. Express X explicitly as a function defined on a sample space.

b. Write down the event "$X = 10$" as a set.

c. Calculate $P(X = a)$ for all positive integers a.

32.3. A fair coin is flipped three times. This is modeled by a sample space (S, P) where S contains the eight lists from HHH to TTT, all with probability $\frac{1}{8}$. Let X denote the number of times we see TAILS.

a. Write X explicitly as a function defined on S.

b. Write the event "X is odd" as a set.

c. Calculate $P(X \text{ is odd})$.

32.4. A pair of dice are rolled. Let X be the (absolute value of the) difference between the numbers on the dice.

a. What is $X[(2, 5)]$?

b. Evaluate $P(X = a)$ for all integers a.

32.5. Two unfair coins are tossed. The first lands HEADS side up with probability p_1, and the second lands HEADS side up with probability p_2. Let X be the random variable that gives the number of HEADS that appear when these two coins are flipped.

Please calculate $P(X = a)$ for $a = 0, 1, 2$.

32.6. A die is rolled ten times. Let X be the number of times the number 1 is rolled. Find $P(X = a)$ for all integers a.

32.7. A coin is flipped ten times. Let X_H be the number of times HEADS is produced and let X_T be the number of times TAILS is produced. Are X_H and X_T independent random variables?

32.8. A coin is flipped ten times. Let X_1 be the number of times we see HEADS immediately before TAILS and let X_2 be the number of times we see TAILS immediately before we see HEADS.

For example, if we flip THHTTHHTHH, then $X_1 = 2$ and $X_2 = 3$ because we have H-T twice and T-H three times in THHTTHHTHH.

Are X_1 and X_2 independent random variables?

32.9. A card is drawn at random from a standard deck of 52 cards. Let X be the value of the card (from 2 to ace) and let Y be the suit of card. Are X and Y independent random variables?

32.10. Two cards are drawn (without replacement) at random from a standard deck of 52 cards. Let X be the value (from 2 to ace) of the first card and let Y be the value of the second card. Are X and Y independent random variables?

32.11. Let X be a random variable defined on a sample space (S, P). Is it possible for X to be independent of itself?

33 Expectation

Most of the random variables we have considered give numerical results such as the number of HEADS in a series of coin flips or the sum of values on a pair of dice. When a random variable yields numerical results, we can ask questions such as: What is the average value this random variable might take? And we might ask: How widely spread are its values?

Not all random variables yield results that are numbers. For example, if a card is drawn at random from a deck, we can define a random variable X as the suit of the card. In this case, the random variable is not real-valued. Rather, its values lie in the set $\{♣, ♢, ♡, ♠\}$.

In this section, we consider the *expected value* of real-valued random variables. The expected value can be interpreted as the average value of a random variable.

Recall the spinner from Examples 29.3 and 31.2. Define the random variable X to be simply the number of the region in which the pointer lands. Thus

$$P(X=1)=\frac{1}{2},\quad P(X=2)=\frac{1}{4},\quad P(X=3)=\frac{1}{8},\quad \text{and}\quad P(X=4)=\frac{1}{8}.$$

What is the average value of X?

A plausible (but incorrect) reply would be the following. The random variable X can take on only four values: 1, 2, 3, and 4. The average of these is $\frac{1+2+3+4}{4} = \frac{10}{4} = \frac{5}{2}$. So the average value of X is $\frac{5}{2}$.

However, the needle lands in region 1 far more often than in region 4. So if we were to spin the pointer many times and average the result, we would be averaging many more 1s and 2s than 3s and 4s. So we would get an average value less than 2.5.

If we were to spin the pointer a huge number N times, we would expect to see (roughly) $\frac{N}{2}$ ones, $\frac{N}{4}$ twos, $\frac{N}{8}$ threes, and $\frac{N}{8}$ fours. If we add these up and divide by N, we get

$$\frac{\frac{N}{2}\times 1+\frac{N}{4}\times 2+\frac{N}{8}\times 3+\frac{N}{8}\times 4}{N}=\frac{1}{2}+\frac{1}{2}+\frac{3}{8}+\frac{1}{2}=\frac{15}{8}=1.875$$

which is less than 2.5.

A straight average of the values of X is not what we want. What we have calculated is a *weighted average* of the values of X. The value a is counted a number of times that is proportional to how often a appears. We call this weighted average of the values of X the *expected value* or *expectation* of X.

Definition 33.1 **(Expectation)** Let X be a real-valued random variable defined on a sample space (S, P). The *expectation* (or the *expected value*) of X is

$$E(X)=\sum_{s\in S} X(s)P(s).$$

The expected value of X is also called the *mean* value of X. The letter μ is often used to denote the expected value of a random variable.

Example 33.2 **(Expected value of the spinner)** Let X be the number that appears on the spinner of Example 29.3. Its expected value is

$$\begin{aligned}
E(X) &= \sum_{a=1}^{4} X(a)P(a)\\
&= X(1)P(1)+X(2)P(2)+X(3)P(3)+X(4)P(4)\\
&= 1\cdot\frac{1}{2}+2\cdot\frac{1}{4}+3\cdot\frac{1}{8}+4\cdot\frac{1}{8}\\
&= \frac{15}{8}.
\end{aligned}$$

Example 33.3 **(Expected value on a die)** A die is tossed. Let X denote the number that we see. What is the expected value of X?

The expected value is

$$\begin{aligned} E(X) &= \sum_{a=1}^{6} X(a)P(a) \\ &= X(1)P(1) + X(2)P(2) + X(3)P(3) \\ &\quad + X(4)P(4) + X(5)P(5) + X(6)P(6) \\ &= 1 \cdot \frac{1}{6} + 2 \cdot \frac{1}{6} + 3 \cdot \frac{1}{6} + 4 \cdot \frac{1}{6} + 5 \cdot \frac{1}{6} + 6 \cdot \frac{1}{6} \\ &= \frac{1+2+3+4+5+6}{6} = \frac{21}{6} = \frac{7}{2} = 3.5. \end{aligned}$$

Suppose we roll a pair of dice. Let X be the sum of the numbers on the two dice. What is the expected value of X? In principle, to calculate $E(X)$, we need to calculate

$$E(X) = \sum_{s \in S} X(s)P(s).$$

However, in this case, there are 36 different outcomes in the set S. This makes the above calculation quite unpleasant. Fortunately, there are alternative methods to calculate expectation. We present two methods that show that $E(X) = 7$.

Imagine that we wrote out all 36 terms in the sum $\sum_{s \in S} X(s)P(s)$. To simplify this mess, we can collect like terms. For example, we could collect all the terms for which $X(s) = 10$. There are three such terms:

$$\cdots + 10P[(4,6)] + 10P[(5,5)] + 10P[(4,6)] + \cdots.$$

Since all three probabilities equal $\frac{1}{36}$, this equals $10 \cdot \frac{3}{36}$. Notice that the outcomes in these three terms are exactly those $s \in S$ for which $X(s) = 10$. So we can rewrite these terms as

$$\cdots + 10P(X = 10) + \cdots.$$

If we collect all like terms, we have

$$E(X) = 2P(X = 2) + 3P(X = 3) + \cdots + 11P(X = 11) + 12P(X = 12).$$

We can use this simplification to complete the calculation of $E(X)$. We have

$$\begin{aligned} E(X) &= 2P(X = 2) + 3P(X = 3) + \cdots + 11P(X = 11) + 12P(X = 12) \\ &= 2 \cdot \frac{1}{36} + 3 \cdot \frac{2}{36} + 4 \cdot \frac{3}{36} + 5 \cdot \frac{4}{36} + 6 \cdot \frac{5}{36} + \\ &\quad + 7 \cdot \frac{6}{36} + 8 \cdot \frac{5}{36} + 9 \cdot \frac{4}{36} + 10 \cdot \frac{3}{36} + 11 \cdot \frac{2}{36} + 12 \cdot \frac{1}{36} \\ &= \frac{2 + 6 + 12 + 20 + 30 + 42 + 40 + 36 + 30 + 22 + 12}{36} \\ &= \frac{252}{36} = 7. \end{aligned}$$

This was still a great deal of work, but better than expanding out 36 terms in the sum $\sum X(s)P(s)$. We shall present an even more efficient technique to find $E(X)$, but first let us generalize what we have learned.

Proposition 33.4 Let (S, P) be a sample space and let X be a real-valued random variable defined on S. Then

$$E(X) = \sum_{a \in \mathbb{R}} a P(X = a).$$

Notice that the summation in Proposition 33.4 is over all real numbers a. This, of course, is ridiculous. It seems we have exchanged a reasonable, finite sum—namely, $\sum_{s \in S} X(s)P(s)$—for an unreasonable, infinite sum. However, because S is finite, there are only finitely many different values that $X(s)$ can actually take. For all other numbers a, $P(X = a)$ is zero, and so we do not need to include them in the sum. So the apparently infinite sum in Proposition 33.4 is, in fact, only a finite sum over just those real numbers a for which $P(X = a) > 0$.

Proof (of Proposition 33.4)

Let X be a real-valued random variable defined on a sample space (S, P). The expected value of X is

$$E(X) = \sum_{s \in S} X(s)P(s).$$

We can rearrange the order of the terms in this sum by collecting those terms with a common value for $X(s)$. We have

$$E(X) = \sum_{a \in \mathbb{R}} \left[\sum_{s \in S : X(s) = a} X(s)P(s) \right].$$

The inner sum is just over those s for which $X(s)$ is a. There are only finitely many values a for which the inner sum is not empty.

The inner sum can be rewritten. Because $X(s) = a$ for all s in the inner sum, we can replace $X(s)$ by a. This gives

$$E(X) = \sum_{a \in \mathbb{R}} \left[\sum_{s \in S : X(s) = a} a P(s) \right] = \sum_{a \in \mathbb{R}} \left[a \sum_{s \in S : X(s) = a} P(s) \right].$$

Notice that we moved a out of the inner sum (by the distributive property).

The inner sum is now simply

$$\sum_{s \in S : X(s) = a} P(s)$$

which is precisely $P(X = a)$. We make this final substitution to yield

$$E(X) = \sum_{a \in \mathbb{R}} \left[a \sum_{s \in S : X(s) = a} P(s) \right] = \sum_{a \in \mathbb{R}} a P(X = a).$$ ■

Example 33.5 In Exercise 32.2 we considered a game in which we spin the spinner from Example 29.3, receiving \$10 for spinning an odd number and \$20 for spinning an even number. Let X be the payout from this game. What is the expected value of X? In other words, how much money do we expect to receive per spin if we play this game many times?

We calculate the answer in two ways. By Definition 33.1, this is

$$\begin{aligned} E(X) &= \sum_{s \in S} X(s)P(s) \\ &= X(1)P(1) + X(2)P(2) + X(3)P(3) + X(4)P(4) \\ &= 10 \cdot \frac{1}{2} + 20 \cdot \frac{1}{4} + 10 \cdot \frac{1}{8} + 20 \cdot \frac{1}{8} \\ &= \frac{110}{8} = 13.75. \end{aligned}$$

Alternatively, we can use Proposition 33.4. In this case, we get

$$\begin{aligned} E(X) &= \sum_{a \in \mathbb{R}} aP(X = a) \\ &= 10 \cdot P(X = 10) + 20 \cdot P(X = 20) \\ &= 10 \cdot \frac{5}{8} + 20 \cdot \frac{3}{8} \\ &= \frac{110}{8} = 13.75. \end{aligned}$$

If we play this game repeatedly, we expect to receive an average of \$13.75 per spin.

Example 33.6 In Exercise 32.4, we defined a random variable X for the pair-of-dice sample space. The value of X is the absolute value of the difference of the numbers on the two dice. What is the expected value of X?

We use Proposition 33.4:

$$\begin{aligned} E(X) &= \sum_{a \in \mathbb{R}} aP(X = a) \\ &= 0 \cdot P(X = 0) + 1 \cdot P(X = 1) + 2 \cdot P(X = 2) \\ &\quad + 3 \cdot P(X = 3) + 4 \cdot P(X = 4) + 5 \cdot P(X = 5) \\ &= 0 \cdot \frac{6}{36} + 1 \cdot \frac{10}{36} + 2 \cdot \frac{8}{36} + 3 \cdot \frac{6}{36} + 4 \cdot \frac{4}{36} + 5 \cdot \frac{2}{36} \\ &= \frac{10 + 16 + 18 + 16 + 10}{36} = \frac{70}{36} = \frac{35}{18} \approx 1.944. \end{aligned}$$

Linearity of Expectation

Suppose X and Y are real-valued random variables defined on a sample space (S, P). We can form a new random variable Z by adding X and Y; that is, $Z = X + Y$. Since X and Y are functions, we need to be precise about what this means. This means that the value of Z evaluated at s is simply the sum of the values $X(s)$ and $Y(s)$.

For example, suppose (S, P) is the pair-of-dice sample space. Define X_1 to be the number on the first die and X_2 to be the number on the second die. Let $Z = X_1 + X_2$. Then Z is simply the sum of the numbers on the two dice. For example, if $s = (3, 4)$, then $X_1(s) = 3$, $X_2(s) = 4$, and $Z(s) = X_1(s) + X_2(s) = 3 + 4 = 7$.

We can perform other operations on random variables. If X and Y are real-valued random variables on a sample space (S, P), then XY is the random variable whose value at s is $X(s)Y(s)$. Likewise we can define $X - Y$ and so on.

If c is a number and X is a real-valued random variable, then cX is the random variable whose value at s is $cX(s)$.

We now address the question: If we know the expected value of X and Y, can we determine the expected value of $X + Y$, XY, or some other algebraic combination of X and Y?

Let us begin with the simplest case: addition. Let (S, P) be the pair-of-dice sample space, $X_1(s)$ the number on the first die, $X_2(s)$ the number on the second die, and $Z = X_1 + X_2$. We previously calculated that $E(X_1) = E(X_2) = \frac{7}{2}$ and $E(Z) = 7$. Notice that $E(Z) = E(X_1) + E(X_2)$. This is not a coincidence.

Proposition 33.7 Suppose X and Y are real-valued random variables defined on a sample space (S, P). Then

$$E(X + Y) = E(X) + E(Y).$$

Proof. Let $Z = X + Y$. We have

$$\begin{aligned} E(Z) &= \sum_{s \in S} Z(s)P(s) \\ &= \sum_{s \in S} [X(s) + Y(s)]P(s) \\ &= \sum_{s \in S} [X(s)P(s) + Y(s)P(s)] \\ &= \sum_{s \in S} X(s)P(s) + \sum_{s \in S} Y(s)P(s) \\ &= E(X) + E(Y). \end{aligned}$$

■

Example 33.8 Let (S, P) be the pair-of-dice sample space and let Z be the random variable giving the sum of the values on the two dice. What is $E(X)$?

Let X_1 be the value on the first die and X_2 the value on the second. Note that $Z = X_1 + X_2$. We know that $E(X_1) = E(X_2) = \frac{7}{2}$, so

$$E(Z) = E(X_1) + E(X_2) = \frac{7}{2} + \frac{7}{2} = 7.$$

Next we apply Proposition 33.7 to a more complicated problem.

A basket holds 100 chips that are labeled with the numbers 1 through 100. Two chips are drawn at random from the basket (without replacement). What is the expected value of their sum, X?

There are three ways we can approach this problem.

First, we can apply the definition of expectation to find $E(X) = \sum_{s \in S} X(s)P(s)$. This summation involves 9900 terms (there are 100 choices for the first chip times 99 choices for the second chip).

Second, we can apply Proposition 33.4 to compute $E(X) = \sum_{a \in \mathbb{R}} aP(X = a)$. The possible sums range from 3 to 199, so this sum has nearly 200 terms.

Third, we can use Proposition 33.7. Let X_1 be the number on the first chip and X_2 the number on the second chip. Note that X_1 can be any value from 1 to 100 and these are all equally likely. Furthermore, X_2 can also be any value from 1 to 100 and these, too, are equally likely. Therefore

The sum of the integers from 1 to 100 is $\binom{101}{2} = \frac{101 \cdot 100}{2} = 5050$. See Proposition 16.5.

$$E(X_1) = E(X_2) = \frac{1 + 2 + \cdots + 100}{100} = \frac{5050}{100} = 50.5.$$

Since $X = X_1 + X_2$, we have $E(X) = E(X_1 + X_2) = E(X_1) + E(X_2) = 50.5 + 50.5 = 101$.

It is important to note that X_1 and X_2 are *dependent* random variables. This does not prevent us from applying Proposition 33.7, which does not require that the random variables in question be independent.

It is also interesting to consider the expected value of the sum of the two chips if we replace the first chip before drawing the second (see Exercise 33.5).

We have seen that the expected value of a sum equals the sum of the expected values. What happens in the case of multiplication? We begin with a special case. Suppose X is a real-valued random variable on a sample space (S, P), and suppose c is a real number. What can we say about $E(cX)$. First, what does cX mean? The symbols cX stand for the random variable whose value at s is $c \cdot X(s)$. We can express this as $(cX)(s) = c[X(s)]$. Now we compute the expected value of cX. It is

$$\begin{aligned} E(cX) &= \sum_{s \in S} (cX)(s)P(s) \\ &= \sum_{s \in S} c[X(s)]P(s) \\ &= c \sum_{s \in S} X(s)P(s) \\ &= cE(X). \end{aligned}$$

We have proved the following:

Proposition 33.9 Let X be a real-valued random variable on a sample space (S, P) and let c be a real number. Then

$$E(cX) = cE(X).$$

Proposition 33.9 can be restated this way: If the average value of X is some number a, then the average value of cX is ca.

We combine Propositions 33.7 and 33.9 into one result as follows:

Theorem 33.10 **(Linearity of expectation)** Suppose X and Y are real-valued random variables on a sample space (S, P) and suppose a and b are real numbers. Then

$$E(aX + bY) = aE(X) + bE(Y).$$

Proof. We have

$$\begin{aligned} E(aX + bY) &= E(aX) + E(bY) && \text{by Proposition 33.7, and} \\ &= aE(X) + bE(Y) && \text{by Proposition 33.9 (twice).} \end{aligned}$$

■

Theorem 33.10 can be extended to apply to a longer sequence of random variables. Suppose $X_1, X_2, \ldots, X_n$ are random variables defined on a sample space (S, P), and $c_1, c_2, \ldots, c_n$ are real numbers. Then it is easy to prove by induction that

$$E\left[c_1X_1 + c_2X_2 + \cdots + c_nX_n\right] = c_1E\left[X_1\right] + c_2E\left[X_2\right] + \cdots c_nE\left[X_n\right].$$

We apply this to the following problem. A coin is tossed 10 times. Let X be the number of times we observe TAILS immediately after seeing HEADS. What is the expected value of X?

To compute $E(X)$, we express X as the sum of other random variables whose expectations are easier to calculate. Let X_1 be the random variable whose value is one if the first two tosses are HEADS-TAILS and is zero otherwise. The random variable X_1 is called an *indicator random variable;* it indicates whether or not some event occurs by taking the value one if the event occurs and the value zero if it does not. Similarly, we let X_2 be the random variable that is one if the second and third tosses come up HEADS-TAILS and is zero otherwise. More generally, let X_k be the random variable defined as follows:

$$X_k = \begin{cases} 1 & \text{if toss } k \text{ is HEADS and toss } k+1 \text{ is TAILS, and} \\ 0 & \text{otherwise.} \end{cases}$$

Then

$$X = X_1 + X_2 + \cdots + X_9.$$

Thus, to calculate $E(X)$, it is enough to calculate $E(X_k)$ for $k = 1, \ldots, 9$. The advantage is that $E(X_k)$ is easy to compute.

The random variable X_k can take on only two values, one and zero, so

$$E(X_k) = 0 \cdot P(X = 0) + 1 \cdot P(X = 1)$$
$$= P(X = 1)$$

and the probability we see HEADS-TAILS in positions k, $k+1$ is exactly $\frac{1}{4}$. Therefore $E(X_k) = \frac{1}{4}$ for each k with $1 \le k \le 9$. Therefore

$$E(X) = E(X_1) + E(X_2) + \cdots + E(X_9) = \frac{9}{4}.$$

Indicator random variables take on only two values: zero and one. Such random variables are often called *zero-one random variables*.

Proposition 33.11 Let X be a zero-one random variable. Then $E(X) = P(X = 1)$.

Example 33.12 **(Fixed points of a random permutation)** Let π be a random permutation of the numbers $\{1, 2, \ldots, n\}$. In other words, the sample space is (S_n, P) where all permutations $\pi \in S_n$ have probability $P(\pi) = \frac{1}{n!}$. Let $X(\pi)$ be the number of values k such that $\pi(k) = k$. (Such a value k is called a *fixed point* of the permutation.) What is the expected value of X?

For k with $1 \le k \le n$, let $X_k(\pi) = 1$ if $\pi(k) = k$ and let $X_k(\pi) = 0$ otherwise. Note that $X = X_1 + X_2 + \cdots + X_n$.

Since X_k is a zero-one random variable, $E(X_k) = P(X_k = 1) = \frac{1}{n}$. Therefore

$$E(X) = E(X_1) + \cdots + E(X_n) = n \cdot \frac{1}{n} = 1.$$

On average, a random permutation has exactly one fixed point.

If the expected values of X and Y are known, we can easily find the expected value of $X + Y$. Next we consider the expected value of XY.

Product of Random Variables

A pair of dice are tossed. Let X be the product of the numbers on the two dice. What is the expected value of X?

We can express X as the product of X_1 (the number on the first die) and X_2 (the number on the second die). We know that $E(X_1) = E(X_2) = \frac{7}{2}$. It seems reasonable to guess that $E(X_1X_2) = E(X_1)E(X_2) = \left(\frac{7}{2}\right)^2$.

We evaluate $E(X)$ by computing $\sum_{a \in \mathbb{R}} aP(X = a)$. The calculations we need are summarized in the following chart.

a	$P(X = a)$	$aP(X = a)$
1	1/36	1/36
2	2/36	4/36
3	2/36	6/36
4	3/36	12/36
5	2/36	10/36
6	4/36	24/36
8	2/36	16/36
9	1/36	9/36
10	2/36	20/36
12	4/36	48/36
15	2/36	30/36
16	1/36	16/36
18	2/36	36/36
20	2/36	40/36
24	2/36	48/36
25	1/36	25/36
30	2/36	60/36
36	1/36	36/36
	Total:	441/36

Therefore $E(X) = \frac{441}{36} = \frac{21}{6} \cdot \frac{21}{6} = \left(\frac{7}{2}\right)^2$. This confirms our guess that $E(X) = E(X_1X_2) = E(X_1)E(X_2)$.

This example emboldens us to conjecture that $E(XY) = E(X)E(Y)$. Unfortunately, this conjecture is incorrect, as the following example shows.

Example 33.13 A fair coin is tossed twice. Let X_H be the number of HEADS and let X_T be the number of TAILS observed. Let $Z = X_H X_T$. What is $E(Z)$?

Note that $E(X_H) = E(X_T) = 1$, so we might guess that $E(Z) = 1$. However,

$$\begin{aligned} E(Z) &= \sum_{a \in \mathbb{R}} aP(Z = a) \\ &= 0 \cdot P(Z = 0) + 1 \cdot P(Z = 1) \\ &= 0 \cdot \frac{2}{4} + 1 \cdot \frac{2}{4} \\ &= \frac{1}{2}. \end{aligned}$$

Therefore $E(X_H X_T) \neq E(X_H)E(X_T)$.

Example 33.13 shows that the conjecture $E(XY) = E(X)E(Y)$ is incorrect. It is therefore surprising that for the dice-rolling example we have $E(X_1X_2) = E(X_1)E(X_2)$. We might wonder why this works for the numbers on the two dice, but a similar equation does not hold for X_H and X_T (the numbers of HEADS and TAILS). Notice that X_1 and X_2 are *independent* random variables, but X_H and X_T are *dependent*. Perhaps the conjectured relationship $E(XY) = E(X)E(Y)$ holds for independent random variables. This revised conjecture is correct.

Theorem 33.14 Let X and Y be independent, real-valued random variables defined on a sample space (S, P). Then

$$E(XY) = E(X)E(Y).$$

Proof. Let $Z = XY$. Then

$$E(Z) = \sum_{a \in \mathbb{R}} aP(Z = a). \tag{40}$$

Let us focus on the term $aP(Z = a)$. Since $Z = XY$, the only way we can have $Z = a$ is to have $X = b$ and $Y = c$ with $bc = a$. So we can write $P(Z = a)$ as

$$P(Z = a) = \sum_{b,c \in \mathbb{R}: bc = a} P(X = b \wedge Y = c). \tag{41}$$

The sum is over all numbers b and c so that $bc = a$. Since X and Y take on at most finitely many values, this sum has only finitely many nonzero terms. Since X and Y are independent, we can replace $P(X = b \wedge Y = c)$ with $P(X = b)P(X = c)$ in Equation (41), which yields

$$P(Z = a) = \sum_{b,c \in \mathbb{R}: bc = a} P(X = b)P(Y = c).$$

We substitute this expression for $P(Z = a)$ into Equation (40) and calculate

$$\begin{aligned}
E(Z) &= \sum_{a \in \mathbb{R}} a \left[\sum_{b,c \in \mathbb{R}: bc = a} P(X = b)P(Y = c) \right] \\
&= \sum_{a \in \mathbb{R}} \left[\sum_{b,c \in \mathbb{R}: bc = a} aP(X = b)P(Y = c) \right] \\
&= \sum_{a \in \mathbb{R}} \left[\sum_{b,c \in \mathbb{R}: bc = a} bcP(X = b)P(Y = c) \right] \\
&= \sum_{b,c \in \mathbb{R}: bc} bcP(X = b)P(Y = c) \\
&= \sum_{b \in \mathbb{R}} \left[\sum_{c \in \mathbb{R}} bP(X = b)cP(Y = c) \right] \\
&= \sum_{b \in \mathbb{R}} bP(X = b) \left[\sum_{c \in \mathbb{R}} cP(Y = c) \right] \\
&= \left[\sum_{b \in \mathbb{R}} bP(X = b) \right] \left[\sum_{c \in \mathbb{R}} cP(Y = c) \right] \\
&= E(X)E(Y).
\end{aligned}$$

■

If X and Y are independent, then $E(XY) = E(X)E(Y)$. Is the converse of this statement true? If X and Y satisfy $E(XY) = E(X)E(Y)$, then may we conclude that X and Y are independent? Surprisingly, the answer is no, as the following example shows.

Example 33.15 Let (S, P) be the sample space with $S = \{a, b, c\}$ in which all three elements have probability $\frac{1}{3}$. Define random variables X and Y according to the following chart.

s	$X(s)$	$Y(s)$
a	1	0
b	0	1
c	-1	0

Note that X and Y are not independent because

$$P(X = 0) = \frac{1}{3},$$
$$P(Y = 0) = \frac{2}{3}, \quad \text{and}$$
$$P(X = 0 \wedge Y = 0) = 0 \neq \frac{2}{9} = P(X = 0)P(Y = 0).$$

Note that for all $s \in S$, we have $X(s)Y(s) = 0$. Therefore

$$E(X) = 0$$
$$E(Y) = \frac{1}{3}$$
$$E(XY) = 0 = E(X)E(Y).$$

Expected Value as a Measure of Centrality

The expected value of a real-valued random variable is in the "middle" of all the values $X(s)$. For example, consider the sample space (S, P) where $S = \{1, 2, \ldots, 10\}$ and $P(s) = \frac{1}{10}$ for all $s \in S$. Define a random variable X by the following chart.

s	$X(s)$	s	$X(s)$
1	1	6	2
2	1	7	8
3	1	8	8
4	1	9	8
5	2	10	8

Note that

$$E(X) = \sum_{a \in \mathbb{R}} aP(X = a) = 1 \times 0.4 + 2 \times 0.2 + 8 \times 0.4 = 4.$$

We illustrate this with a physical model. Imagine a seesaw—a long horizontal plank—along which we place weights. We place a weight at position a provided $P(X = a) > 0$. The weight we place at a is $P(X = a)$ kilograms. For the random variable X described in the table above, we place a total of 0.4 kg at 1 because

$P(X = 1) = 0.4$. We illustrate this in the figure—each circle represents a mass of 100 g.

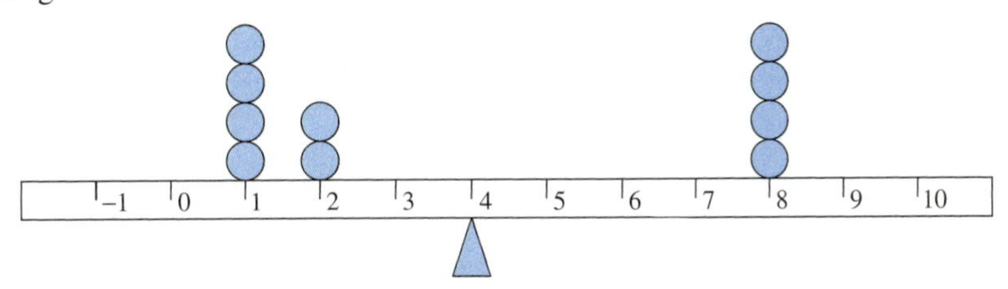

At what point does this device balance (we ignore the mass of the seesaw)? Suppose the seesaw balances at a point ℓ. Masses to the right of ℓ twist the seesaw clockwise, and masses to the left twist it counterclockwise. The greater the distance of a mass from the center, the greater the amount of twist—torque—applied to the seesaw. More precisely, if there is mass m at location x, the amount of torque it applies to the plank is $m(x - \ell)$. The seesaw is in balance if the sum of all the torques is zero. This means we need to solve the equation

$$\sum_{a \in \mathbb{R}} P(X = a)(a - \ell) = 0.$$

This equation can be rewritten as

$$\sum_{a \in \mathbb{R}} a P(X = a) = \ell \sum_{a \in \mathbb{R}} P(X = a)$$

and since $\sum_a P(X = a)$ is 1, we have

$$\ell = \sum_{a \in \mathbb{R}} a P(X = a) = E(X).$$

In the figure, the balancing point is at $\ell = 4$, the expected value of X.

Variance

The expected value of a real-valued random variable is a measure of the centrality of the values $X(s)$. Let us consider three random variables X, Y, and Z. They take on real values as follows:

$$X = \begin{cases} -2 & \text{with probability } \frac{1}{2} \\ 2 & \text{with probability } \frac{1}{2} \end{cases}$$

$$Y = \begin{cases} -10 & \text{with probability } 0.001 \\ 0 & \text{with probability } 0.998 \\ 10 & \text{with probability } 0.001 \end{cases}$$

$$Z = \begin{cases} -5 & \text{with probability } \frac{1}{3} \\ 0 & \text{with probability } \frac{1}{3} \\ 5 & \text{with probability } \frac{1}{3}. \end{cases}$$

Notice that all three of these random variables have an expected value equal to zero; the "centers" of these random variables are all the same. Yet the random variables are quite different. We consider: Which of these is more "spread out"? At first glance, it appears that Y is the most spread out because its values range from -10 to $+10$, whereas X is the most "compact" because its values are restricted to the narrowest range (from -2 to $+2$).

However, Y's extreme values at ± 10 are exceedingly rare. It can be argued that Y is more concentrated near 0 than X because Y is almost always equal to zero, whereas X can be only at ± 2.

To better describe how spread out the values of a random value are, we need a precise mathematical definition. Here is an idea: Let $\mu = E(X)$. Let us calculate how far away each value of X is from μ, but count it only proportional to its probability. That is, we add up $[X(s) - \mu]P(s)$. Unfortunately, this is what happens:

The expression

$$\sum [X(s) - \mu]P(s)$$

measures how far away X is from its mean, μ. It is a weighted average of the distance from X to μ. At first glance, it would appear that when X's values are widely spread out, this weighted average would be large. However, in all cases, it sums to zero.

$$\sum_{s\in S}[X(s) - \mu]P(s) = \left[\sum_{s\in S} X(s)P(s)\right] - \left[\sum_{s\in S} \mu P(S)\right] = E(X) - \mu = 0.$$

The problem is that values to the right of μ are exactly canceled by values to the left. To prevent this cancellation, we can square the distances between X and μ, counting them proportional to their probability. That is, we add up $[X(s) - \mu]^2 P(s)$. We can think of the sum

$$\sum_{s\in S}[X(s) - \mu]^2 P(s)$$

as the expected value of a random variable $Z = (X - \mu)^2$. That is, $Z(s) = [X(s) - \mu]^2$, and the expected value of Z is exactly the measure of "spread" we are creating. This value is called the *variance* of X.

Definition 33.16 **(Variance)** Let X be a real-valued random variable on a sample space (S, P). Let $\mu = E(X)$. The *variance* of X is

$$\text{Var}(X) = E[(X - \mu)^2].$$

Example 33.17 Let X, Y, and Z be the three random variables we introduced at the beginning of this discussion of variance. All three of these random variables have expected value $\mu = 0$. We calculate their variances as follows:

$$\begin{aligned}\text{Var}(X) &= E[(X - \mu)^2] = E(X^2)\\ &= (-2)^2 \cdot 0.5 + 2^2 \cdot 0.5\\ &= 4,\end{aligned}$$

$$\begin{aligned}\text{Var}(Y) &= E[(Y - \mu)^2] = E(Y^2)\\ &= (-10)^2 \cdot 0.001 + 0^2 \cdot 0.998 + 10^2 \cdot 0.001\\ &= 0.2, \quad \text{and}\end{aligned}$$

$$\begin{aligned}\operatorname{Var}(Z) &= E[(Z-\mu)^2] = E(Z^2)\\ &= (-5)^2\cdot\frac{1}{3} + 0^2\cdot\frac{1}{3} + 5^2\cdot\frac{1}{3}\\ &= \frac{50}{3} \approx 16.67.\end{aligned}$$

By this measure, Z is the most spread out and Y is the most concentrated.

Example 33.18 A die is tossed. Let X denote the number that appears on the die. What is the variance of X?

Let $\mu = E(X) = \frac{7}{2}$. Then

$$\begin{aligned}\operatorname{Var}(X) &= E[(X-\mu)^2] = E\left[\left(X-\frac{7}{2}\right)^2\right]\\ &= \left(1-\frac{7}{2}\right)^2\cdot\frac{1}{6} + \left(2-\frac{7}{2}\right)^2\cdot\frac{1}{6} + \left(3-\frac{7}{2}\right)^2\cdot\frac{1}{6}\\ &\quad + \left(4-\frac{7}{2}\right)^2\cdot\frac{1}{6} + \left(5-\frac{7}{2}\right)^2\cdot\frac{1}{6} + \left(6-\frac{7}{2}\right)^2\cdot\frac{1}{6}\\ &= \frac{25}{24} + \frac{3}{8} + \frac{1}{24} + \frac{1}{24} + \frac{3}{8} + \frac{25}{24}\\ &= \frac{35}{12} \approx 2.9167.\end{aligned}$$

The following result gives an alternative method for calculating the variance of a random variable.

Proposition 33.19 Let X be a real-valued random variable. Then

$$\operatorname{Var}(X) = E[X^2] - E[X]^2.$$

Please note that $E[X^2]$ is quite different from $E[X]^2$. The first is the expected value of the random variable X^2, and the second is the square of the expected value of X. These quantities need not be the same.

Proof. Let $\mu = E(X)$. By definition, $\operatorname{Var}(X) = E[(X-\mu)^2]$. We can write $(X-\mu)^2 = X^2 - 2\mu X + \mu^2$. We can think of this as the sum of three random variables: X^2, $-2\mu X$, and μ^2. If we evaluate these at an element s of the sample space, we get $[X(s)]^2$, $-2\mu X(s)$, and μ^2, respectively. Here we are thinking of μ^2 both as a number and as a random variable. As a random variable, its value at every s is simply μ^2. Therefore $E(\mu^2) = \mu^2$. We calculate

$$\begin{aligned}
\operatorname{Var}(X) &= E[(X-\mu)^2] \\
&= E[X^2 - 2\mu X + \mu^2] \\
&= E[X^2] - 2\mu E[X] + E[\mu^2] \qquad \text{by Theorem 33.10} \\
&= E[X^2] - 2\mu^2 + \mu^2 \\
&= E[X^2] - \mu^2 \\
&= E[X^2] - E[X]^2.
\end{aligned}$$

■

Example 33.20 Let X be the number showing on a random toss of a die. What is $\operatorname{Var}(X)$?

We apply Proposition 33.19, $\operatorname{Var}(X) = E[X^2] - E[X]^2$. Note that $E[X]^2 = \left(\frac{7}{2}\right)^2 = \frac{49}{4}$. Also,

$$\begin{aligned}
E[X^2] &= 1^2 \cdot \frac{1}{6} + 2^2 \cdot \frac{1}{6} + 3^2 \cdot \frac{1}{6} + 4^2 \cdot \frac{1}{6} + 5^2 \cdot \frac{1}{6} + 6^2 \cdot \frac{1}{6} \\
&= \frac{1^2 + 2^2 + 3^2 + 4^2 + 5^2 + 6^2}{6} \\
&= \frac{91}{6}.
\end{aligned}$$

Therefore

$$\operatorname{Var}(X) = E[X^2] - E[X]^2 = \frac{91}{6} - \frac{49}{4} = \frac{35}{12}.$$

This agrees with Example 33.18.

The variance of a binomial random variable.

Recall Example 32.5, in which an unfair coin is flipped n times. The coin produces HEADS with probability p and TAILS with probability $1-p$. Let X denote the number of times we see HEADS. We have $E(X) = np$ (see Exercise 33.9). What is the variance of X?

We can express X as the sum of zero-one indicator random variables. Let $X_j = 1$ if the jth flip comes up HEADS and $X_j = 0$ if the jth flip comes up TAILS. Then $X = X_1 + X_2 + \cdots + X_n$.

By Proposition 33.19, $\operatorname{Var}(X) = E[X^2] - E[X]^2$. The term $E[X]^2$ is simple to calculate. Since $E[X] = np$, we have $E[X]^2 = n^2p^2$. The calculation of $E[X^2]$ is more complicated. Since

$$X = X_1 + X_2 + \cdots + X_n$$

we have

$$\begin{aligned}
X^2 &= [X_1 + X_2 + \cdots + X_n]^2 \\
&= X_1X_1 + X_1X_2 + \cdots + X_1X_n + X_2X_1 + \cdots\cdots + X_nX_n.
\end{aligned}$$

There are two kinds of terms in this expansion. There are n terms where the subscripts are the same (e.g., X_1X_1), and there are $n(n-1)$ terms where the subscripts are different (e.g., X_1X_2). We can express this as

$$X^2 = \sum_{i=1}^{n} X_i^2 + \sum_{i \neq j} X_i X_j.$$

To find $E[X^2]$, we apply linearity of expectation. Note that $E[X_i^2] = E[X_i] = p$ (see Proposition 33.11 and Exercise 33.8). If $i \neq j$, then X_i and X_j are independent random variables. Therefore $E[X_i X_j] = E[X_i]E[X_j] = p^2$ (see Proposition 33.14). Therefore

$$\begin{aligned} E[X^2] &= E\left[\sum_{i=1}^{n} X_i^2 + \sum_{i \neq j} X_i X_j\right] \\ &= \sum_{i=1}^{n} E\left[X_i^2\right] + \sum_{i \neq j} E[X_i X_j] \\ &= np + n(n-1)p^2. \end{aligned}$$

We now have that $E[X^2] = np + n(n-1)p^2$ and $E[X]^2 = n^2p^2$. Therefore

$$\begin{aligned} \mathrm{Var}[X] &= E[X^2] - E[X]^2 \\ &= np + n(n-1)p^2 - n^2p^2 \\ &= np + n^2p^2 - np^2 - n^2p^2 \\ &= np - np^2 \\ &= np(1-p). \end{aligned}$$

Recap

The expected value of a real-valued random variable X is the average value of X over many trials. Specifically, $E(X) = \sum_{s \in S} X(s)P(s)$. By rearranging terms, we can write this as $\sum_{a \in \mathbb{R}} aP(X = a)$. If X and Y are real-valued random variables, then $E(X + Y) = E(X) + E(Y)$. If a and b are real numbers, this can be extended to $E(aX+bY) = aE(X)+bE(Y)$. This result is known as linearity of expectation. We can often use linearity of expectation to simplify the calculation of expected values. If X represents the number of times something happens, we can often express X as the sum of indicator random variables whose expectations are easy to calculate. This enables us to calculate $E(X)$. If X and Y are independent random variables, we have $E(XY) = E(X)E(Y)$. We showed how the expected value of X is at the "center" of the values of X, and we introduced the variance as a measure of how spread out the values of X are.

33 Exercises

33.1. Find the expected value of the random variables X, Y, and Z in Exercise 32.1.

33.2. Let (S, P) be the sample space with $S = \{a, b, c\}$ and $P(s) = \frac{1}{3}$ for all $s \in S$. Find the expected value of each of the following random variables:

a. X, where $X(a) = 1$, $X(b) = 2$, and $X(c) = 10$.

b. Y, where $Y(a) = Y(b) = -1$ and $Y(c) = 2$.

c. Z, where $Z = X + Y$.

33.3. A pair of tetrahedral dice are rolled (see Exercise 29.4). Let X be the sum of the two numbers and let Y be the product.

Find $E(X)$ and $E(Y)$.

33.4. You play a game in which you roll a die and you win (in dollars) the square of the number on the die. For example, if you roll a 5, then you win \$25. On average, how much money would you expect to receive per play of this game?

33.5. A basket holds 100 chips that are labeled with the numbers 1 through 100. A chip is drawn at random from the basket, it is replaced, and a second chip is drawn at random (it might be the same chip). Let X be the sum of the numbers on the two chips. What is the expected value of X?

33.6. A coin is flipped 100 times. Let X_H be the number of HEADS and X_T the number of TAILS. Please do the following:

a. Let $Z = X_H + X_T$. What is $Z(s)$? Here s represents an element of the flip-a-coin-one-hundred-times sample space.

b. Evaluate $E(Z)$.

c. Is it true that $X_H = X_T$?

d. Is it true that $E(X_H) = E(X_T)$?

e. Evaluate $E(X_H)$ and $E(X_T)$ using what you have learned from parts (b) and (d).

f. Evaluate $E(X_H)$ by expressing X_H as the sum of 100 indicator random variables.

33.7. Prove Proposition 33.11.

33.8. Suppose X is a zero-one random variable. Prove that $E(X) = E(X^2)$.

33.9. Let X be a binomial random variable as in Example 32.5. Prove that $E(X) = np$.

33.10. Let X and Y be real-valued random variables defined on a sample space (S, P). Suppose $X(s) \le Y(s)$ for all $s \in S$. Prove that $E(X) \le E(Y)$.

33.11. Let (S, P) be a sample space and let $A \subseteq S$ be an event. Define a random variable I_A whose value at $s \in S$ is

$$I_A(s) = \begin{cases} 1 & \text{if } s \in A \text{, and} \\ 0 & \text{otherwise.} \end{cases}$$

The random variable I_A is called an *indicator* random variable because its value indicates whether or not an event occurred.

Prove: $E(X) = P(A)$.

33.12. *Markov's inequality*. Let (S, P) be a sample space and let $X : S \to \mathbb{N}$ be a nonnegative-integer-valued random variable. Let a be a positive integer. Prove that

$$P(X \ge a) \le \frac{E(X)}{a}.$$

A special case of this result is that $P(X > 0) \le E(X)$.

33.13. Find the variance of the random variables X, Y, and Z in Exercise 32.1.

33.14. Let X be the number produced in a toss of a tetrahedral die. Calculate $\text{Var}(X)$.

33.15. Suppose X and Y are independent random variables defined on a sample space (S, P). Prove that $\text{Var}(X + Y) = \text{Var}(X) + \text{Var}(Y)$.

Give an example to show that the hypothesis that the random variables are independent is necessary.

33.16. A pair of dice are tossed. Let X be the sum of the numbers on the two dice. Evaluate $\text{Var}(X)$.

33.17. *Chebyshev's inequality.* Let X be a nonnegative-integer-valued random variable. Suppose $E(X) = \mu$ and $\text{Var}(X) = \sigma^2$. Let a be a positive integer.

Prove:

$$P[|X - \mu| \geq a] \leq \frac{\sigma^2}{a^2}.$$

Chapter 6 Self Test

1. Let (S, P) be a sample space with $S = \{1, 2, 3, \ldots, 10\}$. For $a \in S$, suppose we have

$$P(a) = \begin{cases} x & \text{if } a \text{ is even and} \\ 2x & \text{if } a \text{ is odd.} \end{cases}$$

Find x.

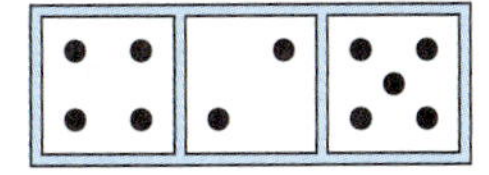

2. Three dice are dropped at random into a frame where they sit snuggly in a row (see the figure). We wish to model this experiment using a sample space, (S, P).
   a. How many outcomes are in S if we think of the dice as being identical?
   b. How many outcomes are in S if we think of the dice as being distinct (e.g., each of the three dice is a different color).
3. Let (S, P) be a sample space where $S = \{1, 2, 3, \ldots, 10\}$ and $P(j) = j/55$ for $1 \leq j \leq 10$. Let A be the event $A = 1, 4, 7, 9$. What is the probability of A?
4. Ten children (five boys and five girls) are standing in line. Assume that all possible ways in which they might line up are equally likely.
   a. What is the probability that they appear in line in alphabetical order by name? Please assume no two of the children have the same name.
   b. What is the probability that all the girls precede all the boys?
   c. What is the probability that between any two girls there are no boys (i.e., the girls stand together in an uninterrupted block)?
   d. What is the probability that they alternate by gender in the line?
   e. What is the probability that neither the boys nor the girls stand together in an uninterrupted block?
5. Thirteen cards are drawn (without replacement) from a standard deck of cards.
   a. What is the probability they are all spades (♤)?
   b. What is the probability they are all black?
   c. What is the probability they are not all of one color?
   d. What is the probability that none of the cards is an ace?
   e. What is the probability that none of the cards is an ace and none is a heart (♡)?

In fact, aces may be taken to have the value 1 or 11, but for this problem we simplify matters by considering only the value 11.

6. In the card game blackjack, each card in the deck has a numerical value. Number cards (2 through 10) have the value printed on the card. Face cards (jacks, queens, and kings) have the value 10, and aces have the value 11.

   Two cards are drawn (without replacement) from a well-shuffled deck.

   a. What is the probability that the sum of the values on the cards is 21?

   b. What is the probability that the sum of the values on the cards is 16 or higher?

   c. What is the probability the second card is a face card given that the first card is an ace?

7. A standard deck of cards is shuffled. What is the probability that the color of the last card is red given that the color of the first card is black? Are the colors of the first and last cards independent; that is, are the events "first card black" and "last card red" independent?

Your answer to problem 8 should begin "Let A be an event in a sample space (S, P). Events A and $\overline{A}$ are independent if and only if . . . ."

8. Let A be an event for a sample space (S, P). Under certain circumstances it is possible for the events A and $\overline{A}$ to be independent. Formulate and prove an if-and-only-if theorem for an event and its complement to be independent.

9. Two squares are chosen (with replacement) from among the 64 squares of a standard chess board; all such choices are equally likely. We consider the following events:
   - R is the event that the two squares are in the same row of the chess board,
   - C is the event that the two squares are in the same column of the chess board, and
   - B is the event that both squares are black.

   Which pairs of these events are independent?

10. Repeat the previous problem, this time assuming the squares are chosen without replacement where all 64×63 possible sequences of choices are equally likely.

11. An unfair coin is tossed twice in a row. What is the probability that the outcome is HEADS and then TAILS, given that the two flips give different results (i.e., not HEADS-HEADS and not TAILS-TAILS)?

12. Let A and B be events for a sample space (S, P). Suppose that $A \subseteq B$ and $P(A) \neq 0$. Prove that $P(A) = P(A|B)P(B)$.

13. Consider the sample space (S, P) where $S = \{a, b, c\}$ and $P(a) = 0.4$, $P(b) = 0.4$, and $P(c) = 0.2$. Let X be a real-valued random variable, and suppose $X(a) = 1$, $X(b) = 2$, and $E(X) = 0$. Find $X(c)$.

14. A card is drawn from a well-shuffled deck. Let X be the blackjack value of the first card in the deck and let Y be the value of the second card. (Recall that face cards are worth 10 and aces are worth 11; see problem 6).

    Please do the following:

    a. Calculate $P(X \text{ is even})$.

    b. Calculate $E(X)$.

    c. Calculate $E(Y)$.

    d. Are X and Y independent? Justify your answer.

    e. Calculate $E(X + Y)$.

    f. Calculate $P(X = Y)$.

    g. Calculate $\text{Var}(X)$.

15. *Simplified stock market.* Suppose there are three kinds of days: GOOD, GREAT, and ROTTEN. The following chart gives the frequency of each of these types of days and the effect on the price of a certain stock on that day.

Type of day	Frequency	Change in stock value
GOOD	60%	+2
GREAT	10%	+5
ROTTEN	30%	−4

The type of a given day is independent of the type of any other day. Let X be the random variable giving the change in value of the stock after five consecutive days.

Please answer:

a. What is the expected change in the stock price? (That is, find $E(X)$.)

b. Calculate $\text{Var}(X)$.

CHAPTER

7 Number Theory

Number theory is one of the oldest branches of mathematics and continues to be a vibrant area of research. It was considered, for some time, to be quintessential pure mathematics—a subject enjoyed for its own sake without any applications. Recently, number theory has become central in the world of cryptography (see Sections 43–45) and computer security.

34 Dividing

Does this sound like you're back in grade school? Sorry! Please bear with us.

Six children find a bag containing 25 marbles. How should they share them?

The answer is that each child should get 4 marbles, and there will be 1 left over. The problem is to divide 25 by 6. The quotient is 4 and the remainder is 1. Here is a formal statement of this process.

Theorem 34.1 **(Division)** Let $a, b \in \mathbb{Z}$ with $b > 0$. There exist integers q and r such that

$$a = qb + r \qquad \text{and} \qquad 0 \le r < b.$$

Moreover, there is only one such pair of integers (q, r) that satisfies these conditions.

The integer q is called the *quotient* and the integer r is called the *remainder*.

Example 34.2 Let $a = 23$ and $b = 10$. Then the quotient is $q = 2$ and the remainder $r = 3$ because

$$23 = 2 \times 10 + 3 \qquad \text{and} \qquad 0 \le 3 < 10.$$

Example 34.3 Let $a = -37$ and $b = 5$. Then $q = -8$ and $r = 3$ because

$$-37 = -8 \times 5 + 3 \qquad \text{and} \qquad 0 \le 3 < 5.$$

Remember: The Well-Ordering Principle states that every nonempty subset of $\mathbb{N}$ contains a least element.

The remainder is the smallest natural number we can form by subtracting multiples of b from a. This observation gives us the key idea in the proof. Consider all natural numbers of the form $a-kb$ and let r be the smallest such natural number. We use the Well-Ordering Principle.

Proof (of Theorem 34.1)

Let a and b be integers with $b > 0$. The first goal is to show that the quotient and remainder exist; that is, there exist integers q and r that satisfy the three conditions

- $a = qb + r$,
- $r \geq 0$, and
- $r < b$.

For example, if $a = 11$ and $b = 3$, then $A = \{\ldots, -4, -1, 2, 5, 8, 11, \ldots\}$ and $B = A \cap \mathbb{N} = \{2, 5, 8, 11, 14, \ldots\}$.

Let

$$A = \{a - bk : k \in \mathbb{Z}\}.$$

We want the remainder to be nonnegative, so we consider only the nonnegative elements of A. Let

$$B = A \cap \mathbb{N} = \{a - bk : k \in \mathbb{Z},\ a - bk \geq 0\}.$$

We want to select the least element of B. Note that the Well-Ordering Principle applies to nonempty subsets of $\mathbb{N}$. Thus we need to check that $B \neq \emptyset$.

In our example, $a = 11 > 0$ so $a \in B = \{2, 5, 8, 11, 14, \ldots\}$.

The simplest thing to do is to choose $k = 0$ in the expression $a - bk$. This shows that $a \in A$ and, if a is nonnegative, then $a \in B$, so $B \neq \emptyset$. But it might be the case that $a < 0$. We know, however, that $b > 0$, so if we take k to be a very negative number, we can certainly make $a - bk$ positive. (As long as we choose k to be any integer less than $\frac{a}{b}$, we know that $a - bk \geq 0$.)

Therefore, regardless of whether a is positive, negative, or zero, the set B is nonempty.

The least element of $B = \{2, 5, 8, 11, \ldots\}$ is $r = 2$. Since $r \in A$, we can express $r = 11 - 3q$, (i.e., when $q = 3$).

Since $B \neq \emptyset$, by the Well-Ordering Principle (Statement 20.6) we can choose r to be the least element of B. Since

$$r \in B \subseteq A = \{a - bk : k \in \mathbb{Z}\}$$

we know that there is an integer, and we call it q, such that $r = a - bq$. This can be rewritten

$$a = qb + r.$$

Since $r \in B \subseteq \mathbb{N}$, we also know that

$$r \geq 0.$$

We still need to show that $r < b$. To prove this, suppose, for the sake of contradiction, that $r \geq b$.

Let's think about this for a moment. We are subtracting multiples of b from a until we reach r, and $r \geq b$. This means we can still subtract another b from r without making a negative result. We have

$$r = a - qb \geq b.$$

Let $r' = (a - qb) - b = r - b \geq 0$, so

$$r' = a - (q + 1)b \geq 0.$$

Therefore $r' \in B$ and $r' = r - b < r$. This contradicts the fact that r is the smallest element of B.$\Rightarrow\Leftarrow$ Therefore $r < b$.

We have proved that the integers q and r exist. We now have to prove that they are unique. Uniqueness is proved by contradiction (see Proof Template 14).

Suppose, for the sake of contradiction, there are two different pairs of numbers (q, r) and (q', r') that satisfy the conditions of the theorem; that is,

$$\begin{array}{ll} a = qb + r & 0 \leq r < b \quad \text{and} \\ a = q'b + r' & 0 \leq r' < b. \end{array}$$

Combining the two equations on the left gives

$$qb + r = q'b + r' \quad \Longrightarrow \quad r - r' = (q' - q)b.$$

This means that $r - r'$ is a multiple of b. But recall that $0 \leq r, r' < b$. The difference of two numbers in $\{0, 1, \ldots, b - 1\}$ can be at most $b - 1$. So the only way that $r - r'$ can be a multiple of b is if $r - r' = 0$ (i.e., $r = r'$).

Now that we know $r = r'$, we turn to q and q'. Since

$$qb + r = a = q'b + r' = q'b + r,$$

we can subtract r from both sides to give

$$qb = q'b,$$

and since $b \neq 0$, we can cancel b from both sides, which yields

$$q = q'.$$

We have shown that these two different pairs of numbers (q, r) and (q', r') have $q = q'$ and $r = r'$, a contradiction.$\Rightarrow\Leftarrow$ Therefore, the quotient and remainder are unique. ■

Armed with Theorem 34.1, we can prove the following:

Corollary 34.4 Every integer is either even or odd, but not both.

Proof. We have previously shown (Proposition 19.3) that no integer can be both even and odd. Thus it remains to show that every integer is one or the other (i.e., there is no integer that is neither).

Let n be any integer. By Theorem 34.1 we can find integers q and r such that $n = 2q + r$ where $0 \leq r < 2$. Note that if $r = 0$ then n is even, and if $r = 1$ then n is odd. ■

Corollary 34.5 Two integers are congruent modulo 2 if and only if they are both even or both odd.

Proof. ($\Rightarrow$) Let a and b be integers and suppose $a \equiv b \pmod 2$. This means that $a - b$ is divisible by 2, say $a - b = 2n$ for some integer n. By Corollary 34.4, a is either even or odd.

- If a is even, then $a = 2k$ for some integer k. Since $a - b = 2n$, we have $b = a - 2n = 2k - 2n = 2(k - n)$ and so b is even.
- If a is odd, then $a = 2k + 1$ for some integer k. Since $a - b = 2n$, we have $b = a - 2n = 2k + 1 - 2n = 2(k - n) + 1$; thus b is odd.

In either case, a and b are either both even or both odd.

($\Leftarrow$) Suppose a and b are integers that are both even or both odd.

If a and b are both even, then $a = 2n$ and $b = 2m$ for some integers n and m. Then $a - b = 2n - 2m = 2(n - m)$ and so $a \equiv b \pmod 2$.

If a and b are both odd, then $a = 2n + 1$ and $b = 2m + 1$ for some integers n and m. Then $a - b = (2n + 1) - (2m + 1) = 2(n - m)$ and so $a \equiv b \pmod 2$.

Thus if a and b are both even or both odd, then $a \equiv b \pmod 2$. ■

Div and Mod

We define two operations associated with the division process. Given a and b, these operations give the quotient and remainder of the division problem. Now it would be quite sensible if mathematicians named these operations with words such as *quot* and *rem,* but we're a mischievous lot; we call them *div* and *mod.* Thus, not only are we guilty of creating new names where perfectly good old names suffice, but we use the word *mod* in two different ways: as an operation and as a relation.

Definition 34.6 **(div and mod)** Let $a, b \in \mathbb{Z}$ with $b > 0$. By Theorem 34.1, there exists a unique pair of numbers q and r with $a = qb + r$ and $0 \le r < b$. We define the operations div and mod by

$$a \text{ div } b = q \qquad \text{and} \qquad a \bmod b = r.$$

Example 34.7 These calculations illustrate the div and mod operations.

$$\begin{array}{cc} 11 \text{ div } 3 = 3 & 11 \bmod 3 = 2 \\ 23 \text{ div } 10 = 2 & 23 \bmod 10 = 3 \\ -37 \text{ div } 5 = -8 & -37 \bmod 5 = 3 \end{array}$$

Pay close attention to the last example. The remainder is never negative. So although $-37 \div 5 = -7.4$, we have $-37 \text{ div } 5 = -8$ and $-37 \bmod 5 = 3$ because $-37 = -8 \times 5 + 3$ and $0 \le 3 < 5$.

A second meaning of mod.

We now need to pay special attention to the overworked word *mod.* We have used this word in two different ways. The two meanings of mod are closely related but different.

When we first introduced the word *mod* (see Definition 14.3) it was used as the name of an equivalence relation. For example,

$$53 \equiv 23 \pmod{10}.$$

The meaning of $a \equiv b \pmod{n}$ is that $a - b$ is a multiple of n. We have $53 \equiv 23 \pmod{10}$ because $53 - 23 = 30$, a multiple of 10.

In the new meaning of this section, mod is a binary operation. For example,

$$53 \bmod 10 = 3.$$

In this context, mod means "divide and take the remainder."

What is the connection between these two meanings of the word *mod?* We have the following result.

Proposition 34.8 Let $a, b, n \in \mathbb{Z}$ with $n > 0$. Then

$$a \equiv b \pmod{n} \iff a \bmod n = b \bmod n.$$

The use of mod on the left is as a relation. The use of mod on the right is as a binary operation.

From the example, notice that $53 \bmod 10 = 3$ and $23 \bmod 10 = 3$.

This if-and-only-if result is not too hard to prove. It sets up as follows:

> Let $a, b, n \in \mathbb{Z}$ with $n > 0$.
> ($\Rightarrow$) Suppose $a \equiv b \pmod{n}$. ... Therefore $a \bmod n = b \bmod n$.
> ($\Leftarrow$) Suppose $a \bmod n = b \bmod n$. ... Therefore $a \equiv b \pmod{n}$. ■

We leave the definition unraveling and the rest of the proof to you (Exercise 34.5).

Recap

We formally developed the process of integer division resulting in quotients and remainders and introduced the binary operations div and mod.

34 Exercises

34.1. For the pairs of integers a, b given below, find the integers q and r such that $a = qb + r$ and $0 \le r < b$.

a. $a = 100, b = 3$.
b. $a = -100, b = 3$.
c. $a = 99, b = 3$.
d. $a = -99, b = 3$.
e. $a = 0, b = 3$.

34.2. For each of the pairs of integers a, b in the previous problem, compute a div b and $a \bmod b$.

34.3. Explain why Theorem 34.1 does not make sense with $b = 0$ or with $b < 0$.

The case $b = 0$ is hopeless. Develop (and prove) an alternative to Theorem 34.1 that allows $b < 0$.

34.4. What is wrong with the following statements? Repair these statements and prove your revised versions.

a. For all integers a, b, we have $b|a$ iff a div $b = \frac{a}{b}$.
b. For all integers a, b, we have $b|a$ iff $a \bmod b = 0$.

34.5. Prove Proposition 34.8.

34.6. Prove that the sum of any three consecutive integers is divisible by 3.

34.7. Many computer programming languages have the mod operation as a built-in feature. For example, the `%` sign in `C` is the mod operation. In `C` the result of `x = 53%10;` is to assign the value 3 to the variable `x`.

Investigate how various languages deal with the mod operation in cases where the second number is zero or negative.

34.8. Computer programming languages allow you to divide two `integer` type numbers and always return an `integer` answer. For example, in `C` the result of `x = 11/5;` is to assign the value 2 to the variable `x`. (Here, `x` is of type `int`.)

Investigate how various languages deal with integer division. In particular, is their implementation of integer division the same as the div operation?

34.9. *Dividing polynomials*. The *degree* of a polynomial is the exponent on the highest power of x. For example, $x^{10} - 5x^2 + 6$ has degree 10, and the degree of $3x - \frac{1}{2}$ is 1. When the polynomial is just a number (there are no x terms), we say the degree is 0. The polynomial 0 is exceptional; we say its degree is -1. If p is a polynomial, we write $\deg p$ to stand for its degree.

You may assume that the coefficients of the polynomials we consider in this problem are rational numbers.

a. Suppose p and q are polynomials. Write a careful definition of what it means for p to *divide* q (i.e., $p|q$).

Please verify that

$$(2x - 6)|(x^3 - 3x^2 + 3x - 9)$$

is true in your definition.

b. Give an example of two polynomials p and q with $p \neq q$ but $p|q$ and $q|p$.

c. What is the relationship between polynomials that divide each other?

d. Prove the following analogue of Theorem 34.1:

Let a and b be polynomials, with b nonzero. Then there exist polynomials q and r so that $a = qb + r$ with $\deg r < \deg b$.

For example, if $a = x^5 - 3x^2 + 2x + 1$ and $b = x^2 + 1$, then we can take $q = x^3 - x - 3$ and $r = 3x + 4$.

e. In this generalized version of Theorem 34.1, are the polynomials q and r uniquely determined by a and b?

35 Greatest Common Divisor

This section deals with the concept of greatest common divisor. The term is virtually self-defining.

Definition 35.1 **(Common divisor)** Let $a, b \in \mathbb{Z}$. We call an integer d a *common divisor* of a and b provided $d|a$ and $d|b$.

For example, the common divisors of 30 and 24 are $-6, -3, -2, -1, 1, 2, 3$, and 6.

Definition 35.2 **(Greatest common divisor)** Let $a, b \in \mathbb{Z}$. We call an integer d the *greatest common divisor* of a and b provided

1. d is a common divisor of a and b and
2. if e is a common divisor of a and b, then $e \le d$.

The greatest common divisor of a and b is denoted $\gcd(a, b)$.

For example, the greatest common divisor of 30 and 24 is 6, and we write $\gcd(30, 24) = 6$. Also $\gcd(-30, -24) = 6$.

Nearly every pair of integers has a greatest common divisor (see Exercise 35.3), and if a and b have a gcd, it is unique (Exercise 35.5). This justifies our use of the definite article when we call $\gcd(a, b)$ *the* greatest common divisor of a and b.

In this section, we explore the various properties of greatest common divisors.

Calculating the gcd

An *algorithm* is a precisely defined computational procedure.

In the foregoing example, we calculated the greatest common divisor of 30 and 24 by explicitly listing all their common factors and choosing the largest. This suggests an *algorithm* for computing gcd. The algorithm is as follows:

- Suppose a and b are positive integers.
- For every positive integer k from 1 to the smaller of a and b, see whether $k|a$ and $k|b$. If so, save that number k on a list.
- Choose the largest number on the list. That number is $\gcd(a, b)$.

This procedure works: Given any two positive integers a and b, it finds their gcd. However, it is a dreadful algorithm because even for moderately large numbers (e.g., $a = 34902$ and $b = 34299883$), the algorithm needs to do many, many divisions. So although correct, this algorithm is terribly slow.

There is a clever way to calculate the greatest common divisor of two positive integers; this procedure was invented by Euclid. It is not only very fast, but it is not difficult to implement as a computer program.

The central idea in Euclid's Algorithm is the following result.

Proposition 35.3 Let a and b be positive integers and let $c = a \bmod b$. Then

$$\gcd(a, b) = \gcd(b, c).$$

In other words, for positive integers a and b, we have

$$\gcd(a, b) = \gcd(b, a \bmod b).$$

Proof. We are given that $c = a \bmod b$. This means that $a = qb + c$ where $0 \le c < b$.

Let $d = \gcd(a, b)$ and let $e = \gcd(b, c)$. Our goal is to prove that $d = e$. To do this, we prove that $d \le e$ and $d \ge e$.

First, we show $d \le e$. Since $d = \gcd(a, b)$, we know that $d|a$ and $d|b$. We can write $c = a - qb$. Since a and b are multiples of d, so is c. Thus d is a common divisor of b and c. However, e is the greatest common divisor of b and c, so $d \le e$.

Next, we show $d \ge e$. Since $e = \gcd(b, c)$, we know that $e|b$ and $e|c$. Now $a = qb + c$, and hence $e|a$ as well. Since $e|a$ and $e|b$, we see that e is a common divisor of a and b. However, d is the greatest common divisor of a and b, so $d \ge e$.

We have shown $d \le e$ and $d \ge e$, and hence $d = e$; that is, $\gcd(a, b) = \gcd(b, c)$. ■

To illustrate how Proposition 35.3 enables us to calculate greatest common divisors efficiently, we compute $\gcd(689, 234)$. The simple, inefficient divide-and-check algorithm we considered first would have us try all possible common divisors from 1 to 234 and select the largest. This implies we would perform $234 \times 2 = 468$ division problems!

689 ↓ 234 ↓ 221 ↓ 13 ↓ 0

Instead, we use Proposition 35.3. To find $\gcd(689, 234)$, let $a = 689$ and $b = 234$. We find $c = 689 \bmod 234$. This requires us to do a division. The result is $c = 221$. To find $\gcd(689, 234)$, it is enough to find $\gcd(234, 221)$ because these two values are the same. Let's record this step here:

$$689 \bmod 234 = 221 \qquad \Rightarrow \qquad \gcd(689, 234) = \gcd(234, 221).$$

Now all we have to do is calculate $\gcd(234, 221)$. We use the same idea. We apply Proposition 35.3 as follows. To find $\gcd(234, 221)$, we calculate $234 \bmod 221 = 13$. Thus $\gcd(234, 221) = \gcd(221, 13)$. Let's record this step (division #2).

$$234 \bmod 221 = 13 \qquad \Rightarrow \qquad \gcd(234, 221) = \gcd(221, 13).$$

Now the problem is reduced to $\gcd(221, 13)$. Notice that the numbers are significantly smaller than the original 689 and 234. We again use Proposition 35.3 and calculate $221 \bmod 13 = 0$. What does that mean? It means that when we divide 221 by 13, there is no remainder. In other words, $13|221$. So clearly the greatest common divisor of 221 and 13 is 13. Let's record this step (division #3).

$$221 \bmod 13 = 0 \qquad \Rightarrow \qquad \gcd(221, 13) = 13.$$

We are finished! We have done three divisions (not 468 ☺), and we found

$$\gcd(689, 234) = \gcd(234, 221) = \gcd(221, 13) = 13.$$

The steps we just performed are precisely the Euclidean algorithm. Here is a formal description:

Euclid's Algorithm for Greatest Common Divisor

Input: Positive integers a and b.
Output: $\gcd(a, b)$.

- Let $c = a \bmod b$.
- If $c = 0$, then we return the answer b and stop.
- Otherwise ($c \neq 0$), we calculate $\gcd(b, c)$ and return this as the answer.

This algorithm for gcd is defined in terms of itself. This is an example of a *recursively* defined algorithm (see Exercise 21.8, where recursion is explored). Let's see how the algorithm works for the integers $a = 63$ and $b = 75$.

- The first step is to calculate $c = a \bmod b$, and we get $c = 63 \bmod 75 = 63$.
- Next we check whether $c = 0$. It's not, so we go on to compute $\gcd(b, c) = \gcd(75, 63)$.

  Very little progress has been made so far! All the algorithm has done is reverse the numbers. The next pass through, however, is more interesting.
- Now we restart the process with $a' = 75$ and $b' = 63$. We calculate $c' = 75 \bmod 63 = 12$. Since $12 \neq 0$, we are told to calculate $\gcd(b', c') = \gcd(63, 12)$.
- We restart again with $a'' = 63$ and $b'' = 12$. We calculate $c'' = 63 \bmod 12 = 3$. Since this is not zero, we need to go on and to calculate $\gcd(b'', c'') = \gcd(12, 3)$.
- We restart yet again with $a''' = 12$ and $b''' = 3$. Now we are told to calculate $c''' = 12 \bmod 3 = 0$. Aha! Now $c''' = 0$, so we return the answer $b''' = 3$ and we are finished.

  The final answer is that $\gcd(63, 75) = 3$.

Here is an overview of the calculation in chart form:

a	b	c
63	75	63
75	63	12
63	12	3
12	3	0

With only four divisions, the answer is produced.

Here is another way to visualize this computation. We create a list whose first two entries are a and b. Now we extend the list by computing mod of the last two entries of the list. When we reach 0, we stop. The next-to-last entry is the gcd of a and b. In this example, the list would be

$$(63, 75, 63, 12, 3, 0).$$

Correctness

Just because someone writes down a procedure to calculate gcd does not make it correct. The point of mathematics is to prove its assertions; the correctness of an algorithm is no exception.

Proposition 35.4 **(Correctness of Euclid's Algorithm for gcd)** Euclid's Algorithm correctly computes $\gcd(a, b)$ for any positive integers a and b.

Proof. Suppose, for the sake of contradiction, that Euclid's Algorithm did not correctly compute gcd. Then there is some pair of positive integers a and b for which it fails. Choose a and b such that $a+b$ is as small as possible. (We are using the smallest-counterexample method.)

It might be the case that $a < b$. If this is so, then the first pass through Euclid's Algorithm will simply interchange the values a and b [as we saw when we calculated $\gcd(63, 75)$] because if $a < b$ then $c = a \bmod b = a$, and Euclid's Algorithm directs us to calculate $\gcd(b, c) = \gcd(b, a)$.

Thus we may assume that $a \geq b$.

The first step of the algorithm is to calculate $c = \gcd(a, b)$. Two outcomes are possible: either $c = 0$ or $c \neq 0$.

In the case $c = 0$, $a \bmod b = 0$, which implies $b|a$. Since b is the largest divisor of b (since $b > 0$ by hypothesis) and since $b|a$, we have b is the greatest common divisor of a and b. In other words, the algorithm gives the correct result, contradicting our supposition that it fails for a and b.

So it must be the case that $c \neq 0$. To get c, we calculated the remainder when dividing a by b. By Theorem 34.1, we have $a = qb + c$ where $0 < c < b$. We also know that $b \leq a$. We add the inequalities:

$$\begin{array}{rr} & c < b \\ + & b \leq a \\ \Rightarrow & b + c < a + b \end{array}$$

Thus b, c are positive integers with $b + c < a + b$.

This means that b and c are not a counterexample to the correctness of Euclid's Algorithm because $b + c < a + b$, and among all counterexamples, a and b was a counterexample with the smallest sum. Thus the algorithm correctly computes $\gcd(b, c)$ and returns its value as the answer. However, by Proposition 35.3, this is the right answer! This contradicts the supposition that Euclid's Algorithm fails on a, b.$\Rightarrow\Leftarrow$ Hence Euclid's Algorithm always returns the greatest common divisor of the positive integers it is given. ■

How Fast?

How many times do we have to divide to calculate the greatest common divisor of two positive integers? We claim that after two rounds of Euclid's Algorithm, the integers with which we are working have decreased by at least 50%. This is the main tool.

Proposition 35.5 Let $a, b \in \mathbb{Z}$ with $a \geq b > 0$. Let $c = a \bmod b$. Then $c < \frac{a}{2}$.

Proof. We consider two cases: (1) $a < 2b$ and (2) $a \geq 2b$.

- **Case (1):** $a < 2b$.

  We know that $2b > a > 0$, so $a > 0$ and $a - b \geq 0$, but $a - 2b < 0$. Hence the quotient when a is divided by b is 1. So the remainder in a divided by b is $c = a - b$.

  Now we can rewrite $a < 2b$ as $b > \frac{a}{2}$, and so

$$c = a - b < a - \frac{a}{2} = \frac{a}{2}$$

  which is what we wanted.

- **Case (2):** $a \geq 2b$, which can be rewritten $b \leq \frac{a}{2}$.

  The remainder, upon division of a by b, is less than b. So $c < b$, and we have $b \leq \frac{a}{2}$, so $c < \frac{a}{2}$.

In both cases, we found $c < \frac{a}{2}$. ■

We may assume that we start Euclid's Algorithm with $a \geq b$; if not, the algorithm reverses a and b on its first pass, and from there on, the numbers come in decreasing order. That is, if the numbers produced by Euclid's Algorithm are listed as

$$(a, b, c, d, e, f, \ldots, 0)$$

then, assuming $a \geq b$, we have

$$a \geq b \geq c \geq d \geq e \geq f \geq \cdots \geq 0.$$

By Proposition 35.5, the numbers c and d are less than half as large as a and b, respectively. Likewise, two steps later, the numbers e and f are less than half as large as c and d, respectively, and less than one-fourth of a and b, respectively. Thus

> *Every two steps of Euclid's Algorithm decreases the integers with which we are working to less than half their current values.*

If we begin with (a, b), then two steps later, the numbers are less than $(\frac{1}{2}a, \frac{1}{2}b)$, and four steps later, less than $(\frac{1}{4}a, \frac{1}{4}b)$, and six steps later, less than $(\frac{1}{8}a, \frac{1}{8}b)$. How large are the numbers after $2t$ passes of Euclid's Algorithm? Since every two steps decrease the numbers by more than a factor of 2, we know that after $2t$ steps the numbers drop by more than a factor of 2^t; that is, the two numbers are less than $(2^{-t}a, 2^{-t}b)$.

Euclid's Algorithm stops when the second number reaches zero. Since the numbers in Euclid's Algorithm are integers, this is the same as when the second number is less than 1. This means that as soon as we have

$$2^{-t}b \leq 1,$$

the second number must have reached zero. Taking base-2 logs of both sides, we have

$$\begin{aligned} \log_2[2^{-t}b] &\leq \log_2 1 \\ -t + \log_2 b &\leq 0 \\ \log_2 b &\leq t. \end{aligned}$$

In other words, once $t \geq \log_2 b$, the algorithm must be finished. So after $2\log_2 b$ passes, the algorithm has completed its work.

How many divisions might this be if, say, a and b were enormous numbers (e.g., 1000 digits each). If $b \approx 10^{1000}$, then the number of steps is bounded by

$$2\log_2(10^{1000}) = 2000\log_2 10 < 2000 \times 3.4 = 6800.$$

(*Note*: $\log_2 10 \approx 3.3219 < 3.4$.) So in under 7000 steps, we have our answer. Compare this to doing 10^{1000} divisions (see Exercise 35.8)!

I hope you do not think I am trying your patience by considering such a ridiculous example. Why on earth would anyone want to compute the gcd of two 1000-digit numbers!? Well, the fact is that this is a practical, important problem with both industrial and military applications. More on this later!

An Important Theorem

The following theorem is central to the study of the greatest common divisor (and beyond).

Theorem 35.6 Let a and b be integers, not both zero. The smallest positive integer of the form $ax + by$, where x and y are integers, is $\gcd(a, b)$.

Let a and b be integers. An *integer linear combination* of a and b is any number of the form $ax + by$ where x and y are also integers. Theorem 35.6 tells us that the smallest positive-integer linear combination of a and b is $\gcd(a, b)$.

For example, suppose $a = 30$ and $b = 24$. We can make a chart of the values $ax + by$ for integers x and y between -4 and 4. We get the following table:

		y								
		−4	−3	−2	−1	0	1	2	3	4
	−4	−216	−192	−168	−144	−120	−96	−72	−48	−24
	−3	−186	−162	−138	−114	−90	−66	−42	−18	6
	−2	−156	−132	−108	−84	−60	−36	−12	12	36
	−1	−126	−102	−78	−54	−30	−6	18	42	66
x	0	−96	−72	−48	−24	0	24	48	72	96
	1	−66	−42	−18	6	30	54	78	102	126
	2	−36	−12	12	36	60	84	108	132	156
	3	−6	18	42	66	90	114	138	162	186
	4	24	48	72	96	120	144	168	192	216

What is the smallest positive value on this chart? We see the number 6 at $x = -3$, $y = 4$ (because $30 \times -3 + 24 \times 4 = -90 + 96 = 6$) and again at $x = 1$, $y = -1$ (because $30 \times 1 + 24 \times -1 = 30 - 24 = 6$).

Now we have shown only a relatively small portion of all the possible values of $ax + by$. Is it possible, if we were to extend this chart, that we might find a smaller positive value for $30x + 24y$? The answer is no. Notice that both 30 and 24 are divisible by 6. Therefore any integer of the form $30x + 24y$ is also divisible by 6 (see Exercise 4.8). So even if we extended this chart out forever, 6 is the smallest positive integer we would find.

Let a and b be any integers (not both zero). It is impossible to find integers x and y with

$$0 < ax + by < \gcd(a, b)$$

because $ax + by$ is divisible by $\gcd(a, b)$. The point of Theorem 35.6 is that we can find integers x and y such that $ax + by = \gcd(a, b)$. Here is the proof:

The set D is the set of all positive integers of the form $ax + by$ (i.e., the set of all positive numbers on the chart we considered above).

Proof (of Theorem 35.6)

Let a and b be integers (not both zero) and let

$$D = \{ax + by : x, y \in \mathbb{Z},\ ax + by > 0\}.$$

We want to examine the smallest member of D (i.e., we are about to invoke the Well-Ordering Principle). First, we must be sure that D is nonempty.

To see that $D \neq \emptyset$, we just have to prove that there is at least one integer in D. Can we select integers x and y to make $ax + by$ positive? If we take $x = a$ and $y = b$, we find that $ax + by = a^2 + b^2$, and this is positive (unless $a = b = 0$, which is forbidden by hypothesis). Therefore $D \neq \emptyset$.

Applying the Well-Ordering Principle to D, a nonempty set of natural numbers, we know that D contains a least element; call that least element d.

Our goal is to show that $d = \gcd(a, b)$. How do we prove that d is the greatest common divisor of a and b? We consult Definition 35.2. We need to show three things: (1) $d|a$, (2) $d|b$, and (3) if $e|a$ and $e|b$, then $e \leq d$. We do each of these in turn.

- **Claim (1):** $d|a$.

  Suppose, for the sake of contradiction, that a is not divisible by d. Then when we divide a by d, we get a nonzero remainder:

  $$a = qd + r \qquad \text{with} \qquad 0 < r < d.$$

  Now $d = ax + by$, so we can solve for r in terms of a and b as follows:

  $$r = a - qd = a - q(ax + by) = a(1 - qx) + b(-qy) = aX + bY$$

  where $X = 1 - qx$ and $Y = -qy$. Notice that $0 < r < d$ and $r = aX + bY$. This means that $r \in D$ and $r < d$, contradicting the fact that d is the least element of D.$\Rightarrow\Leftarrow$ Therefore $d|a$.
- **Claim (2):** $d|b$.

  This proof is analogous to $d|a$.
- **Claim (3):** If $e|a$ and $e|b$, then $e \leq d$.

  Suppose $e|a$ and $e|b$. Then $e|(ax + by)$ (Exercise 4.8). Therefore $e|d$, so $e \leq d$ (because d is positive).

Therefore d is the greatest common divisor of a and b. ■

Example 35.7 Earlier we found that $\gcd(689, 234) = 13$. Note that

$$689 \times -1 + 234 \times 3 = -689 + 702 = 13 = \gcd(689, 234).$$

Here is another example. Note that $\gcd(431, 29) = 1$. And note that

$$431 \times 7 + 29 \times -104 = 3017 - 3016 = 1.$$

Given a and b, how do we find integers x and y such that $ax+by = \gcd(a, b)$? Perhaps it is not too hard to try a few values to guess that $689 \times -1 + 234 \times 3 = 13 = \gcd(689, 234)$, but it seems to be hard to find the right x and y to get $431x + 29y = 1$ (try it!).

The proof of Theorem 35.6 is not of any help. The step that proves that the numbers x and y exist is *nonconstructive*—the Well-Ordering Principle shows that such integers exist but gives us no clue how to find them. The key to finding x and y in $ax + by = \gcd(a, b)$ is to extend Euclid's Algorithm.

Earlier we used Euclid's Algorithm to calculate $\gcd(a, b)$. Each time we did a division, the only information we retained was the remainder (the central computational step is $c = a \bmod b$). By keeping track of the quotients, too, we will be able to find the integers x and y. Here is how this works.

We illustrate this method by finding x and y such that $431x + 29y = \gcd(431, 29) = 1$.

Here are the steps in calculating $\gcd(431, 29)$ by Euclid's Algorithm:

$$\begin{aligned} 431 &= 14 \times 29 + 25 \\ 29 &= 1 \times 25 + 4 \\ 25 &= 6 \times 4 + 1 \\ 4 &= 4 \times 1 + 0. \end{aligned}$$

In all of these equations (except the last), we solve for the remainder (that is, we put the remainders on the left).

$$\begin{aligned} 25 &= 431 - 14 \times 29 \\ 4 &= 29 - 1 \times 25 \\ 1 &= 25 - 6 \times 4. \end{aligned}$$

Now we work from the bottom up. Notice that the last equation has 1 in the form $25x + 4y$. We substitute for 4 using the previous equation:

$$\begin{aligned} 1 &= 25 - 6 \times 4 \\ &= 25 - 6 \times (29 - 1 \times 25) \\ &= -6 \times 29 + 7 \times 25. \end{aligned}$$

Now we use $25 = 431 - 14 \times 29$ to replace the 25 in $1 = -6 \times 29 + 7 \times 25$:

$$\begin{aligned} 1 &= -6 \times 29 + 7 \times 25 \\ &= -6 \times 29 + 7 \times (431 - 14 \times 29) \\ &= 7 \times 431 + [-6 + 7 \times (-14)]29 \\ &= 7 \times 431 + (-104) \times 29. \end{aligned}$$

This is how we found $x = 7$ and $y = -104$ to get $431x+29y = \gcd(431, 29) = 1$.

Pairs of numbers whose greatest common divisor is 1 have a special name.

Definition 35.8 Let a and b be integers. We call a and b *relatively prime* provided $\gcd(a, b) = 1$.

In other words, integers are relatively prime provided the only divisors they have in common are 1 and -1.

Corollary 35.9 Let a and b be integers. There exist integers x and y such that $ax + by = 1$ if and only if a and b are relatively prime.

Theorem 35.6 and its consequence, Corollary 35.9, are extremely useful tools for proving results about gcd and relatively prime numbers. Here is an example. Try proving this without using Theorem 35.6, and then you will appreciate its usefulness.

Proposition 35.10 Let a, b be integers, not both zero. Let $d = \gcd(a, b)$. If e is a common divisor of a and b, then $e|d$.

We know, since $d = \gcd(a, b)$, that $e \le d$, but that does not immediately imply that $e|d$. Here is the proof.

Proof. Let a, b be integers, not both zero, and let $d = \gcd(a, b)$. Suppose $e|a$ and $e|b$. Now, by Theorem 35.6, there exist integers x and y such that $d = ax+by$. Since $e|a$ and $e|b$, we have $e|(ax + by)$ (see Exercise 4.8), and so $e|d$. ■

Recap

In this section we examined the greatest common divisor of a pair of integers. We discussed how to compute the gcd of two integers using Euclid's Algorithm, and we analyzed the efficiency of the Euclidean Algorithm. We showed that for integers a, b (not both zero), the smallest positive value of $ax+by$ (with $x, y \in \mathbb{Z}$) is $\gcd(a, b)$. When two integers' gcd is 1, we call those integers relatively prime.

35 Exercises

35.1. Please calculate:
- **a.** $\gcd(20, 25)$.
- **b.** $\gcd(0, 10)$.
- **c.** $\gcd(123, -123)$.
- **d.** $\gcd(-89, -98)$.
- **e.** $\gcd(54321, 50)$.
- **f.** $\gcd(1739, 29341)$.

35.2. For each pair of integers a, b in the previous problem, find integers x and y such that $ax + by = \gcd(a, b)$.

35.3. Find integers a and b that do not have a greatest common divisor. Prove that the pair you found are the only pair of integers that do not have a gcd.

35.4. Let a and b be positive integers. Find the sum of all the common divisors of a and b.

35.5. Prove that if a and b have a greatest common divisor, it is unique (i.e., they cannot have two greatest common divisors).

35.6. In Proposition 35.3, we did not require that $c \neq 0$. Is Proposition 35.3 (and its proof) correct even in the case $c = 0$?

35.7. Suppose $a \geq b$ and running Euclid's Algorithm yields the numbers (in list form)

$$(a, b, c, d, e, f, \ldots, 0).$$

Prove that

$$a \geq b \geq c \geq d \geq e \geq f \geq \cdots \geq 0.$$

35.8. Suppose we want to compute the greatest common divisor of two 1000-digit numbers on a very fast computer—a computer that can do 1 billion divisions per second. Approximately how long would it take to compute the gcd by the trial division method? (Choose an appropriate unit of time, such as minutes, hours, days, years, centuries, or millennia.)

35.9. We can extend the definition of the gcd of two numbers to the gcd of three or more numbers.

a. Give a careful definition of $\gcd(a, b, c)$ where a, b, c are integers.

b. Prove or disprove: For integers a, b, c, we have $\gcd(a, b, c) = 1$ if and only if a, b, c are pairwise relatively prime.

c. Prove or disprove: For integers a, b, c, we have

$$\gcd(a, b, c) = \gcd(a, \gcd(b, c)).$$

d. Prove that $\gcd(a, b, c) = d$ is the smallest positive integer of the form $ax + by + cz$ where $x, y, z \in \mathbb{Z}$.

e. Find integers x, y, z such that $6x + 10y + 15z = 1$.

f. Is there a solution to part (e) in which one of x, y, or z is zero? Prove your answer.

35.10. Prove that consecutive integers must be relatively prime.

35.11. Let a be an integer. Prove that $2a + 1$ and $4a^2 + 1$ are relatively prime.

35.12. Suppose n and m are relatively prime. Prove that n and $m + jn$ are relatively prime for any integer j.

Conclude that if n and m are relatively prime, and $m' = m \bmod n$, then n and m' are relatively prime.

35.13. Suppose that a and b are relatively prime and that $a|c$ and $b|c$. Prove that $(ab)|c$.

35.14. Suppose $a, b, n \in \mathbb{Z}$ with $n > 0$. Suppose that $ab \equiv 1 \pmod{n}$. Prove that both a and b are relatively prime to n.

35.15. Suppose $a, n \in \mathbb{Z}$ with $n > 0$. Suppose that a and n are relatively prime. Prove that there is an integer b such that $ab \equiv 1 \pmod{n}$.

35.16. Suppose $a, b \in \mathbb{Z}$ are relatively prime. Corollary 35.9 implies that there exist integers x, y such that $ax + by = 1$. Prove that these integers x and y must be relatively prime.

35.17. Let x be a rational number. This means there are integers a and $b \neq 0$ such that $x = \frac{a}{b}$. Prove that we can choose a and b to be relatively prime.

35.18. A class of n children sit in a circle. The teacher walks around the outside of the circle and pats every kth child on the head. Find and prove a necessary and sufficient condition on n and k for every child to receive a pat on the head.

35.7. Suppose $a \geq b$ and running Euclid's Algorithm yields the numbers (in list form)

$$(a, b, c, d, e, f, \ldots, 0).$$

Prove that

$$a \geq b \geq c \geq d \geq e \geq f \geq \cdots \geq 0.$$

35.8. Suppose we want to compute the greatest common divisor of two 1000-digit numbers on a very fast computer—a computer that can do 1 billion divisions per second. Approximately how long would it take to compute the gcd by the trial division method? (Choose an appropriate unit of time, such as minutes, hours, days, years, centuries, or millennia.)

35.9. We can extend the definition of the gcd of two numbers to the gcd of three or more numbers.

a. Give a careful definition of $\gcd(a, b, c)$ where a, b, c are integers.

b. Prove or disprove: For integers a, b, c, we have $\gcd(a, b, c) = 1$ if and only if a, b, c are pairwise relatively prime.

c. Prove or disprove: For integers a, b, c, we have

$$\gcd(a, b, c) = \gcd(a, \gcd(b, c)).$$

d. Prove that $\gcd(a, b, c) = d$ is the smallest positive integer of the form $ax + by + cz$ where $x, y, z \in \mathbb{Z}$.

e. Find integers x, y, z such that $6x + 10y + 15z = 1$.

f. Is there a solution to part (e) in which one of x, y, or z is zero? Prove your answer.

35.10. Prove that consecutive integers must be relatively prime.

35.11. Let a be an integer. Prove that $2a + 1$ and $4a^2 + 1$ are relatively prime.

35.12. Suppose n and m are relatively prime. Prove that n and $m + jn$ are relatively prime for any integer j.

Conclude that if n and m are relatively prime, and $m' = m \bmod n$, then n and m' are relatively prime.

35.13. Suppose that a and b are relatively prime and that $a|c$ and $b|c$. Prove that $(ab)|c$.

35.14. Suppose $a, b, n \in \mathbb{Z}$ with $n > 0$. Suppose that $ab \equiv 1 \pmod{n}$. Prove that both a and b are relatively prime to n.

35.15. Suppose $a, n \in \mathbb{Z}$ with $n > 0$. Suppose that a and n are relatively prime. Prove that there is an integer b such that $ab \equiv 1 \pmod{n}$.

35.16. Suppose $a, b \in \mathbb{Z}$ are relatively prime. Corollary 35.9 implies that there exist integers x, y such that $ax + by = 1$. Prove that these integers x and y must be relatively prime.

35.17. Let x be a rational number. This means there are integers a and $b \neq 0$ such that $x = \frac{a}{b}$. Prove that we can choose a and b to be relatively prime.

35.18. A class of n children sit in a circle. The teacher walks around the outside of the circle and pats every kth child on the head. Find and prove a necessary and sufficient condition on n and k for every child to receive a pat on the head.

35.19. You have two measuring cups. One holds 8 ounces and the other holds 13 ounces. These cups have no marks to show individual ounces. All you can measure is either a full 13 or a full 8 ounces. If you want to measure, say, 5 ounces, you can fill the 13-ounce measuring cup, use it to fill the 8-ounce cup, and you will have 5 ounces left in the larger cup.

a. Show how to use the 13-ounce and 8-ounce cups to measure exactly 1 ounce. You may assume you have a large bowl for holding liquid, but this large bowl has no marks for measuring. At the end, the bowl should contain exactly 1 ounce.

b. Generalize this problem. Suppose the measuring cups hold a and b ounces where a and b are positive integers. Give and prove necessary and sufficient conditions on a and b such that it is possible to measure out exactly 1 ounce using these cups.

35.20. In Exercise 34.9, we considered polynomial division. In this problem, you are asked to develop the concept of polynomial gcd.

Polynomials in this problem may be assumed to have rational coefficients.

a. Let p and q be nonzero polynomials. Write a careful definition for *common divisor* and *greatest common divisor* of p and q.

In this context, *greatest* refers to the degree of the polynomial.

b. Show, by giving an example, that there need not be a unique gcd of two nonzero polynomials.

c. Let d be a greatest common divisor of nonzero polynomials p and q. Prove that there exist polynomials a and b such that $ap + bq = d$.

d. Give a careful definition of *relatively prime* for nonzero polynomials.

e. Prove that two nonzero polynomials p and q are relatively prime if and only if there exist polynomials a and b such that $ap + bq = 1$.

f. Let $p = x^4 - 3x^2 - 1$ and $q = x^2 + 1$. Show that p and q are relatively prime by finding polynomials a and b such that $ap + bq = 1$.

36 Modular Arithmetic

A New Context for Basic Operations

Arithmetic is the study of the basic operations: addition, subtraction, multiplication, and division. The usual contexts for studying these operations are number systems such as the integers, $\mathbb{Z}$, or the rationals, $\mathbb{Q}$.

Division is, perhaps, the most interesting example. In the context of the rational numbers, we can calculate $x \div y$ for any $x, y \in \mathbb{Q}$ except when $y = 0$. In the context of the integers, however, $x \div y$ is defined only when $y \neq 0$ and $y|x$.

The point is that in the two different contexts, $\mathbb{Q}$ and $\mathbb{Z}$, the operation $\div$ takes on slightly different meanings. In this section, we introduce a new context for the symbols $+$, $-$, $\times$, and $\div$ where their meanings are quite different from the traditional context. The difference is so significant, that we use alternative symbols for these operations. We use the symbols $\oplus$, $\ominus$, $\otimes$, and $\oslash$.

Instead of consisting of integers or rationals, the new set in which we perform arithmetic is denoted $\mathbb{Z}_n$ where n is a positive integer. The set $\mathbb{Z}_n$ is defined to be

$$\mathbb{Z}_n = \{0, 1, 2, \ldots, n-1\};$$

that is, $\mathbb{Z}_n$ contains all natural numbers from 0 to $n-1$ inclusive.

We call this number system the *integers mod n*.

To distinguish $\oplus$, $\ominus$, $\otimes$, and $\oslash$ from their uncircled cousins, we refer to these operations as *addition mod n*, *subtraction mod n*, *multiplication mod n*, and *division mod n*.

Mathspeak!

This is a *third* use of the word *mod*! We have *mod* as a relation, as in $13 \equiv 8 \pmod 5$, and we have *mod* as an operation, as in $13 \bmod 5 = 3$. Now we have the integers *mod* n. The three uses are different, but closely related.

Modular Addition and Multiplication

How are the modular operations defined? We begin with $\oplus$ and $\otimes$.

Definition 36.1 **(Modular addition, multiplication)** Let n be a positive integer. Let $a, b \in \mathbb{Z}_n$. We define

$$a \oplus b = (a+b) \bmod n \quad \text{and}$$
$$a \otimes b = (ab) \bmod n.$$

The operations on the left are operations defined for $\mathbb{Z}_n$. The operations on the right are ordinary integer operations.

Example 36.2 Let $n = 10$. We have the following:

$$5 \oplus 5 = 0 \qquad 9 \oplus 8 = 7$$
$$5 \otimes 5 = 5 \qquad 9 \otimes 8 = 2.$$

Notice that the symbols $\oplus$ and $\otimes$ depend on the context. If we are working in $\mathbb{Z}_{10}$, then $5 \oplus 5 = 0$, but if we are working in $\mathbb{Z}_9$, then $5 \oplus 5 = 1$. It might be better to create a more baroque symbol, such as $\overset{n}{\oplus}$ to denote mod n addition, but in most situations, the modulus (n) does not change. We simply must remain vigilant and know the current context.

Notice that if $a, b \in \mathbb{Z}_n$, the results of the operations $a \oplus b$ and $a \otimes b$ are always defined and are elements of $\mathbb{Z}_n$

Given two elements in $\mathbb{Z}_n$, the results of $\oplus$ and $\otimes$ are also in $\mathbb{Z}_n$.

Proposition 36.3 Let $a, b \in \mathbb{Z}_n$. Then $a \oplus b \in \mathbb{Z}_n$ and $a \otimes b \in \mathbb{Z}_n$. (Closure.)

Proof. Exercise 36.7.

The operations $\oplus$ and $\otimes$ exhibit the usual algebraic properties:

Proposition 36.4 Let n be an integer with $n \geq 2$.

- For all $a, b \in \mathbb{Z}_n$, $a \oplus b = b \oplus a$ and $a \otimes b = b \otimes a$. (Commutative.)
- For all $a, b, c \in \mathbb{Z}_n$, $a \oplus (b \oplus c) = (a \oplus b) \oplus c$, and $a \otimes (b \otimes c) = (a \otimes b) \otimes c$. (Associative.)

- For all $a \in \mathbb{Z}_n$, $a \oplus 0 = a$, $a \otimes 1 = a$, and $a \otimes 0 = 0$. (Identity elements.)
- For all $a, b, c \in \mathbb{Z}_n$, $a \otimes (b \oplus c) = (a \otimes b) \oplus (a \otimes c)$. (Distributive.)

The proofs of these are quite similar to each other. We prove only one as an example. Since $a \oplus b = (a + b) \bmod n$, and $a \otimes b = (ab) \bmod n$, the basic step in all of these proofs is to write

$$a \oplus b = a + b + kn \qquad \text{or} \qquad a \otimes b = ab + kn$$

where k is an integer.

Proof. We show that $\oplus$ is associative. Let $a, b, c \in \mathbb{Z}_n$. We want to show that $a \oplus (b \oplus c) = (a \oplus b) \oplus c$.

Now

$$\begin{aligned} & a \oplus (b \oplus c) \in \mathbb{Z}_n, && \text{(by Proposition 36.3)} \\ \text{and} \quad & a \oplus (b \oplus c) = a \oplus (b + c + kn) \\ & \qquad = [a + (b + c + kn)] + jn \\ & \qquad = (a + b + c) + sn \end{aligned}$$

where $k, j, s \in \mathbb{Z}$. Since, obviously,

$$a + b + c + sn \equiv a + b + c \pmod{n},$$

we have $(a+b+c) \bmod n = (a+b+c+sn) \bmod n = (a+b+c+sn)$ because $a + b + c + sn \in \mathbb{Z}_n$. (We used Proposition 34.8.)

In short, $a \oplus (b \oplus c) = (a + b + c) \bmod n$.

By a similar argument, $(a \oplus b) \oplus c = (a + b + c) \bmod n$. Thus $a \oplus (b \oplus c) = (a + b + c) \bmod n = (a \oplus b) \oplus c$.

The rest of this proof is left to you (Exercise 36.8). ■

Modular Subtraction

What is subtraction? We can define ordinary subtraction in a number of different ways. Here is one way based on addition. Let $a, b \in \mathbb{Z}$. We *define* $a - b$ to be the solution to the equation $a = b + x$. We then would prove two things: (1) the equation $a = b + x$ has a solution, and (2) the equation $a = b + x$ has only one solution.

We use the same approach to define modular subtraction. We start by proving that an equation of the form $a = b \oplus x$ has a solution, and only one solution.

Proposition 36.5 Let n be a positive integer, and let $a, b \in \mathbb{Z}_n$. Then there is one and only one $x \in \mathbb{Z}_n$ such that $a = b \oplus x$.

Proof. To show that x exists, let $x = (a - b) \bmod n$. We need to check that $x \in \mathbb{Z}_n$ and that x satisfies the equation $a = b + x$.

By definition of (the binary operation) mod, x is the remainder when we divide $a-b$ by n, so $0 \le x < n$, i.e., $x \in \mathbb{Z}_n$. Note that $x = (a-b) + kn$ for some integer k.

We calculate

$$b \oplus x = (b+x) \bmod n = [b + (a - b + kn)] \bmod n = (a + kn) \bmod n = a$$

because $0 \le a < n$. Therefore x satisfies the equation $a = b \oplus x$.

Now we turn to showing uniqueness (see Proof Template 14). Suppose, for the sake of contradiction, there were two solutions; that is, there exist $x, y \in \mathbb{Z}_n$ (with $x \neq y$) for which $a = b \oplus x$ and $a = b \oplus y$. This means that

$$\begin{aligned} b \oplus x &= (b+x) \bmod n = b + x + kn = a, \text{ and} \\ b \oplus y &= (b+y) \bmod n = b + y + jn = a \end{aligned}$$

for some integers k, j. Combining these, we have

$$\begin{aligned} b + x + kn = b + y + jn \quad &\Rightarrow \quad x = y + (k-j)n \\ &\Rightarrow \quad x \equiv y \pmod{n} \\ &\Rightarrow \quad x \bmod n = y \bmod n \\ &\Rightarrow \quad x = y \end{aligned}$$

because $0 \le x, y < n$. We have shown $x = y$, but $x \neq y$.$\Rightarrow\Leftarrow$ ■

Now that we know that the equation $a = b \oplus x$ has a unique solution, we can use this to define $a \ominus b$.

Definition 36.6 **(Modular subtraction)** Let n be a positive integer and let $a, b \in \mathbb{Z}_n$. We define $a \ominus b$ to be the unique $x \in \mathbb{Z}_n$ such that $a = b \oplus x$.

Alternatively, we could have defined $a \ominus b$ to be $(a-b) \bmod n$. We prove that this would give the same result.

Proposition 36.7 Let n be a positive integer and let $a, b \in \mathbb{Z}_n$. Then $a \ominus b = (a-b) \bmod n$.

Proof. To prove that $a \ominus b = (a-b) \bmod n$, we consult the definition. We need to show (1) that $[(a-b) \bmod n] \in \mathbb{Z}_n$ and (2) that if $x = (a-b) \bmod n$, then $a = b \oplus x$.

Note that (1) is obvious because $(a-b) \bmod n$ is an integer in $\mathbb{Z}_n$.

To show (2), we first note that $x = a - b + kn$ for some integer k. Then

$$b \oplus x = (b + (a - b + kn)) \bmod n = (a + kn) \bmod n = a.$$ ■

We could have used Proposition 36.7 as the *definition* of $\ominus$ and then proved the assertion in Definition 36.6 as a theorem. See Exercise 36.9.

Modular Division

Modular arithmetic is $\frac{3}{4}$ easy. We now come to the difficult $\frac{1}{4}$. Modular division is significantly different from the other modular operations. For example, in ordinary integer arithmetic, we have cancellation laws. If a, b, c are integers with $a \neq 0$, then

$$ab = ac \quad \Longrightarrow \quad b = c.$$

However, in $\mathbb{Z}_{10}$,

$$5 \otimes 2 = 5 \otimes 4 \quad \text{but} \quad 2 \neq 4.$$

Despite the fact that $5 \neq 0$, we cannot cancel, or divide, both sides by 5.

Motivated by the definition of $\ominus$, we might like to define $a \oslash b$ to be the unique $x \in \mathbb{Z}_n$ so that $a = b \otimes x$. This is problematic. Consider $6 \oslash 2$ in $\mathbb{Z}_{10}$. This should be the unique $x \in \mathbb{Z}_{10}$ so that $2 \otimes x = 6$. Is $x = 3$? That would be nice. And we are encouraged by the fact that $2 \otimes 3 = 6$. However, observe that $2 \otimes 8 = 6$. Should we have $6 \oslash 2 = 8$? The problem is that there might not be a unique solution to $6 = 2 \otimes x$.

Example 36.8 Given $a, b \in \mathbb{Z}_{10}$ (with $b \neq 0$), must there be a solution to $a = b \otimes x$? If so, is it unique?

Consider the following three cases.

- Let $a = 6$ and $b = 2$. There are two solutions to $6 = 2 \otimes x$, namely $x = 3$ and $x = 8$.
- Let $a = 7$ and $b = 2$. There are no solutions to $7 = 2 \otimes x$.
- Let $a = 7$ and $b = 3$. There is one and only one solution to $7 = 3 \otimes x$, namely $x = 9$. In this case it makes sense to write $7 \oslash 3 = 9$.

Each of the assertions above can be verified simply by considering all possible values of x; since there are only ten possible values for x, this is not terribly time-consuming.

The situation looks hopelessly muddled. Let's try another approach. In $\mathbb{Q}$, we can define $a \div b$ to be $a \cdot b^{-1}$; that is, division by b is *defined* to be multiplication by b's reciprocal. This explains why division by zero is undefined; zero does not have a reciprocal. Let's be precise about what we mean by reciprocal. The *reciprocal* of a rational number x is a rational number y such that $xy = 1$.

We can use this as our basis for defining division in $\mathbb{Z}_n$. We begin by defining reciprocals.

Definition 36.9 **(Modular reciprocal)** Let n be a positive integer and let $a \in \mathbb{Z}_n$. A *reciprocal* of a is an element $b \in \mathbb{Z}_n$ such that $a \otimes b = 1$. An element of $\mathbb{Z}_n$ that has a reciprocal is called *invertible*.

Let's investigate reciprocals in $\mathbb{Z}_{10}$. Here is the multiplication table for $\mathbb{Z}_{10}$:

$\otimes$	0	1	2	3	4	5	6	7	8	9
0	0	0	0	0	0	0	0	0	0	0
1	0	1	2	3	4	5	6	7	8	9
2	0	2	4	6	8	0	2	4	6	8
3	0	3	6	9	2	5	8	1	4	7
4	0	4	8	2	6	0	4	8	2	6
5	0	5	0	5	0	5	0	5	0	5
6	0	6	2	8	4	0	6	2	8	4
7	0	7	4	1	8	5	2	9	6	3
8	0	8	6	4	2	0	8	6	4	2
9	0	9	8	7	6	5	4	3	2	1

Several comments are in order.

- Element 0 does not have a reciprocal; this is not surprising.
- Elements 2, 4, 5, 6, and 8 do not have reciprocals. This explains why our attempts to divide by 2 were strange.
- Elements 1, 3, 7, and 9 are invertible (have reciprocals). Furthermore, they have only one reciprocal each.
- Notice the elements of $\mathbb{Z}_{10}$ that have reciprocals are precisely those integers in $\mathbb{Z}_{10}$ that are relatively prime to 10.
- The reciprocal of 3 is 7, and the reciprocal of 7 is 3; both 1 and 9 are their own reciprocals.

These observations give us some ideas to develop into theorems.

We observed that not all elements have reciprocals. However, those that do have only one reciprocal. Notice that in Definition 36.9 we used the indefinite article. We wrote "*A* reciprocal of...." We did not write "*The* reciprocal of..." because we had not yet established uniqueness. Let's do that now.

Proposition 36.10 Let n be a positive integer and let $a \in \mathbb{Z}_n$. If a has a reciprocal in $\mathbb{Z}_n$, then it has only one reciprocal.

Proof. Suppose a had two reciprocals, $b, c \in \mathbb{Z}_n$ with $b \neq c$. Consider $b \otimes a \otimes c$. Using the associative property (see Proposition 36.4) for $\otimes$ yields

$$b = b \otimes 1 = b \otimes (a \otimes c) = (b \otimes a) \otimes c = 1 \otimes c = c,$$

contradicting $b \neq c$.$\Rightarrow\Leftarrow$ ■

The overworked superscript -1.

Thus it makes sense to speak of *the* reciprocal of a. We also call the reciprocal of a the *inverse* of a. The notation for the reciprocal of a is a^{-1}. We are overtaxing the superscript -1 and trying your patience as a reader here. The symbol a^{-1} has three different meanings that depend on context. Please be careful! The three meanings are as follows:

- In the context of integers or rational numbers, a^{-1} refers to the rational number $\frac{1}{a}$.

- In the context of relations or functions, R^{-1} stands for the relation formed by reversing all the ordered pairs in R (see Section 13).
- In the context of $\mathbb{Z}_n$, a^{-1} is the reciprocal of a. It is not (and you should *never* write) $\frac{1}{a}$. For example, in the context of $\mathbb{Z}_{10}$, we have $3^{-1} = 7$.

Note that 3 and 7 are reciprocals of each other in $\mathbb{Z}_{10}$. We have the following:

Proposition 36.11 Let n be a positive integer and let $a \in \mathbb{Z}_n$. Suppose a is invertible. If $b = a^{-1}$, then b is invertible and $a = b^{-1}$. In other words, $(a^{-1})^{-1} = a$.

The proof is left to you (Exercise 36.11).

We use reciprocals to define modular division.

Definition 36.12 **(Modular division)** Let n be a positive integer and let b be an invertible element of $\mathbb{Z}_n$. Let $a \in \mathbb{Z}_n$ be arbitrary. Then $a \oslash b$ is defined to be $a \otimes b^{-1}$.

Notice that $a \oslash b$ is defined only when b is invertible; this is analogous to the fact that, for rational numbers, $a \div b$ is defined only when b is invertible—that is, nonzero.

Example 36.13 In $\mathbb{Z}_{10}$, calculate $2 \oslash 7$. Note that $7^{-1} = 3$, so $2 \oslash 7 = 2 \otimes 3 = 6$.

We still have some work to do. We need to address the following issues:

- In $\mathbb{Z}_n$, which elements are invertible?
- In $\mathbb{Z}_n$, given that a is invertible, how do we calculate a^{-1}?

We solved these problems for $\mathbb{Z}_{10}$ by writing out the entire $\otimes$ table for $\mathbb{Z}_{10}$. We would not want to do that for $\mathbb{Z}_{1000}$!

We noticed that the only invertible elements in $\mathbb{Z}_{10}$ are 1, 3, 7, and 9—precisely those elements relatively prime to 10. Does this pattern continue? Let's examine $\mathbb{Z}_9$. Here is the $\otimes$ table for $\mathbb{Z}_9$:

$\otimes$	0	1	2	3	4	5	6	7	8
0	0	0	0	0	0	0	0	0	0
1	0	1	2	3	4	5	6	7	8
2	0	2	4	6	8	1	3	5	7
3	0	3	6	0	3	6	0	3	6
4	0	4	8	3	7	2	6	1	5
5	0	5	1	6	2	7	3	8	4
6	0	6	3	0	6	3	0	6	3
7	0	7	5	3	1	8	6	4	2
8	0	8	7	6	5	4	3	2	1

The invertible elements of $\mathbb{Z}_9$ are 1, 2, 4, 5, 7, and 8 (these are all relatively prime to 9), and the noninvertible elements are 0, 3, and 6 (none of these are relatively prime to 9).

This suggests the following.

Theorem 36.14 **(Invertible elements of $\mathbb{Z}_n$)** Let n be a positive integer and let $a \in \mathbb{Z}_n$. Then a is invertible if and only if a and n are relatively prime.

At first glance, this may seem like a difficult theorem to prove. And if you attempt to prove it by simply unraveling definitions, it is hard. However, we have a power tool for dealing with pairs of numbers that are relatively prime. Corollary 35.9 tells us that a and b are relatively prime if and only if there is an integer solution to $ax + by = 1$. When we are armed with this tool, the proof of Theorem 36.14 almost writes itself.

Here is an outline for the proof.

> Let n be a positive integer and let $a \in \mathbb{Z}_n$.
>
> ($\Rightarrow$) Suppose a is invertible. ... Therefore a and n are relatively prime.
>
> ($\Leftarrow$) Suppose a and n are relatively prime. ... Therefore a is an invertible element of $\mathbb{Z}_n$. ■

For the forward ($\Rightarrow$) direction, we unravel the definition of invertible and keep unraveling.

> Let n be a positive integer and let $a \in \mathbb{Z}_n$.
>
> ($\Rightarrow$) Suppose a is invertible. This means there is an element $b \in \mathbb{Z}_n$ such that $a \otimes b = 1$. In other words, $(ab) \bmod n = 1$. Thus $ab + kn = 1$ for some integer k. ... Therefore a and n are relatively prime.
>
> ($\Leftarrow$) Suppose a and n are relatively prime. ... Therefore a is an invertible element of $\mathbb{Z}_n$. ■

The first part of the proof is 99% done! We have $ab + kn = 1$. We apply Corollary 35.9 to a and n to conclude $\gcd(a, n) = 1$. This finishes the first part of the proof.

> Let n be a positive integer and let $a \in \mathbb{Z}_n$.
>
> ($\Rightarrow$) Suppose a is invertible. This means there is an element $b \in \mathbb{Z}_n$ such that $a \otimes b = 1$. In other words, $(ab) \bmod n = 1$. Thus $ab + kn = 1$ for some integer k. By Corollary 35.9, a and n are relatively prime.
>
> ($\Leftarrow$) Suppose a and n are relatively prime. ... Therefore a is an invertible element of $\mathbb{Z}_n$. ■

For the second half (⇐) of the proof, we start right in with Corollary 35.9.

> Let n be a positive integer and let $a \in \mathbb{Z}_n$.
>
> (⇒) Suppose a is invertible. This means there is an element $b \in \mathbb{Z}_n$ such that $a \otimes b = 1$. In other words, $(ab) \bmod n = 1$. Thus $ab + kn = 1$ for some integer k. By Corollary 35.9, a and n are relatively prime.
>
> (⇐) Suppose a and n are relatively prime. By Corollary 35.9, there are integers x and y such that $ax + ny = 1$. ... Therefore a is an invertible element of $\mathbb{Z}_n$. ■

We have $ax + ny = 1$. This can be rewritten $ax = 1 - ny$. We want b such that $a \otimes b = 1$. The integer x is a likely candidate, but perhaps $x \notin \mathbb{Z}_n$. Of course, we can adjust x up or down by a multiple of n without changing anything important. We can let $b = x \bmod n$. Let's work this into the proof.

> Let n be a positive integer and let $a \in \mathbb{Z}_n$.
>
> (⇒) Suppose a is invertible. This means there is an element $b \in \mathbb{Z}_n$ such that $a \otimes b = 1$. In other words, $(ab) \bmod n = 1$. Thus $ab + kn = 1$ for some integer k. By Corollary 35.9, a and n are relatively prime.
>
> (⇐) Suppose a and n are relatively prime. By Corollary 35.9, there are integers x and y such that $ax + ny = 1$. Let $b = x \bmod n$. So $b = x + kn$ for some integer k. Substituting into $ax + ny = 1$, we have
>
> $$1 = ax + ny = a(b - kn) + ny = ab + (y - ka)n.$$
>
> Therefore $a \otimes b = ab \pmod{n} = 1$. Thus b is the reciprocal of a and therefore a is an invertible element of $\mathbb{Z}_n$. ■

We now know that the invertible elements of $\mathbb{Z}_n$ are exactly those that are relatively prime to n. Also, the proof of Theorem 36.14 gives us a method to calculate inverses.

Let $a \in \mathbb{Z}_n$ and suppose $\gcd(a, n) = 1$. Thus there are integers x and y such that $ax + ny = 1$. To find the numbers x and y, we use back substitution in Euclid's Algorithm (see Section 35).

Example 36.15 In $\mathbb{Z}_{431}$, find 29^{-1}.

Solution. In Section 35, we found integers x and y such that $431x + 29y = 1$, namely $x = 7$ and $y = -104$. Therefore $(-104 \cdot 29) \bmod 431 = 1$.

However, $-104 \notin \mathbb{Z}_{431}$. Instead we can take

$$b = -104 \bmod 431 = 327.$$

Now $29 \otimes 327 = (29 \cdot 327) \bmod 431 = 9483 \bmod 431 = 1$. Therefore $29^{-1} = 327$.

Example 36.16 In $\mathbb{Z}_{431}$, calculate $30 \oslash 29$.

Solution. In the previous example, we found $29^{-1} = 327$. Therefore

$$30 \oslash 29 = 30 \otimes 327 = (30 \cdot 327) \bmod 431 = 9810 \bmod 431 = 328.$$

A Note on Notation

In this book we use different symbols for ordinary addition $+$ and modular addition $\oplus$. This is important because when proving theorems about $\oplus$, we often have both $a+b$ and $a \oplus b$ in the same equation. It would be terribly confusing to write $a+b$ for both.

The good news is that throughout this book, we are consistent in using $\oplus$ for addition in $\mathbb{Z}_n$ and $+$ for addition in $\mathbb{Z}$ or $\mathbb{Q}$. It is still your responsibility to be aware of the modulus (n) currently under discussion.

The bad news is that this $\oplus$ notation is not standard. When mathematicians work in $\mathbb{Z}_n$, they just write $a+b$ or ab in place of $a \oplus b$ and $a \otimes b$, respectively.

The mathematician typically writes a phrase such as "working in $\mathbb{Z}_n$" or "working modulo n" and then uses the conventional operation symbols.

Recap

We introduced the number system $\mathbb{Z}_n$. This is the set $\{0, 1, \ldots, n-1\}$ together with the operations $\oplus$, $\ominus$, $\otimes$, and $\oslash$.

The operations $\oplus$, $\ominus$, and $\otimes$ are similar to $+$, $-$, and $\times$, respectively; one simply operates on the integers in the usual way and then reduces mod n.

The operation $\oslash$ is more subtle. We defined reciprocals in $\mathbb{Z}_n$ and showed that an element of $\mathbb{Z}_n$ is invertible if and only if it is relatively prime to n. We can use Euclid's gcd algorithm to compute reciprocals in $\mathbb{Z}_n$. We then defined $a \oslash b = a \otimes b^{-1}$ just when b is invertible. If b is not invertible, then $a \oslash b$ is undefined.

36 Exercises

36.1. In the context of $\mathbb{Z}_{10}$, please calculate:

a. $3 \oplus 3$.
b. $6 \oplus 6$.
c. $7 \oplus 3$.
d. $9 \oplus 8$.
e. $12 \oplus 4$. [Be careful. The answer is *not* 6.]
f. $3 \otimes 3$.
g. $4 \otimes 4$.
h. $7 \otimes 3$.
i. $5 \otimes 2$.
j. $6 \otimes 6$.
k. $4 \otimes 6$.
l. $4 \otimes 1$.
m. $12 \otimes 5$.
n. $5 \ominus 8$.

o. $8 \ominus 5$.

p. $8 \oslash 7$.

q. $5 \oslash 9$.

36.2. Solve the following equations for x in the $\mathbb{Z}_n$ specified.

a. $3 \otimes x = 4$ in $\mathbb{Z}_{11}$.

b. $4 \otimes x \ominus 8 = 9$ in $\mathbb{Z}_{11}$.

c. $3 \otimes x \oplus 8 = 1$ in $\mathbb{Z}_{10}$.

d. $342 \otimes x \oplus 448 = 73$ in $\mathbb{Z}_{1003}$.

36.3. Solve the following equations for x in the $\mathbb{Z}_n$ specified. *Note*: These are quite different from the previous set of problems. Why? Be sure you find *all* solutions.

a. $2 \otimes x = 4$ in $\mathbb{Z}_{10}$.

b. $2 \otimes x = 3$ in $\mathbb{Z}_{10}$.

c. $9 \otimes x = 4$ in $\mathbb{Z}_{12}$.

d. $9 \otimes x = 6$ in $\mathbb{Z}_{12}$.

36.4. Here are a few more equations for you to solve in the $\mathbb{Z}_n$ specified. Be sure to find all solutions.

a. $x \otimes x = 1$ in $\mathbb{Z}_{13}$.

b. $x \otimes x = 11$ in $\mathbb{Z}_{13}$.

c. $x \otimes x = 12$ in $\mathbb{Z}_{13}$.

d. $x \otimes x = 4$ in $\mathbb{Z}_{15}$.

e. $x \otimes x = 10$ in $\mathbb{Z}_{15}$.

f. $x \otimes x = 14$ in $\mathbb{Z}_{15}$.

The order of operations in $\mathbb{Z}_n$ is the same as in ordinary arithmetic. The expression $x \otimes x \oplus 1$ should be parenthesized as $(x \otimes x) \oplus 1$. In essence, this problem is asking you to determine whether or not there is a $\sqrt{-1}$ in $\mathbb{Z}_p$ for various prime numbers p.

36.5. For some prime numbers p, the equation $x \otimes x \oplus 1 = 0$ has a solution in $\mathbb{Z}_p$. For other primes it does not. For example, in $\mathbb{Z}_{17}$ we have $4 \otimes 4 \oplus 1 = 0$, but in $\mathbb{Z}_{19}$ there is no solution. The equation has a solution for $p = 2$, but this is not a particularly interesting example.

Investigate the first several (say, to 103) odd prime numbers p and divide them into two categories: those for which $x \otimes x \oplus 1 = 0$ has a solution in $\mathbb{Z}_p$ and those for which it does not. I recommend that you write a computer program to do this.

State a conjecture based on your evidence.

36.6. Prove: For all $a, b \in \mathbb{Z}_n$, $(a \ominus b) \oplus (b \ominus a) = 0$.

36.7. Prove that the operations $\oplus$, $\otimes$, and $\ominus$ are *closed*. This means that if $a, b \in \mathbb{Z}_n$, then $a \oplus b$, $a \otimes b$, $a \ominus b$ are all elements of $\mathbb{Z}_n$.

36.8. Prove Proposition 36.4. Why is this proposition restricted to $n \geq 2$?

36.9. Use Proposition 36.7 as the *definition* of $\ominus$ and then prove the assertion in Definition 36.6 as a theorem.

36.10. For ordinary integers, the following is true. If $ab = 0$, then $a = 0$ or $b = 0$. The analogous statement for $\mathbb{Z}_n$ is not necessarily true. For example, in $\mathbb{Z}_{10}$, $2 \otimes 5 = 0$ but $2 \neq 0$ and $5 \neq 0$. However, for some values of n (e.g., $n = 5$) it is true that $a \otimes b = 0$ implies $a = 0$ or $b = 0$.

For which values of $n \geq 2$ does the implication

$$a \otimes b = 0 \qquad \Longleftrightarrow \qquad a = 0 \text{ or } b = 0$$

hold in $\mathbb{Z}_n$?

Prove your answer.

36.11. Prove Proposition 36.11.

36.12. Let n be a positive integer and suppose $a, b \in \mathbb{Z}_n$ are both invertible. Prove or disprove each of the following statements.

a. $a \oplus b$ is invertible.

b. $a \ominus b$ is invertible.

c. $a \otimes b$ is invertible.

d. $a \oslash b$ is invertible.

36.13. Let n be an integer with $n \geq 2$. Prove that in $\mathbb{Z}_n$ the element $n - 1$ is its own inverse.

36.14. *Modular exponentiation*. Let b be a positive integer. The notation a^b means to multiply a by itself repeated, with a total of b factors of a; that is,

$$a^b = \underbrace{a \times a \times \cdots \times a}_{b \text{ times}}.$$

The notation for $\mathbb{Z}_n$ is the same. If $a \in \mathbb{Z}_n$ and b is a positive integer, in the context of $\mathbb{Z}_n$ we define

$$a^b = \underbrace{a \otimes a \otimes \cdots \otimes a}_{b \text{ times}}.$$

Please do the following:

a. In the context of $\mathbb{Z}_n$, prove or disprove: $a^b = a^{b \bmod n}$.

b. Without the aid of a computer or a calculator, find, in $\mathbb{Z}_{100}$, the value of 3^{64}.

The most horrible way to do this problem is to fully calculate 3^{64} and then reduce mod 100 (although this will give the correct answer—why?).

A less horrible way is to multiply 3 by itself 64 times, reducing mod 100 at each stage. This requires you to do 63 multiplication problems.

Try to do this calculation using only 6 multiplications, including the very first $3 \times 3 = 9$.

c. Estimate how many multiplications you need to do to calculate a^b in $\mathbb{Z}_n$.

d. Give a sensible definition for a^0 in $\mathbb{Z}_n$.

e. Give a sensible definition for a^b in $\mathbb{Z}_n$ when $b < 0$. Should you be upset that a^{-1} already has a meaning?

37 The Chinese Remainder Theorem

In this section, we investigate how to solve equations that involve modular equivalences.

Solving One Equation

We start with an easy example.

36.11. Prove Proposition 36.11.

36.12. Let n be a positive integer and suppose $a, b \in \mathbb{Z}_n$ are both invertible. Prove or disprove each of the following statements.

a. $a \oplus b$ is invertible.

b. $a \ominus b$ is invertible.

c. $a \otimes b$ is invertible.

d. $a \oslash b$ is invertible.

36.13. Let n be an integer with $n \geq 2$. Prove that in $\mathbb{Z}_n$ the element $n - 1$ is its own inverse.

36.14. *Modular exponentiation.* Let b be a positive integer. The notation a^b means to multiply a by itself repeated, with a total of b factors of a; that is,

$$a^b = \underbrace{a \times a \times \cdots \times a}_{b \text{ times}}.$$

The notation for $\mathbb{Z}_n$ is the same. If $a \in \mathbb{Z}_n$ and b is a positive integer, in the context of $\mathbb{Z}_n$ we define

$$a^b = \underbrace{a \otimes a \otimes \cdots \otimes a}_{b \text{ times}}.$$

Please do the following:

a. In the context of $\mathbb{Z}_n$, prove or disprove: $a^b = a^{b \bmod n}$.

b. Without the aid of a computer or a calculator, find, in $\mathbb{Z}_{100}$, the value of 3^{64}.

The most horrible way to do this problem is to fully calculate 3^{64} and then reduce mod 100 (although this will give the correct answer—why?).

A less horrible way is to multiply 3 by itself 64 times, reducing mod 100 at each stage. This requires you to do 63 multiplication problems.

Try to do this calculation using only 6 multiplications, including the very first $3 \times 3 = 9$.

c. Estimate how many multiplications you need to do to calculate a^b in $\mathbb{Z}_n$.

d. Give a sensible definition for a^0 in $\mathbb{Z}_n$.

e. Give a sensible definition for a^b in $\mathbb{Z}_n$ when $b < 0$. Should you be upset that a^{-1} already has a meaning?

37 The Chinese Remainder Theorem

In this section, we investigate how to solve equations that involve modular equivalences.

Solving One Equation

We start with an easy example.

Example 37.1 Solve the equation

$$x \equiv 4 \pmod{11}.$$

Solution. This asks for all integers x such that $x - 4$ is a multiple of 11 (i.e., $x - 4 = 11k$ for some integer k). We can rewrite this as $x = 4 + 11k$ where k can be any integer.

So the solutions are $\ldots, -18, -7, 4, 15, 26, \ldots$.

Let's now work on a more complicated example.

Example 37.2 Solve the equation

$$3x \equiv 4 \pmod{11}. \tag{42}$$

Suppose, just for a moment, that we had a solution x_0 to the equation $3x \equiv 4 \pmod{11}$. Now consider the integer $x_1 = x_0 + 11$. If we substitute x_1 for x in Equation (42), we get

$$3x_1 = 3(x_0 + 11) = 3x_0 + 33 \equiv 3x_0 \equiv 4 \pmod{11}$$

so x_1 is also a solution. Thus, if we add or subtract any multiple of 11 to a solution to Equation (42), we obtain another solution to Equation (42). So if there is a solution, then there is a solution in $\{0, 1, 2, \ldots, 10\} = \mathbb{Z}_{11}$. Once we find all the solutions in $\mathbb{Z}_{11}$, we have found all solutions to the equation.

Now there are only 11 possible values of x we need to try, and it might be simplest just to try all the possibilities to find the answer. However, we want to generalize this method to problems where the modulus is a great deal larger than 11.

We seek a number $x \in \mathbb{Z}_{11}$ for which $3x \equiv 4 \ (11)$. But note that

$$3x \equiv 4 \ (11) \quad \Longleftrightarrow \quad (3x) \bmod 11 = 4 \quad \Longleftrightarrow \quad 3 \otimes x = 4$$

where $\otimes$ is modular multiplication in $\mathbb{Z}_{11}$. How do we solve the equation $3 \otimes x = 4$ in $\mathbb{Z}_{11}$? We would like to divide both sides by 3. Do we get $x = \frac{4}{3}$? Nonsense! That is not how we divide in $\mathbb{Z}_{11}$. We multiply both sides of $3 \otimes x = 4$ by 3^{-1}. By the methods of Section 36, we can calculate $3^{-1} = 4$, and so

$$3 \otimes x = 4 \quad \Rightarrow \quad 4 \otimes 3 \otimes x = 4 \otimes 4 \quad \Rightarrow \quad 1 \otimes x = 5 \quad \Rightarrow \quad x = 5$$

(because $12 \bmod 11 = 1$ and $16 \bmod 11 = 5$).

Let's check this answer in Equation (42). We substitute $x = 5$ and calculate

$$3x = 15 \equiv 4 \pmod{11}$$

and so 5 is a solution. Furthermore, there are no other solutions in $\mathbb{Z}_{11}$. If $x' \in \mathbb{Z}_{11}$ were another solution, we would have $3 \otimes x' = 4$, and when we $\otimes$ both sides by 4, we would find $x' = 5$.

Although 5 is the only solution in $\mathbb{Z}_{11}$, it is not the only solution to Equation (42). If we add any multiple of 11 to 5, we get another solution. The full set of solutions is $\{5 + 11k : k \in \mathbb{Z}\} = \{\ldots, -17, -6, 5, 16, 27, \ldots\}$. This completes the solution to Example 37.2.

We summarize what we have learned in the following result.

Proposition 37.3 Let $a, b, n \in \mathbb{Z}$ with $n > 0$. Suppose a and n are relatively prime and consider the equation

$$ax \equiv b \pmod{n}$$

The set of solutions to this equation is

$$\{x_0 + kn : k \in \mathbb{Z}\}$$

where $x_0 = a_0^{-1} \otimes b_0$, $a_0 = a \bmod n$, $b_0 = b \bmod n$, and $\otimes$ is modular multiplication in $\mathbb{Z}_n$. The integer x_0 is the only solution to this equation in $\mathbb{Z}_n$.

We have essentially done the proof by solving Equation (42). Please write out the proof yourself, using our solution to Equation (42) as a guide.

It is not hard to extend Proposition 37.3 to solve equations of the form

$$ax + b \equiv c \pmod{n}$$

where a and n are relatively prime.

Solving Two Equations

Now we solve a pair of congruence equations in different moduli. The type of problem we solve is

$$x \equiv a \pmod{m}, \quad \text{and}$$
$$x \equiv b \pmod{n}.$$

Let's work out the solution to the following problem.

Example 37.4 Solve the pair of equations

$$x \equiv 1 \pmod{7}, \quad \text{and}$$
$$x \equiv 4 \pmod{11}.$$

In other words, we want to find all integers x that satisfy both of these equations.

Let's begin with the first equation. Since $x \equiv 1 \ (7)$, we can write

$$x = 1 + 7k$$

for some integer k. We can substitute $1 + 7k$ for x in the second equation: $x \equiv 4 \ (11)$. This gives

We can check that $7^{-1} = 8$ by calculating $7 \otimes 8 = (7 \cdot 8) \bmod 11 = 56 \bmod 11 = 1$.

$$1 + 7k \equiv 4 \pmod{11} \quad \Rightarrow \quad 7k \equiv 3 \pmod{11}.$$

The problem now reduces to a single equation in k. We apply Proposition 37.3. To solve this equation, we need to $\otimes$ both sides by 7^{-1} working in $\mathbb{Z}_{11}$. In $\mathbb{Z}_{11}$ we find that $7^{-1} = 8$. We calculate, in $\mathbb{Z}_{11}$,

$$7 \otimes k = 3 \quad \Rightarrow \quad 8 \otimes 7 \otimes k = 8 \otimes 3 \quad \Rightarrow \quad k = 2.$$

Furthermore, if we increase or decrease $k = 2$ by any multiple of 11, we again have a solution to $1 + 7k \equiv 4 \;\; (11)$.

We are nearly finished. Let's write down what we have. We know that we want all values of x with

$$x = 1 + 7k$$

and k can be any integer of the form

$$k = 2 + 11j$$

where j is any integer. Combining these two, we have

$$x = 1 + 7k = 1 + 7(2 + 11j) = 15 + 77j \qquad (\forall j \in \mathbb{Z}).$$

In other words, the solution set to the equations in Example 37.4 is $\{x \in \mathbb{Z} : x \equiv 15 \;\; (77)\}$.

To check that this is correct, notice that

$$15 \equiv 1 \pmod 7 \qquad \text{and} \qquad 15 \equiv 4 \pmod{11}.$$

Furthermore, if x is increased or decreased by any multiple of 77, both equations remain valid because 77 is a multiple of both 7 and 11.

Theorem 37.5 **(Chinese Remainder)** Let a, b, m, n be integers with m and n positive and relatively prime. There is a unique integer x_0 with $0 \le x_0 < mn$ that solves the pair of equations

$$x \equiv a \pmod m, \quad \text{and}$$
$$x \equiv b \pmod n.$$

Furthermore, every solution to these equations differs from x_0 by a multiple of mn.

We saw all the steps to prove the Chinese Remainder Theorem when we solved the system in Example 37.4. The general proof follows the method of that example.

Proof. From the equation $x \equiv a \;\; (m)$, we know that $x = a + km$ where $k \in \mathbb{Z}$. We substitute this into the second equation $x \equiv b \;\; (n)$ to get

$$a + km \equiv b \;\; (n) \quad \Rightarrow \quad km \equiv b - a \;\; (n)$$

and we want to solve this for k. Note that adding or subtracting a multiple of n to $b - a$ or to m does not change this equation. So we let

$$m' = m \bmod n, \quad \text{and}$$
$$c = (b - a) \bmod n$$

Since m and n are relatively prime, so are m' and n (see Exercise 35.12). Thus solving $km \equiv b - a \;\; (n)$ is equivalent to solving $km' \equiv c \;\; (n)$. To find a solution in $\mathbb{Z}_n$, we solve, in $\mathbb{Z}_n$,

$$k \otimes m' = c.$$

Since m' is relatively prime to n, we can $\otimes$ both sides by its reciprocal to get

$$k = (m')^{-1} \otimes c.$$

Let $d = (m')^{-1} \otimes c$, so the values for k that we want are $k = d + jn$ for all integers j.

Finally, we substitute $k = d + jn$ into $x = a + km$ to get

$$x = a + km = a + (d + jn)m = a + dm + jnm$$

where $j \in \mathbb{Z}$ is arbitrary. We have shown that the original system of two equations reduces to the single equation

$$x \equiv a + dm \pmod{mn}$$

and the conclusions follow. ■

Example 37.6 Suppose we want to solve a system of three equations. For example, solve for all x:

$$\begin{aligned} x &\equiv 3 \pmod{9}, \\ x &\equiv 5 \pmod{10}, \quad \text{and} \\ x &\equiv 2 \pmod{11}. \end{aligned}$$

Solution: We can solve the first two equations by the usual method

$$\left.\begin{array}{l} x \equiv 3 \ (9) \\ x \equiv 5 \ (10) \end{array}\right\} \quad \Rightarrow \quad x \equiv 75 \ (90).$$

Now we combine this result with the last equation and solve again by the usual method.

$$\left.\begin{array}{l} x \equiv 75 \ (90) \\ x \equiv 2 \ (11) \end{array}\right\} \quad \Rightarrow \quad x \equiv 255 \ (990).$$

Recap

We investigated how to solve equations of the form $ax + b \equiv c \ (n)$ as well as systems of equations of the form $x \equiv a \ (m)$ and $x \equiv b \ (n)$ where m and n are relatively prime.

---

37 Exercises

37.1. Solve the following for all integers x.

a. $3x \equiv 17 \pmod{20}$.

b. $2x + 5 \equiv 7 \pmod{15}$.

c. $10 - 3x \equiv 2 \pmod{23}$.

d. $100x \equiv 74 \pmod{127}$.

37.2. Prove Proposition 37.3.

37.3. Solve the following systems of equations.

a. $x \equiv 4 \ (5)$ and $x \equiv 7 \ (11)$.

b. $x \equiv 34 \ (100)$ and $x \equiv -1 \ (51)$.

c. $x \equiv 3 \ (7)$, $x \equiv 0 \ (4)$, and $x \equiv 8 \ (25)$.

d. $3x \equiv 8 \ (10)$ and $2x + 4 \equiv 9 \ (11)$.

37.4. Explain why it is important for a and n to be relatively prime in the equation $ax \equiv b \ (n)$. Specifically, you should do the following:

a. Create an equation of the form $ax \equiv b \ (n)$ that has no solutions.

b. Create an equation of the form $ax \equiv b \ (n)$ that has more than one solution in $\mathbb{Z}_n$.

37.5. For the pair of equations $x \equiv a \ (m)$ and $x \equiv b \ (n)$, explain why it is important that m and n be relatively prime. Where in the proof of Theorem 37.5 did we use this fact?

Give an example of a pair of equations $x \equiv a \ (m)$ and $x \equiv b \ (n)$ that has *no* solution.

Give an example of a pair of equations $x \equiv a \ (m)$ and $x \equiv b \ (n)$ that has more than one solution in $\mathbb{Z}_{nm}$.

37.6. Consider the system of congruences

$$x \equiv a_1 \pmod{m_1}$$
$$x \equiv a_2 \pmod{m_2}$$

where m_1 and m_2 are relatively prime. Let b_1 and b_2 be integers where

These inverses exist because m_1 and m_2 are relatively prime.

$$b_1 = m_1^{-1} \quad \text{in } \mathbb{Z}_{m_2}$$
$$b_2 = m_2^{-1} \quad \text{in } \mathbb{Z}_{m_1}.$$

Finally, let

$$x_0 = m_1 b_1 a_2 + m_2 b_2 a_1.$$

Please prove that x_0 is a solution to the system of congruences.

37.7. Use the technique of the previous problem to solve the following systems of congruences.

a. $x \equiv 3 \pmod{8}$ and $x \equiv 2 \pmod{19}$.

b. $x \equiv 1 \pmod{10}$ and $x \equiv 3 \pmod{21}$.

38 Factoring

In this section we prove the following well-known fact: Every positive integer can be factored into primes in (essentially) a unique fashion. For example, the integer 60 can be factored into primes as $60 = 2 \times 2 \times 3 \times 5$. It can also be factored as $60 = 5 \times 2 \times 3 \times 2$, but notice that the primes in the two factorizations are exactly the same; the only difference is the order in which we listed them. This is true of all positive integers (we can treat 1 as the empty product of primes—see Section 8). We can consider prime numbers to be already factored into primes: a prime, say 17, is the product of just one prime: 17. Composite numbers are the product of two or more primes.

Theorem 38.1 **(Fundamental Theorem of Arithmetic)** Let n be a positive integer. Then n factors into a product of primes. Furthermore, the factorization of n into primes is unique up to the order of the primes.

The phrase "up to the order of the primes" means that we treat $2 \times 3 \times 5$ the same as $5 \times 2 \times 3$.

A key tool in the proof of this theorem is the following result.

Lemma 38.2 Suppose $a, b, p \in \mathbb{Z}$ and p is a prime. If $p|ab$, then $p|a$ or $p|b$.

Note: If we already had a proof of Theorem 38.1, this lemma would be simple to prove (see Exercise 38.5).

Proof. Let a, b, p be integers with p prime and suppose $p|ab$. Suppose, for the sake of contradiction, that p divides neither a nor b.

Since p is a prime, the only divisors of p are ± 1 and $\pm p$. Since p is not a divisor of a, the largest divisor they have in common is 1. Therefore $\gcd(a, p) = 1$ (i.e., a and p are relatively prime). Thus, by Corollary 35.9, there are integers x and y such that $ax + py = 1$.

Similarly, b and p are relatively prime. By Corollary 35.9, there are integers w and z such that $bz + pw = 1$.

We have found that $ax + py = 1$ and $bz + pw = 1$. Multiplying these two equations together, we get

$$1 = (ax + py)(bz + pw) = abxz + pybz + paxw + p^2yw.$$

Notice that all four of these terms are divisible by p (the first term is a multiple of ab, which in turn is a multiple of p by hypothesis). We have shown that $p|1$, but this is clearly false. $\Rightarrow\Leftarrow$ ■

Lemma 38.3 Suppose $p, q_1, q_2, \ldots, q_t$ are prime numbers. If

$$p|(q_1 q_2 \cdots q_t),$$

then $p = q_i$ for some $1 \le i \le t$.

You can prove Lemma 38.3 by induction on t (or by the smallest-counterexample method). See Exercise 38.6.

Proof (of Theorem 38.1)

Suppose, for the sake of contradiction, that not all positive integers factor into primes. Let X be the set of all positive integers that do not factor into primes. Note that $1 \notin X$ because we can factor 1 into an empty product of primes. Also $2 \notin X$ because 2 is a prime (and factors $2 = 2$).

By the Well-Ordering Principle, there is a least element of X; let's call it x. The integer x is the smallest positive integer that does not factor into primes. Note that $x \neq 1$ (discussed in the previous paragraph). Furthermore, x is not prime, since every prime is the product of just one number (itself). Therefore x is composite.

Since x is composite, there is an integer a with $1 < a < x$ and $a|x$. This means there is an integer b with $ab = x$. Since $a < x$, we may divide both sides of $ab = x$ by a to get $1 < \frac{x}{a} = b$. Because $1 < a$, we may multiply both sides by b to get $b < ab = x$. Thus $1 < b < x$. Therefore a and b are both positive integers less than x. Since x is the least element of X, we know that neither a nor b is in X, so both a and b can be factored into primes. Suppose the factorizations of a and b are

$$a = p_1 p_2 \cdots p_s \quad \text{and} \quad b = q_1 q_2 \cdots q_t$$

where the ps and qs are prime. Then

$$x = ab = (p_1 p_2 \cdots p_s)(q_1 q_2 \cdots q_t)$$

is a factorization of x into primes, contradicting $x \in X$.$\Rightarrow\Leftarrow$ Therefore all positive integers can be factored into primes.

Now we work to show uniqueness. Suppose, for the sake of contradiction, that some positive integers can be factored into primes in two distinct ways. Let Y be the set of all such integers with two (or more) distinct factorizations. Note that $1 \notin Y$ because 1 can be factored only as the empty product of primes. The supposition is that $Y \neq \emptyset$, and therefore Y contains a least element y. Thus y can be factored into primes in two distinct ways:

$$y = p_1 p_2 \cdots p_s \text{ and}$$
$$y = q_1 q_2 \cdots q_t$$

where the ps and qs are primes and the two lists of primes are not rearrangements of one another.

Claim: *The list $(p_1, p_2, \ldots, p_s)$ and the list $(q_1, q_2, \ldots, q_t)$ have no elements in common (i.e., $p_i \neq q_j$ for all i and j).* If the two lists had a prime in common—say, r—then y/r would be a smaller integer (than y) that factors into primes in two distinct ways, contradicting the fact that y is smallest in Y.

Now consider p_1. Notice that $p_1|y$, so $p_1|(q_1 q_2 \cdots q_t)$. However, by Lemma 38.3, p_1 must equal one of the qs, contradicting the claim we just proved.$\Rightarrow\Leftarrow$ ■

Infinitely Many Primes

How many primes are there? At first, it is very easy to find primes; almost every other number is prime: 2, 3, 5, 7, 11, 13, 17, 19, 23, 29, and so on. This suggests that there could be infinitely many primes. However, this pattern does not continue. In Exercise 8.9 you found a sequence of 1001 consecutive composite numbers. Perhaps, after a point, there are no more primes.

Although the prime numbers thin out as we look deeper into the positive integers, they never die out completely. There are infinitely many primes.

Theorem 38.4 **(Infinitude of primes)** There are infinitely many prime numbers.

This proof can also be viewed as an algorithm for generating primes. Given that we have generated primes from 2 to p, the prime factors of $n = 2 \cdot 3 \cdot 5 \cdots p + 1$ must be new primes that we have not previously constructed. In this way, we can build as many primes as we like.

Proof. Suppose, for the sake of contradiction, that there are only finitely many prime numbers. In such a case, we could (in principle) list them all:

$$2, 3, 5, 7, \ldots, p$$

where p is the (alleged) last prime number. Let

$$n = (2 \times 3 \times 5 \times \cdots \times p) + 1.$$

That is, n is the positive integer formed by multiplying together all the prime numbers and then adding 1.

Is n a prime?

The answer is no. Clearly n is greater than the last prime p, so n is not prime. Since n is not prime, n must be composite.

Let q be any prime. Because

$$n = (2 \times 3 \times \cdots \times q \times \cdots \times p) + 1,$$

when we divide n by q, we are left with a remainder of 1. We see that there is no prime number q with $q|n$, contradicting Theorem 38.1.$\Rightarrow\Leftarrow$ ■

A Formula for Greatest Common Divisor

Suppose a and b are positive integers. By Theorem 38.1, we can factor them into primes as

$$a = 2^{e_2} 3^{e_3} 5^{e_5} 7^{e_7} \cdots \qquad \text{and} \qquad b = 2^{f_2} 3^{f_3} 5^{f_5} 7^{f_7} \cdots. \tag{43}$$

For example, if $a = 24$ we would have

$$24 = 2^3 3^1 5^0 7^0 \cdots.$$

Suppose $a|b$. Let p be a prime and suppose it appears e_p times in the prime factorization of a. Since $p^{e_p}|a$ and $a|b$, we have (by Proposition 4.3) $p^{e_p}|b$, and therefore $p^{e_p}|p^{f_p}$. Thus $e_p \le f_p$. In other words, if $a|b$, then the number of factors of p in the prime factorization of a is less than or equal to the number of factors of p in the prime factorization of b.

Thus, if a and b are as in Equation (43) and if $d = \gcd(a, b)$, then

$$d = 2^{x_2} 3^{x_3} 5^{x_5} 7^{x_7} \cdots$$

The notation $\min\{a, b\}$ stands for the smaller of a or b. That is, if $a \le b$, then $\min\{a, b\} = a$; otherwise $\min\{a, b\} = b$.

where $x_2 = \min\{e_2, f_2\}$, $x_3 = \min\{e_3, f_3\}$, $x_5 = \min\{e_5, f_5\}$, and so on. For example,

$$24 = 2^3 3^1 5^0 7^0 \cdots \qquad \text{and} \qquad 30 = 2^1 3^1 5^1 7^0 \cdots$$

and so

$$\gcd(24, 30) = 2^{\min\{3,1\}} 3^{\min\{1,1\}} 5^{\min\{0,1\}} 7^{\min\{0,0\}} \cdots = 2^1 3^1 5^0 7^0 \cdots = 6.$$

Let us summarize what we have observed in the following result.

Theorem 38.5 **(GCD formula)** Let a and b be positive integers with

$$a = 2^{e_2}3^{e_3}5^{e_5}7^{e_7}\cdots \qquad \text{and} \qquad b = 2^{f_2}3^{f_3}5^{f_5}7^{f_7}\cdots.$$

Then

$$\gcd(a, b) = 2^{\min\{e_2, f_2\}}3^{\min\{e_3, f_3\}}5^{\min\{e_5, f_5\}}7^{\min\{e_7, f_7\}}\cdots.$$

Irrationality of $\sqrt{2}$

Is there a square root of 2? In other words, is there a number x such that $x^2 = 2$? This is actually a subtle question. In this section, we show that there is no rational number x such that $x^2 = 2$.

Proposition 38.6 There is no rational number x such that $x^2 = 2$.

In effect, this is asking us to show that the set $\{x \in \mathbb{Q} : x^2 = 2\}$ is empty. To show that something does not exist, we use Proof Template 13.

Proof. Suppose, for the sake of contradiction, that there is a rational number x such that $x^2 = 2$. This means there are integers a and b such that $x = \frac{a}{b}$.

We therefore have $\left(\frac{a}{b}\right)^2 = 2$. This can be rewritten

$$a^2 = 2b^2.$$

Consider the prime factorization of the integer $n = a^2 = 2b^2$. On the one hand, since $n = a^2$, the prime 2 appears an even number (perhaps zero) of times in the prime factorization of n. On the other hand, since $n = 2b^2$, the prime 2 appears an odd number of times in the prime factorization of n. $\Rightarrow\Leftarrow$ Therefore, there is no rational number x such that $x^2 = 2$. ■

There is a real number x that satisfies $x^2 = 2$, but the proof of this fact is complicated. First, we need to define *real number*. Second, we need to define what it means to multiply two real numbers. Finally, we have to show that $x^2 = 2$ has a solution. All of these are a job for *continuous mathematics*, and we do not venture into that realm here.

There are many lovely proofs that $\sqrt{2}$ is irrational. Here is another.

Proof (of Proposition 38.6)

Suppose there is a rational number x such that $x^2 = 2$. Write $x = \frac{b}{a}$. By Exercise 35.17, we may choose a and b to be relatively prime.

Because a and b are relatively prime, there is no prime that divides both.

Since $\frac{b^2}{a^2} = 2$, we have

$$b^2 = 2a^2.$$

Factor both sides of this equation into primes; the two sides of this equation are integers that are greater than or equal to 2. Let p be one of the primes in the factorization. Looking at the left-hand side, we see that the prime factorization of b^2 is simply the prime factorization of b with every prime appearing twice as often. So if $p|b^2$, clearly p is a divisor of b and not a divisor of a. Looking at the right-hand side, we see that p must be a divisor of 2, so $p = 2$. We have shown that the only prime divisor of $b^2 = 2a^2$ is 2. Since $2|b$ and $\gcd(a, b) = 1$, we see that a does not have any prime divisors! Thus $a = \pm 1$ and we have

$$b^2 = 2.$$

In other words, there is an integer b with $b^2 = 2$, and clearly there is no such integer. ■

Here is yet another proof that uses geometry.

Proof (of Proposition 38.6)

Suppose, for the sake of contradiction, there is a rational number x such that $x^2 = 2$. We may assume x is positive, for otherwise we could simply use $-x$ instead [since $(-x)^2 = x^2 = 2$].

Since x is rational, write $x = \frac{b}{a}$ where a and b are both positive and are as small as possible.

Write $x^2 = 2$ as $a^2 + a^2 = b^2$. Construct an isosceles right triangle XYZ (with right angle at Y) whose legs have length a and whose hypotenuse has length b. See the figure.

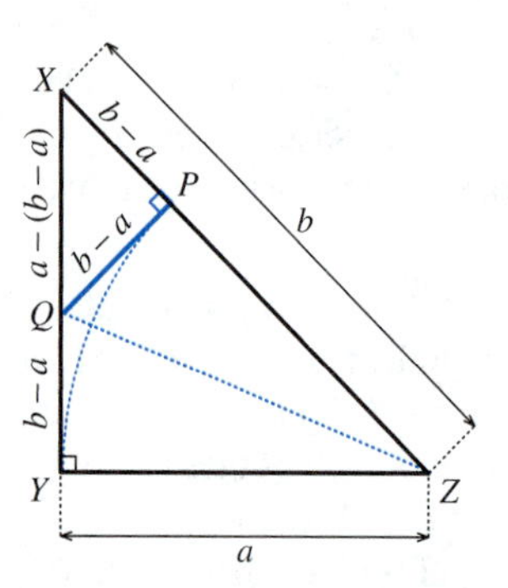

Geometry abbreviations: The HL theorem asserts that given two right triangles, if the hypotenuse and a leg of one triangle are congruent to the hypotenuse and a leg of a second triangle, then the triangles are congruent. The abbreviation CPCTC stands for corresponding parts of congruent triangles are congruent.

Swing an arc centered at Z from Y meeting the hypotenuse at point P. Because the segment ZP has length a (it is a radius of the arc), the segment XP has length $b - a$.

Erect a perpendicular at P that meets leg XY at the point Q. Notice that XPQ is also an isosceles right triangle (angle X is 45°) and so segment PQ has length $b - a$.

Now triangles ZPQ and ZYQ are congruent because they are right triangles with the same hypotenuse (QZ) and congruent legs YZ and PZ (use the HL theorem from geometry). Therefore, by CPCTC, PQ and YQ are congruent. Since the length of PQ is $b - a$, the length of YQ is the same.

Thus, since the length of YQ is $b - a$ and the length of XY is a, the length of XQ is $a - (b - a) = 2a - b$.

- **Claim:** $b > a$, and hence $b - a > 0$.
  This is because $\left(\frac{b}{a}\right)^2 = 2$, and if $b \le a$, we would have $\left(\frac{b}{a}\right)^2 \le 1$. (Also, the length of the hypotenuse of a right triangle is greater than that of its legs.)
- **Claim:** $2a - b > 0$.
  If this were not so, we would have
  $$b \ge 2a \quad \Rightarrow \quad b^2 \ge 4a^2 \quad \Rightarrow \quad \frac{b^2}{a^2} \ge 4$$
  contradicting $\frac{b^2}{a^2} = 2$.
- **Claim:** $(b - a)^2 + (b - a)^2 = (2a - b)^2$.
  This follows by the Pythagorean Theorem applied to triangle XPQ.

Therefore
$$\left(\frac{b'}{a'}\right)^2 = \left(\frac{2a - b}{b - a}\right)^2 = 2$$
where $b' = 2a - b$ and $a' = b - a$. Since triangle XQP is strictly inside triangle XYZ, we have $a' < a$ and $b' < b$, contradicting the choice of a and b as small as possible. ■

Recap

We showed that every positive integer factors uniquely into a product of primes. We proved there are infinitely many primes, and we used prime factorization to develop a formula for the greatest common divisor of two positive integers. We proved that there is no rational number whose square is 2.

38 Exercises

38.1. Suppose you wish to factor a positive integer n. You could write a computer program that tries to divide n by all possible divisors between 1 and n. If n is around one million, this means performing around one million divisions.

Explain why this is not necessary and that it is enough to check all possible divisors from 2 up to (and perhaps including) $\sqrt{n}$.

If n is around one million, then $\sqrt{n}$ is around one thousand.

38.2. Factor the following positive integers into primes.

a. 25.
b. 4200.
c. 10^{10}.
d. 19.
e. 1.

38.3. Let x be an integer. Prove that $2|x$ and $3|x$ if and only if $6|x$.

Generalize and prove.

38.4. Suppose a is a positive integer and p is a prime. Prove that $p|a$ if and only if the prime factorization of a contains p.

38.5. Prove Lemma 38.2 using Theorem 38.1.

38.6. Prove Lemma 38.3 by induction (or Well-Ordering Principle) using Lemma 38.2.

38.7. Suppose we wish to compute the greatest common divisor of two 1000-digit numbers using Theorem 38.5. How many divisions would this take? (Assume we factor using trial division up to the square roots of the numbers.)

How would this compare to using Euclid's Algorithm?

38.8. Let a and b be integers. A *common multiple* of a and b is an integer n for which $a|n$ and $b|n$. We call an integer m the *least common multiple* of n provided (1) m is positive, (2) m is a common multiple of a and b, and (3) if n is any other positive common multiple of a and b, then $n \geq m$.

The notation for the least common multiple of a and b is $\text{lcm}(a, b)$. For example, $\text{lcm}(24, 30) = 120$.

Please do the following:

a. Develop a formula for the least common multiple of two positive integers in terms of their prime factorizations; your formula should be similar to the one in Theorem 38.5.

b. Use your formula to show that if a and b are positive integers, then

$$ab = \gcd(a, b)\text{lcm}(a, b).$$

38.9. Let $a \in \mathbb{Z}$ and suppose a^2 is even. Prove that a is even.

38.10. Generalize the previous exercise. Prove that if $a, p \in \mathbb{Z}$ with p a prime and $p|a^2$, then $p|a$.

38.11. Prove that consecutive perfect squares are relatively prime.

38.12. Let n be a positive integer and suppose we factor n into primes as follows:

$$n = p_1^{e_1} p_2^{e_2} \cdots p_t^{e_t}$$

where the p_js are distinct primes and the e_js are natural numbers.

Find a formula for the number of positive divisors of n. For example, if $n = 18$, then n has six positive divisors: 1, 2, 3, 6, 9, and 18.

38.13. Recall (see Exercise 2.9) that an integer n is called *perfect* if it equals the sum of all its divisors d with $1 \leq d < n$. For example, 28 is perfect because $1 + 2 + 4 + 7 + 14 = 28$.

Let a be a positive integer. Prove that if $2^a - 1$ is prime, then $n = 2^{a-1}(2^a - 1)$ is perfect.

Euler's totient, $\varphi(n)$.

38.14. In this problem we consider the question: How many integers, from 1 to n inclusive, are relatively prime to n? For example, suppose $n = 10$. There are ten numbers in $\{1, 2, \ldots, 10\}$. Of them, the following are relatively prime to 10: $\{1, 3, 7, 9\}$. So there are four numbers from 1 to 10 that are relatively prime to 10.

The notation $\varphi(n)$ stands for the answer to this counting problem; that is, $\varphi(n)$ is the number of integers from 1 to n (inclusive) that are relatively prime to n. From our example, $\varphi(10) = 4$. The symbol φ is a Greek lowercase letter phi. The function φ is known as *Euler's totient*.

Please evaluate the following:

a. $\varphi(14)$.

b. $\varphi(15)$.

c. $\varphi(16)$.
d. $\varphi(17)$.
e. $\varphi(25)$.
f. $\varphi(5041)$. Note: $5041 = 71^2$ and 71 is prime.
g. $\varphi(2^{10})$.
Note: You could do all of these by listing all the possibilities, but the last two would be painful. Try to develop general methods (or see the next problem).

38.15. *Euler's totient, continued.* Suppose p and q are unequal primes. Prove the following:
a. $\varphi(p) = p - 1$.
b. $\varphi(p^2) = p^2 - p$.
c. $\varphi(p^n) = p^n - p^{n-1}$ where n is a positive integer.
d. $\varphi(pq) = pq - q - p + 1 = (p-1)(q-1)$.

38.16. *Euler's totient, continued further.* Suppose $n = p_1 p_2 \cdots p_t$ where the p_is are distinct primes (i.e., no two are the same). For example, $n = 2 \times 3 \times 11 = 66$ is such a number. Prove that

$$\begin{aligned}\varphi(n) = n &- \frac{n}{p_1} - \frac{n}{p_2} - \cdots - \frac{n}{p_t} \\ &+ \frac{n}{p_1 p_2} + \frac{n}{p_1 p_3} + \cdots + \frac{n}{p_{t-1} p_t} \\ &- \frac{n}{p_1 p_2 p_3} - \frac{n}{p_1 p_2 p_4} - \cdots - \frac{n}{p_{t-2} p_{t-1} p_t} \\ &+ \cdots\cdots \pm \frac{n}{p_1 p_2 p_3 \cdots p_t}.\end{aligned}$$

For example,

$$\begin{aligned}\varphi(66) &= 66 - \frac{66}{2} - \frac{66}{3} - \frac{66}{11} + \frac{66}{2 \cdot 3} + \frac{66}{2 \cdot 11} + \frac{66}{3 \cdot 11} - \frac{66}{2 \cdot 3 \cdot 11} \\ &= 66 - 33 - 22 - 6 + 11 + 3 + 2 - 1 \\ &= 20.\end{aligned}$$

Note that this formula simplifies to

$$\varphi(n) = n\left(1 - \frac{1}{p_1}\right)\left(1 - \frac{1}{p_2}\right)\cdots\left(1 - \frac{1}{p_t}\right).$$

For example, $\varphi(66) = 66(1 - \frac{1}{2})(1 - \frac{1}{3})(1 - \frac{1}{11}) = 20$.

38.17. *Again with Euler's totient.* Now suppose n is any positive integer. Factor n into primes as

$$n = p_1^{a_1} p_2^{a_2} \cdots p_t^{a_t}$$

where the p_is are distinct primes and the exponents a_i are all positive integers. Prove that the formulas from the previous problem are valid for this general n.

38.18. Rewrite the second proof of Proposition 38.6 to show the following:

Let n be an integer. If $\sqrt{n}$ is not an integer, then there is no rational number x such that $x^2 = n$.

38.19. Explain why we may assume a and b are both positive in the third proof of Proposition 38.6.

38.20. Prove that $\log_2 3$ is irrational.

38.21. *Sieve of Erasothenes*. Here is a method for finding many prime numbers. Write down all the numbers from 2 to, say, 1000. Notice that the smallest number on this list (2) is a prime. Cross off all multiples of 2 (except 2). The next smallest number on the list is a prime (3). Cross off all multiples of 3 (except 3 itself). The next number on the list is 4, but it's crossed off. The next smallest number on the list that isn't crossed off is 5. Cross off all multiples of 5 (except 5 itself).

a. Prove that this algorithm crosses off all composite numbers on the list but retains all the primes.

b. Implement this algorithm on a computer.

c. Let $\pi(n)$ denote the number of primes that are less than or equal to n. For example, $\pi(19) = 8$ because there are eight primes that are less than or equal to 19—namely 2, 3, 5, 7, 11, 13, 17, and 19.
Use your program from part (b) to evaluate $\pi(1000000)$.

d. The Prime Number Theorem states that $\pi(n) \approx n/\ln n$. How good is this approximation when $n = 1{,}000{,}000$?

38.22. In this and the subsequent problems, you will be working in a different number system. The goal is to illustrate that unique factorization of prime numbers is a special feature of the integers.

We consider all numbers of the form

$$a + b\sqrt{-3}$$

where a and b are integers. For example, $5 - 2\sqrt{-3}$ is a number in this system, but $\frac{1}{2}$ is not.

This number system is denoted $\mathbb{Z}[\sqrt{-3}]$. That is, $\mathbb{Z}[\sqrt{-3}]$ is the set

$$\mathbb{Z}\left[\sqrt{-3}\right] = \left\{a + b\sqrt{-3} : a, b \in \mathbb{Z}\right\}.$$

Please do the following:

a. Prove that if $w, z \in \mathbb{Z}[\sqrt{-3}]$, then $w + z \in \mathbb{Z}[\sqrt{-3}]$.

b. Prove that if $w, z \in \mathbb{Z}[\sqrt{-3}]$, then $w - z \in \mathbb{Z}[\sqrt{-3}]$.

c. Prove that if $w, z \in \mathbb{Z}[\sqrt{-3}]$, then $wz \in \mathbb{Z}[\sqrt{-3}]$.

d. Find all numbers w such that both w and w^{-1} are in $\mathbb{Z}[\sqrt{-3}]$.

38.23. Let $w = a + b\sqrt{-3} \in \mathbb{Z}[\sqrt{-3}]$. Define the *norm* of w to be

$$N(w) = a^2 + 3b^2.$$

Please do the following:

a. Prove that if $w, z \in \mathbb{Z}[\sqrt{-3}]$, then $N(wz) = N(w)N(z)$.

b. Find all $w \in \mathbb{Z}[\sqrt{-3}]$ with $N(w) = 0$, with $N(w) = 1$, with $N(w) = 2$, with $N(w) = 3$, and with $N(w) = 4$.

38.24. Let $w, z \in \mathbb{Z}[\sqrt{-3}]$. We say that w *divides* z provided there is a $q \in \mathbb{Z}[\sqrt{-3}]$ with $wq = z$. In this case, we call w a *factor* of z.

We call $p \in \mathbb{Z}[\sqrt{-3}]$ *irreducible* if and only if (1) $p \neq 1$ and $p \neq -1$ and (2) the only factors of p are ± 1 and $\pm p$. Irreducible elements of $\mathbb{Z}[\sqrt{-3}]$ are much like primes in $\mathbb{Z}$ (only we do not consider negative integers to be prime).

Determine which of the following elements of $\mathbb{Z}[\sqrt{-3}]$ are irreducible.

a. $1 + 2\sqrt{-3}$.
b. $2 + \sqrt{-3}$.
c. 2.
d. $1 + \sqrt{-3}$.
e. 3.
f. 7.
g. -1.
h. 0.

38.25. Let $w \in \mathbb{Z}[\sqrt{-3}]$ with $w \neq 0, \pm 1$. Prove that w can be factored into irreducible elements of $\mathbb{Z}[\sqrt{-3}]$; that is, we can find irreducible elements $p_1, p_2, \ldots, p_t$ with $w = p_1 p_2 \cdots p_t$.

38.26. We have reached the main point of this series of problems about $\mathbb{Z}[\sqrt{-3}]$. Our goal is to make a statement about unique factorization in $\mathbb{Z}[\sqrt{-3}]$.

Suppose we factor a into irreducibles as

$$a = (p_1)(p_2)(p_3)\cdots(p_t)$$

and consider the factorization

$$a = (-p_2)(-p_1)(p_3)\cdots(p_t).$$

We consider these factorizations to be the same. We do not care about the order of the factors (this is the same as for factoring positive integers into primes), and we do not care about stray factors of -1. For example, we consider the following two factorizations of 6 into irreducibles to be the same:

$$6 = (2)(\sqrt{-3})(-\sqrt{-3}) \quad \text{and}$$
$$6 = (-2)(\sqrt{-3})(\sqrt{-3}).$$

These are the same despite the fact that we use 2 in the first factorization and -2 in the second—we do not care about sign changes in the factors.

Thus the following two factorizations of 4 into irreducibles are the same:

$$4 = (2)(2) = (-2)(-2).$$

Here is the surprise and your job for this problem: Find another factorization of 4 into irreducibles.

Therefore, in the number system $\mathbb{Z}[\sqrt{-3}]$, we can factor numbers into irreducibles, but the factorization need not be unique!

Chapter 7 Self Test

1. Find integers q and r such that $23 = 5q + r$ with $0 \leq r < 5$, and calculate 23 div 5 and 23 mod 5.
2. Let a and b be positive integers. Prove that if $b|a$, then a div $b = \frac{a}{b}$.
3. Let $a \geq 2$ and b be positive integers and suppose $a|(b! + 1)$. Prove that $a > b$.
4. Let p and q be primes. Prove that $\gcd(p, q) = 1$ if and only if $p \neq q$.
5. Find integers x and y such that $100x + 57y = \gcd(100, 57)$.

6. Find the reciprocal of 57 in $\mathbb{Z}_{100}$.
7. Prove that consecutive Fibonacci numbers must be relatively prime; that is, $\gcd(F_n, F_{n+1}) = 1$ for all positive integers n.
8. Let p be a prime and let n be a positive integer. Prove that if n is not divisible by p, then $\gcd(n, n + p) = 1$.
9. Let p be a prime and let n be a positive integer. Find, in simplest possible terms, the sum of the positive divisors of p^n.
10. In $\mathbb{Z}_{101}$, please calculate the following:
    a. $55 \oplus 66$.
    b. $55 \ominus 66$.
    c. $55 \otimes 66$.
    d. $55 \oslash 66$.
11. Let n be an integer with $n \geq 2$. Prove that n is prime if and only if all nonzero elements of $\mathbb{Z}_n$ are invertible.
12. Find all integers x that satisfy the following pair of congruences:

$$x \equiv 21 \pmod{64} \quad \text{and}$$
$$x \equiv 12 \pmod{51}.$$

13. Let a and b be positive integers. Prove that $a = b$ if and only if $\gcd(a, b) = \operatorname{lcm}(a, b)$.
14. Let $n = 10^{10}$.
    a. How many positive divisors does n have?
    b. What is $\varphi(n)$?
15. Let n be a positive integer. Prove that n has an odd number of positive divisors if and only if n is a perfect square.
16. Let a, b, c be positive integers. Prove that if $a|bc$ and $\gcd(a, b) = 1$, then $a|c$.
17. Let a be a positive integer. Prove that the sum of a consecutive integers is divisible by a if and only if a is odd.

CHAPTER

8 Algebra

The word *algebra* means various things to different people. On the one hand, *algebra* is a high school subject, often studied in conjunction with trigonometry, in which students learn how to deal with variables and algebraic expressions. An important focus of such a course is solving various types of equations.

The word *algebra* also refers to a more advanced, theoretical subject. Mathematicians often call this subject *abstract algebra* to distinguish it from its more elementary cousin.

This chapter is an introduction to the ideas of abstract algebra. We are primarily concerned with algebraic systems called *groups,* but abstract algebra studies other exotic systems known as rings, fields, vector spaces, and so on.

Abstract algebra has a practical side: We combine ideas from number theory and group theory in our study of public-key cryptography.

39 Groups

Operations

The first operation we learn as children is addition. Later we move on to more complex operations such as division, and in this book, we have investigated more exotic examples, including $\wedge$ and $\vee$ defined on the set {TRUE, FALSE}, $\oplus$ and $\otimes$ defined on $\mathbb{Z}_n$, and $\circ$ defined on S_n.

In this section, we take a broader look at operations defined on sets and their algebraic properties. First, we present a formal definition of *operation.*

Definition 39.1 **(Operation)** Let A be a set. An *operation* on A is a function whose domain contains $A \times A$.

Recall that $A \times A$ is the set of all ordered pairs (two-element lists) whose entries are in A. Thus an operation is a function whose input is a pair of elements from A.

Notice that we write $f(a, b)$, although it would be more proper to write $f[(a, b)]$ because we are applying the function f to the object (a, b). The extra brackets, however, tend to be a distraction. Alternatively, we can think of a function defined on $A \times A$ as a function of two variables.

For example, consider the following function $f : \mathbb{Z} \times \mathbb{Z} \to \mathbb{Z}$ defined by

$$f(a, b) = |a - b|.$$

In words, $f(a, b)$ gives the distance between a and b on a number line.

Although the notation $f(a, b)$ is formally correct, we rarely write the operation symbol in front of the two elements on which we are operating. Rather, we write a symbol for the operation between the two elements of the list. Instead of $f(a, b)$, we write $a\ f\ b$.

Furthermore, we usually do not use a letter to denote an operation. Instead, we use a special symbol such as $+$ or $\otimes$ or $\circ$. The symbols $+$ and $\times$ have preset meanings. A common symbol for a generic operation is $*$. Thus, instead of writing $f(a, b) = |a - b|$, we could write $a * b = |a - b|$.

Example 39.2 Which of the following are operations on $\mathbb{N}$: $+$, $-$, $\times$, and $\div$?

Solution. Certainly addition $+$ is an operation defined on $\mathbb{N}$. Although $+$ is more broadly defined on any two rational (or even real or complex numbers), it is a function whose domain includes any pair of natural numbers. Likewise multiplication $\times$ is an operation on $\mathbb{N}$.

Furthermore, $-$ is an operation defined on $\mathbb{N}$. Note, however, that the result of $-$ might not be an element of $\mathbb{N}$. For example, $3, 7 \in \mathbb{N}$, but $3 - 7 \notin \mathbb{N}$.

Finally, division $\div$ does not define an operation on $\mathbb{N}$ because division by zero is undefined. However, $\div$ is an operation defined on the *positive* integers.

Properties of Operations

Operations may satisfy various properties. For example, an operation $*$ on a set A is said to be *commutative* on A if $a * b = b * a$ for all $a, b \in A$. Addition of integers is commutative, but subtraction is not. Here we present formal definitions of some important properties of operations.

Definition 39.3 **(Commutative property)** Let $*$ be an operation on a set A. We say that $*$ is *commutative* on A provided

$$\forall a, b \in A,\ a * b = b * a.$$

Definition 39.4 **(Closure property)** Let $*$ be an operation on a set A. We say that $*$ is *closed* on A provided

$$\forall a, b \in A,\ a * b \in A.$$

Let $*$ be an operation defined on a set A. Note that Definition 39.1 does not require that the result of $*$ be an element of the set A. For example, $-$ is an operation defined on $\mathbb{N}$, but the result of subtracting two natural numbers might not be a natural number. Subtraction is not closed on $\mathbb{N}$; it is closed on $\mathbb{Z}$.

Definition 39.5 **(Associative property)** Let $*$ be an operation on a set A. We say that $*$ is *associative* on A provided

$$\forall a, b, c \in A,\ (a * b) * c = a * (b * c).$$

For example, the operations $+$ and $\times$ on $\mathbb{Z}$ are associative, but $-$ is not. For example, $(3 - 4) - 7 = -8$, but $3 - (4 - 7) = 6$.

Definition 39.6 **(Identity element)** Let $*$ be an operation on a set A. An element $e \in A$ is called an *identity element* (or *identity* for short) for $*$ provided

$$\forall a \in A,\ a * e = e * a = a.$$

For example, 0 is an identity element for $+$, and 1 is an identity element for $\times$. An identity element for $\circ$ on S_n is the identity permutation ι.

Identity elements must work on both sides of the operation.

Not all operations have identity elements. For example, subtraction of integers does not have an identity element. It is true that $a - 0 = a$ for all integers a, so 0 partially satisfies the requirements of being an identity element for subtraction. However, for 0 to merit the name *identity element for subtraction,* we would need that $0 - a = a$ for all integers, and this is false. Subtraction does not have an identity element.

Is it possible for an operation on a set to have more than one identity element?

Proposition 39.7 Let $*$ be an operation defined on a set A. Then $*$ can have at most one identity element.

Proof. We use Proof Template 14 for proving uniqueness.

Suppose, for the sake of contradiction, that there are two identity elements, e and e', in A with $e \neq e'$.

Consider $e * e'$. On the one hand, since e is an identity element, $e * e' = e'$. On the other hand, since e' is an identity element, $e * e' = e$. Thus we have shown $e' = e * e' = e$, a contradiction to $e \neq e'$.$\Rightarrow\Leftarrow$ ■

Definition 39.8 **(Inverses)** Let $*$ be an operation on a set A and suppose that A has an identity element e. Let $a \in A$. We call element b an *inverse* of a provided $a * b = b * a = e$.

For example, consider the operation $+$ on the integers. The identity element for $+$ is 0. Every integer a has an inverse: The inverse of a is simply $-a$ because $a + (-a) = (-a) + a = 0$.

Now consider the operation $\times$ on the rational numbers. The identity element for multiplication is 1. Most, but not all, rational numbers have inverses. If $x \in \mathbb{Q}$, then $\frac{1}{x}$ is x's inverse, unless, of course, $x = 0$.

Notice that we require that an element's inverse work on both sides of the operation.

Must inverses be unique? Consider the following example.

Example 39.9 Consider the operation $*$ defined on the set $\{e, a, b, c\}$ given in the following chart.

$*$	e	a	b	c
e	e	a	b	c
a	a	a	e	e
b	b	e	b	e
c	c	e	e	c

Notice that e is an identity element. Notice further that both b and c are inverses of a because

$$a * b = b * a = e \qquad \text{and} \qquad a * c = c * a = e.$$

Groups

Example 39.9 is strange. We know that if an operation has an identity element, it must be unique. And it would be quite natural for us to expect "the" inverse of an element be unique. However, we cannot say *the* inverse because we saw that an element might have more than one inverse. For most operations we encounter, elements have at most one inverse. Some examples:

- If $a \in \mathbb{Z}$, there is exactly one integer b such that $a + b = 0$.
- If $a \in \mathbb{Q}$, there is at most one rational number b such that $ab = 1$.
- If $\pi \in S_n$, there is exactly one permutation $\sigma \in S_n$ such that $\pi \circ \sigma = \sigma \circ \pi = \iota$ (see Exercise 26.15).

Most operations we encounter in mathematics are associative, and, as we shall show, associativity implies uniqueness of inverses. Note that the operation in Example 39.9 is not associative (see Exercise 39.4).

This brings us to the notion of a *group*. A group is a common generalization of the following operations and sets:

Mathspeak!

The word *group* is a technical mathematical term. Its meaning in mathematics is far removed from its standard English usage.

- $+$ on $\mathbb{Z}$,
- $\times$ on the positive rationals,
- $\oplus$ on $\mathbb{Z}_n$,
- $\circ$ on S_n, and
- $\circ$ on symmetries of a geometric object.

In each of these cases, we have an operation that behaves nicely; for example, in all these cases, elements have unique inverses. Here is the definition of a *group*:

Definition 39.10 **(Group)** Let $*$ be an operation defined on a set G. We call the pair $(G, *)$ a *group* provided:

(1) The set G is closed under $*$; that is, $\forall g, h \in G, g * h \in G$.
(2) The operation $*$ is associative; that is, $\forall g, h, k \in G, (g * h) * k = g * (h * k)$.

(3) There is an identity element $e \in G$ for $*$; that is, $\exists e \in G, \forall g \in G, g * e = e * g = g$.

(4) For every element $g \in G$, there is an inverse element $h \in G$; that is, $\forall g \in G, \exists h \in G, g * h = h * g = e$.

Notice that a group is a *pair* of objects: a set G and an operation $*$. For example, $(\mathbb{Z}, +)$ is a group. We can pronounce the symbols $(\mathbb{Z}, +)$ aloud as "integers with addition."

Sometimes, however, the operation under consideration is obvious. For example, $(S_n, \circ)$ is a group (we proved this in Proposition 26.4). The only operation on S_n we consider in this book (and virtually the only operation most mathematicians consider on S_n) is composition $\circ$. Thus we may refer to S_n as a group, where we understand that this is shorthand for the pair $(S_n, \circ)$.

Similarly, if we write, "Let G be a group...," we understand that G has a group operation, which, in this book, is denoted by $*$. Please be aware that the symbol $*$ is not customary as the generic group operation. Mathematicians use $\cdot$ or no symbol at all to denote a general group operation. This is the same convention we use for multiplication. To avoid confusion, in this book we use $*$ or $\star$ as the operation symbol of a generic group.

Mathspeak!

Our use of the word *Abelian* honors Niels Henrik Abel, a Norwegian mathematician (1802–1829). Abelian groups are sometimes called *additive* or *commutative*.

The group operation $*$ need not be commutative. For example, we saw in Section 26 that $\circ$ is not a commutative operation on S_n. Groups in which the operation is commutative have a special name.

Definition 39.11 **(Abelian groups)** Let $(G, *)$ be a group. We call this group *Abelian* provided $*$ is a commutative operation on G (i.e., $\forall g, h \in G, g * h = h * g$).

For example, $(\mathbb{Z}, +)$ and $(\mathbb{Z}_{10}, \oplus)$ are Abelian, but $(S_n, \circ)$ is not.

In Example 39.9, we considered an operation in which inverses are not unique. This does not happen in groups; in a group, every element has an inverse, and that inverse is unique.

Proposition 39.12 Let $(G, *)$ be a group. Every element of G has a unique inverse in G.

Proof. We know, by definition, that every element in G has an inverse. At issue is whether or not it is possible for an element of G to have two (or more) inverses.

Suppose, for the sake of contradiction, that $g \in G$ has two (or more) distinct inverses. Let $h, k \in G$ be inverses of g with $h \neq k$. This means

$$g * h = h * g = g * k = k * g = e$$

where $e \in G$ is the identity element for $*$. By the associative property,

$$h * (g * k) = (h * g) * k.$$

Notice that we are using the facts that k and h are inverses of g and the fact that e is an identity element.

Furthermore,

$$h * (g * k) = h * e = h \quad \text{and}$$
$$(h * g) * k = e * k = k.$$

Hence $h = k$, contradicting the fact that $h \neq k$.⇒⇐ ■

The inverse of g in a group $(G, *)$ is denoted g^{-1}.

Proposition 39.12 establishes that if g is an element of a group, then g has a unique inverse. We may speak of *the* inverse of g. The notation for g's inverse is g^{-1}. The superscript -1 notation is in harmony with taking reciprocals in the group of positive rational numbers (with multiplication), or inverse permutations in S_n. It is not a good notation for $(\mathbb{Z}, +)$.

Examples

The concept of a group is rather abstract. It is helpful to have several specific examples. Some of the examples that we present here we have considered before; others are new.

- $(\mathbb{Z}, +)$: Integers with addition is a group.
- $(\mathbb{Q}, +)$: Rational numbers with addition is a group.
- $(\mathbb{Q}, \times)$: Rational numbers with multiplication is not a group. It nearly satisfies Definition 39.10, except that $0 \in \mathbb{Q}$ does not have an inverse. We can repair this example in two ways. First, we can consider only the positive rational numbers: $(\mathbb{Q}^+, \times)$ is a group.

  Another way to repair this example is simply to eliminate the number 0. $(\mathbb{Q} - \{0\}, \times)$ is a group.
- $(S_n, \circ)$ is a group called the *symmetric group*.
- Let A_n be the set of all even permutations in S_n. Then $(A_n, \circ)$ is a group called the *alternating group*.
- The set of symmetries of a square with $\circ$ is a group. This group is called a *dihedral group*.

  In general, if n is an integer with $n \geq 3$, the dihedral group D_{2n} is the set of symmetries of a regular n-gon with the operation $\circ$ (see Section 27).
- $(\mathbb{Z}_n, \oplus)$ is a group for all positive integers n.
- Let $G = \{(0, 0), (0, 1), (1, 0), (1, 1)\}$. Define an operation $*$ on G by

  $$(a, b) * (c, d) = (a \oplus c, b \oplus d)$$

  where $\oplus$ is addition mod 2 (i.e., $\oplus$ in $\mathbb{Z}_2$).

  The $*$ table for this group is this:

*	(0, 0)	(0, 1)	(1, 0)	(1, 1)
(0, 0)	(0, 0)	(0, 1)	(1, 0)	(1, 1)
(0, 1)	(0, 1)	(0, 0)	(1, 1)	(1, 0)
(1, 0)	(1, 0)	(1, 1)	(0, 0)	(0, 1)
(1, 1)	(1, 1)	(1, 0)	(0, 1)	(0, 0)

This group is known as the *Klein 4-group*. Notice that $(0, 0)$ is the identity element and every element is its own inverse.

Δ stands for *symmetric difference* of sets.

- Let A be a set. Then $(2^A, \Delta)$ is a group (Exercise 39.9).
- $(\mathbb{Z}_{10}, \otimes)$ is not a group. The problem is similar to $(\mathbb{Q}, \times)$: Zero does not have an inverse. The remedy in this case is a bit more complicated. We cannot just throw away the element 0. Notice that in $(\mathbb{Z}_{10} - \{0\}, \otimes)$ the operation $\otimes$ is no longer closed. For example, $2, 5 \in \mathbb{Z}_{10} - \{0\}$, but $2 \otimes 5 = 0 \notin \mathbb{Z}_{10} - \{0\}$. Also, elements 2 and 5 do not have inverses.

  In addition to eliminating the element 0, we can discard those elements that do not have inverses. By Theorem 36.14, we are left with the elements in $\mathbb{Z}_{10}$ that are relatively prime to 10; we are left with $\{1, 3, 7, 9\}$.

  These four elements together with $\otimes$ form a group. The $\otimes$ table for them is this:

$\otimes$	1	3	7	9
1	1	3	7	9
3	3	9	1	7
7	7	1	9	3
9	9	7	3	1

The last example is worth exploring in a bit more depth. We observed that $(\mathbb{Z}_{10}, \otimes)$ is not a group and then we eliminated from $\mathbb{Z}_{10}$ those elements that do not have an inverse. We saw, in Theorem 36.14, that the invertible elements of $(\mathbb{Z}_n, \otimes)$ are precisely those that are relatively prime to n.

Definition 39.13 ($\mathbb{Z}_n^*$) Let n be a positive integer. We define

$$\mathbb{Z}_n^* = \{a \in \mathbb{Z}_n : \gcd(a, n) = 1\}.$$

Example 39.14 Consider $\mathbb{Z}_{14}^*$. The invertible elements in $\mathbb{Z}_{14}^*$ (i.e., the elements relatively prime to 14) are 1, 3, 5, 9, 11, and 13. Thus

$$\mathbb{Z}_{14}^* = \{1, 3, 5, 9, 11, 13\}.$$

The $\otimes$ table for $\mathbb{Z}_{14}^*$ is this:

$\otimes$	1	3	5	9	11	13
1	1	3	5	9	11	13
3	3	9	1	13	5	11
5	5	1	11	3	13	9
9	9	13	3	11	1	5
11	11	5	13	1	9	3
13	13	11	9	5	3	1

The inverses of the elements in $\mathbb{Z}_{14}^*$ can be found in this table. We have

$$1^{-1} = 1 \quad 3^{-1} = 5 \quad 5^{-1} = 3$$
$$9^{-1} = 11 \quad 11^{-1} = 9 \quad 13^{-1} = 13.$$

Proposition 39.15 Let n be a positive integer. Then $(\mathbb{Z}_n^*, \otimes)$ is a group.

To prove that $(G, *)$ is a group, we need to check Definition 39.10. We summarize this in Proof Template 23.

Proof Template 23 Proving $(G, *)$ is a group.

To prove that $(G, *)$ is a group:

- Prove that G is closed under $*$: Let $g, h \in G$. ... Therefore $g * h \in G$.
- Prove that $*$ is associative: Let $g, h, k \in G$. ... Therefore $g * (h * k) = (g * h) * k$.
- Prove that G contains an identity element for $*$: Let e be some specific element of G. Let $g \in G$ be arbitrary. ... Therefore $g * e = e * g = g$.
- Prove that every element of G has a $*$-inverse in G: Let $g \in G$. Construct an element h such that $g * h = h * g = e$.

Therefore $(G, *)$ is a group. ■

Proof (of Proposition 39.15)

First, we prove that $\mathbb{Z}_n^*$ is closed under $\otimes$. Let $a, b \in \mathbb{Z}_n^*$. We need to prove that $a \otimes b \in \mathbb{Z}_n^*$. Recall that $a \otimes b = (ab) \bmod n$.

We know that $a, b \in \mathbb{Z}_n^*$. This means that a and b are relatively prime to n. Therefore, by Corollary 35.9, we can find integers x, y, z, w such that

$$ax + ny = 1 \qquad \text{and} \qquad bw + nz = 1.$$

Multiplying these equations gives

$$\begin{aligned} 1 = (ax + ny)(bw + nz) &= (ax)(bw) + (ax)(nz) + (ny)(bw) + (ny)(nz) \\ &= (ab)(wx) + (n)[axz + ybw + ynz] \\ &= (ab)(X) + (n)(Y) \end{aligned}$$

for some integers X and Y. Therefore ab is relatively prime to n. By Exercise 35.12, we may increase or decrease ab by a multiple of n, and the result is still relatively prime to n. Therefore $\gcd(a \otimes b, n) = 1$, and so $a \otimes b \in \mathbb{Z}_n^*$.

Second, we show that $\otimes$ is associative. This was proved in Proposition 36.4.

Third, we show that $(\mathbb{Z}_n^*, \otimes)$ has an identity element. Clearly $\gcd(1, n) = 1$, so $1 \in \mathbb{Z}_n^*$. Furthermore, for any $a \in \mathbb{Z}_n^*$, we have

$$a \otimes 1 = 1 \otimes a = (a \cdot 1) \bmod n = a$$

and therefore 1 is an identity element for $\otimes$.

Fourth, we show that every element in $\mathbb{Z}_n^*$ has an inverse in $\mathbb{Z}_n^*$. Let $a \in \mathbb{Z}_n^*$. We know, by Theorem 36.14, that a has an inverse $a^{-1} \in \mathbb{Z}_n$. The only issue is: Is a^{-1} in $\mathbb{Z}_n^*$? Since a^{-1} is itself invertible, by Theorem 36.14 again, a^{-1} is relatively prime to n, and so $a^{-1} \in \mathbb{Z}_n^*$.

Therefore $(\mathbb{Z}_n^*, \otimes)$ is a group. ■

How many elements are in $\mathbb{Z}_n^*$? This is a problem we have already solved (see Exercises 38.14–17). We recall and record the answer here for later reference.

Proposition 39.16 Let n be an integer with $n \geq 2$. Then

$$\left|\mathbb{Z}_n^*\right| = \varphi(n)$$

where $\varphi(n)$ is Euler's totient.

Recap

We began with a formal description of an operation on a set and listed various properties an operation might exhibit. We then focused on four particular properties: closure, associativity, identity, and inverses. We developed the concept of a group and discussed several examples.

39 Exercises

39.1. Let $(G, *)$ be a group with $G = \{a, b, c\}$. Here is an *incomplete* operation table for $*$:

$*$	a	b	c
a	a	b	c
b	?	?	?
c	?	?	?

Find the missing entries.

39.2. Explain why $(\mathbb{Z}_5, \ominus)$ is not a group. Give at least two reasons.

39.3. Consider the operations $\wedge$, $\vee$, and $\veebar$ defined on the set {TRUE, FALSE}. Which of the various properties of operations do these operations exhibit? (Consider the properties closure, commutativity, associativity, identity, and inverses.)

Which (if any) of these operations define a group on {TRUE, FALSE}?

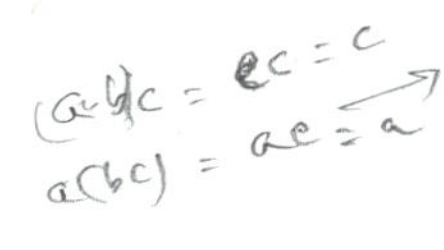

39.4. Show that the operation in Example 39.9 is not associative.

39.5. Prove that if $(G, *)$ is a group and $g \in G$, then $\left(g^{-1}\right)^{-1} = g$.

39.6. Prove that if $(G, *)$ is a group, then $e^{-1} = e$.

39.7. We saw that $(\mathbb{Q}^+, \times)$ is a group (positive rational numbers with multiplication). Is $(\mathbb{Q}^-, \times)$ (negative rationals with multiplication) a group? Prove your answer.

39.8. *This problem is only for those who have studied linear algebra.* Let G be the set of 2×2 real matrices $\begin{bmatrix} a & b \\ c & d \end{bmatrix}$ with $ad - bc \neq 0$. Prove that G, together with the operation of matrix multiplication, is a group.

Note that the set of all 2×2 real matrices do not form a group because some matrices, such as $\begin{bmatrix} 1 & 1 \\ 1 & 1 \end{bmatrix}$, are not invertible. We discard the noninvertible matrices, and what remains is a group. This is analogous to our transformation from $\mathbb{Z}_n$ to $\mathbb{Z}_n^*$.

39.9. Let A be a set. Prove that $(2^A, \Delta)$ is a group.

39.10. Let G be a group and let $a \in G$. Define a function $f : G \to G$ by $f(g) = a * g$. Prove that f is a permutation of G.

39.11. Let G be a group. Define a function $f : G \to G$ by $f(g) = g^{-1}$. Prove that f is a permutation of G.

39.12. Let $*$ be an operation on a finite set G. Form the $*$ operation table. Prove that if $(G, *)$ is a group, then in every row and in every column, each element of G appears exactly once.

Show that the converse of this assertion is false; that is, construct an operation $*$ on a finite set G such that in every row and in every column, each element of G appears exactly once, but $(G, *)$ is not a group.

39.13. Let $(G, *)$ be a group and let $g, h \in G$. Prove that $(g * h)^{-1} = h^{-1} * g^{-1}$.

39.14. Let $(G, *)$ be a group. Prove that G is Abelian if and only if $(g * h)^{-1} = g^{-1} * h^{-1}$ for all $g, h \in G$.

39.15. Let $(G, *)$ be a group. Define a new operation $\star$ on G by

$$g \star h = h * g.$$

Prove that $(G, \star)$ is a group.

39.16. Let $(G, *)$ be a group. Notice that $e^{-1} = e$. Prove that if $|G|$ is finite and even, then there is another element $g \in G$ with $g^{-1} = g$.

Give an example of a finite group with five or more elements in which no element (other than the identity) is its own inverse.

39.17. Let $*$ be an operation defined on a set A. We say that $*$ has *the left cancellation property* on A provided

$$\forall a, b, c \in A,\ a * b = a * c \Longrightarrow b = c.$$

a. Prove that if $(G, *)$ is a group, then $*$ has the left cancellation property on G.

b. Give an example of a set A with an operation $*$ that has the left cancellation property but is not a group.

39.18. *Reverse Polish Notation*. We remarked at the beginning of this section that mathematicians usually put the operation symbol between the two objects (operands) to which the operation applies. There is, however, an alternative notation in which the operation symbol comes *after* the two operands. This notation is called *reverse Polish notation* (RPN for short) or *postfix notation*. For example, in RPN, instead of writing $2 + 3$, we write $2\ 3\ +$.

Consider the RPN expression $2, 3, 4, +, \times$. There are two operation symbols, and each operates on the two operands to its left. What do the $+$ and $\times$ operate on? The $+$ sign immediately follows $3, 4$, so it means to add those two numbers. This reduces the problem to $2, 7, \times$. Now the $\times$ operates on the 2 and the 7 to give 14. Overall, the expression $2, 3, 4, +, \times$ in standard notation is $2 \times (3 + 4)$.

On the other hand, the RPN expression $2, 3, +, 4, \times$ stands for $(2 + 3) \times 4$, which evaluates to 20.

Evaluate each of the following.

a. $1, 1, 1, 1, +, +, +$.

b. $1, 2, 3, 4, \times, +, +$.

c. $1, 2, +, 3, 4, \times, +$.
d. $1, 2, +, 3, 4, +, \times$.
e. $1, 2, +, 3, +, 4, \times$.

39.19. *RPN continued.* Convert the following expressions from standard notation to RPN. Do not evaluate.
a. $(2 + 3) \times (4 + 5)$.
b. $(2 + (3 \times 4)) + 5$.
c. $((2 + 3) \times 4) + 5$.

39.20. *RPN continued.* Suppose we have a list of numbers and operations ($+$ and $\times$) symbols representing an RPN expression. Some such expressions are invalid, such as $2, +, +$ or $+, 3, \times, 4, 4$ or $2, 3, +, 4$.

State and prove a theorem describing when a list of numbers and operation symbols forms a valid RPN expression.

39.21. *RPN continued.* Write a computer program to evaluate RPN expressions.

40 Group Isomorphism

The Same?

What does it mean for two groups to be *the same*?

A simple answer to this question is that $(G, *) = (H, \star)$ provided $G = H$ and $* = \star$ (i.e., $*$ and $\star$ are the same operation). This would certainly be a proper definition for two groups to be *equal,* but we asked a vaguer question.

Consider the following three groups: $(\mathbb{Z}_4, \oplus)$, $(\mathbb{Z}_5^*, \otimes)$, and the Klein 4-group. Their operation tables are as follows:

$\oplus$	0	1	2	3
0	0	1	2	3
1	1	2	3	0
2	2	3	0	1
3	3	0	1	2

$\otimes$	1	2	3	4
1	1	2	3	4
2	2	4	1	3
3	3	1	4	2
4	4	3	2	1

$*$	(0, 0)	(0, 1)	(1, 0)	(1, 1)
(0, 0)	(0, 0)	(0, 1)	(1, 0)	(1, 1)
(0, 1)	(0, 1)	(0, 0)	(1, 1)	(1, 0)
(1, 0)	(1, 0)	(1, 1)	(0, 0)	(0, 1)
(1, 1)	(1, 1)	(1, 0)	(0, 1)	(0, 0)

These three groups are different because they are defined on different sets. However, two of them are, in essence, the same group. Look carefully at the three operation tables and try to distinguish one from the other two.

The *main diagonal* of these tables is the diagonal running from the upper left to the lower right.

The Klein 4-group (right) has a property that the other two don't share. Notice that every element in the Klein 4-group is its own inverse; you can see this by noting the identities running down the main diagonal. However, in the other two groups, there are elements that are not their own inverses. For example, 1 and 3 are inverses of one another in $(\mathbb{Z}_4, \oplus)$, and 2 and 3 are inverses of one another in $(\mathbb{Z}_5^*, \otimes)$. Other than the identity, only 2 is its own inverse in $(\mathbb{Z}_4, \oplus)$ and only 4 is its own inverse in $(\mathbb{Z}_5^*, \otimes)$.

We can superimpose the operation tables for the two groups $(\mathbb{Z}_4, \oplus)$ and $(\mathbb{Z}_5^*, \otimes)$ on top of one another so they look the same. We pair the identity elements in the two groups with one another. We also pair the other elements ($2 \in \mathbb{Z}_4$ and

$4 \in \mathbb{Z}_5^*$) that are their own inverses. Then we have a choice for the other two pairs of elements. Here is a pairing:

$(\mathbb{Z}_4, \oplus)$		$(\mathbb{Z}_5^*, \otimes)$
0	$\longleftrightarrow$	1
1	$\longleftrightarrow$	2
2	$\longleftrightarrow$	4
3	$\longleftrightarrow$	3

Next we superimpose their operation tables.

$\oplus$	$\otimes$	0	1	1	2	2	4	3	3
0	1	0	1	1	2	2	4	3	3
1	2	1	2	2	4	3	3	0	1
2	4	2	4	3	3	0	1	1	2
3	3	3	3	0	1	1	2	2	4

The tables for both $(\mathbb{Z}_4, \oplus)$ and $(\mathbb{Z}_5^*, \otimes)$ are correct [although the table for $(\mathbb{Z}_5^*, \otimes)$ is twisted around a bit because we swapped the rows and columns for elements 3 and 4]. The important thing to note is that every element of $(\mathbb{Z}_4, \oplus)$ (in black) sits next to its mate from $(\mathbb{Z}_5^*, \otimes)$ (color).

More formally, let $f : \mathbb{Z}_4 \to \mathbb{Z}_5^*$ defined by

$$f(0) = 1 \qquad f(2) = 4$$
$$f(1) = 2 \qquad f(3) = 3.$$

Clearly f is a bijection and

$$f(x \oplus y) = f(x) \otimes f(y)$$

where $\oplus$ is mod 4 addition and $\otimes$ is mod 5 multiplication.

In other words, if we rename the elements of $\mathbb{Z}_4$ using the rule f, we get elements in $\mathbb{Z}_5^*$. The operation $\oplus$ for $\mathbb{Z}_4$ and the operation $\otimes$ for $\mathbb{Z}_5^*$ give the exact same results once we rename the elements.

To put it another way, imagine we made a group of four elements $\{e, a, b, c\}$ with the following operation table:

$*$	e	a	b	c
e	e	a	b	c
a	a	b	c	e
b	b	c	e	a
c	c	e	a	b

We then tell you that, in reality, these four elements $\{e, a, b, c\}$ are either (1) aliases for elements of $\mathbb{Z}_4$ with operation $\oplus$ or (2) aliases for elements of $\mathbb{Z}_5^*$ with operation $\otimes$. Would you be able to distinguish case (1) from (2)? No. The relabeling f shows that either group fits the pattern in this table. The groups $(\mathbb{Z}_4, \oplus)$ and $(\mathbb{Z}_5^*, \otimes)$ are, in essence, the same. They are called *isomorphic*.

Definition 40.1 **(Isomorphism of groups)** Let $(G, *)$ and $(H, \star)$ be groups. A function $f : G \to H$ is called a *(group) isomorphism* provided f is one-to-one and onto and satisfies

$$\forall g, h \in G, \ \ f(g * h) = f(g) \star f(h).$$

When there is an isomorphism from G to H, we say *G is isomorphic to H* and we write $G \cong H$.

The is-isomorphic-to relation for groups is an equivalence relation (see Section 13); that is,

- for any group G, $G \cong G$,
- for any two groups G and H, if $G \cong H$, then $H \cong G$, and
- for any three groups G, H, and K, if $G \cong H$ and $H \cong K$, then $G \cong K$.

Cyclic Groups

The groups $(\mathbb{Z}_4, \oplus)$ and $(\mathbb{Z}_5^*, \otimes)$ have everything in common except for the names of their elements. Element 1 of $(\mathbb{Z}_4, \oplus)$ has a special feature; it generates all the elements of the group $(\mathbb{Z}_4, \oplus)$ as follows:

$$\begin{aligned} 1 &= 1 \\ 1 \oplus 1 &= 2 \\ 1 \oplus 1 \oplus 1 &= 3 \\ 1 \oplus 1 \oplus 1 \oplus 1 &= 0. \end{aligned}$$

The element 3 also generates all the elements of $(\mathbb{Z}_4, \oplus)$; please do these calculations yourself.

Of course, because $(\mathbb{Z}_5^*, \otimes)$ is isomorphic to $(\mathbb{Z}_4, \oplus)$, it, too, must have a generator. Since $1 \in \mathbb{Z}_4$ corresponds (according to the isomorphism we found previously) to $2 \in \mathbb{Z}_5^*$, we calculate

$$\begin{aligned} 2 &= 2 \\ 2 \otimes 2 &= 4 \\ 2 \otimes 2 \otimes 2 &= 3 \\ 2 \otimes 2 \otimes 2 \otimes 2 &= 1. \end{aligned}$$

Thus element $2 \in \mathbb{Z}_5^*$ generates the group.

The Klein 4-group does not have an element that generates the entire group. In this group, every element g has the property that $g * g = e = (0, 0)$, so there is no way that $g, g * g, g * g * g, \ldots$ can generate all the elements of the group.

By this pattern, there is no element of $\mathbb{Z}$ that generates $(\mathbb{Z}, +)$. However, we have not formally defined *generator,* so we are going to extend the rules in this case. The element 1 generates all the positive elements of $\mathbb{Z}$: $1, 1+1, 1+1+1$, and so forth. By this system, we never get 0 or the negative integers. If, however, we allow 1's inverse, -1, to participate in the generation process, then we can get 0 [as $1+(-1)$] and all the negative numbers $-1, (-1)+(-1), (-1)+(-1)+(-1)$, etc.

Definition 40.2 **(Generator, cyclic group)** Let $(G, *)$ be a group. An element $g \in G$ is called a *generator* for G provided every element of G can be expressed just in terms of g and g^{-1} using the operation $*$.

If a group contains a generator, it is called *cyclic*.

The special provision for g^{-1} is necessary only for groups with infinitely many elements. If $(G, *)$ is a finite group and $g \in G$, then we can always find a way to write $g^{-1} = g * g * \cdots * g$.

Proposition 40.3 Let $(G, *)$ be a finite group and let $g \in G$. Then, for some positive integer n, we have

$$g^{-1} = \underbrace{g * g * \cdots * g}_{n \text{ times}}.$$

It is inconvenient to write

$$\underbrace{g * g * \cdots * g}_{n \text{ times}}.$$

Instead, we can write g^n; this notation means we $*$ together n copies of g.

Proof. Let $(G, *)$ be a finite group and let $g \in G$. Consider the sequence

$$g^1 = g, \quad g^2, \quad g^3, \quad g^4, \quad \ldots.$$

Since the group is finite, this sequence must, at some point, repeat itself. Suppose the first repeat is at $g^a = g^b$ where $a < b$.

Claim: $a = 1$.

Suppose, for the sake of contradiction, $a > 1$. Then we have

$$\begin{aligned} g^a &= g^b \\ \underbrace{g * g * \cdots * g}_{a \text{ times}} &= \underbrace{g * g * \cdots * g}_{b \text{ times}}. \end{aligned}$$

We operate on the left by g^{-1} to get

$$\begin{aligned} g^{-1} * g^a &= g^{-1} * g^b \\ g^{-1} * (\underbrace{g * g * \cdots * g}_{a \text{ times}}) &= g^{-1} * (\underbrace{g * g * \cdots * g}_{b \text{ times}}) \\ (g^{-1} * g) * (\underbrace{g * g * \cdots * g}_{a-1 \text{ times}}) &= (g^{-1} * g) * (\underbrace{g * g * \cdots * g}_{b-1 \text{ times}}) \\ e * (\underbrace{g * g * \cdots * g}_{a-1 \text{ times}}) &= e * (\underbrace{g * g * \cdots * g}_{b-1 \text{ times}}) \\ \underbrace{g * g * \cdots * g}_{a-1 \text{ times}} &= \underbrace{g * g * \cdots * g}_{b-1 \text{ times}} \\ g^{a-1} &= g^{b-1} \end{aligned}$$

which shows that the first repeat is before $g^a = g^b$, a contradiction. Therefore $a = 1$.

We now know that if we stop at the first repeat, the sequence is

$$g^1, \quad g^2, \quad g^3, \quad \ldots, \quad g^b = g.$$

Notice that since $g = g^b$, if we operate on the left by g^{-1}, we get $e = g^{b-1}$.

It may be the case that $b = 2$, so $g^2 = g$. In this case, $g = e$ and so $g^1 = g^{-1}$, proving the result.

Otherwise, $b > 2$. In this case, we can write

$$e = g^{b-1} = g^{b-2} * g$$

and therefore $g^{b-2} = g^{-1}$. ■

Theorem 40.4 Let $(G, *)$ be a finite cyclic group. Then $(G, *)$ is isomorphic to $(\mathbb{Z}_n, \oplus)$ where $n = |G|$.

Proof. Let $(G, *)$ be a finite cyclic group. Suppose $|G| = n$ and let $g \in G$ be a generator. We claim that $(G, *) \cong (\mathbb{Z}_n, \oplus)$. To this end, we define $f : \mathbb{Z}_n \to G$ by

$$f(k) = g^k$$

where g^k means $g * g * \cdots * g$ (with k copies of g and $g^0 = e$).

To prove that f is an isomorphism, we must show that f is one-to-one and onto and that $f(j \oplus k) = f(j) * f(k)$.

- f is one-to-one.

  Suppose $f(j) = f(k)$. This means that $g^j = g^k$. We want to prove that $j = k$. Suppose that $j \neq k$. Without loss of generality, $0 \leq j < k < n$ (with $<$ in the usual sense of integers). We can $*$ the equation $g^j = g^k$ on the left by $(g^{-1})^j$ to get

  $$(g^{-1})^j * g^j = (g^{-1})^j * g^k$$
  $$e = g^{k-j}.$$

  Since $k - j < n$, this means that the sequence

  $$g, \quad g^2, \quad g^3, \quad \ldots$$

  repeats after $k - j$ steps, and therefore g does not generate the entire group (but only $k - j$ of its elements). However, g is a generator.$\Rightarrow\Leftarrow$ Therefore f is one-to-one.
- f is onto.

  Let $h \in G$. We must find $k \in \mathbb{Z}_n$ such that $f(k) = h$. We know that the sequence

  $$e = g^0, \quad g = g^1, \quad g^2, \quad g^3, \quad \ldots$$

  must contain all elements of G. Thus h is somewhere on this list—say, at position k (i.e., $h = g^k$). Therefore $f(k) = h$ as required. Hence f is onto.

In this calculation, tn might be zero (in which case $g^0 = e$ is fine) or tn might be negative. The meaning of, say, g^{-n} is simply $(g^{-1})^n = (g^n)^{-1}$.

- For all $j, k \in \mathbb{Z}_n$, we have $f(j \oplus k) = f(j) * f(k)$.

  Recall that $j \oplus k = (j + k) \bmod n = j + k + tn$ for some integer t. Therefore

$$\begin{aligned} f(j \oplus k) &= g^{j+k+tn} = g^j * g^k * g^{tn} \\ &= g^j * g^k * g^{tn} = g^j * g^k * (g^n)^t \\ &= g^j * g^k * e^t = g^j * g^k \\ &= f(j) * f(k) \end{aligned}$$

  as required.

Therefore $f : \mathbb{Z}_n \to G$ is an isomorphism, and so $(\mathbb{Z}_n, \oplus) \cong (G, *)$. ■

Recap

In this section we discussed the notion of group isomorphism. Roughly speaking, two groups are isomorphic if they are exactly the same except for the names of their elements. We also discussed the concepts of group generators and cyclic groups.

40 Exercises

40.1. Find an isomorphism from $(\mathbb{Z}_{10}, \oplus)$ to $(\mathbb{Z}_{11}^*, \otimes)$.

40.2. Let $(G, *)$ be the following group. The set G is $\{0, 1\} \times \{0, 1, 2\}$; that is,

$$G = \{(0, 0), (0, 1), (0, 2), (1, 0), (1, 1), (1, 2)\}.$$

The operation $*$ is defined by

$$(a, b) * (c, d) = (a + c \bmod 2, b + d \bmod 3).$$

For example, $(1, 2) * (1, 2) = (0, 1)$.

Find an isomorphism from $(G, *)$ to $(\mathbb{Z}_6, \oplus)$.

40.3. Let $(G, *)$ be the following group. The set G is $\{0, 1, 2\} \times \{0, 1, 2\}$; that is,

$$G = \{(0, 0), (0, 1), (0, 2), (1, 0), (1, 1), (1, 2), (2, 0), (2, 1), (2, 2)\}.$$

The operation $*$ is defined by

$$(a, b) * (c, d) = (a + c \bmod 3, b + d \bmod 3).$$

For example, $(1, 2) * (1, 2) = (2, 1)$.

Show that $(G, *)$ is not isomorphic to $(\mathbb{Z}_9, \oplus)$.

40.4. Suppose $(G, *)$ and $(H, \star)$ are isomorphic groups. Let e be the identity element for $(G, *)$ and let e' be the identity element for $(H, \star)$. Let $f : G \to H$ be an isomorphism.

Prove that $f(e) = e'$.

40.5. Suppose $(G, *)$ and $(H, \star)$ are isomorphic groups. Let $f : G \to H$ be an isomorphism and let $g \in G$.

Prove that $f(g^{-1}) = f(g)^{-1}$.

40.6. We showed that $(\mathbb{Z}_4, \oplus)$ and $(\mathbb{Z}_5^*, \otimes)$ are isomorphic. The isomorphism we found was $f(0) = 1$, $f(1) = 2$, $f(2) = 4$, and $f(3) = 3$. There is another isomorphism (a different function) from $(\mathbb{Z}_4, \oplus)$ to $(\mathbb{Z}_5^*, \otimes)$. Find it.

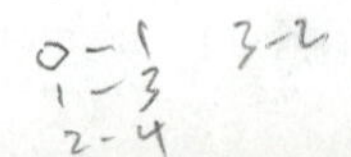

40.7. Let $(G, *)$ and $(H, \star)$ be isomorphic groups. Prove that $(G, *)$ is Abelian if and only if $(H, \star)$ is Abelian.

40.8. The group S_4 (permutations of the numbers $\{1, 2, 3, 4\}$ with the operation $\circ$) has 24 elements. Is it isomorphic to $(\mathbb{Z}_{24}, \oplus)$? Prove your answer.

40.9. Find an isomorphism from the Klein 4-group to the group $(2^{\{1,2\}}, \Delta)$.

40.10. Let $(G, *)$ be a group and let $a \in G$. Define a function $f_a : G \to G$ by $f_a(x) = a * x$. In Exercise 39.10, you showed that the functions f_a are permutations.

Let $H = \{f_a : a \in G\}$. Prove that $(G, *) \cong (H, \circ)$ where $\circ$ is composition.

40.11. Which elements of $\mathbb{Z}_{10}$ are generators of the cyclic group $(\mathbb{Z}_{10}, \oplus)$?

Generalize your answer and prove your result.

40.12. Let $(G, *)$ and $(H, \star)$ be finite cyclic groups and let $f : G \to H$ be an isomorphism. Prove that g is a generator of $(G, *)$ if and only if $f(g)$ is a generator of $(H, \star)$.

40.13. It is an advanced theorem that the group $\mathbb{Z}_p^*$ is a cyclic group for all primes p. Verify this for $p = 5, 7, 11, 13$, and 17 by finding a generator for these $\mathbb{Z}_p^*$.

41 Subgroups

A subgroup is a group within a group. Consider the integers as a group: $(\mathbb{Z}, +)$. Within the set of integers, we find the set of even integers, $E = \{x \in \mathbb{Z} : 2|x\}$. Notice that $(E, +)$ is also a group; it satisfies the four required properties. The operation $+$ is closed on E (the sum of two even integers is again even), addition is associative, E contains the identity element 0, and if x is an even integer, then $-x$ is, too, so every element of E has an inverse in E. We call $(E, +)$ a *subgroup* of $(\mathbb{Z}, +)$.

Definition 41.1 **(Subgroup)** Let $(G, *)$ be a group and let $H \subseteq G$. If $(H, *)$ is also a group, we call it a *subgroup* of $(G, *)$.

Notice that the operation for the group and the operation for its subgroup must be the same. It is incorrect to say that $(\mathbb{Z}_{10}, \oplus)$ is a subgroup of $(\mathbb{Z}, +)$; it is true that $\mathbb{Z}_{10} \subseteq \mathbb{Z}$, but the operations $\oplus$ and $+$ are different.

Example 41.2 **(Subgroups of $(\mathbb{Z}_{10}, \oplus)$)** List all the subgroups of $(\mathbb{Z}_{10}, \oplus)$.
Solution: They are

$$\begin{array}{ll} \{0\} & \{0, 1, 2, 3, 4, 5, 6, 7, 8, 9\} \\ \{0, 5\} & \{0, 2, 4, 6, 8\}. \end{array}$$

In all four cases, the operation is $\oplus$.

Is the solution to Example 41.2 correct? There are two issues to consider:

- For each of the four subsets H we listed, is it the case that $(H, \oplus)$ is a group?
- Are there other subsets $H \subseteq \mathbb{Z}_{10}$ that we neglected to include?

We consider these two questions in turn.

If $(G, *)$ is a group and $H \subseteq G$, how do we determine whether $(H, *)$ is a subgroup?

Definition 41.1 tells us what to do. First, we need to be sure that $H \subseteq G$. Second, we need to be sure that $(H, *)$ is a group. The most direct way to do this is to check that $(H, *)$ satisfies the four conditions listed in Definition 39.10: closure, associativity, identity, and inverses.

To check closure, we need to prove that if $g, h \in H$, then $g * h \in H$. For example, the even integers form a subgroup of $(\mathbb{Z}, +)$, but the odd integers do not—they do not satisfy the closure property. If g and h are odd integers, then $g + h$ is not odd.

Next, we do not have to check associativity. Reread that sentence! We wrote: we do *not* have to check associativity. We know that $(G, *)$ is a group and therefore $*$ is associative on G; that is, $\forall g, h, k \in G,\ g*(h*k) = (g*h)*k$. Since $H \subseteq G$, we must have that $*$ is already associative on H. We get associativity for free!

Next, we check that the identity element is in H. This step is usually easy.

Finally, we know that every element of H has an inverse (because every element of $G \supseteq H$ has an inverse). The issue is as follows: If $g \in H$, show that $g^{-1} \in H$.

These steps for proving that a subset of a group is a subgroup are listed in Proof Template 24.

Proof Template 24 Proving a subset of a group is a subgroup.

Let $(G, *)$ be a group and let $H \subseteq G$. To prove that $(H, *)$ is a subgroup of $(G, *)$:

- Prove that H is closed under $*$ (i.e., $\forall g, h \in H,\ g * h \in H$).
  "Let $g, h \in H$. ... Therefore $g * h \in H$."
- Prove that e (the identity element for $*$) is in H.
- Prove that the inverse of every element of H is in H (i.e., $\forall h \in H,\ h^{-1} \in H$).
  "Let $h \in H$. ... Therefore $h^{-1} \in H$."

We now reconsider the question: Are the four subsets in Example 41.2 truly subgroups of $(\mathbb{Z}_{10}, \oplus)$? We check them all.

- $H = \{0\}$ is a subgroup of $(\mathbb{Z}_{10}, \oplus)$.

  The only element of this set is the identity element for $\oplus$. Since $0 \oplus 0 = 0$, we see that H is closed under $\oplus$, that it contains the identity, and that since 0's inverse is 0, the inverse of every element in H is also in H. Therefore $\{0\}$ is a subgroup.

  In general, if $(G, *)$ is any group, then $H = \{e\}$ is a subgroup (where e is the $*$-identity element).
- $H = \mathbb{Z}_{10} = \{0, 1, 2, 3, 4, 5, 6, 7, 8, 9\}$ is a subgroup of $(\mathbb{Z}_{10}, \oplus)$.
  Since $(\mathbb{Z}_{10}, \oplus)$ is a group, it is a subgroup of itself.
  In general, if $(G, *)$ is any group, then G is a subgroup of itself.

- $H = \{0, 5\}$ is a subgroup of $(\mathbb{Z}_{10}, \oplus)$.

  It is easy to check that H is closed under $\oplus$ since

  $$0 \oplus 0 = 5 \oplus 5 = 0 \quad \text{and} \quad 0 \oplus 5 = 5 \oplus 0 = 5.$$

  Clearly $0 \in H$, and finally 0 and 5 are their own inverses. Therefore H is a subgroup of $(\mathbb{Z}_{10}, \oplus)$.
- $H = \{0, 2, 4, 6, 8\}$ is a subgroup of $(\mathbb{Z}_{10}, \oplus)$.

  Notice that H contains the even elements of $\mathbb{Z}_{10}$. If we add any two even numbers, the result is even, and when we reduce the result mod 10, the answer is still even (Exercise 41.7). We see that $0 \in H$ and the inverses of 0, 2, 4, 6, 8 are 0, 8, 6, 4, 2, respectively. Therefore H is a subgroup of $(\mathbb{Z}_{10}, \oplus)$.

This shows that the four subsets in Example 41.2 are subgroups of $(\mathbb{Z}_{10}, \oplus)$. We now turn to the other issue: Are there other subgroups of $(\mathbb{Z}_{10}, \oplus)$? There are $2^{10} = 1024$ subsets of $\mathbb{Z}_{10}$; we could list them and check them all, but there is a shorter method.

Let $H \subseteq \mathbb{Z}_{10}$ and suppose that $(H, \oplus)$ is a subgroup of $(\mathbb{Z}_{10}, \oplus)$. Since $(H, \oplus)$ is a group, we must have the identity element 0 in H. If the only element of H is 0, we have $H = \{0\}$. Otherwise there must be one, or more, additional elements. We consider them in turn.

- Suppose $1 \in H$.

  Then, by closure, we must also have $1 \oplus 1 = 2$ in H. By closure again, we must also have $1 \oplus 2 = 3$ in H. Continuing in this fashion, we see that $H = \mathbb{Z}_{10}$.

  We have shown that $1 \in H$ implies $H = \mathbb{Z}_{10}$, so now we consider only the cases with $1 \notin H$.
- Suppose $3 \in H$.

  Then $3 \oplus 3 = 6 \in H$ and $3 \oplus 6 = 9 \in H$. Since $9 \in H$, so is its inverse, $1 \in H$. And we know that if $1 \in H$, then $H = \mathbb{Z}_{10}$.

  So we may assume $3 \notin H$.
- Likewise, if $7 \in H$ or if $9 \in H$, then we can show that $1 \in H$, and then $H = \mathbb{Z}_{10}$. (Please verify these for yourself.)

  We may therefore assume that none of 1, 3, 7, or 9 is in H.
- Suppose $5 \in H$.

  We have $H \supseteq \{0, 5\}$. We know that $1, 3, 7, 9 \notin H$. Might an even integer be in H? If $2 \in H$, then $2 \oplus 5 = 7 \in H$, and that leads to $H = \mathbb{Z}_{10}$. Likewise, if any other even number is also in H, then $H = \mathbb{Z}_{10}$.

  So if $5 \in H$, then either $H = \{0, 5\}$ or $H = \mathbb{Z}_{10}$.

  We have exhausted all possible cases in which an odd integer is in H. Henceforth we may assume that all elements in H are even.
- Suppose $2 \in H$. By closure, we have 4, 6, and 8 also in H, so $H = \{0, 2, 4, 6, 8\}$.
- If $4 \in H$, then $4 \oplus 4 \oplus 4 = 2 \in H$, and we're back to $H = \{0, 2, 4, 6, 8\}$.

  By a similar argument, if 6 or 8 is in H, we again arrive at $H = \{0, 2, 4, 6, 8\}$.

In summary, our analysis shows the following: We know that $0 \in H$. If any of 1, 3, 7, or 9 is in H, then $H = \mathbb{Z}_{10}$. If $5 \in H$, then either $H = \{0, 5\}$ or $H = \mathbb{Z}_{10}$. If H contains any of 2, 4, 6, or 8, then $H = \{0, 2, 4, 6, 8\}$ or $H = \mathbb{Z}_{10}$. In all cases, we have that H is one of $\{0\}$, $\mathbb{Z}_{10}$, $\{0, 5\}$, or $\{0, 2, 4, 6, 8\}$, showing that the list in Example 41.2 is exhaustive.

Lagrange's Theorem

In Example 41.2, we found four subgroups of $(\mathbb{Z}_{10}, \oplus)$. The cardinalities of these four subgroups are 1, 2, 5, and 10. Notice that these four numbers are divisors of 10. Here is another example:

Example 41.3 **(Subgroups of S_3)** List all the subgroups of $(S_3, \circ)$.
Solution: Recall that S_3 is the set of all permutations of $\{1, 2, 3\}$; that is,

$$S_3 = \{(1)(2)(3), (12)(3), (13)(2), (1)(23), (123), (132)\}.$$

Its subgroups are

$$\{(1)(2)(3)\}$$

$$\{(1)(2)(3), (12)(3)\} \quad \{(1)(2)(3), (13)(2)\} \quad \{(1)(2)(3), (1)(23)\}$$

$$\{(1)(2)(3), (123), (132)\}$$

$$\{(1)(2)(3), (12)(3), (13)(2), (1)(23), (123), (132)\}.$$

The cardinalities of these subgroups are 1, 2, 3, and 6—all of which are divisors of 6.

Examples 41.2 and 41.3 suggest that if $(G, *)$ is a subgroup of $(H, *)$ (and both are finite), then $|G|$ is a divisor of $|H|$.

Theorem 41.4 **(Lagrange)** Let $(H, *)$ be a subgroup of a finite group $(G, *)$ and let $a = |H|$ and $b = |G|$. Then $a|b$.

The central idea in the proof is to *partition* G into subsets, all of which are the same size as H. Since the parts in a partition are pairwise disjoint, we have divided G into nonoverlapping parts of size $|H|$. This implies that $|H|$ divides $|G|$. (This approach is akin to using Theorem 15.6.)

The partition we create consists of equivalence classes of an equivalence relation which is defined as follows:

Definition 41.5 **(Congruence modulo a subgroup)** Let $(G, *)$ be a group and let $(H, *)$ be a subgroup. Let $a, b \in G$. We say that *a is congruent to b modulo H* if $a * b^{-1} \in H$. We write this as

$$a \equiv b \pmod{H}.$$

This is yet another meaning for the overused word *mod!* We consider an example.

Consider the group $(\mathbb{Z}_{25}^*, \otimes)$. The elements of $\mathbb{Z}_{25}^*$ are

$$\mathbb{Z}_{25}^* = \{1, 2, 3, 4, 6, 7, 8, 9, 11, 12, 13, 14, 16, 17, 18, 19, 21, 22, 23, 24\}.$$

Let $H = \{1, 7, 18, 24\}$. The operation table for $\otimes$ restricted to H is

$\otimes$	1	7	18	24
1	1	7	18	24
7	7	24	1	18
18	18	1	24	7
24	24	18	7	1

Notice that H is closed under $\otimes$, the identity element $1 \in H$, and since

$$1^{-1} = 1 \qquad 7^{-1} = 18 \qquad 18^{-1} = 7 \qquad 24^{-1} = 24,$$

the inverse of every element of H is again a member of H. Therefore H is a subgroup of $\mathbb{Z}_{25}^*$.

For this group and subgroup, do we have $2 \equiv 3 \pmod{H}$? The answer is no. To see why, we calculate

$$2 \otimes 3^{-1} = 2 \otimes 17 = 9 \notin H$$

so $2 \not\equiv 3 \pmod{H}$. (Note that $3^{-1} = 17$ because $3 \otimes 17 = 1$.)

On the other hand, we do have $2 \equiv 11 \pmod{H}$. To see why, we calculate

$$2 \otimes 11^{-1} = 2 \otimes 16 = 7 \in H$$

so $2 \equiv 11 \pmod{H}$. (Note that $11^{-1} = 16$ because $11 \otimes 16 = 176 \bmod 25 = 1$.)

Congruence modulo a subgroup is an equivalence relation on the group.

Lemma 41.6 Let $(G, *)$ be a group and let $(H, *)$ be a subgroup. Then congruence modulo H is an equivalence relation on G.

It is interesting to note that the three portions of this proof correspond precisely to the three conditions we must check to prove a subset of a group is a subgroup (Proof Template 24). The reflexive property follows from the fact that $e \in H$. The symmetry property follows from the fact that the inverse of an element of H must also be in H. And transitivity follows from the fact that H is closed under $*$.

Proof. To check that congruence modulo H is an equivalence relation on G, we need to show that it is reflexive, symmetric, and transitive.

- *Congruence modulo H is reflexive.*

  Let $g \in G$. We need to show that $g \equiv g \pmod{H}$. To do that, we need to show $g * g^{-1} \in H$. Since $g * g^{-1} = e$ and since $e \in H$, we have $g \equiv g \pmod{H}$.
- *Congruence modulo H is symmetric.*

  Suppose $a \equiv b \pmod{H}$. This means that $a * b^{-1} \in H$. Therefore $(a * b^{-1})^{-1} \in H$. Note that

  $$(a * b^{-1})^{-1} = (b^{-1})^{-1} * a^{-1} = b * a^{-1}$$

  and so $b * a^{-1} \in H$. Thus we have $b \equiv a \pmod{H}$.

- *Congruence modulo H is transitive.*

  Suppose $a \equiv b \pmod{H}$ and $b \equiv c \pmod{H}$. Thus $a*b^{-1}, b*c^{-1} \in H$. It follows that

  $$(a * b^{-1}) * (b * c^{-1}) \in H$$

  because H is a subgroup and therefore closed under $*$. Note that

  $$(a * b^{-1}) * (b * c^{-1}) = a * (b^{-1} * b) * c^{-1} = a * c^{-1}$$

  and so $a * c^{-1} \in H$. Therefore $a \equiv c \pmod{H}$.

Therefore congruence modulo H is an equivalence relation on G. ■

Since congruence mod H is an equivalence relation, we may consider the equivalence classes of this relation. Recall the group $(\mathbb{Z}_{25}^*, \otimes)$ and its subgroup $H = \{1, 7, 18, 24\}$ we considered earlier. For the congruence mod H relation, what is the equivalence class [2]? This is the set of all elements of $\mathbb{Z}_{25}^*$ that are related to 2; that is,

$$[2] = \{a \in \mathbb{Z}_{25}^* : a \equiv 2 \pmod{H}\}.$$

We can test all 20 elements of $\mathbb{Z}_{25}^*$ to see which are and which are not congruent to 2 modulo H. We find that

$$[2] = \{2, 11, 14, 23\}.$$

In this manner, we can find all the equivalence classes. They are

$$\begin{aligned}
[1] &= \{1, 7, 18, 24\}, \\
[2] &= \{2, 11, 14, 23\}, \\
[3] &= \{3, 4, 21, 22\}, \\
[6] &= \{6, 8, 17, 19\}, \text{ and} \\
[9] &= \{9, 12, 13, 16\}.
\end{aligned}$$

Several comments are in order.

First, these are all the equivalence classes of congruence mod H. Every element of $\mathbb{Z}_{25}^*$ is in exactly one of these classes. You might ask: Did we neglect the class [4]? The equivalence class [4] is exactly the same as [3] because $4 \equiv 3 \pmod{H}$ (because $3 \otimes 4^{-1} = 3 \otimes 19 = 7 \in H$).

Second, because these are equivalence classes, we know (by Corollary 14.13) that they form a partition of the group (in this case, of $\mathbb{Z}_{25}^*$).

Third, the class [1] equals the subgroup $H = \{1, 7, 18, 24\}$. This is not a coincidence. Let $(G, *)$ be any group and let $(H, *)$ be a subgroup. The equivalence class of the identity element, $[e]$, must equal H. Here's the one-line proof:

$$a \in [e] \iff a \equiv e \pmod{H} \iff a * e^{-1} \in H \iff a \in H.$$

Fourth, the equivalence classes all have the same size (in this example, they all have four elements). This observation is the key step in proving Theorem 41.4, and so we prove it here as a lemma.

Lemma 41.7 Let $(G, *)$ be a group and let $(H, *)$ be a finite subgroup. Then any two equivalence classes of the congruence mod H relation have the same size.

Proof. Let $g \in G$ be arbitrary. It is enough to show that $[g]$ and $[e]$ have the same size. As we noted above, $[e] = H$. To show that $[g]$ and H have the same size, we define a function $f : H \to [g]$ and we prove that f is one-to-one and onto. From this, it follows that $|H| = |[g]|$.

For $h \in H$, define $f(h) = h * g$. Clearly f is a function defined on H, but is $f : H \to [g]$? We need to show that $f(h) \in [g]$. In other words, we need to prove that $f(h) \equiv g \pmod{H}$. This is true because

$$f(h) * g^{-1} = (h * g) * g^{-1} = h * (g * g^{-1}) = h \in H.$$

Therefore f is a function from H to $[g]$.

Next we show that f is one-to-one. Suppose $f(h) = f(h')$. Then $h*g = h'*g$. Operating on the right by g^{-1} gives

$$\begin{aligned}(h * g) * g^{-1} &= (h' * g) * g^{-1}\\ h * (g * g^{-1}) &= h' * (g * g^{-1})\\ h &= h'\end{aligned}$$

and so f is one-to-one.

Finally, we show that f is onto. Let $b \in [g]$. This means that $b \equiv g \pmod{H}$, and so $b * g^{-1} \in H$. Let $h = b * g^{-1}$. Then

$$f(h) = f(b * g^{-1}) = (b * g^{-1}) * g = b * (g * g^{-1}) = b$$

and so f is onto $[g]$.

Therefore H and $[g]$ have the same cardinality and the result is proved. ■

We now have the tools necessary to prove Lagrange's Theorem.

Proof (of Theorem 41.4)

Let $(G, *)$ be a finite group and let $(H, *)$ be a subgroup. The equivalence classes of the is-congruent-to-mod-H relation all have the same cardinality as H. Since the equivalence classes form a partition of G, we know that $|H|$ is a divisor of $|G|$. ■

Recap

In this section, we introduced the notion of a subgroup of a group, and we proved that if H is a finite subgroup of G, then $|H|$ is a divisor of $|G|$.

41 Exercises

41.1. Find all subgroups of $(\mathbb{Z}_6, \oplus)$.

41.2. Find all subgroups of $(\mathbb{Z}_9, \oplus)$.

41.3. Find all subgroups of the Klein 4-group.

41.4. Let $(G, *)$ be a group and suppose H is a nonempty subset of G.

Prove that $(H, *)$ is a subgroup of $(G, *)$ provided that H is closed under $*$ and that for every $g \in H$, we have $g^{-1} \in H$.

This gives an alternative proof strategy to Proof Template 24. You do not need to prove that $e \in H$. You need only prove that H is nonempty.

41.5. Let $(G, *)$ be a group and suppose H is a nonempty subset of G.

Prove that $(H, *)$ is a subgroup of $(G, *)$ provided for every $g, h \in H$, we have $g * h^{-1} \in H$.

This gives yet another alternative to Proof Template 24, although of limited utility.

41.6. Find, with proof, all the subgroups of $(\mathbb{Z}, +)$.

41.7. Prove that if x and y are even, then so is $[(x + y) \bmod 10]$. Conclude that $\{0, 2, 4, 6, 8\}$ is closed under mod 10 addition.

41.8. In $(\mathbb{Z}_{25}^*, \otimes)$ the set $H = \{1, 6, 11, 16, 21\}$ is a subgroup. Find the equivalence classes of the congruence mod H relation.

41.9. Consider the group $(S_3, \circ)$ and the subgroup $H = \{(1)(2)(3), (12)(3)\}$.

Find the equivalence classes of the mod H relation.

41.10. Let $(G, *)$ be a finite group and let $g \in G$.

a. Prove that there is a positive integer k such that

$$g^k = \underbrace{g * g * \cdots * g}_{k \text{ times}} = e.$$

By the Well-Ordering Principle, there is a least positive integer k such that $g^k = e$. We define the *order* of the element g to be the smallest such positive integer.

b. Prove that $\{e, g, g^2, g^3, \ldots\}$ is a subgroup of G whose cardinality is the order of g.

c. Prove that the order of g divides $|G|$.

d. Conclude that $g^{|G|} = e$.

41.11. Let $(G, *)$ be a group and let $(H, *)$ and $(K, *)$ be subgroups. Prove or disprove each of the following assumptions.

a. $H \cap K$ is a subgroup of $(G, *)$.

b. $H \cup K$ is a subgroup of $(G, *)$.

c. $H - K$ is a subgroup of $(G, *)$.

d. $H \Delta K$ is a subgroup of $(G, *)$.

41.12. Why did we reuse the word *mod* for the new equivalence relation in this section? The new relations are a generalization of the more familiar $x \equiv y \pmod{n}$ for integers. Here is the connection:

Consider the group $(\mathbb{Z}, +)$ and let n be a positive integer. Let H be the subgroup consisting of all multiples of n; that is,

$$H = \{a \in \mathbb{Z} : n|a\}.$$

Prove that for all integers x and y,

$$x \equiv y \pmod{H} \iff x \equiv y \pmod{n}.$$

41.13. Let $(G, *)$ be a group and let $(H, *)$ be a subgroup.

Let $a, b, c, d \in G$. We would like to believe that

$$\text{if} \quad a \equiv b \pmod{H} \quad \text{and} \quad c \equiv d \pmod{H},$$

$$\text{then} \quad a * c \equiv b * d \pmod{H}$$

but this is not true. Give a counterexample.

This problem introduces the concept of a *coset*. Given a group $(G, *)$, a subgroup H, and an element $g \in G$, the sets $g * H$ and $H * g$ are called *cosets* of H. More specifically, $g * H$ is called a *left coset* and $H * g$ is called a *right coset*.

41.14. Let $(G, *)$ be a group. Although the operation $*$ operates on two elements of G, in this and the next problem we extend the use of the operation symbol $*$ as follows:

Let $g \in G$ and let $(H, *)$ be a subgroup of G. Define the sets $H * g$ and $g * H$ as follows:

$$H * g = \{h * g : h \in H\}, \quad \text{and}$$
$$g * H = \{g * h : h \in H\}.$$

In other words, $H * g$ is the set of all elements of G that can be formed by operating on an element of H (called h) with g to form $h * g$. If $H = \{h_1, h_2, h_3, \ldots\}$, then

$$H * g = \{h_1 * g, h_2 * g, h_3 * g, \ldots\} \quad \text{and}$$
$$g * H = \{g * h_1, g * h_2, g * h_3, \ldots\}.$$

For example, suppose the group G is S_3 and the subgroup is $H = \{(1)(2)(3), (1, 2, 3), (1, 3, 2)\}$. Let $g = (1, 2)(3)$. Then

$$\begin{aligned} H \circ g &= H \circ (1, 2)(3) \\ &= \{(1)(2)(3) \circ (1, 2)(3), (1, 2, 3) \circ (1, 2)(3), (1, 3, 2) \circ (1, 2)(3)\} \\ &= \{(1, 2)(3), (1, 3)(2), (1)(2, 3)\}. \end{aligned}$$

Please do the following:

a. Prove that $g \in H * g$ and $g \in g * H$.

b. Prove that $g * H = H \iff H * g = H \iff g \in H$.

c. Prove that if $(G, *)$ is Abelian, then $g * H = H * g$.

d. Give an example of a group G, subgroup H, and element g such that $g * H \neq H * g$

See the previous problem for the definition of $g * H$ and $H * g$.

41.15. We call a subgroup $(H, *)$ of $(G, *)$ *normal* provided, for all $g \in G$, we have $g * H = H * g$.

Prove that if H is normal and $a, b, c, d \in G$, the implication

$$\text{if} \quad a \equiv b \pmod{H} \quad \text{and} \quad c \equiv d \pmod{H},$$

$$\text{then} \quad a * c \equiv b * d \pmod{H}$$

is true.

42 Fermat's Little Theorem

This section is devoted to proving the following result.

Theorem 42.1 **(Fermat's Little Theorem)** Let p be a prime and let a be an integer. Then

$$a^p \equiv a \pmod{p}.$$

For example, if $p = 23$, then the powers of 5 taken modulo 23 are

$$\begin{array}{lllll} 5^1 \equiv 5 & 5^2 \equiv 2 & 5^3 \equiv 10 & 5^4 \equiv 4 & 5^5 \equiv 20 \\ 5^6 \equiv 8 & 5^7 \equiv 17 & 5^8 \equiv 16 & 5^9 \equiv 11 & 5^{10} \equiv 9 \\ 5^{11} \equiv 22 & 5^{12} \equiv 18 & 5^{13} \equiv 21 & 5^{14} \equiv 13 & 5^{15} \equiv 19 \\ 5^{16} \equiv 3 & 5^{17} \equiv 15 & 5^{18} \equiv 6 & 5^{19} \equiv 7 & 5^{20} \equiv 12 \\ 5^{21} \equiv 14 & 5^{22} \equiv 1 & 5^{23} \equiv 5 & 5^{24} \equiv 2 & 5^{25} \equiv 10 \end{array}$$

where all congruences are mod 23.

We give three rather different proofs of this lovely result.

First Proof

Proof (of Theorem 42.1)

We first prove (using induction) the result in the special case that $a \geq 0$. We finish by showing that the special case implies the full theorem.

We prove, by induction on a, that if p is prime and $a \in \mathbb{N}$, then $a^p \equiv a \ (p)$.

Basis case: In the case $a = 0$, we have $a^p = 0^p = 0 = a$, so $a^p \equiv a \ (p)$ holds for $a = 0$.

Induction hypothesis: Suppose the result holds for $a = k$; that is, $k^p \equiv k \ (p)$. We need to prove that $(k+1)^p \equiv k+1 \ (p)$.

By the Binomial Theorem (Theorem 16.8), we have

$$(k+1)^p = k^p + \binom{p}{1}k^{p-1} + \binom{p}{2}k^{p-2} + \cdots + \binom{p}{p-1}k + 1. \tag{44}$$

Notice that the intermediate terms (all but the very first and very last) on the right-hand side of Equation (44) are all of the form $\binom{p}{j}k^{p-j}$ where $0 < j < p$. The binomial coefficient $\binom{p}{j}$ is an integer that we can write as (Theorem 16.12):

$$\binom{p}{j} = \frac{p!}{j!(p-j)!} = \frac{p(p-1)!}{j!(p-j)!}. \tag{45}$$

The fraction in Equation (45) is an integer. Imagine we factor the numerator and the denominator of this fraction into primes (by Theorem 38.1). Because this fraction reduces to an integer, every prime factor in the denominator cancels a matching prime factor in the numerator. However, notice that p is a prime factor of the numerator, but p is not a prime factor of the denominator; both j and $p-j$ are less than p (because $0 < j < p$), and so the prime factors in $j!$ and $(p-j)!$ cannot include p. Thus, after we reduce the fraction in Equation (45) to an integer, that integer must be a multiple of p.

Therefore the middle terms in Equation (44) are all multiples of p, so we can write

$$k^p + \binom{p}{1}k^{p-1} + \binom{p}{2}k^{p-2} + \cdots + \binom{p}{p-1}k + 1 \equiv k^p + 1 \pmod{p}. \quad (46)$$

Finally, by induction we know that $k^p \equiv k \;\; (p)$, so combining Equations (44) and (46), we have

$$(k+1)^p \equiv k^p + 1 \equiv k + 1 \pmod{p}$$

completing the induction.

Thus we have proved Theorem 42.1 for all $a \in \mathbb{N}$; we finish by showing that the result also holds for negative integers; that is, we need to prove

$$(-a)^p \equiv (-a) \pmod{p}$$

where $a > 0$. The case $p = 2$ is different from the case for odd primes.

Note that $a \equiv -a \pmod{2}$; see Exercise 14.3.

In the case $p = 2$, we have

$$(-a)^2 \equiv a^2 \equiv a \equiv -a \pmod{2}$$

because $-a \equiv a \;\; (2)$ for all integers a.

In the case $p > 2$ (and therefore p is odd), we have

$$(-a)^p = (-1)^p a^p = -(a^p) \equiv -a \pmod{p}$$

completing the proof. ■

Second Proof

Proof (of Theorem 42.1)

As in the previous proof, we first prove a restricted special case. In this proof, we assume a is a positive integer. The case $a = 0$ is trivial, and the case when $a < 0$ is handled as in the previous proof.

Thus we assume p is a prime and a is a positive integer. We consider the following counting problem.

> How many length-p lists can we form in which the elements of the list are chosen from $\{1, 2, \ldots, a\}$?

The answer to this question is, of course, a^p (see Theorem 7.6).

Next we define an equivalence relation R on these lists. We say that two lists are equivalent if we can get one from the other by cyclically shifting its entries. In a *cyclic shift* we move the last element to the first position on the list. Two lists are related by R if we can form one from the other by performing one (or more) cyclic shifts. For example, the following lists are all equivalent:

$$12334 \; R \; 41233 \; R \; 34123 \; R \; 33412 \; R \; 23341.$$

We now consider a new problem:

> How many nonequivalent length-p lists can we form in which the elements of the list are chosen from $\{1, 2, \ldots, a\}$?

By *nonequivalent* we mean not related by R. In other words, we want to count the number of R-equivalence classes.

Example 42.2 Consider the case $a = 2$ and $p = 3$. There are eight lists we can form: 111, 112, 121, 122, 211, 212, 221, 222. These fall into four equivalence classes:

$$\{111\}, \quad \{222\}, \quad \{112, 121, 211\}, \quad \text{and} \quad \{122, 212, 221\}.$$

Example 42.3 Consider the case $a = 3$ and $p = 5$. There are $3^5 = 243$ possible lists (from 11111 to 33333). There are three equivalence classes that contain just one list, namely

$$\{11111\}, \quad \{22222\}, \quad \text{and} \quad \{33333\}.$$

The remaining lists fall into equivalence classes containing more than one element. For example, the list 12113 is in the following equivalence class:

$$[12113] = \{12113, 31211, 13121, 11312, 21131\}.$$

By experimenting with other lists, please notice that all the equivalence classes with more than one list contain exactly five lists. (We prove this below.)

Thus there are three equivalence classes that contain only one list. The remaining $3^5 - 3$ lists fall into classes containing exactly five lists each; there are $(3^5 - 3)/5$ such lists. Thus, all told, there are

$$3 + \frac{3^5 - 3}{5} = 51$$

different equivalence classes.

The punch line is this: The number $(3^5 - 3)/5$ is an integer. Therefore $3^5 - 3$ is divisible by 5; that is, $3^5 \equiv 3 \pmod 5$.

How do we count the number of equivalence classes in general? If the equivalence classes all had the same size, then we could use Theorem 15.6; we would simply divide the number of lists by the (allegedly) common number of lists in each class. However, as the examples show, the classes might contain different numbers of lists.

Let's explore how many elements an equivalence class might contain. We begin with the simple special case of lists all of whose elements are the same (e.g., $222\cdots 2$ or $aaa\cdots a$); such lists are equivalent only to themselves. There are a equivalence classes that contain exactly one list—namely, $\{111\cdots 1\}$, $\{222\cdots 2\}$, $\ldots$, $\{aaa\cdots a\}$.

Now consider a list with (at least) two different elements, such as 12113. How many lists are equivalent to this list? We saw in Example 42.3 that there are five lists in 12113's equivalence class.

In general, consider the list

$$x_1x_2x_3\cdots x_{p-1}x_p$$

where the elements of the list are drawn from the set $\{1, 2, \ldots, a\}$. The equivalence class of this list contains the following lists:

$$\begin{aligned}
&\text{List } 1: && x_1 x_2 x_3 \cdots x_{p-1} x_p \quad \text{(original)} \\
&\text{List } 2: && x_2 x_3 \cdots x_{p-1} x_p x_1 \\
&\text{List } 3: && x_3 \cdots x_{p-1} x_p x_1 x_2 \\
& && \vdots \\
&\text{List } p: && x_p x_1 x_2 x_3 \cdots x_{p-1}.
\end{aligned}$$

It appears that there are p lists in this equivalence class, but we know this is not quite right; if all the x_is are the same, these p "different" lists are all the same. We need to worry that even in the case where the x_is are not all the same, there still might be a repetition.

We claim: If the elements of the list $x_1 x_2 x_3 \cdots x_{p-1} x_p$ are not all the same, then the p lists above are all different. Suppose, for the sake of contradiction, that two of the lists are the same. That is, there are two lists, say List i and List j, with $1 \le i < j \le p$, with

$$x_i x_{i+1} \cdots x_{i-1} = x_j x_{j+1} \cdots x_{j-1}.$$

What does it mean that these lists are *equal*? It means that, element by element, they are equal; that is,

$$\begin{aligned}
x_i &= x_j \\
x_{i+1} &= x_{j+1} \\
&\vdots \\
x_{i-1} &= x_{j-1}.
\end{aligned}$$

These equations imply the following: If we cyclically shift the list $x_1 x_2 x_3 \cdots x_{p-1} x_p$ by $j - i$ steps, the resulting sequence is identical to the original. In particular, this means that

$$x_1 = x_{1+(j-i)}.$$

If we shift the list another $j - i$ steps, we again return to the original, so

$$x_1 = x_{1+(j-i)} = x_{1+2(j-i)}.$$

We need to be careful. Perhaps the subscript $1 + 2(j - i)$ is larger than p. Although there is no element, say, x_{p+1} (it would be past the end of the list), since we are cyclically shifting we can consider element x_{p+1} to be the same as element x_1. In general, we can always add or subtract a multiple of p so that the subscript on x lies in the set $\{1, 2, \ldots, p\}$. In other words, we consider two subscripts to be the same if they are congruent mod p. Thus the equation $x_1 = x_{1+(j-i)} = x_{1+2(j-i)}$ now makes sense.

We continue the analysis. We have the equation $x_1 = x_{1+(j-i)} = x_{1+2(j-i)}$ by considering two cyclic shifts of the list $x_1 x_2 x_3 \cdots x_{p-1} x_p$ by $j - i$ steps. If we shift another $j - i$ steps, we have

$$x_1 = x_{1+(j-i)} = x_{1+2(j-i)} = x_{1+3(j-i)}.$$

Clearly we have

$$x_1 = x_{1+(j-i)} = x_{1+2(j-i)} = x_{1+3(j-i)} = \cdots = x_{1+(p-1)(j-i)} \tag{47}$$

where subscripts are taken modulo p. We claim that Equation (47) says

$$x_1 = x_2 = \cdots = x_p.$$

To see why, we note that in Equation (47) all subscripts (from 1 to p) appear. This was shown in Exercise 35.18.

It is time to draw these various threads together. We are considering the set of lists equivalent to $x_1x_2x_3 \cdots x_{p-1}x_p$. We know that if all the xs are the same, there is only one list equivalent to $x_1x_2x_3 \cdots x_{p-1}x_p$ (namely, itself). Otherwise, if there are at least two different elements on this list, then there are exactly p different lists equivalent to $x_1x_2x_3 \cdots x_{p-1}x_p$ (if there were any fewer, then we would have $x_1 = x_2 = \cdots = x_p$ by the above analysis).

Thus there are a equivalence classes of size 1, corresponding to the lists $111 \cdots 1$ through $aaa \cdots a$. The remaining $a^p - a$ lists form equivalence classes of size p. Thus, all together, there are

$$a + \frac{a^p - a}{p}$$

different equivalence classes. Since this number must be an integer, we know that $(a^p - a)/p$ must be an integer (i.e., $a^p - a$ is divisible by p). This can be rewritten as $a^p \equiv a \pmod{p}$. ■

Third Proof

Proof (of Theorem 42.1)

For this third proof, we work in the group $(\mathbb{Z}_p^*, \otimes)$. We begin by making some simplifications.

We want to prove $a^p \equiv a \pmod{p}$ where p is a prime and a is any integer. We saw in the previous proofs that we need to prove this result only for $a > 0$; the case $a = 0$ is trivial, and the case $a < 0$ follows from the case when a is positive.

Let us narrow even further the range of values of a we need to consider. First, not only is the case $a = 0$ trivial, it is also easy to prove $a^p \equiv a \ (p)$ when a is a multiple of p (Exercise 42.3).

Second, if we increase (or decrease) a by a multiple of p, there is no change (modulo p) in the value of a^p:

$$\begin{aligned}(a + kp)^p &= a^p + \binom{p}{1}a^{p-1}(kp)^1 + \binom{p}{2}a^{p-2}(kp)^2 + \cdots + \binom{p}{p}a^0(kp)^p \\ &\equiv a^p \pmod{p}\end{aligned}$$

because all the $\binom{p}{j}a^{p-j}(kp)^j$ (with $j > 0$) are multiples of p.

Therefore we may assume that a is an integer in the set $\{1, 2, \ldots, p-1\} = \mathbb{Z}_p^*$. Furthermore the equation $a^p \equiv a \ \ (p)$ is equivalent to

$$\underbrace{a \otimes a \otimes \cdots \otimes a}_{p \text{ times}} = a$$

where the computations are in $\mathbb{Z}_p^*$. This can be rewritten $a^p = a$ where, again, the computations are in $\mathbb{Z}_p^*$. If we $\otimes$ both sides by a^{-1}, we have $a^{p-1} = 1$ (in $\mathbb{Z}_p^*$).

Conversely, if we can prove $a^{p-1} = 1$ in $\mathbb{Z}_p^*$, then our proof of Theorem 42.1 will be complete.

The good news is that you have already solved this problem! Exercise 41.10(d) asserts that for any group G and for any element $g \in G$, we have $g^{|G|} = e$. In our case, the group is $\mathbb{Z}_p^*$, the element is a, and $|\mathbb{Z}_p^*| = p - 1$. Therefore $a^{p-1} = 1$ and we are finished. ■

Euler's Theorem

We can extend the third proof of Fermat's Little Theorem to a broader context. Does the result hold for nonprime moduli? Perhaps we can prove $a^n \equiv a \pmod n$ for any positive integer n. An example shows that this is not the correct extension of Fermat's Little Theorem.

Example 42.4 Does $a^n \equiv a \pmod n$ for nonprime values of n? Consider $n = 9$. We have

$$\begin{array}{lll} 1^9 \equiv 1 & 2^9 \equiv 8 \not\equiv 2 & 3^9 \equiv 0 \not\equiv 3 \\ 4^9 \equiv 1 \not\equiv 4 & 5^9 \equiv 8 \not\equiv 5 & 6^9 \equiv 0 \not\equiv 6 \\ 7^9 \equiv 1 \not\equiv 7 & 8^9 \equiv 8 & 9^9 \equiv 0 \equiv 9 \end{array}$$

where all congruences are modulo 9. The formula $a^p \equiv a \pmod p$ does not extend to nonprime values of p.

Let us return to the inner workings of the third proof. The key was to prove $a^{p-1} = 1$ in $\mathbb{Z}_p^*$. There are two reasons why this equation holds.

First, $a \in \mathbb{Z}_p^*$; if a were a multiple of p, then any power of a would also be a multiple of p, and there is no power of a that would give us 1 modulo p.

Second, the exponent $p - 1$ is the number of elements in $\mathbb{Z}_p^*$. The number of elements in $\mathbb{Z}_n^*$ is not, in general, $n - 1$. Rather, $|\mathbb{Z}_n^*| = \varphi(n)$, Euler's totient. (See Exercises 38.14–17.)

Let us revisit Example 42.4, this time replacing the exponent 9 with the exponent $\varphi(9) = 6$.

Example 42.5 Note that $\mathbb{Z}_9^* = \{1, 2, 4, 5, 7, 8\}$ and $\varphi(9) = 6$. Raising the integers 1 through 9 to the power 6 (mod 9) gives

$$\begin{array}{lll} 1^6 \equiv 1 & 2^6 \equiv 1 & 3^6 \equiv 0 \\ 4^6 \equiv 1 & 5^6 \equiv 1 & 6^6 \equiv 0 \\ 7^6 \equiv 1 & 8^6 \equiv 1 & 9^6 \equiv 0. \end{array}$$

This is much better! For those values of $a \in \mathbb{Z}_9^*$, we have $a^6 = 1$. Of course, if a is increased or decreased by a multiple of 9, the results in Example 42.5 remain the same.

By Exercise 41.10(d), we know that if $a \in \mathbb{Z}_n^*$, then

$$a^{|\mathbb{Z}_n^*|} = 1$$

and since $|\mathbb{Z}_n^*| = \varphi(n)$, this can be rewritten

$$a^{\varphi(n)} = 1$$

where the computations are performed in $\mathbb{Z}_n^*$ (i.e., using $\otimes$). Restated, this says,

$$a^{\varphi(n)} \equiv 1 \pmod{n}$$

with ordinary integer multiplications. The generalization of Fermat's Little Theorem is the following result, which we owe to Euler.

Theorem 42.6 **(Euler's Theorem)** Let n be a positive integer and let a be an integer relatively prime to n. Then

$$a^{\varphi(n)} \equiv 1 \pmod{n}.$$

Proof. We have seen the main steps in this proof already. Let a be relatively prime to n. Dividing a by n, we have

$$a = qn + r$$

where $0 \le r < n$. Since a is relatively prime to n, so is r (see Exercise 35.12). Thus we may assume that $a \in \mathbb{Z}_n^*$.

To show that $a^{\varphi(n)} \equiv 1 \pmod{n}$ is equivalent to showing that $a^{\varphi(n)} = 1$ in $\mathbb{Z}_n^*$, and this follows immediately from Exercise 41.10(d). ■

Primality Testing

Fermat's Little Theorem states that if p is a prime, then $a^p \equiv a \pmod{p}$ for any integer a. We can write this symbolically as

$$p \text{ is a prime} \quad \Rightarrow \quad \forall a \in \mathbb{Z},\ a^p \equiv a \pmod{p}.$$

The contrapositive of this statement is

$$\neg[\forall a \in \mathbb{Z},\ a^p \equiv a \pmod{p}] \quad \Rightarrow \quad p \text{ is not a prime}$$

which can be rewritten

$$\exists a \in \mathbb{Z},\ a^p \not\equiv a \pmod{p} \quad \Rightarrow \quad p \text{ is not a prime.}$$

In other words, if there is some integer a such that $a^p \not\equiv a \pmod{p}$, then p is not a prime. We have the following:

Theorem 42.7 Let a and n be positive integers. If $a^n \not\equiv a \pmod{n}$, then n is not prime.

Example 42.8 Let $n = 3007$. Is n prime? We compute $2^{3007} \bmod 3007$ and the result is 33. If 3007 were prime, we would have $2^{3007} \equiv 2 \pmod{3007}$. Thus 3007 is not prime.

Notice that we have shown that 3007 is not prime without factoring. This may seem a rather complicated way to check whether a number is prime. The number 3007 factors simply as 31×97. Isn't it simpler and faster just to factor 3007 than to compute $2^{3007} \bmod 3007$?

How much effort is involved in factoring 3007? The simplest method is trial division. We can test divisors of 3007 starting from 2 and continuing until just after we pass $\sqrt{3007} \approx 54.8$. This method can, in the worst case, involve around 50 divisions.

On the other hand, computing 2^{3007} seems to demand thousands of multiplications. However, as we saw in Exercise 36.14, the computation $a^b \bmod c$ can be performed very efficiently. The computation $2^{3007} \pmod{3007}$ is accomplished with about 20 multiplications and 20 reductions mod 3007 (i.e., 20 divisions).

The computational efforts of the two methods appear to be roughly the same.

However, suppose we use trial division to see whether a 1000-digit number is prime. Since $n \approx 10^{1000}$, we have $\sqrt{n} \approx 10^{500}$. Thus we would be performing on the order of 10^{500} divisions, and this would take a very long time. (See Exercise 42.4.)

On the other hand, computing $a^n \bmod n$ requires only a few thousand multiplications and divisions; this computation can be done in less than a minute on a desktop computer.

Theorem 42.7 is a terrific tool for showing that an integer is not prime. However, suppose we have positive integers a and n with $a^n \equiv a \pmod{n}$. Does this imply that n is prime? No. Theorem 42.7 only guarantees that certain numbers are not prime.

Thus $a^n \equiv a \pmod{n}$ does not imply n is prime. Computing, say, $2^n \bmod n$ is not a sure-fire way to check whether n is prime. You might wonder, suppose we find that $2^n \bmod n = 2$, $3^n \bmod n = 3$, and $4^n \bmod n = 4$, and so on. Do these imply that n is prime? No. This is explored in Exercise 42.6.

Recap

We presented Fermat's Little Theorem [if p is prime, then $a^p \equiv a \pmod{p}$] and gave three different proofs. We also proved a generalization of this result known as Euler's Theorem. Finally, we showed how Fermat's Little Theorem can be used as a primality test.

42 Exercises

42.1. For all $a \in \mathbb{Z}_{13}$, calculate a^{12} and a^{13}. (hint use Fermat's Thm)

42.2. For all $a \in \mathbb{Z}_{15}^*$, calculate a^{14}, a^{15}, and $a^{\varphi(15)}$.

42.3. Without using Theorem 42.1, prove that if p is a prime and a is a multiple of p, then $a^p \equiv a \pmod{p}$.

42.4. Estimate how long it would take to factor a 1000-digit number using trial divisions. Assume that we try all divisors up to the square root of the number and that we can perform 10 billion trial divisions per second.

Choose a reasonable unit of time for your answer.

42.5. One of the following two integers is prime: 332,461,561 or 332,462,561. Which one is it?

42.6. Find a positive integer n with the following properties:

- n is composite, but
- for all integers a with $1 < a < n$, $a^n \equiv a \pmod{n}$.

Such an integer is called a *Carmichael* number. It always passes our primality test but is not prime.

The point is this: Even if an integer passes our primality test, it is not necessarily prime. However, if it fails the primality test, then it must be composite.

43 Public Key Cryptography I: Introduction

The Problem: Private Communication in Public

This problem is not contrived. Imagine you wish to purchase a product over the World Wide Web. You visit the company's website and place your order. To pay for the order, you enter your credit card number. You do not want anyone else on the Internet to receive your credit card number—only the merchant should receive this sensitive information. When you press the SEND button, your credit card information is sent out over the Internet. On its way to the merchant, it passes through various other computers (e.g., from the computer in your home, the information first passes to your Internet service provider's computer). You want to be sure that an unscrupulous computer operator (between you and the merchant) cannot intercept your credit card number. In this scenario, you (the customer) correspond to Alice, the merchant corresponds to Bob, and the unscrupulous hacker on the Internet is Eve.

Alice wants to tell Bob a secret. The problem is that everything they say to one another is heard by an eavesdropper named Eve. Can Alice tell Bob the secret? Can they hold a private conversation? Perhaps they can create a secret code and converse only in this code. The problem is that Eve can overhear everything they say to each other—including all the details of their secret code! One option is for Alice and Bob to make up their code in private (where Eve can't hear). This option could be impractical, slow, and expensive (e.g., if Alice and Bob live far apart). It seems impossible for Alice and Bob to hold a private conversation while Eve is listening to everything they say. Their attempts to pass private messages could be thwarted by the fact that Eve knows their coding system.

It is therefore an amazing fact that private communication in a public forum is possible! The key is to develop a secret code with the following property: Revealing the encryption procedure does not undermine the secrecy of the decryption procedure. The idea is to find a procedure that is relatively easy to do, but extraordinarily difficult to undo. For example, it is not hard (at least for a computer) to multiply two enormous prime numbers. However, factoring the resulting product (if we don't know the prime factors) is extremely hard.

Factoring

Suppose p and q are large prime numbers—say, around 500 digits each. It is not difficult to multiply these numbers. The result, $n = pq$, is a 1000-digit composite number. On a computer, this computation takes less than a second. Indeed, if you were compelled to multiply two 500-digit numbers with only pencil and paper (lots of paper!), you would be able to do this task in a matter of hours or days.

Suppose that instead of being given the primes p and q, you are given their product $n = pq$. You are asked to factor n to recover the prime factors p and q. You do not know p and q—you know only n. If you try to factor n using trial division, you will need to do about 10^{500} divisions, and this would take an unimaginably long period of time even on a blazingly fast computer (see Exercise 42.4).

There are more sophisticated algorithms for factoring that work much faster than trial division. We do not discuss these more complicated, but faster, methods in this book. The relevant fact is that although these techniques are much faster than trial division, they are not so tremendously fast that they can factor a 1000-digit number in a reasonable period of time (e.g., under a century).

Furthermore, running these techniques on faster computers does not make factoring significantly easier. Instead of using 500-digit primes p and q, we can use 1000-digit primes (so $n = pq$ increases from 1000 to 2000 digits). The time to multiply p and q rises modestly (about 4 times longer). However, the time to factor $n = pq$ increases enormously. The number n is not twice as big as before—it's 10^{1000} times bigger!

The point of this discussion is to convince you that it is extremely difficult to factor large integers. However, this might not be true. All I can say is that to date, there are no efficient factoring algorithms known. Mathematicians and computer scientists believe there are no efficient factoring algorithms, but to date, there is no proof that such an algorithm cannot be created.

Conjecture 43.1 There is no computationally efficient procedure for factoring positive integers.

(We have not defined the term *computationally efficient procedure,* so this conjecture's precise meaning has not been made clear. The imprecise meaning of this conjecture—"Factoring is hard!"—suffices for our purposes.)

This brings us to the second amazing fact for this section. The two techniques we present for sending private messages over public channels are based on this unproven conjecture!

The term *public key* refers to the fact that the encryption procedure is known to everyone, including the eavesdropper.

The security of public-key cryptosystems is based on ignorance, not on knowledge. Both of the public-key systems we present, Rabin's system (Section 44) and the RSA system (Section 45), can be broken by an efficient factoring algorithm. Details follow.

Words to Numbers

Alice's message to Bob will be a large integer. People normally communicate with words, so we need a system for converting a message into a number. Suppose her message is

```
Dear Bob, Do you want to go to the movies tonight? Alice
```

First, Alice converts this message into a positive integer. There is a standard way to convert the Roman alphabet into numbers; this encoding is called the ASCII code. There is nothing secret about this code. It is a standard way to represent the letters A–Z (lower and upper cases), numerals, punctuation, and so on, using numbers in the set $\{0, 1, 2, \ldots, 255\}$. For example, the letter `D` in ASCII is the number 68. The letter `e` is 101. The space character is 32. Alice's message, rendered as numbers, is

```
 D   e   a   r  spc  B   o   b   ,  spc  D   o  spc  y   o   u  ...
068 101 097 114 032 066 111 098 044 032 068 111 032 121 111 117 ...
```

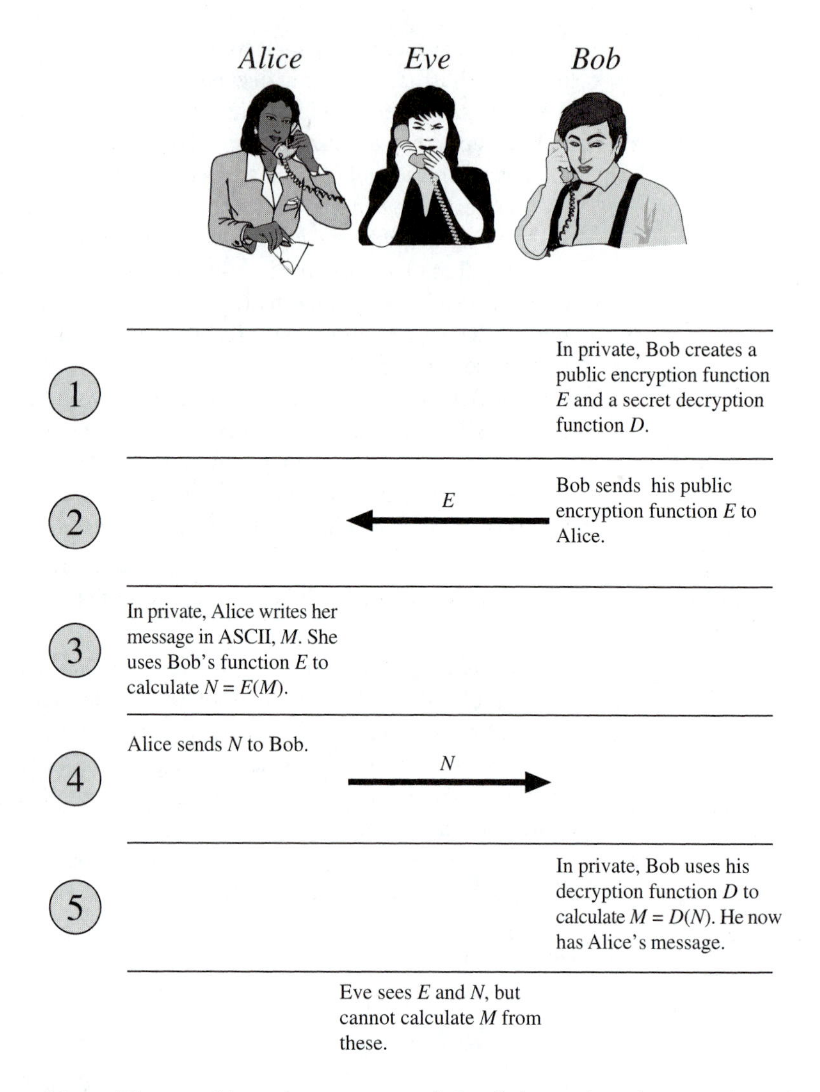

Next, Alice combines these separate three-digit numbers into one large integer, M:

$$M = 68{,}101{,}097{,}114{,}032{,}066{,}111{,}098, \ldots, 099{,}101.$$

Since Alice's original message is about 50 characters long, this message is about 150 digits long. This is how Alice sends her message to Bob:

- In the privacy of his home, Bob creates a pair of functions, D and E; these functions are inverses of one another; that is, $D(E(M)) = M$.
- Bob tells Alice the function E. At this point, Eve gets to see the function E. The function is fairly easy to compute, but it is very hard for Eve to figure out D knowing only E.
- Alice uses Bob's public encryption function E. In the privacy of her own home, she computes $N = E(M)$ (where M is the message she wants to send). She now sends the integer N to Bob. Eve gets to see this integer as well.

- Bob now uses his private decryption function D to compute $D(N)$. The result is

$$D(N) = D(E(M)) = M$$

so now Bob knows the message M. Since Eve does not know D, she cannot figure out what M is.

The challenge is to create functions E and D that work for this protocol. In the next two sections, we present two methods to accomplish this.

Cryptography and the Law

I am most certainly not an expert on law. Nonetheless, let me share some advice about the material in the next two sections.

The techniques in the next two sections are not hard to implement on a computer. Let's suppose you reside in the United States and you write a computer program that implements these cryptographic methods. Indeed, it might be a terrific software package that lots of people would like to use. You realize that since people value your work, they would be willing to pay you for this program. So you sell your program to various people, including individuals outside the United States.

Now, I hope you have an excellent lawyer, because you could be in heap of trouble. You may have violated copyright and patent laws (the RSA system is so protected) as well as U.S. export control laws (because cryptography is of military value, there are export controls restricting its sale).

The point is that you must be careful if you decide to implement the techniques we are about to present. Get knowledgeable legal advice before you start.

Recap

We introduced the central problem in public-key cryptography: How can two people, who have never met, send private messages to each other over a nonsecure channel?

43 Exercises

43.1. Write a computer program to convert ordinary text into ASCII and a sequence of ASCII numbers into ordinary text.

43.2. A message, when converted to ASCII, reads as follows:

```
71  111  111  100  32  119  111  114  107
```

What is the message?

44 Public Key Cryptography II: Rabin's Method

The challenge in public-key cryptography is to create good encryption and decryption functions. The functions should be relatively easy to compute, and (this is the central point) revealing E should not provide enough information about D for Eve to figure D out.

In this section, we present a public-key cryptosystem devised by Michael Rabin. The encryption function is especially simple. Let n be a large (e.g., 200-digit) integer. The encryption function is

$$E(M) = M^2 \bmod n.$$

Decryption involves taking a square root (in $\mathbb{Z}_n$). The integer n needs to be chosen in a special manner (described below). To understand how to decrypt messages and why Rabin's method is secure, we need to understand how to take square roots in $\mathbb{Z}_n$.

Square Roots Modulo n

Most hand-held calculators have a square root button. In the blink of an eye, your calculator can tell you that $\sqrt{17} \approx 4.1231056$. Most calculators, however, cannot give you $\sqrt{17}$ in $\mathbb{Z}_{59}$. What does this mean? When we say that 3 is the square root of 9, we mean that 3 is the root of the equation $x^2 = 9$. Now the use of the word *the* is inappropriate because 9 has two different square roots: $+3$ and -3. However, the positive square root usually enjoys preferential treatment.

In $\mathbb{Z}_{59}$ the situation is similar. When we ask for the square roots of 17, we seek those elements $x \in \mathbb{Z}_{59}$ for which $x^2 = x \otimes x = 17$. The calculator's value of $4.1231056\ldots$ is not of any help here.

There are only 59 different elements in $\mathbb{Z}_{59}$. We can simply square all of them and see which (if any) gives 17 as a result. This is painful to do by hand but fast on a computer. We find that 17 has two square roots in $\mathbb{Z}_{59}$: 28 and 31.

What is $\sqrt{18}$ in $\mathbb{Z}_{59}$? After we try all the possible values, we find that 18 does not have a square root in $\mathbb{Z}_{59}$.

Stranger still, when we search for square roots of 17 in $\mathbb{Z}_{1121}$, we find four answers: 146, 500, 621, and 975.

For this cryptographic application, we need to take square roots modulo numbers that are hundreds of digits long. Trying all the possibilities is not practical! We need a better understanding of square roots in $\mathbb{Z}_n$.

Integers whose square roots are themselves integers are called *perfect squares*. In $\mathbb{Z}_n$ there is a different term.

Definition 44.1 **(Quadratic residue)** Let n be a positive integer and let $a \in \mathbb{Z}_n$. If there is an element $b \in \mathbb{Z}_n$ such that $a = b \otimes b = b^2$, we call a a *quadratic residue modulo* n. Otherwise (there is no such b) we call a a *quadratic nonresidue*.

We do not make a comprehensive study of quadratic residues here. We limit our investigation to those facts that we need to understand the Rabin cryptosystem. We begin by studying square roots in $\mathbb{Z}_p$ where p is a prime.

Proposition 44.2 Let p be a prime and let $a \in \mathbb{Z}_p$. Then a has at most two square roots in $\mathbb{Z}_p$.

Proof. Suppose, for the sake of contradiction, that a has three (or more) square roots in $\mathbb{Z}_p$. Notice that if x is a square root of a, then so is $-x \equiv p - x$ because

$$(p-x)^2 = p^2 - 2px + x^2 \equiv x^2 \equiv a \pmod{p}.$$

Since a has three (or more) square roots, we can choose two square roots, $x, y \in \mathbb{Z}_p$, such that $x \neq \pm y$. Now let's calculate $(x-y)(x+y)$. We get

$$(x-y)(x+y) = x^2 - y^2 \equiv a - a = 0 \pmod{p}.$$

Now the condition $x \neq \pm y$ implies that $x + y \not\equiv 0 \ (p)$ and $x - y \not\equiv 0 \ (p)$ (i.e., neither $x+y$ nor $x-y$ is a multiple of p). This means that p is not a factor of either $x+y$ or $x-y$. Yet p is factor of $(x+y)(x-y)$, contradicting Lemma 38.2.⇒⇐ Therefore a has at most two square roots in $\mathbb{Z}_p$. ■

Lemma 38.2 states that if p is prime and $p|ab$, then $p|a$ or $p|b$.

Proposition 44.3 Let p be a prime with $p \equiv 3 \pmod{4}$. Let $a \in \mathbb{Z}_p$ be a quadratic residue. Then the square roots of a in $\mathbb{Z}_p$ are

$$\left[\pm a^{(p+1)/4}\right] \bmod p.$$

Proof. Let $b = a^{(p+1)/4} \bmod p$. We need to prove that $b^2 = a$.

By hypothesis, a is a quadratic residue in $\mathbb{Z}_p$, so there is an $x \in \mathbb{Z}_p$ such that $a = x \otimes x = x^2$. We now calculate

$$\begin{aligned}
b^2 &\equiv [a^{(p+1)/4}]^2 \\
&\equiv [(x^2)^{(p+1)/4}]^2 \qquad \text{(substitute } a \to x^2\text{)} \\
&\equiv [x^{(p+1)/2}]^2 \\
&\equiv x^{p+1} \\
&\equiv x^p x^1 \\
&\equiv x^2 \\
&\equiv a \pmod{p}.
\end{aligned}$$

The step $x^p x^1 \equiv x^2$ follows from Theorem 42.1 because $x^p \equiv x \ (p)$ for a prime p.

Of course, if $b^2 \equiv a \pmod{p}$, then also $(-b)^2 \equiv a \pmod{p}$. By the proof of Proposition 44.2, there can be no other square roots in $\mathbb{Z}_p$. ■

In reading through this proof, you may have noticed that we did not explicitly use the hypothesis that $p \equiv 3 \pmod{4}$. However, this hypothesis is important and is used implicitly in the proof (see Exercise 44.2).

Example 44.4 Notice that 59 is prime and $59 \equiv 3 \pmod{4}$. In $\mathbb{Z}_{59}$ we have

$$17^{(p+1)/4} = 17^{15} = 28$$

and notice that $28^2 = 28 \otimes 28 = 17$. Also $-28 \equiv 31$, and we have $31^2 = 31 \otimes 31 = 17$.

As we have discussed (see Exercise 36.14), the computation $a^b \bmod c$ can be done efficiently on a computer, so Proposition 44.3 gives us an efficient way to find square roots in $\mathbb{Z}_p$ (for primes congruent to 3 mod 4).

We mentioned earlier that 17 has four square roots in $\mathbb{Z}_{1121}$. This is not a contradiction to Proposition 44.2 because 1121 is not prime; it factors $1121 = 19 \times 59$.

Here we describe how to find the four square roots of 17. But first, some analysis. Suppose x is a square root of 17 in $\mathbb{Z}_{1121}$. This means

$$x \otimes x = 17$$

which can be rewritten

$$x^2 \equiv 17 \pmod{1121}$$

and that's the same as

$$x^2 = 17 + 1121k$$

for some integer k. We can write this (yet again!) in the following two ways:

$$x^2 = 17 + 19(59k) \qquad \text{and} \qquad x^2 = 17 + 59(19k)$$

and so

$$x^2 \equiv 17 \pmod{19} \qquad \text{and} \qquad x^2 \equiv 17 \pmod{59}.$$

This suggests that to solve $x^2 \equiv 17\ (1121)$, we should first solve the two equations

$$x^2 \equiv 17\ (19) \quad \text{and} \quad x^2 \equiv 17\ (59).$$

We have already solved the second equation: In $\mathbb{Z}_{59}$ the square roots of 17 are 28 and 31.

Fortunately, $19 \equiv 3 \pmod 4$, so we can use the formula in Proposition 44.3:

$$17^{(19+1)/4} = 17^5 \equiv 6 \pmod{19}.$$

The other square root is $-6 \equiv 13$.

Let's summarize what we know so far.

- We want to find $\sqrt{17}$ in $\mathbb{Z}_{1121}$.
- We have $1121 = 19 \times 59$.
- In $\mathbb{Z}_{19}$ the square roots of 17 are 6 and 13.
- In $\mathbb{Z}_{59}$ the square roots of 17 are 28 and 31.

Furthermore, if x is square root of 17 in $\mathbb{Z}_{1121}$, then (after we reduce x modulo 59) it is also a square root of 17 in $\mathbb{Z}_{59}$, and (after we reduce x modulo 19) it is also a square root of 17 in $\mathbb{Z}_{19}$. Thus x must satisfy the following:

$$x \equiv 6 \text{ or } 13 \pmod{19} \quad \text{and} \quad x \equiv 28 \text{ or } 31 \pmod{59}.$$

This gives us four problems to solve:

$$\begin{aligned} x &\equiv 6 \pmod{19} & \qquad x &\equiv 6 \pmod{19} \\ x &\equiv 28 \pmod{59} & \qquad x &\equiv 31 \pmod{59} \end{aligned}$$

$$\begin{aligned} x &\equiv 13 \pmod{19} & \qquad x &\equiv 13 \pmod{19} \\ x &\equiv 28 \pmod{59} & \qquad x &\equiv 31 \pmod{59}. \end{aligned}$$

We can solve each of these four problems via the Chinese Remainder Theorem (Theorem 37.5). Here we do one of the calculations. Let us solve the first system of congruences:

$$x \equiv 6 \pmod{19}$$
$$x \equiv 28 \pmod{59}.$$

Since $x \equiv 6\ (19)$, we can write $x = 6 + 19k$ for some integer k. Substituting this into the second congruence $x \equiv 28\ (59)$, we get

$$6 + 19k \equiv 28\ (59) \qquad \Rightarrow \qquad 19k \equiv 22\ (59).$$

We multiply both sides of the latter equation by $19^{-1} = 28$ (in $\mathbb{Z}_{59}$) to get

$$28 \times 19k \equiv 28 \times 22\ (59) \qquad \Rightarrow \qquad k \equiv 26\ (59).$$

Thus we can write $k = 26 + 59j$. Substituting this for k in $x = 6 + 19k$, we have

$$x = 6 + 19k = 6 + 19(26 + 59j) = 500 + 1121j$$

so we find that $x = 500$ is one of the four square roots of 17 (in $\mathbb{Z}_{1121}$).

The other three square roots of 17 are 621, 146, and 975.

Let us recap the steps we took to find the square roots of 17 in $\mathbb{Z}_{1121}$.

- We factored $1121 = 19 \times 59$.
- We found the two square roots of 17 in $\mathbb{Z}_{19}$ (they are 6 and 13) as well as the square roots of 17 in $\mathbb{Z}_{59}$ (they are 28 and 31).

  Because 19 and 59 are congruent to 3 modulo 4, we can use the formula from Proposition 44.3 to compute these square roots.
- We solve four Chinese Remainder Theorem problems corresponding to the four possible pairs of values that $\sqrt{17}$ might take in $\mathbb{Z}_{19}$ and $\mathbb{Z}_{59}$.
- The four answers to these Chinese Remainder Theorem problems are the four square roots of 17 modulo 1121.

Only one of these four steps is computationally difficult: the factoring step. The other steps (finding square roots in $\mathbb{Z}_p$ and using the Chinese Remainder Theorem) may be more novel to you, but they can be done efficiently on a computer.

This procedure can be used to find the square roots of numbers in $\mathbb{Z}_n$ provided the integer n is of the form $n = pq$ where p and q are primes with $p \equiv q \equiv 3 \pmod 4$. However, if p and q are, say, 100-digit primes, then the factoring step makes this procedure utterly impractical.

Does this imply that there is no other procedure for finding square roots? No, but let us show that finding square roots in this context is just as hard as factoring.

Theorem 44.5 Let $n = pq$ where p and q are primes. Suppose $x \in \mathbb{Z}_n$ has four distinct square roots, a, b, c, d. If these four square roots are known, then there is an efficient computational procedure to factor n.

Proof. Suppose $x \in \mathbb{Z}_n$ where $n = pq$ with p, q prime, and suppose x has four distinct square roots. For example,

$$x = a^2 = b^2 = c^2 = d^2$$

in $\mathbb{Z}_n$. Of course, since a is a square root of x, so is $-a$. Because there are four distinct square roots, we may assume that $b = -a$, but $c \neq \pm a$. Notice that

$$(a - c)(a + c) = a^2 - c^2 \equiv x - x \equiv 0 \pmod{n}.$$

This means that $(a-c)(a+c) = kpq = kn$ where k is some integer. Furthermore, since $c \neq \pm a$ (in $\mathbb{Z}_n$), we know that $a - c \not\equiv 0$ and $a + c \not\equiv 0 \ (n)$.

Therefore $\gcd(a-c, n) \neq n$ because $a-c$ is not a multiple of n. Is it possible that $\gcd(a-c, n) = 1$? If so, then neither p nor q is a divisor of $a - c$, and since $(a-c)(a+c) = kpq = kn$, we see that p and q must be factors of $a+c$, but this is a contradiction because $a+c$ is not a multiple of n. If $\gcd(a-c, n) \neq n$ and $\gcd(a-c, n) \neq 1$, what possible values remain for $\gcd(a-c, n)$? The only other divisors of n are p and q, and therefore we must have $\gcd(a-c, n) = p$ or $\gcd(a-c, n) = q$.

Since gcd can be computed efficiently, given the four square roots of x in $\mathbb{Z}_n$, we can efficiently find one of the factors of $n = pq$ and then get the other factor by division into n. ■

Example 44.6 Let $n = 38989$. The four square roots of 25 in $\mathbb{Z}_n$ are $a = 5$, $b = -5 = 38984$, $c = 2154$, and $d = -2154 = 36835$. [Please check these yourself on a computer. For example, verify that $2154^2 \equiv 25 \ (38989)$.] Now we calculate

$$\gcd(a - c, n) = \gcd(-2149, 38989) = 307$$
$$\gcd(a + c, n) = \gcd(2159, 38989) = 127$$

and, indeed, $127 \times 307 = 38989$.

Although there may be other procedures to find square roots in $\mathbb{Z}_{pq}$, an efficient procedure would be a contradiction to Conjecture 43.1. Therefore we believe there is no computationally efficient procedure to find square roots in $\mathbb{Z}_{pq}$.

The Encryption and Decryption Procedures

Alice wants to send a message to Bob. To prepare for this, Bob, in the privacy of his home, finds two large (say, 100 digits each) prime numbers p and q with $p \equiv q \equiv 3 \pmod 4$. He calculates $n = pq$. He then sends the integer n to Alice. Of course, Eve now knows n as well, but because factoring is difficult, neither Alice nor Eve knows the factors p and q.

Next, Alice, in the privacy of her home, forms the integer M by converting her words into ASCII and using the ASCII codes as the digits of her message number M. She calculates $N = M^2 \bmod n$.

Now Alice sends N to Bob. Eve receives the number N as well.

To decrypt, Bob computes the four square roots of N (in $\mathbb{Z}_n$). Because Bob knows the factors of n (namely, p and q), he can compute the square roots. This gives four possible square roots, only one of which is the message M that Alice sent. Presumably, however, only one of the four square roots is the ASCII representation of words; the other three square roots give nonsense.

Eve cannot decrypt because she does not know how to find square roots.

Thus Alice has sent Bob a message that only Bob can decrypt, and all their communication has been in public!

Recap

In this section, we discussed Rabin's public-key cryptosystem. In this system, messages are encrypted by squaring and decrypted by finding square roots. These calculations take place in $\mathbb{Z}_{pq}^*$ where p and q are primes congruent to 3 modulo 4. We explained how to find square roots in this context and noted the connection to factoring.

44 Exercises

44.1. Suppose it takes about 1 second to multiply two 500-digit numbers on a computer. Explain why we should expect it to take about 4 seconds to multiply two 1000-digit numbers.

44.2. Proposition 44.3 includes the hypothesis $p \equiv 3 \pmod 4$. This fact is not explicitly used in the proof. Explain why this hypothesis is necessary and where in the proof we (implicitly) use this condition.

44.3. Find the four square roots of 500 in $\mathbb{Z}_{589}$.

44.4. Find all values of $\sqrt{17985}$ in $\mathbb{Z}_{34751}$.

44.5. The first step in all public-key cryptosystems is to convert the English-language message into a number, M. This is typically done with the ASCII code. In this problem, we use a simpler method.

We write our messages using only the 26 uppercase letters. We use `01` to stand for `A`, `02` to stand for `B`, etc., and `26` to stand for `Z`. The word `LOVE` would be rendered as `12152205` in this encoding.

Suppose Bob's public key is $n = 328419349$. Alice encrypts her message M using Rabin's system as $M^2 \bmod n$. For example, if her message is `LOVE`, this is encrypted as

$$12152205^2 \bmod 328419349 = 27148732$$

and so she transmits `27148732` to Bob.

Alice encrypts four more words to Bob. Their encryptions are as follows:

a. `249500293`.

b. `29883150`.

c. `232732214`.

d. `98411064`.

Decrypt these four words.

44.6. *Long and short messages.* Suppose Bob's public key is a 1000-digit composite number n, and Alice encodes her message M as $E(M) = M^2 \bmod n$.

When Alice wants to send a message containing c characters, she creates an integer with $3c$ digits (using the ASCII code).

a. Suppose $3c > 1000$. What should Alice do?

b. Suppose $3c < 500$. What should Alice be concerned about? What should she do in this situation?

44.7. Let $n = 171121$; this number is the product of two primes.

The four square roots of 56248 in $\mathbb{Z}_n$ are 68918, 75406, 95715, and 102203.

Without using trial division, factor n.

44.8. Let $n = 594752966202352474884l$; this number is the product of two primes.

The four square roots of 574663446180827837l316 in $\mathbb{Z}_n$ are

$$602161451924,$$
$$190932110031878750416\,5,$$
$$403820856170473724467\,6, \quad \text{and}$$
$$594752966142136329691\,7.$$

Factor n.

44.9. The method we presented in this section is a simplified version of Rabin's method. In the complete version, the encryption function is slightly more complicated.

As in the simplified system, Bob chooses two prime numbers p and q with $p \equiv q \equiv 3 \pmod 4$, and he calculates $n = pq$. He also chooses a value $k \in \mathbb{Z}_n$. Bob's encryption function is

$$E(M) = M(M + k) \bmod n.$$

In the simplified version, we took $k = 0$.

a. Explain how Bob decrypts messages sent to him using this encryption function.

b. Suppose $n = 589$ and $k = 321$. If Alice's message is $M = 100$, what value does she send to Bob? Call this number N.

c. Bob receives the value sent by Alice [N from part (b)]. What are the (four) possible messages Alice might have sent?

45 Public Key Cryptography III: RSA

Another public-key cryptosystem is known as the RSA cryptosystem, named after its inventors, R. Rivest, A. Shamir, and L. Adleman. This method is based on Euler's extension (Theorem 42.6) to Fermat's Little Theorem 42.1; we repeat Euler's result here.

Let n be a positive integer and let a be an integer relatively prime to n. Then

$$a^{\varphi(n)} \equiv 1 \pmod n.$$

Here φ is Euler's totient: $\varphi(n)$ is the number of integers from 1 to n that are relatively prime to n. For use with the RSA system, we are especially interested in $\varphi(n)$ with $n = pq$ where p and q are distinct prime numbers. In this case, recall that

$$\varphi(n) = \varphi(pq) = pq - p - q + 1 = (p-1)(q-1)$$

(see Exercise 38.15).

The RSA Encryption and Decryption Functions

We begin our study of the RSA cryptosystem by introducing its encryption and decryption functions. In the privacy of his home, Bob finds two large (e.g., 500-digit) prime numbers p and q and calculates their product $n = pq$. He also finds two integers e and d. The numbers e and d have special properties that we explain below.

The encryption and decryption functions are

$$E(M) = M^e \bmod n \qquad \text{and} \qquad D(N) = N^d \bmod n.$$

These calculations can be done efficiently on a computer (see Exercise 36.14).

Bob tells Alice his encryption function E. In so doing, he reveals the numbers n and e not only to Alice but also to Eve. He keeps the function D secret; that is, he does not reveal the number d.

In the privacy of her home, Alice forms her message M, calculates $N = E(M)$, and sends the result to Bob. Eve gets to see N, but not M.

In the privacy of his home, Bob calculates

$$D(N) = D(E(M)) \stackrel{?}{=} M.$$

For Bob to be able to decrypt the message, it is important that we have $D(E(M)) = M$. Working in $\mathbb{Z}_n$, we want

$$D(E(M)) = D(M^e) = (M^e)^d = M^{ed} \stackrel{?}{=} M.$$

How can we make this work? Euler's theorem helps. Euler's theorem tell us that if $M \in \mathbb{Z}_n^*$, then

$$M^{\varphi(n)} = 1 \qquad \text{in } \mathbb{Z}_n^*.$$

Raising both sides of this equation to a positive integer k gives

$$\left(M^{\varphi(n)}\right)^k = 1^k \qquad \Rightarrow \qquad M^{k\varphi(n)} = 1.$$

If we multiply both sides of the last equation by M, we get

$$M^{k\varphi(n)+1} = M$$

so if $ed = k\varphi(n) + 1$, then we have $D(E(M)) = M^{ed} = M$. In other words, we want

$$ed \equiv 1 \pmod{\varphi(n)}.$$

Now we are ready to explain how to choose e and d.

Bob selects e to be a random value in $\mathbb{Z}_{\varphi(n)}^*$; that is, e is an integer between 1 and $\varphi(n)$ that is relatively prime to $\varphi(n)$. Note that because Bob knows the prime factors of n, he can calculate $\varphi(n)$.

Next he computes $d = e^{-1}$ in $\mathbb{Z}_{\varphi(n)}^*$ (see Section 36). Now we have

$$D(E(M)) = M^{ed} = M^{k\varphi(n)+1} = \left(M^{\varphi(n)}\right)^k \otimes M = 1^k \otimes M = M \qquad \text{in } \mathbb{Z}_n^*$$

and therefore, with this choice of e and d, Bob can decrypt Alice's message.

Example 45.1 Bob picks the prime numbers $p = 1231$ and $q = 337$, and computes $n = pq = 414847$. He can also compute

$$\varphi(n) = (p-1)(q-1) = 1230 \times 336 = 413280.$$

He chooses e at random in $\mathbb{Z}_{413280}^*$—say, $e = 211243$. Finally, he calculates (in $\mathbb{Z}_{413280}^*$)

$$d = e^{-1} = 166147.$$

Let us review the steps in this procedure.

- In the privacy of his home, Bob finds two very large prime numbers, p and q. He calculates $n = pq$ and $\varphi(n) = (p-1)(q-1)$.
- Still in private, Bob chooses a random number $e \in \mathbb{Z}_{\varphi(n)}^*$ and calculates $d = e^{-1}$ where the inverse is in the group $\mathbb{Z}_{\varphi(n)}^*$. He does this using Euclid's Algorithm.
- Bob tells Alice the numbers n and e (but keeps the number d secret). Eve gets to see n and e.
- In the privacy of her home, Alice forms her message M and calculates $N = E(M) = M^e \bmod n$.
- Alice sends the number N to Bob. Eve gets to see this number as well.
- In the privacy of his home, Bob calculates $D(N) = N^d = (M^e)^d = M$ and reads Alice's message.

Note: The decryption assumes M to be relatively prime to n (otherwise Euler's theorem does not apply). See Exercise 45.6 in the case that M is not relatively prime to n.

Example 45.2 (Continued from Example 45.1.) Bob's encryption/decryption functions are

$$E(M) = M^{211243} \bmod 414847 \qquad \text{and} \qquad D(N) = N^{166147} \bmod 414847.$$

Suppose Alice's message is $M = 224455$. In private, she computes,

$$E(M) = 224455^{211243} \bmod 414847 = 376682$$

and sends 376682 to Bob.

In private, Bob calculates

$$D(376682) = 376682^{166147} \bmod 414847 = 224455$$

and recovers Alice's message.

Security

Can Eve decrypt Alice's message? Let's consider what she knows. She knows Bob's public encryption function $E(M) = M^e \bmod n$, but she does not know the two prime factors of n. She also knows $E(M)$ (the encrypted form of Alice's message), but she does not know M.

If Eve can guess the message M, then she can check her guess because she too can compute $E(M)$. If Alice's message is very short (e.g., `Yes`), this might be feasible.

Otherwise Eve can try to break Bob's code. One way she can do this is to factor n. Once she has n, she can compute $\varphi(n)$ and then get $d = e^{-1}$ (in $\mathbb{Z}^*_{\varphi(n)}$). However, our supposition is that factoring is too hard for this to be feasible.

Note that Eve does not really need to know the prime factors of n. She would be happy just knowing $\varphi(n)$, so she can calculate d. This is not practical either.

Proposition 45.3 Let p and q be primes and let $n = pq$. Suppose we are given n, but we do not know p or q. If we are also given $\varphi(n)$, then we can efficiently calculate the prime factors of n.

Proof. We know that

$$n = pq, \qquad \text{and}$$
$$\varphi(n) = (p-1)(q-1).$$

This is a system of two equations in two unknowns (p and q) that we can simply solve. We write $q = n/p$ and substitute this into the second equation, which we solve via the quadratic formula. ■

Thus, if Eve could efficiently calculate $\varphi(n)$ from n, then she could efficiently factor n, contradicting Conjecture 43.1.

Example 45.4 If $n = 414847$, then $\varphi(n) = 413280$. We want to solve

$$pq = 414847 \qquad \text{and} \qquad (p-1)(q-1) = 413280.$$

We substitute $q = 414847/p$ into

$$(p-1)(q-1) = 413280$$

to get

$$(p-1)(414847/p - 1) = 413280$$

which expands to

$$414848 - \frac{414847}{p} - p = 413280$$

which rearranges to

$$p^2 - 1568p + 414847 = 0$$

whose roots are $p = 337$ and 1231 (by the quadratic formula). The prime factors of 414847 are, indeed, 337 and 1231.

Now Eve does not, in point of fact, need $\varphi(n)$. She will be happy just to know d. Is there an efficient procedure for Eve to find d given n and e? Probably not.

Proposition 45.5 Let p, q be large primes and the $n = pq$. Suppose there is an efficient procedure that, given e with $\gcd(e, \varphi(n)) = 1$, produces d with $ed \equiv 1 \pmod{\varphi(n)}$. Then there is an efficient procedure to factor n.

The proof is beyond the scope of this text, but it can be found in more advanced books on cryptology. The point is that if we believe factoring is intractable, then there is no way for Eve to recover the exponent d just from knowing e and n.

This, however, does not completely settle the issue. To break Bob's code, Eve needs to solve the equation

$$M^e \equiv N \pmod{n}$$

where she knows e, N, and n. We have been thinking about the possibility that Eve would recover the decryption function (especially the integer d) and compute M from N the same way Bob might. However, there may be other ways to solve this equation that we have not considered. It is an unsolved problem to prove that breaking RSA is as hard as factoring.

Recap

The RSA cryptosystem is a public-key system. Bob (the recipient) chooses two large primes, p and q, and calculates $n = pq$. He also finds e and d with $ed \equiv 1 \ (\varphi(n))$. He then (publicly) tells Alice his encryption function $E(M) = M^e \bmod n$, while holding confidential his decryption function $D(N) = N^d \bmod n$.

In private, Alice forms her message M, computes $N = E(M)$, and transmits N to Bob.

Finally, in private, Bob takes the value he received, N, and computes $D(N) = D[E(M)] = M$ to recover Alice's message M.

45 Exercises

45.1. Suppose $n = 589 = 19 \times 31$ and let $e = 53$. Bob's encryption function is $E(M) = M^e \bmod n$. What is his decryption function?

45.2. Suppose $n = 589 = 19 \times 31$ and let $d = 53$. Bob's decryption function is $D(N) = N^d \bmod n$. What is his encryption function?

45.3. Suppose Bob's encryption function is $E(M) = M^{53} \pmod{589}$. Alice encrypts a message M, calculates $E(M) = 289$, and sends the value 289 to Bob. What was her message M?

45.4. The first step in all public-key cryptosystems is to convert the English-language message into a number, M. This is typically done with the ASCII code. In this problem, we use a simpler method.

We write our messages using only the 26 uppercase letters. We use `01` to stand for `A`, `02` to stand for `B`, etc., and `26` to stand for `Z`. The word `LOVE` would be rendered as `12152205` in this encoding. (This is the same method as in Exercise 44.5.)

Suppose Bob's RSA public key is $(n, e) = (328419349, 220037467)$. To encrypt the word `LOVE`, Alice calculates

$$12152205^{220037467} \bmod 328419349 = 76010536$$

and sends `76010536` to Bob.

Alice encrypts four more words to Bob. Their encryptions are as follows:

a. `322776966`.

b. `43808278`.

c. `166318297`.

d. `18035306`.

Decrypt these four words.

45.5. Suppose Bob creates two RSA encryption algorithms as follows: First, he picks large primes p and q and calculates $n = pq$. Next he chooses two integers e_1 and e_2 with $\gcd(e_1, \varphi(n)) = \gcd(e_2, \varphi(n)) = 1$ to make two encryption functions:

$$E_1(M) = M^{e_1} \bmod n, \quad \text{and}$$
$$E_2(M) = M^{e_2} \bmod n$$

When Alice puts her message into code, she double-encrypts it by calculating

$$N = E_1(E_2(M))$$

and sends N to Bob.

Please answer the following:

a. How should Bob decrypt the message he receives from Alice?

b. Suppose, by mistake, Alice calculates $N' = E_2(E_1(M))$ and, unbeknownst to Bob, sends N' instead of N to him. What will happen when Bob decrypts N'?

c. How much harder is it, using this double-encryption method, for Eve to decrypt Alice's message (compared to standard single encryption)?

45.6. Let Bob's encryption function be $E(M) = M^e \bmod n$ where $n = pq$ for distinct primes p and q. His decryption function is $D(N) = N^d \bmod n$ where $ed \equiv 1 \pmod{\varphi(n)}$.

Suppose Alice forms a message M (with $1 \leq M < n$) that is not relatively prime to n. You may suppose that M is a multiple of p, but not of q.

Prove that $D(E(M)) = M$.

Chapter 8 Self Test

1. For real numbers x and y, define an operation $x * y$ by

$$x * y = \sqrt{x^2 + y^2}.$$

Please answer the following questions and justify your responses.

a. Evaluate $3 * 4$.

b. Is the operation $*$ closed for real numbers?

c. Is the operation $*$ commutative?

d. Is the operation $*$ associative?

e. Does the operation $*$ have an identity element?

2. List the elements in $\mathbb{Z}_{32}^*$ and find $\varphi(32)$.

3. Consider the group $(\mathbb{Z}_{15}^*, \otimes)$. Find the following subsets of $\mathbb{Z}_{15}^*$:

a. $H = \{x \in \mathbb{Z}_{15}^* : x \otimes x = 1\}$, and

b. $K = \{x \in \mathbb{Z}_{15}^* : x = y \otimes y \text{ for some } y \in \mathbb{Z}_{15}^*\}$.

4. Let $(G, *)$ be an Abelian group. Define the following two subsets of G:

a. $H = \{x \in G : x * x = e\}$, and

b. $K = \{x \in G : x = y * y \text{ for some } y \in G\}$.

Prove that $(H, *)$ and $(K, *)$ are subgroups of $(G, *)$.

Furthermore, give examples to demonstrate that if the requirement that $(G, *)$ be Abelian is deleted, H and K do not necessarily constitute subgroups.

5. Let $(G, *)$ be a group with exactly three elements. Prove that G is isomorphic to $(\mathbb{Z}_3, \oplus)$.

6. Find an isomorphism between $(\mathbb{Z}_{13}^*, \otimes)$ and $(\mathbb{Z}_{12}, \oplus)$.

7. Let $(G, *)$ be a group and let $(H, *)$ and $(K, *)$ be subgroups. Define the set $H * K$ to be the set of all elements for the form $h * k$ where $h \in H$ and $k \in K$; that is,

$$H * K = \{g \in G : g = hk \text{ for some } h \in H \text{ and } k \in K\}.$$

a. In $(\mathbb{Z}_{100}, \oplus)$ let $H = \{0, 25, 50, 75\}$ and $K = \{0, 20, 40, 60, 80\}$. Find the set $H \oplus K$.

b. Prove: If $(G, *)$ is an Abelian group and H and K are subgroups, then $H * K$ is also a subgroup.

c. Show that the result in part (b) is false if the word *Abelian* is deleted.

8. Show that for all elements g of $(\mathbb{Z}_{15}^*, \otimes)$, we have $g^4 = 1$.

Use this to prove that the group $(\mathbb{Z}_{15}^*, \otimes)$ is not cyclic.

9. Without the use of any computational aid, calculate $2^{90} \bmod 89$.

10. Let $n = 38168467$. Use the fact that

$$2^n \equiv 6178104 \pmod{n}$$

to determine whether n is prime or composite.

11. Let $n = 38168467$. Given that $\varphi(n) = 38155320$, calculate (without the assistance of a computational aid)

$$2^{38155321} \bmod 38168467.$$

12. Using only a basic handheld calculator, compute

$$874^{256} \bmod 9432.$$

13. Find all values of $\sqrt{71}$ in $\mathbb{Z}_{883}$.

14. Find all values of $\sqrt{1}$ in $\mathbb{Z}_{440617}$. Note that 440617 factored into primes is 499×883.

15. Let $n = 5460947$. In $\mathbb{Z}_n$ we have

$$1235907^2 = 1842412^2 = 3618535^2 = 4225040^2 = 1010120.$$

Use this information to factor n.

Note: You should find that n is the product of two distinct primes.

16. Alice and Bob communicate using the Rabin public-key cryptosystem. Bob's public key is $n = 713809$.

Alice sends a message to Bob. She first converts her message (a three-letter word) to a number by taking `A` to be `01`, `B` to be `02`, and so on. Then she encrypts her message using Bob's public key and sends the result, 496410, to Bob.

Given that $713809 = 787 \times 907$, decrypt Alice's message.

17. Alice and Bob switch to using the RSA public-key cryptosystem. Alice's public key is $(n, e) = (453899, 449)$. Given that $453899 = 541 \times 839$, find Alice's private decryption exponent, d.

18. Bob sends Alice a message using Alice's RSA public key (as described in the previous problem). Using `A` is `01`, `B` is `02`, etc., Bob converts his message (a three-letter word) into an integer M, and encrypts using Alice's encryption function. The result is $E_A(M) = 105015$.

What was Bob's message?

19. Given that $n = 40119451$ is the product of two distinct primes and $\varphi(n) = 40106592$, factor n.

CHAPTER

9 Graphs

The word *graph* has several meanings. In nonmathematical English, it refers to a method of representing an idea or concept with a picture or in writing. In both mathematics and in English, it often refers to a diagram used to show the relationship of one quantity to another.

In this chapter, we introduce an entirely different mathematical meaning for the word *graph*. For us, a graph is not a picture drawn on x and y axes.

46 Fundamentals of Graph Theory

Before we say just what a graph is or give a formal definition of the word *graph*, we consider some interesting problems.

Map Coloring

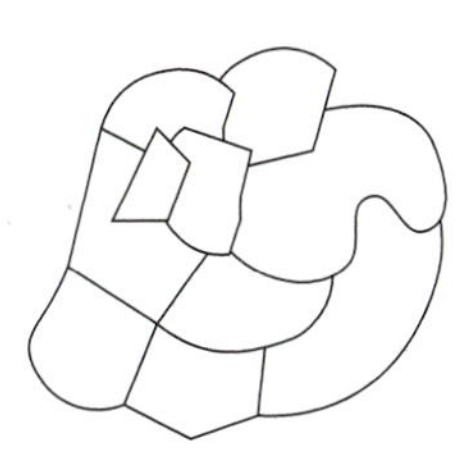

Imagine a map of a mythical continent that has several countries. You are a cartographer charged with designing a map of your continent. To show the different countries clearly, you fill their regions using various colors. However, if you were to make every country a different color, the map would be garish.

To make the map clear, but not gaudy, you decide to use as few colors as possible. However, to maintain clarity, you insist that neighboring countries should not receive the same color.

The question is: What is the smallest number of colors you need to color your map?

The question refers not just to the map in the figure, but to any map that might be drawn. Well, not quite *any* map. We do not allow countries that are disconnected. (For example, Russia includes a region north of Poland and west of Lithuania that is disconnected from the rest of Russia. The United States is in multiple pieces, and the U.S. state Michigan is in two pieces: the upper and lower peninsulas.) Furthermore, regions that touch at just one point need not receive different colors. (For example, the U.S. states Arizona and Colorado may be the same color.)

We can color the map in the figure with just four colors, as shown. This raises a few questions.

- Can this map be colored with fewer than four colors? (Notice that we have only one country that is gray; perhaps if we are clever, we can color this map with only three colors.)
- Is there another map that can be colored with fewer than four colors?
- Is there a map that requires more than four colors?

The answer to the first question is no; this map cannot be colored with fewer than four colors. Can you prove this yourself? We shall return to this specific question in a later section, but try this on your own now.

The answer to the second question is yes. This is an easy question. Try drawing a map that requires only two colors. (Hint: Make your life easy and build a continent with only two countries!)

The third question, however, is notoriously difficult. This problem is known as the *four color map problem*. It was first posed in 1852 by Francis Guthrie and remained unsolved for about a century until, in the mid-1970s, Appel and Haken proved that every map can be colored using at most four colors. We discuss this further in Sections 51 and 52.

Map coloring might seem like a frivolous problem. Instead, let us consider the following: Imagine a university in which there are thousands of students and hundreds of courses. As in most universities, at the end of each term there is an examination period. Each course has a 3-hour final exam. On any given day, the university can schedule two final exams.

Now it would be quite impossible for a student enrolled in two courses to take both final exams if they were held during the same time slot. Recognizing this, the university wishes to devise a final examination schedule with the condition that if a student is enrolled in two courses, these courses must get different examination periods.

A simple solution to this problem is to hold only one examination during any time slot. The problem, of course, is that if the examination period begins in May, it won't end until November!

The solution the university prefers is to have the smallest possible number of examination slots. This way, students (and faculty ☺) can go on their summer vacations as soon as possible.

At first glance, this examination-scheduling problem seems to have little in common with map coloring, but we assert that these problems are essentially the same. In map coloring, we seek the least number of colors, subject to a special condition (countries that share a common border receive different colors). In exam scheduling, we seek the least number of time slots subject to a special condition (courses that share a common student receive different time slots).

Problem	**Map Coloring**	**Exam Scheduling**
Assign	colors	time slots
to	countries	courses
condition	common border $\Rightarrow$ different colors	common student $\Rightarrow$ different slots
objective	fewest colors	fewest time slots

Both problems—map coloring and exam scheduling—have the same basic structure.

Three Utilities

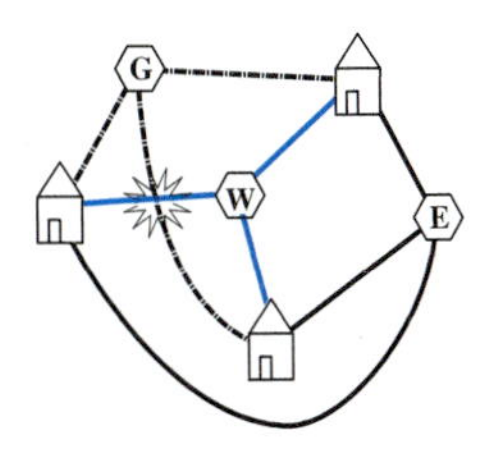

The following is a classic puzzle. Imagine a "city" containing three houses and three utility plants. The three utilities supply gas, water, and electricity. As an urban planner, your job is to run connections from every utility plant to every home. You need to have three electric wires (from the electric plant to each of the three houses), three water pipes (from the water plant to the houses), and three gas lines (from the gas facility to the houses). You may place the houses and the utility plants anywhere you desire. However, you may not allow two wires/pipes/lines to cross! The diagram shows a failed attempt to construct a suitable layout.

I highly recommend you try this problem yourself. After many tries, you may come to believe that no solution is possible. This is correct. It is impossible to construct a gas/water/electric layout to three houses without at least one pair of crossing lines. Later we prove this.

This may seem like a frivolous problem. However, consider the following: A printed circuit board is a flat piece of plastic on which various electronic devices (resistors, capacitors, integrated circuits, etc.) are mounted. Connections between these devices are made by printing bare metal wires onto the surface of the board. If two of these wires were to cross, there would be a short circuit. The problem is: Can we print the various connecting wires onto the board in such a way that there are no crossings?

If there must be crossings, then the circuit board can be constructed in layers, but this is more expensive. Finding a noncrossing layout saves production costs and therefore is worthwhile (especially for a mass-produced device).

The gas/water/electric layout problem is a simplified version of the more complicated "printed circuit board" problem.

Seven Bridges

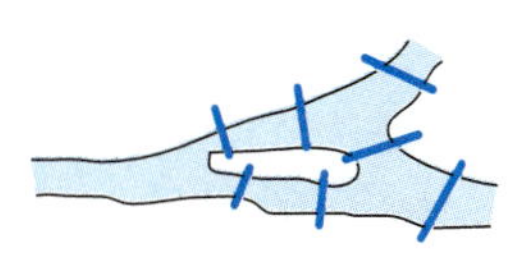

The following is another classic puzzle. In the late 1700s, in the city of Königsburg (now called Kaliningrad) located in the aforementioned disconnected section of Russia, there were seven bridges connecting various parts of the city; these were configured as shown in the figure.

The townspeople enjoyed strolling through their city in the evening. They wondered: Is there a tour we can take through our city so that we cross every bridge exactly once?

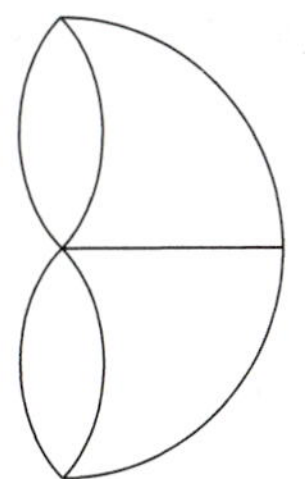

I recommend you try solving this problem yourself. After a number of frustrating false starts, you may decide that no such tour is possible. This is correct. The proof of this fact is attributed to Euler. Euler abstracted the problem into a diagram akin to the one shown in the figure. Each line in the diagram represents a bridge in Königsburg. The problem of walking the seven bridges is now replaced by the problem of drawing the abstract figure without lifting your pencil from the paper and without redrawing a line. Can this figure be so drawn? In the diagram, there are four places where lines come together; at each of these places, the number of lines is odd. We claim: If we could draw this figure in the manner described, a point where an odd number of lines meet must be either the starting point or the finishing point of the drawing. Think about an intermediate point of the drawing—that is, any junction other than where we start or where we end. At this junction, there must be an even number of lines because every time we enter this point along one line, we leave it along another (recall that we are not allowed to retrace a line). So every point in the diagram must be either the first or the last point in the drawing. Of course, this is not possible because there are four such points. Therefore it is impossible to draw the diagram without retracing a line or lifting your pencil, and therefore it is impossible to tour the city of Königsburg and cross each of the seven bridges exactly once.

This is a nice puzzle, but again, it seems a bit frivolous. Here is the same problem again in a more serious setting. Once again, don your urban-planning hat. Now, instead of distributing utility services, you are charged instead with the glamorous job of overseeing garbage collection. Your small city can afford only one garbage truck. Your job is to set the route the garbage truck is to follow. It needs to collect along every street in your city. It would be wasteful if the truck were to traverse the same street more than once. Can you find a route for the garbage truck so that it travels only once down every street?

If your city has more than two intersections where an odd number of roads meet, then such a tour is not possible.

What Is a Graph?

The three problems we considered are modeled best by using the mathematical notion of a *graph*.

Definition 46.1 **(Graph)** A *graph* is a pair $G = (V, E)$ where V is a finite set and E is a set of two-element subsets of V.

This definition is tricky to understand and may appear to have nothing to do with the motivational problems we introduced. Let us study it carefully, beginning with an example.

Example 46.2 Let

$$G = (\{1, 2, 3, 4, 5, 6, 7\}, \{\{1, 2\}, \{1, 3\}, \{2, 3\}, \{3, 4\}, \{5, 6\}\}).$$

Here V is the finite set $\{1, 2, 3, 4, 5, 6, 7\}$ and E is a set containing 5 two-element subsets of V: $\{1, 2\}$, $\{1, 3\}$, $\{2, 3\}$, $\{3, 4\}$, and $\{5, 6\}$. Therefore $G = (V, E)$ is a graph.

The elements of V are called the *vertices* (singular: *vertex*) of the graph, and the elements of E are called the *edges* of the graph. Remember, the elements of E are subsets of V, each of which contains exactly two vertices. The graph in Example 46.2 has seven vertices and five edges.

There is a nice way to draw pictures of graphs. These pictures make graphs much easier to understand. It is vital, however, that you realize that a picture of a graph is not the same thing as the graph itself!

To draw a picture of a graph, we draw a dot for each vertex (element of V). For the graph in Example 46.2, we would draw seven dots and label them with the integers 1 through 7. Each edge in E is drawn as a curve in the diagram. For example, if $e = \{u, v\} \in E$, we draw the edge e as a curve joining the dot for u to the dot for v. The following three pictures all depict the same graph from Example 46.2.

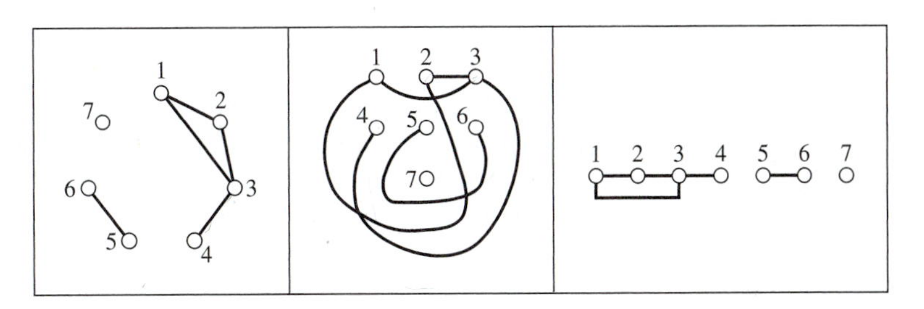

The middle picture is a perfectly valid drawing of the graph. Three pairs of edges cross each other; this is not a problem. The dots in the pictures represent the vertices, and the curves in the pictures represent the edges. We can "read" the pictures and, from them, determine the vertices and edges of the graph. The crossings may make the picture harder to understand, but they do not change the basic information the picture conveys. The first and third pictures are better only because they are clearer and easier to understand.

Adjacency

Definition 46.3 **(Adjacent)** Let $G = (V, E)$ be a graph and let $u, v \in V$. We say that u is *adjacent* to v provided $\{u, v\} \in E$. The notation $u \sim v$ means that u is adjacent to v.

WARNING! DANGER!→

We, most emphatically, do not say that u is "connected" to v. The phrase *is connected to* has an entirely different meaning (discussed later). We may say that u is *joined* to v.

Endpoint.

If $\{u, v\}$ is an edge of G, we call u and v the *endpoints* of the edge. This language is suggestive of the drawing of G: the endpoints of the curve that represents the edge $\{u, v\}$ are the dots that represent the vertices u and v. However, it is important to remember always that an edge of a graph is not a curve or a line segment; it is a two-element subset of the vertex set.

Dropping the curly braces.

It is sometimes cumbersome to write the curly braces for an edge $\{u, v\}$. Provided there is no chance for confusion, it is acceptable to write uv in place of $\{u, v\}$.

Incident.

Suppose v is a vertex and an endpoint of the edge e. We can express this fact as $v \in e$ since e is a two-element set, one of whose elements is v. We also say that v is *incident on* (or *incident with*) e.

Notice that *is-adjacent-to* ($\sim$) is a relation defined on the vertex set of a graph G. Which of the various properties of relations does is-adjacent-to exhibit?

- Is $\sim$ reflexive?

  No.

  This would mean that $u \sim u$ for all vertices in V. This means that $\{u, u\}$ is an edge of the graph. However, by Definition 46.1, an edge is a two-element subset of V. Note that although we have written u twice between curly braces, $\{u, u\}$ is a one-element set. An object either is or is not an element of a set; it cannot be an element "twice."
- Is $\sim$ irreflexive?

  Yes, but.

  By the previous discussion, it is never the case that $\{u, u\}$ is an edge of a graph. Thus a vertex is never adjacent to itself and therefore $\sim$ is irreflexive.

  Then why, you may wonder, did we answer this question "Yes, but"? We were quite emphatic (and remain so) that a vertex can never be considered adjacent to itself. The issue is over the very word *graph*. According to Definition 46.1, an edge of a graph is a two-element subset of V—end of story. However, some mathematicians use the word *graph* in a different way and allow the possibility that a vertex could be adjacent to itself; an edge joining a vertex to itself is called a *loop*. For us, graphs are not allowed to have loops. Some authors also allow more than one edge with the same endpoints; such edges are called *parallel* edges. Again, for us, graphs may not have parallel edges. The set $\{u, v\}$ either is or is not an edge—it can't be an edge "twice."

  When we want to be perfectly clear, we use the term *simple graph*. If we wish to discuss a "graph" that may have loops and multiple edges, we use the word *multigraph*.

Mathspeak!

The word *graph* is not 100% standardized. What we call a *graph* is often called a *simple graph*. There are other, more exotic, forms of graphs.

- Is $\sim$ symmetric?

  Yes.

  Suppose u and v are vertices of a graph G. If $u \sim v$ in G, this means that $\{u, v\}$ is an edge of G. Of course, $\{u, v\}$ is the exact same thing as $\{v, u\}$, so $v \sim u$. Therefore $\sim$ is symmetric.
- Is $\sim$ antisymmetric?

  In general, no.

  Consider the graph from Example 46.2. In this graph, $1 \sim 2$ and $2 \sim 1$ but, of course, $1 \neq 2$. Therefore $\sim$ is not antisymmetric.

  However, it is possible to construct a graph in which $\sim$ is antisymmetric (see Exercise 46.10).
- Is $\sim$ transitive?

  In general, no.

  Consider the graph from Example 46.2. Notice that $2 \sim 3$ and $3 \sim 4$, but 2 is not adjacent to 4.

  However, it is possible to construct a graph in which $\sim$ is transitive (see Exercise 46.10).

A Matter of Degree

Let $G = (V, E)$ be a graph and suppose u and v are vertices of G. If u and v are adjacent, we also say that u and v are *neighbors*. The set of all neighbors of a vertex v is called the *neighborhood* of v and is denoted $N(v)$. That is,

$$N(v) = \{u \in V : u \sim v\}.$$

For the graph in Example 46.2, we have

$$\begin{array}{llll} N(1) = \{2, 3\} & N(2) = \{1, 3\} & N(3) = \{1, 2, 4\} & N(4) = \{3\} \\ N(5) = \{6\} & N(6) = \{5\} & N(7) = \emptyset. & \end{array}$$

The number of neighbors of a vertex is called its *degree*.

Definition 46.4 **(Degree)** Let $G = (V, E)$ be a graph and let $v \in V$. The degree of v is the number of edges with which v is incident. The degree of v is denoted $d_G(v)$ or, if there is no risk of confusion, simply $d(v)$.

Some graph theorists call the degree of a vertex its *valence*. This is a lovely term! The word was chosen because graphs serve as models of organic molecules. The *valence* of an atom in a molecule is the number of bonds it forms with its neighbors.

In other words,

$$d(v) = |N(v)|.$$

For the graph in Example 46.2, we have

$$\begin{array}{llll} d(1) = 2 & d(2) = 2 & d(3) = 3 & d(4) = 1 \\ d(5) = 1 & d(6) = 1 & d(7) = 0. & \end{array}$$

The notation

$$\sum_{v \in V} d(v)$$

means we add the quantity $d(v)$ for all vertices $v \in V$.

Something interesting happens when we add the degrees of the vertices of a graph. For Example 46.2, we have

$$\begin{aligned} \sum_{v \in V} d(v) &= d(1) + d(2) + d(3) + d(4) + d(5) + d(6) + d(7) \\ &= 2 + 2 + 3 + 1 + 1 + 1 + 0 = 10 \end{aligned}$$

which, you might notice, is exactly twice the number of edges in G. This is not a coincidence.

Theorem 46.5 Let $G = (V, E)$. The sum of the degrees of the vertices in G is twice the number of edges; that is,

$$\sum_{v \in V} d(v) = 2|E|.$$

Proof. Suppose the vertex set is $V = \{v_1, v_2, \ldots, v_n\}$. We can create an $n \times n$ chart as follows. The entry in row i and column j of this chart is 1 if $v_i \sim v_j$ and is 0 otherwise. For the graph from Example 46.2, the chart would look like this:

$$\begin{bmatrix} 0 & 1 & 1 & 0 & 0 & 0 & 0 \\ 1 & 0 & 1 & 0 & 0 & 0 & 0 \\ 1 & 1 & 0 & 1 & 0 & 0 & 0 \\ 0 & 0 & 1 & 0 & 0 & 0 & 0 \\ 0 & 0 & 0 & 0 & 0 & 1 & 0 \\ 0 & 0 & 0 & 0 & 1 & 0 & 0 \\ 0 & 0 & 0 & 0 & 0 & 0 & 0 \end{bmatrix}.$$

This chart is called the *adjacency matrix* of the graph.

Our technique for proving Theorem 46.5 is *combinatorial proof* (see Proof Template 9). We ask,

How many 1s are in this chart?

We give two answers to this question.

- *First answer*: Notice that for every edge of G there are exactly two 1s in the chart. For example, if $v_i v_j \in E$, then there is a 1 in position ij (row i, column j) and a 1 in position ji. Thus the number of 1s in this chart is exactly $2|E|$.
- *Second answer*: Consider a given row of this chart—say, the row corresponding to some vertex v_i. There is a 1 in this row exactly for those vertices adjacent to v_i (i.e., there is a 1 in the jth spot of this row exactly when there is an edge from v_i to v_j). Thus, the number of 1s in this row is exactly the degree of the vertex—that is, $d(v_i)$.

  The number of 1s in the entire chart is the sum of the row subtotals. In other words, the number of 1s in the chart equals the sum of the degrees of the vertices of the graph.

Because these two answers are both correct solutions to the question "How many 1s are in this chart?" we conclude that the sum of the degrees of the vertices of G (answer 2) equals twice the number of edges of G (answer 1). ■

Further Notation and Vocabulary

There are many new terms to learn when studying graphs. Here we introduce more terms and notation that are often used in graph theory.

- *Maximum and minimum degree.*

  The maximum degree of a vertex in G is denoted $\Delta(G)$. The minimum degree of a vertex in G is denoted $\delta(G)$. The letters Δ and δ are upper- and lowercase Greek deltas, respectively. For the graph in Example 46.2, we have $\Delta(G) = 3$ and $\delta(G) = 0$.
- *Regular graphs.*

  If all vertices in G have the same degree, we call G *regular*. If a graph is regular and all vertices have degree r, we also call the graph r-regular. The graph in the figure is 3-regular.
- *Vertex and edge sets.*

  Let G be a graph. If we neglect to give a name to the vertex and edge sets of G, we can simply write $V(G)$ and $E(G)$ for the vertex and edge sets, respectively.
- *Order and size.*

  Let $G = (V, E)$ be a graph. The *order* of G is the number of vertices in G—that is, $|V|$. The *size* of G is the number of edges—that is, $|E(G)|$.

  It is customary (but certainly not mandatory) to use the letters n and m to stand for $|V|$ and $|E|$, respectively.

  Various authors invent special symbols to stand for the number of vertices and the number of edges in a graph. Personally, I like the following:

$$\nu(G) = |V(G)| \qquad \text{and} \qquad \varepsilon(G) = |E(G)|.$$

The terms *vertex* and *edge* are not 100% standardized. Some authors refer to vertices as *nodes*, and others call them *points*. Similarly, edges are variously called *arcs*, *links*, and *lines*.

You should think of ν and ε as functions that, given a graph, return the number of vertices and edges, respectively.

The Greek letter ν (nu) corresponds to the Roman letter n (the usual letter for the number of vertices in a graph), and it looks like a v (for vertices). The Greek letter ε (a stylized epsilon) corresponds to the Roman e (for edges).

- *Complete graphs.*

  Let G be a graph. If all pairs of distinct vertices are adjacent in G, we call G *complete*. A complete graph on n vertices is denoted K_n. The graph in the figure is a K_5.

  The opposite extreme is a graph with no edges. We call such graphs *edgeless*.

  A graph with no vertices (and hence no edges) is called an *empty graph*.

Recap

We began by motivating the study of graph theory with three classic problems (and nonfrivolous variations thereof). We then formally introduced the concept of a graph, being careful to distinguish between a graph and its drawing. We studied the adjacency relation, concluding with the result that the sum of the degrees of the vertices in a graph equals twice the number of edges in the graph. Finally, we introduced additional graph theory terminology.

46 Exercises

46.1. The following pictures represent graphs. Please write each of these graphs as a pair of sets (V, E).

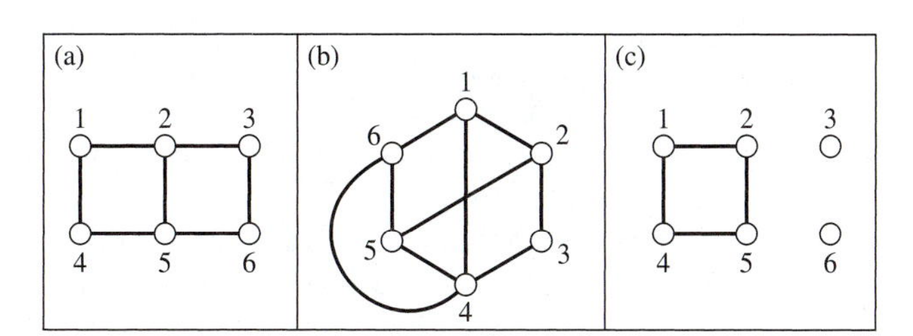

46.2. Draw pictures of the following graphs.
a. $(\{a, b, c, d, e\}, \{\{a, b\}, \{a, c\}, \{a, d\}, \{b, e\}, \{c, d\}\})$.
b. $(\{a, b, c, d, e\}, \{\{a, b\}, \{a, c\}, \{b, c\}, \{b, d\}, \{c, d\}\})$.
c. $(\{a, b, c, d, e\}, \{\{a, c\}, \{b, d\}, \{b, e\}\})$.

46.3. In the map-coloring problem, why do we require that countries be connected (and not in multiple pieces like Russia or Michigan)?

Draw a map, in which disconnected countries are permitted, that requires more than four colors.

46.4. In the map-coloring problem, why do we allow countries that meet at only one point to receive the same color?

Draw a map that requires more than four colors if countries that meet only at one point must get different colors.

46.5. If three countries on a map all border each other, then the map certainly requires at least three colors. (For example, look at Brazil, Venezuela, and Colombia or at France, Germany, and Belgium.)

Devise a map in which no three countries border each other, and yet the map cannot be colored with fewer than three colors.

46.6. Imagine creating a map on your computer screen. This map wraps around the screen in the following way. A line that moves off the right side of the screen instantly reappears at the corresponding position on the left. Similarly, a line that drops off the bottom of the screen instantly reappears at the corresponding position at the top. Thus it is possible to have a country on this map that has a little section on the left and another little section on the right of the screen, but is still in one piece.

Devise such a computer-screen map that requires more than four colors.

Try to create such a map that requires seven colors! (It is possible.)

46.7. *Refer to the previous problem about drawing on your computer screen.* On this screen, can you solve the gas/water/electricity problem? That is, find a way to place the three utilities and the three houses such that the connecting lines don't cross. You may, of course, take advantage of the fact that a pipe can wrap from the left side of the screen across to the exact same point on the right or from top to bottom.

46.8. *Continued from the previous problem.* Suppose now you wish to add a cable television facility to your computer screen city. Can you run three television cables from the cable TV headquarters to each of the three houses without crossing any of the gas/water/electric lines?

46.9. Recall the university examination-scheduling problem. Create a list of courses and students such that more than four final examination periods are required.

46.10. Construct a graph G for which the is-adjacent-to relation, $\sim$, is antisymmetric.

Construct a graph G for which the is-adjacent-to relation, $\sim$, is transitive.

46.11. In Definition 46.4 (degree), we defined $d(v)$ to be the number of edges incident with v. However, we also said that $d(v) = |N(v)|$. Why is this so?

Is $d(v) = |N(v)|$ true for a multigraph?

46.12. Let G be a graph. Prove that there must be an even number of vertices of odd degree. (For example, the graph in Example 46.2 has exactly two vertices of odd degree.)

46.13. Prove that in any graph with two or more vertices, there must be two vertices of the same degree.

46.14. Let G be an r-regular graph with n vertices and m edges. Find (and prove) a simple algebraic relation between r, n, and m.

46.15. Find all 3-regular graphs on nine vertices.

46.16. How many edges are in K_n, a complete graph on n vertices?

46.17. How many different graphs can be formed with vertex set $V = \{1, 2, 3, \ldots, n\}$?

46.18. What does it mean for two graphs to be the same? Let G and H be graphs. We say that G is *isomorphic* to H provided there is a bijection $f : V(G) \to V(H)$ such that for all $a, b \in V(G)$ we have $a \sim b$ (in G) if and only if $f(a) \sim f(b)$ (in H). The function f is called an *isomorphism* of G to H.

We can think of f as renaming the vertices of G with the names of the vertices in H in a way that preserves adjacency. Less formally, isomorphic graphs have the same drawing (except for the names of the vertices).

Please do the following:

a. Prove that isomorphic graphs have the same number of vertices.

b. Prove that if $f : V(G) \to V(H)$ is an isomorphism of graphs G and H and if $v \in V(G)$, then the degree of v in G equals the degree of $f(v)$ in H.

c. Prove that isomorphic graphs have the same number of edges.

d. Give an example of two graphs that have the same number of vertices and the same number of edges but are not isomorphic.

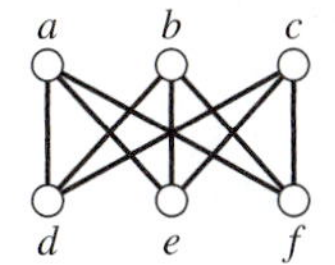

e. Let G be the graph whose vertex set is $\{1, 2, 3, 4, 5, 6\}$. In this graph, there is an edge from v to w if and only if $v - w$ is odd. Let H be the graph in the figure.
Find an isomorphism $f : V(G) \to V(H)$.

47 Subgraphs

Informally, a *subgraph* is a graph contained inside another graph. Here is a careful definition:

Definition 47.1 **(Subgraph)** Let G and H be graphs. We call G a *subgraph* of H provided $V(G) \subseteq V(H)$ and $E(G) \subseteq E(H)$.

Example 47.2 Let G and H be the following graphs:

$$V(G) = \{1, 2, 3, 4, 6, 7, 8\} \qquad V(H) = \{1, 2, 3, 4, 5, 6, 7, 8, 9\}$$

$$E(G) = \{\{1, 2\}, \{2, 3\}, \{2, 6\}, \{3, 6\}, \{4, 7\}, \{6, 8\}, \{7, 8\}\}$$

$$E(H) = \{\{1, 2\}, \{1, 4\}, \{2, 3\}, \{2, 5\}, \{2, 6\}, \{3, 6\}, \{3, 9\}, \{4, 7\}, \{5, 6\}, \{5, 7\}, \{6, 8\}, \{6, 9\}, \{7, 8\}, \{8, 9\}\}$$

Notice that $V(G) \subseteq V(H)$ and $E(G) \subseteq E(H)$, and so G is a subgraph of H. Pictorially, these graphs are

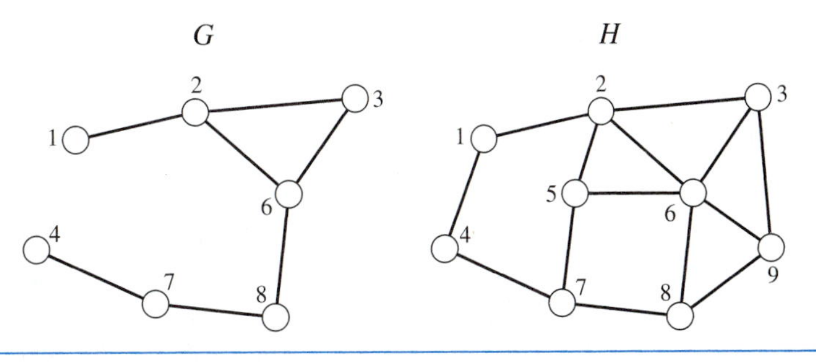

Naturally, if G is a subgraph of H, we call H a *supergraph* of G.

Induced and Spanning Subgraphs

Edge deletion.

We form a subgraph G from a graph H by deleting various parts of H. For example, if e is an edge of H, then removing e from H results in a new graph that we denote $H - e$. Formally, we can write this as

$$V(H-e) = V(H)$$
$$E(H-e) = E(H) - \{e\}.$$

If we form a subgraph of H solely by use of edge deletion, the resulting subgraph is called a *spanning* subgraph of H. Here is another way to express this:

Definition 47.3 **(Spanning subgraph)** Let G and H be graphs. We call G a *spanning subgraph* of H provided G is a subgraph of H, and $V(G) = V(H)$.

When G is a spanning subgraph of H, the definition requires that $V(G) = V(H)$; that is, G and H have all the same vertices. Thus the only allowable deletions from H are edge deletions.

Example 47.4 Let H be the graph from Example 47.2 and let G be the graph with

$$V(G) = \{1, 2, 3, 4, 5, 6, 7, 8, 9\}, \quad \text{and}$$
$$E(G) = \{\{1,2\}, \{2,3\}, \{2,5\}, \{2,6\}, \{3,6\}, \{3,9\}, \{5,7\}, \{6,8\}, \{7,8\}, \{8,9\}\}.$$

Note that G is a subgraph of H and, furthermore, that G and H have the same vertex set. Therefore G is a spanning subgraph of H.

Pictorially, these graphs are

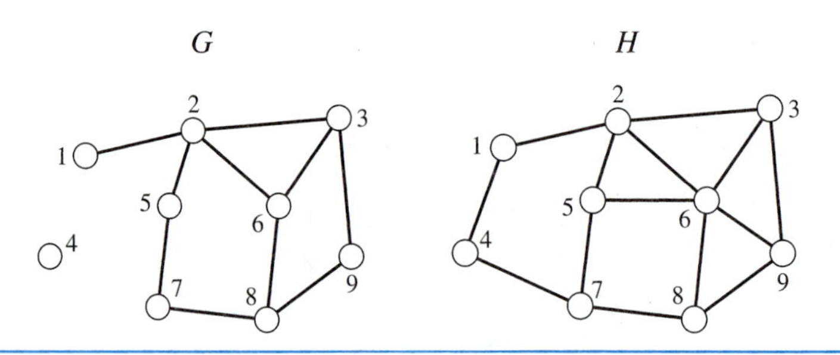

Vertex deletion.

Deleting vertices from a graph is a more subtle process than deleting edges. Suppose v is a vertex of a graph H. How shall we define the graph $H - v$? One idea (incorrect) is to let

$$V(H-v) = V(H) - \{v\}, \quad \text{and}$$
$$E(H-v) = E(H). \qquad \leftarrow \text{WARNING! INCORRECT!!}$$

This looks just like the definition of $H - e$. What is the problem? The problem with this definition is that there may be edges of H that are incident with v. After we delete v from H, it does not make sense to have "edges" in $H - v$ that involve the vertex v. Remember: The edge set of a graph consists of two-element subsets of the vertex set. So an edge with v as an endpoint is not legal in a graph that does not include v as a vertex.

Let's try defining $H - v$ again. When we delete v from H, we must delete all edges that are incident with v; they are not legal to keep once v is gone. Otherwise we retain all the edges that are not incident with v. Here is the correct definition:

$$V(H - v) = V(H) - \{v\}, \quad \text{and}$$
$$E(H - v) = \{e \in E(H) : v \notin e\}.$$

In other words, the vertex set of $H - v$ contains all the vertices of H except v. The edge set of $H - v$ contains all those edges of H that are not incident with v. The notation $v \notin e$ is a terse way to write "v is not incident with e." Recall that e is a two-element set, and $v \notin e$ means v is not an element of e (i.e., not an end point of e).

If we form a subgraph of H solely by means of vertex deletion, we call the subgraph an *induced* subgraph of H.

Definition 47.5 **(Induced subgraph)** Let H be a graph and let A be a subset of the vertices of H; that is, $A \subseteq V(H)$. The *subgraph of H induced on A* is the graph $H[A]$ defined by

$$V(H[A]) = A, \quad \text{and}$$
$$E(H[A]) = \{xy \in E(H) : x \in A \text{ and } y \in A\}.$$

The set A is the set of vertices we keep. The induced subgraph $H[A]$ is the graph whose vertex set is A and whose edges are all those edges of H that are legally possible (i.e., have both end points in A).

When we say that G is an induced subgraph of H, we mean that $G = H[A]$ for some $A \subseteq V(H)$.

The graph $H - v$ is an induced subgraph of H. If $A = V(H) - \{v\}$, then $H - v = H[A]$.

Example 47.6 Let H be the graph from Example 47.2 and let G be the graph with

$$V(G) = \{1, 2, 3, 5, 6, 7, 8\}, \quad \text{and}$$
$$E(G) = \{\{1, 2\}, \{2, 3\}, \{2, 5\}, \{2, 6\}, \{3, 6\}, \{5, 6\}, \{5, 7\}, \{6, 8\}, \{7, 8\}\}.$$

Note that G is a subgraph of H. From H we deleted vertices 4 and 9. We have included in G every edge of H except, of course, those edges incident with vertices 4 or 9. Thus G is an induced subgraph of H and

$$G = H[A] \qquad \text{where} \qquad A = \{1, 2, 3, 5, 6, 7, 8\}.$$

We can also write $G = (H - 4) - 9 = (H - 9) - 4$.

Pictorially, these graphs are

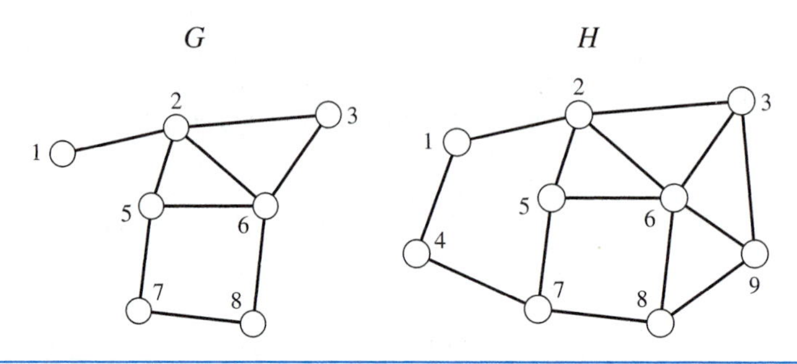

Cliques and Independent Sets

Definition 47.7 **(Clique, clique number)** Let G be a graph. A subset of vertices $S \subseteq V(G)$ is called a *clique* provided any two distinct vertices in S are adjacent.

The *clique number* of G is the size of a largest clique; it is denoted $\omega(G)$.

In other words, a set $S \subseteq V(G)$ is called a clique provided $G[S]$ is a complete graph.

Example 47.8 Let H be the graph from the earlier examples in this section, shown again here.

Mathspeak!

In proper English, *maximum* and *maximal* are closely related, but not interchangeable, words. The difference is that *maximum* is a noun and *maximal* is an adjective. In common usage, people often use *maximum* as both a noun and an adjective. In mathematics, we use both *maximal* and *maximum* as adjectives with slightly different meanings. This difference is explored further in Section 54.

An alternative term for an independent set in a graph is *stable* set, and $\alpha(G)$ is also known as the *stability* number of G.

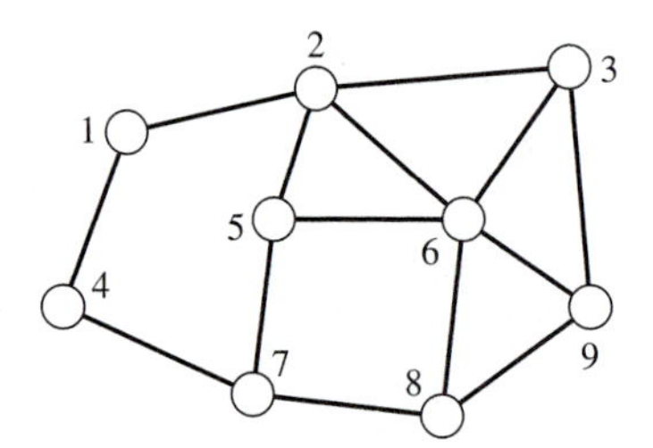

This graph has many cliques. Here we list some of them:

$$\{1,4\} \quad \{2,5,6\} \quad \{9\} \quad \{2,3,6\} \quad \{6,8,9\} \quad \{4\} \quad \emptyset.$$

The largest size of a clique in H is 3, so $\omega(H) = 3$.

The clique $\{1, 4\}$ in the above example is interesting. It only contains two vertices, so it does not have the largest possible size for a clique in H. However, it cannot be extended. It is a *maximal* clique that does not have *maximum* size. By *maximal* we mean "cannot be extended." By *maximum* we mean "largest." Thus $\{1, 4\}$ is a *maximal* clique that is not clique of *maximum* size.

Definition 47.9 **(Independent set, independence number)** Let G be a graph. A subset of vertices $S \subseteq V(G)$ is called an *independent set* provided no two vertices in S are adjacent.

The *independence number* of G is the size of a largest independent set; it is denoted $\alpha(G)$.

In other words, a set $S \subseteq V(G)$ is independent provided $G[S]$ is an edgeless graph.

Example 47.10 Let H be the graph from the earlier examples in this section.

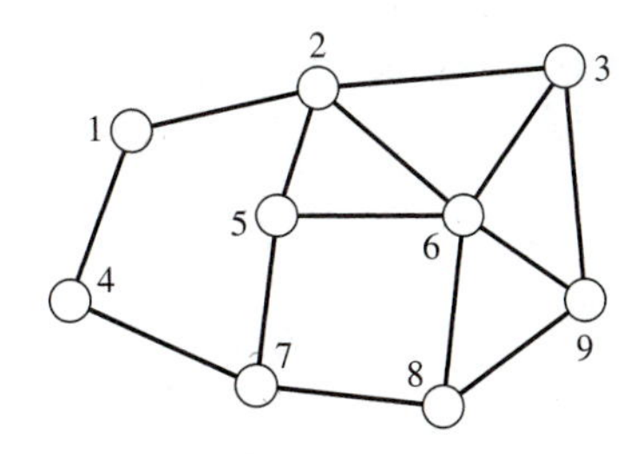

This graph has many independent sets. Here we list some of them:

$$\{1, 3, 5\} \quad \{1, 7, 9\} \quad \{4\} \quad \{1, 3, 5, 8\} \quad \{4, 6\} \quad \{1, 3, 7\} \quad \emptyset.$$

The largest size of an independent set in H is 4, so $\alpha(H) = 4$.

The independent set $\{4, 6\}$ is interesting. It is not a largest independent set, but it is a *maximal* independent set. If you carefully examine the graph H, you should note that each of the other seven vertices is adjacent to vertex 4 or to vertex 6. Thus $\{4, 6\}$ is independent but cannot be extended. It is a maximal independent set that is not of maximum size.

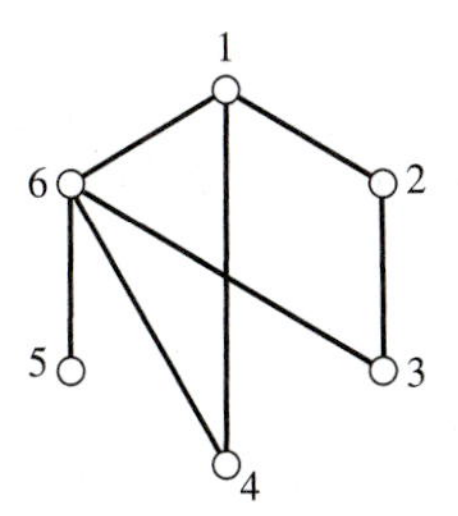

Complements

The two notions of clique and independent sets are flip sides of the same coin; here we discuss what it means to "flip the coin."

The *complement* of a graph G is a new graph formed by removing all the edges of G and replacing them by all possible edges that are not in G. Formally, we state this as follows:

Definition 47.11 **(Complement)** Let G be a graph. The *complement* of G is the graph denoted $\overline{G}$ defined by

$$V(\overline{G}) = V(G), \quad \text{and}$$
$$E(\overline{G}) = \{xy : x, y \in V(G),\ x \neq y,\ xy \notin E(G)\}.$$

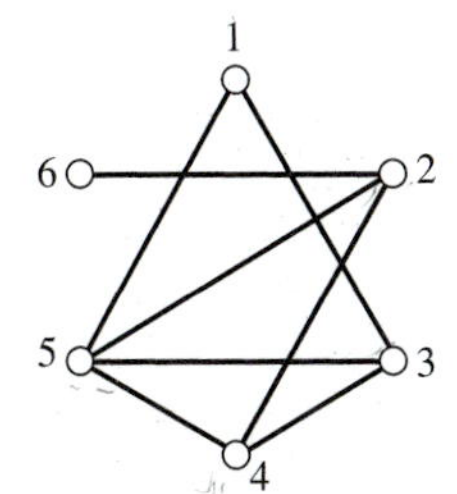

The two graphs in the figure are complements of one another.

The following immediate result makes explicit our assertion that cliques and independent sets are flip sides of the same coin.

Proposition 47.12 Let G be a graph. A subset of $V(G)$ is a clique of G if and only if it is an independent set of $\overline{G}$. Furthermore,

$$\omega(G) = \alpha(\overline{G}) \qquad \text{and} \qquad \alpha(G) = \omega(\overline{G}).$$

Let G be a "very large" graph (i.e., a graph with a great many vertices). A celebrated theorem in graph theory (known as Ramsey's Theorem) implies that either G or its complement, $\overline{G}$, must have a "large" clique. Here we prove a special case of this result; the full statement and general proof of Ramsey's Theorem can be found in more advanced texts. (See also Exercise 47.10.)

Proposition 47.13 Let G be a graph with at least six vertices. Then $\omega(G) \geq 3$ or $\omega(\overline{G}) \geq 3$.

The conclusion may also be written as follows: Then $\omega(G) \geq 3$ or $\alpha(G) \geq 3$.

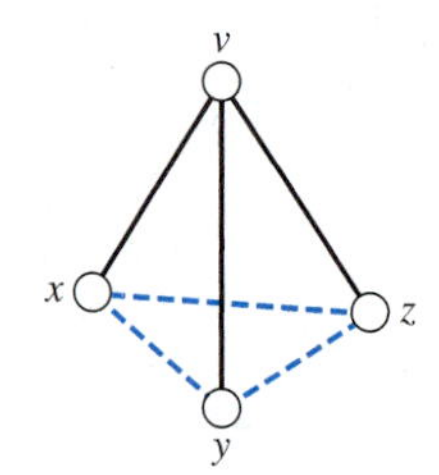

Proof. Let v be any vertex of G. We consider two possibilities: either $d(v) \geq 3$ or else $d(v) < 3$.

Consider first the case $d(v) \geq 3$. This means that v has at least three neighbors: Let x, y, z be three of v's neighbors. See the figure.

If one (or more) of the possible edges xy, yz, or xz is actually an edge of G, then G contains a clique of size 3, and so $\omega(G) \geq 3$.

However, if none of the possible edges xy, yz, or xz is present in G, then all three are edges of $\overline{G}$, and so $\omega(\overline{G}) \geq 3$.

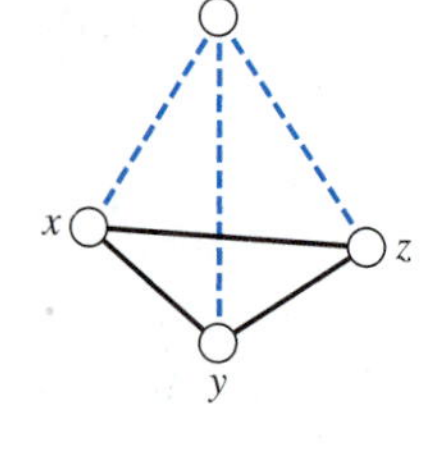

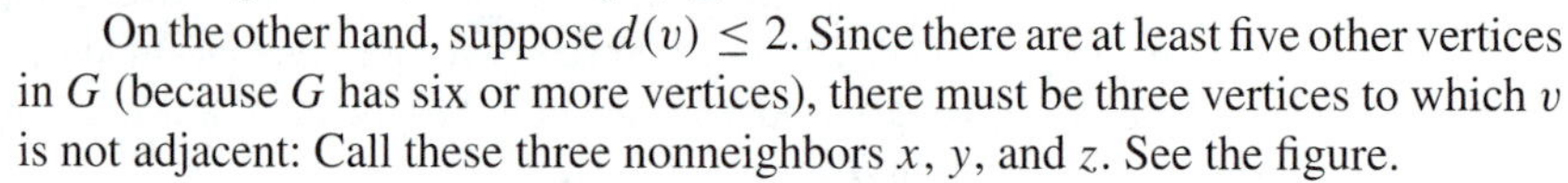

On the other hand, suppose $d(v) \leq 2$. Since there are at least five other vertices in G (because G has six or more vertices), there must be three vertices to which v is not adjacent: Call these three nonneighbors x, y, and z. See the figure.

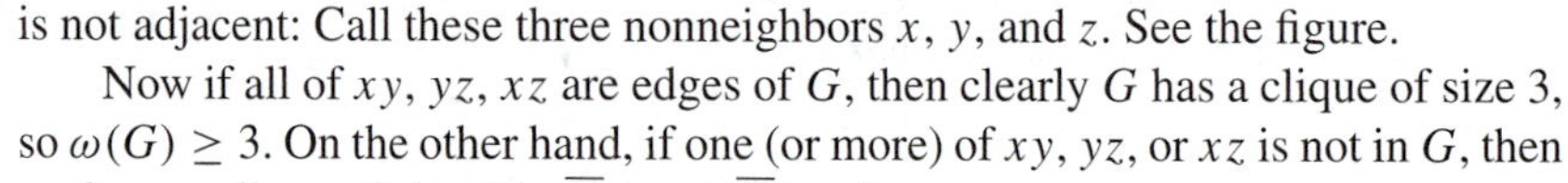

Now if all of xy, yz, xz are edges of G, then clearly G has a clique of size 3, so $\omega(G) \geq 3$. On the other hand, if one (or more) of xy, yz, or xz is not in G, then we have a clique of size 3 in $\overline{G}$, so $\omega(\overline{G}) \geq 3$.

In all, there have been four cases, and in every case, we concluded either $\omega(G) \geq 3$ or $\omega(\overline{G}) \geq 3$. ■

Recap

We introduced the concept of subgraph and the special forms of subgraph: spanning and induced. We discussed cliques and independent sets. We presented the concept of the complement of a graph. Finally, we presented a simplified version of Ramsey's Theorem.

47 Exercises

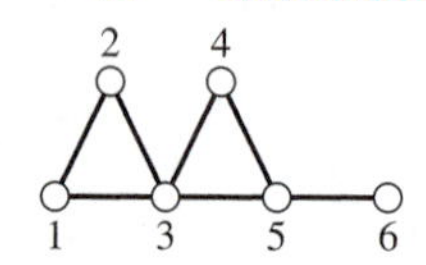

47.1. Let G be the graph in the figure. Draw pictures of the following subgraphs.

a. $G - 1$.
b. $G - 3$.
c. $G - 6$.
d. $G - \{1, 2\}$.
e. $G - \{3, 5\}$.
f. $G - \{5, 6\}$.
g. $G[\{1, 2, 3, 4\}]$.

h. $G[\{2,4,6\}]$.
i. $G[\{1,2,4,5\}]$.

47.2. Which of the various properties of relations does the is-a-subgraph-of relation exhibit? Is it reflexive? Irreflexive? Symmetric? Antisymmetric? Transitive?

47.3. Let G be a complete graph on n vertices.

a. How many spanning subgraphs does G have?
b. How many induced subgraphs does G have?

47.4. Let G and H be the two graphs in the figure.

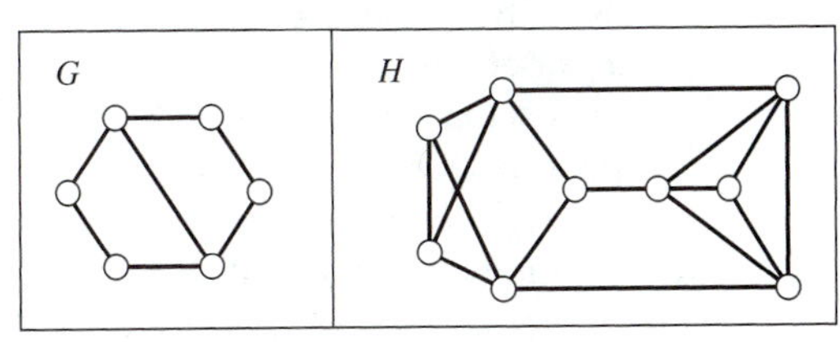

Please find $\alpha(G)$, $\omega(G)$, $\alpha(H)$, and $\omega(H)$.

47.5. Find a graph G with $\alpha(G) = \omega(G) = 5$.

47.6. Suppose that G is a subgraph of H. Prove or disprove:

a. $\alpha(G) \le \alpha(H)$.
b. $\alpha(G) \ge \alpha(H)$.
c. $\omega(G) \le \omega(H)$.
d. $\omega(G) \ge \omega(H)$.

47.7. *Self-complementary graphs.* Recall the definition of graph isomorphism from Exercise 46.18. We call a graph G *self-complementary* if G is isomorphic to $\overline{G}$.

a. Show that the graph $G = (\{a,b,c,d\}, \{ab, bc, cd\})$ is self-complementary.
b. Find a self-complementary graph with five vertices.
c. Prove that if a self-complementary graph has n vertices, then $n \equiv 0 \pmod 4$ or $n \equiv 1 \pmod 4$.

47.8. Find a graph G on five vertices for which $\omega(G) < 3$ and $\omega(\overline{G}) < 3$. This shows that the number six in Proposition 47.13 is best possible.

47.9. Let G be a graph with at least two vertices. Prove that $\alpha(G) \ge 2$ or $\omega(G) \ge 2$.

47.10. *Ramsey arrow notation.* Let $n, a, b \ge 2$ be integers. The notation $n \to (a, b)$ is an abbreviation for the following sentence:

Every graph G on n vertices has $\alpha(G) \ge a$ or $\omega(G) \ge b$.

For example, Proposition 47.13 says that if $n \ge 6$, then $n \to (3,3)$ is true. However, Exercise 47.8 asserts that $5 \to (3,3)$ is false.

Please prove the following:

a. If $n \ge 2$, then $n \to (2,2)$.
b. For any integer $n \ge 2$, $n \to (n, 2)$.
c. If $n \to (a,b)$ and $m \ge n$, then $m \to (a,b)$.
d. If $n \to (a,b)$, then $n \to (b,a)$.
e. The least n such that $n \to (3,3)$ is $n = 6$.
f. $10 \to (3,4)$.

g. Suppose $a, b \geq 3$. If $n \to (a-1, b)$ and $m \to (a, b-1)$, then $(n+m) \to (a, b)$.

h. $20 \to (4, 4)$.

48 Connection

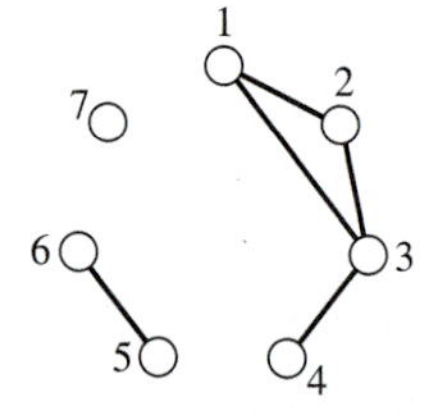

Graphs are useful in modeling communication and transportation networks. The vertices of a graph can represent major cities in a country, and the edges in the graph can represent highways that link them. A fundamental question is: For a given pair of sites in the network, can we travel from one to the other?

For example, in the United States, we can travel by interstate from Baltimore to Denver, but we cannot get to Honolulu from Chicago, even though both of these cities are serviced by interstates. (Some so-called "interstate" highways actually reside entirely within one state, such as I-97 in Maryland and H-1 in Hawaii.)

In this section, we consider what it means for a graph to be connected and related issues. The intuitive notion is clear. The graph in Example 46.2 (reproduced in the figure) is not connected, but it does contain three connected components. These ideas are made explicit next.

Walks

Definition 48.1 **(Walk)** Let $G = (V, E)$ be a graph. A *walk* in G is a sequence (or list) of vertices, with each vertex adjacent to the next; that is,

$$W = (v_0, v_1, \ldots, v_\ell) \quad \text{with} \quad v_0 \sim v_1 \sim v_2 \sim \cdots \sim v_\ell.$$

The *length* of this walk is ℓ. Note that we started the subscripts at zero and that there are $\ell + 1$ vertices on the walk.

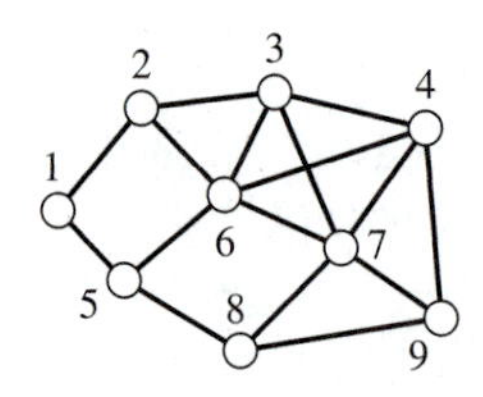

For example, consider the graph in the figure. The following sequences of vertices are walks:

- $1 \sim 2 \sim 3 \sim 4$.

  This is a walk of length three. It starts at vertex 1 and ends at vertex 4, and so we call it a $(1, 4)$-walk.

  In general, a (u, v)-*walk* is a walk in a graph whose first vertex is u and whose last vertex is v.

- $1 \sim 2 \sim 3 \sim 6 \sim 2 \sim 1 \sim 5$.

  This is a walk of length six. There are seven vertices on this walk (counting vertices 1 and 2 twice, because they are visited twice by this walk).

  We are permitted to visit a vertex more than once on a walk.

- $5 \sim 1 \sim 2 \sim 6 \sim 3 \sim 2 \sim 1$.

  This is also a walk of length six. Notice that this sequence is exactly the reverse of that of the previous example.

  If $W = v_0 \sim v_1 \sim \cdots \sim v_{\ell-1} \sim v_\ell$, then its *reversal* is also a walk (because $\sim$ is symmetric). The reversal of W is $W^{-1} = v_\ell \sim v_{\ell-1} \sim \cdots \sim v_1 \sim v_0$.

- 9.

  This is a walk of length zero. A singleton vertex is considered a walk.
- $1 \sim 5 \sim 1 \sim 5 \sim 1$.

  This is a walk of length four. This walk is called *closed* because it begins and ends at the same vertex.

However, the sequence $(1, 1, 2, 3, 4)$ is not a walk because 1 is not adjacent to 1. Likewise the sequence $(1, 6, 7, 9)$ is not a walk because 1 is not adjacent to 6.

Definition 48.2 **(Concatenation)** Let G be a graph. Suppose W_1 and W_2 are the following walks:

$$W_1 = v_0 \sim v_1 \sim \cdots \sim v_\ell$$
$$W_2 = w_0 \sim w_1 \sim \cdots \sim w_k$$

and suppose $v_\ell = w_0$. Their *concatenation*, denoted $W_1 + W_2$, is the walk

$$v_0 \sim v_1 \sim \cdots \sim (v_\ell = w_0) \sim w_1 \sim \cdots \sim w_k.$$

Continuing the example from above, the concatenation of the walks $1 \sim 2 \sim 3 \sim 4$ and $4 \sim 7 \sim 3 \sim 2$ is the walk $1 \sim 2 \sim 3 \sim 4 \sim 7 \sim 3 \sim 2$.

Paths

Definition 48.3 **(Path)** A *path* in a graph is a walk in which no vertex is repeated.

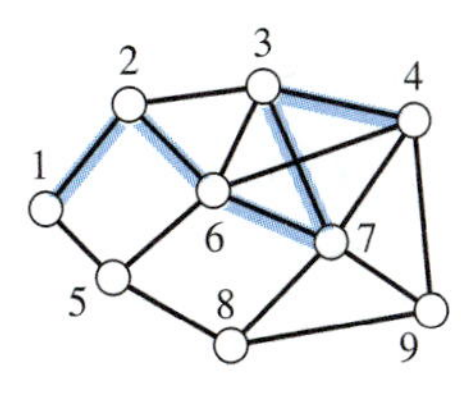

For example, for the graph in the figure, the walk $1 \sim 2 \sim 6 \sim 7 \sim 3 \sim 4$ is a path. It is also called a $(1, 4)$-path because it begins at vertex 1 and ends at vertex 4. In general, a (u, v)*-path* is a path whose first vertex is u and whose last vertex is v.

Note that the definition of path explicitly requires that no vertex of the graph be repeated. Implicit in this condition is that no edge be used twice on the path. What do we mean by *using* an edge? If a walk (or path) is of the form $\cdots \sim u \sim v \sim \cdots$, then we say that the walk *used* or *traversed* the edge uv.

Proposition 48.4 Let P be a path in a graph G. Then P does not traverse any edge of G more than once.

Proof. Suppose, for the sake of contradiction, that some path P in a graph G traverses the edge $e = uv$ more than once. Without loss of generality, we have

$$P = \cdots \sim u \sim v \sim \cdots \sim u \sim v \sim \cdots \qquad \text{or}$$
$$P = \cdots \sim u \sim v \sim \cdots \sim v \sim u \sim \cdots.$$

In the first case, we clearly have repeated both vertices u and v, contradicting the fact that P is a path. In the second case, it is conceivable that the first and second

v we wrote are really one in the same; that is, the path is of the form

$$P = \cdots \sim u \sim v \sim u \sim \cdots$$

but as in the previous case, we have repeated vertex u, contradicting the fact that P is a path. Therefore P does not traverse any edge more than once. ■

Thus a path of length k contains exactly $k + 1$ (distinct) vertices and traverses exactly k (distinct) edges. The word *path* in graph theory has an alternative meaning. Properly speaking, a *path* is a sequence of vertices. However, we often think of a path as a graph or as a subgraph of a given graph.

Definition 48.5 **(Path graph)** A *path* is a graph with vertex set $V = \{v_1, v_2, \ldots, v_n\}$ and edge set

$$E = \{v_i v_{i+1} : 1 \le i < n\}.$$

A path on n vertices is denoted P_n.

A P_5 graph:

Given a sequence of vertices in G constituting a path, we can also view that sequence as a subgraph of G; the vertices of this subgraph are the vertices of the path, and the edges of this subgraph are the edges traversed by the path.

Note that P_n stands for a path with n vertices. Its length is $n - 1$.

We use paths to define what it means for one vertex to be *connected* to another.

Definition 48.6 **(Connected to)** Let G be a graph and let $u, v \in V(G)$. We say that *u is connected to v* provided there is a (u, v)-path in G (i.e., a path whose first vertex is u and whose last vertex is v).

Is-connected-to is reflexive...

Note that the is-connected-to relation is different from the is-adjacent-to relation. For example, a vertex is always connected to itself: If v is a vertex, then the path (v)—yes, one vertex by itself makes a perfectly legitimate path—is a (v, v)-path, so v is connected to v. However, a vertex is never adjacent to itself. In the language of relations, is-connected-to is reflexive, and is-adjacent-to is irreflexive.

The is-connected-to relation is reflexive. What other properties does it exhibit? It is not hard to check that is-connected-to is not (in general) irreflexive or antisymmetric. (See Exercise 48.8.)

... and symmetric ...

Is the is-connected-to a relation symmetric? Suppose, in a graph G, vertex u is connected to vertex v. This means there is a (u, v)-path in G; call this path P. Its reversal, P^{-1}, is a (v, u)-path, and so v is connected to u. Thus is-connected-to is a symmetric relation.

... and transitive.

Is the is-connected-to relation transitive? Suppose, in a graph G, we know that x is connected to y and that y is connected to z. We want to prove that x is connected to z.

Since x is connected to y, there must be an (x, y)-path; let's call it P. And since y is connected to z, there must be a (y, z)-path. Let's call it Q. Notice that the last vertex of P is the same as the first vertex of Q (it's y). Therefore we can form the concatenation $P + Q$, which is an (x, z)-path. Therefore x is connected to z.

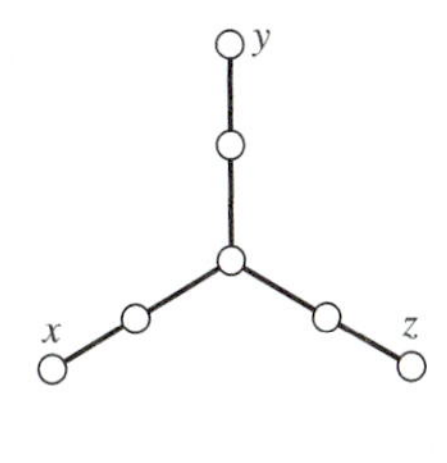

Nice proof, huh? Not really. The above proof is incorrect! What went wrong? Try to figure out the difficulty yourself. The figure gives you a big hint.

The problem with the proof is that although P and Q are paths, and it is true that the last vertex of P and the first vertex of Q are the same, we do not know that $P + Q$ is a path. All we can say for certain is that $P + Q$ is an (x, y)-walk.

To complete our argument that is-connected-to is transitive, we need to prove that the existence of an (x, y)-walk implies the existence of an (x, y)-path. Let's state this formally and prove it.

Lemma 48.7 Let G be a graph and let $x, y \in V(G)$. If there is an (x, y)-walk in G, then there is an (x, y)-path in G.

The truth of this lemma is not too hard to see. If there is a walk and if this walk contains a repeated vertex, we can shorten the walk by removing the portion of the walk between the repeated vertex. Of course, this might not be a walk, so we may need to do this operation again. This analysis can lead to a mushy proof. Here is a crisp way to express the same basic idea.

Proof. Suppose there is an (x, y)-walk in a graph G. Note that the length of an (x, y)-walk is a natural number. Thus, by the Well-Ordering Principle, there is a shortest (x, y)-walk, P.

There may be more than one shortest (x, y)-walk; let P be any one of them.

We claim that P is, in fact, an (x, y)-path. Suppose, for the sake of contradiction, that P is not an (x, y)-path. If P is not a path, then there must be some vertex, u, that is repeated on the path. In other words,

$$P = x \sim \cdots \sim ? \sim u \underbrace{\sim \cdots \sim u} \sim ?? \sim \cdots \sim y.$$

Note: We do not rule out the possibility that $u = x$ and/or $u = y$. We only assume that vertex u appears at least twice, so the second (colored) u appears later in the sequence than the first. Form a new walk P' by deleting the portion of the walk marked in color. Note that this results in a new walk. Note that vertices ? and ?? are both adjacent to u, so the shortened sequence P' is still an (x, y)-walk. However, by construction P is a shortest (x, y)-walk, contradicting the fact that P' is an even shorter (x, y)-walk.$\Rightarrow\Leftarrow$

Therefore P is an (x, y)-path. ■

We return to where we left off before we proved this lemma. We were trying to show that the relation is-connected-to is transitive. Let's try the proof again. Suppose, in a graph G, we know that x is connected to y and that y is connected to z. By definition, this means there are an (x, y)-path P and a (y, z)-path Q. Form the walk $W = P + Q$. This is an (x, z)-walk, so by Lemma 48.7, there must be an (x, z)-path in G. Therefore x is connected to z.

We have shown that is-connected-to is reflexive, symmetric, and transitive. In other words, we have proved the following:

Theorem 48.8 Let G be a graph. The is-connected-to relation is an equivalence relation on $V(G)$.

Whenever we have an equivalence relation, we also have a partition: the equivalence classes of the relation. What can we say about the equivalence classes of the is-connected-to relation?

Let u and v be vertices of a graph G. If u and v are in the same equivalence class of the is-connected-to relation, then there is a path joining them (from u to v, as well as its reversal, from v to u). On the other hand, if u and v are in different equivalence classes, then u and v are not related by the is-connected-to relation. In this case, we know there is no path joining u to v, or vice versa.

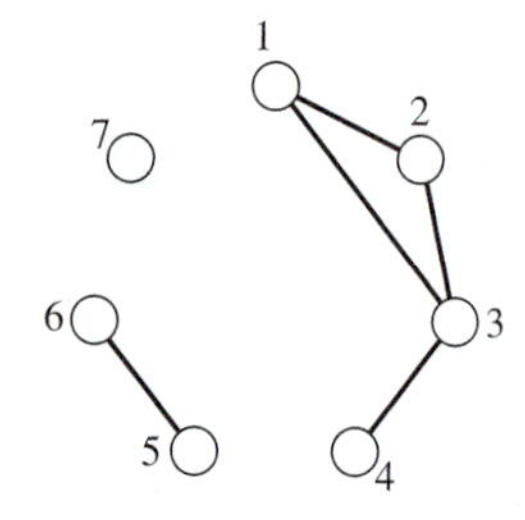

Consider the graph in the figure (the same graph from Example 46.2). The equivalence classes of the is-connected-to relation on this graph are

$$\{1, 2, 3, 4\}, \quad \{5, 6\}, \quad \text{and} \quad \{7\}.$$

The equivalence classes of is-connected-to decompose a graph into what we call *components*.

Definition 48.9 **(Component)** A *component* of G is a subgraph of G induced on an equivalence class of the is-connected-to relation on $V(G)$.

In other words, we partition the vertices; two vertices are in the same part exactly when there is a path from one to the other. For each part of this partition, there is a component of the graph. The component is the subgraph formed by taking all vertices in one of these parts and all edges of the graph that involve those vertices.

The graph we have been considering (from Example 46.2) has three components:

$$G[\{1, 2, 3, 4\}], \quad G[\{5, 6\}], \quad \text{and} \quad G[\{7\}].$$

The first component has four vertices and four edges. The second component has two vertices and one edge. And the third component has just one vertex and no edges.

If a graph is edgeless, then each of its vertices forms a component unto itself. At the other extreme, it is possible that there is only one component. In this case, we call the graph *connected*. Here is another way to state this:

Definition 48.10 **(Connected)** A graph is called *connected* provided each pair of vertices in the graph are connected by a path; that is, for all $x, y \in V(G)$, there is an (x, y)-path.

Disconnection

Definition 48.11 **(Cut vertex, cut edge)** Let G be a graph. A vertex $v \in V(G)$ is called a *cut vertex* of G provided $G - v$ has more components than G.

Similarly, an edge $e \in E(G)$ is called a *cut edge* of G provided $G - e$ has more components than G.

In particular, if G is a connected graph, a cut vertex v is a vertex such that $G - v$ is disconnected. Likewise e is a cut edge if $G - e$ is disconnected. The graph in the figure has two cut edges and four cut vertices (highlighted).

Theorem 48.12 Let G be a connected graph and suppose $e \in E(G)$ is a cut edge of G. Then $G - e$ has exactly two components.

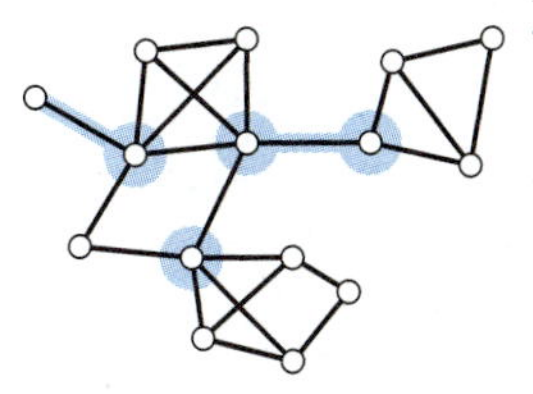

Proof. Let G be a connected graph and let $e \in E(G)$ be a cut edge. Because G is connected, it has exactly one component. Because e is a cut edge, $G - e$ has more components than G (i.e., $G - e$ has at least two components). Our job is to show that it does not have more than two components.

Suppose, for the sake of contradiction, $G - e$ has three (or more) components. Let a, b, and c be three vertices of $G - e$, each in a separate component. This implies that there is no path joining any pair of them.

Let P be an (a, b)-path in G. Because there is no (a, b)-path in $G - e$, we know P must traverse the edge e. Suppose x and y are the endpoints of the edge e, and without loss of generality, the path P traverses e in the order x, then y; that is,

$$P = a \sim \cdots \sim x \sim y \sim \cdots \sim b.$$

Similarly, since G is connected, there is a path Q from c to a that must use the edge $e = xy$. Which vertex, x or y, appears first on Q as we travel from c to a?

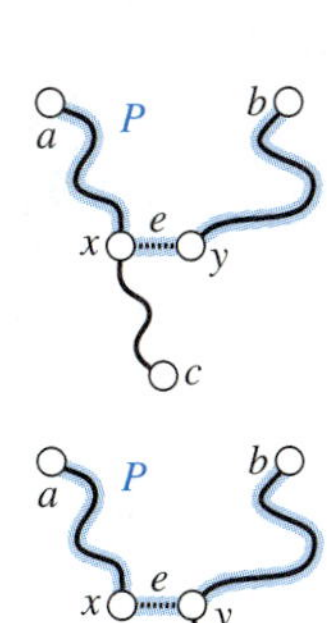

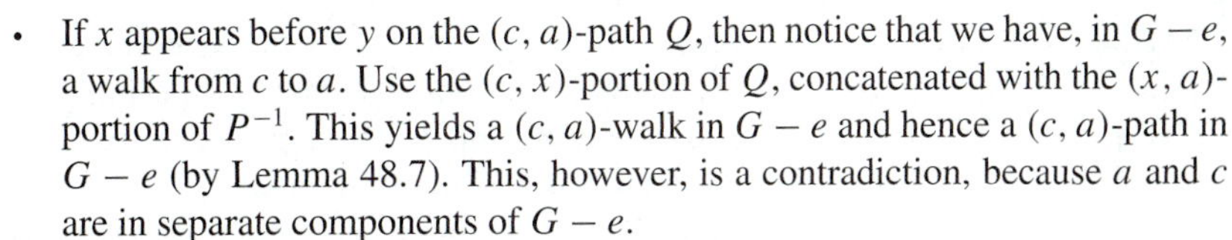

- If x appears before y on the (c, a)-path Q, then notice that we have, in $G - e$, a walk from c to a. Use the (c, x)-portion of Q, concatenated with the (x, a)-portion of P^{-1}. This yields a (c, a)-walk in $G - e$ and hence a (c, a)-path in $G - e$ (by Lemma 48.7). This, however, is a contradiction, because a and c are in separate components of $G - e$.

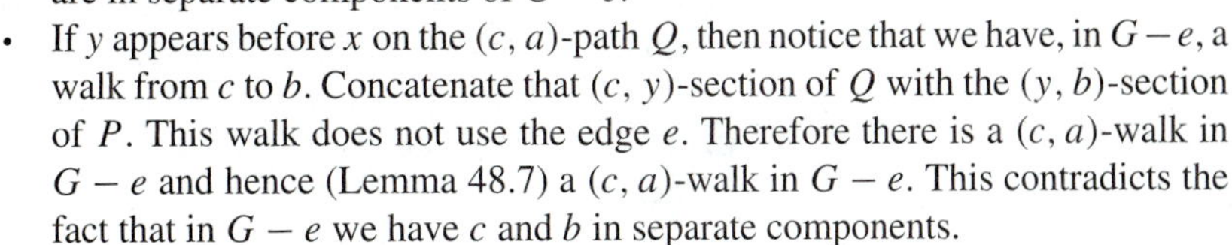

- If y appears before x on the (c, a)-path Q, then notice that we have, in $G - e$, a walk from c to b. Concatenate that (c, y)-section of Q with the (y, b)-section of P. This walk does not use the edge e. Therefore there is a (c, a)-walk in $G - e$ and hence (Lemma 48.7) a (c, a)-walk in $G - e$. This contradicts the fact that in $G - e$ we have c and b in separate components.

Therefore $G - e$ has at most two components. ∎

Recap

We began with the concepts of walk and path. From there, we defined what it means for a graph to be connected and what its connected components are. We discussed cut vertices and cut edges.

48 Exercises

48.1. Let G be the graph in the figure.

a. How many different paths are there from a to b?

b. How many different walks are there from a to b?

48.2. Is concatenation a commutative operation?

48.3. Prove that K_n is connected.

48.4. Let $n \geq 2$ be an integer. Form a graph G_n whose vertices are all the two-element subsets of $\{1, 2, \ldots, n\}$. In this graph we have an edge between distinct vertices $\{a, b\}$ and $\{c, d\}$ exactly when $\{a, b\} \cap \{c, d\} = \emptyset$.

Please answer:

a. How many vertices does G_n have?

b. How many edges does G_n have?

c. For which values of $n \geq 2$ is G_n connected? Prove your answer.

48.5. Consider the following (incorrect) restatement of the definition of *connected*: "A graph G is connected provided there is a path that contains every pair of vertices in G."

What is wrong with this sentence?

48.6. Let G be a graph. A path P in G that contains all the vertices of G is called a *Hamiltonian path*. Prove that the following graph does not have a Hamiltonian path.

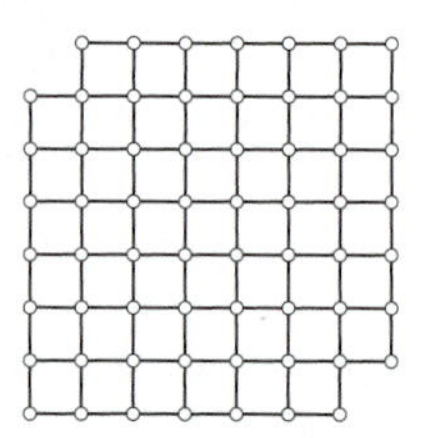

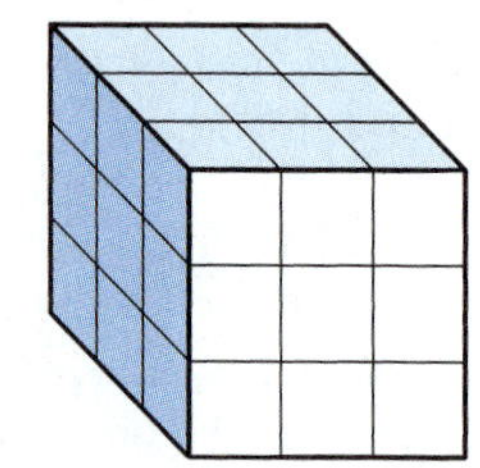

48.7. *Mouse and cheese*. A block of cheese is made up of $3 \times 3 \times 3$ cubes as in the figure. Is it possible for a mouse to tunnel its way through this block of cheese by (a) starting at a corner, (b) eating its way from cube to adjacent cube, (c) never passing though any cube twice, and, finally, (d) finishing at the center cube? Prove your answer.

48.8. Consider the is-connected-to relation on the vertices of a graph. Show that is-connected-to need not be irreflexive or antisymmetric.

48.9. Let G be a graph. Prove that G or $\overline{G}$ (or both) must be connected.

48.10. Let G be a graph with $n \geq 2$ vertices. Prove that if $\delta(G) \geq \frac{1}{2}n$, then G is connected.

48.11. Let G be a graph with $n \geq 2$ vertices.

a. Prove that if G has at least $\binom{n-1}{2} + 1$ edges, then G is connected.

b. Show that the result in (a) is best possible; that is, for each $n \geq 2$, prove there is a graph with $\binom{n-1}{2}$ edges that is not connected.

48.12. *For those who have studied linear algebra*. Let A be the adjacency matrix of a graph G. That is, we label the vertices of G as $v_1, v_2, \ldots, v_n$. The matrix A is an $n \times n$ matrix whose i, j-entry is 1 if $v_i v_j \in E(G)$ and is 0 otherwise.

Let $k \in \mathbb{N}$. Prove that the i, j-entry of A^k is the number of walks of length k from v_i to v_j.

48.13. Let n and k be integers with $1 \le k < n$. Form a graph G whose vertices are the integers $\{0, 1, 2, \ldots, n-1\}$. We have an edge joining vertices a and b provided

$$a - b \equiv \pm k \pmod{n}.$$

For example, if $n = 20$ and $k = 6$, then vertex 2 would be adjacent to vertices 8 and 16.

a. Find necessary and sufficient conditions on n and k such that G is connected.

b. Find a formula involving n and k for the number of connected components of G.

49 Trees

Perhaps the simplest family of graphs are the *trees*. Graph theory problems can be difficult. Often, a good way to begin thinking about these problems is to solve them for trees. Trees are also the most basic connected graph. What are trees? They are connected graphs that have no cycles. We begin by defining the term *cycle*.

Cycles

Definition 49.1 **(Cycle)** A *cycle* is a walk of length at least three in which the first and last vertex are the same, but no other vertices are repeated.

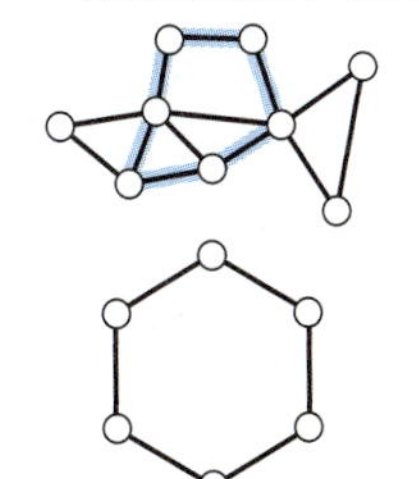

The term *cycle* also refers to a (sub)graph consisting of the vertices and edges of such a walk. In other words, a cycle is a graph of the form $G = (V, E)$ where

$$V = \{v_1, v_2, \ldots, v_n\}, \quad \text{and}$$
$$E = \{v_1v_2, v_2v_3, \ldots, v_{n-1}v_n, v_nv_1\}.$$

A cycle (graph) on n vertices is denoted C_n.

In the upper figure we see a cycle of length six as a walk in a graph. The lower figure shows the graph C_6.

Forests and Trees

Definition 49.2 **(Forest)** Let G be a graph. If G contains no cycles, then we call G *acyclic*. Alternatively, we call G a *forest*.

The term *acyclic* is more natural and (almost) does not need a definition—its standard English meaning is a perfect match for its mathematical usage. The term *forest* is widely used as well. The rationale for this word is that here, just as in real life, a forest is a collection of trees.

Definition 49.3 **(Tree)** A *tree* is a connected, acyclic graph.

In other words, a *tree* is a connected forest.

The forest in the figure contains four connected components. Each component of a forest is a tree.

Note that a single isolated vertex (e.g., the graph K_1) is a tree; it is the simplest tree possible.

There is only one possible structure for a tree on two vertices: Since a tree on two vertices must be connected, there must be an edge joining the two vertices. This is the only possible edge in the graph, and a graph on two vertices cannot have a cycle (a cycle requires at least three distinct vertices). Therefore any tree on two vertices must be a K_2.

There is also only one possible structure for a tree on three vertices. Since the graph is connected, there certainly must be at least one edge—say, joining vertices a and b. However, if there were only one edge, then the third vertex, c, would not be connected to either a or b, and so the graph would not be connected. Thus there must be at least one more edge—without loss of generality, let us say that it is the edge from b to c. So far we have $a \sim b \sim c$, but $ac \notin E$. Now the graph is connected. Might we also add the edge ac? If we do, the graph is connected, but it is no longer acyclic, as we would have the cycle $a \sim b \sim c \sim a$. Any tree on three vertices must be a P_3.

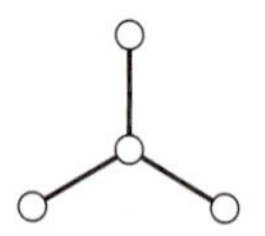

However, on four vertices, we can have two different sorts of trees. We can have the path P_4 and we can have a *star*: a graph of the form $G = (V, E)$ where

$$V = \{a, x, y, z\} \quad \text{and} \quad E = \{ax, ay, az\}.$$

Properties of Trees

Trees have a number of interesting properties. Here we explore several of them.

Theorem 49.4 Let T be a tree. For any two vertices a and b in $V(T)$, there is a unique (a, b)-path.

Conversely, if G is a graph with the property that for any two vertices u, v, there is exactly one (u, v)-path, then G must be a tree.

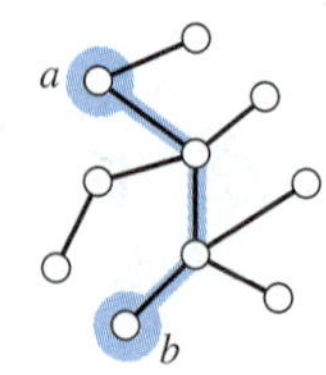

Proof. This is an if-and-only-if style theorem. It can be rephrased: A graph is a tree if and only if between any two vertices there is a unique path.

($\Rightarrow$) Suppose T is a tree and let $a, b \in V(T)$. We need to prove that there is a unique (a, b)-path in T. We have two things to prove:

- *Existence*: The path exists.
- *Uniqueness*: There can be only one such path.

The first task is easy. There exists an (a, b)-path because (by definition) trees are connected.

The second task is more complicated. To prove uniqueness, we use Proof Template 14.

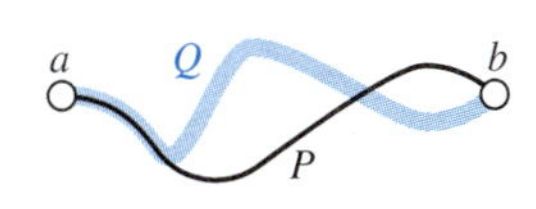

Suppose, for the sake of contradiction, there were two (or more) different (a, b)-paths in T; let us call them P and Q. It would be tempting at this point to reason as follows: "Follow that path P from a to b and then the path Q from b to a; this gives a cycle—contradiction! Therefore there can be only one (a, b)-path." However, this reasoning is incorrect. As the figure suggests, the paths P and Q might overlap or cross each other; we cannot say that $P + Q^{-1}$ is a cycle. We need to be more careful.

Since P and Q are different paths, we know that at some point one of them traverses a different edge than the other. Let us say that from a to x the paths are the same (perhaps $a = x$) but then they traverse different edges; that is,

$$P : a \sim \cdots \sim x \sim y \sim \cdots \sim b$$
$$Q : a \sim \cdots \sim x \sim z \sim \cdots \sim b.$$

This implies that xy is an edge of P and not an edge of Q (because Q cannot repeat vertices—it's a path!—the vertex x does not appear again on Q and so there is no opportunity to see the edge xy on Q).

Now consider the graph $T - xy$ (delete the edge xy from T). We claim there is an (x, y)-path in $T - xy$. Why? Notice that there is an (x, y)-walk in $T - xy$: Start at x, follow P^{-1} from x to a, follow Q from a to b, and then follow P^{-1} from b to y. Notice that on this walk we never traverse the edge xy. Thus there is an (x, y)-walk in $T - xy$. Therefore, by Lemma 48.7, there is an (x, y)-path in $T - xy$; let us call this path R. The path R must contain at least one vertex in addition to x and y because R does not use the edge xy to get from x to y. Now, if we add the edge xy to the path R, we have a cycle (traverse R from x to y and then back to x along the edge yx). This, at long last, is the contradiction we sought: a cycle in the tree T.$\Rightarrow\Leftarrow$ Therefore there can be at most one (a, b)-path.

($\Leftarrow$) Let G be a graph with the property that between any two vertices there is exactly one path. We must prove that G is a tree. We leave this for you (Exercise 49.5). ■

Theorem 49.4 gives an alternative *characterization* of trees. We can prove that a graph is a tree directly by the definition: show that it is connected and acyclic. Alternatively, we can prove that a graph is a tree by showing that between any two vertices of G there is a unique path. The next theorem gives yet another characterization of trees.

Theorem 49.5 Let G be a connected graph. Then G is a tree if and only if every edge of G is a cut edge.

Proof. Let G be a connected graph.

($\Rightarrow$) Suppose G is a tree. Let e be any edge of G. We must prove that e is a cut edge. Suppose the endpoints of e are x and y. To prove that e is a cut edge, we must prove that $G - e$ is disconnected.

Notice that in G there is an (x, y)-path—namely, $x \sim y$ (traverse just the edge e). By Theorem 49.4, this path is unique—there can be no other (x, y)-paths.

Thus, if we delete the edge $e = xy$ from G, there can be no (x, y)-paths (i.e., $G - e$ is disconnected). Therefore e is a cut edge.

($\Leftarrow$) Suppose every edge of G is a cut edge. We must prove that G is a tree. By assumption, G is connected, so we must show that G is acyclic.

Suppose, for the sake of contradiction, that G contains a cycle C. Let $e = xy$ be an edge of this cycle. Notice that the vertices and other edges of C form an (x, y)-path, which we call P.

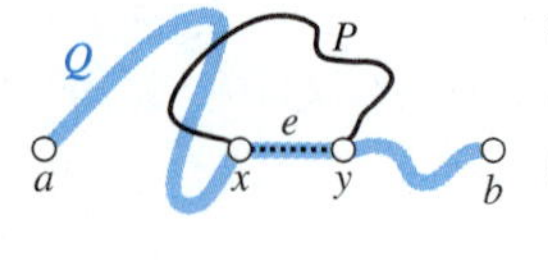

Since e is a cut edge of G, we know that $G - e$ is disconnected. This means there exist vertices a, b for which there is no (a, b)-path in $G - e$. However, in G, there is an (a, b)-path Q; hence Q must traverse the edge e. Without loss of generality, we traverse e from x to y as we step along Q:

$$Q = a \sim \cdots \sim x \sim y \sim \cdots \sim b.$$

We are nearly finished. Notice that in $G - e$ there is an (a, b)-walk. We traverse Q from a to x, then P from x to y, and then Q from y to b (see the figure). By Lemma 48.7, this implies that in $G - e$ there is an (a, b)-path, contradicting the fact that there is no such path.$\Rightarrow\Leftarrow$

Thus G has no cycles and is therefore a tree. ■

Leaves

In biology, a *leaf* is a part of the tree that hangs at the "ends" of the tree. We use the same word in graph theory to convey a similar idea.

Definition 49.6 **(Leaf)** A *leaf* of a graph is a vertex of degree 1.

Leaves are also called *end vertices* or *pendant vertices*. The tree in the figure has four leaves (marked).

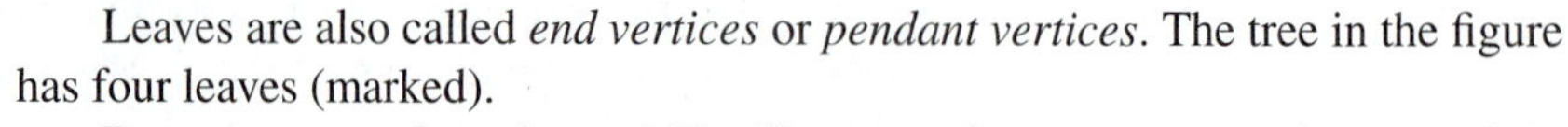

Does every tree have leaves? No. However, the counterexamples are a little silly. The empty graph and the graph K_1 are trees, and they have no vertices of degree 1. However, other than these, every tree has a leaf.

Theorem 49.7 Every tree with at least two vertices has a leaf.

Proof. Let T be a tree with at least two vertices. Let P be a longest path in T (i.e., P is a path in T and there are no paths in T that are longer). Since T is connected and contains at least two vertices, P has two or more vertices. Say,

$$P = v_0 \sim v_1 \sim \cdots \sim v_\ell$$

where $\ell \geq 1$.

We claim that the first and last vertices of P (v_0 and v_ℓ) are leaves of T.

Suppose, for the sake of contradiction, that v_0 is not a leaf. Since v_0 has at least one neighbor (v_1), we have that $d(v_0) \geq 2$. Let x be another neighbor of v_0 (i.e., $x \neq v_1$).

Note that x is not a vertex on P, for otherwise we would have a cycle:

$$v_0 \sim v_1 \sim \cdots \sim x \sim v_0.$$

Thus we can prepend x to the path P to form the path Q:

$$Q = x \sim \underbrace{v_0 \sim v_1 \sim \cdots \sim v_\ell}_{P}.$$

However, notice that Q is a path in T that is longer than P.$\Rightarrow\Leftarrow$ Therefore v_0 is a leaf.

Likewise v_ℓ is a leaf. Therefore T has at least two leaves. ■

In fact, we proved that a tree with at least two vertices must have two (or more) leaves.

Next we prove that plucking a leaf off a tree leaves behind a smaller tree.

Proposition 49.8 Let T be a tree and let v be a leaf of T. Then $T - v$ is a tree.

A converse of this statement is also true; we leave the proof of the converse to you as an exercise (Exercise 49.7).

Proof. We need to prove that $T - v$ is a tree. Clearly $T - v$ is acyclic: If $T - v$ contained a cycle, that cycle would also exist in T. Thus we must show that $T - v$ is connected.

Let $a, b \in V(T - v)$. We must show there is an (a, b)-path in $T - v$. We know, since T is connected, that there is an (a, b)-path P in T. We claim that P does not include the vertex v. Otherwise we would have

$$P = a \sim \cdots \sim v \sim \cdots \sim b$$

and since v is neither the first nor the last vertex on this path, it has two distinct neighbors on the path, contradicting the fact that $d(v) = 1$. Therefore P is an (a, b)-path in $T - v$, and so $T - v$ is connected and a tree. ■

Proposition 49.8 forms the basis of a proof technique for trees. Many proofs about trees are by induction on the number of vertices. Proof Template 25 gives the basic form for such a proof.

We demonstrate this proof technique for the following result.

Theorem 49.9 Let T be a tree with $n \geq 1$ vertices. Then T has $n - 1$ edges.

Proof Template 25 Proving theorems about trees by leaf deletion.

To prove: Some theorem about trees.

Proof. We prove the result by induction on the number of vertices in T.

Basis case: Claim the theorem is true for all trees on $n = 1$ vertices. (This should be easy!)

Induction hypothesis: Suppose the theorem is true for all trees on $n = k$ vertices.

Let T be a tree on $n = k + 1$ vertices. Let v be a leaf of T. Let $T' = T - v$. Note that T' is a tree with k vertices, so by induction T' satisfies the theorem.

Now we use the fact that the theorem is true for T' to somehow prove that the conclusion of the theorem holds for T. (This might be tricky.)

Thus the result is proved by induction. ■

We use Proof Template 25 to prove this result.

Proof. We prove Theorem 49.9 by induction on the number of vertices in T.

Basis case: Claim the theorem is true for all trees on $n = 1$ vertices. If T has only $n = 1$ vertex, then clearly it has $0 = n - 1$ edges.

Induction hypothesis: Suppose Theorem 49.9 is true for all trees on $n = k$ vertices.

Let T be a tree on $n = k + 1$ vertices. We need to prove that T has $n - 1 = k$ edges.

Let v be a leaf of T and let $T' = T - v$. Note that T' is a tree with k vertices, so by induction T' satisfies the theorem (i.e., T' has $k - 1$ edges).

Since v is a leaf of T, we have $d(v) = 1$. This means that when we deleted v from T, we deleted exactly one edge. Therefore T has one more edge than T'; that is, T has $(k - 1) + 1 = k$ edges.

Thus the result is proved by induction. ■

Spanning Trees

Trees are, in a sense, minimally connected graphs. By definition, they are connected, but (see Theorem 49.5) the deletion of any edge disconnects a tree.

Definition 49.10 **(Spanning tree)** Let G be a graph. A *spanning tree* of G is a spanning subgraph of G that is a tree.

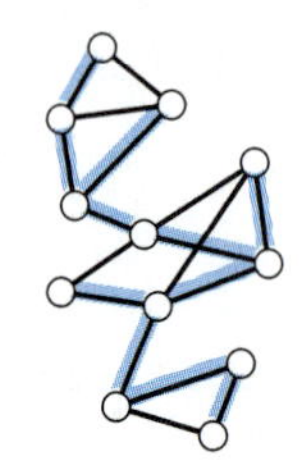

(Recall that a *spanning subgraph* of G is a subgraph that has the same vertices as G. See Definition 47.3.)

The definition appears not to say anything because the words *spanning tree* are perfectly descriptive. A spanning tree of G is a tree subgraph of G that includes all the vertices of G. For the graph in the figure, we have highlighted one of its many spanning trees.

Theorem 49.11 A graph has a spanning tree if and only if it is connected.

Note: G is, itself, a spanning connected subgraph of G. Thus there is at least one such subgraph. Among all spanning connected subgraphs, we choose one with the least number of edges and we call it T.

Proof. ($\Rightarrow$) Suppose G is connected. Let T be a spanning connected subgraph of G with the least number of edges.

We claim that T is a tree. By construction, T is connected. Furthermore, we claim that every edge of T is a cut edge. Otherwise, if $e \in E(T)$ were not a cut edge of T, then $T - e$ would be a smaller spanning connected subgraph of G.$\Rightarrow\Leftarrow$ Therefore every edge of T is a cut edge. Hence (Theorem 49.5) T is a tree, and so G has a spanning tree.

($\Leftarrow$) Suppose G has a spanning tree T. We want to show that G is connected. Let $u, v \in V(G)$. Since T is spanning, we have $V(T) = V(G)$, and so $u, v \in V(T)$. Since T is connected, there is a (u, v)-path P in T. Since T is a subgraph of G, P is a (u, v)-path of G. Therefore G is connected. ■

We can use this result to provide yet another characterization of trees.

Theorem 49.12 Let G be a connected graph on $n \geq 1$ vertices. Then G is a tree if and only if G has exactly $n - 1$ edges.

Proof. ($\Rightarrow$) This was shown in Theorem 49.9.

($\Leftarrow$) Suppose G is a connected graph with n vertices and $n - 1$ edges. By Theorem 49.11, we know that G has a spanning tree T; that is, T is a tree, $V(T) = V(G)$, and $E(T) \subseteq E(G)$. Note, however, that

$$|E(T)| = |V(T)| - 1 = |V(G)| - 1 = |E(G)|$$

so we actually have $E(T) = E(G)$. Therefore $G = T$ (i.e., G is a tree). ■

Recap

We introduced the notions of cycle, forest, and tree. We proved that the following statements about a graph G are equivalent:

- G is a tree.
- G is connected and acyclic.
- G is connected and every edge of G is a cut edge.
- Between any two vertices of G there is a unique path.
- G is connected and $|E(G)| = |V(G)| - 1$.

We also introduced the concept of spanning tree and proved that a graph has a spanning tree if and only if it is connected.

49 Exercises

49.1. Let G be a graph in which every vertex has degree 2. Is G necessarily a cycle?

49.2. Let T be a tree. Prove that the average degree of a vertex in T is less than 2.

49.3. Let $d_1, d_2, \ldots, d_n$ be $n \geq 2$ positive integers (not necessarily distinct). Prove that $d_1, \ldots, d_n$ are the degrees of the vertices of a tree on n vertices if and only if $\sum_{i=1}^{n} d_i = 2n - 2$.

49.4. Let e be an edge of a graph G. Prove that e is not a cut edge if and only if e is in a cycle of G.

49.5. Complete the proof of Theorem 49.4. That is, prove that if G is a graph in which any two vertices are joined by a unique path, then G must be a tree.

49.6. Why is the empty graph a tree?

49.7. Prove the following converse to Proposition 49.8:

Let T be a tree with at least two vertices and let $v \in V(T)$. If $T - v$ is a tree, then v is a leaf.

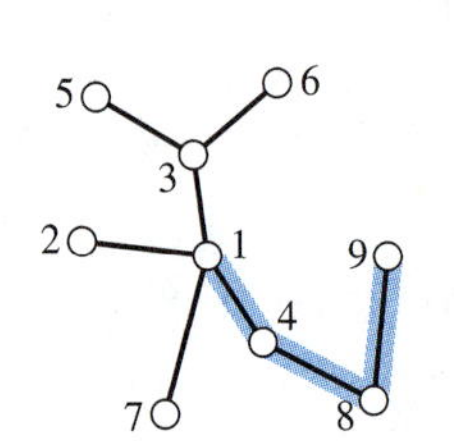

49.8. Let T be a tree whose vertices are the integers 1 through n. We call T a *recursive tree* if it has the following special property. Let P be any path in T starting at vertex 1. Then, as we move along the path P, the vertices we encounter come in increasing numerical order. The tree in the figure is an example of a recursive tree. Notice that all paths starting at vertex 1 encounter the vertices in increasing order. For example, the highlighted path encounters the vertices $1 < 4 < 8 < 9$.

Please do the following:

a. Prove: If T is a recursive tree on n vertices, then vertex n is a leaf (provided $n > 1$).

b. Prove: If T is a recursive tree on $n > 1$ vertices, then $T - n$ (the tree T with vertex n deleted) is also a recursive tree (on $n - 1$ vertices).

c. Prove: If T is a recursive tree on n vertices and a new vertex $n + 1$ is attached as a leaf to any vertex of T to form a new tree T', then T' is also a recursive tree.

d. How many different recursive trees on n vertices are there? Prove your answer.

49.9. Let G be a forest with n vertices and c components. Find and prove a formula for the number of edges in G.

49.10. Prove that a graph is a forest if and only if all of its edges are cut edges.

49.11. In this problem, you will develop a new proof that every tree with two or more vertices has a leaf. Here is an outline for your proof.

a. First prove, using strong induction and the fact that every edge of a tree is a cut edge (Theorem 49.5), that a tree with n vertices has exactly $n - 1$ edges.

Please note that our previous proof of this fact (Theorem 49.9) used the fact that trees have leaves; that is why we need an alternative proof.

b. Use (a) to prove that the average degree of a vertex in a tree is less than 2.

c. Use (b) to prove that every tree (with at least two vertices) has a leaf.

49.12. Let T be a tree with $u, v \in V(T)$, $u \neq v$, and $uv \notin E(T)$. Prove that if we add the edge e to T, the resulting graph has exactly one cycle.

49.13. Let G be a connected graph with $|V(G)| = |E(G)|$. Prove that G contains exactly one cycle.

49.14. Prove:

a. Every cycle is connected.

b. Every cycle is 2-regular.

c. Conversely, every connected, 2-regular graph must be a cycle.

49.15. Let e be an edge of a graph G. Prove that e is a cut edge if and only if e is not in any cycle of G.

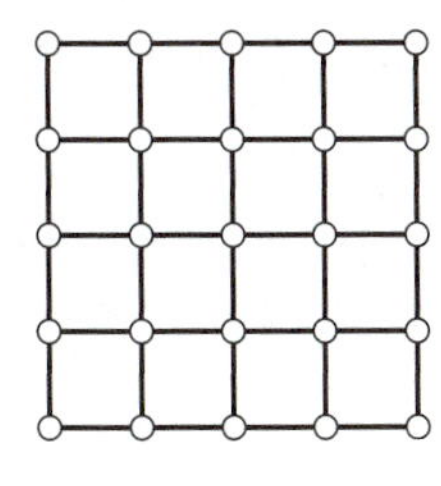

49.16. Let G be a graph. A cycle of G that contains all the vertices in G is called a *Hamiltonian cycle*.

a. Show that if $n \geq 5$, then $\overline{C_n}$ has a Hamiltonian cycle.

b. Prove that the graph in the figure does not have a Hamiltonian cycle.

49.17. Consider the following algorithm.

- **Input:** A connected graph G.
- **Output:** A spanning tree of G.

(1) Let T be a graph with the same vertices as G, but with no edges.

(2) Let $e_1, e_2, \ldots, e_m$ be the edges of G.

(3) For $k = 1, 2, \ldots, m$, do:

(3a) **If** adding edge e_k to T does not form a cycle with edges already in T, **then** add edge e_k to T.

(4) Output T.

Prove that this algorithm is correct. In other words, prove that whenever the input to this algorithm is a connected graph, the output of this algorithm is a spanning tree of G.

49.18. Consider the following algorithm.

- **Input:** A connected graph G.
- **Output:** A spanning tree of G.

(1) Let T be a copy of G.

(2) Let $e_1, e_2, \ldots, e_m$ be the edges of G.

(3) For $k = 1, 2, \ldots, m$, do:

(3a) **If** edge e_k is not a cut edge of T, **then** delete e_k from T

(4) Output T.

Prove that this algorithm is correct. In other words, prove that whenever the input to this algorithm is a connected graph, the output of this algorithm is a spanning tree of G.

50 Eulerian Graphs

Earlier (in Section 46) we presented the classic Seven Bridges of Königsburg problem. We explained that it is impossible to walk all seven bridges without retracing a bridge (or taking a swim across the river) because the multigraph that represents the bridges has more than two vertices of odd degree.

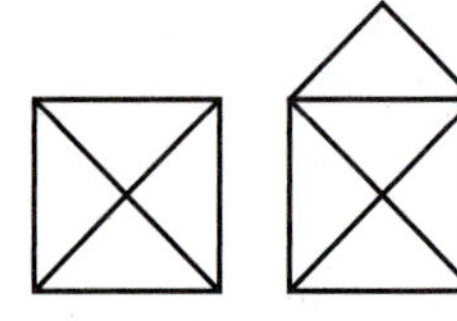

Consider the two figures shown. The figure on the left has four corners where an odd number of lines meet. Therefore, it is impossible to draw this figure without lifting your pencil or redrawing a line. The odd corners must be the first or last points on such a drawing.

The figure on the right, however, has only two corners with an odd number of lines (the lower two). These points must be the first/last points in a drawing. Can this figure be drawn without lifting your pencil or retracing a line? Try it! You have an important hint. You must start at one of the lower two corners. With that hint, it is simple to draw this figure.

In this section, we recast this bridge-walking/figure-drawing problem as a graph theory problem.

Definition 50.1 **(Eulerian trail, tour)** Let G be a graph. A walk in G that traverses every edge exactly once is called an *Eulerian trail*. If, in addition, the trail begins and ends at the same vertex, we call the walk an *Eulerian tour*. Finally, if G has an Eulerian tour, we call G *Eulerian*.

The problems we consider are the following: Which graphs have Eulerian trails? Which graphs have Eulerian tours (i.e., are Eulerian)? In this section, we give a complete answer.

Necessary Conditions

If a graph G has an Eulerian trail, then it is (almost) necessary that G be connected. If the graph has two (or more) components, it would be impossible for the trail to visit more than one component, so there is no way we can traverse all the edges of the graph. Impossible, that is, unless those additional components did not have any edges to traverse! This can happen if all (but one) of the components consist of just a single isolated vertex.

An *isolated* vertex is a vertex of degree 0.

Let us call a component of a graph *trivial* if it contains only one vertex. Otherwise we call the component *nontrivial*. Thus the first necessary condition for the existence of an Eulerian trail is the following:

- If G is Eulerian, then G has at most one nontrivial component.

We revisit the degree conditions. Suppose v is a vertex of a graph G in which there is an Eulerian trail W. If v is neither the first nor the last vertex on this trail, then we observe that v must have even degree:

$$W = \text{first} \sim \cdots \sim ? \sim v \sim ? \sim \cdots \sim ? \sim v \sim ? \sim \cdots \sim ? \sim v \sim ? \sim \cdots \sim \text{last}.$$

Since every edge of the graph is traversed exactly once, and since for every edge entering v on this tour there is another edge exiting v, it must be the case that $d(v)$ is even.

We therefore have the following:

- If G has an Eulerian trail, then it has at most two vertices of odd degree.

What can we say about the degrees of the first and last vertices on the trail? Suppose that the first and last vertices on the trail are different. The degree of the first vertex on the trail must be odd by the following reasoning. There is one edge traversed from this vertex when the trail begins. Then, every other time we visit the first vertex, an entering edge is paired with an exiting edge. Therefore, its degree must be odd. The same is true for the last vertex on the trail; its degree must be odd.

- If G has an Eulerian trail that begins at a vertex a and ends a vertex b (with $a \neq b$), then vertices a and b have odd degree.

Another reason $d(a)$ is even: If $d(a)$ were odd, it would be the only vertex of odd degree, contradicting Exercise 46.12.

If the trail begins and ends at the same vertex a, we observe that $d(a)$ must be even. We have one edge exiting a at the start of the tour which matches the final edge entering a at the end of the tour. Every other time we visit a, entering and exiting edges pair up, and so, all told, the number of edges incident with a must be even. We therefore have the following:

- If G has an Eulerian tour (i.e., if G is Eulerian), then all vertices in G have even degree.

We have one last remark to make about Eulerian tours before we present the main theorems for this section. Suppose we have an Eulerian tour in a connected graph that begins and ends at a vertex a, and suppose b is the second vertex on this tour:

$$W = a \sim b \sim \cdots\cdots \sim a.$$

We can, instead, begin the tour at b, follow the original tour until we get to the last visit to a, and finish at b; that is,

$$W' = b \sim \cdots\cdots \sim a \sim b$$

is also an Eulerian tour starting/ending at b. If we shift the tour repeatedly, we see that we can begin an Eulerian tour at any vertex we choose.

- If G is a connected Eulerian graph, then G has an Euler tour that begins/ends at any vertex.

Main Theorems

The necessary conditions we just delineated motivate what we seek to prove.

Theorem 50.2 Let G be a connected graph all of whose vertices have even degree. For every vertex $v \in V(G)$, there is an Eulerian tour that begins and ends at v.

Theorem 50.3 Let G be a connected graph with exactly two vertices of odd degree: a and b. Then G has an Eulerian trail that begins at a and ends at b.

A traditional way to prove these results is first to prove Theorem 50.2 and then use it to prove Theorem 50.3. We take a different, more interesting approach. We establish these two theorems with a single proof! The proof is by induction on the number of edges in the graph. To prove the two results at the same time, we require a more elaborate induction hypothesis, but this makes the induction easier—an example of induction loading.

Proof. We prove both Theorems 50.2 and 50.3 by induction on the number of edges in G.

Basis case: Suppose G has 0 edges. Then G consists of just 1 isolated vertex, v. The walk (v)—remember: a single vertex by itself is a walk—is an Eulerian trail of G.

(This is a perfectly valid basis case, but it is so simple we do one more unnecessary basis step to make sure nothing strange is happening here. It also appears to have nothing to do with Theorem 50.3.)

Another basis case: Suppose G has one edge. Since G is connected, the graph must consist of just two vertices, a and b, and a single edge joining them. Now G has exactly two vertices of odd degree, and clearly $a \sim b$ is an Eulerian trail starting at one and ending at the other.

Induction hypothesis: Suppose a connected graph has m edges. If all of its vertices have even degree, then there is an Eulerian tour beginning/ending at any vertex. If exactly two of its vertices have odd degree, then there is an Eulerian trail that begins at one of these vertices and ends at the other.

Let G be a connected graph with $m + 1$ edges.

- **Case 1:** All of G's vertices have even degree.

  In this case, we must show that we can form an Eulerian tour starting at any vertex of G. Let v be an arbitrary vertex of G.

  Let w be any neighbor of v. Consider the graph $G' = G - vw$. Notice that in G' all vertices have exactly the same degree that they had in G, except for v and w; their degrees have decreased by exactly 1. Thus G' has exactly two vertices of odd degree.

  We also assert that G' is connected. We defer this part of the proof to Lemma 50.4 (see the "Unfinished business" section), which assures us that if all vertices in a graph have even degree, then no edge is a cut edge.

  Here is the lovely part: Since G' is connected and has exactly two vertices of odd degree, it has (by induction) an Eulerian trail that begins at w and ends at v.

  If we add the edge vw to the beginning of W, the result is an Eulerian tour of G that begins/ends at v!
- **Case 2:** Exactly two of G's vertices, a and b, have odd degree.

  We must show there is an Eulerian trail that begins at a and ends at b.

  - **Subcase 2a:** Suppose $d(a) = 1$.

    In this case, a has exactly one neighbor, x. It is possible that $x = b$ or $x \neq b$. We check both possibilities.

    Let $G' = G - a$; that is, delete vertex a (and the one edge incident thereon) from G. Notice that $d(x)$ drops by 1, while all other vertices have the same degree as before. Also note that G' has m edges and is connected (see the proof of Proposition 49.8).

If $x = b$, then all vertices in G' have even degree (a is gone and b's degree has changed by 1). Therefore, by induction, G' has an Eulerian tour W that begins and ends at vertex b. If we insert the edge ab at the beginning of W, we have constructed an Eulerian trail that begins at a and ends at b.

If $x \neq b$, then G' has exactly two vertices of odd degree (the degree of x in G' is now odd, and b still has odd degree). Therefore, by induction, there is an Eulerian trail W that begins at x and ends at b. If we prepend the edge ax to W, we have an Eulerian trail in G that begins at a and ends at b.

- **Subcase 2b:** Suppose $d(a) > 1$.
  Since $d(a)$ is odd, we have $d(a) \geq 3$. We claim that at least one of the edges incident with a is not a cut edge (this is proved in Lemma 50.5 in "Unfinished business,").

  Let ax be an edge incident with a that is not a cut edge of G. Let $G' = G - ax$. Notice that, just as in subcase 2a, we might have $x = b$ or $x \neq b$.

  In the case $x = b$, then, just as before, all vertices of G' have even degree, and we can form, by induction, an Eulerian tour in G' that begins/ends at b and then prepend the edge ab to form an Eulerian trail in G that begins at a and ends at b, as required.

  In the case $x \neq b$, then, just as before, we have exactly two vertices of odd degree in G', namely, x and b. By induction, we form, in G', an Eulerian trail that starts at x and ends at b. We prepend the edge ax to yield the requisite Euler trail in G.

In all cases, we find the required Eulerian trail/tour in G. ■

The proof of Theorems 50.2 and 50.3 implicitly gives an algorithm for finding Eulerian trails in graphs. The algorithm can, rather imprecisely, be expressed as follows: Don't make any blatant mistakes. What do we mean by this?

First, if the graph has two vertices of odd degree, you must begin the trail at one of these vertices.

Second, imagine you are part way through drawing the graph. You are currently at vertex v, and let us suppose H represents the subgraph of the original graph consisting of those edges you have not yet traversed. Which edge from v should you take? The proof shows that you can take any edge you like, just as long as it is not a cut edge. Of course, if there is only one edge of H incident with v, you must take it, but this isn't a problem; you will never need to revisit that vertex again!

Unfinished Business

The proof of Theorems 50.2 and 50.3 used the following two results.

Lemma 50.4 Let G be a graph all of whose vertices have even degree. Then no edge of G is a cut edge.

Proof. Suppose, for the sake of contradiction, $e = xy$ is a cut edge of such a graph. Notice that $G - e$ has exactly two components (by Theorem 48.12), and each of these components contains exactly one vertex of odd degree, contradicting Exercise 46.12. ■

Lemma 50.5 Let G be a connected graph with exactly two vertices of odd degree. Let a be a vertex of odd degree and suppose $d(a) \neq 1$. Then at least one of the edges incident with a is not a cut edge.

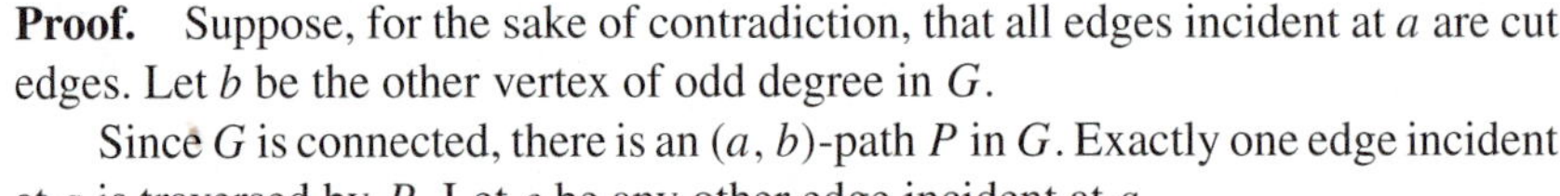

Proof. Suppose, for the sake of contradiction, that all edges incident at a are cut edges. Let b be the other vertex of odd degree in G.

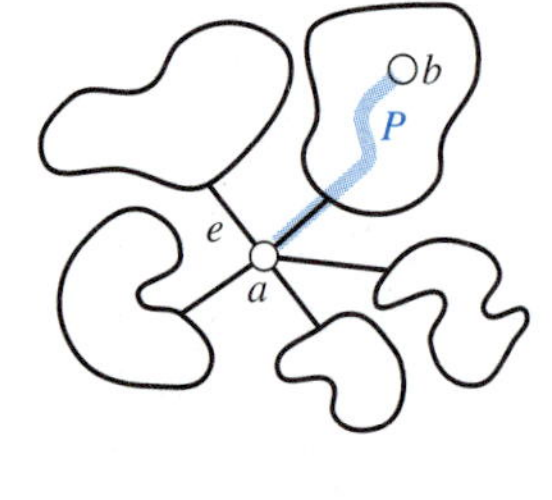

Since G is connected, there is an (a, b)-path P in G. Exactly one edge incident at a is traversed by P. Let e be any other edge incident at a.

Now consider the graph $G' = G - e$. This graph has exactly two components (Theorem 48.12). Since the path P does not use the edge e, vertices a and b are in the same component. Notice also that, in G', vertex a has even degree, and all other vertices in its component have not changed degree. This means that, in G', the component containing vertex a has exactly one vertex of odd degree, contradicting Exercise 46.12. ■

Recap

Motivated by the Seven Bridges of Königsburg problem, we defined Eulerian trails and tours in graphs. We showed that every connected graph with at most two vertices of odd degree has an Eulerian trail. If there are no vertices of odd degree, it has an Eulerian tour.

50 Exercises

50.1. We noticed that a graph with more than two vertices of odd degree cannot have an Eulerian trail, but connected graphs with zero or two vertices of odd degree do have Eulerian trails. The missing case is connected graphs with exactly one vertex of odd degree. What can you say about those graphs?

50.2. A *domino* is a 2×1 rectangular piece of wood. On each half of the domino is a number, denoted by dots. In the figure, we show all $\binom{5}{2} = 10$ dominoes we can make where the numbers on the dominoes are all pairs of values chosen from $\{1, 2, 3, 4, 5\}$ (we do not include dominoes where the two numbers are the same). Notice that we have arranged the ten dominoes in a ring such that, where two dominoes meet, they show the same number.

For what values of $n \geq 2$ is it possible to form a domino ring using all $\binom{n}{2}$ dominoes formed by taking all pairs of values from $\{1, 2, 3, \ldots, n\}$? Prove your answer.

Note: In a conventional box of dominoes, there are also dominoes both of whose squares have the same number of dots. You may either ignore these "doubles" or explain how they can easily be inserted into a ring made with the other dominoes.

50.3. Let G be a connected graph that is not Eulerian. Prove that it is possible to add a single vertex to G, together with some edges from this new vertex to some old vertices such that the new graph is Eulerian.

50.4. Let G be a connected graph that is not Eulerian. In G there must be an even number of odd-degree vertices (see Exercise 46.12). Let $a_1, b_1, a_2, b_2, \ldots, a_t, b_t$ be the vertices of odd degree in G.

If we add edges $a_1b_1, a_2b_2, \ldots, a_tb_t$ to G, does this give an Eulerian graph?

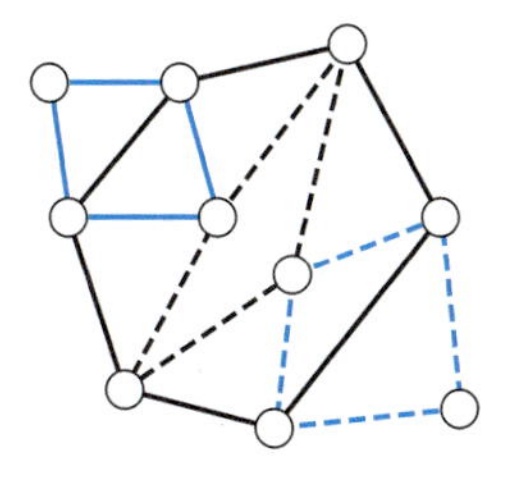

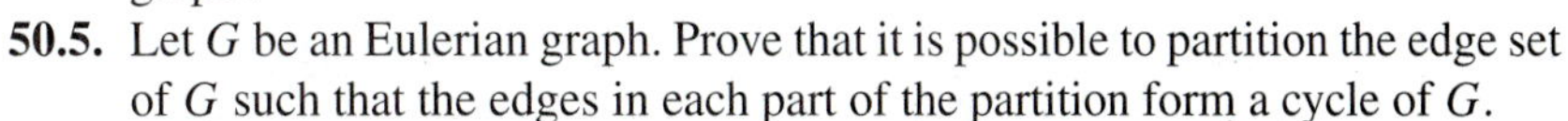

50.5. Let G be an Eulerian graph. Prove that it is possible to partition the edge set of G such that the edges in each part of the partition form a cycle of G.

The figure shows such a partition in which the edges from different parts of the partition are drawn in different colors and line styles.

50.6. Is it possible to walk the seven bridges of Königsburg so that you cross every bridge exactly twice, once in each direction?

51 Coloring

The four color map problem and the exam-scheduling problem are both examples of *graph-coloring* problems. The general problem is as follows: Let G be a graph. To each vertex of G, we wish to assign a color. The restriction is that adjacent vertices must receive different colors. Of course, we could give every vertex its own color, but this is not terribly interesting and not relevant to applications. The objective is to use as few colors as possible.

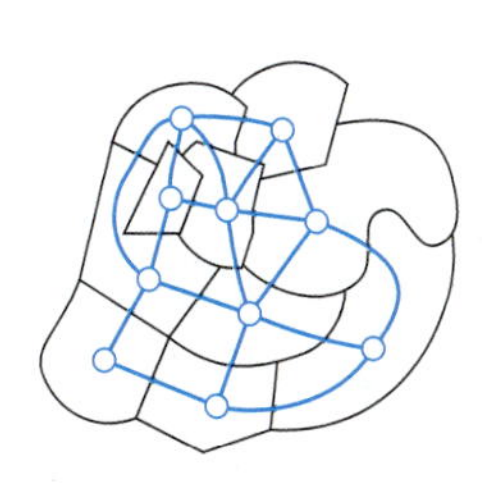

For example, consider the map-coloring problem from Section 46. We can convert this problem into a graph-coloring problem by representing each country as a vertex of a graph. Two vertices in this graph are adjacent exactly when the countries they represent share a common border. Thus coloring the countries on the map corresponds exactly to coloring the vertices of the graph.

We can also convert the exam-scheduling problem into a graph-coloring problem. The vertices of this graph represent the courses at the university. Two vertices are adjacent when the courses they represent have a common student enrolled. The colors on the vertices represent the different examination time slots. Minimizing the number of colors assigned to the vertices corresponds to minimizing the number of exam periods.

Core Concepts

Colors are phenomena of the physical world, and graphs are mathematical objects. It is mildly illogical to speak of applying colors (physical pigments) to vertices (abstract elements).

The careful way to define graph coloring is to give a mathematical definition of *coloring*.

Definition 51.1 **(Graph coloring)** Let G be a graph and let k be a positive integer. A *k-coloring* of G is a function

$$f : V(G) \to \{1, 2, \ldots, k\}.$$

We call this coloring *proper* provided

$$\forall xy \in E(G), \ \ f(x) \neq f(y).$$

If a graph has a proper k-coloring, we call it *k-colorable*.

The central idea in the definition is the function f. To each vertex $v \in V(G)$, the function f associates a value $f(v)$. The value $f(v)$ is the color of v. The palette of colors we use is the set $\{1, 2, \ldots, k\}$; we are using positive integers as "colors." Thus $f(v) = 3$ means that vertex v is assigned color 3 by the coloring f.

The condition $\forall xy \in E(G), \ f(x) \neq f(y)$ means that whenever vertices x and y are adjacent (form an edge of G), then $f(x) \neq f(y)$ (the vertices must get different colors). In a proper coloring, adjacent vertices are not assigned the same color.

Notice what the definition does not require: It does not say that all the colors must be used; that is, it does not require f to be onto. The number k refers to the size of the palette of colors available—it is not a demand that all k colors be used. If, say, a graph is five-colorable, then it is also six-colorable. We can simply add color 6 to the palette and then not use it.

Although the formal definition of coloring specifies that the colors we use are integers, we often refer to real colors when describing graph coloring.

The goal in graph coloring is to use as few colors as possible.

Definition 51.2 **(Chromatic number)** Let G be a graph. The smallest positive integer k for which G is k-colorable is called the *chromatic number* of G. The chromatic number of G is denoted $\chi(G)$.

The symbol χ is not an x. It is a lowercase Greek chi.

Example 51.3 Consider the complete graph K_n. We can properly color K_n with n colors by giving every vertex a different color. Can we do better? No. Since every vertex is adjacent to every other vertex in K_n, no two vertices may receive the same color, and so n colors are required. Therefore $\chi(K_n) = n$.

Notice that for any graph G with n vertices, we have $\chi(G) \leq n$ because we can always give each vertex a separate color. This means that among all graphs with n vertices, K_n has the largest chromatic number. We can say a little bit more.

Proposition 51.4 Let G be a subgraph of H. Then $\chi(G) \leq \chi(H)$.

Proof. Given a proper coloring of H, we can simply copy those colors to the vertices of G to achieve a proper coloring of G. So if we used only $\chi(H)$ colors to color the vertices of H, we have used at most $\chi(H)$ colors in a proper coloring of G. ■

Proposition 51.5 Let G be a graph with maximum degree Δ. Then $\chi(G) \leq \Delta + 1$.

Proof. Suppose the vertices of G are $\{v_1, v_2, \ldots, v_n\}$ and we have a palette of $\Delta + 1$ colors. We color the vertices of G as follows:

To begin, no vertex in G is assigned a color. Assign any color from the palette to vertex v_1. Next we color vertex v_2. We take any color we wish from the palette, as long as the coloring is proper. In other words, if $v_1 v_2$ is an edge, we may not assign the same color to v_2 that we gave to v_1. We continue in exactly this fashion through all the vertices. That is, when we come to vertex v_j, we assign to vertex v_j any color from the palette we wish, just making certain that the color on vertex v_j is not the same as any of its already-colored neighbors.

The issue is whether there are sufficiently many colors in the palette so that this procedure never gets stuck (i.e., we never reach a vertex where there is no legal color left to choose). Since every vertex has at most Δ neighbors and since there are $\Delta + 1$ colors in the palette, we can never get stuck. Thus this procedure produces a proper $\Delta + 1$-coloring of the graph. Hence $\chi(G) \leq \Delta + 1$. ■

Example 51.6 What is the chromatic number of the cycle C_n? If n is even, then we can alternate colors (black, white, black, white, etc.) around the cycle. When n is even, this yields a valid coloring. However, if n is odd, then vertex 1 and vertex n would both be black if we alternated colors around the cycle. See the figure. Thus, for n-odd, C_n is not two-colorable. It is, however, three-colorable. We can alternately color vertices 1 through $n - 1$ with black and white and then color vertex n with, say, blue. This gives a proper three-coloring of C_n. [Also, by Proposition 51.5, we have $\chi(C_n) \leq \Delta(C_n) + 1 = 2 + 1 = 3$.] Thus

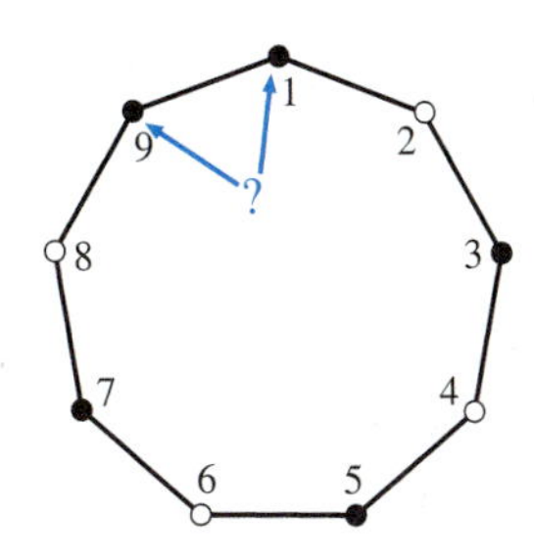

$$\chi(C_n) = \begin{cases} 2 & \text{if } n \text{ is even, and} \\ 3 & \text{if } n \text{ is odd.} \end{cases}$$

Note the following interesting point about this example: The chromatic number of C_9 is 3, but C_9 does not contain K_3 as a subgraph.

Bipartite Graphs

Which graphs are one-colorable? That is, can we describe the class of all graphs G for which $\chi(G) = 1$?

Notice that $\chi(G) = 1$ means that we can properly 1-color the graph G. This means that if we assign all vertices the same color, this is a proper coloring. How can this be? It implies that both endpoints of any edge in G are the same color, which is a blatant violation! The answer is: There can be no edges in G. In other words, we have the following:

Proposition 51.7 A graph G is one-colorable if and only if it is edgeless.

That was easy! Let's move on to characterizing two-colorable graphs—that is, graphs G for which $\chi(G) \le 2$. These graphs have a special name.

Definition 51.8 **(Bipartite graphs)** A graph G is called *bipartite* provided it is 2-colorable.

Here is another useful way to describe bipartite graphs. Let $G = (V, E)$ be a bipartite graph and select a proper two-coloring. Let X be the set of all vertices that receive one of the two colors, and let Y be the set of all vertices that receive the other color. Notice that $\{X, Y\}$ forms a partition of the vertex set V. Furthermore, if e is any edge of G, then e has one of its endpoints in X and its other endpoint in Y.

The partition of V into the sets X and Y such that every edge of G has one end in X and one end in Y is called a *bipartition* of the bipartite graph. When writing about bipartite graphs, it is customary to write sentences such as the following: Let G be a bipartite graph with bipartition $V = X \cup Y \ldots$. This means that X and Y are the two parts of the bipartition. The sets X and Y are called by some authors (this author not included) the *partite* sets of the bipartite graph.

The problem we address here is: Which graphs are bipartite? For example, on the basis of Example 51.6, we conclude that even cycles are bipartite, but odd cycles are not. The following result gives another wide class of examples.

Proposition 51.9 Trees are bipartite.

We prove this using the method in Proof Template 25.

Proof. The proof is by induction on the number of vertices in the tree.

Basis case: Clearly a tree with only one vertex is bipartite. Indeed, $\chi(K_1) = 1 \le 2$.

Induction hypothesis: Every tree with n vertices is bipartite.

Let T be a tree with $n + 1$ vertices. Let v be a leaf of T and let $T' = T - v$. Since T is a tree with n vertices, by induction T' is bipartite. Properly color T' using the two colors black and white.

Now consider v's neighbor—call it w. Whatever color w has, we can give v the other color (e.g., if w is white, we color v black).

Since v has only one neighbor, this gives a proper two-coloring of T. ■

Trees and even cycles are bipartite. What other graphs are bipartite? Here is another important class of bipartite graphs:

Definition 51.10 **(Complete bipartite graphs)** Let n, m be positive integers. The *complete bipartite graph* $K_{n,m}$ is a graph whose vertices can be partitioned $V = X \cup Y$ such that

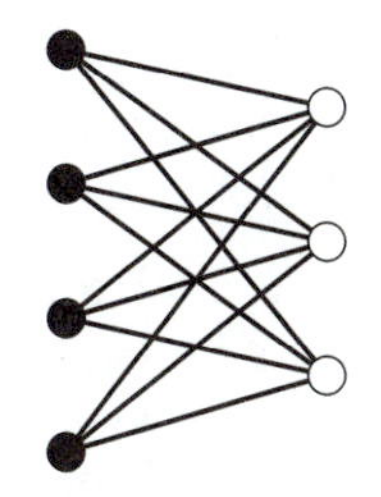

- $|X| = n$,
- $|Y| = m$,
- for all $x \in X$ and for all $y \in Y$, xy is an edge, and
- no edge has both its endpoints in X or both its endpoints in Y.

The graph in the figure is $K_{4,3}$.

The following theorem describes precisely which graphs are bipartite.

Theorem 51.11 A graph is bipartite if and only if it does not contain an odd cycle.

This is an example of a *characterization theorem.*

The proof of this result is a bit complicated. We present it in a moment, but first, we explain why this is a wonderful theorem.

Suppose I have a graph and I want to convince you that it is bipartite. I can do this by coloring the vertices and then showing you my coloring. You can patiently inspect each edge and notice that the two endpoints of every edge have different colors. You will be certain that the graph is bipartite.

On the other hand, suppose I present you with a complicated graph that is not bipartite. The following argument is not terribly persuasive: "I tried for days to two-color this graph, and I really worked quite hard. Trust me! There is no *way* this graph can be two-colored."

Theorem 51.11 guarantees that I will always be able to present a much better and simpler argument. I can find an odd cycle in the graph and show it to you, and then you will be convinced that the graph is not bipartite.

The proof of Theorem 51.11 requires the following concept.

Definition 51.12 **(Distance)** Let G be a graph and let x, y be vertices of G. The *distance* from x to y in G is the length of a shortest (x, y)-path. In cases where there is no such path, we may either say that the distance is undefined or ∞.

The distance from x to y is denoted $d(x, y)$.

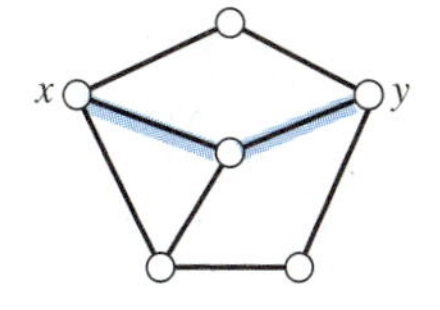

For the graph in the figure, there are several (x, y)-paths; the shortest among them have length 2. Thus $d(x, y) = 2$.

Proof (of Theorem 51.11)

($\Rightarrow$) Let G be a bipartite graph. Suppose, for the sake of contradiction, that G contains an odd cycle C as a subgraph. By Proposition 51.4, we have

$$3 = \chi(C) \le \chi(G) \le 2,$$

a contradiction. Therefore G does not contain an odd cycle.

($\Leftarrow$) Next we show that if G does not contain an odd cycle, then G is bipartite. We begin by proving a special case of this result. We show that if G is connected and does not contain an odd cycle, then G is bipartite.

Suppose G is connected and does not contain an odd cycle. Let u be any vertex in $V(G)$. Define two subsets of $V(G)$ as follows:

$$X = \{x \in V(G) : d(u, x) \text{ is odd}\}, \quad \text{and}$$
$$Y = \{y \in V(G) : d(u, y) \text{ is even}\}.$$

In words, X and Y contain those vertices in G that are at odd and even distance from u, respectively. Note that $u \in Y$ because $d(u, u) = 0$. Also note that $V(G) = X \cup Y$ (every vertex is some finite distance from u because, by hypothesis, G is connected) and $X \cap Y = \emptyset$ (because the distance from a given vertex to u cannot be both odd and even).

We color the vertices in X black and the vertices in Y white. We claim that this gives a proper two-coloring of G. To prove this, we must show that there are no two vertices in X that are adjacent and no two vertices in Y that are adjacent.

Suppose, for the sake of contradiction, there are two vertices $x_1, x_2 \in X$ with $x_1 \sim x_2$. Let P_1 be a shortest path from u to x_1. Because $x_1 \in X$, we know that $d(u, x_1)$ is odd, so the length of P_1 is odd. Likewise let P_2 be a shortest (u, x_2)-path; its length is also odd.

It is tempting (but incorrect!) to conclude as follows: Concatenate

$$P_1 + (x_1 \sim x_2) + P_2^{-1}.$$

That is, traverse P_1 from u to x_1 (odd distance), go from x_1 to x_2 (odd distance), and, finally, go back to u along P_2 (odd again). The total distance is odd, so we have an odd cycle.

The error is that $P_1 + (x_1 \sim x_2) + P_2^{-1}$ might not be a cycle (see the figure). The paths P_1 and P_2 might have vertices and edges in common.

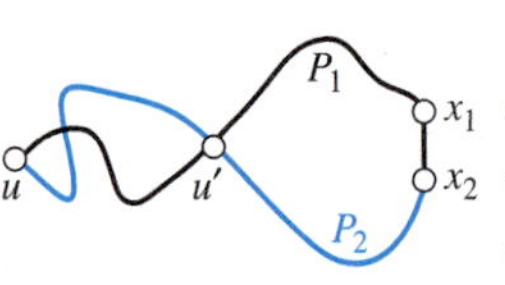

To fix this problem, let u' denote the last vertex that P_1 and P_2 have in common. That is, as we traverse P_1 from u to x_1, we know that P_1 and P_2 have at least one vertex in common—namely, u. Perhaps they have other vertices in common. In any case, since P_1 ends at x_1 and P_2 ends at x_2, eventually along P_1 we reach the last vertex these two paths have in common. After u', there are no further P_2 vertices on P_1. Therefore, if we traverse P_1 from u' to x_1, then traverse the edge x_1x_2, and finally return to u' along P_2^{-1}, we have a cycle. The question is: Is this an odd cycle?

We note that the section of P_1 from u to u' is as short as possible. Otherwise, if there were a shorter path Q from u to u', then we could concatenate Q with the (u', x_1)-section of P_1 and achieve a (u, x_1)-walk that is shorter than P_1, from which we could construct a (u, x_1)-path that is shorter than P_1; this is a contradiction. So the (u, u')-section of P_1 is as short as possible. Likewise the (u, u')-section of P_2 is as short as possible. Hence the (u, u')-sections of P_1 and P_2 must have the same length.

Now consider the (u, x_1)- and (u, x_2)-sections of P_1 and P_2, respectively. We know that P_1 and P_2 both have odd length. From them, we delete the same length: their (u, u')-sections. Thus the two sections that remain are either both odd or both even—they have the same parity.

We now conclude that the cycle C is an odd cycle. The cycle consists of the edge x_1x_2 (length 1) and the two sections from u' of P_1 and P_2 (same parity). Since $1 + \text{odd} + \text{odd}$ and $1 + \text{even} + \text{even}$ are both odd, we conclude that C is an odd cycle. But by hypothesis, G has no odd cycles.$\Rightarrow\Leftarrow$ Therefore there is no edge in G both of whose endpoints are in X.

Might there be an edge with both ends in Y? No. The argument is exactly the same as before. The only fact we used about the paths P_1 and P_2 is that their lengths had the same parity; it didn't really matter that they were both odd. If they were both even, the exact same argument applies. There are no edges between any pair of vertices of Y.

Therefore we have a proper two-coloring of G, and hence G is bipartite.

To finish the proof, we need to consider the case when G is disconnected. Suppose G is a disconnected graph that contains no odd cycles. Let $H_1, H_2, \ldots, H_c$ be its connected components. Note that since G does not contain an odd cycle, neither do any of its components. Hence, by the argument above, they are bipartite. Let $X_i \cup Y_i$ be a bipartition of $V(H_i)$ (with $1 \le i \le c$). Finally, let

$$X = X_1 \cup X_2 \cup \cdots \cup X_c \quad \text{and}$$
$$Y = Y_1 \cup Y_2 \cup \cdots \cup Y_c.$$

We claim that $X \cup Y$ is a bipartition of $V(G)$.

Please observe that X and Y are pairwise disjoint and their union is $V(G)$. There can be no edge between two vertices in X_i because $X_i \cup Y_i$ is a bipartition, and there can be no edge between vertices of X_i and X_j (with $i \ne j$) because these vertices are in separate components of G. Therefore no edge has both ends in X. Similarly, no edge has both ends in Y. Therefore $X \cup Y$ is a bipartition of $V(G)$, and so G is bipartite. ■

The Ease of Two-Coloring and the Difficulty of Three-Coloring

The proof of Theorem 51.11 gives us a method for determining whether or not a graph is bipartite, and the statement itself gives us an efficient way to convince others that we have correctly determined whether or not a graph is bipartite.

We begin with a graph, all of whose vertices are uncolored. We arbitrarily color one vertex white. Then we color all its neighbors black. Now we color all neighbors of black vertices white, and then all neighbors of white vertices black.

At some point in this procedure, we may color two adjacent vertices the same color. If we do, we can retrace our steps and find an odd cycle, proving the graph is not bipartite.

We may also find that this coloring procedure finds no new vertices to color, but yet, there remain uncolored vertices. In this case, we realize the graph is not connected, and we restart this procedure in another component.

If, after doing this procedure in every component, we never find adjacent vertices with the same color, then we have found a bipartition of the graph.

This procedure is simple and efficient. We know that once we color a vertex, say, black, all its neighbors must be white. There is no choice in this matter because there are only two colors.

The situation for three-coloring graphs is more complicated. Let's suppose the three colors are red, blue, and green. We color one vertex red. Now, what shall we color its neighbors? We have choices, and in this case, choices complicate our lives.

We do not have a result akin to Theorem 51.11 for three-colorable graphs. If I have a three-coloring for a graph G, I can convince you that G is three-colorable simply by showing you the coloring. However, if G is not three-colorable, how can I readily convince you that no such coloring is possible? There is no known answer to this problem.

We ask:

Is it difficult to three-color graphs?

This question itself is difficult! Most computer scientists and mathematicians believe that it is difficult to color a graph properly with three colors or to show that no such coloring exists. However, there is no proof that this is a hard problem.

Computer scientists have identified a wide collection of problems that are on a par with graph coloring. That is, they have shown that if any one problem in this special collection has an efficient solution, then they all do. Problems in this category are known as NP-*complete*. A full description of what it means for a problem to be in this category is beyond the realm of this book. Our point is that there are no known efficient procedures to determine whether or not a graph is three-colorable (or k-colorable for any fixed value of $k > 2$), and so there is no known efficient procedure for calculating $\chi(G)$. There are, however, heuristic and approximate methods that often give good results.

Recap

We introduced the concepts of a proper coloring of a graph and the chromatic number. We analyzed the class of bipartite (two-colorable) graphs and characterized such graphs by the fact that they do not contain odd cycles.

51 Exercises

51.1. Let G and H be the graphs in the following figure.

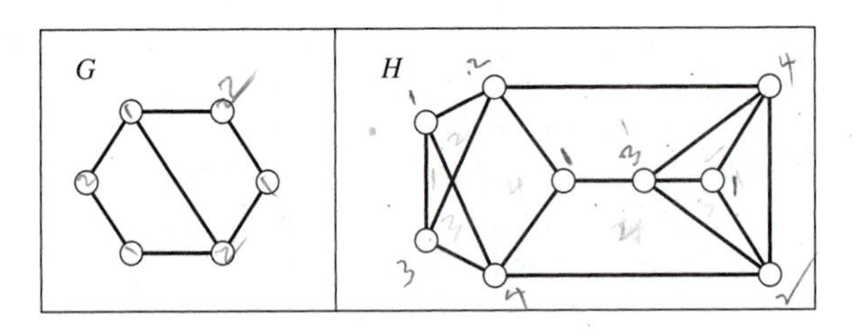

Please find $\chi(G)$ and $\chi(H)$.

51.2. Let G be a graph with just one vertex. It is correct to say that G is three-colorable. How can this be if G has only one vertex?

51.3. Let G be a properly colored graph and let us suppose that one of the colors used is red. The set of all red-colored vertices have a special property. What is it?

Graph coloring can be thought of as partitioning $V(G)$ into subsets with this special property.

51.4. Let G be a graph with n vertices that is not a complete graph. Prove that $\chi(G) < n$.

51.5. Let G be a graph with n vertices. Prove that $\chi(G) \geq \omega(G)$ and $\chi(G) \geq n/\alpha(G)$.

51.6. Let $G = K_{n,m}$. Determine $|V(G)|$ and $|E(G)|$.

51.7. Let G be a graph with n vertices. Prove that $\chi(G)\chi(\overline{G}) \geq n$.

51.8. Let G be the graph in the figure. Prove that $\chi(G) = 4$.

51.9. Let G be a graph with exactly one cycle. Prove that $\chi(G) \leq 3$.

51.10. Let n be a positive integer. The *n-cube* is a graph, denoted Q_n, whose vertices are the 2^n possible length-n lists of 0s and 1s. For example, the vertices of Q_3 are 000, 001, 010, 011, 100, 101, 110, and 111.

Two vertices of Q_n are adjacent if their lists differ in exactly one position. For example, in Q_4, vertices 1101 and 1100 are adjacent (they differ only in their fourth element) but 1100 and 0110 are not adjacent (they differ in positions 1 and 3).

Please do the following:

a. Show that Q_2 is a four-cycle.

b. Draw a picture of Q_3 and explain why this graph is called a *cube*.

c. How many edges does Q_n have?

d. Prove that Q_n is bipartite.

51.11. Suppose G has maximum degree $\Delta > 1$, but it has only one vertex of degree Δ. Prove that $\chi(G) \leq \Delta$.

51.12. Let G be a graph with the property that $\delta(H) \leq d$ for all induced subgraphs H of G. Prove that $\chi(G) \leq d + 1$.

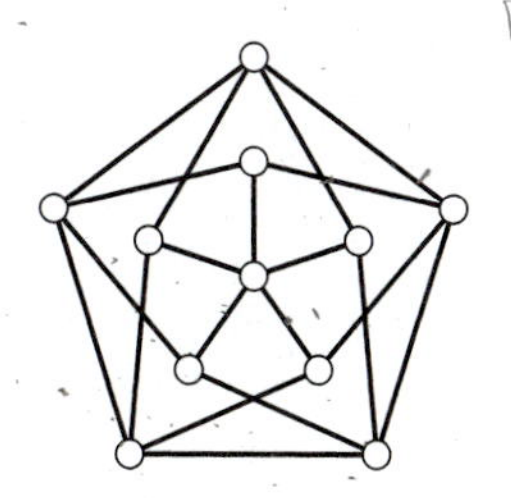

51.13. Consider the graph in the figure. Notice that it does not contain K_3 as a subgraph. Please do the following:

a. Show that this graph is four-colorable.

b. Show that this graph has chromatic number equal to 4.

c. Show that if we delete any edge from this graph, the resulting graph has chromatic number 3.

51.14. Suppose G is a graph with 100 vertices. One way to determine whether G is three-colorable is to examine all possible three-colorings of G. If a computer can check 1 million colorings per second, about how long would it take to check all possible three-colorings?

52 Planar Graphs

In this section, we study graph *drawings*. We are especially interested in graphs that can be drawn without crossing edges.

Dangerous Curves

A graph and its drawing are very different objects. A *graph* is, by Definition 46.1, a pair of finite sets (V, E) that satisfy certain properties. Its *drawing* is ink on paper; it is notational shorthand that is often easier to grasp than writing out the two sets V and E in full.

In this section, we take a different approach. We study not only graphs, but their drawings as well. A drawing is ink on paper—it is not a mathematical object. (A picture of a circle is not a circle.) Thus our first order of business ought to be a careful *mathematical* definition of a graph drawing. Unfortunately, this is complicated. The difficulty lies primarily in defining just what we mean by a curve in the plane. The precise definition of *curve* requires concepts from continuous mathematics that we have not developed and are beyond the scope of this book.

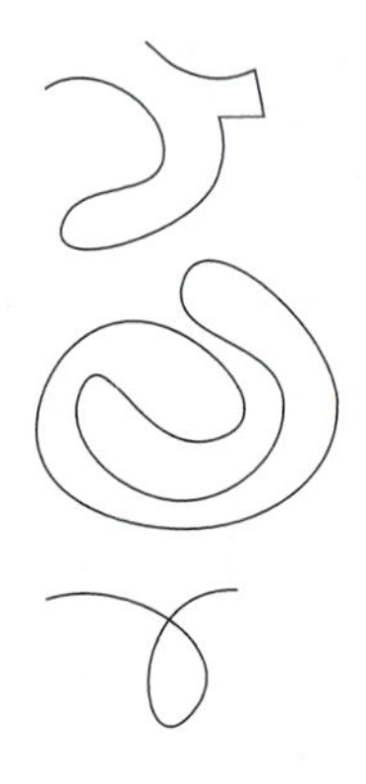

Instead, we shall just live dangerously. We proceed with our intuitive understanding of what a curve is. Note that a curve may have corners and straight sections. Indeed, a line segment is a curve. It must, however, be all in one piece. The figure in the margin shows three separate curves. A *simple curve* is a curve that joins two distinct points in the plane and does not cross itself. The top curve in the figure is simple; the other two are not.

If a curve returns to its starting point, we call the curve *closed*. If the first/last point of the curve is the only point on the curve that is repeated, then we call the curve a *simple closed* curve. The middle curve in the diagram is a simple closed curve. The third curve is neither simple nor closed.

Before we get to work on planar graphs, we need to present a word of warning. Some of the proofs in this section are not rigorous. We shall be honest with you concerning where we are not using full rigor. The problem is that fully proving these results requires a deep understanding of curves, and we have not even given a proper definition of curve. For example, we use (implicitly) the following theorem.

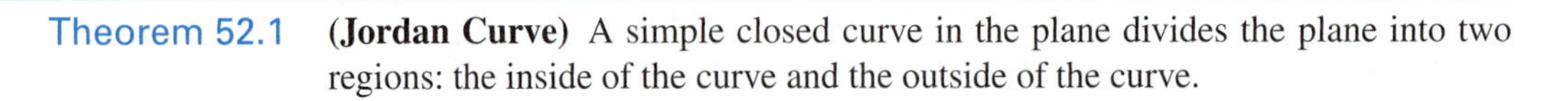

Theorem 52.1 **(Jordan Curve)** A simple closed curve in the plane divides the plane into two regions: the inside of the curve and the outside of the curve.

Many students' reaction to the Jordan Curve Theorem is that it is so obvious that it does not require a proof. Ironically, this "simple" and "obvious" statement is difficult to prove. We shall accept it and use it nevertheless.

Embedding

A *drawing* is a diagram made of ink on paper. The mathematical abstraction of a drawing is called an *embedding*. An embedding of a graph is a collection of points and curves in a plane that satisfies the following conditions:

- Each vertex of the graph is assigned a point in the plane; distinct vertices receive distinct points (i.e., no two vertices share the same point).
- Each edge of the graph is assigned a curve in the plane. If the edge is $e = xy$, then the endpoints of the curve for e are exactly the points assigned to x and y. Furthermore, no other vertex point is on this curve.

If all the curves are simple (do not cross themselves) and if the curves from two edges do not intersect (except at an endpoint if they both are incident with the same vertex), then we call the embedding *crossing-free*.

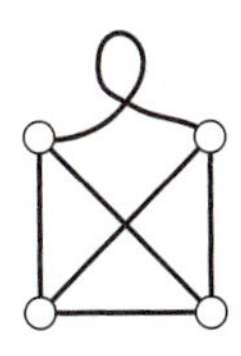
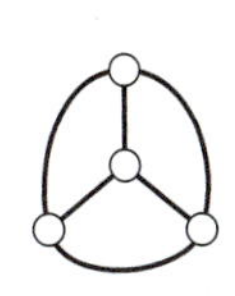

The figure shows two embeddings of the graph K_4. Note that we greatly exaggerated the points, drawing them as large round dots. The drawing on the right represents a crossing-free embedding on K_4.

Not all graphs have crossing-free embeddings in the plane. Those that do have a special name.

Definition 52.2 **(Planar graph)** A *planar graph* is a graph that has a crossing-free embedding in the plane.

For example, the graph K_4 is planar. However, the graph K_5 is not planar. How do we know? We can try to find a crossing-free drawing of K_5 and not succeed, but that is not much of a proof. Alternatively, we study properties of planar graphs and use that knowledge to prove that K_5 is not planar. The first step toward this goal is a classic result of Euler.

Euler's Formula

Let G be a planar graph and consider a crossing-free embedding of G, as in the figure. In this drawing, we see the points and curves of the embedding. We also see another feature: *faces*. A *face* is a portion of the plane cut off by the drawing. Imagine the graph drawn on a physical piece of paper. If we cut along the curves representing the edges of G, the paper falls apart into various pieces. Each of these pieces is called a *face* (or *region*) of the embedding.

This definition of *face* is not rigorous.

The drawing of the graph in the figure has five faces. Yes, five is the correct number. There are four *bounded* faces (faces with only finite area) and one *unbounded* face that surrounds the graph.

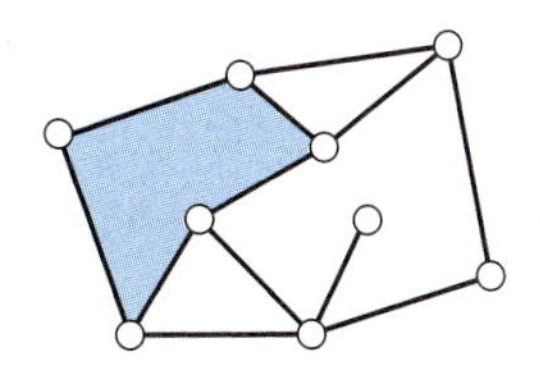

The graph in this figure has $n = 9$ vertices, $m = 12$ edges, and $f = 5$ faces. I encourage you to make a number of other crossing-free drawings of connected planar graphs and, for each, record how many vertices, edges, and faces each drawing has. Stare at your numbers and see whether you discover the following result (don't peek).

Theorem 52.3 **(Euler's formula)** Let G be a connected planar graph with n vertices and m edges. Choose a crossing-free embedding for G, and let f be the number of faces in the embedding. Then

$$n - m + f = 2.$$

Please note that the hypothesis *connected* is important. An extension to this result covers the cases when the graph is not connected (see Exercise 52.3).

This proof is not 100% rigorous. There are no untrue statements, but some of our claims are unsupported. In particular, when we delete a noncut edge from the graph, we assert, but do not prove, that two faces collapse into a single face.

Proof. This proof is by induction on the number of edges in the planar graph G.

Suppose G has n vertices. The basis case for this proof is when the number of edges is $n - 1$ since a connected graph with n vertices must have at least $n - 1$ edges (see Section 49).

Basis case: Since G is connected and has $m = n - 1$ edges, we know that G is a tree. In a drawing of a tree, there is only one face (the unbounded face) because there are no cycles to enclose additional faces. Thus $f = 1$. We therefore have

$$n - m + f = n - (n - 1) + 1 = 2$$

as required.

Induction hypothesis: Suppose all connected planar graphs with n vertices and m edges satisfy Euler's formula.

Let G be a planar graph with n vertices and $m + 1$ edges. Choose a crossing-free embedding of G and let f be the number of faces in this embedding. We need to prove that $n - (m + 1) + f = 2$.

Let e be an edge of G that is not a cut edge. Because G has more than $n - 1$ edges, it is not a tree, and therefore (Theorem 49.5) not all of its edges are cut edges. Therefore $G - e$ is connected.

If we erase e from the drawing of G, we have a crossing-free embedding of $G - e$, and so $G - e$ is planar. Notice that $G - e$ has n vertices and $(m + 1) - 1 = m$ edges. The drawing, we claim, has $f - 1$ faces. The edge we deleted causes the two faces on either side of it to merge into a single face, so $G - e$'s drawing has one less face than G's.

Now, by induction, we have

$$n - m + (f - 1) = 2$$

which rearranges to

$$n - (m + 1) + f = 2$$

which is what we needed to prove. ■

Let G be a connected planar graph with n vertices and m edges. We can solve the equation $n - m + f = 2$ for f and we get $f = 2 - n + m$. This has an important consequence. The number of vertices and edges are quantities that depend only on the graph G—they have nothing to do with how the graph is drawn in the plane. On the other hand, the quantity f is the number of faces in a particular crossing-free drawing of G. There may be many different ways to draw G without crossings. The implication of Euler's formula is that regardless of how we draw the graph, the number of faces is always the same.

For example, consider the two drawings of the graph in the figure. In both cases, the graph has $f = 2 - n + m = 2 - 9 + 12 = 5$ faces.

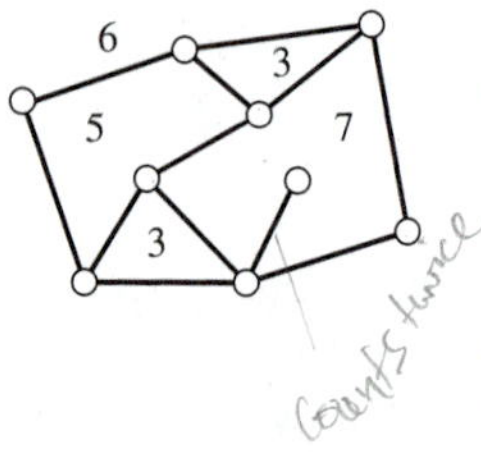

Notice that we wrote a number inside each face. This indicates the number of edges that are on the boundary of that face; it is called the *degree* of the face. In the upper figure, the face with degree equal to 7 is noteworthy. Observe there are only six edges that touch that face. Why, then, do we say this face has degree 7? The edge to the leaf has both sides on the boundary of the face; therefore this edge counts twice when we calculate the degree. The concept of *side* of an edge has no meaning whatsoever when we are considering only graphs. However, it makes sense when we consider a graph's embedding.

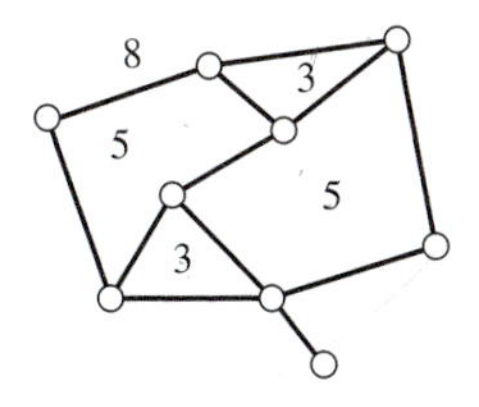

Since every edge has two sides, it contributes a total value of 2 to the degrees of the faces it touches. If an edge only touches one face, then it counts twice toward that face's degree. If it touches two faces, it counts once toward each of the two faces' degrees. Therefore, if we add the degrees of all the faces in the embedding, we get twice the number of edges in the graph. We have shown the following:

Proposition 52.4 Let G be a planar graph. The sum of the degrees of the faces in a crossing-free embedding of G in the plane equals $2|E(G)|$.

How small can the degree of a face be? If the graph is simply K_1, then the embedding is just one point, and there is just one face (the entire plane minus the one point). This face is bounded by zero edges, so it has degree equal to 0.

If the graph has just one edge, then, as before, there is only one face. The "boundary" of this face is just the one edge—it counts twice to the degree, and so this face has degree 2.

As soon as a planar graph has two (or more) edges, then all faces have degree 3 or greater. (Technically, we should prove this, but we are taking a less than rigorous approach to planar graphs just for this section. Draw pictures to convince yourself of this fact.)

We use the face-degree concept to prove the following corollary to Euler's formula.

Corollary 52.5 Let G be a planar graph with at least two edges. Then

$$|E(G)| \leq 3|V(G)| - 6.$$

Furthermore, if G does not contain K_3 as a subgraph, then

$$|E(G)| \leq 2|V(G)| - 4.$$

Proof. First note that, without loss of generality, G is connected. If G is not connected, we can add single edges between components to make it connected, and the resulting graph is still planar with more edges than the original graph. If the larger graph satisfies the inequality $|E(G)| \leq 3|V(G)| - 6$, so does the original graph.

Let G be a connected planar graph with at least two edges. Pick a crossing-free embedding of G; this embedding has f faces. By Euler's formula, $f = 2 - |V(G)| + |E(G)|$.

We calculate the sum of the degrees of the faces in this embedding.

On the one hand, by Proposition 52.4, the sum of the face degrees is $2|E(G)|$.

On the other hand, every face has degree at least 3, so the sum of the face degrees is at least $3f$. Therefore we have

$$2|E(G)| \geq 3f$$

which we can rearrange to read $f \leq \frac{2}{3}|E(G)|$.

Substituting this into Euler's formula, we get

$$2 - |V(G)| + |E(G)| = f \leq \frac{2}{3}|E(G)|,$$

which rearranges to $2 - |V(G)| + \frac{1}{3}|E(G)| \leq 0$, which yields

$$|E(G)| \leq 3|V(G)| - 6.$$

The proof of the second inequality is left for you in Exercise 52.4. ■

Here is another consequence of Euler's formula:

Corollary 52.6 Let G be a planar graph with minimum degree δ. Then $\delta \leq 5$.

Proof. Let G be a planar graph. If G has fewer than two edges, clearly $\delta \leq 5$. So we may assume that G has at least two edges.

Thus, by Corollary 52.5, we have $|E(G)| \leq 3|V(G)| - 6$.

The minimum degree δ cannot be greater than the average degree. Let $\bar{d}$ denote the average degree in G. So $\delta \leq \bar{d}$.

We now calculate

$$\delta \leq \bar{d} = \frac{\sum_{v \in V(G)} d(v)}{|V(G)|} = \frac{2|E(G)|}{|V(G)|} \leq \frac{2(3|V(G)| - 6)}{|V(G)|} = 6 - \frac{12}{|V(G)|} < 6$$

but since δ is an integer, we have $\delta \leq 5$. ■

Nonplanar Graphs

A graph that is not planar is called *nonplanar*. We can use Corollary 52.5 to prove that certain graphs are nonplanar.

Proposition 52.7 The graph K_5 is nonplanar.

Corollary 52.5 is not an if-and-only-if result. The graph in the figure satisfies the inequality $|E| \leq 3|V| - 6$ but is not planar.

Proof. Suppose, for the sake of contradiction, that K_5 were planar. By Corollary 52.5, we would have

$$10 = |E(G)| \leq 3|V(G)| - 6 = 3 \times 5 - 6 = 9,$$

a contradiction.⇒⇐ Therefore K_5 is nonplanar. ■

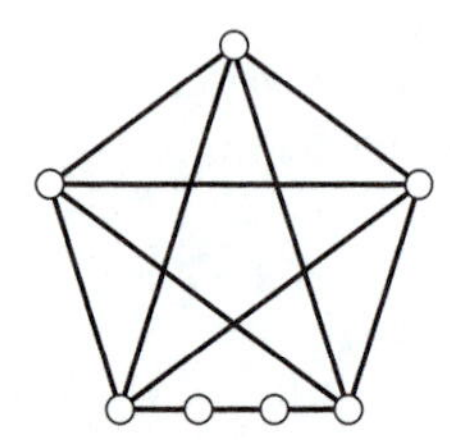

Consider the graph in the figure: Is it planar? Note that it has 7 vertices and 12 edges. Does it satisfy the formula $|E(G)| \leq 3|V(G)| - 6$? Yes: Note that $12 \leq 15 = 3 \times 7 - 6$.

We claim the graph in the figure is nonplanar. Suppose it were planar. Then it would have a crossing-free embedding. Given such an embedding, we can ignore the two vertices of degree 2. The path between the lower left and lower right vertices is represented by a three-section curve that we can think of as a single curve. Thus, if the graph in the figure had a crossing-free planar embedding, so would K_5. However, since K_5 has no such embedding, neither does the graph in the figure.

The graph in the figure is an example of a *subdivision* of K_5. A *subdivision* of G is formed from G by replacing edges with paths. Clearly if a graph is planar, so are its subdivisions. And the converse of this statement is also true: If a graph is nonplanar, then all of its subdivisions are also nonplanar. Therefore any subdivision of K_5 is nonplanar.

Moreover, any graph that contains a subdivision of K_5 as a subgraph must also be nonplanar.

Next let us consider the complete bipartite graph $K_{3,3}$. It has six vertices and nine edges, and so it satisfies the inequality $9 = |E(G)| \leq 3|V(G)| - 6 = 3 \times 6 - 6 = 12$. However, because $K_{3,3}$ is bipartite, it contains no odd cycles. In particular, it does not contain K_3 as a subgraph. We can therefore consider the stronger inequality $|E(G)| \leq 2|V(G)| - 4$ in Corollary 52.5.

Proposition 52.8 The graph $K_{3,3}$ is nonplanar.

Proof. Suppose, for the sake of contradiction, that $K_{3,3}$ were planar. Since it does not contain K_3 as a subgraph, we would have

$$9 = |E(G)| \leq 2|V(G)| - 4 = 2 \times 6 - 4 = 8$$

which is a contradiction.$\Rightarrow\Leftarrow$ Therefore $K_{3,3}$ is nonplanar. ■

This solves the gas/water/electricity problem from Section 46. It is impossible to run noncrossing utility lines between the three utilities and the three homes—if we could, we would have a crossing-free embedding of $K_{3,3}$, and that is impossible!

Not only is $K_{3,3}$ nonplanar, but so is any subdivision graph we can form from $K_{3,3}$. Furthermore, any graph that contains a subdivision of $K_{3,3}$ as a subgraph must be nonplanar as well.

The following remarkable result of Kuratowski says that K_5 and $K_{3,3}$ are the "only" nonplanar graphs. Here is what we mean:

Theorem 52.9 **(Kuratowski)** A graph is planar if and only if it does not contain a subdivision of K_5 or $K_{3,3}$ as a subgraph.

We have shown the easier half of Kuratowski's Theorem. If G contains a subdivision of K_5 or $K_{3,3}$ as a subgraph, then G cannot be planar—if G were planar, we would be able to create a crossing-free embedding of K_5 or $K_{3,3}$, and that's impossible.

The more difficult part of this result is to prove that if a graph does not contain a subdivision of K_5 or $K_{3,3}$ as a subgraph, then the graph must be planar. For the proof, please see an advanced text on graph theory.

Kuratowski's Theorem is a marvelous characterization of planarity. If a graph is planar, I can convince you of this fact by presenting you with a crossing-free drawing. On the other hand, if a graph is nonplanar, I can convince you of this fact by finding a subdivision of K_5 or $K_{3,3}$ as a subgraph of my graph.

Coloring Planar Graphs

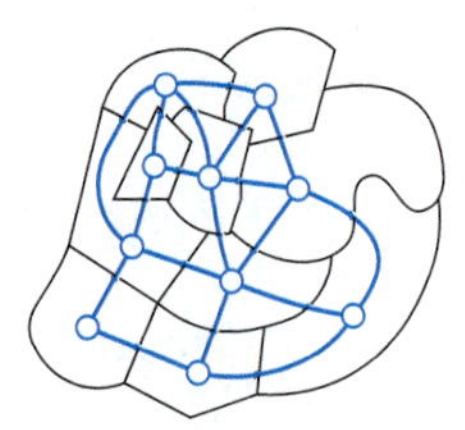

We return to the map-coloring problem of Section 46. As we discussed in Section 51, the problem of coloring a map is equivalent to the problem of coloring a graph. What we did not consider previously is that the graph that arises from a map has a special property: It must be planar. To see why, we begin with a map. We locate one vertex for each country at the capital city of that country. From that capital city, we draw curves out to its various borders. These curves fan out in a starlike pattern and do not cross each other. We send each curve to the midpoint of the border where it connects to the curve emanating from the capital city of its neighbor. In this way, we have constructed a planar embedding of the graph we want to color.

Thus the map-coloring problem becomes: Is every planar graph four-colorable? The answer is yes. This was proved in the 1970s by Appel and Haken.

Theorem 52.10 **(Four color)** If G is a planar graph, then $\chi(G) \leq 4$.

This theorem is best possible in the sense that the number 4 cannot be replaced by a smaller value. The graph K_4 is planar, and $\chi(K_4) = 4$ (Example 51.3).

The proof of the Four Color Theorem is long and complicated. One of the interesting aspects of the proof is that it requires a large amount of computation. Roughly speaking, Appel and Haken showed how to reduce the four color problem to about 2000 cases. They also proved how each case can be checked by a computer program. They then created and ran the necessary programs to check each of these cases.

In this section, we prove a simpler version of the Four Color Theorem. We show that every planar graph is five-colorable. We start by proving that every planar graph is six-colorable.

Proposition 52.11 **(Six color)** If G is a planar graph, then $\chi(G) \leq 6$.

Proof. The proof is by induction on the number of vertices in the graph.

Basis case: The theorem is obviously true for all graphs on six or fewer vertices, because we can give each vertex a separate color.

Induction hypothesis: Suppose the theorem is true for all graphs on n vertices (i.e., all planar graphs with n vertices are six-colorable).

Let G be a planar graph with $n + 1$ vertices. By Corollary 52.6, G contains a vertex, v, with $d(v) \leq 5$. Let $G' = G - v$. Notice that G' is planar and has n vertices. By induction, G' is six-colorable. Properly color the vertices of G' using just six colors. We can extend this coloring to G by giving v a color. Notice that v has at most five neighbors, and so there is some other color that we can assign to v that is different from the colors of its neighbors. This yields a proper six-coloring of G, and so $\chi(G) \leq 6$. ■

The overall logic in proving that $\chi(G) \leq 5$ for planar graphs is similar. The difficult part comes when there are five neighbors of vertex v, and they all have different colors.

Theorem 52.12 **(Five color)** If G is a planar graph, then $\chi(G) \leq 5$.

Proof. The proof is by induction on the number of vertices in the graph.

Basis case: The theorem is obviously true for all graphs on five or fewer vertices, because we can give each vertex a separate color.

Induction hypothesis: Suppose the theorem is true for all graphs on n vertices (i.e., all planar graphs with n vertices are five-colorable).

Let G be a planar graph with $n+1$ vertices. By Corollary 52.6, G contains a vertex, v, with $d(v) \leq 5$. Let $G' = G - v$. Notice that G' is planar and has n vertices. By induction, G' is five-colorable. Properly color the vertices of G' using just five colors.

We want to extend this coloring to G by giving v a color. Consider the neighbors of v. If among the neighbors of v there are only four different colors, then there is a left over color that we can assign to v. This yields a proper five-coloring of G.

The problem has been reduced to the case where $d(v) = 5$ and all five of its neighbors are different colors. There is no way to extend this coloring to v; whatever color we might choose for v would be the same color as one of its neighbors. So to extend the coloring to vertex v, we need to recolor some vertices.

Since G is planar, choose a crossing-free embedding of G. Every vertex of G, except v, has been colored with colors from the set $\{1, 2, 3, 4, 5\}$. Let $u_1, u_2, \ldots, u_5$ be the five neighbors of v in clockwise order, and, without loss of generality, let us assume that u_i has color i (for $i = 1, 2, \ldots, 5$).

The basic idea is to change the color on one of v's neighbors. Let's change the color of u_1 from 1 to 3. Now we can simply color v with color 1 and celebrate. The problem, however, is that u_1 might have a neighbor that has color 3; in that case, changing u_1 to color 3 creates an edge both of whose endpoints have the same color, and so the coloring would not be proper (see the figure).

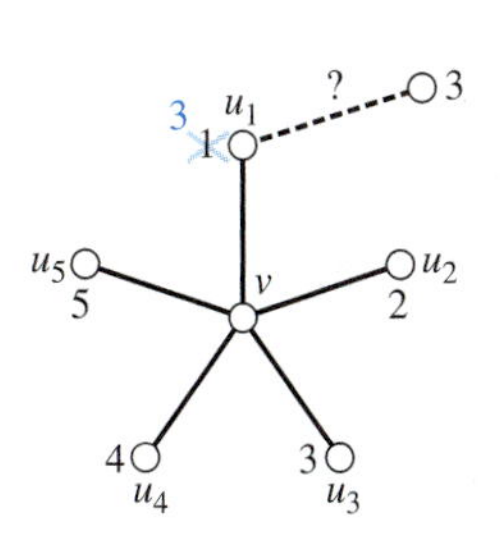

Simply changing the color of u_1 from 1 to 3 does not solve this problem. We need to be more aggressive!

Let $H_{1,3}$ be the subgraph of G induced by all vertices with color 1 or 3. In other words, we take only those vertices with color 1 or 3, and all edges that join such vertices, and call that subgraph $H_{1,3}$. Notice that if in one component of $H_{1,3}$, we exchange colors 1 and 3, then we still have a proper coloring of G' (remember: v is not colored yet).

We therefore exchange colors 1 and 3 in the component of $H_{1,3}$ that contains vertex u_1. This color exchange results in a proper coloring of G' in which vertex u_1 has color 3. We are all set to color vertex v with color 1. The problem, however, is that vertex u_3 might also be in the same component of $H_{1,3}$ as vertex u_1. Then, despite a 1-for-3 color exchange, v still has all five colors present on its neighbors.

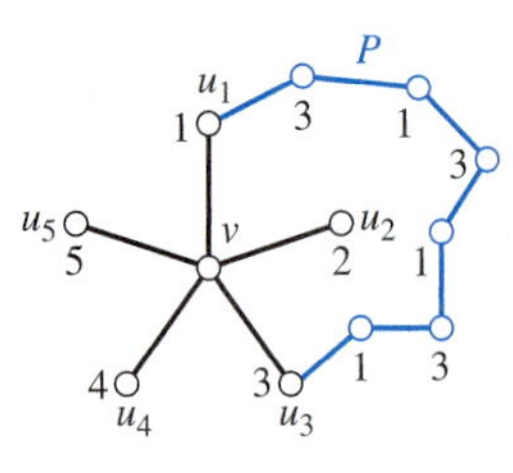

If u_1 and u_3 are in separate components of $H_{1,3}$, then the 1-for-3 color exchange works fine. We exchange colors 1 and 3 in the component of $H_{1,3}$ that includes u_1 (but not u_3). This gives a modified (but proper) five-coloring of G' in which color 3 is not present on any of v's neighbors, and so we may color v with color 1.

It remains to consider the case where u_1 and u_3 are in the same component of $H_{1,3}$ (i.e., there is a path P in $H_{1,3}$ from u_1 to u_3 as in the figure).

If u_1 and u_3 are in the same component of $H_{1,3}$, we proceed as follows: We argue as before, but now we attempt to recolor vertex u_2 with color 4. Let $H_{2,4}$ denote the subgraph of G induced on the vertices of color 2 or color 4. If u_2 and u_4 are in separate components of $H_{2,4}$, then we can recolor u_2's component, exchanging colors 2 and 4. The resulting modified coloring is a proper five-coloring of G' in which no neighbor of v has color 2. In this case, we can simply give vertex v color 2 and have a proper five-coloring of G.

The problem, as before, is that perhaps u_2 and u_4 are in the same component of $H_{2,4}$. We claim, however, that this cannot happen! Suppose there is a path, Q, from u_2 to u_4. Note that the vertices along Q are colored with colors 2 and 4, and the vertices on P are colored with colors 1 and 3. Thus P and Q have no vertices in common. Furthermore, path P, together with vertex v, forms a cycle. This cycle becomes a simple closed curve in the plane. Notice that vertices u_2 and u_4 are on different sides of this curve! Therefore the path Q from u_2 to u_4 must pass from the inside of this simple closed curve to the outside, and where it does, there is an edge crossing. However, by construction, this embedding has no edge crossings! Therefore vertices u_2 and u_4 must be in separate components of $H_{2,4}$, and the 2-for-4 recoloring technique may be used. Finally, we color vertex v with color 2, giving a proper five-coloring of G. ■

Recap

We introduced the concept of planar graphs: graphs that can be drawn in the plane without edges crossing. We presented Euler's formula that relates the number of vertices, edges, and faces of a connected planar graph and used it to find bounds on the number of edges in a planar graph. We showed that K_5 and $K_{3,3}$ are nonplanar and discussed Kuratowski's Theorem, which says, in essence, that these two graphs are the only "fundamental" nonplanar graphs. We then discussed the Four Color Theorem and proved the simpler result that all planar graphs are five-colorable.

52 Exercises

52.1. Give an example of a curve that is closed but not simple.

52.2. Each of the graphs in the figure is planar. Redraw these graphs without crossings.

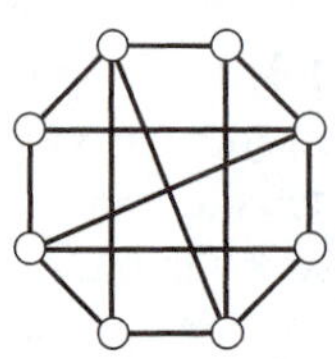
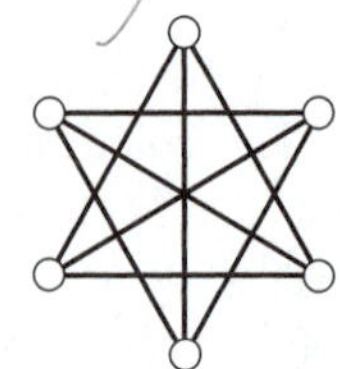
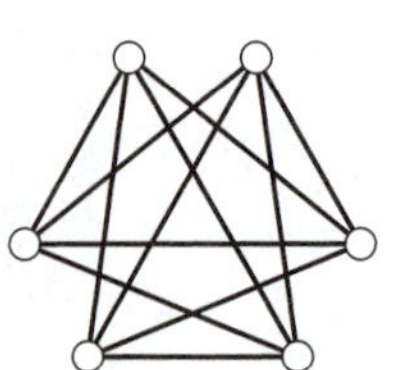

52.3. Let G be a planar graph with n vertices, m edges, and c components. Let f be the number of faces in a crossing-free embedding of G. Prove that

$$n - m + f - c = 1.$$

52.4. Complete the proof of Corollary 52.5. That is, prove that if G is planar, has at least two edges, and does not contain K_3 as a subgraph, then $|E(G)| \leq 2|V(G)| - 4$.

52.5. Let G be a graph with 11 vertices. Prove that G or $\overline{G}$ must be nonplanar. use cor 52.6

52.6. Let G be a 5-regular graph with ten vertices. Prove that G is nonplanar.

52.7. For which values of n is the n-cube Q_n planar? (See Exercise 51.10.) Prove your answer.

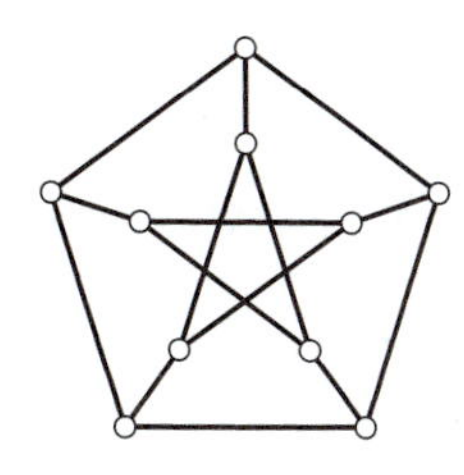

52.8. The graph in the figure is known as *Petersen's graph*. Prove that it is nonplanar by finding either a subdivision of K_5 or a subdivision of $K_{3,3}$ as a subgraph.

52.9. Let $G = (V, E)$ be a planar graph in which every cycle has length 8 or greater.

a. Prove that $|E| \leq \frac{4}{3}|V| - \frac{8}{3}$. (You may assume the graph has at least one cycle.)

b. Prove that $\delta(G) \leq 2$.

c. Prove that $\chi(G) \leq 3$.

52.10. A *Platonic* graph is a connected planar graph in which all vertices have the same degree r (with $3 \leq r \leq 5$) and in whose crossing-free embedding all faces have the same degree s (with $3 \leq s \leq 5$). Let G be a Platonic graph with v vertices, e edges, and f faces.

a. Prove that $vr = fs$. How is this quantity related to e?

b. Prove that if $r = s = 3$, then $v = f = 4$. Conclude that K_4 is the only Platonic graph with $r = s = 3$.

c. Prove that

$$e = \frac{2}{\frac{2}{r} + \frac{2}{s} - 1}.$$

d. In all, there are nine ordered pairs (r, s) with $3 \leq r, s \leq 5$. Use the equation in part (c) to rule out the existence of Platonic graphs with some of these values.

e. For the pairs (r, s) that were not ruled out in part (d), find a Platonic graph with vertex degree r and face degree s.

52.11. A soccer ball is formed by stitching together pieces of material that are regular pentagons and regular hexagons. The lengths of the sides of these polygons are all the same, so the edges match up exactly. Each corner of a polygon is the meeting place for exactly three polygons.

Prove that there must be exactly 12 pentagons.

Chapter 9 Self Test

1. Draw a picture of the following graph:

$$(\{1, 2, 3, 4, 5\}, \{\{1, 2\}, \{1, 3\}, \{3, 4\}\}).$$

2. Find a graph on ten vertices whose degrees are 6, 5, 5, 5, 4, 4, 4, 4, 3, and 3, or prove that no such graph exists.
3. Let G be a graph with 100 vertices. The vertex set of G can be partitioned into ten sets of ten vertices each; thus,

$$V(G) = W_1 \cup W_2 \cup \cdots \cup W_{10}.$$

The W_is are pairwise disjoint and all have cardinality 10.

In G there are no edges between vertices in the same W_i, but between W_i and W_j (with $i \neq j$) all possible edges are present.

How many edges does G have?
4. Let G be a graph with 10 vertices and 15 edges.
   a. How many induced subgraphs does G have?
   b. How many spanning subgraphs does G have?
5. Let a and b be distinct vertices in a complete graph on ten vertices, K_{10}. How many paths of length 5 are there from a to b?
6. Let a and b be distinct vertices in a complete graph on ten vertices, K_{10}. How many walks of length 5 are there from a to b?

   This question is more difficult than the one posed in Problem 5. To assist you in answering this question, use the following steps:
   a. Define $f(k)$ to be the number of length-k walks between distinct vertices in K_{10} and $g(k)$ to be the number of length-k walks in K_{10} from a vertex back to itself.

   Deduce the values of $f(0)$, $g(0)$, $f(1)$, and $g(1)$.
   b. Suppose $k > 1$. Express $f(k)$ in terms of $f(k-1)$ and $g(k-1)$.
   c. Suppose $k > 1$. Express $g(k)$ in terms of $f(k-1)$ and $g(k-1)$.
   d. Use your answers to the previous parts to work out $f(5)$.
7. Let G be a graph with n vertices. Suppose $\delta(G) \geq n/2$. Prove that G is connected.
8. Among the various subgraphs of K_5, how many are cycles?

   Note: Since you are asked to count subgraphs, do not consider the orientation or the starting vertex of the cycle.
9. Let G be a connected graph in which the average degree of a vertex is less than 2. Prove that G is a tree.

   Note: This is the converse of Exercise 49.2.
10. Suppose that T_1 and T_2 are trees on a common vertex set; that is, $V(T_1) = V(T_2)$. Suppose further that for any vertex v, the degree of v in the two trees is the same (i.e., $d_{T_1}(v) = d_{T_2}(v)$).

    Please answer, with proof, the following question: Is it the case that T_1 and T_2 must be isomorphic graphs?
11. What is the maximum number of edges that a disconnected graph on ten vertices can have?

12. Recall that a Hamiltonian path of a graph is a path that includes all the vertices of the graph. Show that the edges of K_8 can be partitioned into Hamiltonian paths, but the edges of K_9 cannot be so partitioned.

Note: A partition of $E(K_8)$ into Hamiltonian paths is a collection of paths that includes each of the edges of K_8 exactly once.

13. Let T be a tree containing three distinct vertices a, b, and c. By Theorem 49.4, there is a unique path from a to b (call it P), a unique path from b to c (call it Q), and a unique path from a to c (call it R).

Prove that P, Q, and R have exactly one vertex in common.

14. Let G be a graph. Prove that G is Eulerian if and only if for every partition of $V(G) = A \cup B$ (with $A \cap B = \emptyset$ and A and B nonempty), the number of edges with one end in A and one end in B is even but not zero.

15. Let G be the graph in the following figure.

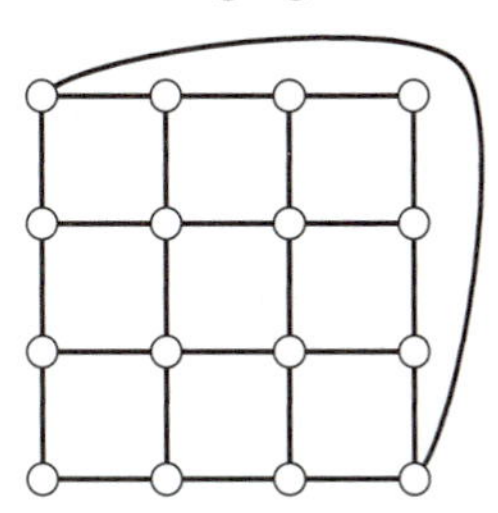

Find, with proof, $\chi(G)$.

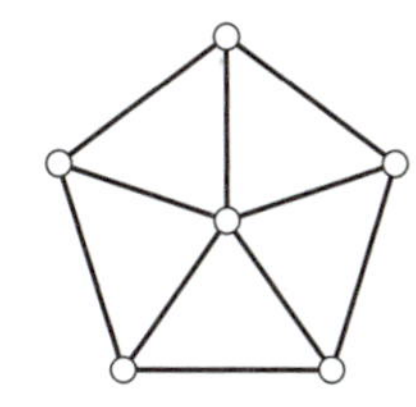

16. A *wheel* is a graph formed from a cycle by the addition of a new vertex that is adjacent to all the vertices on the cycle. A wheel with n vertices is denoted W_n; the graph W_6 is shown in the figure. Note that W_6 is based on a 5-cycle plus an additional vertex.

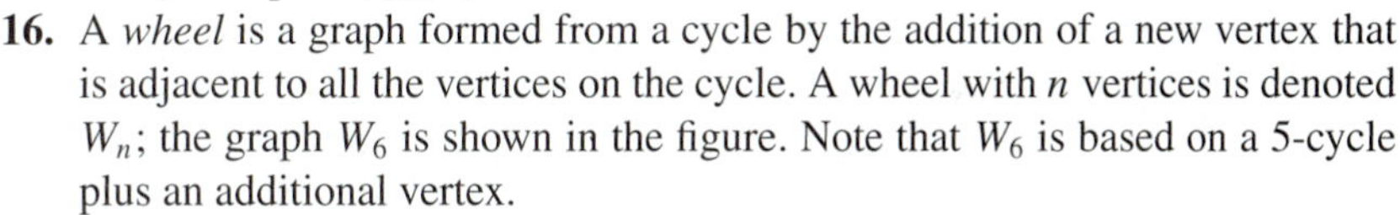

For $n \geq 3$, find, with proof, $\chi(W_n)$.

17. Let n be an integer with $n \geq 4$. Find, with proof, $\chi(\overline{C_n})$.

18. Let G be a graph and let k be a positive integer. We write $\chi(G, k)$ to stand for the number of proper k-colorings of G. For example, if $G = K_3$, then $\chi(G, k) = k(k-1)(k-2)$ because there are k choices for coloring vertex 1, and for each such choice, $k - 1$ choices for coloring vertex 2, and, finally, for each choice of colors for vertices 1 and 2, there are $k - 2$ choices for vertex 3.

a. Prove that $\chi(G) \geq k$ if and only if $\chi(G, k) > 0$.

b. Prove that if T is a tree with n vertices, then $\chi(T, k) = k(k-1)^{n-1}$.

19. From the graph K_6 delete three edges that have no endpoints in common. That is, if $V(K_6) = \{1, 2, 3, 4, 5, 6\}$, delete the edges 12, 34, and 56. Show that the resulting graph is planar.

20. Prove that the graphs $\overline{C_7}$ and $\overline{C_8}$ are nonplanar.

21. A planar graph has vertices only of degree 5 and 7. If there are 10 vertices of degree 7, prove that there are at least 22 vertices of degree 5.

CHAPTER

10 Partially Ordered Sets

We have studied various kinds of relations in this book: equivalence relations, function relations, and adjacency relations (for graphs). In this final chapter, we study another important class of relations: partial orders.

An equivalence relation R on a set A is a relation that satisfies three conditions: It is reflexive, symmetric, and transitive (see Section 14). In graph theory, the adjacency relation ($\sim$) on the vertex set of a graph is irreflexive and symmetric (see Section 46). Now we explore a new class of relations that satisfies a different suite of relation properties. We study relations that are reflexive, antisymmetric, and transitive.

53 Fundamentals of Partially Ordered Sets

What Is a Poset?

Consider the following relations defined on sets:

- the less-than-or-equal-to relation $\leq$ defined on the integers, $\mathbb{Z}$,
- the divides relation $|$ defined on the natural numbers, $\mathbb{N}$, and
- the is-a-subset-of relation $\subseteq$ defined on 2^A for some set A.

Please review Section 13, where the concepts of reflexive, antisymmetric, and transitive are introduced.

In all three cases, the relation R captures the flavor of *is smaller than* for the elements of the set X on which it is defined. Notice also that all three relations are reflexive, antisymmetric, and transitive on the sets on which they are defined. A *partially ordered set* is a set together with a relation that satisfies these three conditions.

Definition 53.1 **(Partially ordered set, poset)** A *partially ordered set* is a pair $P = (X, R)$ where X is a set and R is a relation on X that satisfies the following:

- R is reflexive: $\forall x \in X,\ x\ R\ x$,
- R is antisymmetric: $\forall x, y \in X$, if $x\ R\ y$ and $y\ R\ x$, then $x = y$, and
- R is transitive: $\forall x, y, z \in X$, if $x\ R\ y$, and $y\ R\ z$, then $x\ R\ z$.

The set X is called the *ground set* of P. The elements of X are simply called *elements* of the partially ordered set. The relation R is called a *partial order* relation.

The term *poset* is an abbreviation for *partially ordered set.*

Example 53.2 Let $P = (X, R)$ where $X = \{1, 2, 3, 4\}$ and

$$R = \{(1,1), (1,2), (1,3), (1,4), (2,2), (3,3), (3,4), (4,4)\}.$$

It is not hard to see that R is reflexive [all of $(1, 1)$ through $(4, 4)$ are in R] and antisymmetric [the only time we have both (x, y) and (y, x) in R is when $x = y$]. Checking transitivity is tedious. The only interesting case is that we have both $1\ R\ 3$ and $3\ R\ 4$, and note that we also have $(1, 4) \in R$.

Thus P is a poset.

This is a diagram depicting the poset from Example 53.2.

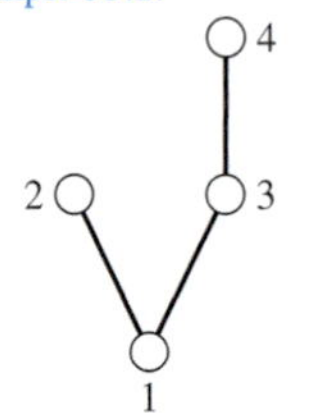

Although poset diagrams (called *Hasse diagrams*) look much like drawings of graphs, they represent rather different mathematical objects.

The poset in Example 53.2 is nearly incomprehensible. It is difficult to understand relations when they are written out as a list of ordered pairs. It is often easier to understand mathematical concepts when we can draw pictures of them.

The figure shows a diagram for the poset in Example 53.2. Each element of X, the ground set of the poset, is represented by a dot in the diagram. If $x\ R\ y$ in the poset, then we draw x's dot below y's and draw a line segment (or curve) from x to y. For example, in the figure, we position 1's dot below 2's dot, and we draw a line between them because $1\ R\ 2$.

We do not need to draw a curve from a dot to itself. We know that partial order relations are reflexive; we don't need the diagram to remind us of this fact.

If you look carefully at the figure, it appears that we have neglected to draw one of the connecting lines. Notice that $(1, 4) \in R$, but we did not draw a line from 1's dot to 4's.

The relationships $(1, 3)$ and $(3, 4)$ are explicit in the figure. The relationship $(1, 4)$ is implicit. Because partial order relations are transitive, we can infer $1\ R\ 4$ from the diagram. We can read this in the diagram by following an upward path from 1 through 3 to 4. By not drawing a curve from 1 to 4, we keep the diagram less cluttered and easier to read.

These diagrams of posets are known as *Hasse diagrams*.

For better or for worse, Hasse diagrams look exactly like (pictures of) graphs. It is important to remember, however, that posets and graphs are different mathematical objects. Their pictures look remarkably similar, but these pictures are just notational shorthand for the true underlying mathematical structures. Also, in a graph drawing, the geometric positions of the vertices are irrelevant. However, in a Hasse diagram, the vertical positioning of the dots is important.

Example 53.3 **Problem:** Draw the Hasse diagram of the poset whose ground set is $\{1, 2, 3, 4, 5, 6\}$ and whose relation is $|$ (divides).
Solution:

4 6
2 3 5
1

Example 53.4 **Problem:** Draw the Hasse diagram for the poset whose ground set is $2^{\{1,2,3\}}$ and whose relation is $\subseteq$.
Solution:

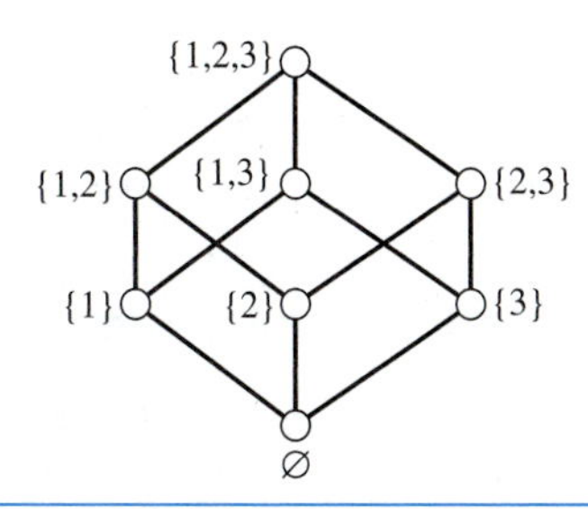

There is a natural way in which we can partially order the partitions of a set (see Section 15).

Definition 53.5 **(Refinement)** Let $\mathcal{P}$ and $\mathcal{Q}$ be partitions of a set A. We say that $\mathcal{P}$ *refines* $\mathcal{Q}$, if every part in $\mathcal{P}$ is a subset of some part in $\mathcal{Q}$. We also say that $\mathcal{P}$ is *finer* than $\mathcal{Q}$.

For example, let $A = \{1, 2, 3, 4, 5, 6, 7\}$, and let

$$\mathcal{P} = \{\{1, 2\}, \{3\}, \{4\}, \{5, 6\}, \{7\}\}, \quad \text{and,}$$
$$\mathcal{Q} = \{\{1, 2, 3, 4\}, \{5, 6, 7\}\}.$$

Notice that every part of $\mathcal{P}$ is a subset of a part of $\mathcal{Q}$. Thus we say that $\mathcal{P}$ is a refinement of $\mathcal{Q}$ or that $\mathcal{P}$ is finer than $\mathcal{Q}$.

It is not hard to see that every partition of a set is finer than itself (since every part of $\mathcal{P}$ is a subset of itself). Thus *refines* is reflexive. Furthermore, *refines* is antisymmetric, because if every part of $\mathcal{P}$ is contained in a part of $\mathcal{Q}$ and vice versa, you can prove (Exercise 53.6) that they must contain exactly the same parts (i.e., $\mathcal{P} = \mathcal{Q}$). Furthermore, *refines* is transitive. Therefore, *refines* is a partial order on the set of all partitions of A.

Example 53.6 **(Partitions poset) Problem:** Draw the Hasse diagram of the *refines* partial order on all partitions of $\{1, 2, 3, 4\}$.

Solution: It is convenient to write $1/2/34$ in lieu of $\{\{1\}, \{2\}, \{3, 4\}\}$. Here is the Hasse diagram:

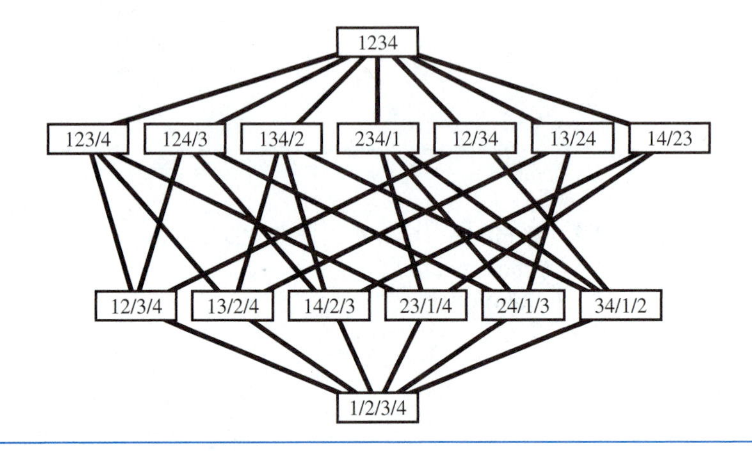

Notation and Language

A partially ordered set is a pair $P = (X, R)$ where X is a set and R is a relation. Mathematicians rarely use the letter R to stand for a poset's relation. For some posets, there is a natural symbol to use. For the poset in Example 53.4, it is natural to use the symbol $\subseteq$ to denote the partial order relation.

However, for a general poset such as the one in Example 53.2, the symbol most often used for the partial order relation is $\leq$. The use of this symbol is both good and bad. It is bad because the symbol $\leq$ already has a meaning: ordinary less than or equal to. We need to infer from context what meaning $\leq$ has: the ordinary or some partial order relation. However, there are some good features to this notation. A partial order relation is a generalization of ordinary $\leq$. We may also use the symbols $<$, $\geq$, and $>$ as follows: Let $P = (X, \leq)$ be a poset (now we are using $\leq$ to stand for a generic partial order relation). We define the following:

- $x < y$ means $x \leq y$ and $x \neq y$,
- $x \geq y$ means $y \leq x$, and
- $x > y$ means $y \leq x$ and $y \neq x$.

We may also put a slash through any of these symbols to mean that the given relationship does not hold. For example, $x \not\geq y$ means $y \leq x$ is false.

When we read the symbols such as $\leq$ aloud, it is awkward to pronounce $\leq$ as "less than or equal to." Further, we want to distinguish poset $\leq$ from ordinary $\leq$. One comfortable way to pronounce the symbol $\leq$ is to read it as "is below." For the other symbols, we read $<$ as "is strictly below," $\geq$ as "is above," and $>$ as "is strictly above."

Some mathematicians use a different-shaped $\leq$ symbol for partial orders, such as $\preceq$. This is a reasonable approach in printed work, but it can be annoying when writing mathematics by hand.

There is only one symbol $\leq$, and we may have occasion to discuss two different posets at once. We cannot use the same symbol for both partial orders! One solution is to attach various decorations to the $\leq$ symbol, such as a prime mark $\leq'$ or subscripts $\leq_2$.

Why do we need separate symbols for $<$ and for $\ngeq$? Don't these two mean the same thing?

For ordinary less than or equal to, $x < y$ is true if and only if $x \ngeq y$ is true. So in that context, the symbols $<$ and $\ngeq$ carry the same meaning.

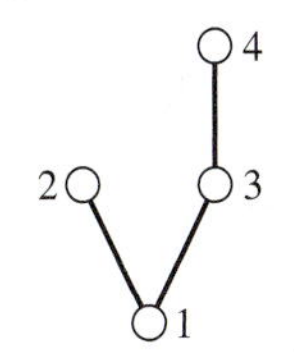

However, for a poset, $<$ and $\ngeq$ mean different things. For the poset in Example 53.2 (see the figure), we have $2 \ngeq 4$ is true (since 2 is not above 4) but $2 < 4$ is false (since 2 is not strictly below 4).

For the poset in this example, all three of the following are false: $2 < 4$, $2 = 4$, and $2 > 4$. This cannot happen for ordinary $\leq$. Elements 2 and 4 cannot be compared by the relation $\leq$. Neither $2 \leq 4$ nor $4 \leq 2$ is true. We call such a pair of elements *incomparable*.

Definition 53.7 **(Comparable, incomparable)** Let $P = (X, \leq)$ be a poset. Let $x, y \in X$. We call elements x and y *comparable* provided $x \leq y$ or $y \leq x$.

We call the elements x and y *incomparable* if $x \nleq y$ and $y \nleq x$.

In the example poset, elements 2 and 4 are incomparable, whereas elements 1 and 4 are comparable.

Definition 53.8 **(Chain, antichain)** Let $P = (X, \leq)$ be a poset and let $C \subseteq X$. We call C a *chain* of P provided every pair of elements in C are comparable.

Let $A \subseteq X$. We call A an *antichain* of P provided every pair of distinct elements of A are incomparable.

Consider the poset P from Example 53.2. The following sets are some of the chains of P:

$$\{1\}, \quad \{1, 2\}, \quad \{1, 4\}, \quad \{1, 3, 4\}, \quad \emptyset.$$

Note that in the Hasse diagram for this poset, elements 1 and 4 are not joined by a line. Nonetheless, $\{1, 4\}$ is a chain because 1 and 4 are comparable.

The following sets are some of the antichains of P:

$$\{3\}, \quad \{2, 3\}, \quad \{2, 4\}, \quad \emptyset.$$

Definition 53.9 **(Height, width)** Let P be a poset. The *height* of P is the maximum size of a chain.

The *width* of P is the maximum size of an antichain.

The largest chain in the poset of Example 53.2 is $\{1, 3, 4\}$, so this poset has height equal to 3.

The largest antichains in this poset are $\{2, 3\}$ and $\{2, 4\}$, this poset has width equal to 2.

Recap

We introduced the concept of partially ordered set (or poset for short) and gave several examples. We often use the symbol $\leq$ for the partial order relation despite the fact that it also stands for ordinary less than or equal to.

We showed how to draw a picture of a poset. We introduced a number of terms, including comparable/incomparable, chain/antichain, and height/width.

53 Exercises

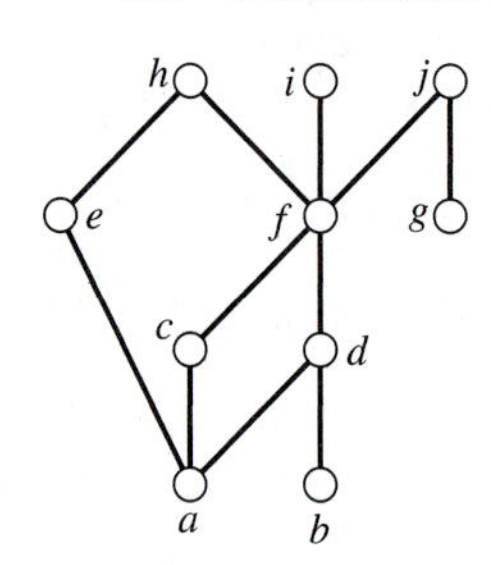

53.1. Let P be the poset in the figure. For each pair of elements x, y listed below, determine whether $x < y$, $y < x$, or x and y are incomparable.

a. a, b.
b. a, c.
c. c, g.
d. b, h.
e. c, i.
f. h, d.

53.2. For the poset from the previous problem, please find the following:

a. The height of the poset and a chain of largest size.
b. The width of the poset and an antichain of largest size.
c. A chain containing three elements that cannot be extended to a larger chain.
d. A chain containing two elements that cannot be extended to a larger chain.
e. An antichain containing three elements that cannot be extended to a larger antichain.

53.3. Let P_n denote the set of all positive divisors of the positive integer n ordered by divisibility. In other words, $|$ is the partial order relation.

Draw the Hasse diagram of P_n for the following values of n:

a. $n = 6$.
b. $n = 10$.
c. $n = 12$.
d. $n = 16$.
e. $n = 18$.

53.4. For each of the posets in the previous problem, find a largest chain, a largest antichain, the height of the poset, and the width.

See Definition 13.4 for the definition of the inverse of a relation.

53.5. Suppose $P = (X, R)$ is a partially ordered set. Prove that $\hat{P} = (X, R^{-1})$ is also a partially ordered set. We call $\hat{P}$ the *dual* of P.

If the partial order relation is denoted by $\leq$, what is a better way to write $\leq^{-1}$?

53.6. Prove that *refines* is a partial order relation on the set of all partitions of a set A.

53.7. Let x and y be elements of a poset. Prove that we cannot have both $x < y$ and $x > y$.

53.8. *True or false*: Please label each of the following statements as either true or false and then give a proof.

a. Let x and y be elements of a poset. It must be the case that exactly one of the following is true: $x < y$, $x = y$, or $x > y$.

b. Let x and y be elements of a poset and suppose there is a chain that contains both x and y. Then it must be the case that exactly one of the following is true: $x < y$, $x = y$, or $x > y$.

c. Let C and D be chains in a poset. Then $C \cup D$ is also a chain.

d. Let C and D be chains in a poset. Then $C \cap D$ is also a chain.

e. Let A and B be antichains in a poset. Then $A \cup B$ is also an antichain.

f. Let A and B be antichains in a poset. Then $A \cap B$ is also an antichain.

g. Let A be an antichain and C be a chain in a poset. Then $A \cap C$ must be empty.

h. Two points in a Hasse diagram (representing two elements of a poset) can never be joined by a horizontal line segment.

i. Let A be a set of elements in a poset. If no two elements of A are joined by a curve in the Hasse diagram, then A is an antichain.

j. Let A be a set of elements in a poset. If A is an antichain, then no two elements of A are joined by a curve in the Hasse diagram.

53.9. Which of the various properties of relations does *is comparable to* exhibit? That is, determine (with proof) whether or not it is always reflexive, irreflexive, symmetric, antisymmetric, and/or transitive.

53.10. Which of the various properties of relations does *is incomparable to* exhibit? That is, determine (with proof) whether or not it is always reflexive, irreflexive, symmetric, antisymmetric, and/or transitive.

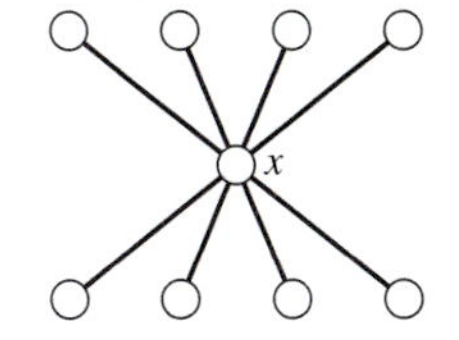

53.11. What does it mean to delete an element from a poset? Let $P = (X, \leq)$ and let $x \in X$. Create a sensible definition for $P - x$.

Let P be the poset in the figure. Draw the Hasse diagram of $P - x$.

54 Max and Min

In this section, we discuss various notions of *largest* and *smallest* in partially ordered sets.

Definition 54.1 **(Maximum, minimum)** Let $P = (X, \leq)$ be a partially ordered set. An element $x \in X$ is called *maximum* if, for all $a \in X$, we have $a \leq x$.

We call x *minimum* if, for all $b \in X$, we have $x \leq b$.

In other words, x is maximum if all other elements of the poset are below x, and x is minimum if all other elements of the poset are above x.

For example, consider the poset consisting of the positive divisors of 36 ordered by divisibility (see the figure on the left). In this poset, element 1 is minimum because it is strictly below all other elements of the poset. Element 36 is maximum because it is strictly above all other elements.

However, consider the poset consisting of the integers 1 through 6 ordered by divisibility (see the figure on the right). In this poset, element 1 is minimum, but there is no maximum element.

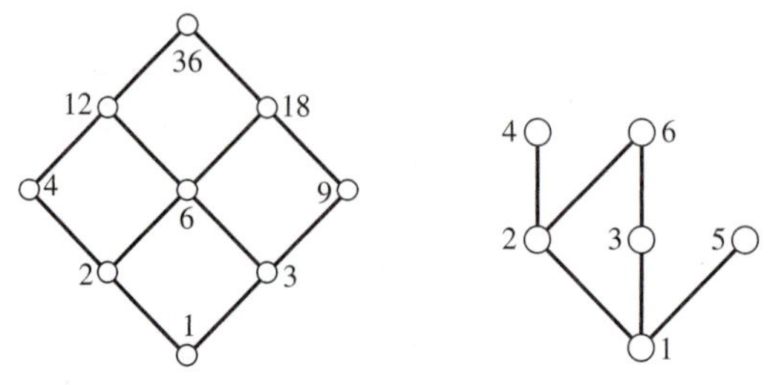

It is possible to construct an example of a poset that has neither a maximum nor a minimum element (Exercise 54.3).

An alternative concept of *largest* (or *smallest*) is presented in the next definition.

Definition 54.2 **(Maximal, minimal)** Let $P = (X, \leq)$ be a partially ordered set. An element $x \in X$ is called *maximal* if there is no $b \in X$ with $x < b$.

Element x is called *minimal* if there is no $a \in X$ with $a < x$.

In other words, x is maximal if there is no element strictly above x and minimal if there is no element strictly below it. In the poset consisting of the integers 1 through 6 ordered by divisibility (lower figure), elements 4, 5, and 6 are maximal, and element 1 is minimal.

The concepts of maxim*um* and minim*um* are similar to, but not the same as, those of minim*al* and maxim*al*. Use the following chart to help you remember the definitions.

Term	Meaning
maximum	all other elements are below
maximal	no other element is above
minimum	all other elements are above
minimal	no other element is below

It is also helpful to have an interpretation of *not maximal* and *not minimal*. Element x is *not maximal* if there is some other element y with $y > x$. Likewise, element x is *not minimal* if there is another element *strictly below* x.

We have seen an example of a poset that has no maximum element; instead, it has three maximal elements. Is it possible for a poset to have no maximal elements? Yes! Consider the poset $(\mathbb{Z}, \leq)$—the integers ordered by ordinary less than or equal to. This poset has no maximal and no minimal elements. However, finite posets must have maximal (and minimal) elements.

Proposition 54.3 Let $P = (X, \leq)$ be a finite, nonempty poset. Then P has maximal and minimal elements.

When we say that P is *finite* and *nonempty*, we mean that X is a finite set and $X \neq \emptyset$.

The value $u(x)$ is called the *up-degree* of x.

Proof. Let x be any element of P. Let us write $u(x)$ to stand for the number of elements of P that are strictly above x; that is,

$$u(x) = |\{a \in X : a > x\}|.$$

Because P is finite, $u(x)$ is a natural number (i.e., is finite).

Choose an element m such that $u(m)$ is as small as possible (since P is nonempty, there must be such an element). We claim that m is a maximal element of P.

Suppose, for the sake of contradiction, that m is not maximal. This means that there is an element a with $m < a$. By transitivity, every element that is strictly above a is also strictly above m. Furthermore, a is strictly above m, so $u(m) \geq u(a) + 1$, so $u(m) > u(a)$. However, m was selected to have smallest up-degree.$\Rightarrow\Leftarrow$ Therefore m is maximal.

A similar argument shows that every finite, nonempty poset has a minimal element. ■

Recap

We introduced the concepts of maximum, maximal, minimum, and minimal elements in a poset. We proved that every finite, nonempty poset must have maximal and minimal elements.

54 Exercises

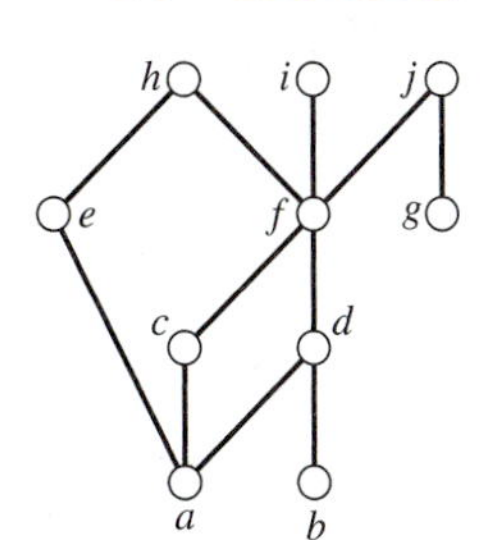

54.1. Let P be the poset in the figure. Determine which elements are maximal, maximum, minimal, and minimum.

54.2. For each of the following partially ordered sets, determine which elements are maximum, maximal, minimum, and minimal.

a. The integers $\{1, 2, 3, 4, 5\}$ ordered by ordinary less than or equal to, $\leq$.

b. The integers $\{1, 2, 3, 4, 5\}$ ordered by divisibility, $|$.

c. $(2^{\{1,2,3\}}, \subseteq)$, that is, the set of all subsets of $\{1, 2, 3\}$ ordered by is-a-subset-of (see Example 53.4).

d. Let $X = \{n \in \mathbb{Z} : n \geq 2\}$. Let $P = (X, |)$; that is, P is the poset of all integers that are greater than 1, ordered by divisibility.

e. Let X be the set of all people who are currently living. Form a partial order on X with $a < b$ provided a is a descendant of b. (In other words, a is the child, grandchild, or great grandchild, etc. of b.)

54.3. Find a poset that has neither a maximum nor a minimum element.

54.4. Prove or disprove each of the following statements.

a. If a poset has a maximum element, then it must be unique.

b. It is possible for a poset to have an element that is both maximum and minimum.

c. It is possible for a poset to have an element that is both maximal and minimal but is neither maximum nor minimum.

d. If a poset has exactly one maximal element, then it must be a maximum.

e. If x is a minimal element in a poset and y is a maximal element in a poset, then $x \leq y$.

f. If x and y are incomparable, then neither is a minimum.

g. Distinct (i.e., unequal) maximal elements must be incomparable.

54.5. Let P be a finite, nonempty poset. We know (Proposition 54.3) that P must have a minimal and a maximal element. Prove the following stronger statement.

Let P be a finite, nonempty poset. Prove that P must contain a minimal element x and a maximal element y with $x \leq y$.

55 Linear Orders

Partially ordered sets can contain *incomparable* elements. This feature makes the order relation $\leq$ *partial:* Only some of the elements can be compared using $\leq$.

There are two ways we can think about incomparable elements: On the one hand, it may not make sense to say which is "bigger" for a given pair of objects. For example, in terms of divisibility, we cannot compare 10 and 12: Neither is a divisor of the other. Another example comes from psychology in the study of preference. We may be able to say that we prefer going to the movies to going to the dentist, but there may be pairs of activities (say, movie-going versus eating a candy bar) where we might not have a clear preference.

On the other hand, two objects may be incomparable because we cannot determine which is larger. We might want to rank-order sports teams, and at some point we might ask, "Which team is better: the Baltimore Orioles (baseball) or the Baltimore Ravens (American football)?" A reasonable answer is that they cannot be compared because they play different sports. Or we might not be able to compare some objects simply because we do not have enough information.

In this section, we consider *total* (or *linear*) orders: These are partially ordered sets that do not have incomparable elements.

Definition 55.1 **(Total/linear order)** Let $P = (X, \leq)$ be a partially ordered set. We call P a *total* or *linear* order provided P does not contain incomparable elements.

For example, $(\mathbb{Z}, \leq)$ is a total order.

If x and y are elements of a total order, then we must have either $x \leq y$ or $y \leq x$. Another way to state this is that total orders satisfy the *trichotomy* rule: For all x and y in the poset, exactly one of the following is true:

- $x < y$,
- $x = y$, or
- $x > y$.

Example 55.2 Let P be the poset $(\{1, 2, 3, 4, 5\}, \leq)$—that is, the integers 1 through 5 ordered by ordinary less than or equal to. This is a total order whose Hasse diagram looks like this:

Let Q be the partially ordered set consisting of the positive divisors of 81 ordered by divisibility. In other words, the elements of Q are 1, 3, 9, 27, and 81, and they are totally ordered $1|3|9|27|81$. Notice that this poset has the same Hasse digram as P.

This example is interesting because we have two different total orders that, in essence, are "the same." A few moments thinking and doodling might convince you that all total orders on five elements are "the same." This is correct. Let us pause to consider the precise meaning of "the same." The proper term is that these posets are *isomorphic*.

Definition 55.3 **(Isomorphism of posets)** Let $P = (X, \leq)$ and $Q = (Y, \leq')$ be posets. A function $f : X \to Y$ is called a (poset) *isomorphism* provided f is a bijection and

$$\forall a, b \in X,\ a \leq b \iff f(a) \leq' f(b).$$

In the case when there is an isomorphism from P to Q, we say that P is *isomorphic* to Q and we write $P \cong Q$.

Compare this definition with the definitions of group isomorphism (Definition 40.1) and graph isomorphism (Exercise 46.18).

The condition

$$a \leq b \iff f(a) \leq' f(b)$$

means that the function f is *order-preserving*; that is, whatever order relation holds between a and b in P, we must have the corresponding relation between $f(a)$ and $f(b)$ in Q (see Exercise 55.4).

We now show that any two finite total orders with the same number of elements are isomorphic. We do this by showing that they are isomorphic to a common reference poset.

Theorem 55.4 Let $P = (X, \preceq)$ be a finite total order containing n elements. Let $Q = (\{1, 2, \ldots, n\}, \leq)$ (the integers 1 through n in their standard order). Then $P \cong Q$.

Proof. The proof is by induction on n. The basis case $n = 0$ is trivial (as is the basis case $n = 1$ in case you dislike empty posets).

We assume that the result is true for $n = k$ and suppose $P = (X, \preceq)$ is a total order on $k + 1$ elements. Let $Q = (\{1, 2, \ldots, k + 1\}, \leq)$. We must show that P is isomorphic to Q.

By Proposition 54.3, we know that P has a maximal element x. Let P' be the poset $P - x$, the poset formed by deleting x from P (see Exercise 53.11). Let Q' be the poset $(\{1, 2, \ldots, k\}, \leq)$.

By induction, P' is isomorphic to Q' so we can find an order-preserving bijection f' between their ground sets.

We define $f : X \to \{1, 2, \ldots, k + 1\}$ by

$$f(a) = \begin{cases} f'(a) & \text{if } a \neq x, \\ k + 1 & \text{if } a = x. \end{cases}$$

We must show that f is a bijection and is order-preserving.

To show that f is a bijection, we first check that f is one-to-one. Suppose $f(a) = f(b)$.

- If neither a nor b equals x, then $f(a) = f'(a)$ and $f(b) = f'(b)$, so $f'(a) = f'(b)$. Since f' is one-to-one, we have $a = b$.
- If both a and b are x, then clearly $a = b$.
- Finally, note that if $f(a) = f(b)$, it is impossible for one of a or b to be x and the other one not x; in this case, one of $f(a)$ or $f(b)$ evaluates to $k + 1$ and the other does not.

Therefore f is one-to-one.

Next we check that f is onto. Let $b \in \{1, 2, \ldots, k + 1\}$, the ground set of Q.

- If $b = k + 1$, then note that $f(x) = b$.
- If $b \neq k + 1$, then (since f' is onto $\{1, \ldots, k\}$) we can find $a \in X - \{x\}$ with $f'(a) = b$. But then $f(a) = f'(a) = b$ as required.

Thus f is onto.

Therefore f is a bijection.

Next we need to show that f is order-preserving; that is, for all $a, b \in X$,

$$a \preceq b \iff f(a) \leq f(b).$$

($\Rightarrow$) Suppose $a, b \in X$ and $a \preceq b$. We must show that $f(a) \leq f(b)$.

- If neither a nor b is equal to x, then $f(a) = f'(a)$ and $f(b) = f'(b)$. Since $f'(a) \leq f'(b)$ (because f' is order-preserving), we have $f(a) \leq f(b)$.
- If both a and b equal x, then $f(a) = f(b) = k + 1$, so clearly $f(a) \leq f(b)$.
- If $a \neq x$ and $b = x$, then $f(a) = f'(a) \leq k < k+1 = f(b)$, so $f(a) \leq f(b)$.
- Finally, we cannot have $a = x$ and $b \neq x$ because that would give $x \prec b$, and x is maximal in P.

Thus, in all possible cases, we have $a \preceq b \Rightarrow f(a) \leq f(b)$.

($\Leftarrow$) Suppose $f(a) \leq f(b)$. We must show that $a \preceq b$.

- If neither a nor b is x, then $f(a) = f'(a)$ and $f(b) = f'(b)$. Thus $f'(a) \leq f'(b)$ and so $a \preceq b$ (because f' is order-preserving).
- If both a and b are x, then $a \preceq b$.
- Note that we cannot have $a = x$ and $b \neq x$ because then $k + 1 = f(a) \leq f(b) \leq k$, which is a contradiction.
- Thus the only remaining case is $a \neq x$ and $b = x$. Since $b = x$ is maximal, we know that $a \not\succ b$. Since P is a total order, we must have $a \preceq b$.

Note that this is the first (and only) place in the proof where we use the fact that P is a total order.

Thus, in all cases, we have $a \preceq b$.

Therefore f is an order-preserving bijection between P and Q, and therefore f is an isomorphism, and P and Q are isomorphic. ■

Recap

We defined the notions of total (linear) orders and isomorphism of posets. We showed that any two finite total orders on n elements must be isomorphic; indeed, they are isomorphic to the poset $(\{1, 2, \ldots, n\}, \leq)$.

55 Exercises

55.1. What is the width of a nonempty total order?

55.2. Let n be a positive integer.

a. How many different (unequal) linear orders can be formed on the elements $\{1, 2, \ldots, n\}$?

b. How many different (nonisomorphic) linear orders can be formed on the elements $\{1, 2, \ldots, n\}$?

55.3. Prove that a minimal element of a total order is a minimum element. (Likewise, a maximal element of a total order is maximum.)

55.4. Suppose f is an isomorphism between posets P and Q, and let x and y be elements of P. Prove that x and y are incomparable (in P) if and only if $f(x)$ and $f(y)$ are incomparable (in Q).

55.5. Let P and Q be isomorphic posets and let f be an isomorphism. Let x be an element of the ground set of P. Please prove:

a. x is minimum in P iff $f(x)$ is minimum in Q.

b. x is maximum in P iff $f(x)$ is maximum in Q.

c. x is minimal in P iff $f(x)$ is minimal in Q.

d. x is maximal in P iff $f(x)$ is maximal in Q.

55.6. Prove that $(\mathbb{N}, \leq)$ and $(\mathbb{Z}, \leq)$ are not isomorphic.

Note: This exercise shows that infinite total orders need not be isomorphic; there can be no analogue to Theorem 55.4 if the posets are not finite. Furthermore, these two posets have the same size (transfinite cardinality): $\aleph_0$.

56 Linear Extensions

There are two ways to think about a partially ordered set. On the one hand, there may truly be incomparabilities among the elements of the set—we cannot compare 8 and 11 with respect to divisibility. On the other hand, we can think of a partially ordered set as representing partial information about an ordered set.

For example, consider the poset in the left portion of the figure. We see that a is a minimum element, e is a maximum element, and we have $a < b < c < e$ and $a < d < e$. However, d is—so far—incomparable to b and c. We can imagine that we simply do not yet know the order relation between b and d (or c and d).

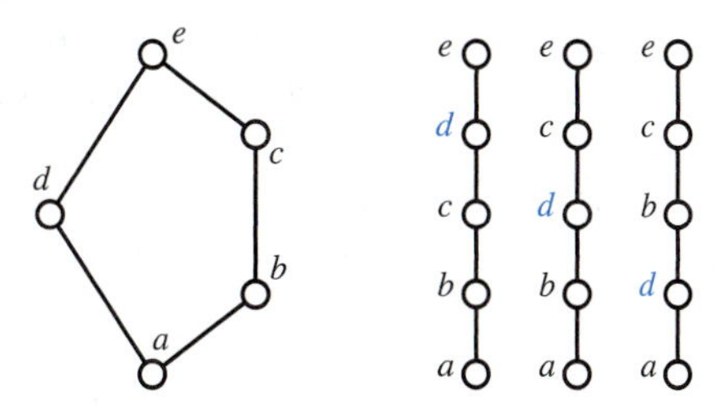

Given that elements $\{a, b, c, d, e\}$ are partially ordered, we can ask: What linear orders are consistent with the partial ordering already given on these elements?

For consistency, we must have a below all the other elements and e above all the other elements. We also must have $b < c$. The figure on the right shows the three possibilities: d might be above both b and c, d might be between b and c, or d might be below both b and c. The three linear orderings on the right are called *linear extensions* of the poset.

Definition 56.1 **(Linear extension)** Let $P = (X, \preceq)$ be a partially ordered set. A *linear extension* of P is a linear order $L = (X, \leq)$ with the property that

$$\forall x, y \in X,\ x \preceq y \Longrightarrow x \leq y.$$

It is important to notice three things about a linear extension L of a poset P:

- The posets P and L have the same ground set, X. That is, they are both partial orders on the same set of elements.
- The poset L is a linear (total) order.
- The poset L is an *extension* of P. This means that if $x \preceq y$ in P (if x and y are related in P), then $x \leq y$ (then they must also be related in L).

  No claim is made about incomparable elements of P. If x and y are incomparable in P, we might have either $x < y$ or $x > y$ in L. (We cannot have x and y incomparable in L because L is a total order.)

  The condition $x \preceq y \Rightarrow x \leq y$ can be written in the following interesting way:

$$\preceq \subseteq \leq.$$

Remember: The relations $\preceq$ and $\leq$ are relations and, as such, are sets of ordered pairs. The condition "$\preceq \subseteq \leq$" means "if $(x, y) \in \preceq$, then $(x, y) \in \leq$," which is more sensibly written "if $x \preceq y$, then $x \leq y$."

Example 56.2 Let $P = (X, \leq)$ be an antichain containing n elements. Then all possible linear orders on those n elements are linear extensions of P. Thus there are $n!$ possible linear extensions of P.

We now consider the following problem: Does every poset have a linear extension? We prove that every finite poset has a linear extension. We actually prove a stronger result.

If P is a linear order, then it is already its own linear extension. Otherwise, suppose x and y are incomparable in a finite poset P. Then we can find a linear extension L in which $x < y$ (and another linear extension L' in which $y <' x$).

Theorem 56.3 Let P be a finite partially ordered set. Then P has a linear extension. Moreover, if x and y are incomparable elements of P, then there is a linear extension L of P in which $x < y$.

Proof. Let $P = (X, \preceq)$ where X is a finite set. If P is a total order, then P is its own linear extension. Henceforth, we assume P is not a total order.

Suppose x and y are incomparable elements of P. We define a new relation, $\preceq'$, on X as follows. The basic idea is to "add" the relation (x, y) to $\preceq$.

For example, consider the poset on the left in the figure. Notice that elements x and y are incomparable. We now wish to extend $\preceq$ so that x is below y.

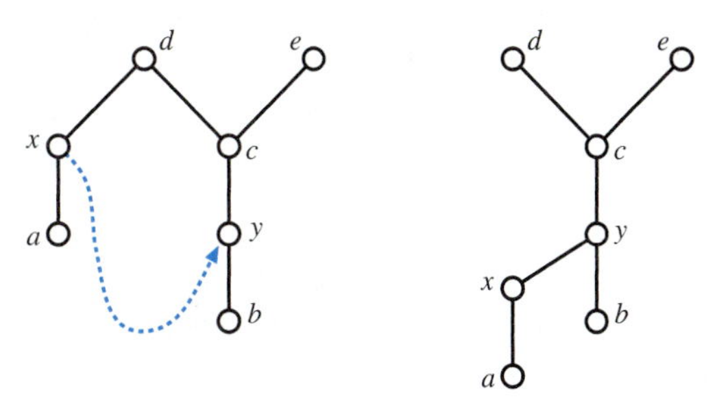

We cannot simply add the pair (x, y) to $\preceq$ because the resulting relation might not be a partial order. In particular, since $a \preceq x$, if we add the pair (x, y), we also need to add the pair (a, y). Thus we want $\preceq'$ to do three things:

- $\preceq'$ should extend $\preceq$ (i.e., if $u \preceq v$ then $u \preceq' v$),
- (x, y) should be in $\preceq'$ (i.e., $x \preceq' y$), and
- $\preceq'$ should be a partial order on X.

To this end, we define $\preceq'$ as follows. Let $s, t \in X$. We have $s \preceq' t$ provided either of the following conditions holds:

(A) $s \preceq t$ or
(B) $s \preceq x$ and $y \preceq t$.

The poset on the right in the figure above shows the relation $\preceq'$ formed from $\preceq$ (on the left).

Condition (A) guarantees that $\preceq'$ extends $\preceq$: If two elements of P are related by $\preceq$, then they are also related by $\preceq'$. Condition (B) guarantees that $x \preceq' y$ because we can take $s = x$ and $t = y$ in the definition; since $x \preceq x$ and $y \preceq y$, we have $x \preceq' y$.

Now we check that $\preceq'$ is a partial order. To do this, we need to show that $\preceq'$ is reflexive, antisymmetric, and transitive.

- $\preceq'$ is reflexive.

  Let $a \in X$ be any element of the poset P. Since $a \preceq a$ (because $\preceq$ is reflexive), we have, by condition (A) $a \preceq' a$. Therefore $\preceq'$ is reflexive.
- $\preceq'$ is antisymmetric.

  Suppose $a \preceq' b$ and $b \preceq' a$. We must prove that $a = b$. There are two possible ways we might have $a \preceq' b$: either by condition (A) or by condition (B). Likewise there are two ways we might have $b \preceq' a$. This gives four cases.
  - Suppose $a \preceq' b$ because $a \preceq b$ (A), and $b \preceq' a$ because $b \preceq a$ (A). Since $\preceq$ is antisymmetric, and because we have $a \preceq b$ and $b \preceq a$, we have $a = b$.

- Suppose $a \preceq' b$ because $a \preceq b$ (A), and $b \preceq' a$ because $b \preceq x$ and $y \preceq a$ (B).
  We claim this case cannot happen! Notice that we have $y \preceq a \preceq b \preceq x$, implying that $y \preceq x$. However, x and y are incomparable in P.$\Rightarrow\Leftarrow$ Therefore this case cannot arise.
- Suppose $a \preceq' b$ because $a \preceq x$ and $y \preceq b$ (B), and $b \preceq' a$ because $b \preceq a$. This case is just like the previous case and cannot occur.
- Finally, suppose $a \preceq' b$ because $a \preceq x$ and $y \preceq b$ (B), and $b \preceq' a$ because $b \preceq x$ and $y \preceq a$ (B).
  In this case, we have $y \preceq b \preceq x$, contradicting the fact that x and y are incomparable.$\Rightarrow\Leftarrow$ Therefore this case cannot occur.

Therefore, in all possible cases, we have $a \preceq' b$ and $b \preceq' a$ imply that $a = b$. Thus $\preceq'$ is antisymmetric.

- $\preceq'$ is transitive.

  Suppose $a \preceq' b$ and $b \preceq' c$. We must show that $a \preceq' c$. As in the demonstration of antisymmetry, there are two possible cases for $a \preceq' b$ and two possible cases for $b \preceq' c$. This gives us four cases to consider.

  - Suppose $a \preceq' b$ because $a \preceq b$ (A), and $b \preceq' c$ because $b \preceq c$ (A).
    Then $a \preceq c$ (since $\preceq$ is transitive) and so $a \preceq' c$ by (A).
  - Suppose $a \preceq' b$ because $a \preceq b$ (A), and $b \preceq' c$ because $b \preceq x$ and $y \preceq c$ (B).
    In this case, we have $a \preceq b \preceq x$, so $a \preceq x$. We also have $y \preceq c$, so $a \preceq' c$ by (B).
  - Suppose $a \preceq' b$ because $a \preceq x$ and $y \preceq b$ (B), and $b \preceq' c$ because $b \preceq c$ (A).
    In this case, we have $y \preceq b \preceq c$, so $y \preceq c$. Since $a \preceq x$, we have $a \preceq' c$ by (B).
  - Finally, suppose $a \preceq' b$ because $a \preceq x$ and $y \preceq b$ (B), and $b \preceq' c$ because $b \preceq x$ and $y \preceq c$ (B).
    We claim this case cannot occur. Notice that we have $y \preceq b \preceq x$, and so $y \preceq x$. However, x and y are incomparable.$\Rightarrow\Leftarrow$ Thus this case cannot occur.

  In all cases, have $a \preceq' c$, and so $\preceq'$ is transitive.

Therefore $P' = (X, \preceq')$ is a poset. It has the following properties. First, $a \preceq b \implies a \preceq' b$ for all $a, b \in X$. Second, $x \preceq' y$, but x and y are incomparable in P.

Thus the number of pairs of elements related by $\preceq'$ is *strictly greater* than the number of pairs of elements related by $\preceq$.

It is conceivable that $\preceq'$ is a linear order. In this case, P' is the desired linear extension of P. However, if P' is not a linear order, then it contains incomparable elements x' and y'. We can extend $\preceq'$ to form $\preceq''$ in precisely the same way as before. The relation $\preceq''$ will include all relations in $\preceq'$ and will also have the relation $x' \preceq'' y'$.

In this way, we create a sequence of partial order relations each containing more pairs than the previous: $\preceq, \preceq', \preceq'', \preceq''', \ldots$.

Because X is finite, this process cannot go on forever. Eventually, we will reach a relation in this sequence that is a total order. Let that relation be $\leq$. Since

$x \preceq' y$, and all subsequent relations are extensions of $\preceq'$, we see that $x \le y$ (see the figure).

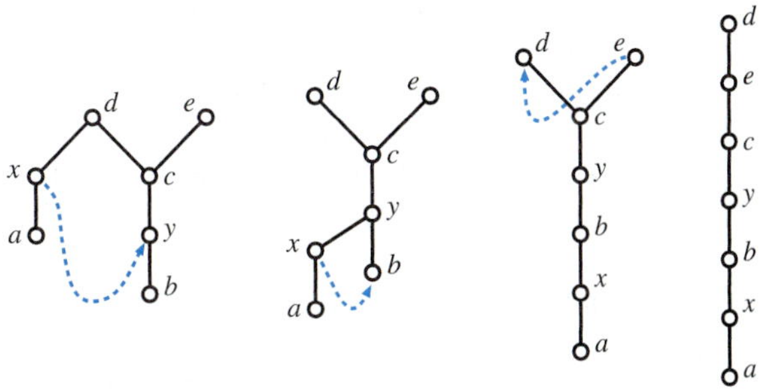

Thus we have constructed a linear extension of P in which $x \le y$. ■

Sorting

A data *record* is a collection of data about one object. In a company's personnel database, one data record might include the employee's name, Social Security number, salary, phone number, age, etc. Each of these categories is called a *field*. The goal of a sorting algorithm is to arrange the records in the natural order of one of its fields (e.g., numerically by age).

The term *sorting* refers to the process of taking a collection of data and placing it in numerical or alphabetical order. For example, imagine a company with many employees. We can create various lists of the employees. A phone roster might list all employees alphabetically by name. The accountant might list all the employees numerically by Social Security number or by salary. The telecommunications department might want a list sorted by the employees' telephone numbers. And the security department might want a roster ordered by office number.

There are a variety of techniques for sorting data. The typical methods involve making comparisons between the various data records and, from there, placing the records in their proper order.

When such an algorithm begins, the computer has no information about the order of any of the records. It starts by comparing two records. Then it compares another pair of records, and then another, and then another, and so on. On the basis of these comparisons, the computer places the records in their proper order.

The question we address here is: How many comparisons do we need to make in order to sort the data?

How does $\binom{n}{2}$ compare to $n \log_2 n$? When $n = 1000$ (a modest-sized collection of data), $\binom{n}{2}$ is about 500,000 and $n \log_2 n$ is about 10,000, or about $\frac{1}{50}$ the size.

For example, we could compare every record to every other. If there are n data records, this method takes $\binom{n}{2}$ comparisons. But this does not mean that $\binom{n}{2}$ comparisons are necessary. Indeed, there are a variety of sorting algorithms that require only $n \log_2 n$ comparisons.

We might wonder whether it is possible to develop a sorting algorithm that uses fewer than $n \log_2 n$ comparisons. For example, a sorting algorithm might begin by checking whether the n records are already sorted. If they are, the algorithm is finished after only $n - 1$ comparisons (check record 1 against 2, then 2 against 3, etc.). However, such an algorithm is not guaranteed to complete its work with only $n - 1$ comparisons. We want to know: Is there a sorting algorithm that can sort n records with fewer than $n \log_2 n$ comparisons in all cases? The answer is no. Here is the analysis.

The state of a sorting algorithm can be modeled as a partially ordered set. The elements of the poset correspond to the data records. The partial order contains all the order relations between the records that we have tested or that we can deduce from our tests.

In the beginning (when the algorithm starts), the computer has no information on the order of the records. We can represent this state of knowledge as a poset all of whose elements are incomparable to each other. The first thing the computer does is to compare two records to see which is larger. Then it compares another pair, and another, and so on. At each stage of the algorithm, the knowledge the computer has of the order of the record is partial. We can represent this information as a poset! At each stage of the sorting procedure, there is a poset P representing all we know about the relative order of the records. The linear extensions of P are all the possible ways the records might be sorted based on what we know so far.

At the start of the algorithm, all $n!$ linear extensions are feasible: We have no information (yet) about the order of the records, and so none of the $n!$ linear extensions can be ruled out.

There is no reason to compare comparable elements because we already know their relative order.

At each stage of the algorithm, we have a poset P based on our partial knowledge of the order, and all linear extensions of P are possible outcomes of the sorting algorithm. At the next step of the sorting algorithm, the computer compares two records x and y. These records correspond to incomparable elements of P. When we compare x and y, we may learn either that $x < y$ or that $x > y$. If $x < y$, some of the linear extensions of P (those in which $x < y$) will remain feasible, and the others (those in which $x > y$) will become infeasible. Conversely, if $x > y$, then the situation is reversed—those linear extensions with $x > y$ are feasible, and the others are not.

In short, there are linear extensions of P with $x < y$ and linear extensions with $x > y$; both are consistent with what we know so far. If P has k linear extensions, then there are at least $k/2$ possibilities with one order for x and y (and at most $k/2$ with the other order). If we take a worst-case outlook, the comparison of x and y yields a new poset that still has at least $k/2$ linear extensions.

In other words, each comparison in the sorting algorithm might rule out only half (or fewer) of the possible linear extensions. Since we begin with $n!$ linear orders possible at the start of the algorithm, after c comparisons, there can still be $n!/2^c$ (or more) linear orders feasible. Note that if $n!/2^c > 1$, then the sorting algorithm has not completed its work—there is more than one possible order, and so we do not yet know the actual order of the records. Thus the algorithm cannot be guaranteed to finish unless we have $n!/2^c \leq 1$.

We can solve the inequality $n!/2^c \leq 1$ for c as follows. First, we rewrite the inequality as

$$2^c \geq n!$$

and take base-2 logarithms of both sides to get

$$c \geq \log_2(n!).$$

Next, we substitute Stirling's formula (see Exercise 8.6) $n! \approx \sqrt{2\pi n}\, n^n e^{-n}$ for the $n!$ term and we have

$$c \geq \log_2\left[\sqrt{2\pi n}\, n^n e^{-n}\right],$$

which, by the rules of logarithms, gives

$$c \geq \log_2\left(\sqrt{2\pi}\right) + \frac{1}{2}\log_2 n + n\log_2 n - n\log_2 e.$$

The dominant term in this expression is $n \log_2 n$. Indeed, we can write this as

$$c \geq n\log_2 n + O(n).$$

[See Section 28 for an explanation of the $O(n)$ term.]

Since c is the number of comparisons we need to make in order to find the true order of the records, we see that we need $n \log_2 n$ comparisons to sort the data.

Linear Extensions of Infinite Posets

We proved that every finite partially ordered set has a linear extension. We now consider the same issue for infinite posets: Must they have linear extensions as well? The bizarre answer to this question is yes and no.

How is this possible? Surely the statement "Every poset has a linear extension" is either true or false—it can't be both!

Recall the Pythagorean Theorem (Theorem 3.1). In Exercise 3.7, we noted that right triangles on the surface of a sphere do not observe the Pythagorean Theorem. This does not undermine the truth of the Pythagorean Theorem because right triangles on the surface of the sphere are not the sort of right triangles to which the Pythagorean Theorem applies.

Thus the Pythagorean Theorem is true for some sorts of right triangles (the "real" right triangles in the plane) and not for others (the "fake" right triangles on the sphere). The Pythagorean Theorem is definitive once we are precise about the term *right triangle*.

The situation for linear extensions of infinite posets is similar. The truth of the statement "Every poset has a linear extension" depends on the precise meaning of the word *set*. In this book, we have been deliberately vague about what a set is. We rely on our readers' intuition that a set is a "collection of things." It is not necessary, however, to work with a vague notion of sets. A branch of mathematics, known as set theory, directly addresses the issue of what is a set. Set theory provides the foundation for all of mathematics.

Surprisingly, there is no single, unequivocal concept of set. In laying down the defining properties of sets, there are various conditions, called *axioms,* that we demand be satisfied by sets. For example, one axiom states that if X and Y are sets, then there is a set that contains all the elements in X and all the elements in Y. In essence, this axiom says that if X and Y are sets, so is $X \cup Y$.

A more exotic axiom is known as the Axiom of Choice. There are a number of different ways to state this axiom. One way is as follows: Given a collection of pairwise disjoint sets, there is another set X that contains exactly one element from each set in the collection.

If one accepts this axiom as part of the definition of set, then one can prove that every poset (finite or infinite) has a linear extension. However, if one denies the Axiom of Choice, then there are posets that do not have linear extensions.

Does this mean that the statement "Every poset has a linear extension" is both true and false? No. It is true or false depending on what we mean by set. The strange issue here is that there is more than one way to define set, and, depending on which definition you choose, different mathematical results follow.

The Axiom of Choice is (mostly) a nonissue in discrete mathematics. Results about finite collections of finite sets do not depend upon it. Thus all of the theorems in this book are true irrespective of which concept of set we use. It is only when we consider infinite sets, or infinite collections of sets, that these issues come into play.

Recap

We proved that every finite partially ordered set has a linear extension. Indeed, we showed that if P contains incomparable elements x and y, then P has a linear extension in which x is below y and another linear extension in which x is above y. We then used linear extensions to discuss the number of comparisons necessary to sort n data records. Finally, we considered the issue of whether or not infinite posets have linear extensions and discussed the fact that the answer to this question depends on our fundamental notion of precisely what a set is.

56 Exercises

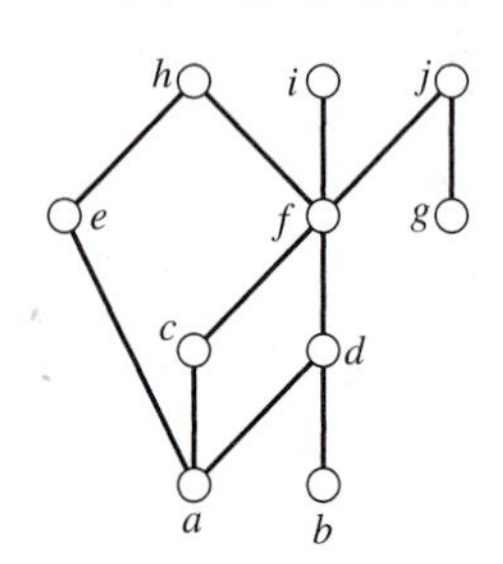

56.1. Let P be the poset in the figure. Which, if any, of the following are linear extensions of P?

a. $a < b < c < d < e < f < g < h < i < j$.

b. $b < a < e < g < d < c < f < j < i < h$.

c. $a < c < f < j$.

d. $a < b < c < e < f < h < i < j < h < g$.

56.2. Find the number of linear extensions of each of these three posets.

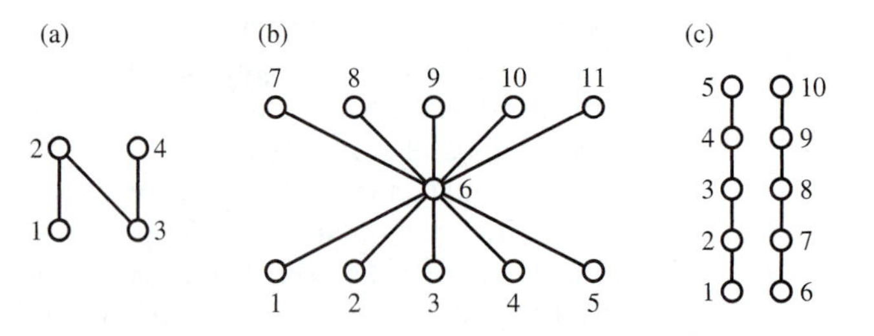

56.3. Let $P = (X, \leq)$ be a poset with incomparable elements x and y. Show that the relation $\leq'$ defined by

$$\leq' = \leq \cup \{(x, y)\}$$

must be reflexive and antisymmetric.

Give an example of a poset $P = (X, \leq)$ with incomparable elements x and y where $\leq'$ (as defined above) is not a partial order relation.

56.4. Let $P = (X, \leq)$ be a finite poset that is not a total order. Prove that P contains incomparable elements x and y such that

$$\leq' = \leq \cup \{(x, y)\}$$

is a partial order relation.

Such a pair of elements is called a *critical pair*.

56.5. Find all critical pairs in the poset from Exercise 56.1 that include the element g.

57 Dimension

Realizers

We return to the example at the beginning of the previous section. We examined the following partially ordered set and its linear extensions.

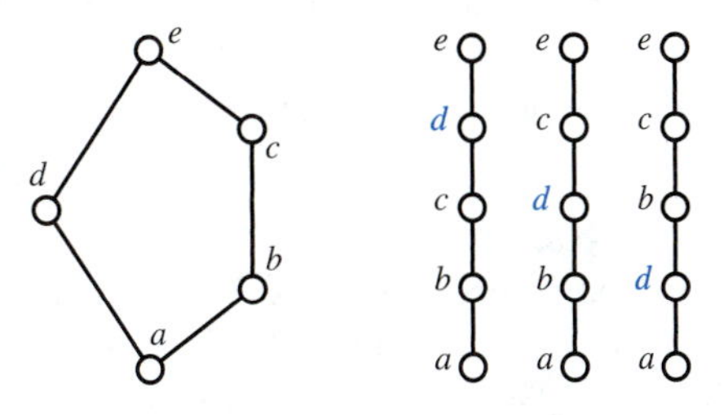

We make the following claim: The three linear extensions of the poset P contain enough information to reconstruct the poset. Consider elements b and c. Notice that $b < c$ in all three linear extensions. By Theorem 56.3, this can happen only if $b < c$ in P itself. On the other hand, consider elements b and d. In the first linear extension, we have $b < d$, but in the third, we have $b > d$. Were it the case that $b < d$ in P, then we would have $b < d$ in all linear extensions. So we can deduce that b and d are incomparable in P.

We formalize these remarks as follows:

Corollary 57.1 Let P be a finite partially ordered set, and let x and y be distinct elements of P. If $x < y$ in all linear extensions of P, then $x < y$ in P. Conversely, if $x < y$ in one linear extension, but $x > y$ in another, then x and y are incomparable in P.

The proof is left to you (Exercise 57.2).

This observation gives us a way to store a partially ordered set in a computer. We can save, as lists, the linear extensions of P. To see whether $x < y$ in P, we simply check that x is below y in all of the linear extensions.

However, some partially ordered sets have a large number of linear extensions. For example, consider an antichain on ten elements (see Example 56.2). It contains 10! (over 3 million) linear extensions. However, we do not need all 10! linear extensions to represent this antichain in our computer. Instead, we can use just the two linear orders:

$$1 < 2 < 3 < 4 < 5 < 6 < 7 < 8 < 9 < 10, \quad \text{and}$$
$$10 < 9 < 8 < 7 < 6 < 5 < 4 < 3 < 2 < 1.$$

Notice that for any two elements x and y of the antichain, we have $x < y$ in one of the orders and $x > y$ in the other.

The same idea works for the five-element poset we considered earlier. We do not need all three of its linear extensions to serve as a representation. Consider just the first and third linear extensions:

$$a < b < c < d < e \quad \text{and} \quad a < d < b < c < e.$$

Notice that if $x < y$ in the poset, then we have $x < y$ in both of these linear extensions, but if x and y are incomparable (e.g., $x = b$ and $y = d$), then we have $x < y$ in one extension and $x > y$ in another. So it is enough just to hold these two linear extensions in the computer.

Let us be more precise. A set of linear extensions that captures all the information in a poset is called a *realizer,* and this is the proper definition:

Definition 57.2 **(Realizer)** Let $P = (X, \leq)$ be a partially ordered set. Let $\mathcal{R}$ be a set of linear extensions of P. We call $\mathcal{R}$ a *realizer* of P, provided that for all $x, y \in X$ we have $x \leq y$ in P if and only if $x \leq y$ in all linear extensions in $\mathcal{R}$.

We say that $\mathcal{R}$ *realizes* P.

Another way to express Corollary 57.1 is as follows: *Let P be a finite partially ordered set and let $\mathcal{R}$ be the set of all linear extensions of P. Then $\mathcal{R}$ is a realizer of P.*

If $\mathcal{R} = \{L_1, L_2, \ldots, L_t\}$ is a realizer for a poset P, then we know that $x \leq y \iff x \leq_i y$ for all $i = 1, 2, \ldots, t$. Half of this statement (the $\Rightarrow$ implication) always holds by virtue of the fact that the L_i are linear extensions. If $x \leq y$ in P, then, because the L_i are linear extensions of P, we must have $x \leq_i y$ for all i.

Here the notation $x \leq_i y$ means $x \leq y$ in L_i.

The other implication (the $\Leftarrow$ half) is the important feature. This says that if $x \not\leq y$, then we do not have $x \leq_i y$ for all i. Of course, if $y < x$, this is obvious, for then we have $y <_i x$ for all i. The interesting case is when x and y are incomparable. Since $x \not\leq y$, there is an i with $x >_i y$. And since $y \not\leq x$, there is a j with $x <_j y$.

We have the following:

Proposition 57.3 Let P be a poset and let $\mathcal{R} = \{L_1, L_2, \ldots, L_t\}$ be a set of linear extensions of P. Then $\mathcal{R}$ is a realizer of P if and only if for all pairs of incomparable elements of P, x and y, there are indices i and j so that $x <_i y$ and $x >_j y$.

We gave virtually the entire proof in the preceding discussion, and we leave it to you to write this out carefully (Exercise 57.3).

Example 57.4 Let P be the poset whose Hasse diagram is shown here:

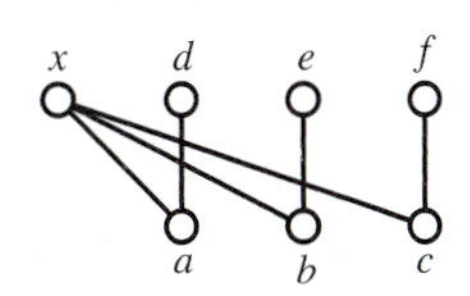

Let L_1, L_2, and L_3 be the following linear extensions of P:

$$L_1 : b < c < e < f < a < x < d,$$
$$L_2 : a < c < d < f < b < x < e, \quad \text{and}$$
$$L_3 : a < b < d < e < c < x < f.$$

Let $\mathcal{R} = \{L_1, L_2, L_3\}$. We claim that $\mathcal{R}$ is a realizer of P.

Checking that $\mathcal{R}$ is a realizer for the poset in Example 57.4 is tedious.

First, we need to make sure that all three L_i are linear extensions of P (i.e., if $u < v$ in P, then we must have $u < v$ in all three L_i). Observe that $a < x$ and $a < d$ in all three L_i. Then check that $b < x$ and $b < e$ in all three. Finally, note that $c < x$ and $c < f$ in all three.

Second, we check that if u and v are incomparable in P, then $u < v$ in one linear extension and $u > v$ in another. There are several cases, but we can check these systematically as well. Consider first the incomparabilities among a, b, and c. Note that we have $a < b$ in L_3 and $a > b$ in L_1. The incomparabilities between a and c and between b and c are checked in the same way.

We also see that $d < e$ in L_2 and $d > e$ in L_1. The other incomparabilities among $\{d, e, f\}$ are checked in the same way.

Next, $x < d$ in L_1 and $x > d$ in L_2. The other incomparabilities involving x are checked in the same manner.

Finally, notice that $a < e$ in L_2 and $a > e$ in L_1. The incomparabilities a-f, b-d, b-f, c-d, and c-e are checked in a similar manner.

Therefore $\mathcal{R}$ is a realizer.

Dimension

Let P be an antichain on ten elements. We can form a realizer of P using all 10! linear extensions, and we can also form a realizer of P using just two linear extensions. Clearly the latter is more efficient (especially if we wish to use linear extensions to store a poset in a computer).

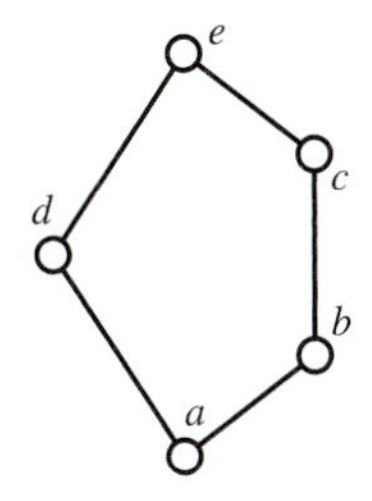

It is not difficult to realize a poset when we use all its linear extensions. The tricky (and interesting) problem is to realize a poset with as few linear extensions as possible. For example, the poset at the beginning of this section (see the figure) can be realized using all three of its linear extensions or with just two.

Can we realize this poset with just one linear extension? No. Because this poset has incomparable elements (call them x and y), we need at least two linear extensions: one in which $x < y$ and another in which $x > y$. This poset can be realized with two linear extensions, but no fewer.

The technical terminlogy that applies here is that the poset has *dimension* equal to 2.

Definition 57.5 **(Dimension)** Let P be a finite poset. The smallest size of a realizer of P is called the *dimension* of P. The dimension of P is denoted $\dim P$.

An antichain on ten elements and the poset in the figure both have dimension equal to 2.

Recall the poset P from Example 57.4. We showed that this poset has a realizer containing three linear extensions. Because P is not a linear order, it cannot be realized by a single linear extension. The question becomes: Can P be realized using just two linear extensions? We claim that it cannot.

Suppose, for the sake of contradiction, that P (the poset in Example 57.4) can be realized with just two linear extensions L' and L''. Consider the pairwise incomparable elements a, b, and c. By symmetry, and without loss of generality, we have $a < b < c$ in L' and $a > b > c$ in L''. Since x is above all of a, b, and c, we also know that x is above them in L' and L''. So far we have

$$a < b < c < x \quad \text{in } L' \text{ and}$$
$$c < b < a < x \quad \text{in } L''.$$

Now consider element e. We know that e and x are incomparable, so $e < x$ in one of L' or L'' and $e > x$ in the other. Since the situation is still symmetrical, we assume $e > x$ in L' (so in L' we have $a < b < c < x < e$). In L'' we know that $e < x$, but we also know that $e > b$ (because $e > b$ in P). So in L'' we have $c < b < e < x$. The point is that in both L' and L'' we have $c < e$, despite the fact that c and e are incomparable. Therefore $\{L', L''\}$ is not a realizer for P, and so there can be no realizer of size 2. In Example 57.4, we presented a realizer of size 3. Therefore $\dim P = 3$.

Here is another family of posets whose dimension we calculate:

Example 57.6 **(Standard example)** Let n be an integer with $n \geq 2$ and let P_n denote the following poset. The ground set of P_n consists of $2n$ elements: $\{a_1, a_2, \ldots, a_n, b_1, b_2, \ldots, b_n\}$. The only strict order relations in P_n are those of the form $a_i < b_j$ where $i \neq j$. The poset P_4 is shown in the figure.

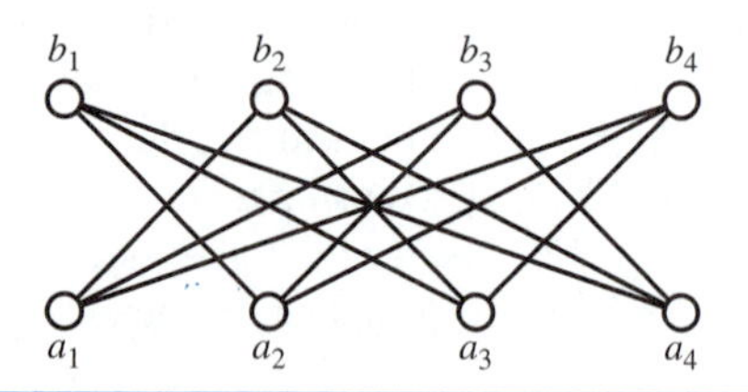

Proposition 57.7 Let n be an integer with $n \geq 2$ and let P_n be the poset defined in Example 57.6. The dimension of P_n is n.

The proof has two parts. First, we show that P_n has a realizer of size n. Second, we show that P_n cannot have a realizer with fewer than n linear extensions.

Proof. Let i be an integer with $1 \leq i \leq n$. Let L_i be a linear order on the ground set of P_n of the following form:

$$\text{(other } as) < b_i < a_i < \text{(other } bs).$$

The "other as" means we place all a_j (except a_i) before b_i in this linear order. Similarly, the "other bs" means we place all b_j (except b_i) after a_i. We claim that regardless of how we arrange the "other as" and "other bs," L_i is a linear extension of P_n. We just need to check that $a_j < b_k$ whenever $j \neq k$. Indeed, we have $a_j < b_k$ for all j and k except for $j = k = i$. Thus L_i is a linear extension (for each $i = 1, 2, \ldots, n$).

Let $\mathcal{R} = \{L_1, L_2, \ldots, L_n\}$. We claim that $\mathcal{R}$ is a realizer for P_n. There are three types of incomparable pairs in P_n: two as, two bs, and a_i-b_i for some i.

- Incomparable pairs of the form a_i-a_j: Notice that $a_i < a_j$ in L_j and $a_i > a_j$ in L_i.
- Incomparable pairs of the form b_i-b_j: Notice that $b_i < b_j$ in L_i and $b_i > b_j$ in L_j.
- Incomparable pairs of the form a_i-b_i: Notice that $a_i > b_i$ in L_i, but $a_i < b_i$ in any other L_k $(k \neq i)$.

Therefore $\mathcal{R}$ is a realizer of P_n.

We now show that P_n cannot have a realizer with fewer than n linear extensions. Suppose, for the sake of contradiction, there is a realizer $\mathcal{R}$ of P_n with $|\mathcal{R}| < n$. For each k (with $1 \leq k \leq n$), there must be a linear extension $L \in \mathcal{R}$ in which $a_k > b_k$ (because a_k and b_k are incomparable). There are n such incomparable pairs, but at most $n - 1$ linear extensions in $\mathcal{R}$. Therefore (by the Pigeonhole Principle—see Section 24), there must be a linear extension L and two distinct indices i and j such that $a_i > b_i$ and $a_j > b_j$ in L. Since $b_j > a_i$ and $b_i > b_j$ in P_n, we must also have these relations in L. Thus in L we have

$$b_j > a_i > b_i > a_j > b_j \Rightarrow b_j > b_j$$

which is impossible.$\Rightarrow\Leftarrow$ Therefore $\mathcal{R}$ is not a realizer of P_n, and so we cannot realize P_n with fewer than n linear extensions.

Therefore $\dim P_n = n$. ■

Embedding

Hasse diagrams are helpful geometric representations for partially ordered sets. In this section, we consider an alternative geometric representation.

Every point in the plane can be represented by a pair of real numbers: the (x, y)-coordinates of the point. This is why the plane is often referred to as $\mathbb{R}^2$. Likewise, every point in three-dimensional space can be described as an ordered triple: (x, y, z). We write $\mathbb{R}^3$ to stand for three-dimensional space. We do not need to stop at three dimensions. Four-dimensional space is simply the set of all points of the form (x, y, z, w) and we denote this set as $\mathbb{R}^4$. In general $\mathbb{R}^n$ stands for the set of all ordered n-tuples of real numbers and represents n-dimensional space.

The symbol $\mathbb{R}^n$ stands for n-dimensional space.

The goal of this section is to show the connection between the two uses (geometry and posets) of the word *dimension*.

Let $\mathbf{p}$ and $\mathbf{q}$ be two points in n-dimensional space $\mathbb{R}^n$. We say that $\mathbf{p}$ *dominates* $\mathbf{q}$ provided each coordinate of $\mathbf{p}$ is greater than or equal to the corresponding coordinate of $\mathbf{q}$. In other words, if the coordinates of $\mathbf{p}$ and $\mathbf{q}$ are

$$\mathbf{p} = (p_1, p_2, \ldots, p_n)$$
$$\mathbf{q} = (q_1, q_2, \ldots, q_n)$$

then $p_1 \geq q_1$, $p_2 \geq q_2$, $\ldots$, $p_n \geq q_n$. Let us write $\mathbf{p} \succeq \mathbf{q}$ in the case where $\mathbf{p}$ dominates $\mathbf{q}$. We also write $\mathbf{q} \preceq \mathbf{p}$, and we say that $\mathbf{q}$ is dominated by $\mathbf{p}$.

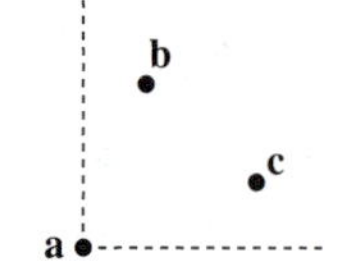

For example, suppose $\mathbf{p}$ and $\mathbf{q}$ are points in the plane. If $\mathbf{p} \succeq \mathbf{q}$, then both of $\mathbf{p}$'s coordinates are at least as large as those of $\mathbf{q}$. Thus $\mathbf{q}$ must lie to the "northeast" of $\mathbf{p}$. In the figure, $\mathbf{a}$ is dominated by both $\mathbf{b}$ and $\mathbf{c}$ (i.e., $\mathbf{a} \preceq \mathbf{b}$ and $\mathbf{a} \preceq \mathbf{c}$), but $\mathbf{b}$ and $\mathbf{c}$ are incomparable.

Definition 57.8 **(Embedding in $\mathbb{R}^n$)** Let $P = (X, \leq)$ be a poset and let n be a positive integer. An *embedding* of P in $\mathbb{R}^n$ is a one-to-one function $f : X \to \mathbb{R}^n$ such that $x \leq y$ (in P) if and only if $f(x) \preceq f(y)$ (in $\mathbb{R}^n$).

Example 57.9 The following figure shows a poset on the left and an embedding in $\mathbb{R}^2$ on the right.

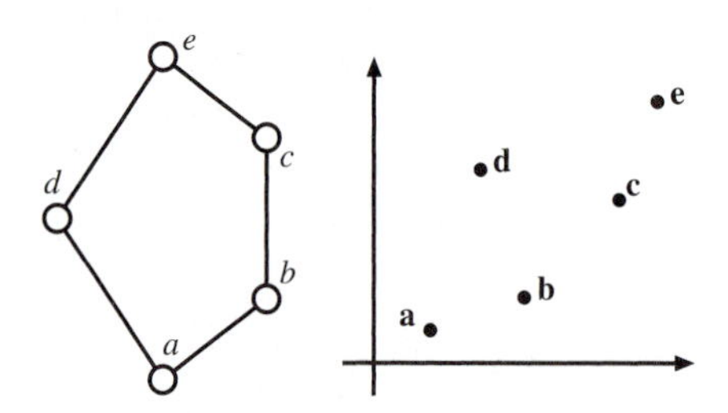

The embedding is $a \mapsto \mathbf{a}$, $b \mapsto \mathbf{b}$, $c \mapsto \mathbf{c}$, $d \mapsto \mathbf{d}$, and $e \mapsto \mathbf{e}$. Notice that the chain $a < b < c < e$ corresponds to the sequence of points $\mathbf{a}$, $\mathbf{b}$, $\mathbf{c}$, $\mathbf{e}$ where each point is to the northeast of the previous point. Also note that since b and d are incomparable, their points $\mathbf{b}$ and $\mathbf{d}$ are also incomparable in the dominance ($\preceq$) order.

Theorem 57.10 Let P be a finite poset and let n be a positive integer. Then P has a realizer of size n if and only if P embeds in $\mathbb{R}^n$. Thus $\dim P$ is the least positive integer n such that P embeds in $\mathbb{R}^n$.

Equivalently, we know that L_i is a finite linear order and thus that it is isomorphic to $\{1, 2, \ldots, |X|\}$ ordered by ordinary $\leq$ (see Theorem 55.4). The function h_i is simply the poset isomorphism from L_i to $\{1, 2, 3, \ldots, |X|\}$.

Proof. ($\Rightarrow$) Suppose that $P = (X, \leq)$ has a realizer of size n—say, $\mathcal{R} = \{L_1, L_2, \ldots, L_n\}$. For $x \in X$, let $h_i(x)$ denote the *height* of x in L_i; that is, $h_i(x)$ is the number of elements less than or equal to x in L_i. Thus $h_i(x) = 1$ if x is the least element of L_i, $h_i(x) = 2$ if it is next to bottom, and so on.

Let $f : P \to \mathbb{R}^n$ be defined by

$$f(x) = (h_1(x), h_2(x), \ldots, h_n(x)).$$

Clearly f is one-to-one: If $x \neq y$, then $h_1(x) \neq h_1(y)$ (because x and y are at different heights in L_1), and so $f(x) \neq f(y)$.

We must show that $x \leq y$ (in P) iff $f(x) \preceq f(y)$.

- Suppose $x \leq y$ in P. Then $h_i(x) \leq h_i(y)$ (because $x \leq y$ in all the linear extensions, L_i). Hence $f(x)$ is, coordinate by coordinate, less than or equal to $f(y)$, and so $f(x) \preceq f(y)$.
- Suppose $f(x) \preceq f(y)$. This means that $h_i(x) \leq h_i(y)$ for all i. Thus $x \leq y$ in all linear extensions L_i, and so (by definition of realizer) $x \leq y$ in P.

($\Leftarrow$) Suppose $P = (X, \leq)$ can be embedded in $\mathbb{R}^n$. This means there is a one-to-one mapping $f : X \to \mathbb{R}^n$ so that for all $x, y \in X$ we have $x \leq y \iff f(x) \preceq f(y)$.

Let i be an integer with $1 \leq i \leq n$. We define a linear extension L_i on P as follows: Let $f_i(x)$ be the ith coordinate of $f(x)$. We form L_i by arranging the elements of X in increasing order of f_i. That is, we have $x \leq_i y$ provided $f_i(x) \leq f_i(y)$. This would give a total order on the elements X were it not for the annoying problem of elements with equal ith coordinate. We break such ties as follows: Suppose $f_i(x) = f_i(y)$ for some $x \neq y$. Since f is a one-to-one function, there must be some other coordinate j where $f_j(x) \neq f_j(y)$. In this case, we declare the order of x and y in L_i to be determined by lowest index j where $f_j(x) \neq f_j(y)$ (see Example 57.11).

We claim that L_i is a linear extension of P. Clearly L_i is a linear order. Suppose $x < y$ in P. Then $f(x) \prec f(y)$ and so $f_i(x) \leq f_i(y)$. In case $f_i(x) = f_i(y)$ and $x < y$, we note that for all j, $f_j(x) \leq f_j(y)$ and for some indices j, the inequality is strict. Thus $x < y$ in P implies $x <_i y$, and so L_i is a linear extension of P.

Now we claim that $\mathcal{R} = \{L_1, \ldots, L_n\}$ is a realizer. We must show that if x and y are incomparable, then there are indices i and j with $x <_i y$ and $x >_j y$. Since $f(x)$ is incomparable to $f(y)$ (by definition of embedding in $\mathbb{R}^n$), we know that there are indices i and j with $f_i(x) < f_i(y)$ and $f_j(x) > f_j(y)$, and this gives $x <_i y$ and $x >_j y$. ■

Example 57.11 Let P be the poset in the figure (left) and let $a \mapsto \mathbf{a}$, $b \mapsto \mathbf{b}$, ..., $f \mapsto \mathbf{f}$ (on the right) be an embedding of P in $\mathbb{R}^2$.

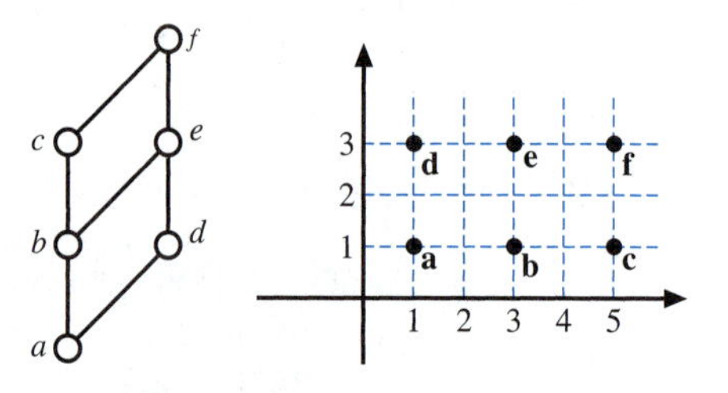

For example, d is embedded at $\mathbf{d} = (1, 3)$.

The two linear extensions we extract from this embedding are

$$L_1 : a < d < b < e < c < f$$
$$L_2 : a < b < c < d < e < f.$$

We found L_1 by sorting the six points by their first coordinate (and breaking ties using the second coordinate). Likewise we found L_2 by sorting the points by their second coordinate (and breaking ties using the first coordinate).

Observe that $\mathcal{R} = \{L_1, L_2\}$ is a realizer for P.

Recap

We introduced the notion of a realizer of a partially ordered set. We defined the dimension of a poset to be the size of a smallest realizer. We showed that the concept of a poset dimension is closely linked to the geometric concept of dimension by studying embeddings of posets in $\mathbb{R}^n$.

57 Exercises

57.1. Let P be the poset in the figure.

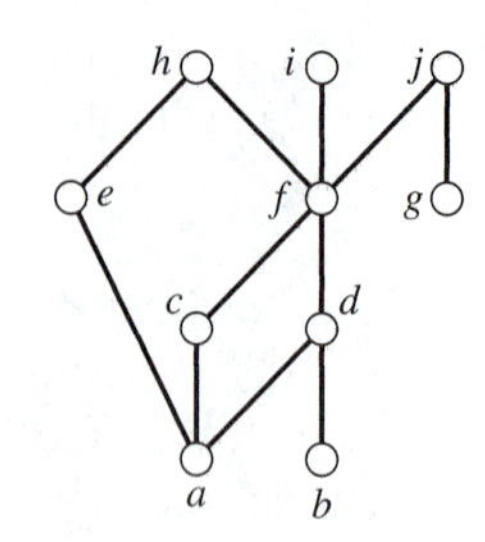

a. Find $d = \dim P$.

b. Find a realizer of P containing d linear extensions.

c. Give an embedding of P in $\mathbb{R}^d$ (either via a picture or by specifying coordinates).

57.2. Prove Corollary 57.1.

57.3. Prove Proposition 57.3.

57.4. Let P be a subposet of Q. This means that the elements of P are a subset of the elements of Q, and the relation between elements of P are exactly the same as their relation in Q (i.e., $x \le y$ in P iff $x \le y$ in Q).

Prove that $\dim P \le \dim Q$.

57.5. Let n be an integer with $n \ge 3$. A *fence* is a poset on $2n$ elements $a_0, a_1, \ldots, a_{n-1}, b_0, b_1, \ldots, b_{n-1}$ where the only strict relations are of the form $a_i < b_i$ and $a_i < b_{i+1}$ (where subscript addition is modulo n, so $a_{n-1} < b_0$). The figure shows a fence with $n = 5$.

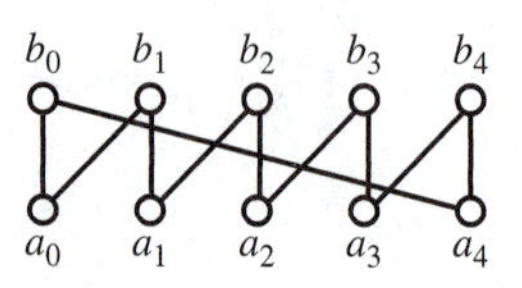

Prove that fences have dimension equal to 3.

58 Lattices

We have seen that subset ($\subseteq$), ordinary less than or equal to ($\le$), and divides ($|$) share three essential features: They are reflexive, antisymmetric, and transitive, and hence they are partial order relations.

In this section, we show that the operations $\cap$ (intersection), $\wedge$ (Boolean and), and gcd (greatest common divisor) are similarly related.

Meet and Join

The usual way to define the intersection of two sets, A and B, is to say that $A \cap B$ is the set of all elements that are in both A and B. Now consider the following challenge: Can we describe the intersection of two sets, $A \cap B$, without using the word *element?*

Notice first that $A \cap B$ is a subset of both A and B. Of course, there may be many sets X with $X \subseteq A$ and $X \subseteq B$, so this does not uniquely specify $A \cap B$. However, among subsets of both A and B, we know that $A \cap B$ is the "biggest." By this we mean that if $X \subseteq A$ and $X \subseteq B$, then we must have $X \subseteq A \cap B$.

Let's show this rigorously.

Proposition 58.1 Let A and B be sets. Let Z be a set with the following properties:

- $Z \subseteq A$ and $Z \subseteq B$ and
- if $X \subseteq A$ and $X \subseteq B$, then $X \subseteq Z$.

Then $Z = A \cap B$.

Proof. First, suppose $x \in Z$. Since $Z \subseteq A$, we have $x \in A$. Likewise $Z \subseteq B$ implies $x \in B$. Therefore $x \in A \cap B$.

Second, suppose $x \in A \cap B$. This means that $x \in A$ and $x \in B$, and so $X = \{x\}$ is a subset of both A and B. Therefore $X = \{x\}$ is a subset of Z (by the second property). Thus $x \in Z$.

We have shown that $x \in Z \iff x \in A \cap B$, and so $Z = A \cap B$. ■

A similar result holds for the greatest common divisor of two positive integers.

Proposition 58.2 Let a, b be positive integers. Let d be a positive integer with the following properties:

- $d|a$ and $d|b$, and
- if $e \in \mathbb{N}$ with $e|a$ and $e|b$, then $e|d$.

Then $d = \gcd(a, b)$.

The proof is left for you (Exercise 58.4).

These Propositions suggest an alternative way to define intersection and greatest common divisor. We can define $A \cap B$ to be the largest set that is below both

A and B where *largest* and *below* are in terms of the subset ($\subseteq$) partial order. Similarly, we can define $\gcd(a, b)$ to be the largest positive integer that is below both a and b where *largest* and *below* are in terms of the divisibility order ($|$).

We extend these ideas to other posets.

Definition 58.3 **(Lower and upper bounds)** Let $P = (X, \leq)$ be a poset and let $a, b \in X$.

We say that $x \in X$ is a *lower bound* for a and b provided $x \leq a$ and $x \leq b$.

Similarly, we say that $x \in X$ is an *upper bound* for a and b provided $a \leq x$ and $b \leq x$.

The lower bound concept is an extension of the common divisor concept: Let $a, b \in \mathbb{N}$. In the poset $(\mathbb{N}, |)$, the lower bounds of a and b are precisely the common divisors of a and b.

Next, we define the notions of *greatest lower bound* and *least upper bound*.

Definition 58.4 **(Greatest lower bound/least upper bound)** Let $P = (X, \leq)$ be a poset and let $a, b \in X$.

Some authors abbreviate *greatest lower bound* as glb and *least upper bound* as lub.

We say that $x \in X$ is a *greatest lower bound* for a and b provided (1) x is a lower bound for a and b and (2) if y is lower bound for a and b, then $y \leq x$.

Similarly, we say that $x \in X$ is a *least upper bound* for a and b provided (1) x is an upper bound for a and b and (2) if y is an upper bound for a and b, then $y \geq x$.

Example 58.5 Let P be the following poset.

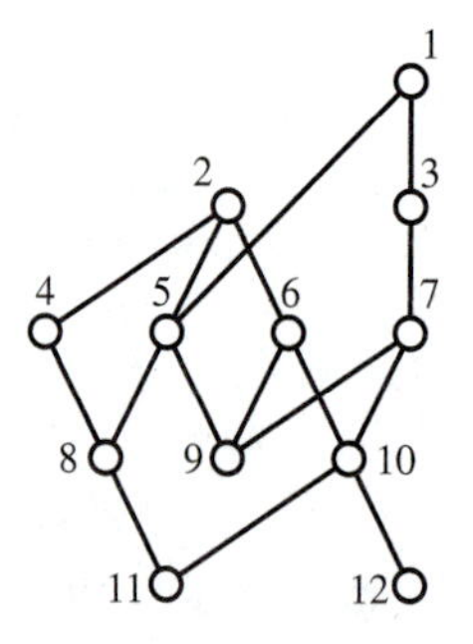

- Consider elements 8 and 9. Notice that 1, 2, and 5 are upper bounds for 8 and 9. Since $5 < 1$ and $5 < 2$, we have that 5 is the least upper bound of 8 and 9. On the other hand, 8 and 9 have no lower bounds and consequently no greatest lower bound.
- Elements 4 and 7 have 11 as their only lower bound; thus 11 is their greatest lower bound. Elements 4 and 7 have no upper bound and hence no least upper bound.
- Elements 5 and 6 have 2 as the least (and only) upper bound. They have incomparable lower bounds 9 and 11, so they do not have a greatest lower bound.

- Elements 9 and 10 have no greatest lower bound and no least upper bound.
- Elements 4 and 5 have 2 as their least upper bound and 8 as their greatest lower bound.

Greatest lower bounds and least upper bounds, if they exist, are unique.

If a pair of elements of a poset have a greatest lower bound, it must be unique. Suppose x and y are both greatest lower bounds of a and b. We have $x \leq y$ because y is greatest and we have $y \leq x$ because x is greatest. Therefore $x = y$. Likewise, if a and b have a least upper bound, it must be unique.

There are alternative terms for least upper bound and greatest lower bound and a special notation for them as well.

Definition 58.6 **(Meet and join)** Let $P = (X, \leq)$ be a poset and let $a, b \in X$.

If a and b have a greatest lower bound, it is called the *meet* of a and b, and it is denoted $a \wedge b$.

If a and b have a least upper bound, it is called the *join* of a and b, and it is denoted $a \vee b$.

We use the symbols $\wedge$ and $\vee$ for the meet and join operations because $\wedge$ is an abstraction of $\cap$ and $\vee$ is an abstraction of $\cup$. Unfortunately, we have used the symbols $\wedge$ and $\vee$ in two different ways. In Section 6 these symbols stand for the Boolean operations *and* and *or*. Here they stand for the poset operations *meet* and *join*. Fortunately, we can reach a peaceful resolution to this crisis. Consider the poset P whose ground set is {TRUE, FALSE}. We make the mathematical (as well as ethical) decision to place truth above falsehood; that is, we have FALSE < TRUE in this poset—see the figure.

T

F

Notice that in this poset we have $\mathsf{T} \wedge \mathsf{F} = \mathsf{F}$ because FALSE is the greatest (and only) lower bound for TRUE and FALSE. Indeed, all of the following are true:

$$\begin{array}{llll} \mathsf{T} \wedge \mathsf{T} = \mathsf{T} & \mathsf{T} \wedge \mathsf{F} = \mathsf{F} & \mathsf{F} \wedge \mathsf{T} = \mathsf{F} & \mathsf{F} \wedge \mathsf{F} = \mathsf{F} \\ \mathsf{T} \vee \mathsf{T} = \mathsf{T} & \mathsf{T} \vee \mathsf{F} = \mathsf{T} & \mathsf{F} \vee \mathsf{T} = \mathsf{T} & \mathsf{F} \vee \mathsf{F} = \mathsf{F}. \end{array}$$

Therefore the operations $\wedge$ and $\vee$ on $\{\mathsf{T}, \mathsf{F}\}$ are exactly the same whether we interpret them as *and* and *or* or as *meet* and *join*.

Example 58.7 The results from Example 58.5 can be expressed as follows:

- $8 \wedge 9$ is undefined and $8 \vee 9 = 5$.
- $4 \wedge 7 = 11$ and $4 \vee 7$ is undefined.
- $5 \wedge 6$ is undefined and $5 \vee 6 = 2$.
- Both $9 \wedge 10$ and $9 \vee 10$ are undefined.
- $4 \wedge 5 = 8$ and $4 \vee 5 = 2$.

Lattices

Note that for some pairs of elements, meet or join might be undefined. However, in some posets, meet and join are defined for all pairs of elements. There is a special name for such posets.

Definition 58.8 **(Lattice)** Let P be a poset. We call P a *lattice* provided that, for all elements x and y of P, $x \wedge y$ and $x \vee y$ are defined.

Let us look at some examples of lattices.

Example 58.9 Let P be the poset in the figure. The $\wedge$ and $\vee$ operation tables are given as well.

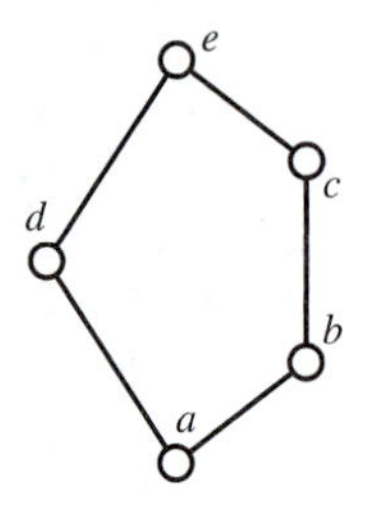

$\wedge$	a	b	c	d	e
a	a	a	a	a	a
b	a	b	b	a	b
c	a	b	c	a	c
d	a	a	a	d	d
e	a	b	c	d	e

$\vee$	a	b	c	d	e
a	a	b	c	d	e
b	b	b	c	e	e
c	c	c	c	e	e
d	d	e	e	d	e
e	e	e	e	e	e

Since $\wedge$ and $\vee$ are defined for every pair of elements, this poset is a lattice.

Example 58.10 **(Subsets of a set)** Let A be a set and let $P = (2^A, \subseteq)$; that is, P is the poset of all subsets of A ordered by containment. In this poset we have, for all $x, y \in 2^A$,

$$x \wedge y = x \cap y \qquad \text{and} \qquad x \vee y = x \cup y.$$

Therefore P is a lattice.

Example 58.11 **(Natural numbers/positive integers ordered by divisibility)** Consider the poset $(\mathbb{N}, |)$ (i.e., the set of natural numbers ordered by divisibility). Let $x, y \in \mathbb{N}$. Then $x \wedge y$ is the greatest common divisor of x and y, and $x \vee y$ is their least common multiple. However, $(\mathbb{N}, |)$ is not a lattice because $0 \wedge 0 = \gcd(0, 0)$ is not defined.

However, the poset $(\mathbb{Z}^+, |)$ is a lattice. Here $\mathbb{Z}^+$ stands for the set of positive integers which we order by divisibility. In this case, $\wedge$ and $\vee$ (gcd and lcm) are defined for all pairs of positive integers, and so $(\mathbb{Z}^+, |)$ is a lattice.

Example 58.12 **(Linear orders)** Let $P = (X, \leq)$ be a linear (total) order. Note that for any $x, y \in X$,

$$x \wedge y = \begin{cases} x & \text{if } x \leq y \\ y & \text{if } x \geq y. \end{cases}$$

We can rewrite this as $x \wedge y = \min\{x, y\}$ where $\min\{x, y\}$ stands for the smaller of x and y.

Likewise $x \vee y = \max\{x, y\}$ (i.e., the larger of the pair). Thus all linear orders are lattices.

Look at the $\wedge$ and $\vee$ tables in Example 58.9, and look at the diagonal entries running from the upper left to the lower right.

What algebraic properties do $\wedge$ and $\vee$ exhibit? For example, it is easy to see that $x \wedge x = x$. Let us prove this. First, x is a lower bound of both x and x because $x \leq x$ and $x \leq x$. Second, if y is any other lower bound of x and x, we have $y \leq x$ (because y is a lower bound!). Therefore x is the greatest lower bound of x and x. Likewise, $x \vee x = x$.

Also, $\wedge$ and $\vee$ are commutative operations: $x \wedge y = y \wedge x$ and $x \vee y = y \vee x$. The following result covers the significant algebraic properties exhibited by meet and join.

Theorem 58.13 Let $P = (X, \leq)$ be a lattice. For all $x, y, z \in X$, the following hold:

- $x \wedge x = x \vee x = x$.
- $x \wedge y = y \wedge x$ and $x \vee y = y \vee x$. (Commutative)
- $(x \wedge y) \wedge z = x \wedge (y \wedge z)$ and $(x \vee y) \vee z = x \vee (y \vee z)$. (Associative)
- $x \wedge y = x \iff x \vee y = y \iff x \leq y$.

Proof. The first property was shown earlier, and the second and fourth are easy to prove (we leave them for you).

Here we prove that $\wedge$ is associative. The proof that $\vee$ is associative is similar.

Let $a = (x \wedge y) \wedge z$ and let $b = x \wedge (y \wedge z)$. We must show that $a = b$. To this end, we first prove $a \leq b$.

Since $a = (x \wedge y) \wedge z$, we know that a is a lower bound for $x \wedge y$ and for z. Thus $a \leq x \wedge y$ and $a \leq z$. Since $a \leq x \wedge y$ and since $x \wedge y \leq x$ and $x \wedge y \leq y$, we have that $a \leq x$ and $a \leq y$. Thus a is below x, y, and z.

Symbolically, the argument of the preceding paragraph can be written as follows:

$$
\begin{array}{ccccc}
a = (x \wedge y) \wedge z & \Longrightarrow & a \leq x \wedge y & \text{and} & a \leq z \\
 & & \Downarrow & & \\
 & & a \leq x \text{ and } a \leq y. & &
\end{array}
$$

Since $a \leq y$ and $a \leq z$, we see that a is a lower bound for y and z. Therefore $a \leq y \wedge z$ since $y \wedge z$ is the greatest lower bound of y and z.

Since $a \leq x$ and $a \leq y \wedge z$, we see that a is a lower bound for x and $y \wedge z$. But b is the greatest lower bound for x and $y \wedge z$, so $a \leq b$.

By an identical argument, we have $b \leq a$, and so $a = b$—that is, $(x \wedge y) \wedge z = x \wedge (y \wedge z)$. ∎

Recap

We introduced the concepts of lower bound, greatest lower bound, upper bound, and least upper bound. The greatest lower bound of two elements is called their meet

($\wedge$), and the least upper bound is called their join ($\vee$). Meet and join are abstract versions of intersection and union (and of gcd and lcm). Finally, we presented the notion of a lattice and discussed some of the algebraic properties of meet and join.

58 Exercises

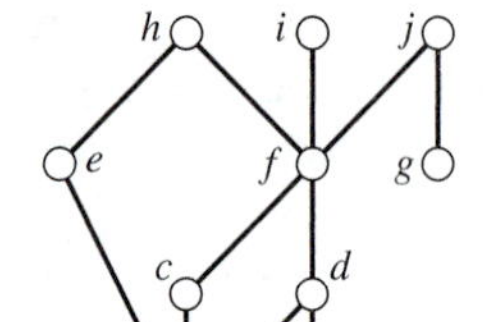

58.1. Let P be the poset in the figure. Please calculate:

a. $a \wedge b$.
b. $a \vee b$.
c. $c \wedge i$.
d. $c \vee i$.
e. $e \wedge d$.
f. $e \vee d$.
g. $(c \wedge d) \vee g$.
h. $c \wedge (d \vee g)$.

Is this poset a lattice?

58.2. Consider the poset $(\mathbb{Z}, \leq)$ (ordinary less than or equal to). For $x, y \in \mathbb{Z}$, explain in simple terms what $x \wedge y$ and $x \vee y$ are.

58.3. Let $P = (X, \leq)$ be a lattice. Prove that P is a linear order if and only if $\{x \wedge y, x \vee y\} = \{x, y\}$ for all $x, y \in X$.

58.4. Prove Proposition 58.2.

58.5. The following statement is false: Every lattice has a maximum element and a minimum element. Show, by presenting a counterexample, that this statement is false. However, by inserting one word into the statement, we can make it a true statement. Show how to repair the statement, and prove the true version.

58.6. Let $P = (X, \leq)$ be a lattice and let m be an element of the lattice. Prove that m is maximum in P if and only if $\forall x \in X, x \vee m = m$ if and only if $\forall x \in X, x \wedge m = x$.

What is the analogous statement for a minimum element?

58.7. In Theorem 11.3, we showed that $\cup$ and $\cap$ satisfy the distributive properties:

$$A \cup (B \cap C) = (A \cup B) \cap (A \cup C), \text{ and}$$
$$A \cap (B \cup C) = (A \cap B) \cup (A \cap C).$$

These equations can be rewritten with $\wedge$ in place of $\cap$ and $\vee$ in place of $\cup$:

$$a \vee (b \wedge c) = (a \vee b) \wedge (a \vee c), \text{ and}$$
$$a \wedge (b \vee c) = (a \wedge b) \vee (a \wedge c).$$

Give an example of a lattice for which the distributive laws are false.

58.8. Consider the following infinite poset P. The elements of P are various subsets of the plane. These subsets are (a) the entire plane itself, (b) all lines in the plane, (c) all single points in the plane, and (d) the empty set. The partial order is containment. This poset is a lattice. Explain, in geometric terms, the effect of the meet and join operations in this lattice.

58.9. Let P be a lattice with minimum element b and maximum element t.

a. What is the identity element for $\wedge$?
b. What is the identity element for $\vee$?
c. Show, by means of an example, that elements of P need not have inverses for either $\wedge$ or $\vee$.

Chapter 10 Self Test

1. Let $P = (\{1, 2, 3, \ldots, 20\}, |)$; that is, P is the poset whose elements are the integers from 1 to 20 ordered by divisibility.
   a. Draw a Hasse diagram of P.
   b. Find a largest chain in P.
   c. Find a largest antichain in P.
   d. Find the set of all maximal elements of P.
   e. Find the set of all minimal elements of P.
   f. Find the set of all maximum elements of P.
   g. Find the set of all minimum elements of P.
2. Let C be a chain and A be an antichain of a poset $P = (X, \leq)$. Prove that $|C \cap A| \leq 1$.
3. Let $P = (X, \leq)$ be a poset. Suppose there are chains C_1 and C_2 in P such that $X = C_1 \cup C_2$. Prove that the width of P is at most 2.
4. Let $P = (X, \leq)$ be a poset. Prove that P is an antichain if and only if every element of X is both maximal and minimal.
5. Let $P = (X, \leq)$ be a finite poset. We say that P is a *weak order* if we can partition X into disjoint antichains

$$X = A_1 \cup A_2 \cup \cdots \cup A_h$$

such that for all $x \in A_i$ and $y \in A_j$, if $i < j$ then $x < y$. One may think of the A_is as "levels" in the weak order; two elements on the same level must be incomparable, but an element on a lower-numbered level must be less than an element on a higher-numbered level.
   a. Show that (finite) chains and antichains are weak orders.
   b. Prove that a poset is a weak order if and only if it does not contain the subposet shown in the figure.

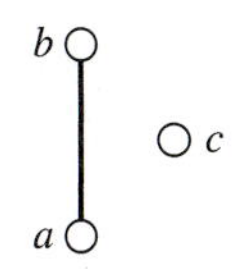

   c. Suppose $P = (X, \leq)$ is a weak order in which $X = A_1 \cup \cdots \cup A_h$ where all of the antichains A_i have k elements. Thus X has hk elements in total. How many linear extensions does P have?
   d. Prove that if P is a weak order, then the dimension of P is at most 2.
6. Let $P = (X, \leq)$ be a poset. We say that P is an *interval order* provided we can assign to each element $x \in X$ a real interval $[a_x, b_x]$ such that $x < y$ in P if and only if the interval $[a_x, b_x]$ is entirely to the left of $[a_y, b_y]$ (i.e., $b_x < a_y$). Note that this implies that if x and y are incomparable, then $[a_x, b_x]$ and $[a_y, b_y]$ must overlap (have nonempty intersection).
   a. Show that finite chains and antichains are interval orders.
   b. Prove that weak orders are interval orders (see Problem 5 for the definition of a weak order).

For real numbers $a < b$, the interval $[a, b]$ is the set of all real numbers between a and b inclusive. That is, $[a, b] = \{x \in \mathbb{R} : a \leq x \leq b\}$.

c. Show that the poset in the following figure is not an interval order.

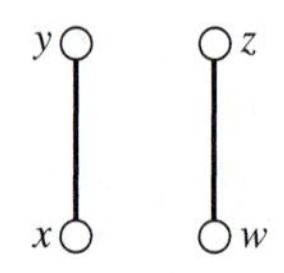

d. Show that the poset in the following figure is an interval order, but the lengths of the intervals used to represent this poset cannot all be the same.

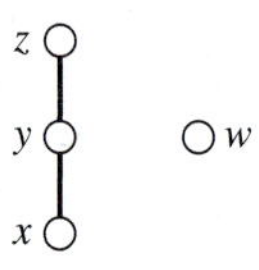

7. Let P be the poset whose Hasse diagram is shown in the figure.

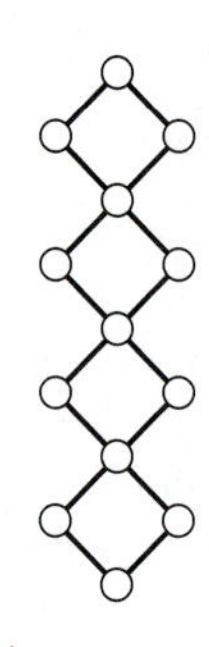

How many linear extensions does P have?

8. Let P be the poset whose Hasse diagram is shown in the figure.

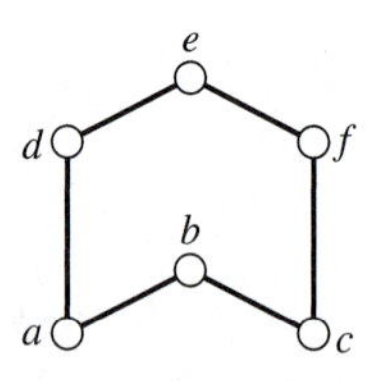

Please do the following:

a. List all pairs of elements that are incomparable in P.

b. Find three linear extensions of P that form a realizer of P.
Verify that your solution is correct by finding, for each incomparable pair $\{x, y\}$, one extension in which $x < y$ and another in which $y < x$.

c. Prove that there can be no linear extension of P in which $f < a$ and $d < c$.
Prove that there can be no linear extension of P in which $b < f$ and $f < a$.
Prove that there can be no linear extension of P in which $b < d$ and $d < c$.

Prove that there can be no linear extension of P in which $e < b$ and $b < d$. Prove that there can be no linear extension of P in which $e < b$ and $b < f$.

d. In a realizer of P, there must be linear extensions in which $f < a$, $d < c$, $b < d$, $e < b$, and $b < f$. Show that no more than two of these can hold in a single linear extension.

e. Show that $\dim P = 3$; that is, show that P does not have a realizer of size 2.

9. Let $P = (X, \leq)$ be a lattice, and suppose that for all $x, y \in X$, we have $x \wedge y = x \vee y$. Prove that P contains at most one element.

10. Recall from Definition 53.5 and Example 53.6 that the set of all partitions of a given set, together with the refines relation, forms a poset. Please answer the following:

a. Let $\mathcal{P} = \{\{1, 2, 3, 4\}, \{5, 6, 7, 8, 9\}\}$ and $\mathcal{Q} = \{\{1, 3, 5, 7, 9\}, \{2, 4, 6, 8\}\}$. Calculate $\mathcal{P} \wedge \mathcal{Q}$ and $\mathcal{P} \vee \mathcal{Q}$.

b. Let $\mathcal{P}$, $\mathcal{Q}$, and $\mathcal{R}$ be partitions of an n-element set for which

$$\mathcal{P} = \{X_1, X_2, \ldots, X_p\},$$
$$\mathcal{Q} = \{Y_1, Y_2, \ldots, Y_q\}, \quad \text{and}$$
$$\mathcal{R} = \mathcal{P} \wedge \mathcal{Q} = \{Z_1, Z_2, \ldots, Z_r\}.$$

Show that every Z_k in $\mathcal{R}$ is of the form $X_i \cap Y_j$.

11. Let $P = (X, \leq)$ be a lattice. Let $a, x_1, x_2, \ldots, x_n \in X$ and suppose that $a \leq x_i$ for all $1 \leq i \leq n$. Prove that $a \leq x_1 \wedge x_2 \wedge \cdots \wedge x_n$.

12. Let $P = (X, \leq)$ be a finite poset. Let $a, b \in X$ and define $U(a, b) = \{x \in X : a \leq x \text{ and } a \leq y\}$; that is, $U(a, b)$ is the set of all elements above both a and b.

Prove the following: If $a \vee b$ is defined and $U(a, b)$ is nonempty with $U(a, b) = \{u_1, u_2, \ldots, u_n\}$, then $a \vee b = u_1 \wedge u_2 \wedge \cdots \wedge u_n$.

Appendices

A Lots of Hints and Comments; Some Answers

1.1 Sorry, there is no hint for this problem; that would utterly defeat the purpose! Trust yourself and keep thinking about this. You will succeed, and when you have the answer, you will be absolutely sure you are right—you won't need the back of the book to tell you!

2.1 To determine whether $a|b$ is true, see if you can find an *integer* x such that $ax = b$.

2.2 In the previous problem there are integers a and b where $b|a$ is true but $\frac{a}{b}$ is *not* an integer.

2.3 Read Definition 2.6 carefully. Check each number to see if it fulfills all the requirements set forth in the definition.

2.4 Your definition for $\leq$ (less than or equal to) should look like this:

Let x and y be integers. We say that x is *less than or equal to* y (written $x \leq y$) provided ....

where the ... represents a condition involving x, y, and the natural numbers.

Once you define $\leq$, you may use this concept to define $<$, $>$, and $\geq$.

2.5 You need to do two things:

(1) Explain why integers are rational numbers. You must explain why if x is an integer, then you can find integers a and b such that $x = \frac{a}{b}$. The integers a and b depend on x, and you can find *simple* values for these. Beware not to choose $b = 0$.

(2) Explain why some rational numbers are not integers. All that is required is that you find a rational number that is not an integer.

2.6 The number 169 is square. How would you convince someone this is true? You would need to tell them about the number 13.

Here is the full answer:

An integer x is called a *perfect square* provided there is an integer y such that $x = y^2$.

2.7 Use the notation $d(A, B)$ to denote the distance between the points A and B. Determine a relation between $d(A, B)$, $d(B, C)$, and $d(A, C)$ that determines whether C is between A and B.

2.8 For small numbers, the easiest thing to do is simply write out all the possibilities. For larger numbers, try to develop a better method. Try factoring the numbers into primes. Factoring and prime numbers are discussed at length in Section 38.

2.9 The answer to (a) is best found by starting with 2 and checking each number. You should find the answer fairly quickly.

The hardest part for (b) is to write a subroutine to check whether, given integers a and b, $a|b$ is true. One way to do this is to calculate a/b and then round down to the nearest integer giving the integer c. Then check whether $ac = b$. Beware that this idea works well when a and b are positive, and that is sufficient for the problem at hand. However, if you plan to use this subroutine in other projects, it is worth your while to write a subroutine that will work correctly for any pair of integers a and b.

3.1 An answer to (a) is: If x is an odd integer and y is an even integer, then xy is an even integer.

Note that statement (d) is false.

3.2 There are many possible correct answers. For example, the statement "If (A) an animal is a cat, then (B) it is a mammal" is true, but "If (B) an animal is a mammal, then (A) it is a cat" is false.

Now create your own example.

3.3 The statement "If A, then B" is true unless A is true and B is false.

An "or" statement is true unless both conditions are false. This tells you when "(not A) or B" is true and when it is false. Compare to "If A, then B."

3.6 To what kind of triangles does the Pythagorean Theorem apply?

3.7 To what kind of triangles does the Pythagorean Theorem apply?

3.8 A distance is a number and lines are infinite. In your rewrite, use the term *line segment.*

3.9 Check out guinea pig anatomy.

Michael Newman/Photo Edit, Inc.

3.10 *Lemmas* is one, and there is another.

4.1 Here is a full answer:

Converted to if-then form, the problem asks you to prove: If x and y are odd integers, then $x + y$ is even.

Proof. Let x and y be odd integers. By the definition of *odd*, there is an integer a such that $x = 2a+1$. Likewise, there is an integer b such that $y = 2b + 1$. Therefore

$$\begin{aligned} x + y &= (2a + 1) + (2b + 1) \\ &= 2a + 2b + 2 \\ &= 2(a + b + 1). \end{aligned}$$

Since a and b are integers, so is $a + b + 1$. Therefore, by the definition of *divisible,* $x + y$ is divisible by 2. Therefore, by the definition of *even,* $x + y$ is even.

4.2 The first line of the proof is: Let x be an odd integer and let y be an even integer.

The last line of the proof is: Therefore $x + y$ is odd.

4.3 One line, in the middle of this proof, is

$$xy = (2a)(2b) = 4ab = 2(2ab).$$

4.5 $(2a + 1)(2b + 1) = 2(2ab + a + b) + 1.$

4.6 It may be more work, but if you do Exercise 4.8 first, you can derive this as a corollary.

4.7 See the previous hint.

4.8 If you did the previous two problems without first doing this problem, you can use them to do this.

4.10 ($\Rightarrow$) Suppose x is odd.... Therefore $x + 1$ is even.

($\Leftarrow$) Suppose $x + 1$ is even.... Therefore x is odd.

Algebraic trick: $2b - 1 = 2(b - 1) + 1$.

4.12 The smallest positive integer is 1, and $a < b$ implies $b - a > 0$. Be sure you prove both halves of this if-and-only-if statement.

4.13 $2a + 1 = (a) + (a + 1)$.

4.14 Construct a statement of the form "If A or B, then C" that is false, but "If A, then C" is true.

4.15 Are we sure now that whenever A is true, so is B, and whenever B is true, so is A?

5.1 Negative integers.

5.3 Don't choose a to be a prime number.

5.5 There is no very small value of n for which $n^2 + n + 41$ is not prime. You have to take n to be modestly large. If you choose the correct n, you won't need to do any calculations to see that $n^2 + n + 41$ is composite.

6.1 (b) TRUE. (c) FALSE.

6.3 Make a truth table for $(x \wedge y) \vee (x \wedge \neg y)$ and check that the column for x exactly matches the column for $(x \wedge y) \vee (x \wedge \neg y)$.

6.4 Make a truth table for both and make sure they are the same.

6.9 More than 1000. Try some smaller examples first.

6.10 Answer to (b): Take x = FALSE and y = TRUE. Then $x \to y$ evaluates to TRUE, whereas $x \leftrightarrow y$ evaluates to FALSE.

6.11 To show that (a) is a tautology, we construct the following truth table.

x	y	$x \vee y$	$x \vee \neg y$	$(x \vee y) \vee (x \vee \neg y)$
T	T	T	T	T
T	F	T	T	T
F	T	T	F	T
T	F	F	T	T

6.12 Here is a truth table for part (a):

x	y	$x \vee y$	$x \vee \neg y$	$\neg x$	$(x \vee y) \wedge (x \vee \neg y) \wedge \neg x$
T	T	T	T	F	F
T	F	T	T	F	F
F	T	T	F	T	F
F	F	F	T	T	F

6.13 ($\Rightarrow$) Suppose A is logically equivalent to B. ... Therefore $A \leftrightarrow B$ is a tautology.

($\Leftarrow$) Suppose $A \leftrightarrow B$ is a tautology. ... Therefore A is logically equivalent to B.

6.15 For (c) we show $x \underline{\vee} y = (x \vee y) \wedge (\neg(x \wedge y))$ via the following truth table.

x	y	$x \underline{\vee} y$	$x \vee y$	$\neg(x \wedge y)$	$(x \vee y) \wedge (\neg(x \wedge y))$
T	T	F	T	F	F
T	F	T	T	T	T
F	T	T	T	T	T
F	F	F	F	T	F

6.17 If necessary, you can write down all possible tables and find ways to express each. However, there is a mechanical way to convert an arbitrary binary Boolean operation using $\wedge$, $\vee$, and $\neg$.

7.1 2^k.

7.6 Try a smaller version of this problem first. For instance, show that there are 36 ways to place pairwise nonattacking rooks on a 3×3 chess board.

7.9 (a) 10^9. (c) 5^9. (e) 9^9.

7.10 Break this problem into 8 cases depending on the length of the name, and total your answer.

7.13 The answer is $20 \times 19 \times 18 \times \cdots \times 2 \times 1$.

7.14 The answer is not $(10 \times 9 \times \cdots \times 2 \times 1)^2$.

8.1 Here is an answer to a question we didn't ask. There are $6! \times 8! \times 5!$ ways to place the books on the shelf if the French books must be to the left, the Russian books in the middle, and the Spanish books to the right.

8.2 The point of this discussion is that the product of a list containing just one number ought to be the number on the list. No actual multiplication takes place.

8.3 Try to use the formula $(n)_k = \frac{n!}{(n-k)!}$ to calculate $(3)_6$.

8.5 $2^{100} = (2^4)^{25} = 16^{25}$.

8.6 Approximate error is computed as

$$\frac{\text{approximate value} - \text{true value}}{\text{true value}}.$$

8.7 (a) 945; (b) 0.

8.8 The answer is *not* 20.

8.9 The last two on this list work slightly differently from the others.

8.10 $n! = n \times (n-1)!$.

9.1 (a) $\{0, 3, 6, 9\}$.
(f) $\{-10, 10, -20, 20, -50, 50, -100, 100\}$.

9.2 (a) 21.

9.3 (a) $2 \in \{1, 2, 3\}$.
(b) $\{2\} \subseteq \{1, 2, 3\}$.
(c) $\{2\} \in \{\{1\}, \{2\}, \{3\}\}$.

9.7 Let $x \in C = \{x \in \mathbb{Z} : x|12\}$, so x is a divisor of 12; i.e., $12 = xa$ for some integer a. Multiply both sides by 3 and we have $36 = 3xa = (3a)x$. Therefore x is a divisor of 36 and so $x \in D$. Therefore $C \subseteq D$.

9.10 You need to find a triple (a, b, c) that is in one of the sets, but not the other. Since $T \subseteq P$, you should try to find a Pythagorean triple $(a, b, c) \in P$ for which $(a, b, c) \notin T$.

As an extra hint: What can you say about the middle term of the triples $(p, q, r) \in T$?

10.1 (a) $\forall x \in \mathbb{Z}$, x is prime.
(g) $\forall x \in \mathbb{Z}, \exists y \in \mathbb{Z}, xy = 1$.

10.2 (a) $\exists x \in \mathbb{Z}$, x is not prime. "There is an integer that is not prime."
(g) $\exists x \in \mathbb{Z}, \forall y \in \mathbb{Z}, xy \neq 1$. "There is an integer x such that no matter what integer we multiply x by, the answer is never 1."

10.4 (a) False. (g) True.

10.5 (a) $\exists x \in \mathbb{Z}, x \not< 0$: There is an integer that is not negative.
(g) $\exists x \in \mathbb{Z}, \forall x \in \mathbb{Z}, x + y \neq 0$: There is an integer x with the property that for any integer y, the sum of x and y is not zero.

11.1 (b) $\{4, 5\}$. (c) $\{1, 2, 3\}$. (e) $\{1, 2, 3, 6, 7\}$.

11.5 This is false. Find a counterexample.

11.7 This is true. Here is an outline for your proof:

($\Rightarrow$) Suppose $A \cup B = A \cap B$. We want to prove $A = B$.

Suppose $x \in A$. ... Therefore $x \in B$.
Suppose $x \in B$. ... Therefore $x \in A$.
Therefore $A = B$.

($\Leftarrow$) Suppose $A = B$. ... Therefore $A \cup B = A \cap B$. ■

11.12 ($\Leftarrow$) If $A - B = \emptyset$, then if $x \in A$, then x must also be in B (otherwise $A - B$ wouldn't be empty), so $A \subseteq B$.

($\Rightarrow$) On the other hand, if $A \subseteq B$, clearly there are no elements in A that are not in B, so $A - B = \emptyset$.

11.15 Use DeMorgan's Law from Boolean algebra.

11.16 Most of these are false. Venn diagrams will help you figure out which are true and which are false. Then construct small counterexamples (for the false ones).

11.18 One way to do this: Start with a standard Venn diagram with three circles (for sets A, B, and C) and then add a complicated shape for D.

Note that there must be at least 16 regions in the final figure.

11.19 Let $X = A \cup B$. Apply Equation (4) to $|X \cup C|$. You now have

$$|A \cup B \cup C| = |X \cup C| = |X| + |C| - |X \cap C|.$$

Apply what you know to find $|X|$ and $|X \cap C|$ and substitute into the above equation to finish the proof.

11.20 Be sure you have done Exercise 6.15.

11.21 Union ($\cup$) is commutative.

11.22 Be sure you have done Exercise 11.20 and use properties of the corresponding idea from Boolean algebra.

11.24 Use the Multiplication Principle (Theorem 7.2).

11.25 Here is a template for a proof of part (a).

To show that two sets are equal, we use Proof Template 5.

> Suppose first that $x \in A \times (B \cup C)$. ... Therefore $x \in (A \times B) \cup (A \times C)$.
>
> Suppose second that $x \in (A \times B) \cup (A \times C)$. ... Therefore $x \in A \times (B \cup C)$.
>
> Therefore $A \times (B \cup C) = (A \times B) \cup (A \times C)$. ■

We expand this a little as follows.

> Suppose first that $x \in A \times (B \cup C)$. This means $x = (a, z)$ where $a \in A$ and $z \in B \cup C$.... Therefore $x \in (A \times B) \cup (A \times C)$.
>
> Suppose second that $x \in (A \times B) \cup (A \times C)$. Thus either $x \in A \times B$ or $x \in A \times C$.... Therefore $x \in A \times (B \cup C)$.
>
> Therefore $A \times (B \cup C) = (A \times B) \cup (A \times C)$. ■

And we can expand still further to give the following structure for you to complete.

> Suppose first that $x \in A \times (B \cup C)$. This means $x = (a, z)$ where $a \in A$ and $z \in B \cup C$. We have two cases:
> - If $z \in B$, then ...
> - If $z \in C$, then ...
>
> ... Therefore $x \in (A \times B) \cup (A \times C)$.

> Suppose second that $x \in (A \times B) \cup (A \times C)$. Thus either $x \in A \times B$ or $x \in A \times C$.
> We have two cases:
> - If $x \in A \times B$, then ...
> - If $x \in A \times C$, then ...
>
> ...Therefore $x \in A \times (B \cup C)$.
>
> Therefore $A \times (B \cup C) = (A \times B) \cup (A \times C)$. ■

12.1 Use the following question: How many length-n lists can we form using the elements 0 and 1 (repetition allowed) in which the elements are not all zero?

12.2 For the first part, the expression should simplify to $x^n - 1$.

For the second part, let $x = 2$.

12.3 For the first part, use the question "How many length-n lists can we form using the elements in $\{1, 2, 3\}$ in which the elements are not all 3?"

For the second part, note that $99 + 1 = 100$, $999 + 1 = 1000$, and so on.

12.4 Create two patterns of dots. In the first pattern, the dots are laid out in a rectanglular grid of $a-b$ rows and $a+b$ columns. Clearly this pattern has $(a-b)(a+b)$ dots. Now find a rearrangement of the dots that clearly has $a^2 - b^2$ dots.

12.5 Do not use algebra! Give two different answers to the question "How many length-2 lists can we make from n elements?"

13.1 (a) reflexive, symmetric, antisymmetric, transitive

(b) irreflexive, antisymmetric

13.2 (d) is true. Here is a proof: Suppose $x\ R\ y$. Then $|x-y| \le 2$. Note that $|x-y| = |y-x|$, so $|y - x| \le 2$. Therefore $y\ R\ x$.

(f) is false. You should find three numbers a, b, and c with $a\ R\ b$, $b\ R\ c$, but $a \not R\ c$.

13.3 (a) $R^{-1} = \{(2, 1), (3, 2), (4, 3)\}$.

(c) $R^{-1} = \{(x, y) : x, y \in \mathbb{Z},\ x-y = -1\}$ or $R^{-1} = \{(x, y) : x, y \in \mathbb{Z},\ y-x = 1\}$.

13.4 Remember that R, S, R^{-1}, and S^{-1} are sets. To prove that two sets are equal, use Proof Template 5.

13.5 This is false. Find a counterexample with $A = \{1, 2\}$.

13.6 All proofs and counterexamples for this problem are quite short. For example, here is the proof that the "has-the-same-size-as" relation R is transitive.

Let A, B, and C be finite sets of integers and suppose $A\ R\ B$ and $B\ R\ C$. This means that $|A| = |B|$ and $|B| = |C|$. Therefore $|A| = |C|$ and so $A\ R\ C$. Therefore R is transitive.

13.7 For one part of this problem, Exercise 9.4 is useful.

13.9 For part (b): Does this seem impossible? It isn't. Perhaps you are reasoning as follows:

Let $x \in A$. In order for R to be reflexive, we have to have $x\ R\ x$. In order for R to be irreflexive, we have to have $x \not R\ x$. We can't have it both ways (x either *is* or *is not* related to itself).

The mistake in this reasoning is in the first sentence.

13.10 Here is a proof template for this problem.

> ($\Rightarrow$) Suppose R is symmetric. To show that $R = R^{-1}$, we need to prove that the two sets, R and R^{-1} are the same. We use Proof Template 5.
>
> Suppose $(x, y) \in R$. ... $(x, y) \in R^{-1}$.
>
> Suppose $(x, y) \in R^{-1}$. ... $(x, y) \in R$. Therefore $R = R^{-1}$.
>
> ($\Leftarrow$) Suppose $R = R^{-1}$. We must prove that R is symmetric. Suppose $x\ R\ y$. ... Therefore $y\ R\ x$, so R is symmetric. ■

14.1 (a) Yes. (f) No. (g) Yes.

14.2 Here is the proof for the first statement:

Suppose x and y are both odd. By definition, we can find integers a and b such that $x = 2a + 1$ and $y = 2b + 1$. Now $x - y = (2a + 1) - (2b + 1) = 2a - 2b = 2(a - b)$, so $2|(x - y)$. Therefore $x \equiv y \pmod 2$.

14.3 What is $a - (-a)$?

14.4 Be sure you did Exercise 4.6.

14.5 (a) $[1] = \{1, 2\}$.
(e) [you] is the set containing all people born on your birthday.

14.6 You may use Proof Template 5 (the default manner to prove sets are equal), but you'll have an easier time if you apply Proposition 14.12.

14.10 Here is a useful notation for this problem. Write $[a]_R$ for the equivalence class of R with respect to the relation R and $[a]_S$ for the equivalence class with respect to S. That is,

$$[a]_R = \{x \in A : x \; R \; a\}$$
$$[a]_S = \{x \in A : x \; S \; a\}.$$

14.13 It is painful to write out an equivalence relation as a full set of ordered pairs. For example, consider the relation

$$R = \{(1, 1), (1, 2), (2, 1), (2, 2), (3, 3), (3, 4), (4, 3), (4, 4)\}.$$

It is simpler just to write out its equivalence classes: $\{1, 2\}$ and $\{3, 4\}$. A convenient shorthand for this is just to write 12/34.

15.1 Use the notation 1/23 to stand for the partition $\{\{1\}, \{2, 3\}\}$, etc.

The partitions of $\{1, 2, 3\}$ are 1/2/3, 1/23, 2/13, 3/12, and 123.

There are 15 partitions of $\{1, 2, 3, 4\}$.

15.2 Answer to (d): $\frac{7!}{2! \times 3!}$.

15.5 Here is an outline for this proof.

Part (1): Let $[a]$ be an equivalence class of $\overset{\mathcal{P}}{\equiv}$. Prove that there is a part $P \in \mathcal{P}$ such that $[a] = P$ (you will need to prove that two sets are equal here).

Part (2): Let P be a part of $\mathcal{P}$, i.e. $P \in \mathcal{P}$. Prove there is an element $a \in A$ such that $[a] = P$.

You have shown that every equivalence class of $\overset{\mathcal{P}}{\equiv}$ is a part of $\mathcal{P}$ and, conversely, that every part of $\mathcal{P}$ is an equivalence class of $\overset{\mathcal{P}}{\equiv}$.

15.7 Imagine the 12 people are arranged around a clock face. In how many ways can you locate them around the clock? [Answer: 12!.] Of course, if everyone moves one position clockwise, the arrangement is equivalent. Develop an equivalence relation on the set of arrangements, and figure out the size of the equivalence classes.

15.8 Be careful. It is easy to be off by a factor of 2 in this problem.

15.9 It's easy to be off by a factor of 2 in this problem as well.

15.10 12926008369442488320000, but this is an awful way to report the answer.

15.11 Imagine the 20 people first line up. In how many ways can this be done? [Answer: 20!.] The first 10 people on the line form a team, and the last 10 people on the line form the other team. Consider two line-ups equivalent if they result in the same two teams being formed. Count the size of the equivalence classes and figure out the number of classes. Note that if we switch all players on both teams, we have not really changed anything at all.

Test your answer by considering the number of ways to divide 6 people into two teams of 3. The answer should be 10: 123/456, 124/356, 125/346, 126/345, 134/256, 135/246, 136/245, 145/236, 146/235, and 156/234.

15.16 Yes. Find the set A.

16.1 The second answer is twice the first.

16.4 The answer to (b) is $\binom{50}{10}$.

16.5 The chart looks like this:

$$\{1, 2, 3\} \leftrightarrow \{4, 5, 6, 7\}$$
$$\vdots$$
$$\{5, 6, 7\} \leftrightarrow \{1, 2, 3, 4\}$$

16.6 For (b), consider the cases with zero, with one, and with two doubles separately.

16.7 *Warning*: There's a little trap waiting for you. Be sure you don't step into it.

16.9 Expand both $\binom{n}{k}$ and $\binom{n}{n-k}$ in terms of factorials, and then use algebra to show they are equal.

16.10 The question is "How many subsets does an n-element set have?"

16.11 Expand $(1 - 1)^n$.

The equation means that the number of subsets of an n-element set with an even number of elements equals the number of subsets with an odd number of elements.

16.15 Consider a set of $2n+2$ people with two weirdos.

16.16 Stirling's formula is $n! \approx \sqrt{2\pi n}n^n e^{-n}$. For the second part, note that $4^n = 2^{2n}$.

16.17 Put everything over a common denominator and don't lose your courage.

16.18 The question is "How many 3-element subsets does $\{1, 2, 3, \ldots, n\}$ have?"

And consider, how many of those 3-element subsets have largest element 3? ...largest element 4?... largest element n?

16.21 Think of a classroom containing n girls and n boys.

16.23 The answer is $\binom{n}{k}$. We can think of this labeling process as selecting the k elements of A that receive the "good" label (and the remainder are assigned "bad"). So there is a one-to-one correspondence between assigning labels and selecting a k-element subset of A.

16.24 (a) If in doubt, write out all the possibilities. There are not that many.

(b) Note that $1+2+5 \neq 10$, so there are not enough labels to go around. The answer to this problem is a number, not the word "impossible."

(d) Since there are no labels of Type 3 available, this reduces to the previous problem; the answer is a binomial coefficient.

(e) Changing the names of the labels doesn't change the number of ways to distribute them.

16.25 For (a), think of the labeling process as proceeding in two stages. First we assign the labels of Type 1 (in how many ways?) and then we assign the labels of Type 2 (in how many ways?).

For (b), you may use (a) and expand the binomial coefficients into factorials, but there is a combinatorial proof. Place the n elements in a repetition-free list (in how many ways?). Then give the first a elements in the list Type 1 labels, the next b elements Type 2 labels, and the last c elements Type 3 labels. Call two lists equivalent if they result in the same distribution of labels, and count the number of equivalence classes.

16.26 The proof is similar to that of Theorem 16.8.

16.27 Answer: $\binom{52}{5}$.

16.28 (a) 13×48. (d) $13 \times \binom{4}{3} \times 12 \times \binom{4}{2}$.

16.30 Write down the first six or seven rows of Pascal's triangle. In a different color, record next to each entry how many additions it takes to calculate that value. Notice that the 1s on the ends of each row take 0 calculations.

Now, to compute an interior value, such as $\binom{n}{k}$, you calculate $\binom{n-1}{k-1}$ (that takes a certain number of additions) and $\binom{n-1}{k}$ (that takes a certain number of additions). Finally, to calculate $\binom{n}{k}$ you do one more calculation.

Do you see a pattern? This will enable you to find the number of additions to compute $\binom{100}{30}$.

17.1 For $\left(\!\binom{3}{2}\!\right)$: We list all six 2-element multisets we can form with the elements in $\{1, 2, 3\}$. They are $\langle 1, 1\rangle$, $\langle 1, 2\rangle$, $\langle 1, 3\rangle$, $\langle 2, 2\rangle$, $\langle 2, 3\rangle$, $\langle 3, 3\rangle$.

Theorem 17.8 gives $\left(\!\binom{3}{2}\!\right) = \binom{3+2-1}{2} = \binom{4}{2} = 6$.

Give a similar answer for $\left(\!\binom{2}{3}\!\right)$.

17.2 For $\left(\!\binom{3}{2}\!\right)$: $**||$, $*|*|$, $*||*$, $|**|$, $|*|*$, and $||**$.

17.3 You can check your answers using the chart. The point of this problem is to verify the first row and the first column of the chart.

17.4 $\langle 1, 4, 4, 4\rangle$.

17.5 Convert to a binomial coefficient.

17.6 You can resort to factorials if you must. Here's a better idea: $* \leftrightarrow |$.

17.8 This calls for a combinatorial proof. The question is "How many k-element multisets can we form using the integers 1 through n?" The first answer is $\left(\!\binom{n}{k}\!\right)$. The second answer depends on the multiplicity of element n in the multiset.

This problem can also be solved by conversion to binomial coefficients.

17.9 This problem calls for a combinatorial proof. You must find a question that is answered by both the left- and right-hand sides of the equation. The question should be "How many k-element multisets can be formed using integers chosen from $\{1, 2, \ldots, n\}$?"

The left-hand side of the equation is clearly one such answer. Try to figure out how the right-hand side is also an answer. And to help you with that, answer the following:

How many multisets have size 10, with elements that are chosen from $\{1, 2, \ldots, 99\}$, whose largest element is 23? Answer: $\left(\!\binom{23}{9}\!\right)$.

18.1 Call the four groups of people A_1, A_2, A_3, and A_4. The problem gives you the sizes of these sets and their various intersections. You need to find $|A_1 \cup A_2 \cup A_3 \cup A_4|$.

18.2 This is true, so don't try to disprove it. Start from

$$\begin{aligned}|A \cup B \cup C| = |A| + |B| + |C| \\ - |A \cap B| - |A \cap C| \\ - |B \cap C| + |A \cap B \cap C|\end{aligned}$$

and cancel $|A \cup B \cup C| = |A| + |B| + |C|$.

18.3 First count "bad" words and subtract from 26^5.

Let B_1 denote the set of words whose first two letters are the same. Let B_2 denote the set of words whose second two letters are the same. And so on.

Figure out the sizes of the various intersections and apply inclusion-exclusion.

18.4 For (a), write $9^n = [10 + (-1)]^n$.

For (b), the combinatorial proof, you need the right question. Here is a good way to start your question: How many lists of length n can we make using the standard digits 0 through 9 in which ...?

19.1 (a) If x^2 is not odd, then x is not odd.
(d) If a parallelogram is not a rhombus, then its diagonals are not perpendicular.

19.2 Remember: $\neg\neg B = B$.

19.4 (b) Let a and b be negative integers. Suppose, for the sake of contradiction, that $a + b$ is nonnegative.
(d) Let p and q be primes for which $p + q$ is also prime. Suppose, for the sake of contradiction, that neither p nor q is equal to 2.

19.5 Your proof should begin as follows:

Let x and $x + 1$ be consecutive integers. Suppose, for the sake of contradiction, that x and $x + 1$ are both even....

19.8 Suppose $(A - B) \cap (B - A) \neq \emptyset$. This means there is an element x in both $A - B$ and $B - A$. Argue from here to a contradiction.

19.9 This is an if-and-only-if style theorem; be sure to prove both halves. Both halves can be proved by *contrapositive*.

19.10 A direct proof here is possible. The point of this problem is to introduce the term *converse*.

19.11 Answer for (a): We say that x is a *smallest element* of A provided (1) $x \in A$ and (2) if $y \in A$, then $x \leq y$.

First sentence for (b): Suppose, for the sake of contradiction, E contains a smallest element x.

Comment for (c): This is quite obvious, but please write a careful proof by contradiction using Proof Template 14. Here is a good start for your proof: "Let A be a subset of the integers with a smallest element. Suppose, for the sake of contradiction, A contains two distinct smallest elements a and b..."

20.1 There is no such thing.

20.5 It it helpful to use a computer to generate the first several Fibonacci numbers and the values of 1.6^n, although a hand-held calculator will suffice.

You should find that the inequality holds for all $n \geq 29$.

20.6 Here is a handy chart to get you started.

n	F_n	$F_0 + \cdots + F_n$
0	1	1
1	1	2
2	2	4
3	3	7
4	5	12
5	8	20
6	13	33
7	21	54
8	34	88
9	55	143
10	89	232

Compare the numbers in the third column to those in the second.

20.8 Expressed as a theorem, you need to show: For every $n \in \mathbb{N}$, the nth row of Pascal's triangle is

$$\binom{n}{0}, \binom{n}{1}, \binom{n}{2}, \ldots \binom{n}{n-1}, \binom{n}{n}.$$

21.3 Here is a complete answer to (a).

Proof. (by induction on n):

Basis case $n = 1$. Both sides of the equation evaluate to 1, so the basis case is true.

Induction hypothesis: Suppose the result is true when $n = k$.

That is, we have

$$1+4+7+\cdots+(3k-2) = \frac{k(3k-1)}{2} \quad (*).$$

We want to show

$$1+4+7+\cdots+(3k-2)+[3(k+1)-2] = \frac{(k+1)[3(k+1)-1]}{2}.$$

Add $3(k+1) - 2 = 3k+1$ to both sides of $(*)$ to get

$$\begin{aligned} 1+4+7+\cdots&+(3k-2)+(3k+1) \\ &= \frac{k(3k-1)}{2} + (3k+1) \\ &= \frac{(3k^2-k)+(6k+2)}{2} \\ &= \frac{3k^2+5k+2}{2} \\ &= \frac{(k+1)(3k+2)}{2} \\ &= \frac{(k+1)[3(k+1)-1]}{2}. \end{aligned}$$

For (c), notice that this is a fancy generalization of the fact that $999 = 1000 - 1$.

21.5 This fact is rather obvious. The point here is to get practice writing proofs by induction. Let n be the number of people on the line. In the basis case, $n = 2$.

21.6 The proof is by induction on the number of disks.

21.8 This is a full answer to (a).

The next three terms of the sequence are $a_4 = 31$, $a_5 = 63$, and $a_6 = 127$.

To prove: $a_n = 2^{n+1} - 1$.
Basis case: When $n = 0$, we just need to notice that $a_0 = 1 = 2^{0+1} - 1 = 2 - 1$, as required.

Induction hypothesis: Suppose $a_k = 2^{k+1} - 1$.
We need to prove that $a_{k+1} = 2^{(k+1)+1} - 1$. Notice that

$$\begin{aligned} a_{k+1} &= 2a_k + 1 \quad \text{by definition} \\ &= 2[2^{k+1} - 1] + 1 \quad \text{by induction} \\ &= 2^{k+2} - 2 + 1 = 2^{k+2} - 1 \end{aligned}$$

as required. ■

21.9 Let a_n denote the number of possible solutions. Find a recurrence relation for a_n.

21.10 Use strong induction. If n is a Fibonacci number, there is nothing to prove. If n is not a Fibonacci number, let F_k be the largest Fibonacci number less than n. You will want to show that $n - F_k < F_k$.

21.11 We use induction on $n =$ `last-first`.

The basis case is when $n = 0$. In this case, `first` equals `last`, so the program returns `array[first]`, which is the only value under consideration.

Induction hypothesis: Assume the result is true for all values of `last-first` that are less than n.

Suppose the program is called with `last-first` $= n$. Note that `mid` is between `first` and `last`, and we have `mid` $<$ `last`, so, by induction, the line

```
a = findMax(array,first,mid);
```

sets the variable `a` to the largest value in the array from index `first` to index `last`. Also, `mid+1` is greater than `first`, so, by induction, the line

```
b = findMax(array,mid+1,last);
```

sets `b` to the largest value in the array from index `mid+1` to index `last`.

Finally, the last two lines of the program return the larger of `a` and `b`, which must be the largest value in the array from index `first` to index `last`.

21.13 The point of this problem is that you should *not* be able to do this problem! The full, correct answer to this problem is *"I give up!"*

In the proof in this book, we used the following induction hypothesis:

Induction hypothesis 1: Every triangulated polygon with at most k sides has at least *two* exterior triangles.

Your job is to try to work from this induction hypothesis:

Induction hypothesis 2: Every triangulated polygon with at most k sides has at least *one* exterior triangle.

Hypothesis 1 is easier to use because it gives you more leverage. This is known as induction loading.

21.14 The proof is much like that of Theorem 21.2.

21.15 Here are the first few lines of a proof to get you started.

The basis case is 0. In this case we can write 0 as an empty sum.

Let n be a positive integer and suppose the result has been shown for all natural numbers less than n. Let k be the largest natural number such that $2^k \le n$. (There are only finitely many natural numbers $\le n$ and $2^0 \le n$; so k exists.)

21.16 Here is the statement you need to prove: If A is a nonempty subset of $\mathbb{N}$, then A contains a least element.

To prove this by induction, consider the following alternative. Let $A \subseteq \mathbb{N}$. If n is a natural number in A, then A contains a least element.

Now induct on n. We recommend strong induction.

22.1 Here is a complete solution for (a):

$$\begin{aligned} a_0 &= 1 \qquad \text{(given)} \\ a_1 &= 2a_0 + 2 = 2 \cdot 1 + 2 = 4 \\ a_2 &= 2a_1 + 2 = 2 \cdot 4 + 2 = 10 \\ a_3 &= 2a_2 + 2 = 2 \cdot 10 + 2 = 22 \\ a_4 &= 2a_3 + 2 = 2 \cdot 22 + 2 = 46 \\ a_5 &= 2a_4 + 2 = 2 \cdot 46 + 2 = 92. \end{aligned}$$

For the other parts, here is a_5: (b) 20, (c) 11, (d) 0, (e) 15, and (f) 16.

22.2 (a) $a_n = 4(\frac{2}{3})^n$. $a_9 = 2^{11}/3^9 = 2048/19683$.
(e) $a_n = \frac{19}{2}3^n + \frac{1}{2}$. $a_9 = 186989$.
(h) $a_n = 2 \cdot 2^n - 2$. $a_9 = 1022$.
(n) $a_n = 5(-1)^n - 6n(-1)^n$. $a_9 = 49$.
(o) $a_n = \frac{3}{2}(1+\sqrt{3})^n + \frac{3}{2}(1-\sqrt{3})^n$. $a_9 = 12720$.

22.3 Here is a complete solution to (b). We write down the sequences a_n, Δa_n, $\Delta^2 a_n$, and so on until we reach the all-zeros sequence:

```
6     5     6     9    14    21    30
   -1     1     3     5     7     9
       2     2     2     2     2
          0     0     0     0
```

We then use the first terms from each row and apply Theorem 22.17 to give

$$a_n = 6\binom{n}{0} + (-1)\binom{n}{1} + 2\binom{n}{2}$$
$$= 6 - n + n(n-1) = n^2 - 2n + 6.$$

22.4 The difference operator applies to sequences, not to individual numbers. The notation $(\Delta a)_n$ means the nth term of the sequence Δa; this is the intended meaning. The notation $\Delta(a_n)$ is not defined since a_n

is a number and we have not assigned a meaning to Δ applied to a single number.

22.5 Let k be a positive integer and let $a_n = \binom{n}{k}$. We know that $\Delta a_n = \Delta\binom{n}{k} = \binom{n}{k-1}$. Repeating this, we see that $\Delta^j a_n = \binom{n}{k-j}$. So we have

$$\begin{aligned} a_0 &= \binom{0}{k} = 0 \\ \Delta a_0 &= \binom{0}{k-1} = 0 \\ \Delta^2 a_0 &= \binom{0}{k-2} = 0 \\ &\vdots \\ \Delta^{k-1} a_0 &= \binom{0}{k-(k-1)} = \binom{0}{1} = 0, \end{aligned}$$

but $\Delta^k a_0 = \binom{0}{0} = 1$.

22.6 We can express a_n in the form $c_1 r_1^n + c_2 r_2^n$ where r_1, r_2 are the roots of a quadratic equation. Find r_1, r_2 first. Then set up two equations and two unknowns to find c_1, c_2.

22.8 Let $a_n = 1^t + 2^t + \cdots + n^t$. Note that $\Delta a_n = (n+1)^t$. Apply Δ another $t+1$ times. What do you get? Use Theorem 22.17 to conclude that a_n can be written as a polynomial expression.

22.10 For example, when $s = 3$, we need to solve $a_n = 3\Delta a_n$. Remember that $\Delta a_n = a_{n+1} - a_n$, so what we really want to solve is $a_n = 3(a_{n+1} - a_n)$, which can be rearranged to $a_{n+1} = \frac{4}{3} a_n$. This recurrence is not quite in standard form but is equivalent to $a_n = \frac{4}{3} a_{n-1}$.

22.11 $\Delta^2 a_n = a_{n+2} - 2a_{n+1} + a_n$.

22.13 (a) The answer is $a_n = 3^n - 2^n + 1$.

(d) The form of the answer is $a_n = c_1 5^n + c_2 + c_3 n$.

(e) The form of the answer is $a_n = c_1 3^n + c_2 n 3^n + c_3$.

(f) a_n is given by a quadratic polynomial.

22.14 Here is a complete solution to (a). From the recurrence relation $a_n = 4a_{n-1} - a_{n-2} - 6a_n$, we form the associated cubic equation $x^3 - 4x^2 + x + 6 = 0$. This factors $(x-2)(x-3)(x+1) = 0$; hence the roots are 2, 3, and -1. We therefore expect a_n to be of the form $c_1 2^n + c_2 3^n + c_3(-1)^n$.

We now use the values for a_0, a_1, and a_2 to solve for c_1, c_2, c_3:

$$\begin{aligned} a_0 &= 8 = c_1 + c_2 + c_3 \\ a_1 &= 3 = 2c_1 + 3c_2 - c_3 \\ a_2 &= 27 = 4c_1 + 9c_2 + c_3. \end{aligned}$$

This gives $c_1 = 1$, $c_2 = 2$, and $c_3 = 5$. Therefore

$$a_n = 2^n + 2 \cdot 3^n + 5(-1)^n.$$

22.15 Implement this program in your favorite language. At the start of the procedure, add a debugging statement that prints out the argument. Something like this:

```
print 'Calling get_term with argument ' n
```

Now call get_term(10) and see what happens.

Note that to compute a_0 or a_1, only one call to get_term is generated. To compute a_2, three calls are generated (the original call get_term(2) plus the two embedded calls.

To calculate a_3, get_term is called five times: once for the original call get_term(3), and then it calls get_term(2) (three calls to do that) and get_term(1) (one call for that). The first few values of b_n are 1, 1, 3, 5, 9, 15, 25.

22.16 For (a), use the recurrence to generate the values a_1, a_2, a_3, a_4, but don't perform the actual multiplications.

The answer to (b) is $a_n = 2^{*2^n)}$.

For (c), write out the first several values. You will note that a_n does not exactly fit the pattern you should observe. It is fine to report your answer in the form

$$a_n = \begin{cases} 1 & \text{if } n = 0, \\ \text{a formula} & \text{if } n > 0. \end{cases}$$

Part (d) also has the difficulty that a_0 does not fit the pattern of the subsequent terms. Try to find a second-order recurrence of the form $a_n = s_1 a_{n-1} + s_2 a_{n-2}$ that works once $n \geq 3$ and solve that.

Part (e) is an unsolved problem. The values a_n are called *chaotic,* and no reasonable formula can be expected to exist.

23.1 A complete answer to (a): f is a function, $\operatorname{dom} f = \{1, 3\}$, and $\operatorname{im} f = \{2, 4\}$. f is one-to-one and $f^{-1} = \{(2, 1), (4, 3)\}$.

23.2 There are 2^3 such functions, and none of them is one-to-one.

One of the functions is $\{(1, 4), (2, 4), (3, 4)\}$; it is neither one-to-one nor onto.

23.3 There are 3^2 such functions, and none of them is onto B.

One of the functions is $\{(1, 3), (2, 3)\}$. It is neither one-to-one nor onto B.

23.4 Here is a complete answer.

Function	One-to-one?	Onto?
$\{(1, 3), (2, 3)\}$	no	no
$\{(1, 3), (2, 4)\}$	yes	yes
$\{(1, 4), (2, 3)\}$	yes	yes
$\{(1, 4), (2, 4)\}$	no	no

23.7 In (a), there is no explicit set B to which the definition applies. In particular, every function f is onto if we think of B as being the image of f.

In (b), the notation $f : A \to B$ establishes a context for the phrase "f is onto." In this context, the issue is: Does $\operatorname{im} f$ equal B?

23.9 Here is a complete answer to (a).

First, f is one-to-one. Proof: We need to show that if $f(a) = f(b)$, then $a = b$. So, suppose we have integers a, b with $f(a) = f(b)$. By definition of f, we have $2a = 2b$. Dividing both sides by 2 gives $a = b$. Therefore f is one-to-one.

Second, f is not onto. Proof: We claim that $1 \in \mathbb{Z}$, but there is no $x \in \mathbb{Z}$ with $f(x) = 1$. Suppose, for the sake of contradiction, there is an integer x such that $f(x) = 1$. Then $2x = 1$, and so $x = \frac{1}{2}$. However, $\frac{1}{2}$ is not an integer, so there is no integer x with $f(x) = 1$. Therefore f is not onto.

23.11 This problem requires you to write *three* proofs:

1. If (a) and (b), then (c).
2. If (a) and (c), then (b).
3. If (b) and (c), then (a).

To this end, Proposition 23.24 (the Pigeonhole Principle) is quite helpful.

23.14 How many subsets of A have exactly k elements?

23.15 See Exercises 16.24 and 16.25.

24.1 How many different patterns of $<$s and $>$s are possible in a sequence of five distinct integers?

24.3 Create six categories of integers based on their ones digits. Because there are seven integers in the set, two of these must be in the same category.

24.4 Apply the Pigeonhole Principle by making pigeonholes in the square.

24.5 Think about the parity of the coordinates.

24.6 The number 9 should figure in your proposition.

24.7 Here is a length-nine sequence with no monotone subsequence of length four.

3 2 1 6 5 4 9 8 7.

Try to generalize this and use the Pigeonhole Principle in your proof that the sequence does not contain a monotone subsequence of length $n + 1$.

24.8 If the sequence has length n, then it has 2^n subsequences. Even for moderate values of n, it is highly inefficient to try to scan through all the sequences. Instead, use the labeling scheme in the proof of Theorem 24.3.

25.1 (a) $g \circ f = \{(1, 1), (2, 1), (3, 1)\}$ and $f \circ g = \{(2, 2), (3, 2), (4, 2)\}$; $g \circ f \neq f \circ g$.

(c) $g \circ f = \{(1, 1), (2, 5), (3, 3)\}$ but $f \circ g$ is undefined.

(h) $(g \circ f)(x) = x + 1$ and $(f \circ g)(x) = x - 1$; $g \circ f \neq f \circ g$.

25.8 What are the domain and image of $f \circ f^{-1}$?

25.11 The answer to both is yes if the set, A, is finite. However, ...

25.12 Part (a) was already dealt with in Exercise 25.9. Part (b) is false; find a counterexample. Part (c) is true; use Exercise 25.7.

26.2 The answer to (a) is $(1,2,4)(3,6,5)$.

26.3 For $n=3$ the answer is two: $(1,2,3)$ and $(1,3,2)$. For $n=4$ the answer is six.

26.4 This is a deranged problem.

26.5 The answer to (a) is $(1,4,7,6,9,3,2,5,8)$, and the answer to (b) is different. The answer to (d) is $(1)(2,5,4,3)(6,9,8,7)$, although this may also be written $(1)(5,4,3,2)(9,8,7,6)$.

26.6 This is false.

26.8 This was dealt with in a problem in Section 25.

26.11 Note that for any transposition τ, we have $\tau \circ \tau = \iota$. Therefore $\tau^{-1} = \tau$.

To prove that two permutations are inverses of one another, just compose them and show that the answer must be ι.

26.14 We are given $\pi \circ \sigma = \sigma$. Composing on the right by σ^{-1} gives

$$\begin{aligned} \pi \circ \sigma &= \sigma \\ (\pi \circ \sigma) \circ \sigma^{-1} &= \sigma \circ \sigma^{-1} \\ \pi \circ (\sigma \circ \sigma^{-1}) &= \iota \\ \pi \circ \iota &= \iota \\ \pi &= \iota. \end{aligned}$$

26.16 The answer to (a) is that $\pi = (1,2)(2,3)(3,4)(4,5)$, it has four inversions, and it is even.

26.17 A big hint: Draw a left-to-right arrow picture of the permutation and its inverse, and count crossings.

26.20 Imagine that the blank space carries the number 16. Then several moves of the puzzle result in a permutation of the numbers 1 through 16. In particular, a single move of the Fifteen Puzzle is a transposition.

27.1 Please note that

$$\begin{aligned} R_{90} &= (1,2,3,4), \\ F_H &= (1,2)(3,4), \quad \text{and} \\ F_\backslash &= (1)(2,4)(3). \end{aligned}$$

Now calculate $(1,2)(3,4) \circ (1,2,3,4)$.

27.2 You should find four symmetries of a rectangle.

27.4 There are six symmetries of an equilateral triangle.

27.5 There are two symmetries.

27.7 There are ten symmetries: an identity, four rotations, and five flips.

27.8 The answer to (c): The difference is that the first 24 symmetries involve rotating the cube about. The second collection of 24 are the mirror images of the first 24.

27.9 A rotation through an angle θ can be represented by the matrix $\left[\begin{smallmatrix} \cos\theta & -\sin\theta \\ \sin\theta & \cos\theta \end{smallmatrix}\right]$.

28.1 For (b), expand 1.1^n using the Binomial Theorem:

$$\begin{aligned} 1.1^n &= (1+0.1)^n \\ &= 1^n + \\ &\qquad \binom{n}{1} 1^{n-1}(0.1)^1 + \\ &\qquad \binom{n}{2} 1^{n-2}(0.1)^2 + \cdots \end{aligned}$$

and throw away the terms you don't need.

28.2 (c) is false. For example, $\lfloor 0.7 + 0.8 \rfloor = \lfloor 1.5 \rfloor = 1$, but $\lfloor 0.7 \rfloor + \lfloor 0.8 \rfloor = 0 + 0 = 0$.

28.3 Here is a complete proof.

Since $f(n)$ is $O(g(n))$, there is a positive number A such that, with at most finitely many exceptions,

$$|f(n)| \le A|g(n)|.$$

Similarly, since $g(n)$ is $O(h(n))$, there is a positive number B such that, with at most finitely many exceptions,

$$|g(n)| \le B|h(n)|.$$

Combining these two inequalities, we have, with at most finitely many exceptions,

$$|f(n)| \le A|g(n)| \le AB|h(n)|$$

and so $f(n)$ is $O(h(n))$.

28.5 Use the identity

$$\log_a n = (\log_b a)(\log_a n).$$

28.7 The answer is either $\lceil x - \frac{1}{2} \rceil$ or $\lfloor x + \frac{1}{2} \rfloor$.

28.8 The answer to this problem would be quite easy if you were allowed to use the mod function; it would be just $n \bmod 10$.

29.1 $x = 0.6$.

29.3 An outcome of this experiment can be recorded as (a, b) where a is either H or T (the result of the coin flip) and b is an integer with $1 \le b \le 6$ (the up-face of the die). Thus

$$S = \{(\text{H}, 1), (\text{H}, 2), (\text{H}, 3), (\text{H}, 4), (\text{H}, 5), (\text{H}, 6), (\text{T}, 1), (\text{T}, 2), (\text{T}, 3), (\text{T}, 4), (\text{T}, 5), (\text{T}, 6)\}.$$

All of these $2 \times 6 = 12$ outcomes are equally likely, so $P : S \to \mathbb{R}$ is given by $P(s) = \frac{1}{12}$ for every $s \in S$.

29.4 For (a): The sample space is (S, P) where $S = \{1, 2, 3, 4\}$ and $P(s) = \frac{1}{4}$ for all $s \in S$.

29.5 A complete answer to (b): The set S consists of all 5-element subsets of the set $\{1, 2, \ldots, 20\}$. Thus $|S| = \binom{20}{5}$. All of these outcomes are equally likely, so $P(s) = 1 \Big/ \binom{20}{5}$ for all $s \in S$.

29.6 Here is the answer for region 3: $P(3) = \frac{5}{16}$.

Explanation: The total area of the target (all four regions together) is 16π. The area of region 3 is $9\pi - 4\pi = 5\pi$. So region 3 covers $\frac{5}{16}$ of the total area.

29.7 Let $S = \{1, 2, 3\}$ and let $P(1) = 1$, $P(2) = 0$, and $P(3) = 0$.

29.8 Let $S = \{1\}$ and let $P(1) = 1$.

Note that if a sample space (S, P) has two (or more) elements, we cannot have $P(s) = 1$ for all $s \in S$; if $|S| > 1$, then

$$\sum_{s \in S} P(s) = \sum_{s \in S} 1 = |S| > 1$$

which is forbidden.

29.9 See the discussion "Much ado about 0!" in Section 8.

30.1 Here is the answer for $k = 4$. We have $A_4 = \{(1, 3), (2, 2), (3, 1)\}$ and $P(A_4) = \frac{3}{16}$.

30.2 $A = \{\text{HHTT, HTHT, HTTH, THHT, THTH, TTHH}\}$, and $P(A) = \frac{6}{16} = \frac{3}{8}$.

Notice that $|A| = \binom{4}{2} = 6$.

30.5 (a) $A = \{\text{HTHTHTHTHT, THTHTHTHTH}\}$.
(b) $P(A) = 2/2^{10} = 2^{-9} = 1/512$.

30.6 $A = \{(2, 6), (3, 5), (4, 4), (5, 3), (6, 2)\}$.

30.9 The set A contains $1+2+3+4+5$ outcomes.

30.10 Call the boxes $1, 2, 3, \ldots, 10$. The sample space S contains all length-2 lists of boxes without repetition. So $|S| = (10)_2 = 90$. Let us assume box 1 is the least valuable, and so on up to box 10 being the most valuable. Now this problem is just like the previous problem.

30.11 To compare dice 1 and 2 we make a chart. The rows of the chart are indexed by the numbers on die 1 and the columns by the numbers on die 2. We place a $\star$ for each combination where die 1 beats die 2.

	2	3	4	15	16	17
5	⋆	⋆	⋆			
6	⋆	⋆	⋆			
7	⋆	⋆	⋆			
8	⋆	⋆	⋆			
9	⋆	⋆	⋆			
18	⋆	⋆	⋆	⋆	⋆	⋆

Notice that there are 21 ways in which 1 beats 2, so the probability that die 1 beats die 2 is $\frac{21}{36} = \frac{7}{12} \approx 58.33\%$.

30.12 Here is a complete answer to (b). There are 13 choices for which value will be used in the triple, and for each such value, $\binom{4}{3} = 4$ choices for which cards will be used in that triple. Given the choice of the triple, there are 12 choices for which value will be used in the pair. Given the value, there are $\binom{4}{2} = 6$ choices for which cards we use in the pair. Thus, there are $13 \times 4 \times 12 \times 6 = 3744$ different full houses. Therefore, the probability of choosing a full house is

$$\frac{3744}{\binom{52}{5}} = \frac{3744}{2598960} = \frac{6}{4165} \approx 0.14\%.$$

The approximate numerical answers for the other parts are as follows:

(a) 2.11%, (c) 42.26%, (d) 4.75%, and (e) 0.198%.

30.13 By convention, an empty sum has value 0, so $P(\emptyset) = 0$.

That $P(S) = 1$ follows from the definition of sample space.

If $A \cap B = \emptyset$, then $P(A \cup B) = P(A) + P(B) - P(A \cap B) = P(A) + P(B) - P(\emptyset) = P(A) + P(B)$.

30.14 (a) $\binom{10}{5}/2^{10} = \frac{63}{256} \approx 24.61\%$.

(b) $2^7/2^{10} = 2^{-3} = \frac{1}{8} = 12.5\%$.

(c) $\binom{7}{2}/2^{10} = \frac{21}{1024} \approx 2.05\%$.

(d) By Proposition 30.7, the probability is

$$\frac{\binom{10}{5}}{2^{10}} + \frac{2^7}{2^{10}} - \frac{\binom{7}{2}}{2^{10}} = \frac{359}{1024} \approx 35.06\%.$$

30.15 This problem can get confusing, so it helps to have some good notation. Let A be the event that we see at least one 1, and let B be the event that we see at least one 2. The parts of this problem ask for the following:

(a) $P(\overline{A})$.

(b) $P(A) = 1 - P(\overline{A})$.

(c) $P(B)$ (which is the same as $P(A)$).

(d) $P(\overline{A} \cap \overline{B})$. Note that this is the same as $P(\overline{A \cup B})$.

(e) $P(A \cup B) = 1 - P(\overline{A \cup B})$.

(f) $P(A \cap B) = P(A) + P(B) - P(A \cup B)$.

30.16 Note that $(A \cap B) \cap (A \cap \overline{B}) = \emptyset$. Also note that $(A \cap B) \cup (A \cap \overline{B}) = A$. Therefore

$$\begin{aligned} P(A) &= P[(A \cap B) \cup (A \cap \overline{B})] \\ &= P(A \cap B) + P(A \cap \overline{B}) - P(\emptyset) \\ &= P(A \cap B) + P(A \cap \overline{B}). \end{aligned}$$

30.17 Here is the proof. Note that

$$P(A) = \sum_{s \in A} P(s) \quad \text{and} \quad P(B) = \sum_{s \in B} P(s).$$

Since $A \subseteq B$, every term in the first sum is also present in the second. Since probabilities are nonnegative, this implies that the second sum is at least as large as the first; that is, $P(B) \geq P(A)$. ■

30.18 Use proof by contradiction.

30.19 Use Proposition 30.8 and induction.

30.20 $P(A \cap \overline{A}) = P(\emptyset) = 0$. Interpretation: It is impossible for an event both to occur and not to occur.

30.21 $k = 57$.

31.1 Complete answer to (a): $P(A|B) = P(A \cap B)/P(B) = P(\{2,3\})/P(\{2,3,4\}) = 0.3/0.5 = 3/5 = 60\%$.

31.2 Here is a complete answer. Let A be the event that neither die shows a 2, and let B be the event that they sum to 7. Note that $P(B) = \frac{6}{36} = \frac{1}{6}$. Furthermore, $A \cap B = \{(1,6),(3,4),(4,3),(6,1)\}$, so $P(A \cap B) = \frac{4}{36} = \frac{1}{9}$. Thus $P(A|B) = \frac{P(A \cap B)}{P(B)} = \frac{1/9}{1/6} = \frac{6}{9} = \frac{2}{3}$.

31.3 This problem is not the same as the previous problem and has a different answer. In this problem you need to find $P(B|A)$, whereas in the previous problem you found $P(A|B)$. The answer is $P(B|A) = \frac{4}{25}$.

31.5 Nominally, you need to prove (1) $\iff$ (2), (1) $\iff$ (3), and (2) $\iff$ (3). However, it is enough to prove (1) $\Rightarrow$ (2) $\Rightarrow$ (3) $\Rightarrow$ (1). Or simpler yet, just prove (1) $\iff$ (3) because (2) $\iff$ (3) has an identical proof. These two imply (1) $\iff$ (2).

31.6 Disjoint events are not, in general, independent. For example, consider the roll of a die. Let A be the event we roll an even number and let B be the event we roll an odd number. Then $P(A \cap B) = 0 \neq P(A)P(B) = \frac{1}{4}$.

31.9 Two hints: First, $A \cap B$ and $A \cap \overline{B}$ are disjoint events, so $P(A \cap B) + P(A \cap \overline{B}) = P[(A \cap B) \cup (A \cap \overline{B})]$. Second, $(A \cap B) \cup (A \cap \overline{B}) = A \cap (B \cup \overline{B})$ by the distributive property.

31.10 The answer is yes in both cases. Use the formulas $P(A|B) = P(A \cap B)/P(B)$ and $P(B|A) = P(A \cap B)/P(A)$ to show why.

31.11 Yes. Suppose $P(A|B) > 0$. This says that $P(A \cap B)/P(B) > 0$ so $P(A \cap B) > 0$. Since $A \cap B \subseteq A$, we have (see Exercise 30.17) $P(A) \geq P(A \cap B) > 0$. ■

31.12 For the equation to make sense, we need the fact that $P(A) \neq 0$ (otherwise $P(B|A)$ is undefined). This follows from Exercise 30.17 because $A \subseteq A \cap B$, so $P(A) \geq P(A \cap B) > 0$. Now just use the definition of conditional probability.

31.13 (a) $P(A) = \frac{13}{52} = \frac{1}{4}$.
(b) $P(B) = \frac{4}{52} = \frac{1}{13}$.
(c) $P(A \cap B) = \frac{1}{52}$.
(d) Yes, because $P(A \cap B) = P(A)P(B)$.

31.14 You need to calculate $P(A)$, $P(B)$, and $P(A \cap B)$ and check if $P(A)P(B) = P(A \cap B)$. You should find that $P(A \cap B) = \frac{1}{221}$.

31.15 In principle, you need to calculate $P(A)$, $P(B)$, and $P(A \cap B)$ and check if $P(A)P(B) = P(A \cap B)$. However, for this problem, notice that $P(A \cap B) = 0$, but $P(A)P(B) \neq 0$. Therefore the events are *not* independent.

31.17 Both statements are true. For (a) use Exercise 30.16. For (b), use (a).

31.19 All three are *false*! Here is a counterexample for (a). Suppose the sample space is the pair-of-dice sample space of Example 29.4. Let the three events be as follows:
- A, the dice sum to 2; i.e., $A = \{(1,1)\}$,
- B, the dice sum to 17; i.e., $B = \emptyset$, and
- C, the dice sum to 12; i.e., $C = \{(6,6)\}$.

Note that A and B are independent (because $P(B) = 0$—see Exercise 31.18) and B and C are independent (again, because $P(B) = 0$). However, A and C are not independent because

$$P(A \cap C) = 0 \neq \frac{1}{36^2} = P(A)P(C).$$

31.20 $S^2 = \{(1,1), (1,2), (1,3), (1,4), (2,1), (2,2), (2,3), (2,4), (3,1), (3,2), (3,3), (3,4), (4,1), (4,2), (4,3), (4,4)\}$.

Their probabilities are as follows:

$$P[(1,1)] = \frac{1}{4} \qquad P[(1,2)] = \frac{1}{8}$$
$$P[(1,3)] = \frac{1}{16} \qquad P[(1,4)] = \frac{1}{16}$$
$$P[(2,1)] = \frac{1}{8} \qquad P[(2,2)] = \frac{1}{16}$$
$$P[(2,3)] = \frac{1}{32} \qquad P[(2,4)] = \frac{1}{32}$$
$$P[(3,1)] = \frac{1}{16} \qquad P[(3,2)] = \frac{1}{32}$$
$$P[(3,3)] = \frac{1}{64} \qquad P[(3,4)] = \frac{1}{64}$$
$$P[(4,1)] = \frac{1}{16} \qquad P[(4,2)] = \frac{1}{32}$$
$$P[(4,3)] = \frac{1}{64} \qquad P[(4,4)] = \frac{1}{64}$$

31.21 Let A be the event that the two spins sum to 6. As a set, $A = \{(2,4), (3,3), (4,2)\}$. Therefore

$$P(A) = \frac{1}{4} \cdot \frac{1}{8} + \frac{1}{8} \cdot \frac{1}{8} + \frac{1}{8} \cdot \frac{1}{4} = \frac{5}{64}.$$

31.23 (a) A = {HHTTT, HTHTT,HTTHT, HTTTH, THHTT, THTHT, THTTH, TTHHT, TTHTH, TTTHH}.
(b) $P(A) = 10p^2(1-p)^3$.

31.24 The set A contains $\binom{n}{h}$ sequences, all of which have the same probability.

31.25 Answer to (a): $p(1-p)$.
For (c), remember that $P(A|A \cup B) = \frac{P[A \cap (A \cup B)]}{P(A \cup B)}$.

31.26 Olivia.

31.27 (a) $a_0 = 0$ and $a_{2n} = 1$.
(b) $a_k = pa_{k+1} + qa_{k-1}$ (where $q = 1-p$).
(c) There is a formula for a_k of the form $a_k = c_1 + c_2 s^k$ where c_1, c_2, and s are specific numbers and $s \neq 1$.
Use part (b) to find s and use part (a) to find c_1, c_2.

32.1 Complete answer for (a): "$X > 3$" is the set

$$\{s \in S : X(s) > 3\} = \{c, d\}$$

and $P(X > 3) = 0.7$.

Guidance for (c): The event "$X > Y$" is the set $\{s \in S : X(s) > Y(s)\}$. Which of a, b, c, and d are in this set?

Hint for (f): Is it true that $P(X = m \wedge Y = n) = P(X = m)P(Y = n)$ for all integers m and n?

32.2 Here is a complete solution.

(a) Let (S, P) be the sample space for the spinner, so $S = \{1, 2, 3, 4\}$. Then $X : S \to \mathbb{Z}$ is defined by $X(1) = 10$, $X(2) = 20$, $X(3) = 10$, and $X(4) = 20$.

(b) The event "$X = 10$" is the set $\{1, 3\}$.

(c) $P(X = 10) = \frac{1}{2} + \frac{1}{8} = \frac{5}{8}$ and $P(X = 20) = \frac{3}{8}$. For all other integers a (i.e., $a \neq 10$ and $a \neq 20$), we have $P(X = a) = 0$.

32.3 The answer to (c) is $\frac{4}{8} = \frac{1}{2}$.

32.4 The answer to (a) is 3. $P(X = 1) = \frac{10}{36} = \frac{5}{18}$. $P(X = -2) = 0$.

32.6 If $a < 0$ or $a > 10$ then $P(X = a)$ is zero. Otherwise, this is just like a binomial random variable where the probability of success is $\frac{1}{6}$.

32.7 No. Note that $P(X_H = 1) = P(X_T = 1) > 0$, but $P(X_T = 1 \wedge X_H = 1) = 0 \neq P(X_H = 1)P(X_H = 1)$.

32.8 Calculate $P(X_1 = 5)$, $P(X_2 = 5)$, and $P(X_1 = 5 \text{ and } X_2 = 5)$.

32.9 Yes. Let a be any value in the set

$$\{2, 3, 4, 5, 6, 7, 8, 9, 10, J, Q, K, A\}$$

and let b be any value in the set $\{\clubsuit, \diamondsuit, \heartsuit, \spadesuit\}$. Note that $P(X = a) = \frac{1}{13}$ and $P(Y = b) = \frac{1}{4}$. Finally,

$$\begin{aligned} P(X = a \wedge Y = b) &= \frac{1}{52} \\ &= \frac{1}{13} \times \frac{1}{4} \\ &= P(X = a)P(Y = b). \end{aligned}$$

Therefore X and Y are independent random variables.

32.10 Calculate $P(X = 2)$, $P(Y = 2)$, and $P(X = Y = 2)$.

33.1 $E(X) = 1 \times 0.1 + 3 \times 0.2 + 5 \times 0.3 + 8 \times 0.4 = 5.4$.

33.2 $E(X) = \frac{13}{3}$, $E(Y) = 0$, and $E(Z) = \frac{13}{3}$.

33.4 Let (S, P) be the sample space for a single die; i.e., $S = \{1, 2, 3, 4, 5, 6\}$. Let $X(s) = s^2$. Find $E(X)$.

33.5 Let X_1 be the number on the first chip and X_2 be the number on the second. So $X = X_1 + X_2$. Note that $E(X_1) = E(X_2) = (1 + 2 + \cdots + 100)/100 = 50.5$, so $E(X) = 101$.

33.6 Answer to (d): By symmetry, yes.

Answer to (e): Since $100 = E(Z) = E(X_H + X_T) = E(X_H) + E(X_T)$ and since $E(X_H) = E(X_T)$, we clearly have $E(X_H) = E(X_T) = 50$.

33.7 Let X be a zero-one random variable. Then $E(X) = 0 \cdot P(X = 0) + 1 \cdot P(X = 1) = P(X = 1)$.

33.8 Note that $1^2 = 1$ and $0^2 = 0$.

33.9 Express X as the sum of n zero-one indicator random variables and apply linearity of expectation.

33.11 Apply Proposition 33.4.

33.13 Earlier we showed that $E(X) = 5.4$. We can calculate

$$\begin{aligned} \text{Var}(X) &= E[(X - 5.4)^2] \\ &= (1 - 5.4)^2 \cdot 0.1 + (3 - 5.4)^2 \cdot 0.2 \\ &\quad + (5 - 5.4)^2 \cdot 0.3 + (8 - 5.4)^2 \cdot 0.4 \\ &= (-4.4)^2 \cdot 0.1 + (-2.4)^2 \cdot 0.2 \\ &\quad + (-0.4)^2 \cdot 0.3 + (2.6)^2 \\ &= 5.84. \end{aligned}$$

Alternatively, we can use the formula $\text{Var}(X) = E(X^2) - E(X)^2$. We have

$$\begin{aligned} E(X^2) &= 1^2 \cdot 0.1 + 3^2 \cdot 0.2 + 5^2 \cdot 0.3 + 8^2 \cdot 0.4 \\ &= 35. \end{aligned}$$

and so $\text{Var}(X) = E(X^2) - E(X)^2 = 35 - 5.4^2 = 5.84$.

33.15 Use the formula $\text{Var}(Z) = E(Z^2) - E(Z)^2$. For the second part, see Exercise 33.13.

33.16 Use Exercise 33.15.

33.17 Use Markov's inequality (Exercise 33.12).

34.1 The answer to (b) is $q = -34$ and $r = 2$.

34.2 The answer to (b) is $-100 \text{ div } 3 = -34$ and $-100 \bmod 3 = 2$.

34.4 Read carefully the first sentence of Definition 34.6.

34.5 This is the more difficult half of the proof. (⇒) Suppose $a \equiv b \ (n)$. This means that $n|(a-b)$, or, equivalently, $a - b = kn$ for some integer k. If we divide a and b by n, we get

$$a = qn + r$$
$$b = q'n + r'$$

with $0 \le r, r' < n$. Note that $r = a \bmod n$ and $r' = b \bmod n$.

If we subtract these equations, we get

$$a - b = (q - q')n + (r - r')$$

and since $a - b = kn$, we can rewrite this as

$$kn = (q - q')n + (r - r')$$
$$\Rightarrow \quad r - r' = (k - q + q')n$$

so $r - r'$ is a multiple of n. But r and r' are between 0 and $n - 1$ so their difference is no more than $n - 1$. Thus we must have $r - r' = 0$; i.e., $r = r'$. Since $r = a \bmod n$ and $r' = b \bmod n$, we have

$$a \bmod n = b \bmod n.$$

34.6 Bad idea: Call the three consecutive integers a, b, and c.

Good idea: Call the three consecutive integers a, $a + 1$, and $a + 2$.

34.9 Here is a good definition for part (a): Let p and q be polynomials. We say that p divides q (and we write $p|q$) provided there is a polynomial r such that $q = pr$.

35.1 The answer to (d) is $\gcd(-89, -98) = 1$.

35.2 The answer to (d) is $(-89)(11) + (-98)(-10) = 1$.

35.4 Try a few examples.

35.5 Use Proof Template 14.

35.6 Yes, they are still correct. Explain the equality $\gcd(a, b) = \gcd(b, c)$ in this case.

35.7 It is enough to prove that if $a \ge b > 0$, then $b \ge a \bmod b$.

35.9 (a) A complete answer: The greatest common divisor of three integers, a, b, and c, is an integer d with the following two properties: (1) $d|a$, $d|b$, and $d|c$, and (2) if $e|a$, $e|b$, and $e|c$, then $e \le d$.

(b) The phrase a, b, c are *pairwise relatively prime* means that $\gcd(a, b) = \gcd(a, c) = \gcd(b, c) = 1$.

35.10 Use Corollary 35.9.

35.11 Try to find integers X and Y such that $X(2a + 1) + Y(4a^2 + 1) = 1$; the integers X and Y will depend on a.

35.12 Use proof by contradiction.

35.13 Use the fact that we can find integers x, y such that $ax + by = 1$. Therefore $c = cax + cby$.

35.14 Use Corollary 35.9.

35.15 Use Corollary 35.9.

35.16 Reverse the roles of a, b and x, y.

35.17 Explain why we can take $b > 0$ and then choose b to be as small as possible (thereby invoking the Well-Ordering Principle).

35.18 Number the children from 0 to $n - 1$ and imagine the teacher starts by patting child 0's head first.

Note that if $k = 4$ and $n = 10$, the teacher will never pat the heads of the odd-numbered children.

However, if $k = 3$ and $n = 10$, then the children will be patted in the order 0, 3, 6, 9, 2, 5, 8, 1, 4, and then 7, so all children will be patted.

35.19 $5 \times 13 - 8 \times 8 = 1$.

36.1 Some answers: (a) 6. (g) 6. (n) 7.

36.2 The order of operations for modular arithmetic is the same as that of ordinary arithmetic, so we do $\otimes$ and $\oslash$ before $\oplus$ and $\ominus$.

Although the first three of these can be done by the guess-and-check method, the fourth is not amenable to such a brute force attack. In each case, you need to compute a reciprocal in $\mathbb{Z}_n$. You do this using the extended Euclidean Algorithm.

36.3 Because the coefficient of x in each of these problems is noninvertible, the normal method for solving these equations won't work. For these, I recommend you resort to guess-and-check. It is possible that there are no solutions.

36.4 Use guess-and-check. The answer to (d) is 2, 7, 8, and 13.

36.6 Use the facts that

$$a \oplus b = (a + b) \bmod n, \quad \text{and}$$
$$a \ominus b = (a - b) \bmod n.$$

The first is from Definition 36.1, and the second is from Proposition 36.7.

36.7 Use Theorem 34.1.

36.9 You should assume that $a \ominus b = (a-b) \bmod n$, and you need to prove that $b \oplus (a \ominus b) = a$.

36.10 The answer is that

$$a \otimes b = 0 \iff a = 0 \text{ or } b = 0$$

is a theorem if and only if n is prime.

The structure of the proof is a bit complicated.

First, suppose that n is prime and then prove that

$$a \otimes b = 0 \iff a = 0 \text{ or } b = 0.$$

Second, suppose n is not prime and prove that

$$a \otimes b = 0 \iff a = 0 \text{ or } b = 0$$

is false.

36.11 This is, actually, an easy problem. You need to prove that the inverse of a^{-1} is a. Read Definition 36.9 slowly and carefully.

36.12 Here is a complete answer to (a). False. Counterexample: Note that in $\mathbb{Z}_5$ both 2 and 3 are invertible ($2^{-1} = 3$ and $3^{-1} = 2$), however, $2 \oplus 3 = 0$ is *not* invertible.

36.14 To calculate 3^{32}, you can first find 3^{16} and then calculate $3^{32} = 3^{16} \otimes 3^{16}$.

37.3 Here is a complete solution to (a).

We know first that $x \equiv 4 \ (5)$. This means we can write

$$x = 4 + 5k$$

where k is an integer. We substitute this into the second equation $x \equiv 7 \ (11)$ and we have

$$4 + 5k \equiv 7 \ (11) \quad \Rightarrow \quad 5k \equiv 3 \ (11).$$

To solve $5k \equiv 3 \ (11)$, we multiply both sides by 5^{-1} in $\mathbb{Z}_{11}$. Using the extended GCD method, we have $5^{-1} = 9$. So we multiply both sides by 9:

$$9 \otimes 5 \otimes k = 9 \otimes 3 = 5$$

so $k \equiv 5 \ (11)$. This means we can write k as

$$k = 5 + 11j$$

for some integer j. Substituting this back into $x = 4 + 5k$ we get

$$x = 4 + 5k = 4 + 5(5 + 11j) = 29 + 55j$$

and so we see that $x \equiv 29 \ (55)$.

For (c), solve the first two equivalences to obtain an intermediate answer of the form $x \equiv ? \ (28)$ and then solve the system

$$x \equiv ? \ (28) \quad \text{and} \quad x \equiv 8 \ (25)$$

by the usual method.

For (d), first simplify the two equations so they are both of the form $x \equiv ? \ (?)$.

37.7 Here is a complete solution for (a).

Let $b_1 = 8^{-1} = 12$ in $\mathbb{Z}_{19}$. Let $b_2 = 19^{-1} = 3^{-1} = 3$ in $\mathbb{Z}_8$. Thus

$$\begin{aligned} x_0 &= m_1 b_1 a_2 + m_2 b_2 a_1 \\ &= 8 \cdot 12 \cdot 2 + 19 \cdot 3 \cdot 3 \\ &= 363. \end{aligned}$$

Note that $363 \bmod 8 = 3$ and $363 \bmod 19 = 2$, as required.

Since $8 \times 19 = 152$, we can reduce x_0 modulo 152 to give $363 \bmod 152 = 59$. Note that $59 \bmod 8 = 3$ and $59 \bmod 19 = 2$.

A complete answer to this problem is $x = 59 + 152k$ where $k \in \mathbb{Z}$. However, we can also write $x = 363 + 152k$ where $k \in \mathbb{Z}$. This is exactly the same answer; it is just expressed in a different form.

38.1 Here is a good way to begin your proof: "Suppose, for the sake of contradiction, there is a composite integer n all of whose factors (other than 1) are greater than $\sqrt{n}$...."

38.2 Answer to (b): $4200 = 2^3 \cdot 3 \cdot 5^2 \cdot 7$.

38.3 The generalization is: Let p and q be unequal primes. Then $p|x$ and $q|x$ if and only if $(pq)|x$.

38.5 Begin by factoring a and b (uniquely) into primes.

38.8 Note that for any two numbers s and t, we have

$$s + t = \min[s, t] + \max[s, t].$$

38.11 Call the consecutive perfect squares a^2 and $(a+1)^2$ (where a is an integer) and suppose (for the sake of contradiction) that there is a prime p that divides them both.

38.12 Notice that $18 = 2^1 \cdot 3^2$, so every positive divisor of 18 is of the form $2^a 3^b$ where $0 \le a \le 1$ and $0 \le b \le 2$. Hence there are 2 choices for a and 3 choices for b giving $2 \times 3 = 6$ positive divisors.

38.13 The divisors of n are 2^k and $2^k(2^a - 1)$ for all $0 \le k < a$.

38.14 It is easier to figure out how many numbers between 1 and n are *not* relatively prime to n.

The answer to (f) is $\varphi(5041) = 4970$. To see why, note that the only numbers between 1 and 71^2 that are *not* relatively prime to 71^2 are the multiples of 71: 71, $2 \times 71, 3 \times 71, \ldots, 71 \times 71$. So there are $5041 - 71 = 4970$ integers from 1 to 71^2 that *are* relatively prime to 71^2.

38.16 Use inclusion-exclusion. Let A_i denote the set of multiples of p_i between 1 and n.

38.18 The following sentence is useful: If n is not a perfect square, then there must be a prime p that appears an odd number of times in n's prime factorization.

38.20 Let $x = \log_2 3$. This means that $2^x = 3$. Suppose $x = \frac{a}{b}$ for some integers a and b, and argue to a contradiction.

38.22 Here is a complete answer for (c).

Suppose $w, z \in \mathbb{Z}[\sqrt{-3}]$. This means that $w = a + b\sqrt{-3}$ and $z = c + d\sqrt{-3}$ where $a, b, c, d \in \mathbb{Z}$. Notice that

$$\begin{aligned} wz &= (a + b\sqrt{-3})(c + d\sqrt{-3}) \\ &= (ac - 3bd) + (ad + bc)\sqrt{-3} \end{aligned}$$

and since $ac - 3bd$ and $ad + bc$ are integers, we have $wz \in \mathbb{Z}[\sqrt{-3}]$.

Here is a hint for (d). If $w = a + b\sqrt{-3}$, then

$$w^{-1} = \frac{a}{a^2 + 3b^2} + \frac{-b}{a^2 + 3b^2}\sqrt{-3}.$$

Try to deduce: For which integers a and b are $\frac{a}{a^2+3b^2}$ and $\frac{-b}{a^2+3b^2}$ also integers?

38.23 For (a), let $w = a + b\sqrt{-3}$ and $z = c + d\sqrt{-3}$. To prove that $N(wz) = N(w)N(z)$, just expand everything in terms of a, b, c, and d, and be careful with the algebra.

Here is a partial answer for (b).

There are no $w \in \mathbb{Z}[\sqrt{-3}]$ with $N(w) = 2$. Proof: Suppose $N(a+b\sqrt{-3}) = a^2 + 3b^2 = 2$. If $b \neq 0$ we have $N(w) \ge 3$, so $b = 0$. This leaves us with $a^2 = 2$, which is impossible since $a \in \mathbb{Z}$.

There are exactly six possible values for w with $N(w) = 4$.

38.24 First prove the following lemma: *If $N(w)$ is prime, then w is irreducible.*

38.25 If the statement were false, we could find a counterexample w with $N(w)$ as small as possible.

38.26 We want to write $4 = ab$ with $a, b \neq \pm 2, \pm 1$. Taking norms of both sides of $4 = ab$, we get $16 = N(4) = N(ab) = N(a)N(b)$. We cannot have $N(a)N(b) = 2 \times 8$ because there is no element with norm 2. So we must have $N(a) = N(b) = 4$.

39.3 Answer for $\wedge$: The operation $\wedge$ is closed, commutative, and associative, and TRUE is an identity element. However, FALSE does not have an inverse. Therefore, $(\{\text{TRUE}, \text{FALSE}\}, \wedge)$ is not a group.

39.4 Evaluate: $a * b * c$.

39.5 To show that X and Y are inverses, you just need to show that $X*Y = Y*X = e$. For this problem, you need to show that the inverse of g^{-1} is g. Your answer should be very short.

39.8 Dust off your linear algebra text, and reread the material on the determinant of a matrix.

39.9 Remember that 2^A stands for the set of all subsets of A and that Δ is the symmetric difference operation.

To prove that $(2^A, \Delta)$ is a group, prove the four properties: closure, associativity, identity, and inverses. Note: One of these has already been proved.

39.10 This means that you must prove that $f: G \to G$ is one-to-one and onto. Here is a skeleton of the proof.

First we prove that f is one-to-one. Suppose $f(g) = f(k)$ for some $g, k \in G$. ... Therefore $g = k$, and so f is one-to-one.

Second we prove that f is onto. Let $b \in G$. Let $x \in G$ be defined by .... Therefore $f(x) = b$, and so f is onto.

Thus f is a permutation.

39.11 See the hint for Exercise 39.10.

39.12 Exercise 39.10 is useful here. Note that the row of the $*$ table corresponding to element a contains all elements of the form $a * g$ where g is an arbitrary member of G.

39.13 Remember: To prove that X and Y are inverses, show that $X * Y = Y * X = e$.

39.15 Check that $\star$ satisfies the four requisite properties. Notice that identity and inverses for $*$ and $\star$ are the same.

39.16 Use proof by contradiction. Let G be a group with an even number of elements, and suppose e is the only element that is its own inverse....

39.17 This exercise is similar to Exercise 39.12.

39.18 The answer to (c) is 15; here's why. The expression we need to evaluate is $1, 2, +, 3, 4, \times, +$. The $1, 2, +$ portion evaluates to 3 and the $3, 4, \times$ portion evaluates to 12. The expression has been reduced to $3, 12, +$ and that evaluates to 15.

39.19 The answer to (a) is $2, 3, +, 4, 5, +, \times$.

39.20 Here is a theorem you should prove:

Let ℓ be a list of numbers and operations. We claim that ℓ is a valid RPN expression if and only if the following two conditions hold: (1) the number of operations in ℓ is one less than the number of numbers, and (2) the sublists of ℓ, starting from the beginning of ℓ and including all members of ℓ up to any point in the list, must contain more numbers than operations.

Prove both directions of this if-and-only-if theorem by induction.

40.1 You need to find a one-to-one and onto function

$$f : \{0, 1, 2, \ldots, 9\} \to \{1, 2, 3, \ldots, 11\}$$

with the property that

$$(x + y) \bmod 10 = z \iff [f(x) \times f(y)] \bmod 11 = f(z).$$

Begin with $f(1) = 2$. From there you can work out $f(1 + 1)$, etc.

40.2 Let $f[(1, 1)] = 1$.

40.3 One of these groups is cyclic; the other is not.

40.4 Consider $f(e * e)$.

40.5 To prove that $f(g)$ and $f(g^{-1})$ are inverses, $\star$ them together and hope you get the identity.

40.7 Here is an outline of the proof. Fill in the blanks.

Let $f : G \to H$ be an isomorphism.

($\Rightarrow$) Suppose $(G, *)$ is Abelian. Let $x, y \in H$ be arbitrary.... Therefore $x \star y = y \star x$, and so $(H, \star)$ is Abelian.

($\Leftarrow$) Suppose $(H, \star)$ is Abelian.... Therefore $(G, *)$ is Abelian.

40.9 The Klein 4-group is defined in Section 39. It has four elements: $(0, 0)$, $(0, 1)$, $(1, 0)$, and $(1, 1)$.

Recall that the set $2^{\{1,2\}}$ is the set of all subsets of $\{1, 2\}$ and that Δ is symmetric difference.

40.10 Define a map $F : G \to H$ by $F(a) = f_a$. Prove that F is a bijection and that $F(a * b) = F(a) \circ F(b)$.

40.11 The generators of $(\mathbb{Z}_{10}, \oplus)$ are 1, 3, 7, and 9. Notice that these are exactly the elements of $\mathbb{Z}_{10}$ that are relatively prime to 10.

If you can prove your answer, the teacher will give you a pat on the head.

40.13 Here is the answer for $\mathbb{Z}_5^*$. In $\mathbb{Z}_5^*$, note that $2^1 = 2$, $2^2 = 4$, $2^3 = 3$, and $2^4 = 1$. Therefore 2 is a generator for $\mathbb{Z}_5^*$.

41.1 The subgroups of $(\mathbb{Z}_6, \oplus)$ are $\{0\}$, $\{0, 3\}$, $\{0, 2, 4\}$, and $\{0, 1, 2, 3, 4, 5\}$.

41.2 There are three subgroups.

41.3 There are five subgroups.

41.4 Don't miss a key hypothesis: $H \neq \emptyset$.

You are given
(1) H is closed under $*$.
(2) $H \neq \emptyset$.
(3) For all $g \in H$, $g^{-1} \in H$.
You need to show
(a) H is closed under $*$.
(b) $e \in H$.
(c) For all $g \in H$, $g^{-1} \in H$.
Proving (a) and (c) is trivial. So you need to show how (b) follows from (1), (2), and (3).

41.5 You are given that (a) $H \neq \emptyset$ and (b) for all $g, h \in H$, $g * h^{-1} \in H$.

You should prove: (1) H is closed under $*$, (2) $e \in H$, and (3) if $g \in H$, then $g^{-1} \in H$.

Prove these in the order (2), then (3), then (1).

41.6 Let H be a subgroup of $(\mathbb{Z}, +)$. Think about the least positive element of H (if any).

41.8 The easiest equivalence class to find is $[1] = [e] = H$.

To find other equivalence classes, use the idea from the proof of Lemma 41.7. Choose an element $g \in G$ and define a function $f : H \to [g]$ by $f(h) = h*g$. To compute $[g]$, just compute $f(h)$ for all $h \in H$.

For this problem, $[2] = \{1, 6, 11, 16, 21\}$. There are three other equivalence classes.

41.9 See the previous hint.

41.10 See Proposition 40.3.

41.11 Only one of these is true.

41.12 Remember that in a general group, $x \equiv y \pmod{H}$ means $x * y^{-1} \in H$. In this problem, the operation $*$ is ordinary addition of integers, and the inverse of $y \in \mathbb{Z}$ is simply $-y$.

41.13 Do *not* use an Abelian group.

41.14 The hardest part of this problem is swallowing the definition of $g * H$. Remember that $g * H$ is a *set*. Also, $x \in g * H$ *means* that $x = g * h$ for some $h \in H$. (Likewise, $x \in H * g$ means that there is an $h \in H$ such that $x = h * g$.)

For part (b), start by proving $g * H = H \iff g \in H$. The forward ($\Rightarrow$) direction is not hard [use part(a)]. For the reverse ($\Leftarrow$) direction, you need to show that two sets ($g * H$ and H) are equal, so use Proof Template 5.

For part (d), consult your answer to the previous exercise.

42.1 I recommend using a calculator or a computer. You should find that $a^{13} = a$ for all $a \in \mathbb{Z}_{13}$.

42.5 If you do trial division, it will take tens of thousands of divisions to find the prime factors of the composite number. Compute $2^n \bmod n$ for both values of n, and see what you get.

There is a technical difficulty in working with these large integers. Expressed in binary, they are about 30 bits long; these numbers can fit nicely in your computer. However, multiplying two 30-bit numbers gives a 60-bit product; it may be difficult—using an ordinary computer programming language—for you to deal with numbers this large.

42.6 You will need to write a computer program to solve this problem. The smallest answer to this problem is under 1000.

43.1 Many computer languages have built-in functions for converting text to and from ASCII.

44.2 Try to use this formula for a prime $p \not\equiv 3 \pmod{4}$, such as $p = 17$. What goes wrong?

44.3 To check yourself, the answers are 33, 157, 556, and 432.

44.4 There are eight answers. Note that $34751 = 19 \times 31 \times 59$ and $19 \equiv 31 \equiv 59 \equiv 3 \pmod{4}$.

44.5 Factoring n into primes gives $n = 45343 \times 7243$. Note that both of these primes are congruent to 3 modulo 4. Let $M = 249500293$.

In $\mathbb{Z}_{45343}$ we have

$$249500293^{(45343+1)/4} = 12690.$$

So in $\mathbb{Z}_{45343}$, $\sqrt{M} = \pm 12690 = 12690$ or 32653.

In $\mathbb{Z}_{7243}$ we have

$$249500293^{(7243+1)/4} = 2663$$

so in $\mathbb{Z}_{7243}$ we have $\sqrt{M} = \pm 2663 = 2663$ or 4580.

We solve the following four problems using the Chinese Remainder Theorem.

$$x \equiv 12690 \pmod{45343}$$
$$x \equiv 2663 \pmod{7243}$$

$$x \equiv 32653 \pmod{45343}$$
$$x \equiv 2663 \pmod{7243}$$

$$x \equiv 12690 \pmod{45343}$$
$$x \equiv 4580 \pmod{7243}$$

$$x \equiv 32653 \pmod{45343}$$
$$x \equiv 4580 \pmod{7243}$$

The solutions to these four problems are $x \equiv 111103040$, $x \equiv 7151504$, $x \equiv 321267845$, and $x \equiv 217316309$ (all mod n). The second one gives `07 15 15 04`, which spells `GOOD`.

44.6 For (a) send more than one message. The concern for (b) is that $M^2 \bmod n = M^2$ (there is no "wrapping" modulo n, and so Eve can find M by taking an ordinary square root). Figure out strategies for dealing with this.

44.7 $\gcd(75406 - 68918, 171121) = \gcd(6488, 171121) = 811$.

44.9 (a) Hint: quadratic formula. (b) Answer: 281.

45.1 Answer: $D(N) = N^{377} \bmod 589$.

45.2 Remember: $d = e^{-1}$ in $\mathbb{Z}^*_{\phi(n)}$.

45.3 The answer is $M = 100$.

45.4 The decryption exponent is $d = e^{-1} \pmod{\varphi(n)}$. The answer to (a) is `PIGS`.

46.1 Answer to (a): $(\{1,2,3,4,5,6\}, \{\{1,2\}, \{1,4\}, \{2,3\}, \{2,5\}, \{3,6\}, \{4,5\}, \{5,6\}\})$.

46.2 Answer for (a):

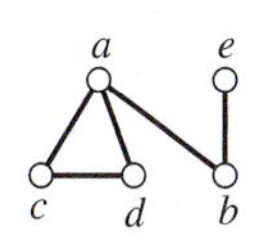

46.4 If you are having trouble with this one, I recommend you take a break and order a pizza. Before you eat, take a good look at the slices.

46.9 There is a solution with only one student.

46.10 Do you think these are impossible? I assure you they are not. What is wrong with the following "proof" that in no graph is the is-adjacent-to relation antisymmetric?

Let G be a graph. Let uv be an edge of G. Then $u \sim v$ and $v \sim u$, but $u \neq v$. Therefore $\sim$ is not antisymmetric. Thus, in no graph is $\sim$ antisymmetric. "■"

Read this "proof" carefully. When you spot the error, you will know how to answer this problem.

46.12 Use Theorem 46.5.

46.13 Use proof by contradiction. If the vertices all have different degrees, what must they be?

46.15 Use the previous problem.

46.16 See Section 16.

46.17 Note that the following are two different graphs:

$$G_1 = (\{1,2,3,4\}, \{12, 13, 14\}) \quad \text{and}$$
$$G_2 = (\{1,2,3,4\}, \{12, 23, 24\})$$

even though their drawings look very much the same. (They have the same vertex sets, but different edge sets.)

I recommend you begin with the special cases $n = 0$, $n = 1$, $n = 2$, $n = 3$ and, $n = 4$ before you dive into the general case. Try writing down all the possibilities. Note, however, that for $n = 4$, you should find 64 different graphs, so you will need to be organized.

46.18 For (a), use Proposition 23.25.

Here is a complete answer to (b): Let $v \in V(G)$. Note that u is adjacent to v if and only if $f(u)$ is adjacent to $f(v)$. Since f is a bijection, this gives a one-to-one correspondence between the neighbors of v and the neighbors of $f(v)$. Therefore, v and $f(v)$ have the same degree (in their respective graphs).

47.1 The graphs for (a), (d), and (g) are shown here:

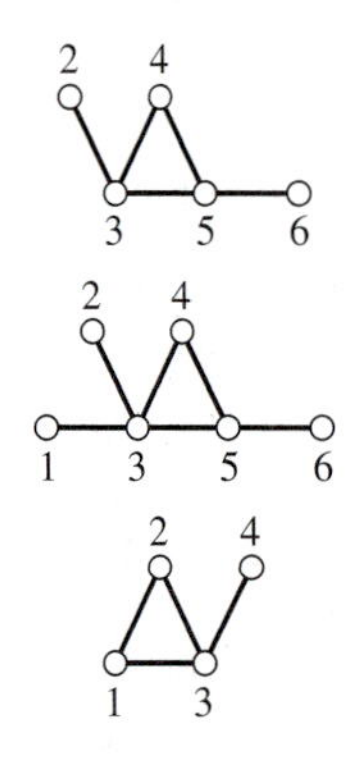

47.2 Here is a complete proof that is-a-subgraph-of is antisymmetric.

Suppose G is a subgraph of H and that H is a subgraph of G. The first implies that $V(G) \subseteq V(H)$ and $E(G) \subseteq E(H)$ and the second implies the reverse containments. Thus $V(G) = V(H)$ and $E(G) = E(H)$, and therefore $G = H$. Thus is-a-subgraph-of is antisymmetric.

47.3 For example, the graph K_2 has two spanning subgraphs and 4 induced subgraphs. We list them all out here. Let a and b be the vertices of K_2.

The spanning subgraphs of K_2 are the following two graphs:

$$(\{a, b\}, \{ab\}) \quad \text{and} \quad (\{a, b\}, \emptyset).$$

The induced subgraphs of K_2 are the following four graphs:

$$\begin{array}{ll} (\{a, b\}, \{ab\}) & (\{a\}, \emptyset) \\ (\{b\}, \emptyset) & (\emptyset, \emptyset). \end{array}$$

If you thought there were only three induced subgraphs of K_2, your answer would be wrong, but not terrible. The graphs $(\{a\}, \emptyset)$ and $(\{b\}, \emptyset)$ have exactly the same drawing (just one dot), but these are not the exact same graphs. (In one case, the sole vertex is a, and in the other case the sole vertex is b.) This fussiness actually makes this problem *easier* to solve. There is a sense in which the graphs $(\{a\}, \emptyset)$ and $(\{b\}, \emptyset)$ are "the same"; see Exercise 46.18.

For K_3, there are eight spanning and eight induced subgraphs.

47.4 $\alpha(G) = 3, \omega(G) = 2$.

47.6 Exactly one of these statements is true. Prove the true statement and find counterexamples for the other three.

47.7 For (b), see the following figure.

Self-Complementary Graph
Artist: Laura Tateosian

For (c), note that if G is self-complementary, then we know that G and $\overline{G}$ must have the same number of edges.

47.10 (d) Let G be any graph on n vertices. Note that $\overline{G}$ is also a graph on n vertices. Since $n \to (a, b)$, we know that $\alpha(\overline{G}) \geq a$ or $\omega(\overline{H}) \geq b$. By Proposition 47.12,

$$\alpha(G) = \omega(\overline{G}) \geq a, \quad \text{or}$$
$$\omega(G) = \alpha(\overline{G}) \geq b.$$

(e) By Proposition 47.13, if $n \geq 6$, then $n \to (3, 3)$. By Exercise 47.8, $5 \not\to (3, 3)$. If $n < 5$ and $n \to (3, 3)$, then by part (c) we would have $5 \to (3, 3)$.$\Rightarrow\Leftarrow$ Therefore 6 is the least positive integer n such that $n \to (3, 3)$.

(f) Let G be a graph on ten vertices. Let v be any vertex of G.

If $d(v) \geq 6$, then among v's six (or more) neighbors we can either find (a) a clique of size 3 or (b) an independent set of size 3 (by Proposition 47.13). In the first case (a), $\omega(G) \geq 4$ because v, together with its three pairwise adjacent neighbors, forms a clique of size 4. In the second case (b), we clearly have $\alpha(G) \geq 3$.

Otherwise ($d(v) \not\geq 6$), we have $d(v) \leq 5$. This means there are (at least) four vertices w, x, y, z to which v is *not* adjacent. If they form a clique, we have $\omega(G) \geq 4$. But otherwise, some pair of them are not adjacent and (together with v) form an independent set of size 3, so $\alpha(G) \geq 3$.

In every case, $\alpha(G) \geq 3$ or $\omega(G) \geq 4$. Therefore $10 \to (3, 4)$.

(g) Hint: Consider an arbitrary vertex v. Either $d(v) \geq m$ or v has at least n vertices to which it is not adjacent.

48.4 The answer to (a) is $\binom{n}{2}$. For (b), note that all the vertices have the same degree.

For (c), to prove that a G_n is connected, pick any two arbitrary vertices and prove there is a path that connects them.

48.5 $\forall\exists$ is not the same as $\exists\forall$.

48.6 Here's a hint:

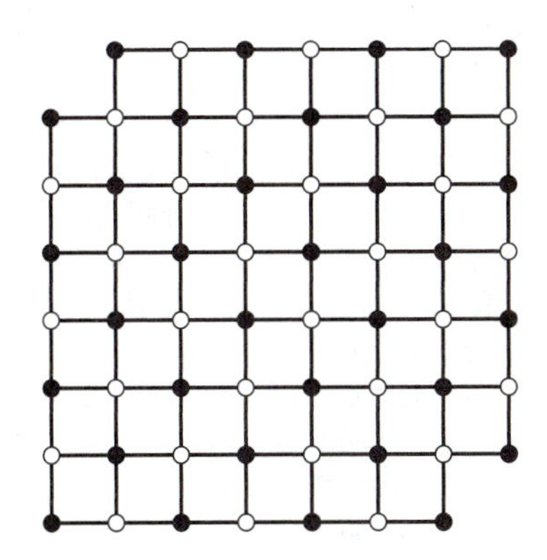

48.9 Here is a detailed outline of a proof. Fill in the blanks.

Suppose G is disconnected. We must show that $\overline{G}$ is connected.

Because G is disconnected, there exist vertices x and y in different components of G.

Now consider any two vertices a and b in $\overline{G}$. We consider cases depending on which component(s) of G contain(s) a and b.

- *a and b are both in x's component of G.*
  ... therefore, there is an (a, b)-path in $\overline{G}$.
- *One of a or b is in x's component, and the other is not.*
  ... therefore, there is an (a, b)-path in $\overline{G}$.
- *Neither a nor b is in x's component.*
  ... therefore, there is an (a, b)-path in $\overline{G}$.

In all cases, there is an (a, b)-path in $\overline{G}$. Since a and b were arbitrarily chosen vertices, $\overline{G}$ is connected.

48.10 Begin your proof thus: Suppose, for the sake of contradiction, G is not connected. Let H be a component of G with the fewest vertices.

48.12 Use induction on k. The i, j-entry of A^{t+1} can be expressed as

$$\sum_{k=1}^{n} [A^t]_{i,k} A_{k,j}.$$

49.3 Since this problem asks you to prove an if-and-only-if statement, you have two jobs. First, suppose $d_1, \ldots, d_n$ are the degrees of the vertices of some tree T. You must show that $d_1 + \cdots + d_n = 2n - 2$; this is fairly easy.

Your second, and more challenging, task is the following: Suppose you are given positive integers $d_1, \ldots, d_n$ for which $d_1 + \cdots + d_n = 2n - 2$. You must prove there is a tree on n vertices whose degrees are precisely $d_1, \ldots, d_n$. For this, we recommend induction. Begin by showing that at least one of the d_i is equal to 1.

49.5 Here is an outline for your proof.

> Let G be a graph in which every pair of vertices is joined by a unique path. We want to show that G is a tree. By Definition 49.3, we must show that G is *connected* and *acyclic*.
> - Claim: G is connected. *Prove that G is connected by direct proof.*
> - Claim: G is acyclic. *Prove that G is acyclic by contradiction.*
>
> Therefore G is a tree. ■

49.6 Your answer should use the word *vacuously*.

49.7 This problem is best done as a proof by contrapositive (Proof Template 11):

Suppose v is not a leaf.... Therefore $T - v$ is not a tree.

49.8 The answer to (d) is $(n-1)!$. Prove this by induction.

49.9 The formula is $n - c$.

49.10 Use Exercise 49.4.

49.11 For (a), you also should use Theorem 48.12.

49.12 Use Theorem 49.4.

49.13 Use Theorems 49.9 and 49.11 and the previous problem.

49.14 For (c), use Exercise 49.13.

49.15 This is an if-and-only-if statement, so be sure to do both directions. Here is half of the proof.

($\Rightarrow$) Let $e = xy$ be a cut edge of G. Suppose, for contradiction, e is contained in a cycle C. Since e is a cut edge of G, there must be vertices a and b that are connected in G but not connected in $G - e$. Let P be an (a, b)-path in G. Necessarily, P contains the edge e. Without loss of generality, vertex x precedes vertex y as we traverse P from a to b.

Notice that in $G - e$ there is an (a, b)-walk: Start at a, traverse P up to x, traverse $C - e$ to b, and then traverse P to y. By Lemma 48.7, there must be an (a, b)-path in $G - e$.$\Rightarrow\Leftarrow$ Therefore e is not contained in any cycle of G.

49.17 Prove that at step (4), the graph T is connected and acyclic.

49.18 Just as for the previous problem, prove that the output graph (the final T) is connected and acyclic. Exercise 49.15 will help.

50.2 Does K_n have an Euler tour?

50.3 Add edges from this new vertex in a way that changes all vertices to even degree. Then check that the new graph you created is connected.

50.4 This question is not so simple.

51.2 The condition that G is 3-colorable means that G can be properly colored using *at most* three colors. We are not required to use all three.

51.3 See Section 47.

51.4 If G has n vertices and is not complete, then $n > 1$. Furthermore, G must contain two vertices that are not adjacent to each other.

51.7 Given a proper coloring of G with a colors and a proper coloring of $\overline{G}$ with b colors, show how to construct a proper coloring of K_n with ab colors.

51.8 First show that G is properly 4-colorable (this is easy). Next, suppose that G is 3-colorable and argue to a contradiction. Give a color to vertex 1 and then discuss the colors of the other vertices in numerical order.

51.11 Color the large-degree vertex first.

51.12 Prove this by induction on the number of vertices in G.

51.13 For (a), color the vertices with four colors.

For (b), please note that you need to prove the graph is *not* 3-colorable. To do this, you should use proof by contradiction. It is also helpful to give sensible names to the vertices. Call the vertex in the center of the picture u. Call its five neighbors a_1 through a_5 and the corresponding five vertices on the outer rim of the picture b_1 through b_5.

For (c), although this graph has many edges, you should use symmetry to reduce this problem to just a few cases. Color each of those graphs with just three colors.

52.1 There are ∞-ly many answers.

52.3 You may use Euler's formula (Theorem 52.3) and/or mimic its proof.

52.4 If G does not contain K_3 as a subgraph, then every face must have degree at least 4.

52.5 Count edges (use Corollary 52.5).

52.8 Although the graph looks a bit like K_5, it does not contain a subdivision of K_5 as a subgraph. (If it did, it would have a vertex of degree at least 4.) Find a subdivision of $K_{3,3}$ as a subgraph.

52.9 For (a), mimic the proof of Corollary 52.5.

For (c), use induction.

52.10 Part (a) has already been proved in the text. The quantity vr is the sum of the degrees and equals $2e$. The quantity fs is the sum of the degrees of the faces and also equals $2e$.

For part (b) and (c), use Euler's formula (Theorem 52.3).

For part (d), note that e must be a positive integer.

Part (e) is a bit trickier. The case $(3, 3)$ was done in part (b). The cases $(3, 4)$ and $(4, 3)$ are not too bad. Try $(3, 5)$ next; remember, the unbounded face also has degree 5. You should be able to calculate how many degree-5 faces you need (answer: 12). Finally, the case $(5, 3)$ is the most complicated. You need to fit 20 triangles (degree-3 faces) together with 5 triangles meeting at every vertex. Good luck!

52.11 Draw a graph on the surface of the soccer ball. Place one vertex in each polygon and join vertices by an edge if their polygons abut each other. Notice that this is a planar graph in which every face is a 3-cycle and all vertices have degree 5 or 6.

Suppose there are a vertices of degree 5 and b vertices of degree 6. Count the number of edges in two ways to derive $a = 12$.

53.1 (a) a and b are incomparable. (e) $c < i$.

53.2 For (a), the height is 4 and there are several chains containing four elements, including $\{b, d, f, i\}$ and $\{a, c, f, j\}$.

For (c), note that $\{a, c, i\}$ is a chain containing three elements, but it can be extended to $\{a, c, f, i\}$. So $\{a, c, i\}$ is not a correct answer for (c).

53.5 Prove that R^{-1} is reflexive, antisymmetric, and transitive.

53.7 Here is the full proof. Suppose, for the sake of contradiction, that there are two elements x and y with $x < y$ and $x > y$. Unraveling these definitions, we have (1) $x \leq y$, (2) $y \leq x$, and (3) $x \neq y$. However, $x \leq y$ and $y \leq x$ imply (by antisymmetry) that $x = y$, contradicting (3).$\Rightarrow\Leftarrow$ Therefore we cannot have both $x < y$ and $x > y$.

53.11 Here is a good definition for $P - x$. Let $P = (X, \leq)$. Let $P - x$ be the poset with ground set $X - \{x\}$ and relation $\leq'$ where $a \leq' b$ if and only if $a \leq b$ for all $a, b \in X - \{x\}$.

54.1 There are no maximum or minimum elements. There are three maximals and three minimals.

54.2 Answer to (b): 1 is minimum and minimal. 3, 4, and 5 are maximal. There is no maximum.

54.4 Statement (d) is false.

Statement (g) is true. Here is a complete proof:

Suppose x and y are distinct maximal elements. Suppose, for the sake of contradiction, that they are not incomparable. Then either $x < y$ or $y < x$. If $x < y$, then x is not maximal, and if $y < x$, then y is not maximal.$\Rightarrow\Leftarrow$ Therefore x and y are incomparable.

55.2 For (a), note that $1 < 2 < 3$ and $1 < 3 < 2$ are different linear orders on $\{1, 2, 3\}$, but they are isomorphic.

55.3 Here is a template for your proof.

Let a be a minimal element of a total order P. Let x be any other element of P.... Therefore $a < x$, and so a is a minimum element.

55.5 Here is *half* the proof of (a):

($\Rightarrow$) Suppose x is minimum in P. To show that $f(x)$ is minimum in Q, we need to show that if b is any element of Q, then $f(x) \leq b$ in Q. Since f is onto, there is an a in P with $f(a) = b$. Since x is minimum in P, $a \geq x$. Thus $b = f(a) \geq f(x)$ (since f is order-preserving). Therefore $f(x)$ is minimum in Q.

55.6 Use part (a) of the previous problem.

56.2 The answers are (a) 4, (b) 14400, and (c) 252. You must supply the explanations.

56.4 By the previous problem, if (x, y) is *not* a critical pair, then $\leq'$ is not transitive. This happens because there is an $a \leq x$ and a $b \geq y$ (other than $a = x$ and $b = y$) with $a \not\leq b$.

Let $u(a)$ denote the number of elements strictly above a and let $\ell(a)$ denote the number of elements strictly below a.

Prove that an incomparable pair (x, y) with $\ell(x) + u(y)$ as small as possible is critical.

56.5 Note that (c, g) is *not* a critical pair because $a < c$ but $a \not< g$. Also (g, c) is *not* a critical pair because $c < f$ but $g \not< f$. However, (a, g) is a critical pair because any $x < a$ is also below g (vacuously) and any $y > g$ (namely, $y = j$) is also above a.

57.1 The dimension is 2, but finding a realizer takes some work.

57.2 Use Theorem 56.3.

57.4 Here is a complete proof.

Suppose P is a subposet of Q. Choose a realizer $\mathcal{R} = \{L_1, L_2, \ldots, L_t\}$ of Q of minimum size. (Thus $t = \dim Q$.)

Let L'_i be the suborder of L_i restricted to the elements of P. Note that for all elements x and y of P we have $x \leq y$ (in P) iff $x \leq y$ (in Q) iff $x \leq_i y$ (in L_i for all i) iff $x \leq_i y$ (in L'_i for all i). Thus $\mathcal{R}' = \{L'_1, L'_2, \ldots, L'_t\}$ is a realizer for P, and so $\dim P \leq t = \dim Q$. ■

57.5 This is a difficult problem. First show that $\dim P \leq 3$ by constructing a realizer using 3 linear extensions. It helps to work out the special cases $n = 3$ and $n = 4$ first and then to look for a general pattern.

To show that $\dim P > 2$ is tricky. Here is a technique that uses counting: Show that there are $n(n-2)$ incomparable pairs of the form a_i-b_j. Then show that a linear extension can have $a_i > b_j$ at most $\binom{n-1}{2}$ times. So if $\dim P \leq 2$, we would need to have $n(n-2) \leq 2\binom{n-1}{2}$, and this leads to a contradiction.

58.1 (b) d. (d) i. (e) a.

58.3 Prove the $\Rightarrow$ part by observing that for any x, y in a linear order we have one of $x < y$, $x = y$, or $x > y$. For the $\Leftarrow$ direction, use contradiction and consider an incomparable pair x, y.

58.4 Note that this result requires all numbers involved to be in $\mathbb{N}$—that is, no negative integers allowed. Therefore $x|y \Rightarrow x \leq y$.

58.5 The poset $(\mathbb{Z}, \leq)$ is a lattice with no maximum and no minimum element. The poset $(\mathbb{N}, |)$ is a lattice with no maximum element (but it does have a minimum element, 1).

To make the sentence true, insert the word *finite*.

58.7 Show the following is not distributive.

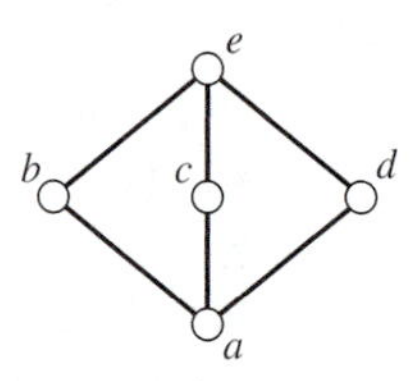

58.8 Start with the meet and join of two distinct points. Their meet is the empty set, and their join is the unique line that contains both of them. Now continue to consider the meet and join of two lines, or of a point and a line.

B Solutions to Self Tests

Chapter 1

1. False. The positive integer 1 is neither prime nor composite.
2. $x|(x+2)$ for x equal to -2, -1, 1, and 2.
3. The $|$ notation means that the number on the left divides the number on the right. The expression $(a|b)+1$ is nonsense because $a|b$ is a statement and 1 is a number; it is not possible to add a statement and an integer!
4. "If an integer is perfect, then it is even" or "if x is a perfect integer, then x is even."
5. "If you will marry me, then you love me."
6. (a) false, (b) false, (c) true, (d) false, (e) true, and (f) true (vacuously).
7. First line: "Suppose M is a graphic matroid."

   Last line: "Therefore, M is representable." ■
8. (a) There are many possible counterexamples, including $x=3$, $y=2$, and $z=-6$. Note that $x>y$ but $xz=-18<-12=yz$.

   (b) If we require z to be positive, the conclusion will follow. The edited statement should read, "If x, y, and z are integers, $x>y$, and $z>0$, then $xz>yz$.
9. (a) This is false. For example, $2|10$ and $5|10$, but $7=2+5$ does not divide 10.

   (b) This is true, and here is a proof. Suppose $a|b$. Then there is an integer x such that $ax=b$. Multiplying both sides of this equation by c gives $axc=bc$, which can be rewritten $(ac)x=bc$. Therefore $ac|bc$. ■
10. Most two-digit numbers are counterexamples to the proposition. For example, $15^2=225$ but $51^2=2601\neq 522$. Therefore the proposition is false.

    The mistake in the proof is that we neglected the effect of carrying in arithmetic. It is true that $(10a+b)^2=(a^2)\times 100+(2ab)\times 10+(b^2)\times 1$, but that does not imply that the digits of $(10a+b)^2$ are $a^2, 2ab, b^2$. For example, if $b>3$, then $b^2>10$ and so the ones digit of $(10a+b)$ cannot be b^2.
11. The proof is incorrect because it assumes what we wish to prove and then works to an obvious known fact (i.e., $0=0$). The approach should be the opposite.

    Here is, essentially, the same proof used to show $a=b$ for any two numbers a and b.

    Proof. Start with
    $$a=b$$
    from which we also have
    $$b=a.$$
    Multiplying these equations together gives
    $$ab=ab$$
    and then canceling ab from both sides gives
    $$0=0$$
    which is correct. ■

    From such a "proof" follows the result that $3=4$. Clearly this is incorrect.
12. The two expressions are not logically equivalent. Consider the following truth table:

x	y	$x\to\neg y$	$\neg(x\to y)$
T	T	F	F
T	F	T	T
F	T	T	F
F	F	T	F

    Since the columns for $x\to\neg y$ and $\neg(x\to y)$ are not identical, the two expressions are not logically equivalent.
13. The expression $(x\to y)\vee(x\to\neg y)$ is a tautology. Consider the following truth table:

x	y	$x\to y$	$x\to\neg y$	$(x\to y)\vee(x\to\neg y)$
T	T	T	F	T
T	F	F	T	T
F	T	T	T	T
F	F	T	T	T

Since the formula evaluates to T for all possible values of its variables, it is a tautology.

14. Let $a, a+1$, and $a+2$ be three consecutive integers. Their sum is $a + (a+1) + (a+2) = 3a + 6$. Note that $3a + 6 = 3(a+2)$. Since $a+2$ is an integer, $3|(3a+6)$. Thus the sum of any three consecutive integers is divisible by 3. ■

15. Let a be a positive integer. The sum of a consecutive integers is divisible by a if and only if a is odd.

16. Let $a \geq 3$ be an integer. Multiplying both sides by a gives $a^2 \geq 3a$. Note that $3a = 2a + a > 2a + 1$ because $a > 1$. Thus $a^2 \geq 3a > 2a + 1$. ■

17. Suppose a is a perfect square and $a \geq 9$. Because a is a perfect square, there is an integer b with $a = b^2$. We may assume that $b > 0$. In order for $a \geq 9$, we must have $b \geq 3$. Observe that $a - 1 = b^2 - 1 = (b-1)(b+1)$. Since $b \geq 3$, we know that $b - 1 \geq 2 > 1$ and $b + 1 \geq 4 > 1$, and so these factors of $a - 1$ are both greater than 1 hence $a - 1$ is composite. ■

18. The definition of square mates applies only to positive integers. Although $10 + (-1)$ is a perfect square, -1 is not positive. Therefore 10 and -1 are not square mates.

19. Let x be a positive integer. Let $y = x^2 + x + 1$; clearly $y > x$ because we have added a positive quantity $(x^2 + 1)$ to x.

    Note that $x + y = x + (x^2 + x + 1) = x^2 + 2x + 1 = (x+1)^2$. Since $x + 1$ is an integer, $x + y$ is a perfect square, and therefore x and y are square mates.

20. For $x = 5, 6, 7, 8, 9$ it is easy to find square mates smaller than x, to wit:

$$(5, 4),\ (6, 3),\ (7, 2),\ (8, 1),\ \text{and } (9, 7).$$

Thus it is enough to prove the result for $x > 9$.

As allowed by the problem statement, choose a positive integer a such that $a^2 \leq x < (a+1)^2$. Since $x > 9$, clearly $a \geq 3$.

Let $y = (a+1)^2 - x$. Clearly $x + y$ is a perfect square and $y > 0$ because $(a+1)^2 > x$. Thus x and y are square mates. It remains to show that $y < x$. To this end we calculate

$$\begin{aligned} y &= (a+1)^2 - x \\ &< (a+1)^2 - a^2 \\ &= 2a + 1 \\ &< a^2 \qquad \text{by Problem 16} \\ &\leq x. \end{aligned}$$

■

Chapter 2

1. There are $2 \times 26 \times 26 = 1352$ ways to form a 3-letter call sign and $2 \times 26 \times 26 \times 26 = 35152$ ways to form a 4-letter call sign. Adding these gives 36,504 possible call signs.

2. Notice that we can choose a and b arbitrarily and then choose c carefully so that $a+b+c$ is even. More specifically, there are 10 choices for a and, for each such choice, 10 choices for b and then, once a and b have been chosen, exactly 5 choices for c. This gives $10 \times 10 \times 5 = 500$ possible choices.

    Alternatively, without the restriction on the sum, there are $10^3 = 1000$ choices for a, b, c exactly half of which have even sum, giving $1000 \div 2 = 500$ choices.

3. There are 10^3 ways to choose a, b, c regardless of their product. In order for abc to be odd, all three of a, b, c must be odd. There are $5^3 = 125$ ways that might happen. Thus there are $1000 - 125 = 875$ ways to choose a, b, c such that abc is even.

4.
$$\frac{20!}{17! \cdot 3!} = \frac{20 \cdot 19 \cdot 18}{3 \cdot 2 \cdot 1} = 20 \cdot 19 \cdot 3 = 1140.$$

5. There are 13! ways to arrange the cards within a given suit, so there are $13!^4$ possible ways to order the cards with suits. Then there are 4! ways to order the suits. All together, this gives $4! \times 13!^4$ possible arrangements.

    There is no need to calculate this any further, but if you did, you should get the following result.

3608548172171337597466673456087040000000

6. The ten couples may appear in 10! orders. For each such order, there are 2 choices per couple, depending on whether a wife is in front of her husband or vice versa. This gives a total of $10!2^{1}0$ possible arrangements.

7. The answer is 0 since the first term in the product is $\frac{0^2}{0+1} = 0$.

8. We can write A as $\{-9, -8, \ldots, 8, 9\}$ and so $|A| = 19$.

9. (a) is TRUE and (b)–(e) are FALSE.

10. (a) TRUE. *Proof:* Suppose $X \in 2^{A \cap B}$. Then $X \subseteq A \cap B$, and hence $X \subseteq A$ and $X \subseteq B$. Therefore $X \in 2^A$ and $X \in 2^B$, and so $X \in 2^{A \cap B}$.
On the other hand, suppose $X \in 2^{A \cap B}$. Then $X \subseteq A \cap B$, and so $X \subseteq A$ and $X \subseteq B$. Therefore $X \in 2^A$ and $X \in 2^B$, so $X \in 2^A \cap 2^B$.
Because we have shown $X \in 2^{A \cap B} \iff X \in 2^A \cap 2^B$, we have $2^{A \cap B} = 2^A \cap 2^B$. ■

(b) FALSE. *Counterexample:* Let $A = \{1, 2\}$ and $B = \{3, 4\}$. Note that $2^{A \cup B} = 2^{\{1,2,3,4\}}$ contains 16 elements, but $2^A \cup 2^B$ contains $4 + 4 = 8$ elements. So $2^{A \cup B} \neq 2^A \cup 2^B$.

(c) FALSE. *Counterexample:* Let A and B be any sets. We know that $\emptyset \in 2^{A \Delta B}$. However, since $\emptyset \in 2^A$ and $\emptyset \in 2^B$, we have that $\emptyset \notin 2^A \Delta 2^B$.

11. (a) FALSE, (b) TRUE, (c) TRUE, and (d) TRUE.

12. (a) FALSE, (b) FALSE, (c) TRUE, and (d) TRUE.

13. Statement (a) is not necessarily true; for example, if $p(x, y)$ is always true, this would be false.
Statement (b) must be true based on the rules for negating quantified statements (and the fact that $\neg[\neg p(x, y)]$ is logically equivalent to $p(x, y)$).
Statement (c) must also be true. If the statement $\exists y,\ p(x, y)$ is true for all possible integers x, then certainly it is true for some integer x.

14. Given that $A \times B = \{(1, 2), (1, 3), (2, 2), (2, 3)\}$, it must be the case that $A = \{1, 2\}$ and $B = \{2, 3\}$. Then $A \cup B = \{1, 2, 3\}$, $A \cap B = \{2\}$, and $A - B = \{1\}$.

15. The following figure gives a Venn diagram illustration of $(A - C) \cup (B - C) = (A \cup B) - C$. Notice that if we combine the shaded regions $A - C$ (upper left) and $B - C$ (upper right), we have the shaded region $(A \cup B) - C$ (bottom).

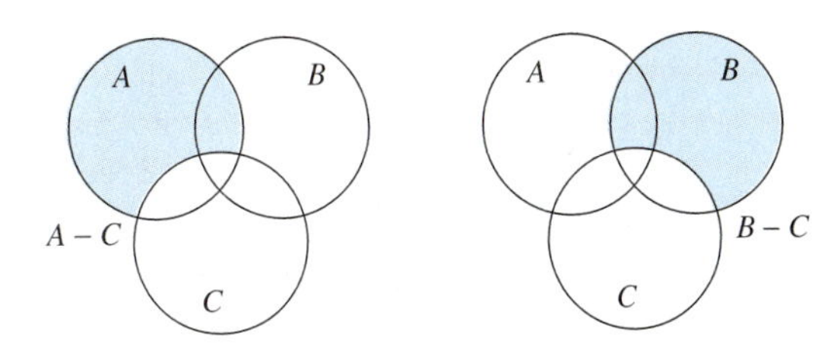

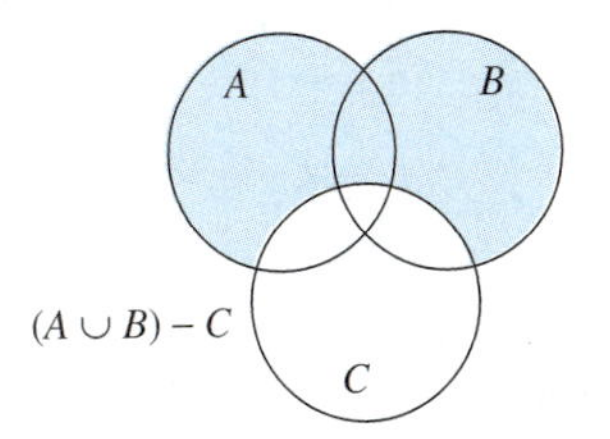

Now for a standard proof.
Suppose $x \in (A - C) \cup (B - C)$. This means that $x \in A - C$ or $x \in B - C$. If $x \in A - C$, then $x \in A$ and $x \notin C$. Since $x \in A$, we certainly have $x \in A \cup B$. And as $x \notin C$, we may conclude $x \in (A \cup B) - C$. Likewise, if $x \in B - C$, we conclude $x \in (A \cup B) - C$. Thus if $x \in (A - C) \cup (B - C)$, then $x \in (A \cup B) - C$.
On the other hand, suppose $x \in (A \cup B) - C$. This means $x \in A \cup B$ and $x \notin C$. Because $x \in A \cup B$, we know that $x \in A$ or $x \in B$. In case $x \in A$, since $x \notin C$, we have $x \in A - C$, and so $x \in (A - C) \cup (B - C)$. Likewise, if $x \in B$ we derive that $x \in (A - C) \cup (B - C)$. Therefore, if $x \in (A \cup B) - C$, then $x \in (A - C) \cup (B - C)$.
Since $x \in (A - C) \cup (B - C)$ iff $x \in (A \cup B) - C$, we have that $(A - C) \cup (B - C) = (A \cup B) - C$. ■

16. From $|A| + |B| = |A \cup B| + |A \cap B|$ we have the equation $10 + |B| = 15 + 3$, whence $|B| = 8$.

17. $((A \cup B) - (A - B)) - (B - A)$.

18. We ask: How many length-3 lists can be formed using n elements?

On the one hand, there are n^3 such lists.

On the other hand, the three elements on the list might be (a) all different, (b) two the same and one different, or (c) all the same. In case (a), there are $(n)_3 = n(n-1)(n-2)$ such lists. In case (b), there are n choices for the repeated element, $(n-1)$ choices for the nonrepeated element, and 3 choices for the slot the nonrepeated element can occupy, for a total of $3n(n-1)$ lists. Finally, there are n lists in which all three elements are the same. Summing these, we find the answer to the question is $n(n-1)(n-2)+3n(n-1)+n$.

Since these are both correct answers to the same question, we must have

$$n^3 = n(n-1)(n-2) + 3n(n-1) + n.$$

■

Chapter 3

1. (a) This is the set containing your children.
   (b) This is the set containing your parents.
   (c) R is irreflexive (no one is their own parent) and antisymmetric (vacuously since $x\ R\ y$ and $y\ R\ x$ cannot hold for any x, y). The other properties (reflexive, symmetric, and transitive) do not hold.
   (d) R^{-1} is the is-the-child-of relation.

2. The relations in (a) and (b) are equivalence relations; it is easy to see they are transitive, symmetric, and reflexive.

The relation in (c) is not an equivalence relation. Although it is reflexive and symmetric, it is not transitive. For example, suppose Alice and Bob have a son George (g), Bob and Cindy have a son Harry (h), and Cindy and Dave have a daughter Inga (i). Then $g\ R\ h$ and $h\ R\ i$ hold, but $h\ R\ i$ is false.

3. A relation R on A is a subset of $A \times A$. In other words, $R \subset A \times A$ or, equivalently, $R \in 2^{A \times A}$. Since $|A \times A| = 4 \cdot 4 = 16$, the cardinality of $2^{A \times A}$ is 2^{16}, and that is our answer. There are 2^{16} relations defined on A.

4. No. For example, take $x = 2$ and $y = 112$. Then $x \equiv y$ in both mod 10 and 11.

5. (a) R is reflexive: if x is any integer, then clearly $|x| = |x|$. R is symmetric: if $x\ R\ y$, then $|x| = |y|$, hence $|y| = |x|$ and so $y\ R\ x$. R is transitive: if $x\ R\ y$ and $y\ R\ z$, then $|x| = |y|$ and $|y| = |z|$ and so $|x| = |z|$. Therefore $x\ R\ z$. Therefore R is an equivalence relation.
   (b) $[5] = \{-5, 5\}$, $[-2] = \{-2, 2\}$, and $[0] = \{0\}$.

6. The equivalence classes are $[1] = [2] = [3] = \{1, 2, 3\} = A$ and $[4] = [5] = \{4, 5\} = B$.

7. There are 6 equivalence classes depending on the cardinality of the sets (from 0 to 5).

8. Suppose x and y are integers. Then $x \stackrel{\mathcal{P}}{\equiv} y$ if and only if x and y have the same sign.

Note that the sign of an integer x is often denoted sgn x and is defined by

$$\operatorname{sgn} x = \begin{cases} 1 & \text{if } x > 0, \\ 0 & \text{if } x = 0, \text{ and} \\ -1 & \text{if } x < 0. \end{cases}$$

9. The answer is $10!2^{10}/20$; here's why. Imagine the couples first stand in line. There are $10!2^{10}$ ways for them to do this in which husbands and wives are next to their respective spouses. See Problem 6 from Chapter Test 2 (page 81).

Once they are lined up, they sit around the table (say, in clockwise order). Two of these seating arrangements are equivalent if one is a rotation of the other. Each equivalence class has 20 seating patterns thus there are $10!2^{10}/20$ equivalence classes.

10. The number of ordinary anagrams of ELECTRICITY is $11!/(2^3)$ because the word is eleven letters long and includes two each of E, C, and T. For each such anagram, there are 10 ways to insert a space to create a two-word anagram. Hence the answer is $10 \cdot 11!/8$.

11. Let us call the three types of squares on a tic-tac-toe board *corner, side,* and *center*. We count the possibilities depending on the first player's move.
   - The first player puts an X in a corner square.
     In this case the second player has five distinct responses: near corner, far corner, near edge, far edge, center.
   - The first player puts an X in an edge square.
     In this case again the second player has five distinct responses: near corner, far corner, near edge, far edge, center.
   - The first player puts an X in the center.
     In this case the second player has two distinct responses: corner, edge.

   Therefore there are 12 distinct (inequivalent) opening pair of moves in tic-tac-toe.

12. The answer is $21!/(2^{10} \cdot 10!)$; here's why. Imagine the students stand in line, and then the teacher picks the lab partners by choosing two students at a time from the line. The last student in line works alone. Two arrangements of the line that yield the same pairing are considered equivalent.

   There are $21!$ different ways for the students to line up. The size of all the equivalence classes is $10! \cdot 2^{10}$ because the first ten pairs may be rearranged and the students within a pair may swap their positions.

   Therefore there are $21!/(2^{10} \cdot 10!)$ inequivalent ways for the students to line up, and this gives the number of pairings.

13. The answer is $\binom{50}{10}$ because we are choosing a 10-element subset of $A' = \{1, 3, 5, \ldots, 99\}$ and $|A'| = 50$.

14. By the binomial theorem, the x^{17} term is $\binom{50}{17}x^{17}2^{50-17}$, so the answer is $\binom{50}{17}2^{33}$.

15. The problem can be rewritten

$$(n+0) + (n+1) + (n+2) + \cdots + (n+n)$$

   which equals $n^2 + \binom{n+1}{2}$.

   Optionally, this may be further simplified to $(3n^2 + n)/2$.

16. The team can be chosen in $\binom{200}{15}$ ways, and for each such choice, there are $\binom{15}{2}$ ways to pick the co-captains, for a total of $\binom{200}{15}\binom{15}{2}$ possible ways to pick the team and co-captains.

17. *Combinatorial proof:* Let N be a finite set with $|N| = n + 2$ and suppose two elements of N are considered weirdos. How many $k+2$-element subsets of N can be formed?

   On the one hand, the answer is simply $\binom{n+2}{k+2}$.

   On the other hand, we can consider how many weirdos are in the set: zero, one, or two. There are $\binom{n}{k}$ ways to choose a k-element subset that contains both weirdos, $2\binom{n}{k+1}$ ways with one weirdo, and $\binom{n}{k+2}$ ways with neither weirdo, for a total of $\binom{n}{k} + 2\binom{n}{k+1} + \binom{n}{k+2}$.

   Therefore, $\binom{n+2}{k+2} = \binom{n}{k} + 2\binom{n}{k+1} + \binom{n}{k+2}$.

   Proof via Pascal's Identity: Applying Pascal's Identity to $\binom{n+2}{k+2}$ we have $\binom{n+2}{k+2} = \binom{n+1}{k+1} + \binom{n+1}{k+2}$. Applying Pascal's Identity to each of these gives

$$\binom{n+1}{k+1} = \binom{n}{k} + \binom{n}{k+1}$$
$$\binom{n+1}{k+2} = \binom{n}{k+1} + \binom{n}{k+2}$$

   and then adding these two equations yields $\binom{n+2}{k+2} = \binom{n}{k} + 2\binom{n}{k+1} + \binom{n}{k+2}$.

18. For (a) the answer is $\binom{10}{4}$ and for (b) the answer is $\left(\binom{10}{4}\right)$.

19. The answer is 4^n. Each potential element j (with $1 \le j \le n$) may appear in the multiset 0, 1, 2, or 3 times. Let m_j be the multiplicity of element j. Instead of counting multisets directly, we can count lists of the form $(m_1, m_2, \ldots, m_n)$ where each $m_j \in \{0, 1, 2, 3\}$. Thus there are 4 choices for each element of the list, for a total of 4^n lists.

20. Let A_j be the set of colorings in which row j is entirely of one color. The set of "bad" colorings is $A_1 \cup \cdots \cup A_4$. The number of "good" colorings is $2^{16} - |A_1 \cup \cdots \cup A_4|$.

We evaluate this as follows:

$$\begin{aligned}\text{answer} &= 2^{16} - |A_1 \cup \cdots \cup A_4| \\ &= 2^{16} - \sum_i |A_i| + \sum_{i<j} |A_i \cap A_j| \\ &\quad - \sum_{i<j<k} |A_i \cap A_j \cap A_k| \\ &\quad + |A_1 \cap A_2 \cap A_3 \cap A_4| \\ &= 2^{16} - \binom{4}{1} \cdot 2 \cdot 2^{12} + \binom{4}{2} \cdot 4 \cdot 2^8 \\ &\quad - \binom{4}{3} \cdot 8 \cdot 2^4 + \binom{4}{4} \cdot 16.\end{aligned}$$

This equals $\binom{4}{0}16^4(-2)^0 + \binom{4}{1}16^3(-2)^1 + \binom{4}{2}16^2(-2)^2 + \binom{4}{3}16^1(-2)^3 + \binom{4}{4}16^0(-2)^4$, which simplifies to $(16-2)^4$ by the Binomial Theorem.

Alternatively, there are $16 - 2$ ways to color each row, so there are 14^4 possible colorings.

Chapter 4

1. Suppose, for the sake of contradiction, that $x^2 + 1 = 0$ has a real root, a. Since a is a real number, we must have one of $a < 0$, $a = 0$, or $a > 0$, but in every case $a^2 \geq 0$. Therefore $a^2 \neq -1$, so $a^2 + 1 \neq 0$.$\Rightarrow\Leftarrow$ ■

2. Suppose, for the sake of contradiction, there are four consecutive integers, $a, a+1, a+2, a+3$, whose sum is divisible by 4. That is, $a+(a+1)+(a+2)+(a+3) = 4a+6$ is divisible by 4. Therefore, there is an integer b such that $4b = 4a + 6$, giving $b - a = \frac{6}{4}$ or $b - a - 1 = \frac{1}{2}$. Note that $b - a - 1$ is an integer but $\frac{1}{2}$ is not.$\Rightarrow\Leftarrow$ ■

3. We are given that $a|b$ and $b|a$. So there exist integers x and y with $ax = b$ and $by = a$. Multiplying these together gives $abxy = ab$. Since $a, b > 0$, we know that $ab \neq 0$, so dividing by ab gives $xy = 1$. The only pairs of integers that multiply to give 1 are $(x, y) = (1, 1)$ and $(x, y) = (-1, -1)$. The latter is impossible because if $a = (-1)b$, then a and b cannot both be positive. Therefore $(x, y) = (1, 1)$ and so $a = b$. ■

4. Sets (b), (c), (d), and (f) are well-ordered, and the others are not.

5. The proof is by induction on n. The case $n = 1$ (basis case) is obvious because both sides evaluate to 1.

   Suppose (induction hypothesis) that the result is true when $n = k$; that is,

$$1 + 4 + 7 + \cdots + (3k - 2) = \frac{3k^2 - k}{2}.$$

   Adding $3(k+1) - 2 = 3k + 1$ to both sides gives

$$\begin{aligned}1 + 4 + \cdots + (3k - 2) + (3k + 1) \\ = \frac{3k^2 - k}{2} + 3k + 1.\end{aligned}$$

   Note that

$$\begin{aligned}\frac{3k^2 - k}{2} + 3k + 1 &= \frac{3k^2 - k + 6k + 2}{2} \\ &= \frac{3k^2 + 5k + 2}{2}\end{aligned}$$

$$\begin{aligned}\frac{3(k+1)^2 - (k+1)}{2} &= \frac{3k^2 + 6k + 3 - k - 1}{2} \\ &= \frac{3k^2 + 5k + 2}{2}\end{aligned}$$

   and so

$$1 + 4 + \cdots + [3(k+1) - 2] = \frac{3(k+1)^2 - (k+1)}{2}$$

   as required. ■

6. The proof is by induction on n. The basis case, $n = 0$, holds because both sides of the inequality evaluate to 1.

   Assume (induction hypothesis) that the inequality has been proved for $n = k$; that is, $0! + 1! + \cdots + k! \leq (k+1)!$. Adding $(k+1)!$ to both sides gives $0! + 1! + \cdots + k! + (k+1)! \leq 2 \cdot (k+1)!$.

   Note that $(k+2)! = (k+2) \cdot (k+1)! \geq 2 \cdot (k+1)!$ (because $k + 2 \geq 2$ since $k \geq 0$). Therefore $0! + 1! + \cdots + (k+1)! \leq (k+2)!$ as required. ■

7. The proof is by induction on n. The basis case, $n = 0$, is true because $a_0 = 1$ and $(2 \cdot 4^0 + 1)/3 = 3/3 = 1$.

Suppose the result has been proved when $n = k$ (i.e., $a_k = (2 \cdot 3^k + 1)/3$). Now consider a_{k+1}. We know that $a_{k+1} = 4a_k - 1$, and so

$$a_{k+1} = 4a_k - 1 = 4\left[\frac{2 \cdot 4^k + 1}{3}\right] - 1$$
$$= \frac{4 \cdot 2 \cdot 4^1 + 4 - 3}{3} = \frac{2 \cdot 4^{k+1} + 1}{3}$$

as required. ■

8. The proof is by induction on n. The basis case, $n = 0$, is true since $0 < 2^0$.

   Suppose (induction hypothesis) that $k < 2^k$. We must show that $k+1 < 2^{k+1}$. To this end, we add 1 to both sides of $k < 2^k$ to find $k + 1 < 2^k + 1 \le 2^k + 2^k = 2^{k+1}$. ■

9. *Proof by contradiction:* Suppose P is a finite set of points in which any three points are collinear but, for the sake of contradiction, the points do not all lie on a common line. Choose a line L that includes two points, say x and y, in P. Since L does not contain all the points in P, there is a third point $z \in P$ that is not on L. But then x, y, z are three points of P that are not collinear. $\Rightarrow\Leftarrow$ Therefore, all points in P lie on a common line. ■

   Proof by induction: The proof is by induction on the number of points (i.e., $|P|$). In the case where the set has only 3 points, the result is obvious.

   Suppose the proposition has been proved for all sets of k points. Let P be a set of $k+1$ points that satisfies the hypothesis of the proposition. Let a be any point in P, and let $P' = P - \{a\}$. Then P' also satisfies the hypothesis of the proposition, and so, by induction, the points in P' lie on a common line L. Let x and y be two distinct points (other than a) in P; note that x, y, a lie on a common line, and since L contains x and y, all three lie on L. Hence all points in P lie on L. ■

10. The proof is by induction on n. The basis case, $n = 1$, is true because both sides of the inequality evaluate to 1.

    Suppose (induction hypothesis) that the result is true when $n = k$; that is,

$$\sqrt{1} + \sqrt{2} + \cdots + \sqrt{k} \le k\sqrt{k}.$$

To show that the result is true when $n = k + 1$, we add $\sqrt{k+1}$ to both sides of this inequality to get

$$\sqrt{1} + \sqrt{2} + \cdots + \sqrt{k} + \sqrt{k+1} \le k\sqrt{k} + \sqrt{k+1}.$$

Because $\sqrt{k} < \sqrt{k+1}$, we have

$$k\sqrt{k} + \sqrt{k+1} < k\sqrt{k+1} + \sqrt{k+1}$$
$$= (k+1)\sqrt{k+1}$$

and so $\sqrt{1} + \sqrt{2} + \cdots + \sqrt{k} + \sqrt{k+1} \le (k+1)\sqrt{k+1}$ as required. ■

11. The proof is by induction on n. For the basis case, $n = 0$, we note that $(x+y)^0 = 1$ and also $\sum_j \binom{n}{j} x^j y^{n-j} = \binom{0}{0} x^0 y^0 = 1$.

    Assume (induction hypothesis) that the result has been proved for $n = k$. We seek to prove the case $n = k + 1$. Note that $(x+y)^{k+1} = (x+y)(x+y)^k$ and we can expand the latter, giving

$$(x+y)^{k+1} = (x+y)\sum_{j=0}^{k}\binom{k}{j}x^j y^{k-j}$$
$$= \sum_{j=0}^{k}\binom{k}{j}x^{j+1}y^{k-j}$$
$$+ \sum_{j=0}^{k}\binom{k}{j}x^j y^{k+1-j}.$$

The first term of the second sum is $x^0 y^{k+1}$ and the last term of the first sum is $x^{k+1}y^0$. Otherwise, we can collect like terms from the two sums and note that the coefficient of $x^j y^{k+1-j}$ (with $1 \le j \le k$) is $\binom{k}{j-1} + \binom{k}{j}$, which, by Pascal's Identity (Theorem 16.10), equals $\binom{k+1}{j}$.

Therefore

$$(x+y)^{k+1} = \sum_{j}^{k+1}\binom{k+1}{j}x^j y^{k+1-j}$$

as required. ■

12. The proof is by induction on n. In case $n = 1$, the single line divides the plane into two regions and $\binom{1}{0} + \binom{1}{1} + \binom{1}{2} = 1 + 1 + 0 = 2$, verifying the basis case.

Suppose (induction hypothesis) the result is true for collections of k lines. Consider a collection of $k + 1$ lines that satisfies the hypothesis of the result. Let L be any one of these lines.

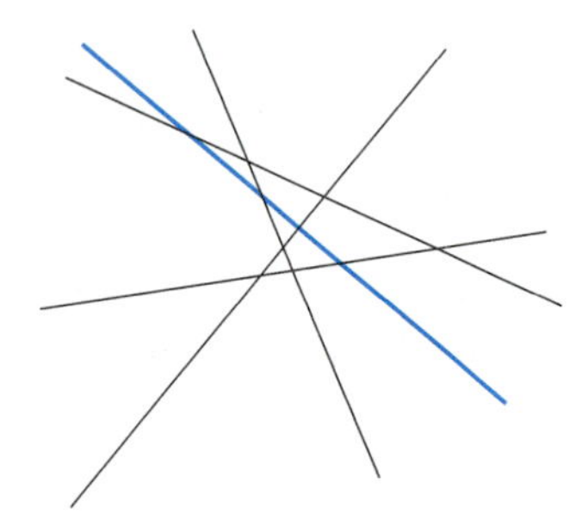

Observe that the k other lines divide the plane into $\binom{k}{0} + \binom{k}{1} + \binom{k}{2}$ regions. Line L intersects each of the three k lines and cuts through $k+1$ regions. Thus line L creates an additional k region; therefore the $k + 1$ lines cut the plane into $\binom{k}{0} + \binom{k}{1} + \binom{k}{2} + (k + 1)$ regions. This can be rewritten as

$$\binom{k}{0} + \left[\binom{k}{1} + 1\right] + \left[\binom{k}{2} + k\right].$$

Note that

$$\binom{k}{0} = \binom{k+1}{0}$$

$$\binom{k}{1} + 1 = \binom{k+1}{1}$$

$$\binom{k}{2} + k = \frac{k(k-1)}{2} + k = \frac{(k+1)k}{2} = \binom{k+1}{2}$$

and so the $k + 1$ lines cut the plane into $\binom{k+1}{0} + \binom{k+1}{1} + \binom{k+2}{2}$ regions. ■

13. The proof is by strong induction on n. The basis case, $n = 0$, is clear because both sides of the equation evaluate to 3. The equation is also true for $n = 1$ because both sides evaluate to 5.

Assume (strong induction hypothesis) that the equation has been shown for $n = 0, 1, 2, \ldots, k$ (where $k \geq 1$). In particular, we have

$$F_k + 2F_{k+1} = F_{k+4} - F_{k+2}$$

$$F_{k-1} + 2F_k = F_{k+3} - F_{k+1}.$$

Adding these equations together gives

$$F_{k+1} + 2F_{k+2} = F_{k+5} - F_{k+3}$$

which is precisely the $n = k + 1$ case of the result. ■

14. We prove this by induction on n. For the basis case, $n = 0$, we note that $F_0 = F_1 = 1$, so the only positive divisor of both is 1.

Suppose (induction hypothesis) that the only positive divisor of F_k and F_{k+1} is 1. We must show that 1 is the only positive divisor of F_{k+1} and F_{k+2}. Suppose, for the sake of contradiction, that there is an integer $d > 1$ with $d|F_{k+2}$ and $d|F_{k+1}$. Because $F_{k+2} = F_k + F_{k+1}$, we have $F_k = F_{k+2} - F_{k+1}$, and so if d divides both F_{k+2} and F_{k+1}, we see that $d|F_k$. But then $d|F_k$ and $d|F_{k+1}$, but $d > 1$.⇒⇐ Therefore the only positive divisor of F_{k+1} and F_{k+2} is 1. ■

15. (a) Consider the first tile. If the tile is a 1×1, then there are 2 choices for the color of that tile, and the remainder of the stripe can be completed in a_{n-1} ways. If the tile is a 1×2, then there are 3 choices for its colors and a_{n-2} ways to tile the rest of the stripe. Thus there are $2a_{n-1} + 3a_{n-2}$ ways to tile the stripe.

(b) We prove this by strong induction on n. In the case $n = 1$, there are only 2 ways to tile the stripe, and $(3^2 + (-1)^1)/4 = (9 - 1)/4 = 2$. In the case $n - 2$, there are 2×2 ways to tile using two 1×1 tiles and 3 ways to tile using a single 1×2 tile, for a total of 7 possible tilings.

Let us assume (strong induction hypothesis) that the result has been proved for $n = 1, 2, \ldots, k$ (where $k \geq 2$). In particular, we know that

$$a_{k-1} = \frac{3^k + (-1)^{k-1}}{4} \quad \text{and}$$
$$a_k = \frac{3^{k+1} + (-1)^k}{4}.$$

Using the identity we proved in (a), we have

$$\begin{aligned} a_{k+1} &= 2a_k + 3a_{k-1} \\ &= 2 \cdot \frac{3^{k+1} + (-1)^k}{4} + 3 \cdot \frac{3^k + (-1)^{k-1}}{4} \\ &= \frac{(2 \cdot 3^{k+1} + 3 \cdot 3^k) + (2 \cdot (-1)^k + 3 \cdot (-1)^{k-1})}{4} \\ &= \frac{3 \cdot 3^{k+1} + (2-3) \cdot (-1)^k}{4} \\ &= \frac{3^{k+2} + (-1)^{k+1}}{4} \end{aligned}$$

as required. ■

16. **Proof.**

Existence. The proof is by the method of smallest counterexample. [Alternatively, we could write this proof using strong induction.] Suppose the result is false and let x be a smallest counterexample. Note that $x \neq 1$ since we can write $1 = 2^0 \cdot 1$. Also observe that x is not odd, for then we could write $x = 2^0 \cdot x$. Thus we may assume x is even. In this case $x/2$ is a smaller positive integer (and therefore not a counterexample), so we can write $\frac{x}{2} = 2^a \cdot b$ where b is odd. But then $x = 2^{a+1} \cdot b$, undermining our supposition that x is a counterexample.⇒⇐ Therefore every positive integer n can be expressed in the form $n = 2^a b$ where b is odd.

Uniqueness. Suppose, for the sake of contradiction, there is an integer n and distinct pairs of nonnegative integers (a, b) and (c, d) such that b and d are odd and

$$n = 2^a b = 2^c d.$$

Were it the case that $a = c$, then $2^a b = 2^c d \Rightarrow b = d$, contradicting the assertion that (a, b) and (c, d) were different pairs. Thus $a \neq c$.

Without loss of generality, $a < c$. Therefore

$$2^a b = 2^c d \qquad \Rightarrow \qquad b = 2^{c-a} d$$

where $c - a > 0$. Therefore b is even, but it is also odd.⇒⇐ Therefore there is only a pair of nonnegative integers (a, b) such that $n = 2^a b$ and b is odd. ■

17. (a) We prove this by induction on the size of A. In the case that A has only one element, t, then clearly $t|t$, so the basis case is true.

Assume (induction hypothesis) that the result is true for all sets of positive integers with k elements. Let A be a set with $k + 1$ elements that satisfies the condition (i.e., $\forall r \in A$, $\forall s \in A$, $(r|s$ or $s|r)$.) Let x be any element of A and let $A' = A - \{x\}$. Note that A' is a set of k positive integers that satisfies the condition. So, by induction, A' contains an element t' for which $a|t'$ for all $a \in A'$. Now, since $x \in A$, either $x|t'$ or $t'|x$. In the first case, note that all elements of A are divisors of t', and we are finished. Otherwise $(t'|x)$, and since all elements of A' are divisors of t' and $t'|x$, it follows that all elements of A are divisors of x. ■

(b) Suppose A contains two different elements t and s with the property that all elements in A are divisors of s and also of t. This implies that $s|t$ and $t|s$. Since $s, t > 0$, it follows (see Problem 3) that $s = t$.⇒⇐ Therefore A contains a unique element that is a multiple of all elements in A. ■

(c) Let $A = \{-2, -1, 1, 2\}$. It is easy to check that for all a, b in A, either $a|b$ or $b|a$. However, A has two distinct elements, -2 and 2, that are multiples of all.

18. (a) $a_n = \frac{5}{2}(-3)^n + \frac{3}{2}(2)^n$.

(b) $a_n = -4 \cdot 2^n + 8$.

(c) $a_n = 6^n - \frac{2}{3}n6^n$.

19. We apply Δ repeatedly to this sequence and find the following:

5		26		67		146		281		490
	21		41		79		135		209	
		20		38		56		74		
			18		18		18			
				0		0				

Therefore

$$\begin{aligned} a_n &= 5\binom{n}{0} + 21\binom{n}{1} + 20\binom{n}{2} + 18\binom{n}{3} \\ &= 5 + 21n + 10n(n-1) + 3n(n-1)(n-2) \\ &= 3n^3 + n^2 + 17n + 5. \end{aligned}$$

Chapter 5

1. (a) $f(2) = 3$.
   (b) $f(4)$ is undefined.
   (c) $\operatorname{dom} f = \{1, 2, 3\}$.
   (d) $\operatorname{im} f = \{2, 3, 4\}$.
   (e) $f^{-1} = \{(2, 1), (3, 2), (4, 3)\}$.
   (f) $g^{-1} = \{(1, 2), (1, 3), (2, 4)\}$ is not a function because it contains two distinct ordered pairs of the form $(1, ?)$. (Also, g is not one-to-one.)
   (g) $g \circ f = \{(1, 1), (2, 1), (3, 2)\}$.
   (h) $f \circ g = \{(2, 2), (3, 2), (4, 3)\}$.
2. (a) True. (b) True. (c) True. (d) False. For (d), note that $f : A \to B$ requires only that $\operatorname{im} f \subseteq B$.
3. (a) $4^3 = 64$. (b) $(4)_3 = 4 \times 3 \times 2 = 24$. (c) none.
4. No, f is not necessarily onto. For example, let $f : \mathbb{Z} \to \mathbb{Z}$ by $f(x) = 2x$, and let g be the same function. Note that both f and g are one-to-one, but neither is onto.

   However, if A and B are finite sets, then it would follow that f is onto.
5. (a) First, f is not one-to-one. For example, $f(-2) = |-2| = 2$ and $f(2) = |2| = 2$, so $f(2) = f(-2)$ but, of course, $2 \neq -2$.
   (b) Second, f is onto. Let $x \in \mathbb{N}$. Since $\mathbb{N} \subseteq \mathbb{Z}$, certainly $x \in \mathbb{Z}$, and $f(x) = |x| = x$ (since x is nonnegative). Therefore f is onto.
6. (a) f is one-to-one. We offer two proofs.
   Proof 1: We claim that f is an increasing function; that is, if $x < y$, then $f(x) < f(y)$. To see why, we consider three cases:
   * x and y are both nonnegative (i.e., $0 \le x < y$). In this case, $x < y$ implies $f(x) = f^3 = x \cdot x^2 < y \cdot x^2 < y \cdot y^2 = y^3 = f(y)$.
   * x and y are both negative (i.e., $x < y < 0$). Since $x < y$ and both are negative, $x^2 > y^2 > 0$ and so $x^3 < y^3$, so $f(x) < f(y)$.
   * x is negative and y is nonnegative. In this case, $f(x) = x^3 < 0 \le y^3 = f(y)$.

   In all cases, $x < y \Longrightarrow f(x) < f(y)$. Thus if $x \neq y$, we certainly have $f(x) \neq f(y)$, and so f is one-to-one.
   Proof 2: Suppose $f(x) = f(y)$. Thus $x^3 = y^3$, and so $x^3 - y^3 = (x - y)(x^2 + xy + y^2)$.
   We claim that $x^2 + xy + y^2$ cannot equal zero. From the quadratic formula,

$$x = \frac{-y \pm \sqrt{y^2 - 4y^2}}{2} = \frac{-y \pm y\sqrt{-3}}{2}$$

   and because $\sqrt{-3}$ is imaginary, there can be no pair of integers x and y for which $x^2 + xy + y^2 = 0$.
   Since $x^2 + xy + y^2 \neq 0$ and $0 = x^3 - y^3 = (x - y)(x^2 + xy + y^2) = 0$, it follows that $x - y = 0$, so $x = y$. Therefore f is onto.
   (b) f is not onto because there is no integer x such that $f(x) = x^3 = 2$.
7. The only function that is an equivalence relation on $\{1, 2, 3, 4, 5\}$ is $\{(1, 1), (2, 2), (3, 3), (4, 4), (5, 5)\}$.
8. There are 64 positions for a 2×2 block, but only $2^4 = 16$ different ways to color the squares in the block. Therefore, by the Pigeonhole Principle, two blocks must be identically colored.

   Moreover, suppose (for contradiction) that each of the 16 possible colorings occurs at most 3 times. This can happen only if there are $3 \times 16 = 48$ or fewer 2×2 blocks. However, there are 64 such blocks.$\Rightarrow\Leftarrow$ Therefore some 2×2 pattern must repeat four times. ■

9. Because $h(1) = 3$ and $h(1) = f[g(1)]$, we must have $g(1) = 2$ or 4. Similarly, from $h(2) = f[g(2)] = 3$, we must have $g(2) = 2$ or 4. From $h(3) = 2 = f[g(3)]$, we must have $g(3) = 1$. From $h(4) = 5 = f[g(4)]$, we must have $g(4) = 5$. From $h(5) = 3 = f[g(5)]$, we must have $g(5) = 2$ or 4.

   Therefore we know that $g = \{(1, ?), (2, ?), (3, 1), (4, 5), (5, ?)\}$ where each ? may be either a 2 or a 4, giving eight possible answers.

10. Note that

$$\begin{aligned}(f \circ g)(x) &= f(3x+2)\\ &= (3x+2)^2 + (3x+2) - 1\\ &= 9x^2 + 15x + 5 \qquad \text{and}\end{aligned}$$

$$\begin{aligned}(g \circ h)(x) &= g(x^2 + x - 1)\\ &= 3(x^2 + x - 1) + 2\\ &= 3x^2 + 3x - 1 \qquad \text{and so}\end{aligned}$$

$$\begin{aligned}(f \circ g)(x) - (g \circ f)(x) &= (9x^2 + 15x + 5)\\ &\quad - (3x^2 + 3x - 1)\\ &= 6x^2 + 12x + 6\\ &= 6(x+1)^2.\end{aligned}$$

11. Note that

$$\begin{aligned}(f \circ g \circ h)(x) &= f[g(2x+1)] = f[a(2x+1) + b]\\ &= f[a(2x+1) + b]\\ &= f(2ax + a + b)\\ &= 3(2ax + a + b) - 4\\ &= 6ax + 3a + 3b - 4\\ &= 6x + 5.\end{aligned}$$

   From this it follows that $a = 1$ and $b = 2$. Therefore

$$\begin{aligned}(h \circ g \circ f)(x) &= h[g(3x-4)]\\ &= h[(3x-4) + 2] = h[3x-2]\\ &= 2(3x-2) + 1 = 6x - 3.\end{aligned}$$

12. A discrete mathematician would be comfortable with setting $0^0 = 1$.

   First, we can consider 0^0 to be an empty product and therefore equal to 1 (as in Section 8).

   Second, we can consider 0^0 to be the number of functions from the empty set to itself. There is exactly one such function—namely, the empty set. Indeed, the empty set is a function because it satisfies the definition of function, albeit vacuously. The domain and image of the empty set (as a function) are both the empty set. This is the only possible function from the empty set to itself.

13. The assertion $f = g^{-1}$ is false, as the following counterexample shows. Let $A = \mathbb{Z}$, let $g(x) = 2x$ for all $x \in \mathbb{Z}$, and let

$$f(x) = \begin{cases} \frac{x}{2} & \text{if } x \text{ is even, and} \\ 0 & \text{if } x \text{ is odd.} \end{cases}$$

   Notice that for any integer x, $(f \circ g)(x) = f[g(x)] = f[2x] = x$, so $f \circ g = \mathrm{id}_{\mathbb{Z}}$. However, $f \neq g^{-1}$. For example, $(5, 0) \in f$, but $(0, 5) \notin g$, or, in customary notation, $f(5) = 0$ but $g(0) \neq 5$.

14. (a) $\{(1,3), (2,9), (3,2), (4,6), (5,5), (6,7), (7,4), (8,1), (9,8)\}$.
   (b) $\pi = (1,3,2,9,8)(4,6,7)(5)$.
   (c) $\pi^{-1} = (1,8,9,2,3)(4,7,6)(5)$.
   (d) $\pi \circ \pi = (1,2,8,3,9)(4,7,6)(5)$.
   (e) $\pi = (1,8) \circ (1,9) \circ (1,2) \circ (1,3) \circ (4,7) \circ (4,6)$ and so π is even.

15. There are only $n!$ elements of S_n, so the sequence $\pi = \pi^{(1)}, \pi^{(2)}, \pi^{(3)}, \ldots,$ must repeat itself eventually. Let j be the smallest index such that $\pi^{(j)} = \pi^{(k)}$ for some $k > j$.

   We claim that $j = 1$. Suppose, for the sake of contradiction, we have $\pi^{(j)} = \pi^{(k)}$ with $1 < j < k$. Composing both sides of this equation on the left by π^{-1} gives

$$\begin{aligned}\pi^{-1} \circ \pi^{(j)} &= \pi^{-1} \circ \pi^{(k)}\\ \Rightarrow \quad \pi^{(j-1)} &= \pi^{(k-1)}\end{aligned}$$

   contradicting the fact that j was the first index of a repeated element.$\Rightarrow\Leftarrow$ Therefore $j = 1$.

Thus $\pi = \pi^{(k)}$ for some $k > 1$. Composing on the left by π^{-1} gives $\pi^{-1} \circ \pi = \pi^{-1} \circ \pi^{(k)}$ whence $\iota = \pi^{(k-1)}$ where $k - 1 > 0$.

If $k - 1 = 1$, that means $\iota = \pi$ and so $\pi = pi^{-1} = \pi^{(1)}$. Otherwise (i.e., $k-1 \geq 2$), we have $\iota = \pi^{(k-1)} = \pi \circ \pi^{(k-2)}$. So $\pi^{(k-2)} = \pi^{-1}$ and $k - 2$ is positive. ■

16. The sum evaluates to 0 since both the positive terms and the negative terms (in different orders) rearrange to $1 + 2 + \cdots + n$.

17. (a) We know we can write π as a composition of transpositions (see Theorem 26.11) as follows:

$$\pi = \tau_1 \circ \tau_2 \circ \cdots \circ \tau_b.$$

If $\tau_i = (1, x_i)$, or, equivalently, $(x_i, 1)$, leave it alone. If $\tau_i = (x_i, y_i)$ with $x_i, y_i > 1$, then we can replace it with $(1, x_i) \circ (1, y_i) \circ (1, x_i)$ since $(x_i, y_i) = (1, x_i) \circ (1, y_i) \circ (1, x_i)$. After these substitutions, we have expressed π as a composition of transpositions of the form $(1, x)$.

(b) In S_3 observe that

$$\iota = (1, 2) \circ (1, 3) \circ (1, 2) \circ (1, 3) \circ (1, 2) \circ (1, 3).$$

Note that both $(1, 2)$ and $(1, 3)$ appear three times.

18. *Proof 1*: Suppose π has ℓ inversions, so $\operatorname{sgn} \pi = (-1)^\ell$. Notice that in every factor of $\prod_{1 \leq i < j \leq n} (x_j - x_i)$, the larger subscripted term precedes the smaller. But in $\prod_{1 \leq i < j \leq n} (x_{\pi(j)} - x_{\pi(i)})$ there are exactly ℓ factors in which the smaller subscripted term precedes the larger, so to restore equality, we can multiply by $(-1)^\ell$. ■

Proof 2: Decompose π as the composition of transpositions, $\pi = \tau_1 \circ \cdots \circ \tau_a$. Starting with the original product $\prod_{1 \leq i < j \leq n} (x_j - x_i)$, replace subscripts i with $\tau_a(i)$ for $i = 1, \ldots, n$. If $\tau_a = (p, q)$, then $n - 2$ of the terms of the form $\pm(x_p - x_j)$ become $\pm(x_a - x_j)$, and another $n - 2$ terms of the form $\pm(x_q - x_j)$ become $\pm(x_q - x_j)$; there is no effect on the product as the result of these changes. However, the term $\pm(x_p - x_q)$ becomes $\pm(x_q - x_p)$, resulting in a change of sign for the entire product. As each subsequent transposition is applied, the sign changes, so in the end we have changed signs a times. Therefore the resulting product is $(-1)^a = \operatorname{sgn} \pi$ times the original. ■

19. (a) Starting from a home position, vertex 1 can be moved to any of the four corners. Then vertex 2 can be rotated into any of three positions, and finally, vertex 3 may be reflected into any of 2 positions, so there are $4 \cdot 3 \cdot 2 = 24$ distinct symmetries. Hence the set of symmetries is all of S_4.

(b) There are only 12 possible symmetries of the tetrahedron (when we omit reflections), and we can explicitly list them.

$(1)(2,3,4)$	$(1)(2,4,3)$	$(2)(1,3,4)$	$(2)(1,4,3)$
$(3)(1,2,4)$	$(3)(1,4,2)$	$(4)(1,2,3)$	$(4)(1,3,2)$
$(1,2)(3,4)$	$(1,3)(2,4)$	$(1,4)(2,3)$	$(1)(2)(3)(4)$

Notice that these are precisely the even permutations of S_4.

20. We can conclude that x must be an integer.

21. To show that 2^n is $O(3^n)$, it is enough to note that $|2^n| \leq |3^n|$ for all positive integers n.

Suppose, for the sake of contradiction, that 3^n is $O(2^n)$. Then there is a positive number M such that $|3^n| \leq M|2^n|$ for all but finitely many positive integers n. We may drop the absolute value bars because 2^n and 3^n are always positive, and so we have

$$3^n \leq M \cdot 2^n \quad \Rightarrow \quad \left(\frac{3}{2}\right)^n \leq M$$

for all but finitely many n. However, the values $\left(\frac{3}{2}\right)^n$ get larger and larger as n grows, exceeding any specific number.$\Rightarrow\Leftarrow$ Therefore 3^n is not $O(2^n)$.

Chapter 6

1. The sum of $P(a)$ over all $a \in S$ must be 1. In S, there are five even numbers, for which $P(a) = x$, and five odd numbers, for which

$P(a) = 2x$. Therefore $5(x) + 5(2x) = 1$ so $15x = 1$ and therefore $x = \frac{1}{15}$.

2. (a) 6^3. (b) 6^4.

3. $P(A) = P(1) + P(4) + P(7) + P(9) = \frac{1}{55} + \frac{4}{55} + \frac{7}{55} + \frac{9}{55} = \frac{21}{55}$.

4. (a) There are 10! ways in which the children may line up, and all are equally likely. There is only one ordering in which the children appear alphabetically by name, so the probability is $1/10!$.
   (b) There are $5! \cdot 5!$ ways in which the children may line up so that all the girls precede all the boys. Therefore the probability is $5! \cdot 5!/10! = 1/252$.
   (c) The first girl may be in position 1 through 6 on the line. Once the position for the first girl is set, there are $5! \cdot 5!$ ways for the children to take their places, giving a total of $6 \cdot 5! \cdot 5!$ successful outcomes. Therefore the probability is $6 \cdot 5! \cdot 5!/10! = 1/42$.
   (d) There are $2 \cdot 5! \cdot 5!$ ways in which the children may stand so that they alternate by gender. Therefore the probability is $2 \cdot 5! \cdot 5!/10! = 1/126$.
   (e) Let B be the event that the boys are in a contiguous block, and let G be the corresponding event for the girls. We seek $P(\overline{B} \cap \overline{G})$. Note that

$$\begin{aligned} P(\overline{B} \cap \overline{G}) &= P(\overline{B \cup G}) \\ &= 1 - P(B \cup G) \\ &= 1 - [P(B) + P(G) - P(B \cap G)]. \end{aligned}$$

From part (c) we have $P(G) = P(B) = 1/42$. To calculate $P(B \cap G)$, note there are $2 \cdot 5! \cdot 5!$ ways in which the children might stand in which all the boys are together and all the girls are together, so $P(B \cap G) = 2 \cdot 5! \cdot 5!/10! = 1/126$. Therefore

$$P(\overline{B} \cap \overline{A}) = 1 - \frac{2}{42} + \frac{1}{126} = \frac{5}{126}.$$

5. (a) There are $\binom{52}{13}$ ways to choose the hand, each of which is equally likely. There is only one way to select the cards if all 13 are spades. Thus the probability is $1/\binom{52}{13}$.
   (b) There are $\binom{26}{13}$ ways to choose the cards such that all are black, so the probability is $\binom{26}{13}/\binom{52}{13}$.
   (c) Let B be the event that all the cards are black and R be the event that all the cards are red; we seek $P(\overline{R} \cap \overline{B})$. We can rewrite this as

$$\begin{aligned} P(\overline{R} \cap \overline{B}) &= P(\overline{R \cup B}) \\ &= 1 - P(R \cup B) \\ &= 1 - [P(R) + P(B) - P(R \cap B)]. \end{aligned}$$

From part (b), $P(R) = P(B) = \binom{26}{13}/\binom{52}{13}$ and $P(R \cap B) = 0$ since there is no way to choose the cards such that they are all black and all red. Therefore

$$\begin{aligned} P(\overline{R} \cap \overline{B}) &= 1 - \frac{2\binom{26}{13}}{\binom{52}{13}} \\ &= \frac{580008}{580027} \approx 0.999967. \end{aligned}$$

   (d) Let A be the event that one (or more) of the cards is an ace. There are $\binom{52-4}{13} = \binom{48}{13}$ ways to choose an aceless hand, so $P(\overline{A}) = \binom{48}{13}/\binom{52}{13}$. This evaluates to approximately 30%.
   (e) There are $52 - 13 - 4 + 1 = 36$ cards in the deck that are neither hearts nor aces. Therefore the probability of drawing 13 cards none of which is an ace or a heart is $\binom{36}{13}/\binom{52}{13}$.

6. (a) The only way to draw 21 is to pick an ace together with a face card or 10. There are 52×51 ways to pick two cards (in sequence) from the deck. Of these, there are $2 \times 4 \times 16$ sequences in which one of the cards is an ace and the other is a ten or face card. So the probability is $(2 \cdot 4 \cdot 16)/(52 \cdot 51) = 32/663$, or roughly 4.8%.
   (b) Of the $52 \times 51 = 2652$ ways to draw a face card, the following chart gives the number of ways to draw the sum 16 or higher, depending on the first card.

First card	Choices for 2nd card	Total
2, 3, or 4	0	0
5	4 (aces only)	16
6	20 (10 or higher)	80
7	24 (9 or higher)	96
8	28 (8 or higher)	112
9	32 (7 or higher)	128
10 or face	36 (6 or higher)	576
ace	40 (5 or higher)	160
	Total number of ways:	1168

Therefore, the probability that the two cards drawn sum to 16 or higher is $1168/2652 = 292/663$, or approximately 44%.

(c) Let A be the event that the first card is an ace, and let F be the event that the second card is a face card. We seek $P(F|A)$. This equals $P(F \cap A)/P(A)$. The numerator equals $(4 \cdot 12)/(52 \cdot 51)$ and the denominator equals $4/52$. Therefore

$$P(F|A) = \frac{(4 \cdot 12)/(52 \cdot 51)}{4/52} = \frac{4}{17}.$$

7. Let FB be the event that the first card is black and LR be the event that the last card is red. We seek $P(LR|FB)$, which equals $P(LR \cap FB)/P(FB)$. The numerator equals $(26 \cdot 26)/(52 \cdot 51)$ and the denominator equals $26/52 = 1/2$. Therefore

$$P(LR|FB) = \frac{(26 \cdot 26)/(52 \cdot 51)}{1/2} = 26/51.$$

The two events, FB and LR, are not independent. Above we showed that $P(FB \cap LR) = (26 \cdot 26)/(52 \cdot 51) = 13/51$, but $P(FB) \cdot P(LR) = \frac{1}{2} \cdot \frac{1}{2} = \frac{1}{4}$.

8. Let A be an event for a sample space (S, P). Events A and $\overline{A}$ are independent if and only if $P(A) = 0$ or $P(A) = 1$.

(Alternatively, the theorem may conclude "... if and only if $P(A) = 0$ or $P(\overline{A}) = 0$," etc.)

Proof. ($\Rightarrow$) Suppose A and $\overline{A}$ are independent. Then $P(A)P(\overline{A}) = P(A \cap \overline{A})$, and these must equal 0 since $A \cap \overline{A} = \emptyset$. Therefore $P(A) = 0$ or $P(\overline{A}) = 0$ and the latter is equivalent to $P(A) = 1$.

($\Leftarrow$) Suppose $P(A) = 0$ or $P(A) = 1$. In either case, $P(A)P(\overline{A}) = 0 = P(\emptyset) = P(A \cap \overline{A})$ and so A and $\overline{A}$ are independent. ■

9. For events R and C, we have

$$P(R) = \frac{64 \cdot 8}{64^2} = \frac{1}{8},$$
$$P(C) = \frac{1}{8}, \quad \text{and}$$
$$P(R \cap C) = \frac{64 \cdot 1}{64^2} = \frac{1}{64} = P(R) \cdot P(C).$$

Therefore R and C are independent.

For events R and B, we have

$$P(R) = \frac{1}{8},$$
$$P(B) = \frac{32^2}{64^2} = \frac{1}{4}, \quad \text{and}$$
$$P(R \cap B) = \frac{32 \cdot 4}{64^2} = \frac{1}{32} = P(R) \cdot P(B).$$

Therefore R and B are independent. Likewise, C and B are independent.

10. For events R and C, we have

$$P(R) = \frac{64 \cdot 7}{64 \cdot 63} = \frac{1}{9}$$
$$P(C) = \frac{1}{9}, \quad \text{and}$$
$$P(R \cap C) = 0 \neq P(R) \cdot P(C).$$

Therefore R and C are not independent.

For events R and B, we have

$$P(R) = \frac{1}{9},$$
$$P(B) = \frac{32 \cdot 31}{64 \cdot 63} = \frac{31}{126} \quad \text{and}$$
$$P(R \cap B) = \frac{32 \cdot 3}{64 \cdot 63} = \frac{1}{42} \neq \frac{31}{5292} = P(R) \cdot P(B).$$

Therefore R and B are not independent. Likewise, C and B are not independent.

Alternative analysis: Instead of choosing the squares in sequence, we can choose them as a pair in $\binom{64}{2}$ ways, all of which are equally likely.
For the events R and C, we have

$$P(R) = \frac{8\binom{8}{2}}{\binom{64}{2}} = \frac{1}{9},$$
$$P(C) = \frac{1}{9}, \quad \text{and}$$
$$P(R \cap C) = 0 \neq P(R) \cdot P(C).$$

For the events R and B, we have

$$P(R) = \frac{1}{9},$$
$$P(B) = \frac{\binom{32}{2}}{\binom{64}{2}} = \frac{31}{126}, \quad \text{and}$$
$$P(R \cap B) = \frac{8\binom{4}{2}}{\binom{64}{2}} = \frac{1}{42} \neq \frac{31}{5292} = P(R) \cdot P(B).$$

11. Suppose the coin produces HEADS with probability p and TAILS with probability $1 - p$. Let A be the event that we flip HEADS-TAILS, and let B be the events that the two flips are different. We calculate as follows:

$$P(A|B) = \frac{P(A \cap B)}{P(B)} = \frac{p(1-p)}{p(1-p) + (1-p)p} = \frac{1}{2}.$$

12. **Proof.** We are given that $A \subseteq B$ and $P(A) \neq 0$. Therefore $P(B) \neq 0$. Note that

$$P(A|B) = \frac{P(A \cap B)}{P(B)} = \frac{P(A)}{P(B)}$$

with the last equality because $A \cap B = A$ since $A \subseteq B$. The result now follows by multiplying the displayed equation through by $P(B)$. ■

13. Since

$$0 = E(X) = X(a)P(a) + X(b)P(b) + X(c)P(c)$$
$$= (1)(0.4) + (2)(0.4) + X(c)(0.2)$$
$$= 1.2 + 0.2X(c)$$

it follows that $X(c) = -1.2/0.2 = -6$.

14. (a) There are 32 cards with even value (four each of 2, 4, 6, 8, 10, jack, queen, and king). Therefore $P(X \text{ is even}) = 32/52 = 8/13$.

(b) Each sort of card appears with probability $1/13$ so the expected value of X is

$$E(X) = \frac{2+3+4+5+6+7+8+9+4\times 10+11}{13} = \frac{95}{13}.$$

(c) The expectation of Y is the same as that of X; that is, $E(Y) = 95/13$.

(d) The random variables X and Y are not independent. For example, consider the probability that both are equal to 2. We have

$$P(X = 2 \wedge Y = 2) = \frac{4 \times 3}{52 \times 51} = \frac{1}{221},$$
$$P(X = 2) = P(Y = 2) = \frac{4}{52} = \frac{1}{13}, \quad \text{but}$$
$$P(X = 2) \cdot P(Y = 2) = \frac{1}{169} \neq P(X = 2 \wedge Y = 2).$$

(e) By linearity of expectation, $E(X+Y) = E(X) + E(Y) = \frac{95}{13} + \frac{95}{13} = \frac{190}{13}$.

(f) The probability $X = Y$ can be calculated by

$$P(X=2 \wedge Y=2) + \cdots + P(X=11 \wedge Y=11).$$

We calculate each summand as follows:

$$P(X = 2 \wedge Y = 2) = \frac{4 \times 3}{52 \times 51} = \frac{1}{122}$$
$$\vdots$$
$$P(X = 9 \wedge Y = 9) = \frac{1}{122}$$
$$P(X = 10 \wedge Y = 10) = \frac{16 \times 15}{52 \times 51} = \frac{20}{221}$$

$$P(X = 11 \wedge Y = 11) = \frac{1}{221}$$

$$\sum_{j=2}^{11} P(X = j \wedge Y = j) = \frac{29}{221}.$$

(g) We use the formula $\text{Var}(X) = E(X^2) - E(X)^2$.

$$E(X^2) = \frac{2^2 + 3^2 + \cdots + 9^2 + 4 \times 10^2 + 11^2}{13} = \frac{805}{13}$$

$$E(X)^2 = \left(\frac{95}{13}\right)^2 = \frac{9025}{169}$$

$$\text{Var}(X) = E(X^2) - E(X)^2 = \frac{805}{13} - \frac{9025}{169} = \frac{1440}{169}.$$

15. The analysis is simplest if we write $X = X_1 + X_2 + \cdots + X_5$ where X_j is the change in stock price on the jth day.
   (a) Note that $E(X) = E(X_1) + \cdots + E(X_5)$. Each summand is given by $E(X_j) = (0.6)(2) + (0.1)(5) + (0.3)(-4) = 0.5$. Therefore $E(X) = 5(0.5) = 2.5 = \frac{1}{2}$. We expect the stock value to rise \$2.50.
   (b) Recall that $\text{Var}(X) = E(X^2) - E(X)^2$. Using the fact that the X_is are independent, we calculate as follows:

$$\begin{aligned} E(X^2) &= E[(X_1 + \cdots + X_5)^2] \\ &= E\big(X_1^2 + \cdots + X_5^2 + 2X_1X_2 \\ &\quad + \cdots + 2X_4X_5\big) \\ &= 5E\big(X_1^2\big) + 20E(X_1X_2) \\ &= 5E\big(X_1^2\big) + 20E(X_1)E(X_2) \\ &= 5(0.6(2)^2 + 0.1(5)^2 \\ &\quad + 0.3(-4)^2) + 20(0.5)^2 \\ &= 53.5 \end{aligned}$$

$$E(X)^2 = (2.5)^2 = 6.25$$

$$\text{Var}(X) = E(X^2) - E(X)^2 = 53.5 - 6.25 = 47.25.$$

Alternatively, we could use the fact that $\text{Var}(X) = \text{Var}(X_1) + \cdots + \text{Var}(X_5)$ (because the X_is are independent).

Chapter 7

1. $q = 4$ and $r = 3$, so 23 div 5 = 4 and 23 mod 5 = 3.

2. **Proof.** Suppose $b|a$. Then there is an integer q such that $a = qb$. So we can write $a = qb + 0$, so by definition of div, we have $q = a$ div b. Since $q = \frac{a}{b}$, the result follows. ■

3. **Proof.** We are given that a, b are positive integers with $a \geq 2$ and $a|(b!+1)$. Suppose, for the sake of contradiction, that $a \leq b$, and so $a|b!$. Since a divides both $b! + 1$ and $b!$, it divides their difference $(b!+1) - b! = 1$, and so $a = 1$; but $a \geq 2$.⇒⇐ Therefore, $a > b$. ■

4. (⇒) Suppose $\gcd(p, q) = 1$ but, for the sake of contradiction, $p = q$. But then $\gcd(p, q) = p = q > 1$.⇒⇐ So $p \neq q$.

   (⇐) Suppose $p \neq q$ but, for the sake of contradiction, $\gcd(p, q) = d > 1$. Then $d|p$ and $d|q$. Since $d > 1$, this is possible only if $d = p$ and $d = q$, whence $p = q$.⇒⇐ Therefore $\gcd(p, q) = 1$. ■

5. $x = 4$ and $y = -7$. Other answers are possible, so long as $100x + 57y = 1$.

6. From the previous problem, we know that $100 \times 4 + 57 \times (-7) = 1$, so $-7 \times 57 \equiv 1 \pmod{100}$. Since $-7 \equiv 93 \pmod{100}$, we have that the reciprocal of 57 is 93. Checking: $57 \times 93 = 5301 \equiv 1 \pmod{100}$, so $57 \otimes 93 = 1$ in $\mathbb{Z}_{100}$.

7. **Proof.** We prove that $\gcd(F_n, F_{n+1}) = 1$ by induction on n.

   The basis case, $n = 1$, is simple because $\gcd(F_1, F_2) = \gcd(1, 2) = 1$.

   Suppose (induction hypothesis) that F_k and F_{k+1} are relatively prime; we must prove that F_{k+1} and F_{k+2} are also relatively prime.

   Suppose, for the sake of contradiction, that $\gcd(F_{k+1}, F_{k+2}) = d > 1$. So $d|F_{k+1}$

and $d|F_{k+2}$. Note that $F_{k+2} = F_k + F_{k+1}$, which can be rewritten as

$$F_k = F_{k+2} - F_{k+1}.$$

Because d divides both terms on the right, $d|F_k$. Therefore d is a common divisor of both F_k and F_{k+1}.$\Rightarrow\Leftarrow$

Therefore F_{k+1} and F_{k+2} are relatively prime. ■

Alternatively, we can complete the proof as follows.

Since F_k and F_{k+1} are relatively prime, there exist integers a and b such that $aF_k + bF_{k+1} = 1$. Substituting $F_k = F_{k+2} - F_{k+1}$ gives

$$\begin{aligned} 1 &= aF_k + bF_{k+1} \\ &= a(F_{k+2} - F_{k+1}) + bF_{k+1} \\ &= (b-a)F_{k+1} + aF_{k+2}. \end{aligned}$$

Therefore F_{k+1} and F_{k+2} are relatively prime by Corollary 35.9. ■

8. Let n, p be as given in the problem and let $d = \gcd(n, n+p)$. Since d is a common divisor of $n+p$ and n, we know that $d|n+p-n$; i.e., $d|p$. Thus either $d = 1$ or $d = p$. The latter is impossible because we are given that p does not divide n. ■

9. Because p is prime, the sum of the positive divisors of p^n is the (finite) geometric series

$$1 + p + p^2 + \cdots + p^n$$

which simplifies to

$$\frac{p^{n+1} - 1}{p - 1}.$$

10. (a) 20, (b) 90, (c) 95, and (d) 85.

11. **Proof.** We rely on Theorem 36.14 that $a \in \mathbb{Z}_n$ is invertible if and only if a and n are relatively prime.

($\Rightarrow$) Suppose n is prime. If $1 \le a \le n-1$, then a and n can have no common factor and hence are relatively prime. Thus a is invertible in $\mathbb{Z}_n$.

($\Leftarrow$) Suppose all nonzero elements of $\mathbb{Z}_n$ are invertible but, for the sake of contradiction, n is not prime. Then there exists an integer a such that $1 < a < n$ and $a|n$. This means $\gcd(a, n) = a$, so a is not invertible and yet is a nonzero element of $\mathbb{Z}_n$.$\Rightarrow\Leftarrow$ Therefore n is prime. ■

12. The congruences are satisfied by all integers x such that $x \equiv 981 \pmod{3264}$.

13. **Proof.** ($\Rightarrow$) This is trivial.

($\Leftarrow$) Suppose $\gcd(a, b) = \text{lcm}(a, b) = d$. This implies that $d|a$, $a|d$, $b|d$, and $d|b$. By Problem 3 in the Self Test for Chapter 4 (page 190), we have $a = d$ and $b = d$, and so $a = b$. ■

14. (a) Because $n = 10^{10} = 2^{10}5^{10}$, we see that n has $11 \times 11 = 121$ positive divisors.
    (b) Of the n integers between 1 and n, there are $n/2$ that share a factor of 2 with n and $n/5$ that share a factor of 5 with n. Of these, we have double-counted the multiples of both 2 and 5, and there are $n/10$ of those. Therefore

$$\begin{aligned} \varphi(n) &= \varphi(10^{10}) \\ &= 10^{10} - \frac{10^{10}}{2} - \frac{10^{10}}{5} + \frac{10^{10}}{10} \\ &= 4 \times 10^9. \end{aligned}$$

15. **Proof.** Let n be a positive integer. Factoring n into primes gives

$$n = p_1^{a_1} p_2^{a_2} \cdots p_t^{a_t}$$

where the p_j are distinct primes and the a_j are natural numbers. The number of divisors of n is

$$D = (a_1 + 1)(a_2 + 1) \cdots (a_t + 1).$$

(See Exercise 38.12.)

We now see that D is odd if and only if $(a_j + 1)$ is odd for all j if and only if a_j is even for all j if and only if n is a perfect square. ■

16. **Proof.** Let a, b, c be positive integers and suppose that $a|bc$ and $\gcd(a, b) = 1$. Factor

a, b, c into primes as follows:

$$a = 2^{x_1}3^{x_2}5^{x_3}7^{x_4}\cdots$$
$$b = 2^{y_1}3^{y_2}5^{y_3}7^{y_4}\cdots$$
$$c = 2^{z_1}3^{z_2}5^{z_3}7^{z_4}\cdots$$

Since $a|bc$, we have $x_j \le y_j + z_j$. Since $\gcd(a, b) = 1$, we have $x_j > 0 \Rightarrow y_j = 0 \Rightarrow x_j \le z_j$. Of course, if $x_j = 0$, then $x_j \le z_j$. Therefore $x_j \le z_j$ for all j, and so $a|c$. ■

17. Note that the sum of a consecutive integers beginning with x is

$$x+(x+1)+(x+2)+\cdots+(x+a-1) = ax + \binom{a}{2}.$$

The term ax is clearly divisible by a, so the sum is divisible by a if and only if $a|\binom{a}{2}$.

Note that $\binom{a}{2} = \frac{a(a-1)}{2}$ is an integer.

($\Leftarrow$) If a is odd, then $a-1$ is even, so $2|(a-1)$. Therefore $\binom{a}{2} = a \times \frac{a-1}{2}$, and because $\frac{a-1}{2}$ is an integer, $a|\binom{a}{2}$.

($\Rightarrow$) If a is not odd (i.e., a is even), then $a-1$ is odd. Since a is even, $a = 2^b \times$ powers of odd primes.

So if we factor $\binom{a}{2} = \frac{a}{2} \times (a-1)$ into primes, we note that

$$\binom{a}{2} = 2^{b-1} \times \text{powers of odd primes}$$
$$a = 2^b \times \text{powers of odd primes}$$

and so a cannot divide $\binom{a}{2}$.

Chapter 8

1. (a) $3 * 4 = \sqrt{3^2+4^2} = \sqrt{25} = 5$.
   (b) The operation $*$ is closed for real numbers. If x and y are real numbers, then x^2+y^2 is a nonnegative real number, and so $x*y = \sqrt{x^2+y^2}$ is a real number.
   (c) The operation is commutative because
   $$x*y = \sqrt{x^2+y^2} = \sqrt{y^2+x^2} = y*x.$$
   (d) The operation $*$ is associative because
   $$x*(y*z) = x*\sqrt{y^2+z^2}$$
   $$= \sqrt{x^2+\left(\sqrt{y^2+z^2}\right)^2}$$
   $$= \sqrt{x^2+y^2+z^2}$$
   and, by a similar analysis, $(x*y)*z = \sqrt{x^2+y^2+z^2}$. Therefore $x*(y*z) = (x*y)*z$.
   (e) The operation $*$ does not have an identity element. Suppose, for the sake of contradiction, that e is an identity element. Then $(-1)*e = -1$, but
   $$(-1)*e = \sqrt{(-1)^2+e^2} = \sqrt{1+e^2} > 0$$
   and so $(-1)*e \ne -1$.$\Rightarrow\Leftarrow$

2. $\mathbb{Z}_{32}^* = \{1, 3, 5, 7, 9, 11, 13, 15, 17, 19, 21, 23, 25, 27, 29, 31\}$. In other words, $\mathbb{Z}_{32}^*$ is the set of odd integers between 0 and 32. Thus $\varphi(32) = 16$.

3. (a) $H = \{1, 4, 11, 14\}$
   (b) $K = \{1, 4\}$.

4. Suppose that $(G, *)$ is Abelian and that H and K are defined as in the statement of the problem.
   (a) We need to prove that H is closed under $*$ and inverses.

   Suppose $a, b \in H$. Then $a*a = b*b = e$. To show that $a*b \in H$, we note that $(a*b)*(a*b) = a*a*b*b = e*e = e$ (valid because $*$ is commutative by hypothesis).

   Suppose $a \in H$. Note that $a*a = e = a*a^{-1}$, from which we have $a = a^{-1}$, and so $a^{-1} \in H$.

   Therefore $(H, *)$ is a subgroup of $(G, *)$. ■
   (b) We need to prove that K is closed under $*$ and inverses.

   Suppose $a, b \in K$. Then there exist $x, y \in G$ such that $a = x*x$ and $b = y*y$. Note that $a*b = (x*x)*(y*y) = (x*y)*(x*y)$ (where we use the fact that $*$ is commutative). Therefore we know that $a*b = z*z$ for some $z \in G$ (namely, $z = x*y$) and so $a*b \in K$.

   Suppose $a \in H$. Then $a = x*x$ for some $x \in G$. Let $b = x^{-1}*x^{-1}$; clearly

$b \in K$ by definition of K. Observe that $a * b = x * x * x^{-1} * x^{-1} = e$, and so $b = a^{-1}$. Therefore $b^{-1} \in K$.

Thus $(K, *)$ is a subgroup of $(G, *)$. ■

Next we present the counterexamples for non-Abelian groups.

(a) To show that H is not necessarily a subgroup when $(G, *)$ is not Abelian, we let $(G, *) = (S_4, \circ)$.

Observe that $(1, 2)$ and $(2, 3)$ are in H because $(1, 2) \circ (1, 2) = (2, 3) \circ (2, 3) = \iota$. However, consider $\pi = (1, 2) \circ (2, 3) = (1, 2, 3)$. Note that $\pi \circ \pi = (1, 2, 3) \circ (1, 2, 3) = (1, 3, 2) \neq \iota$. Therefore $\pi \notin K$. Therefore K is not closed under the group operation $\circ$ and so is not a subgroup.

(b) To show that K is not necessarily a subgroup when $(G, *)$ is not Abelian, we let $(G, *) = (A_4, \circ)$ where A_4 is the set of all even permutations in S_4. The twelve elements of A_4 are listed here.

$(1)(2)(3)(4)$	$(1)(2, 3, 4)$	$(1)(2, 4, 3)$	$(2)(1, 3, 4)$
$(2)(1, 4, 3)$	$(3)(1, 2, 4)$	$(3)(1, 4, 2)$	$(4)(1, 2, 3)$
$(4)(1, 3, 2)$	$(1, 2)(3, 4)$	$(1, 3)(2, 4)$	$(1, 4)(2, 3)$

We form the set K by computing $\pi \circ \pi$ for every $\pi \in A_4$. When we do this, we get the following results.

$(1)(2)(3)(4)$	$(1)(2, 4, 3)$	$(1)(3, 4, 2)$	$(2)(1, 4, 3)$
$(2)(1, 3, 4)$	$(3)(1, 4, 2)$	$(3)(1, 2, 4)$	$(4)(1, 3, 2)$
$(4)(1, 2, 3)$	$(1)(2)(3)(4)$	$(1)(2)(3)(4)$	$(1)(2)(3)(4)$

Thus, not counting duplicates, K has nine elements. We claim that $(K, \circ)$ cannot be a subgroup of $(A_4, \circ)$ for otherwise, by Lagrange's Theorem (Theorem 41.4), we would have $9|12.\Rightarrow\Leftarrow$ Therefore K does not constitute a subgroup.

5. We know that G must have an identity element, which we may call e; let a and b be the other two elements.

Since e is the identity, we must have $e * e = e$, $a * e = e * a = a$, and $b * e = e * b = b$.

Next, we work out the value of $a * b$. There are only three possibilities: e, a, and b. We show that $a * b$ must equal e by ruling out the other two possibilities.

If $a * b = b$, then operating on the right by b^{-1} gives $a*b*b^{-1} = b*b^{-1} \Rightarrow a = e$, which is false. Likewise, $a * b = a$ leads to $b = e$, which is also false. Therefore, $a * b = e$.

By a similar analysis, $b * a = e$.

Now we consider $a * a$; it might be e, a, or b. If $a * a = a$, then operating by a^{-1} would give $a = e$, a contradiction. If $a * a = e$, then operating on both sides by b gives

$$a * a * b = e * b$$
$$a * e = b$$
$$a = b,$$

another contradiction. Therefore $a * a = b$.

Likewise $b * b = a$.

Thus we have deduced the $*$ operation table.

$*$	e	a	b
e	e	a	b
a	a	b	e
b	b	e	a

It is now easy to see that $e \mapsto 0$, $a \mapsto 1$, and $b \mapsto 2$ is an isomorphism of $(G, *)$ to $(\mathbb{Z}_3, \oplus)$. ■

6. Consider the powers of 2 in $(\mathbb{Z}_{13}^*, \otimes)$:

2^0	2^1	2^2	2^3	2^4	2^5
1	2	4	8	3	6

2^6	2^7	2^8	2^9	2^{10}	2^{11}
12	11	9	5	10	7

Therefore the following function $f : \mathbb{Z}_{13}^* \to \mathbb{Z}_{12}$ is an isomorphism:

$1 \mapsto 0$	$2 \mapsto 1$	$3 \mapsto 4$	$4 \mapsto 2$
$5 \mapsto 9$	$6 \mapsto 5$	$7 \mapsto 11$	$8 \mapsto 3$
$9 \mapsto 8$	$10 \mapsto 10$	$11 \mapsto 7$	$12 \mapsto 6$

7. (a) $H \oplus K = \{0, 5, 10, 15, \ldots, 95\}$.

(b) **Proof.** Suppose that $(G, *)$ is an Abelian group and H and K are subgroups. To prove that $H * K$ is also a subgroup, we must show that $H * K$ is closed under $*$ and inverses.

To show that $H * K$ is closed under $*$, let $x, y \in H * K$. This means there exist $h_1, h_2 \in H$ and $k_1, k_2 \in K$ such that $x = h_1 * k_1$ and $y = h_2 * k_2$. Note that

$$x * y = (h_1 * k_1) * (h_2 * k_2) = (h_1 * h_2) * (k_1 * k_2)$$

and because H and K are subgroups, $h_1 * h_2 \in H$ and $k_1 * k_2 \in K$. Therefore $x * y \in H * K$.

To show that $H * K$ is closed under inverses, let $x \in H * K$. Then $x = h * k$ for some $h \in H$ and $k \in K$. Note that

$$x^{-1} = (h * k)^{-1} = k^{-1} * h^{-1} = h^{-1} * k^{-1}.$$

Because H and K are subgroups, we have that $h^{-1} \in H$ and $k^{-1} \in K$. Therefore $x^{-1} \in H * K$.

Thus $H * K$ is a subgroup of $(G, *)$. ■

(c) Let $(G, *) = (S_3, \circ)$, and let $H = \{\iota, (1, 2)\}$ and $K = \{\iota, (1, 3)\}$. Note that H and K are subgroups of $(S_3, \circ)$.

Then the elements of $H \circ K$ are $\iota \circ \iota = \iota$, $\iota \circ (1, 3) = (1, 3)$, $(1, 2) \circ \iota = (1, 2)$, and $(1, 2) \circ (1, 3) = (1, 3, 2)$.

Observe that although $(1, 3, 2) \in H \circ K$, its inverse $(1, 3, 2)^{-1} = (1, 2, 3)$ is not in $H \circ K$. Therefore $H \circ K$ is not a subgroup of $(S_3, \circ)$.

8. The elements of $\mathbb{Z}_{15}^*$ are $\{1, 2, 4, 7, 8, 11, 13, 14\}$. We calculate g^4 for each:

g	g^4	$g^4 \pmod{15}$
1	1	1
2	16	1
4	256	1
7	2401	1
8	4096	1
11	14641	1
13	28561	1
14	38416	1

Suppose, for the sake of contradiction, that $(\mathbb{Z}_{15}^*, \otimes)$ is cyclic. Then there is an element $g \in \mathbb{Z}_{15}^*$ such that the powers of g generate all the elements in $\mathbb{Z}_{15}^*$. But because $g^4 = 1$, the sequence

$$1 \quad g \quad g^2 \quad g^3 \quad g^4 \quad g^5 \quad \cdots$$

must repeat after four (or fewer) steps and therefore cannot include all eight elements of $\mathbb{Z}_{15}^*$. $\Rightarrow\Leftarrow$ Therefore $(\mathbb{Z}_{15}^*, \otimes)$ is not cyclic. ■

9. Because 89 is prime, Fermat's Little Theorem (Theorem 42.1) asserts that $2^{89} \bmod 89 = 2$. Since $2^{90} = 2^{89} \times 2$, we have $2^{90} \bmod 89 = 4$.

10. If n were prime, then Fermat's Little Theorem (Theorem 42.1) implies that $2^n \equiv 2 \pmod{n}$, but we are given that $2^n \not\equiv 2 \pmod{n}$. $\Rightarrow\Leftarrow$ Therefore n is composite.

11. By Euler's Theorem (Theorem 42.6), if a and n are relatively prime, $a^{\varphi(n)} \equiv 1 \pmod{n}$. Since 2 and 38168467 are clearly relatively prime, $2^{\varphi(n)} = 2^{38155320} \equiv 1 \pmod{n}$. However, we are asked to evaluate $2^{\varphi(n)+1} = 2^{\varphi(n)} \times 2$ modulo n, and so

$$2^{38155321} \bmod 38168467 = 2$$

12. $874^{256} \bmod 9432 = 1296$.

The calculation can be done by squaring 874 ten times, reducing modulo 9432 after each squaring.

$$\begin{array}{rl}
a = 874 & \text{square} \\
\to 763876 & \text{mod } 9432 \\
\to a^2 = 9236 & \text{square} \\
\to 85303696 & \text{mod } 9432 \\
\to a^4 = 1077 & \text{square} \\
\to 1159929 & \text{mod } 9432 \\
\to a^8 = 9103 & \text{square} \\
\to 82864609 & \text{mod } 9432 \\
\to a^{16} = 5137 & \text{square} \\
\to 26388769 & \text{mod } 9432 \\
\to a^{32} = 4668 & \text{square} \\
\to 21790224 & \text{mod } 9432
\end{array}$$

$$\begin{aligned}
\to a^{64} &= 9427 && \text{square} \\
&\to 88868329 && \text{mod } 9432 \\
\to a^{128} &= 36 && \text{square} \\
&\to 1296 && \text{mod } 9432 \\
\to a^{256} &= 1296
\end{aligned}$$

13. Since $p = 883$ is prime and $p \equiv 3 \pmod{4}$, we can find the square roots of 71 using Proposition 44.3:

$$\sqrt{71} = \pm 71^{(883+1)/4} \bmod 883 = \pm 707$$

and so the square roots of 71 are 707 and $-707 = 176$.

14. Let $p = 499$, $q = 883$, and $n = pq = 440617$. We want to find all $x \in \mathbb{Z}_n$ such that $x \otimes x = 1$.

In $\mathbb{Z}_p$, we have $x \equiv \pm 1 \pmod{p}$ and likewise in $\mathbb{Z}_q$. This gives rise to the following congruences.

$$\begin{array}{ll}
x \equiv 1 \pmod{p} & x \equiv 1 \pmod{p} \\
x \equiv 1 \pmod{q} & x \equiv -1 \pmod{q}
\end{array}$$

$$\begin{array}{ll}
x \equiv -1 \pmod{p} & x \equiv -1 \pmod{p} \\
x \equiv 1 \pmod{q} & x \equiv -1 \pmod{q}
\end{array}$$

These can be solved using the Chinese Remainder Theorem method, but note that the first gives $x \equiv 1 \pmod{n}$ and the last gives $x \equiv -1 \equiv n - 1 \pmod{n}$. Also, the solutions to the second and third are simply negatives of each other.

Solving $x \equiv 1 \pmod{p}$ and $x \equiv -1 \pmod{q}$ gives $x \equiv 429139 \pmod{440617}$. Therefore, the four square roots of 1 in $\mathbb{Z}_n$ are 1, $-1 = 440616$, 429139, and $-429139 = 11478$.

15. We are given four square roots of 1010120 in $\mathbb{Z}_n$ (with $n = 5460947$): $s, -s, t$, and $-t$. Calculating $\gcd(s+t, n)$ (or $\gcd(s-t, n)$) will reveal a prime factor of n.

$$\begin{aligned}
&\gcd(1235907 + 1842412, 5460947) \\
&= \gcd(3078319, 5460947) \\
&= 4547.
\end{aligned}$$

Note that 4547 is prime and $5460947/4547 = 1201$, so

$$5460947 = 4547 \times 1201.$$

16. Let m stand for the plaintext message. We have that $m^2 = 496410$ in $\mathbb{Z}_{713809}$. We find the four square roots of m^2 in $\mathbb{Z}_n$, and they are

$$160907 \quad 356083 \quad 357726 \quad 552902.$$

Of these, only the first corresponds to a word, and the word is PIG.

17. Since $n = pq = 453899 = 541 \times 839$, we have that $\varphi(n) = (p-1)(q-1) = 540 \times 838 = 452520$. Then d is e^{-1} in $\mathbb{Z}_{452520}$, which is 345689.

18. Using the decryption exponent $d = 345689$ we found in Problem 17, we find

$$\begin{aligned}
D_A(105015) &= 105015^{345689} \pmod{453899} \\
&= 190625.
\end{aligned}$$

The number 190625 is the word SPY.

19. Solving the pair of equations

$$\begin{aligned}
pq &= 40119451 \\
(p-1)(q-1) &= 40106592
\end{aligned}$$

for p and q gives the factors 5323 and 7537.

Chapter 9

1. One possible picture:

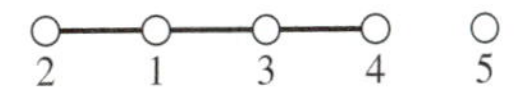

2. No such graph exists. If there were such a graph G, then the sum of the degrees of the vertices would be 43, an odd number, but the sum must equal twice the number of edges, an even number.$\Rightarrow\Leftarrow$

3. We offer two solutions.

First solution. Between W_i and W_j there are $10 \times 10 = 100$ edges. Since there are $\binom{10}{2} = 45$ ways to select the i and j, there are 4500 edges in G.

Second solution. Each vertex is adjacent to 90 others (the 9×10 in other W_js). Therefore, the sum of the degrees of the vertices in G is 9000, which is twice the number of edges. Therefore there are 4500 edges.

4. (a) 2^{10}. (b) 2^{15}.

5. A length-5 path from a to b is of the form

$$a \sim x_1 \sim x_2 \sim x_3 \sim x_4 \sim b$$

where we can choose the x_is without repetition from among the vertices in K_{10} other than a and b. There are $8 \cdot 7 \cdot 6 \cdot 5 = 1680$ possible paths.

6. (a) $f(0) = 0$, $g(0) = 1$, $f(1) = 1$, and $g(1) = 0$.

(b) Consider (a, b)-paths with $a \neq b$. Starting from a, the next vertex we choose might or might not be b. If the next vertex is b, then there are $g(k-1)$ ways to complete the path. If the next vertex is not b (and there are 8 possible ways in which that can happen), then there are $f(k-1)$ ways to complete the path. Therefore

$$f(k) = 8f(k-1) + g(k-1).$$

(c) Starting from a, there are 9 choices for the next vertex on an (a, a) path, and then there are $f(k-1)$ ways to complete the path. Therefore

$$g(k) = 9f(k-1).$$

(d) Using this information, we can build a chart of values for $f(k)$ and $g(k)$.

k	$f(k)$	$g(k)$
0	0	1
1	1	0
2	8	9
3	73	72
4	656	657
5	5905	5904

Therefore, there are 5905 length-5 paths from a to b (with $a \neq b$) in K_{10}.

7. Suppose, for the sake of contradiction, that G is not connected. Then G has two (or more) components. Let A be the vertices in a smallest component of G. Because A is a smallest component of G, it has no more than $n/2$ vertices. Choose $x \in A$. Note that x is adjacent only to other vertices in A (and not to itself) and so $d(x) < n/2$, contradicting the hypothesis that $\delta(G) \geq n/2$.⇒⇐ Therefore G is connected. ■

8. The number of 3-cycles is $\binom{5}{3} = 10$.

There are $4!/(4 \cdot 2) = 3$ different ways to make a cycle on four vertices, so the number of 4-cycles is $\binom{5}{4} \cdot 3 = 15$.

The number of 5-cycles on 5 vertices is $5!/(5 \cdot 2) = 12$.

Therefore the total number of cycles is $10 + 15 + 12 = 37$.

9. Let G be a connected graph with n vertices. Suppose the average degree of a vertex $\bar{d}$ is less than 2. Note that

$$\bar{d} = \frac{1}{n} \sum_{v \in V(G)} d(v) = \frac{2|E(G)|}{n}$$

The number of edges in G is therefore $\frac{1}{2} n\bar{d} < n$. Because G is connected, G has at least $n-1$ edges, and since G has fewer than n edges, it must be the case that G has exactly $n-1$ edges. Therefore, by Theorem 49.12, G is a tree. ■

10. The following two trees are not isomorphic, but the degrees of their vertices are the same.

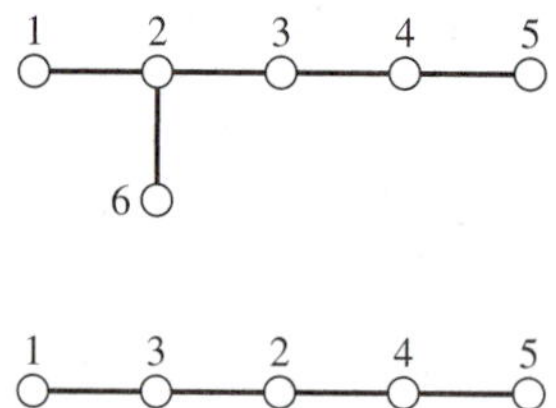

11. A disconnected graph on 10 vertices can have up to 36 edges, but no more. Here is why:

If G has 10 vertices and is disconnected, then G has two (or more) components. Let

A be the vertices in one component and let B be the remaining vertices. If we add edges to either A or B, we will not create a connected graph (since there are no edges between A and B). Thus we may assume that $G[A]$ and $G[B]$ are complete graphs.

Suppose $|A| = a$ where $1 \le a \le 9$. Then G has $\binom{a}{2} + \binom{10-a}{2}$ edges. If we evaluate this expression with various values of $a = 1, 2, \ldots, 9$, we find the largest value when $a = 1$ or $a = 9$; in this case, we find that G has 36 edges.

12. The following is a partition of $E(K_8)$ into four Hamiltonian paths.

$$1 \sim 8 \sim 2 \sim 7 \sim 3 \sim 6 \sim 4 \sim 5$$
$$2 \sim 1 \sim 3 \sim 8 \sim 4 \sim 7 \sim 5 \sim 6$$
$$3 \sim 2 \sim 4 \sim 1 \sim 5 \sim 8 \sim 6 \sim 7$$
$$4 \sim 3 \sim 5 \sim 2 \sim 6 \sim 1 \sim 7 \sim 8$$

Each of the $\binom{8}{2} = 28$ edges of K_8 is on one of these four paths.

However, no such partition of K_9 is possible. The graph K_9 contains $\binom{9}{2} = 36$ edges, and a Hamiltonian path of K_9 contains 8 edges. Were a partition of K_9 into p Hamiltonian paths possible, we would have $8p = 36$, but 36 is not divisible by 8. Therefore no such partition is possible.

13. *Existence*. We know that P and R have vertices in common since a is on both paths. Let x be the last vertex on P (as we traverse from a to b) that is also on R. Note that x is also the last vertex on R as we traverse from a to c (if there were a vertex after x on R that is also on P—call it y—then there would be two different (x, y)-paths in T).

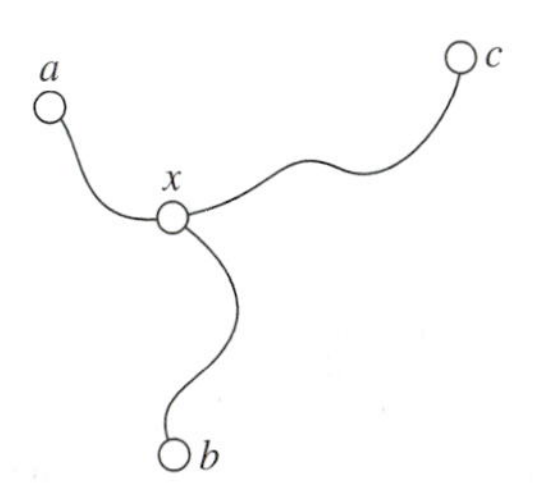

Thus the (b, x)-segment of P and the (c, x)-segment of Q have only x in common. If we concatenate these segments, we have a (b, c)-path S (since the two segments have no vertices in common other than x). Since there is one and only one (b, c)-path in T, namely R, it must be the case that $R = S$. Since x is a vertex of S, it must be a vertex of R, and so x is on all three paths: P, Q, and R.

Uniqueness. Suppose P, Q, and R have two (or more) vertices in common; let us call them x and y. Without loss of generality, as we traverse P from a to b we encounter x before y.

In T, there are a unique (a, x)-path and a unique (x, y)-path, and these must be the corresponding segments of P. Also, there is a unique (a, y)-path, and that path must contain x.

Since R also contains a, x, and y, the (a, x) and (x, y) segments of R must be identical to those of P, and x must be between a and y on R. See the figure.

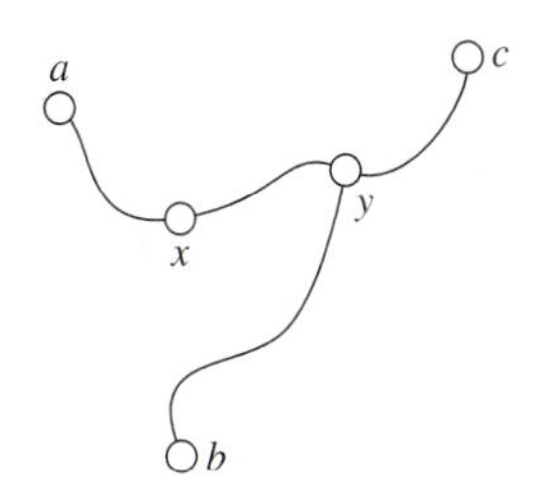

We now observe that the (b, y)-segment of P does not contain x and the (y, c)-segment of R does not contain x. Hence there is a (b, c)-walk that does not contain x. Therefore, there is a shortest (b, c)-walk that does not contain x, and that, necessarily is a (b, c)-path, which must be R. Therefore R does not contain x.$\Rightarrow\Leftarrow$ Therefore the paths P, Q, and R have exactly one vertex common to all three. ■

14. $(\Rightarrow)$ Suppose G is Eulerian. Let $A \cup B$ be a partition of $V(G)$ into disjoint, nonempty sets. Because G is Eulerian, it must be

connected, and so there must be at least one edge from A to B.

If we consider any Eulerian tour W, the number of times W takes an edge from A to B must equal the number of times W takes an edge from B to A (else the tour would not begin and end at the same vertex). Hence, the number of edges between A and B must be even.

($\Leftarrow$) Suppose G is a graph such that for every partition $A \cup B = V(G)$, the number of edges between A and B is even but not zero. Note that this implies that G is connected, for otherwise we could take A to be the vertex set of one component of G and $B = V(G) - A$; in this case there would be no edges between A and B.

Let v be any vertex of G. Let $A = \{v\}$ and $B = V(G) - A$. Note that the number of edges from A to B is exactly $d(v)$, and so $d(v)$ is even. Since G is connected and all vertices have even degree, G is Eulerian. ■

15. It is not hard to find a proper three-coloring of G (for example, color the vertices in a checkerboard fashion, except use a third color for the rightmost vertex in the last row). Therefore $\chi(G) \leq 3$. It is also not hard to find an odd cycle in G; hence G is not bipartite, so $\chi(G) > 2$. Therefore $\chi(G) = 3$.

16. For $n \geq 3$, we have

$$\chi(W_n) = \begin{cases} 3 & \text{if } n \text{ is odd, and} \\ 4 & \text{if } n \text{ is even.} \end{cases}$$

Proof. If n is odd, then the cycle in W_n contains an even number of vertices. These can be colored alternately using two colors, leaving a third color for the additional vertex; therefore $\chi(W_n) \leq 3$. At least three colors are required because W_n contains a complete graph on three vertices (any two consecutive vertices on the cycle plus the additional vertex).

If n is even, then the cycle in W_n contains an odd number of vertices. This odd cycle can be colored with three colors, leaving a fourth color for the additional vertex; therefore $\chi(W_n) \leq 4$. We claim that W_n cannot be colored with fewer than four colors. Suppose, for the sake of contradiction, that such a coloring is possible. The additional vertex (which is adjacent to all others) receives some color. Therefore none of the other vertices can use that color, leaving at most two colors for the vertices of the cycle. Since that cycle has an odd number of vertices, it cannot be colored with only two colors.$\Rightarrow\Leftarrow$ Therefore $\chi(W_n)$ is not less than 4. ■

17. Suppose the vertices of C_n are named, in order, $1, 2, 3, \ldots, n$.

Note that vertices $1, 3, 5, \ldots, \lfloor n/2 \rfloor$ form a clique in $\overline{C_n}$, and so $\chi(\overline{C_n}) \geq \lfloor n/2 \rfloor$.

In the case that n is even, we can color $\overline{C_n}$ properly with $n/2$ colors by assigning color 1 to vertices 1 and 2, color 2 to vertices 3 and 4, and so on. Thus, for n even, $\chi(\overline{C_n}) = n/2$.

In the case that n is odd, the color scheme described above will use $(n+1)/2$ colors (vertex n will be not be paired with another vertex of the same color). Therefore, if n is odd, $\chi(\overline{C_n}) \leq (n+1)/2$.

Can C_n (for n odd) be colored with fewer colors? If that were possible, there would be (at most) $(n-1)/2$ colors available, and so three distinct vertices would receive the same color. Since $n \geq 4$, three vertices of C_n cannot be pairwise adjacent and so cannot be given the same color in $\overline{C_n}$. Thus $\chi(G) > (n-1)/2$.

In conclusion, for $n \geq 4$,

$$\chi(\overline{C_n}) = \begin{cases} n/2 & \text{in case } n \text{ is even, and} \\ (n+1)/2 & \text{in case } n \text{ is odd.} \end{cases}$$

This can be written more concisely as $\chi(\overline{C_n}) = \lceil \frac{n}{2} \rceil$.

18. (a) ($\Rightarrow$) Suppose $\chi(G) \geq k$. This implies that G has at least one proper k-coloring, and hence $\chi(G, k) > 0$.

($\Leftarrow$) Suppose $\chi(G, k) > 0$. This implies that G has at least one proper

k-coloring, and so G is k-colorable. Therefore $\chi(G) \geq k$.

(b) We use Proof Template 25.

The proof is by induction on n. The basis case is when $n = 1$—that is, the tree has just one vertex. In this case, if there are k colors available, there are k different ways to color the sole vertex. Hence $\chi(G, k) = k = k(k-1)^0$, as required.

Suppose (induction hypothesis) that the result has been proved for all trees with ℓ vertices. Let T be a tree with $n = \ell + 1$ vertices. We must prove that $\chi(T, k) = k(k-1)^{n-1} = k(k-1)^{\ell}$.

Let v be a leaf of T and let $T' = T - v$. Note that T' is a tree on ℓ vertices. Therefore, by induction, $\chi(T', k) = k(k-1)^{\ell-1}$.

We now count proper k-colorings of T. Note that given a proper coloring of T, if we ignore vertex v, we have a proper coloring of T'. There are $\chi(T', k)$ ways to k-color T' properly. For each such coloring, there are $k-1$ ways to color v since v may be any color except the color assigned to its sole neighbor.

Thus $\chi(T, k) = \chi(T', k) \times (k-1) = k(k-1)^{\ell-1}(k-1) = k(k-1)^{\ell}$, as required. ■

19. The following drawing demonstrates that the graph is planar.

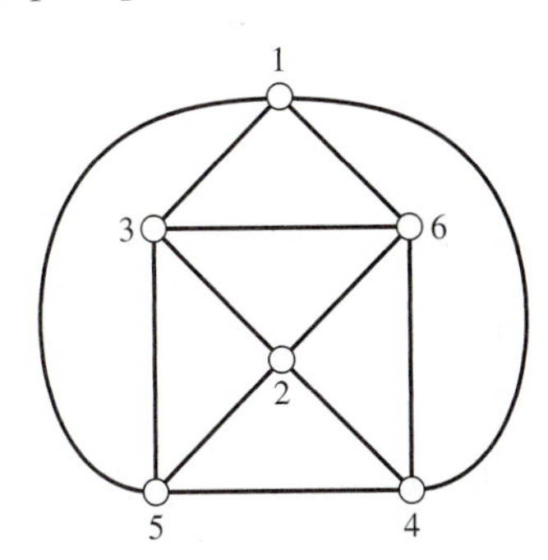

20. The graph $\overline{C_7}$ contains a subdivision of $K_{3,3}$ as illustrated in the following diagram.

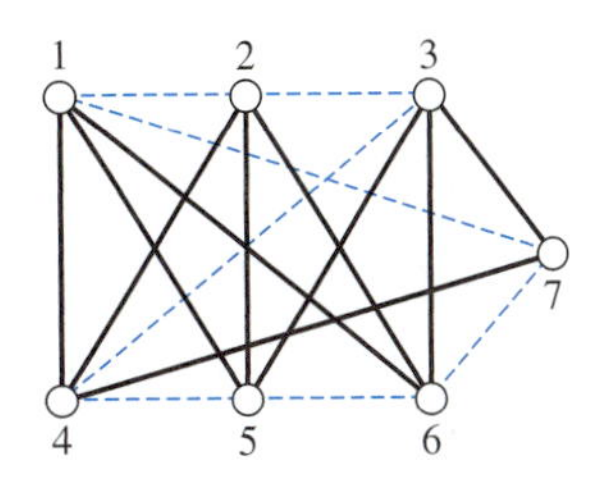

Notice that $\{1, 2, 3\}$ and $\{4, 5, 6\}$ form the two parts of the complete bipartite graph $K_{3,3}$. All edges of this $K_{3,3}$ are present in $\overline{C_7}$ except that the edge from 3 to 4 appears as the two-step path $3 \sim 7 \sim 4$. (To verify that all the edges shown belong to $\overline{C_7}$ the edges of C_7 are shown as colored, broken lines.)

Because $\overline{C_7}$ contains a subdivision of $K_{3,3}$ as a subgraph, it is nonplanar (see Theorem 52.9).

The graph $\overline{C_8}$ has $n = 8$ vertices and $m = \binom{8}{2} - 8 = 20$ edges. If $\overline{C_8}$ were planar, we would have $m \leq 3n - 6$ (see Corollary 52.5). However, $3n - 6 = 3 \times 8 - 6 = 18$ and $20 \not\leq 18$. Therefore $\overline{C_8}$ is nonplanar.

Alternatively, it is not difficult to show that $K_{3,3}$ is a subgraph of $\overline{C_8}$.

21. Suppose there are a vertices of degree 5. We know there are $a + 10$ vertices in this graph and $(5a + 7 \times 10)/2 = \frac{5}{2}a + 35$ edges. By Corollary 52.5, we have

$$\frac{5}{2}a + 35 \leq 3(a + 10) - 6 = 3a + 24$$

which simplifies to $11 \leq \frac{1}{2}a$ and so $a \geq 22$.

Chapter 10

1. (a) The following figure gives the Hasse diagram of P.

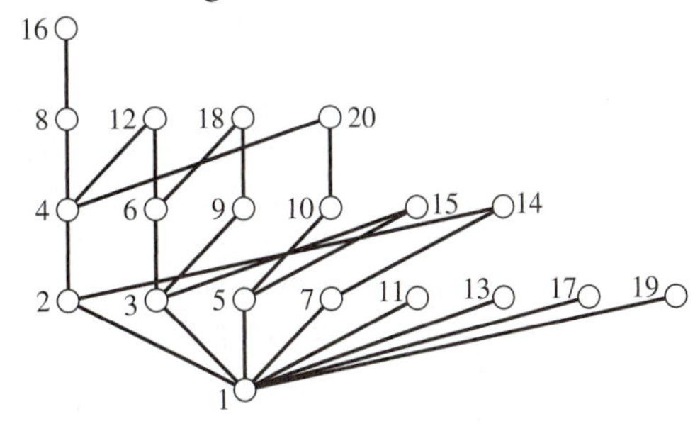

(b) The largest chain in P is $\{1, 2, 4, 8, 16\}$.
(c) The largest antichain in P is

$$\{11, 12, 13, 14, 15, 16, 17, 18, 19, 20\}.$$

(d) The set of maximal elements of P is

$$\{11, 12, 13, 14, 15, 16, 17, 18, 19, 20\}.$$

(e) The set of minimal elements of P is $\{1\}$.
(f) The set of maximum elements of P is $\emptyset$.
(g) The set of minimum elements of P is $\{1\}$.

2. Suppose, for the sake of contradiction, that $C \cap A$ contains two distinct elements x and y. Because $x, y \in C$, we know that $x < y$ or $y < x$; that is, x and y are comparable. However, because $x, y \in A$, we know that x and y are incomparable.$\Rightarrow\Leftarrow$ Therefore $C \cap A$ cannot contain two (or more) elements, and so $|C \cap A| \leq 1$.

3. Let A be an antichain of P. We know (from Problem 2) that A can have at most one element in C_1 and at most one element in C_2. Therefore A can have at most two elements, and so the maximum size of an antichain of P cannot be greater than 2.

4. ($\Rightarrow$) Suppose $P = (X, \leq)$ is an antichain. Let $x \in X$. Since there is no element y such that $y < x$, we have that x is minimal. Likewise, x is maximal. Therefore all elements of X are both maximal and minimal.

($\Leftarrow$) Suppose every element of X is both maximal and minimal. We claim that P is an antichain. If not, there would be elements $x \neq y$ in X with $x < y$. But then x is not maximal and y is not minimal.$\Rightarrow\Leftarrow$ Therefore P is an antichain. ■

5. (a) Let $P = (X, \leq)$ be a finite chain. By Theorem 55.4, we may assume that $X = \{1, 2, \ldots, n\}$ and $\leq$ is ordinary less than or equal to. For j between 1 and n, we let $A_j = \{j\}$. Because the A_js contain only one element, they are antichains. Note that $X = A_1 \cup \cdots \cup A_n$ and if $x \in A_i$ and $y \in A_j$ with $i < j$, then we must have $x < y$ (indeed, $x = i$ and $y = j$). Therefore P is a weak order.

Let $P = (X, \leq)$ be an antichain. Then simply letting $A_1 = X$ gives the required partition.

(b) Let $P = (X, \leq)$ be a finite poset and let Q be the three-element poset depicted in the figure for this problem.

($\Rightarrow$) Suppose P is a weak order but, for the sake of contradiction, contains Q as a subposet. Since P is a weak order, we can partition X into antichains $X = A_1 \cup \cdots \cup A_h$ such that for all $x \in A_i$ and $y \in A_j$, if $i < j$ then $x < y$. Now consider elements a, b, c of Q. Since $a < b$, we must have that $a \in A_i$ and $b \in A_j$ where $i < j$. Suppose $c \in A_k$. Note that since $i < j$ we must have that either $k < j$ or $k > i$. If $k < j$ we would have $c < b$, and if $k > i$ we would have $c > a$; but c is incomparable to both a and b.$\Rightarrow\Leftarrow$ Therefore Q is not a subposet of P.

($\Leftarrow$) Suppose P is a finite poset that does not contain Q as a subposet. Let A_1 be the set of all minimal elements of P. Let A_2 be the set of all minimal elements of $P - A_1$ (that is, the poset formed by deleting the elements of A_1 from P). Let A_3 be the set of all minimal elements in $P - A_1 - A_2$. We continue in this fashion, choosing A_t to be the set of all minimal elements of $P - A_1 - A_2 - \cdots - A_{t-1}$ until there are no elements left. Note that each A_j is an antichain (since it is the set of minimal elements of some subposet of P) and the A_js partition X. It remains to show that if $x \in A_i$ and $y \in A_j$ and $i < j$, then $x < y$.

Suppose, for the sake of contradiction, that $x \not< y$. Note that x is a minimal element of the poset $P - A_1 - \cdots - A_{i-1}$, and so we cannot have $x > y$. Therefore x and y are incomparable. Also, y is not a minimal element of $P - A_1 - \cdots - A_{i-1}$ (because $y \in A_j$ and $j > i$), and so there is some element z in $P - A_1 - \cdots - A_{i-1}$ with $y > z$.

It cannot be the case that $z < x$ because x is minimal, and it cannot be the case that $z \geq x$ for then we would have $y > z \geq x$, implying that x and y are comparable. Therefore z and x are incomparable. That is, we have x incomparable to both y and z, and $z < y$. Therefore we have a copy of Q (with $a = z, b = y$, and $c = x$) as a subposet of P.$\Rightarrow\Leftarrow$ Therefore $x < y$ and so P is a weak order. ■

(c) In a linear extension of P we must have all elements in A_i below all elements of A_j (where $i < j$), but it does not matter in what order the elements of A_i (or A_j) appear among themselves. There are $k!$ ways to arrange the elements of each A_i, and each of the h antichains can be arranged in any way irrespective of the others. Therefore there are $k!^h$ linear extensions of P.

(d) Let $P = (X, \leq)$ be a weak order. To show that $\dim P \leq 2$, we find two linear extensions $L_1 = (X, \leq')$ and $L_2 = (X, \leq'')$ such that $\mathcal{R} = \{L_1, L_2\}$ is a realizer of P.

Since P is a weak order it can be partitioned into antichains by $X = A_1 \cup \cdots \cup A_h$. Let n_i be the number of elements in A_i and let us name the elements of A_i as follows:

$$A_i = \{a_{i,1}, a_{i,2}, \ldots a_{i,n_i}\}.$$

We define L_1 and L_2 as follows:
The order L_1 is given by

$$\begin{aligned} &a_{1,1} <' a_{1,2} <' \cdots <' a_{1,n_1} <' \\ &\quad <' a_{2,1} <' a_{2,2} <' \cdots <' a_{2,n_2} <' \\ &\quad <' \cdots\cdots <' \\ &\quad <' a_{h,1} <' a_{h,2} <' \cdots <' a_{h,n_n} \end{aligned}$$

and the order L_2 is given by

$$\begin{aligned} &a_{1,n_1} <'' a_{1,n_1-1} <'' \cdots <'' a_{1,1} <'' \\ &\quad <'' a_{2,n_2} <'' a_{2,n_2-1} <'' \cdots <'' a_{2,1} \\ &\quad <'' \cdots\cdots <'' \\ &\quad <'' a_{h,n_h} <'' a_{h,n_n-1} <''< \cdots <'' a_{h,1}. \end{aligned}$$

Note that both L_1 and L_2 are linear extensions of P since, for $i < j$, all elements of A_i are $<'$ or $<''$ all elements of A_j. Thus if $x \leq y$ in P, it must be the case that $x \leq' y$ and $x \leq'' y$. Conversely, if x and y are incomparable in P, then $x, y \in A_j$ for some j. In this case we see that $x <' y$ and $x >'' y$ (or vice versa). Therefore $\mathcal{R}$ is a realizer of P and so $\dim P \leq 2$.

6. (a) Let $P = (X, \leq)$ be a finite chain. By Theorem 55.4, we may assume that $X = \{1, 2, \ldots, n\}$ and $\leq$ is ordinary less than or equal to. Let $[a_j, b_j] = [j, j + \frac{1}{2}]$ and note that if $i < j$, then $[a_i, b_i]$ is entirely to the left of $[a_j, b_j]$ as required. Therefore P is an interval order.

Let $P = (X, \leq)$ be an antichain. For all $x \in X$, let $[a_x, b_x] = [0, 1]$. Note that for all $x \neq y$ in X, the elements x and y are incomparable and the intervals $[a_x, b_x]$ and $[a_y, b_y]$ intersect as required. Therefore P is an interval order.

Note: Part (a) of this problem also follows as a corollary of part (b).

(b) Let $P = (X, \leq)$ be a weak order. Then we can partition X into antichains $A_1 \cup \cdots \cup A_h$ so that for all $x \in A_i$ and $y \in A_j$ with $i < j$, we have $x < y$. To show that P is an interval order, we assign intervals as follows:

For $x \in A_j$ let $[a_x, b_x] = [j, j+\frac{1}{2}]$.

Note that if $x, y \in A_j$, then x and y are incomparable, and $[a_x, b_x]$ intersects $[a_y, b_y]$, as required. However, if $x \in A_i$ and $y \in A_j$ with $i < j$, then $x < y$, and note that $[a_x, b_x] = [i, i+\frac{1}{2}]$ is entirely to the left of $[a_y, b_y] = [j, j + \frac{1}{2}]$, as required.

(c) Let P be the poset in the figure. Suppose, for the sake of contradiction, that P is an interval order and so there are intervals $[a_x, b_x]$, $[a_y, b_y]$, $[a_w, b_w]$, and $[a_z, b_z]$ with the properties that

1. $[a_x, b_x]$ is to the left of $[a_y, b_y]$,
2. $[a_w, b_w]$ is to the left of $[a_z, b_z]$,
3. $[a_x, b_x]$ intersects both $[a_w, b_w]$ and $[a_z, b_z]$, and
4. $[a_y, b_y]$ intersects both $[a_w, b_w]$ and $[a_z, b_z]$.

Because (by 2) $[a_w, b_w]$ is to the left of $[a_z, b_z]$ and (by 3) the interval $[a_x, b_x]$ must intersect both $[a_w, b_w]$ and $[a_z, b_z]$; therefore the interval $[a_x, b_x]$ must completely span the gap between the two intervals $[a_w, b_w]$ and $[a_z, b_z]$.

$[a_x, b_x]$

$[a_w, b_w]$ $[a_z, b_z]$

Similarly, interval $[a_y, b_y]$ must also span the gap between $[a_w, b_w]$ and $[a_z, b_z]$, and therefore $[a_x, b_x]$ must intersect $[a_y, b_y]$, a contradiction (to 1).

(d) It is not hard to see that the poset in the figure is an interval order; use the following intervals:

$$x \leftrightarrow [0, 1] \qquad y \leftrightarrow [2, 3]$$
$$z \leftrightarrow [4, 5] \qquad w \leftrightarrow [0.5, 4.5].$$

Suppose, for the sake of contradiction, that there is some choice of the intervals such that they all have the same length. Note that we must have that $[a_x, b_x]$ to the left of $[a_y, b_y]$, which in turn is to the left of $[a_z, b_z]$, but $[a_w, b_w]$ must intersect the other three; see the figure.

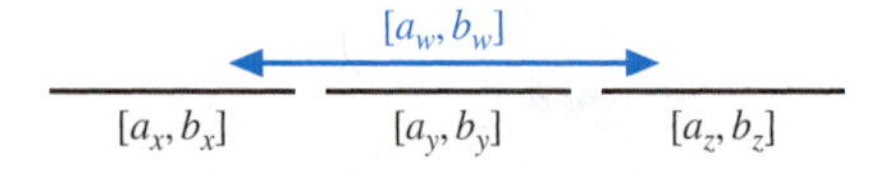

In order for $[a_w, b_w]$ to intersect both $[a_x, b_x]$ and $[a_z, b_z]$, it must completely contain $[a_y, b_y]$ as well as the gaps between $[a_x, b_x]$, $[a_y, b_y]$, and $[a_z, b_z]$. Therefore the length of $[a_w, b_w]$ must be greater than that of $[a_y, b_y]$.⇒⇐

Therefore P cannot be represented by intervals that all have the same length.

7. The poset has 16 linear extensions. For each antichain of size 2, there are two choices for the order of the elements in the linear extension, and this choice can be made for each of the four size-2 antichains, for a total of $2^4 = 16$ linear extensions.

8. (a) The pairs of elements that are incomparable are $\{a, c\}$, $\{a, f\}$, $\{c, d\}$, $\{b, d\}$, $\{b, f\}$, $\{b, e\}$, and $\{d, f\}$.

(b) Here are three linear orders whose intersection gives P:

$$L_1: \quad c < f < a < b < d < e$$
$$L_2: \quad a < b < d < c < f < e$$
$$L_3: \quad a < c < f < d < e < b$$

The following chart shows, for each incomparable pair $\{x, y\}$, the extensions in which $x < y$ and $y < x$.

$\{x, y\}$	$x < y$	$y < x$
$\{a, c\}$	2,3	1
$\{a, f\}$	2,3	1
$\{c, d\}$	1,3	2
$\{b, d\}$	1,2	3
$\{b, f\}$	2	1,3
$\{b, e\}$	1,2	3
$\{d, f\}$	2	1,3

(c) Suppose there were a linear extension L in which $f < a$ and $d < c$. Since $a < d$ in P, we must have that $a < d$ in L, and so, in L, $f < a < d < c$, which implies that $f < c$ in L as well. But this contradicts the fact that $c < f$ in P.⇒⇐ Thus we cannot have both $f < a$ and $d < c$.

Suppose there were a linear extension in L in which $b < f$ and $f < a$. Then $b < a$ in L, but $a < b$ in P.⇒⇐ Thus we cannot have both $b < f$ and $f < a$.

Likewise, we cannot have $b < d < c$ in a linear extension L, for then $b < c$ in L, but $c < b$ in P.

Likewise, we cannot have $e < b < d$ in a linear extension L, for then $e < d$ in L, but $d < e$ in P.

Finally, we cannot have $e < b < f$ in a linear extension L, for then $e < f$ in L, but $f < e$ in P.

(d) If any two of the relations $f < a, d < c, b < d, e < b$, and $b < f$ held in a linear extension L, then we would contradict one of the statements in part (c) of this problem.

(e) From part (b) we know that $\dim P \leq 3$; hence it remains to show that $\dim P > 2$. Suppose, for the sake of contradiction, that $\dim P \leq 2$, so P has a realizer with (at most) two linear extensions. Then one of those linear extensions would satisfy at least three of $f < a$, $d < c$, $b < d$, $e < b$, and $b < f$, contradicting part (d).$\Rightarrow\Leftarrow$ Therefore $\dim P = 3$.

9. Suppose, for the sake of contradiction, that P contains two distinct elements x and y. Note that $x \wedge y \leq x \leq x \vee y$, and since $x \wedge y = x \vee y$, we have that $x \wedge y = x$. Likewise $x \wedge y = y$, and so $x = y$.$\Rightarrow\Leftarrow$ Therefore P has at most one element. ■

10. (a) $\mathcal{P} \wedge \mathcal{Q} = \{\{1,3\},\{2,4\},\{5,7,9\},\{6,8\}\}$ and $\mathcal{P} \vee \mathcal{Q} = \{\{1,2,3,4,5,6,7,8,9\}\}$.

(b) Let $Z_k \in \mathcal{R}$. Because $\mathcal{R} = \mathcal{P} \wedge \mathcal{Q}$, it refines both $\mathcal{P}$ and $\mathcal{Q}$. This means that every part of $\mathcal{R}$ is a subset of a part of $\mathcal{P}$, and likewise of $\mathcal{Q}$. In particular, there is an $X_i \in \mathcal{P}$ such that $Z_k \subseteq X_i$ and a $Y_j \in \mathcal{Q}$ such that $Z_k \subseteq Y_j$. Thus $Z_k \subseteq X_i \cap Y_j$.

Suppose, for the sake of contradiction, that $Z_k \neq X_i \cap Y_j$. Then there must be some other part of $\mathcal{R}$, say $Z_{k'}$ that intersects $X_i \cap Y_j$. This implies that $Z_{k'}$ must be a subset of both X_i and Y_j because every part of $\mathcal{R}$ must be a subset of a part of $\mathcal{P}$ and of a part of $\mathcal{Q}$.

We can now form a new partition from $\mathcal{R}$ simply by combining Z_k and $Z_{k'}$ into a single part $Z_k \cup Z_{k'}$; call this new partition $\mathcal{R}'$. Note that $\mathcal{R}$ is strictly finer than $\mathcal{R}'$, and $\mathcal{R}'$ refines both $\mathcal{P}$ and $\mathcal{Q}$. However, this contradicts the fact that $\mathcal{R} = \mathcal{P} \wedge \mathcal{Q}$ (i.e., $\mathcal{R}$ is the coarsest common refinement of $\mathcal{P}$ and $\mathcal{Q}$).$\Rightarrow\Leftarrow$ Therefore $Z_k = X_i \cap Y_j$.

11. The proof is by induction on n. The basis case, $n = 1$, is obvious. Suppose (induction hypothesis) this result has been proved for $n = k$. Let $a, x_1, \ldots, x_{k+1}$ be given such that $a \leq x_j$ for all $1 \leq j \leq k+1$. We know (by induction) that $a \leq x_1 \wedge \cdots \wedge x_k$ and we know that $a \leq x_{k+1}$ (by hypothesis). This means that a is a lower bound for $x_1 \wedge \cdots \wedge x_k$ and x_{k+1}, so the a is below the greatest lower bound of $x_1 \wedge \cdots \wedge x_k$ and x_{k+1}; that is, $a \leq (x_1 \wedge \cdots \wedge x_k) \wedge x_{k+1}$, as required.

12. To prove that $a \vee b = u_1 \wedge u_2 \wedge \cdots \wedge u_n$, we prove $a \vee b \leq u_1 \wedge u_2 \wedge \cdots \wedge u_n$ and $a \vee b \geq u_1 \wedge u_2 \wedge \cdots \wedge u_n$.

To show $a \vee b \leq u_1 \wedge u_2 \wedge \cdots \wedge u_n$: We know, using Problem 11, that $a \leq u_1 \wedge u_2 \wedge \cdots \wedge u_n$ and $b \leq u_1 \wedge u_2 \wedge \cdots \wedge u_n$. Therefore $u_1 \wedge u_2 \wedge \cdots \wedge u_n$ is an upper bound for a and b. Since $a \vee b$ is the least upper bound of a and b, we have $a \vee b \leq u_1 \wedge u_2 \wedge \cdots \wedge u_n$.

To show $a \vee b \geq u_1 \wedge u_2 \wedge \cdots \wedge u_n$: We know that $a \leq a \vee b$ and $b \leq a \vee b$. Therefore $a \vee b \in U(a, b)$. Without loss of generality, $u_1 = a \vee b$. So the expression $u_1 \wedge u_2 \wedge \cdots \wedge u_n$ can be rewritten $(a \vee b) \wedge (u_2 \wedge \cdots \wedge u_n)$. Note that this latter expression is the greatest lower bound of $a \vee b$ and $u_2 \wedge \cdots \wedge u_n$, and so $(a \vee b) \wedge (u_2 \wedge \cdots \wedge u_n) \leq a \vee b$. Therefore $u_1 \wedge u_2 \wedge \cdots \wedge u_n \leq a \vee b$.

C Glossary

This glossary provides quick reminders of concepts presented in the main text. Please consult the index for pages in the main text that contain a more thorough and rigorous presentation.

A

A See **the.**

Abelian group A group whose operation is commutative; i.e., $g * h = h * g$ for all g and h in the group.

Acyclic Having no cycles. See **forest.**

Adjacent v and w are adjacent if vw is an edge. Notation: $v \sim w$.

Algorithm A precisely defined sequence of calculations.

And The statement "A and B" is true exactly when both A and B are true. In Boolean algebra, $a \wedge b$.

Antichain A subset of a poset all of whose elements are incomparable to each other.

Antisymmetric A relation R is antisymmetric means that for all a and b, if $a\ R\ b$ and $b\ R\ a$, then $a = b$.

Arbitrary Without any restrictions, completely general, generic.

Argument A proof.

Associative property $a * (b * c) = (a * b) * c$ for all a, b, c.

B

Basis step Part of a proof by induction in which the truth of the result is established in the smallest allowable case.

Bernoulli trial A sample space with exactly two outcomes, often called *success* and *failure*.

Bijection A one-to-one and onto function.

Binomial coefficient The number of k-element subsets of an n-element set; denoted $\binom{n}{k}$.

Binomial random variable The number of successes in a finite sequence of independent Bernoulli trials;

$$P(X = a) = \binom{n}{a} p^a (1 - p)^{n-a}$$

where $n, a \in \mathbb{N}$ and $0 \le p \le 1$.

Binomial Theorem For $n \in \mathbb{N}$,

$$(x + y)^n = \sum_{k=0}^{n} \binom{n}{k} x^k y^{n-k}.$$

Bipartite Two-colorable.

Birthday problem What is the probability that among n randomly chosen people some pair of people have the same birthday?

Boolean algebra Calculations and expressions involving the values TRUE and FALSE and the operations $\wedge$, $\vee$, $\neg$, etc.

C

$\mathbb{C}$ The complex numbers.

Cardinality The size of a set; i.e., the number of elements in that set. The cardinality of A is denoted $|A|$.

Carmichael number A positive integer n that is not prime, but $a^n \equiv a \pmod{n}$ for all integers a with $1 \le a < n$.

Cartesian product $A \times B$ is the set of all ordered pairs of the form (a, b) where $a \in A$ and $b \in B$.

Ceiling The ceiling of x is the least integer greater than or equal to x; denoted $\lceil x \rceil$. See also **floor.**

Chain A subset of a poset all of whose elements are comparable to each other.

Characterization theorem An if-and-only-if theorem that gives an alternative description of a mathematical concept.

Chinese Remainder Theorem Technique for solving a pair of modular congruences.

Chromatic number The least k such that G is k-colorable; denoted $\chi(G)$.

Claim A statement proved during the course of a proof.

Clique A set of pairwise adjacent vertices.

Clique number Maximum size of a clique; denoted $\omega(G)$.

Colorable A graph is k-colorable if it has a proper k-coloring.

Coloring A k-coloring of G is a function $f: V(G) \to \{1, 2, \ldots, n\}$, which is *proper* if $xy \in E(G) \Rightarrow f(x) \neq f(y)$.

Combinatorial proof A proof by counting.

Common divisor A common divisor of $a, b \in \mathbb{Z}$ is an integer d with $d|a$ and $d|b$.

Commutative property $a * b = b * a$ for all a, b.

Comparable Elements x and y in a poset for which $x \leq y$ or $y \leq x$.

Complement (graph) $\overline{G}$ is the graph with the same vertices as G in which distinct vertices are adjacent iff they are not adjacent in G.

Complement (set) $\overline{A}$ is the set of elements not in A.

Complete bipartite graph A graph $V(G) = A \cup B$, with $A \cap B = \emptyset$ and $E(G) = \{ab : a \in A, b \in B\}$. Denoted $K_{a,b}$ where $a = |A|$ and $b = |B|$.

Complete graph A graph in which every pair of distinct vertices is adjacent; denoted K_n.

Complex number A number of the form $a + bi$ where $a, b \in \mathbb{R}$ and $i^2 = -1$.

Component A maximal connected subgraph.

Composite A positive integer equal to the product of two smaller positive integers.

Composition $(g \circ f)(x) = g[f(x)]$.

Concatenation Merging two lists together to form a longer list. In particular, the concatenation of walks in a graph is a new walk formed by combining the two walks.

Conclusion The *then* part of an if-then statement.

Conditional probability The probability of one event given another; $P(A|B) = P(A \cap B)/P(B)$.

Congruent (mod n) $a \equiv b \pmod{n}$ means $a - b$ is divisible by n.

Conjecture A statement believed to be true, but for which no proof or counterexample has been found.

Connected Vertex u is connected to vertex v means there is a (u, v)-path in the graph. The graph is connected means every pair of vertices is connected.

Contradiction A statement that is blatantly false. A Boolean expression that yields FALSE for all values of its variables.

Contrapositive The contrapositive of "If A, then B" is "If not B, then not A."

Converse The converse of "If A, then B" is "If B, then A."

Corollary A statement that can be proved readily from another theorem.

Counterexample An example that demonstrates that a statement is false.

Cryptography The art of concealing messages in secret codes.

Cube A graph whose vertices are all length-n lists of 0s and 1s in which two vertices are adjacent iff their lists disagree in exactly one location. Also called a **hypercube.**

Cut edge An edge e of G such that $G - e$ has more components than G.

Cut vertex A vertex v of G such that $G - v$ has more components than G.

Cycle A walk with at least three vertices in which the only repeated vertex is the first/last. Also, a graph of this form, C_n.

Cycle notation A notation for writing permutations as parenthesized collections of elements.

Cyclic group A group generated by a single element.

D

Definition A precise statement creating a new mathematical concept.

Degree (face) The number of edges bounding a face in a planar embedding of a graph; if both sides of an edge are on the face, that edge counts twice.

Degree (polynomial) The highest power on the variable.

Degree (vertex) $d(v)$ is the number of edges incident with v.

Derangement A permutation π with the property that $\pi(x) \neq x$ for all x.

Difference (set) $A - B$ is the set of all elements of A that are not in B.

Dimension The dimension of a poset is the smallest size of a realizer for that poset.

Direct proof A proof technique that proceeds from the hypothesis to the conclusion.

Disjoint Having nothing in common; i.e., $A \cap B = \emptyset$. See also **pairwise disjoint.**

Distance The length of a shortest path between a specified pair of vertices.

Distinct Unequal. If we say "Let x, y, and z be distinct numbers," we mean that $x \neq y$, $x \neq z$, and $y \neq z$.

Div a div b is the quotient when we divide a by b.

Divides $a|b$ means there is an integer c with $b = ac$.

Domain The set of first elements of the ordered pairs in a function; denoted dom f.

E

Edgeless Having no edges.

Element A member of a set. $x \in A$ means x is an element of A.

Empty set The set with no elements; denoted $\emptyset$.

Equivalence class $[a] = \{x : xRa\}$ where R is an equivalence relation. That is, $[a]$ is the set of all elements equivalent to a by the relation R.

Equivalence relation A relation that is reflexive, symmetric, and transitive.

Equivalent statements Two (or more) statements are equivalent if each implies the other(s).

Euclid's Algorithm A method to find the gcd of two integers. Extended version is useful for finding modular reciprocals.

Euler's formula (graph theory) If a planar graph with n vertices, m edges, and c components is drawn in the plane with f faces, then $n - m + f - c = 1$.

Euler's Theorem (number theory) $a^{\varphi(n)} \equiv 1 \pmod{n}$. See also **Fermat's Little Theorem** and **totient.**

Eulerian An Eulerian *trail* is a walk in a graph that traverses each edge exactly once. An Eulerian *tour* is such a walk that begins and ends at the same vertex. An Eulerian *graph* is a graph in which there is an Eulerian tour.

Even (integer) An integer that is divisible by 2.

Even (permutation) A permutation equal to the composition of an even number of transpositions.

Event A subset of a sample space.

Exactly Compare the following sentences:

- There are three numbers with property X.
- There are *exactly* three numbers with property X.

The first sentence may (and often is) interpreted to mean that there are three or more different numbers with property X. However, the second sentence means that there are three (no more, no fewer) different numbers that satisfy property X.

Exclusive or $a \veebar b$, which is true exactly when a or b, but not both, is true. Also written **xor.**

Existential quantifier $\exists$, meaning *there is* or *there exists.*

Expected value The weighted average of a random variable; $E(X) = \sum_s X(s)P(s)$.

F

Fact A simple theorem.

Factorial $n! = n(n-1)(n-2)\cdots 3 \cdot 2 \cdot 1$. Also: $0! = 1$.

Fermat's Little Theorem If p is a prime, then $a^p \equiv a \pmod{p}$.

Fibonacci numbers The sequence 1, 1, 2, 3, 5, 8, 13, ... in which each term equals the sum of the two previous terms.

Finer See **refine.**

Floor The floor of x is the greatest integer less than or equal to x; denoted $\lfloor x \rfloor$. See also **ceiling.**

Forest An acyclic graph.

Four Color Theorem If G is planar, then $\chi(G) \leq 4$.

Function A *function* is a set of ordered pairs f with the property that if $(x, y) \in f$ and $(x, z) \in f$, then $y = z$. $(x, y) \in f$ is usually written $y = f(x)$.

Fundamental Theorem of Arithmetic Every positive integer can be uniquely represented as a product of primes.

G

gcd The greatest common divisor.

glb The greatest lower bound.

Graph A pair (V, E) where V is a finite set and E is a set of two-element subsets of V.

Greatest common divisor The largest common divisor (factor) of a pair of integers. Abbreviated **gcd.**

Group A set with an operation that is closed, is associative, has an identity, and every element of which has an inverse.

Guinea pig A cute rodent of the genus *Cavia* having no tail to speak of.

H

Hamiltonian path, cycle, graph A path [cycle] of a graph that contains all the vertices in the graph. A Hamiltonian graph is a graph with a Hamiltonian cycle.

Hasse diagram A diagram representing a poset.

Height The size of a largest chain.

Hypercube See **cube.**

Hypothesis The *if* part of an if-then statement.

I

Identity element (group) An element e of a group $(G, *)$ with the property that $g * e = e * g = g$ for all $g \in G$.

Identity function, permutation A function $f : A \to A$ given by $f(x) = x$ for all $x \in A$; denoted id_A in general and ι in the context of permutations.

Iff If and only if.

Image The set of all possible outputs of a function; if $f : A \to B$, the image of f is $\{f(a) : a \in A\} \subseteq B$.

Incident Vertex v and edge e are incident provided $v \in e$; i.e., v is an endpoint of e.

Inclusion-exclusion A counting technique for finding the cardinality of a union of sets based on the sizes of the various intersections of these sets.

Incomparable Not comparable; i.e., elements x and y for which $x \not\preceq y$ and $y \not\preceq x$.

Independence number The maximum size of an independent set; denoted $\alpha(G)$.

Independent Events A and B are independent means that $P(A \cap B) = P(A)P(B)$. Random variables X and Y are independent means that the events $X = a$ and $Y = b$ are independent for all a, b.

Independent set A set of vertices no two of which are adjacent. Also called a **stable** set.

Indicator random variable A random variable whose value is 1 if a given event occurs and is 0 otherwise.

Indirect proof See **proof by contradiction.**

Induced subgraph A subgraph formed by vertex deletion.

Induction A proof technique described in Section 21. See Proof Templates 17 and 18.

Induction hypothesis An assumption in a proof by induction that the result is true for a certain case size; it is used to establish the result for the next case size.

Injection A one-to-one function.

Integers $\mathbb{Z} = \{\ldots, -3, -2, -1, 0, 1, 2, 3, \ldots\}$.

Intersection $A \cap B$ is the set of all elements in both A and B.

Inverse (function) If $f : A \to B$ is a bijection, then the inverse relation f^{-1} is also function, $f^{-1} : B \to A$. See **inverse (relation).**

Inverse (group element) If $(G, *)$ is a group and $g \in G$, then h is the inverse of G provided $g * h = h * g = e$ where e is the identity element. The inverse is denoted g^{-1}.

Inverse (number theory) See **reciprocal.**

Inverse (permutation) If π is a permutation, it is a bijection from some set to itself. Thus the inverse function π^{-1} is also a permutation on that set. Also, π^{-1} is the group inverse of π in the symmetric group. Thus $\pi \circ \pi^{-1} = \pi^{-1} \circ \pi = \iota$.

Inverse (relation) R^{-1} is the relation formed from R by replacing each ordered pair (x, y) with (y, x); i.e., $R^{-1} = \{(y, x) : (x, y) \in R\}$.

Inverse (statement) The inverse of "If A, then B" is "If not A, then not B."

Inversion Given a permutation π of $\{1, 2, \ldots, n\}$, an inversion is a pair of values $i < j$ for which $\pi(i) > \pi(j)$.

Invertible Having an inverse.

Irrational A number that is not a rational number.

Irreflexive A relation R is irreflexive if $x\,R\,x$ is always false.

Isolated vertex A vertex of degree 0.

Isometry A distance-preserving function.

Isomorphism (graphs) A bijection f between the vertex sets of two graphs such that xy is an edge iff $f(x)f(y)$ is an edge.

Isomorphism (group) A bijection f between two groups such that $f(g * h) = f(g) \star f(h)$.

Isomorphism (posets) A bijection f between two posets such that $x \le y$ iff $f(x) \le f(y)$.

J

Join $a \vee b$ is the greatest lower bound of a and b.

K

Kuratowski's Theorem A graph is planar iff it does not contain a subdivision of K_5 or $K_{3,3}$ as a subgraph.

L

Lagrange's Theorem The size of a finite group is divisible by the size of any of its subgroups.

Lattice A poset in which the meet and join of every pair of elements are defined.

lcm The least common multiple.

Leaf A vertex of degree 1.

Lemma A theorem chiefly used to prove another, more "important" theorem.

LHS The left-hand side.

Linear extension A total order $L = (X, \le)$ is a linear extension of a poset $P = (X, \preceq)$ provided for all $x, y \in X$, $x \preceq y \Rightarrow x \le y$.

Linear order A poset in which all pairs of elements are comparable. Also called a **total order.**

Linearity of expectation If X, Y are real-valued random variables defined on a sample space and if $a, b \in \mathbb{R}$, then $E(aX + bY) = aE(X) + bE(Y)$.

List An ordered sequence of items.

Logically equivalent Two statements, A and B, such that $A \iff B$ is true. Two Boolean expressions whose values are the same for each possible substitution of its variables.

lub The least upper bound.

M

Map A synonym for function.

Maximal (general) Unextendable; cannot be made larger.

Maximal (posets) x is maximal means there is no y with $x < y$.

Maximum (general) Of largest possible size.

Maximum (posets) x is maximum means for all y, $y \le x$.

Mean A synonym for expected value.

Meet $a \wedge b$ is the least upper bound of a and b.

Minimal (general) Unshrinkable; cannot be made smaller.

Minimal (posets) x is minimal means there is no y with $y < x$.

Minimum (general) Of smallest possible size.

Minimum (posets) x is minimum means for all y, $x \le y$.

Mod (operation) $a \bmod b$ is the remainder when we divide a by b.

Mod (relation on a group) If $(H, *)$ is a subgroup of $(G, *)$, then $a \equiv b \pmod{H}$ means $a * b^{-1} \in H$.

Mod (relation on integers) See **congruent (mod n).**

Modular arithmetic Arithmetic in the number system $\mathbb{Z}_n$.

Multichoose $\left(\!\binom{n}{k}\!\right)$ is the number of k-element multisets we can form whose elements are taken from an n-element set.

Multiplication Principle A counting theorem that asserts that the number of two-element lists we can form in which there are a choices for the first element of the list, and, for each such choice, b choices for the second element of the list, is ab.

Multiplicity The number of times an element is present in a multiset.

Multiset A generalization of a set in which an object may be present in the collection more than once.

N

$\mathbb{N}$ The natural numbers.

Nand A Boolean algebra operation $a \barwedge b$ equivalent to $\neg(a \wedge b)$.

Natural numbers $\mathbb{N} = \{0, 1, 2, 3, \ldots\}$. Some authors do not consider 0 to be a natural number.

Necessary Condition A is necessary for condition B means $B \Rightarrow A$.

Neighbors Adjacent vertices

Not The statement "not A" is true exactly when A is false. In Boolean algebra, $\neg a$.

O

Odd (integer) An integer of the form $2a+1$ where a is an integer.

Odd (permutation) A permutation equal to the composition of an odd number of transpositions.

One and only one Exactly one. See **exactly.**

One-to-one A function is one-to-one means $f(a) = f(b) \Rightarrow a = b$.

Onto A function $f : A \to B$ is onto B means that for all $b \in B$, there is an $a \in A$ with $f(a) = b$. Equivalently, im $f = B$.

Or The statement "A or B" is true exactly when one or both of A and B are true. In Boolean algebra, $a \vee b$. See also **exclusive or.**

Order (graph) The number of vertices in a graph.

Outcome An element of a sample space.

P

Pairwise disjoint A collection of sets no two of which have a common element.

Parity Even or odd. For example, the parity of 3 is odd, and the parity of 0 is even. Two integers with the same parity are either both even or both odd.

Part A member set of a partition.

Partial order A relation that is reflexive, antisymmetric, and transitive.

Partially ordered set $(X, \leq)$ where X is a set and $\leq$ is a relation on X that is reflexive, antisymmetric, and transitive. Also called a **poset.**

Partition A partition of A is a set of nonempty, pairwise disjoint subsets of A whose union is A.

Pascal's triangle A triangular chart of numbers whose entry in the nth row and kth diagonal is $\binom{n}{k}$.

Path A walk without a repeated vertex. Also a graph of that form, P_n.

Perfect number A positive integer equal to the sum of its positive divisors (other than itself).

Perfect square An integer of the form n^2 where n is an integer. See also **quadratic residue.**

Permutation A bijection from a set to itself.

Pigeonhole Principle If $f : A \to B$ with $|A| > |B|$, then f is not one-to-one.

Planar Can be drawn in the plane without edges crossing.

Poset Partially ordered set.

Power set The set of all subsets of a given set; denoted 2^A.

Prime An integer, greater than 1, whose only positive divisors are 1 and itself.

Probability A measure of likelihood, specifically the function P in a sample space (S, P) and its extension to events.

Proof A precise, incontrovertible essay establishing a mathematical truth.

Proof by contradiction A proof that starts with the hypothesis and the negation of the conclusion and proceeds to a contradiction. Also known as *indirect proof* and *reductio ad absurdum*.

Proper coloring A coloring in which adjacent vertices receive different colors. See **coloring.**

Proposition A theorem of lesser generality or importance.

Public-key cryptography Cryptography in which the method for putting messages into code is completely revealed, but the method for decryption is held secret.

Q

$\mathbb{Q}$ The rational numbers.

Quadratic residue The square of an element of $\mathbb{Z}_n$. See also **perfect square.**

Quantifier The symbols $\forall$ (universal) and $\exists$ (existential).

Quod erat demonstrandum Literally, "that which is to be proved." Written at the end of proofs to assert that the proof is complete. Often abbreviated QED.

R

$\mathbb{R}$ The real numbers.

Random variable A function whose domain is the set of outcomes of a sample space.

Rational A number of the form a/b where $a, b \in \mathbb{Z}$ and $b \neq 0$. $\mathbb{Q}$ is the set of all rational numbers.

Realizer A set of linear extensions $\{L_1, \ldots, L_t\}$ is a realizer of a poset $P = (X, \leq)$ provided that for all $x, y \in X$, $x \leq y$ if and only if $x \leq_i y$ for all $i = 1, \ldots, t$.

Reciprocal A multiplicative inverse. For $a \in \mathbb{Z}_n$, its reciprocal b satisfies $a \otimes b = 1$; denoted a^{-1}.

Recurrence relation Given a sequence of numbers, $a_0, a_1, a_2, \ldots$, a recurrence relation is a rule that shows how to calculate a_n in terms of earlier elements of the sequence.

Reductio ad absurdum Proof by contradiction.

Refine If $\mathcal{P}$ and $\mathcal{Q}$ are partitions of a set, we say $\mathcal{P}$ refines (or is finer than) $\mathcal{Q}$ if every part of $\mathcal{P}$ is a subset of some part in $\mathcal{Q}$.

Reflexive A relation R on a set A is reflexive means $\forall a \in A,\ a\,R\,a$.

Regular graph A graph in which all vertices have the same degree. In a k-regular graph, all vertices have degree k.

Relation A set of ordered pairs.

Relatively prime A pair of integers whose greatest common divisor is 1.

Result A theorem.

Reverse Polish notation Notation in which operations appear after their operands. Abbreviated RPN.

RHS The right-hand side.

S

Sample space A pair (S, P) where S is a finite set and P is a function that gives the probability of each element in S.

Sequence A list, typically of numbers.

Set An unordered collection of objects.

Sign (permutation) The sign of π is 1 if π is an even permutation and -1 if π is an odd permutation. Denoted $\operatorname{sgn} \pi$.

Size (graph) The number of edges in a graph.

Size (set) The number of elements in the set; denoted $|A|$. See **cardinality.**

Sorting Placing in order, such as in ascending numerical order or in alphabetical order.

Spanning subgraph A subgraph formed by deleting edges.

Spanning tree A subgraph that is spanning and a tree.

Stable See **independent.**

Stirling's formula An approximation for factorial: $n! \approx \sqrt{2\pi n}\, n^n e^{-n}$.

Strong induction A variant form of induction

using a more extensive induction hypothesis that assumes the result for all possible cases up to a given size.

Subgraph A graph contained in another graph.

Subgroup A group contained in another group.

Subset $A \subseteq B$ means that every element of A is also an element of B.

Sufficient Condition A is sufficient for condition B means $A \Rightarrow B$.

Superset $A \supseteq B$ means that every element of B is also an element of A.

Surjection An onto function.

Symmetric A relation R is symmetric means $a\,R\,b \Rightarrow b\,R\,a$.

Symmetric difference $A \,\Delta\, B$ is the set of all elements in A or B, but not both.

Symmetric group S_n is the set of all permutations of $\{1, 2, \ldots, n\}$.

Symmetry A motion of a geometric object that does not change the appearance of the object.

T

Tautology A Boolean expression that evaluates to TRUE for all possible values of its variables. Informally, something that is true just by definition.

The The definite article, suggesting uniqueness. Use *a* or *an* when there may be more than one possibility. "Let x be *the* solution to..." implies there is one and only one solution. "Let x be *a* solution to..." allows for the possibility of multiple solutions.

Theorem A provable statement about mathematics.

Total order A poset in which all pairs of elements are comparable. Also called a **linear order.**

Totient The number of integers from 1 to n that are relatively prime to n, denoted $\varphi(n)$.

Transitive A relation R is transitive means that for all x, y, z if $x\,R\,y$ and $y\,R\,z$, then $x\,R\,z$.

Transposition A permutation τ for which $\tau(a) = b$, $\tau(b) = a$, $a \neq b$, and for all other elements c, $\tau(c) = c$.

Tree A connected, acyclic graph.

Triangle inequality $|a + b| \leq |a| + |b|$.

Tuple A list of numbers; e.g., $(1, 1, 3, 7)$ is a 4-tuple.

U

Union $A \cup B$ is the set of all elements that are in A or B (or both).

Unique Exactly one.

Universal quantifier $\forall$, meaning *for all* or *for every*.

V

Vacuous An if-then statement whose hypothesis (if clause) is always false. Such statements are regarded as true.

Venn diagram A pictorial representation in which sets are represented by circles or other shapes.

W

Walk A sequence of vertices, each adjacent to the next.

Well-Ordering Principle Every nonempty subset of $\mathbb{N}$ contains a least element.

Width Size of a largest antichain.

Without loss of generality When there is more than one case in a proof, but the proofs in these cases are all the same, we can elect to prove just one of the cases. We announce this by declaring that the choice of this case is "without loss of generality." For example, if a proof involves two different numbers, x and y, and there are no further restrictions on x and y, we might want to break the proof into the cases $x < y$ and $x > y$. Since x and y are, so far, arbitrary, we may assume without loss of generality that $x < y$. Sometimes abbreviated wlog or wolog.

Xor See **exclusive or.**

Z

$\mathbb{Z}$ The integers.

D Fundamentals

This appendix presents various properties of numbers, operations, and relations that you may use freely in any proof.

Numbers

The *natural numbers* consist of zero and the positive whole numbers. The set of natural numbers is denoted by $\mathbb{N}$, so

$$\mathbb{N} = \{0, 1, 2, 3, 4, \ldots\}.$$

The *integers* are the positive and negative whole numbers, and zero. The set of integers is denoted by $\mathbb{Z}$, so

$$\mathbb{Z} = \{\ldots, -3, -2, -1, 0, 1, 2, 3, \ldots\}.$$

The *rational numbers* are all fractions of the form $\frac{a}{b}$ where a and b are integers and $b \neq 0$. The set of all rational numbers is denoted $\mathbb{Q}$.

Two rational numbers $\frac{a}{b}$ and $\frac{c}{d}$ are *equal* exactly when $ad = bc$.

There are many ways to express rational numbers. For example, the rational number $\frac{3}{2}$ is equal to all of the following: $\frac{6}{4}$, $\frac{-3}{-2}$, $1\frac{1}{2}$, 150%, and 1.5.

A precise definition of *real number* is beyond the scope of this book. Informally, real numbers are those that can be expressed as follows: Begin with an integer and append a decimal point and either a finite or an infinite sequence of digits. For example, the following are real numbers:

$$-1.4444444444\ldots$$

$$3$$

$$3.1415926535\ldots$$

$$-99.013$$

The set of real numbers is denoted $\mathbb{R}$.

Real numbers are mentioned on occasion in this book, but in nearly all cases, little is lost by allowing only rational numbers.

Every natural number is an integer, every integer is a rational number, and every rational number is a real number. This can be expressed in symbols as follows:

$$\mathbb{N} \subseteq \mathbb{Z} \subseteq \mathbb{Q} \subseteq \mathbb{R}.$$

In every case, the subset relation is strict (i.e., no two of the sets listed above are equal).

Operations

The fundamental operations of arithmetic are addition (+) and multiplication ($\times$). Basic calculations, such as $3 + 4 = 7$ and $7 \times 3 = 21$, do not require proof.

If we assume that we know how to add and multiply integers, we can define addition and multiplication for rational numbers. If $\frac{a}{b}$ and $\frac{c}{d}$ are rational numbers (where b and d are nonzero), we have

$$\frac{a}{b} + \frac{c}{d} = \frac{ad + bc}{bd}$$

and

$$\frac{a}{b} \cdot \frac{c}{d} = \frac{ac}{bd}.$$

You may assume the following properties of addition and multiplication. Below, the unqualified word *number* may refer either to a rational number or to a real number; the statements are correct in either context.

- *Closure property*: If x and y are integers, then so are $x + y$ and xy.

  Likewise, if x and y are natural/rational/real numbers, then so are $x + y$ and xy.
- *Commutative property*: For any numbers x and y, we have $x + y = y + x$ and $xy = yx$.
- *Associative property*: For any numbers x, y, and z, we have $x + (y + z) = (x + y) + z$ and $x(yz) = (xy)z$.
- *Identity elements*: For any number x, $x + 0 = x$ and $x \cdot 1 = x$.
- *Inverses*: For any number x, there is a number $-x$ with the property that $x + (-x) = 0$. Furthermore, if x is an integer, so is $-x$.

  For any nonzero number x, there is a number x^{-1} with the property that $x \cdot x^{-1} = 1$.

  Consequently, if x and y are nonzero numbers, then xy is also nonzero.

- *Distributive property*: For any numbers x, y, and z, we have $x(y+z) = xy + xz$.

The operations of subtraction ($-$) and division ($\div$) are defined in terms of addition and multiplication. We define $a - b$ to be $a + (-b)$, and for $b \neq 0$, we define $a \div b$ to be $a \cdot b^{-1}$.

Ordering

The *less-than* relation places an ordering on numbers. The expression $x < y$ means that x is less than y. We also have the symbol $\leq$, which stands for *less than or equal to*. When we write $x \leq y$, this means that x is less than or equal to y.

Similarly, we have the symbols $>$ and $\geq$ which stand for *greater than* and *greater than or equal to*, respectively.

We call a number x *positive* provided $x > 0$. We call x *negative* if $x < 0$. We call x *nonnegative* provided $x \geq 0$.

The following are basic properties of $<$ (and its relatives) that you may use without proof.

- *Trichotomy property*: Let x and y be numbers. Then exactly one of the following is true: $x < y$, $x = y$, or $x > y$.

  Consequently, $a \leq b$ if and only if $b \geq a$. Similarly, $a < b$ if and only if $b > a$.
- If $a < b$ and $c < d$, then $a + c < b + d$. Likewise for $\leq$, $>$, and $\geq$.

  Consequently, $a < b$ if and only if $b - a$ is positive.

  Furthermore, if a and b are positive, then so is $a + b$.
- Let x be a positive number. Then $a < b$ if and only if $ax < bx$. Likewise for $\leq$, $>$, and $\geq$.

  Furthermore, if a and b are positive, then so is ab.
- Let a and b be positive numbers. Then $a < b$ if and only if $a^{-1} > b^{-1}$. Likewise for $\leq$ / $\geq$. Likewise if a and b are both negative.
- $a < b$ if and only if $-a > -b$. Likewise for $\leq$ / $\geq$.

  Consequently, if a and b are negative, then ab is positive.
- *Transitive property*: If $x \leq y$ and $y \leq z$, then $x \leq z$. Likewise for $\leq$, $>$, and $\geq$.
- *Well-Ordering Principle*: If A is a nonempty subset of $\mathbb{N}$, then A contains a least element.

Complex Numbers

Complex numbers are an extension of the real numbers. They are formed by defining a new number i with the property that $i^2 = -1$. The set of all complex numbers is denoted $\mathbb{C}$ and contains all numbers of the form $a + bi$ where a and b are real. The usual operations are defined for complex numbers. Let $a + bi$ and $c + di$ be complex numbers (where $a, b, c, d \in \mathbb{R}$); we have the following:

$$
\begin{aligned}
(a+bi) + (c+di) &= (a+c) + (b+d)i \\
(a+bi) - (c+di) &= (a-c) + (b-d)i \\
(a+bi) \times (c+di) &= (ac-bd) + (ad+bc)i \\
|a+bi| &= \sqrt{a^2+b^2} \\
(a+bi)^{-1} &= \frac{a}{a^2+b^2} + \frac{-b}{a^2+b^2}i \\
(a+bi)/(c+di) &= (a+bi) \times [(c+di)^{-1}].
\end{aligned}
$$

Of course, reciprocal and division are defined only in the case where $c + di \neq 0$. In this book, complex numbers are needed only in Section 22 on recurrence relations.

Substitution

The following observation is, perhaps, beyond obvious, but we mention it anyway. When we say two mathematical objects are *equal,* we mean that they are exactly the same. Thus, if a statement involving a mathematical entity x is true, and if $x = y$, then a new statement formed from the first statement by replacing some (or all) occurrences of x with y is also true.

Index

Page numbers in **boldface** refer to concepts presented in a numbered definition.
Page numbers in *italics* refer to enteries in the glossary.

D

E

F

G

H

I

S

T

U

V

W

X

Y